Radiography Essentials
FOR LIMITED PRACTICE

Radiography Essentials

FOR LIMITED PRACTICE

Margaret Mary Hunkele, RT (R)(M)

Clinical Coordinator, Practical Technology in Radiography
The Bryman School of Phoenix
Phoenix, Arizona;
Director of Operations
Med-Teq VIII Mobile Stereotactic Breast Biopsy Service
Founder, The American Society of Practical Technologists in Radiography
Founder, The American Registry of Practical Technologists in Radiography
Scottsdale, Arizona

with 1050 illustrations

W.B. SAUNDERS COMPANY
A Harcourt Health Sciences Company
Philadelphia   London   New York   St. Louis   Sydney   Toronto

W.B. Saunders Company
A Harcourt Health Sciences Company

The Curtis Center
Independence Square West
Philadelphia, PA 19106

Library of Congress Cataloging-in-Publication Data

Hunkele, Margaret Mary.
Radiography essentials for limited practice / Margaret Mary Hunkele.

p. cm.

Includes bibliographical references and index.

ISBN 0–7216–8212-X

1. Radiography, Medical.   I. Title.
   [DNLM: 1. Radiography—methods. 2. Allied Health Personnel. 3. Technology,
Radiologic. WN 200 H937r 2002]

RC78 .H86 2002   616.078572—dc21   2001043386

Publishing Director: Andrew Allen
Acquisitions Editor: Jeanne Wilke
Developmental Editor: Linda Woodard
Project Manager: Linda McKinley
Production Editors: Peggy Fagen, Jennifer Furey
Designer: Julia Ramirez
Medical Illustrator: Jeanne Robertson

Radiography Essentials for Limited Practice   0-7216-8212-x

Printed in the United States of America.

Last digit is the print number: 9   8   7   6   5   4   3   2   1

Contributors

Ruth Ann Ehrlich, RT (R)
Senior Instructor, Radiology
Western States Chiropractic College
Portland, Oregon

Dennis M. Hoyer, MT (ASCP), DC
Associate Professor of Clinical Sciences
Western States Chiropractic College
Portland, Oregon

Reviewers

Debra L. Caldwell, MA, RT (R)(QM)(CV)
Radiologic Technology Program Director
York Technical College
Rock Hill, South Carolina

John M. Gray, MA, BS, RT
Executive Director
Radiologic Technology Board of Examiners
Arizona Radiation Regulatory Agency
Phoenix, Arizona

LaVerne T. Gurley, PhD, BA, RT (R)(T)
Professor Emeritus
Shelby State Community College
Memphis, Tennessee

Henry C. Hirsh, RT (R)
Chief Technologist, Instructor
Western States Chiropractic College
Portland, Oregon

Debbie Hornbacher, BS, RT (R)
Program Director
Trinity Hospital School of Radiologic Technology
Minot, North Dakota

Joseph A. Melanson, CRT (R)(N), ARRT
Campus Director
Central California School of Continuing Education—Corona
Corona, California

Patricia A. Stoddard, MS, RT (R), MT, CMA
Instructor
Medical Assisting
Western Business College
Vancouver, Washington

To the Hunkele family, East, West, and across the pond.

Preface

Education in radiography is presently growing in two different directions. On one hand there is a need for professional imaging specialists with a knowledge base that prepares them for careers in major medical centers, dealing with critically ill or injured patients and using highly sophisticated computerized equipment for a variety of imaging modalities. The trend in this area is toward an ever-increasing degree of specialization. The greatest value is placed on the highest levels of technical specialty. On the other hand, there is a need in many smaller health care facilities for personnel who are trained in a variety of basic clinical skills, including routine radiography. The trend in these areas is toward a greater degree of generalization, with the greatest value placed on cross-training and a wide variety of clinical abilities.

Limited or practical radiographers are personnel who are not certified as registered radiologic technologists but who are trained to perform the radiographic procedures commonly needed in outpatient clinics, physicians' offices, and urgent care centers. A nationwide shortage of registered radiologic technologists has created an increasing demand for practical radiographers who can provide a limited scope of radiographic services in outpatient facilities. In some geographic areas, hospitals have created positions and job descriptions that utilize the skills of practical radiographers. Nearly half the states in the United States officially recognize and regulate the practice of limited radiography.

Education programs for practical radiographers have traditionally used textbooks written for college courses in radiologic technology. These textbooks are devoted exclusively to one subject such as physics, anatomy, radiographic positioning, or patient care. Their scope of content is greater than that needed by practical radiographers, and the comprehension level is designed for the educational background of the college scholar. Clearly, there was a need for a comprehensive textbook to meet the specific requirements of students and instructors in practical radiography. *Radiography Essentials for Limited Practice* was written to fill that need. It will also serve as a valuable reference after the course is completed and for those already involved in the limited practice of radiography.

ORGANIZATION

The scope of this text is designed to provide in one book all of the essential information needed by practical radiographers to pass certification examinations, meet the requirements of their job descriptions, and provide exemplary patient care.

Part I explains the role of the limited radiographer in the health care system and introduces the equipment and procedures involved in radiography.

Part II is devoted to the principles of science that the radiographer uses to operate x-ray equipment safely, set exposures correctly, process x-ray film, make technical judgments, and correct technical errors. The effects of each aspect of the radiographic process on the finished radiographic image are explained in detail.

Part III covers anatomy, positioning, and radiographic pathology. Following a general introductory chapter, these topics are integrated and addressed with respect to a specific part of the body. The content of these chapters is consistent with the categories often used by states to certify proficiency in limited radiography: upper extremity, lower extremity, spine, chest and abdomen, and skull. This section also contains information on pediatric and geriatric radiography and a chapter on film critique.

Part IV contains information on professionalism and patient care. These chapters emphasize principles of practice, communication, and judgment to ensure patient safety and quality care while enhancing job satisfaction and minimizing risks of legal liability.

Part V provides instruction in ancillary clinical skills that increase the value and versatility of the radiographer in the clinical setting. This section includes medication administration, medical laboratory skills, and diagnostic procedures such as weighing and measuring patients, conducting visual screening examinations, and performing electrocardiography and spirometric pulmonary function testing.

FEATURES

The comprehension level of this text is designed to be easily understood by students without a high level of background education and by those who are returning to school after a long absence. For this reason it is not only appropriate for students entering the study of practical radiography, it may also be helpful to students in professional radiologic technology programs that are struggling to master basic concepts. The general vocabulary level is simple, and the technical vocabulary is defined clearly within the text. Key terms are listed at the beginning of each chapter and are bolded in the text. A glossary is included for easy reference to terminology. Discussions of mathematical formulas and calculations are accompanied by clear examples. Chapter 26 provides detailed instruction, practice problems, and practical applications of the mathematical skills needed by radiographers. It is especially designed to elevate the understanding of students who have difficulty relating to the measurements, formulas, and calculations in the preceding chapters. Students who are not proficient in mathematics may find it helpful to study the first section of Chapter 26, "Fundamental Mathematical Principles," before proceeding with Chapter 3.

Learning objectives at the beginning of each chapter provide a preview of the chapter's content and help to establish study goals. More than 1000 illustrations support the text by clarifying concepts and adding interest. The text is also supplemented with charts and tables. Boxes are often used to summarize content or to emphasize key concepts. Positioning tips are included in the positioning chapters. Summaries and questions at the ends of chapters aid in review and assessment of comprehension. The index allows quick location of topics and ensures that this book will be useful as a reference.

Radiography Essentials for Limited Practice was conceived in the hope of assisting its readers to perform basic radiographic procedures with both competence and confidence and to inspire them to strive for excellence. It is dedicated to that end.

Margaret Mary Hunkele

Acknowledgments

I would first like to acknowledge Ruth Ann Ehrlich, RT (R), without whose hard work and dedication this publication would not have been possible. In addition to her direct chapter contributions, Ann oversaw all rewrites, photographs, and radiographs contained within this text. Next, I would like to thank Deborah A. Shuga, RT (R), BSMR, Program Director of the Practical Technology in Radiography department of the Bryman School of Phoenix, for her technical expertise, personal encouragement, and patience during this process. I would like to acknowledge Eric DeVito, PTR, administrative assistant in the Practical Technology in Radiography department of the Bryman School of Phoenix, for his technical expertise and ability to handle my overflow of work as Clinical Coordinator. I would also like to acknowledge John Gray, Administrator of the Medical Radiologic Technology Board of Examiners, state of Arizona, who has served as my mentor since my arrival in Arizona in 1992. I would like to thank Dr. James Lash at Western Arizona Regional Medical Center for providing radiographs for this book. Lastly, I would like to acknowledge all the students I have had the pleasure to instruct, particularly Laverne Abbas, PTR; Richard Cole, RT (R); Richard Shipman, PTR; and especially Karen McBride, MA, whose compassion and dedication exemplified the best our profession has to offer.

Ann Ehrlich would like to thank the clinical staff and radiology department of Western States Chiropractic College, who extended their support to this project with consultation and advice, assisted in the search for radiographic illustrations, and provided the use of facilities for photography. Special thanks to Dr. Beverly Harger, Dr. Elaine Johnston, Dr. Chad Warshel, Dr. Rui Domingues, Dr. Sharon Grant, and Chief Technologist Hank Hirsh. Facilities for photography were also provided by Kaiser Sunnyside Medical Center.

She would also like to thank medical photographer Jeff Watson, whose vision and expertise translated vague ideas into clear and effective illustrations, as well as the photographic models, most of whom were drafted from the student body at WSCC, whose cooperation and assistance made these illustrations possible. Many thanks to Jeanne Robertson, who created the anatomy illustrations that give this book a distinctive artistic style and are well designed to meet the specific needs of its readers.

Experts chosen by the publisher reviewed the submitted manuscript, contributed valuable ideas, and identified errors. Their expertise and insight added significantly to the text, and their work is sincerely appreciated. Thanks also to Ellen D. McCloskey, Richard Kirkland, and Jeff Watson for reviewing major portions of the manuscript prior to submission and for offering excellent suggestions.

The thousands of details involved in planning, encouraging, organizing, designing, editing, and publishing were handled by the highly skilled and professional staff at Harcourt. The dedication, patience, and gracious cooperation of the following members of the editorial staff are acknowledged with gratitude: Jeanne Wilke, Carolyn Kruse, Linda Woodard, Peggy Fagen, and Melissa Kuster.

Margaret Mary Hunkele
Ruth Ann Ehrlich

Contents

Part V   Ancillary Clinical Skills

Appendices

Photo Credits, 593
Glossary, 595

Radiography Essentials

FOR LIMITED PRACTICE

PART 1

Introduction to Limited Radiography

Role of the Radiographer in Limited Practice

Learning Objectives

At the conclusion of this chapter the student will be able to:

* Compare the role of the limited radiographer with that of the registered radiologic technologist
* Identify the discoverer of x-rays and the date of the discovery
* Explain the primary purposes of ASRT, ARRT, and ASPTR
* Determine the legal requirements for the practice of radiography in his or her state
* Describe the typical work environment of the limited radiographer
* Describe in a general way the duties of a limited radiographer

Key Terms

American Registry of Radiologic Technologists (ARRT)
American Society for Practical Technologists in Radiography (ASPTR)
American Society of Radiologic Technologists (ASRT)

back office
front office
limited radiography
medical assistant
radiograph
radiologist
reciprocity

Welcome to the fascinating field of radiography! You are beginning a study of the art and science needed to create images of the internal structures of the human body. The images you create will aid physicians in diagnosis and will help patients receive treatment needed to promote or regain health. This is a vital role in the health care delivery system, one that requires knowledge, skill, judgment, integrity, and dedication.

RADIOGRAPHY AND LIMITED RADIOGRAPHY

X-rays were discovered in 1895 by Wilhelm Conrad Roentgen (Fig. 1-1) at the University of Würzburg in Germany. The first radiographers were scientists and physicians who experimented with primitive x-ray

Fig. 1-1 Wilhelm Conrad Roentgen (1845-1923). He discovered x-rays on November 8, 1895.

apparatus to make x-ray images of the human body (Fig. 1-2). Soon these pioneers trained their assistants to make these "x-ray pictures," now called **radiographs,** and the profession of radiography was born.

Early radiographers soon began meeting to share their knowledge. The organization now called the **American Society of Radiologic Technologists (ASRT)** was founded in Chicago in 1920. It is the world's oldest and largest radiologic science organization. ASRT provides many services to its members, including continuing education, a professional journal, a newsletter, guidelines and assistance for radiography educators, and an annual national meeting. Through the efforts of this society, the **American Registry of Radiologic Technologists (ARRT)** was formed in 1922 to establish standards and examinations necessary to certify radiographers. Radiographers certified by ARRT use the initials "R.T." (Registered Technologist) after their names.

The profession of radiography has expanded to include a variety of imaging and treatment modalities, including those listed in Box 1-1. Many of these modalities require specialized training beyond that needed for certification by ARRT in radiography [R.T.(R)].

ARRT certification in radiography requires at least 2 years of education in an approved program that includes comprehensive academic course work in sciences. These programs are affiliated with acute general hospitals to provide extensive clinical experience in the care of patients who are severely ill or injured. As a result of this background, radiographers who are

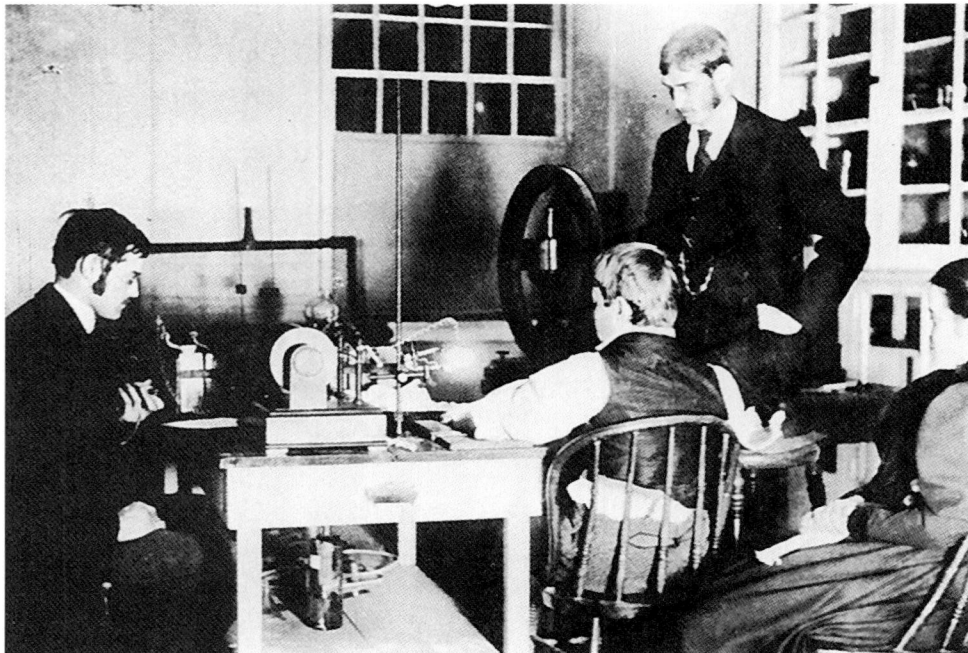

Fig. 1-2 First clinical radiograph in the United States was made at Dartmouth College in 1896.

registered technologists may be overqualified for positions that involve radiography in the typical outpatient clinic or physician's office. On the other hand, they may lack certain skills needed in this type of setting.

Limited radiography is practiced primarily in clinics and physicians' offices (Fig. 1-3). In some areas, however, limited radiographers are employed in hospitals, and this practice is expanding. There is currently a nationwide shortage of radiographers, creating more opportunities for individuals qualified in limited radiography.

Limited radiography, sometimes called "practical radiography," developed as nurses, medical assistants, chiropractic assistants, and other health care office personnel were trained to perform basic radiography in addition to their primary duties. It is called "limited" because, compared to registered technologists, the scope of practice is restricted. Limited practice does not usually involve the use of contrast media for the imaging of blood vessels and abdominal organs.* Additional restrictions may be applied, since the scope of practice varies among the states. In con-

---

*One exception to this statement is the granting of limited permits in California that allow qualified limited radiographers to use contrast media for examinations of the gastrointestinal tract and the urinary system.

BOX 1-1

Imaging and Treatment Modalities in Radiology

Angiography: imaging of blood vessels with the injection of special compounds called contrast media

Computed tomography (CT): computerized x-ray system that provides axial images (transverse "slices") of all parts of the body

Fluoroscopy: real-time viewing of x-ray images in motion

Magnetic resonance imaging (MRI): computerized imaging system that uses a powerful magnetic field and radiofrequency pulses to produce images of all parts of the body

Mammography: x-ray imaging of the breast using a special x-ray machine

Nuclear medicine: the injection or ingestion of radioactive materials and the recording of their uptake in the body using a gamma camera

Positron emission tomography (PET): a highly sophisticated computerized form of nuclear medicine imaging

Radiation therapy: treatment of malignant diseases using radiation

Sonography: imaging of soft tissue structures using sound echoes

---

trast to registered technologists, limited radiographers are often educated as **medical assistants** and perform a variety of office procedures that do not involve imaging, such as drawing blood samples, performing diagnostic tests, and assisting the physician with treatments and patient care procedures. Information about some of these duties is included in the final chapters of this text.

CREDENTIALS

Certification in radiography by ARRT has been accepted as the minimum credential for radiographers in hospital practice for many years, but no *legal* requirements applied to the practice of radiography until early in the 1970s, when licensure laws for radiologic technologists were passed in New York and New Jersey. California passed licensing legislation shortly thereafter. Today, 41 states have laws requiring some type of credential to practice radiography, and 21 of these include provision for limited practice. Currently, depending on jurisdiction, limited radiography may be prohibited, permitted without restriction, or regulated quite specifically. For example, New York prohibits limited practice, while North Dakota and South Dakota freely permit limited practice with no requirements for credentials. Oregon issues eight different categories of limited permit, based on the specific procedures for which the radiographer is qualified.

Legal requirements are subject to change, so it will be necessary for you to inquire directly of the appropriate agency in your state to determine whether limited radiography is permitted, whether its practice is regulated, and how to proceed to obtain the necessary credentials. Appendix A lists the names and addresses of these agencies.

Most states that regulate limited practice also regulate the education needed to qualify. The student

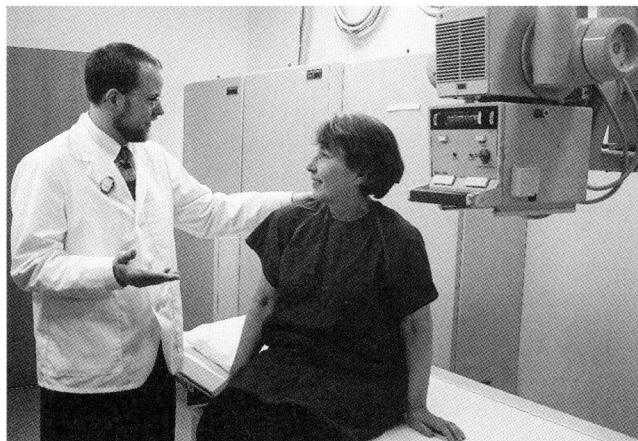

Fig. 1-3 Radiographer at work.

who plans to obtain a limited license or permit must be certain that the planned education meets state requirements. These requirements may include the need for clinical experience and clinical supervision. Some states also have continuing education requirements for renewal of licenses or permits.

Each state that regulates limited practice has established standards for the scope of practice that is allowed. There are serious sanctions for practicing radiography outside the boundaries defined by state laws and regulations. Practicing without a valid license or permit, or practicing outside the scope of one's credentials, may result in fines, imprisonment, or both. In addition, a license or permit may be suspended or permanently revoked. Employers may also be penalized if their employees practice radiography in violation of regulations. *All radiographers must be aware of the legal standards that apply to them and take care that their practice conforms to these standards.* This information also is available from the agencies listed in Appendix A.

The practice of radiography involves a variety of knowledge and skills. While the *scope of practice* may be limited, there is no restriction on the knowledge and skill needed for the practice that is allowed. In other words, the limited radiographer is held to the same high standards as a registered radiologic technologist in performing procedures within the permitted scope of practice.

AMERICAN SOCIETY FOR PRACTICAL TECHNOLOGISTS IN RADIOGRAPHY

One hallmark of a profession is self-regulation by means of an organization that sets minimum standards for credentials and establishes a code of ethical conduct. (Ethical conduct and codes of ethics are discussed in Chapter 20.) Limited radiographers are now beginning the exciting process of becoming true professionals. **The American Society for Practical Technologists in Radiography (ASPTR)** was founded in 1997 in Arizona by a group of committed radiography educators. The membership is still small, but growing. Efforts are underway to organize groups in Florida, Minnesota, Tennessee, and Alaska. The primary purposes of ASPTR are to promote and provide continuing education for practical technologists in radiography and to promote national standards for the practice of limited radiography, including a national certification process. The ASPTR Mission Statement is printed in Box 1-2.

The ASPTR is committed to the pursuit of **reciprocity,** that is, recognition of credentials acquired in one state by other states. Reciprocity facilitates the ability of limited radiographers to qualify for practice when moving from state to state.

The ARRT provides Limited Scope Examinations with several categories of content for state licensing purposes only. Some states that allow and regulate limited radiography use the ARRT Limited Scope Examination to qualify applicants for licenses or permits to practice. The minimum score for certification varies among the states that use this examination. Other states write and administer their own examinations for this purpose. The ASPTR is working to establish a national registry for limited radiographers that would serve the same purpose as the ARRT registry for registered technologists. The planned ASPTR registry would be based on the ARRT Limited Scope Examination, and would greatly simplify the process of instituting nationwide uniform standards.

The Arizona Chapter of ASPTR has taken preliminary steps to become affiliated with the Arizona Society of Radiologic Technologists, a chapter of ASRT. This is just the beginning of a long and uncertain process that could eventually lead to an affiliation between ASPTR and ASRT at the national level. This affiliation would unite all radiographers for greater effectiveness in providing education, professional growth, and political strength. It would also provide a forum for reconciling present differences in viewpoint between members of the two groups.

These are ambitious goals for this struggling, new organization, and they can only be accomplished with the support of limited radiographers who are

---

BOX 1-2

Mission Statement of The American Society of Practical Technologists in Radiography

The mission of The American Society of Practical Technologists in Radiography is to establish nationally recognized standards to promote exemplary patient care, continuing education, and reciprocity among the States in which Practical/Limited Scope Technology is recognized.

Supporting Statements
In support of this mission, the ASPTR will:
1. Require the following for admission:
   a. A minimum 70% weighted score on the ARRT Limited Scope Examination.
   b. Two favorable letters of reference from an M.D. or R.T. describing patient care medical ethic and work ethic of the individual applying for membership.
2. Offer continuing education classes and newsletters to its members consistent with the requirements to be developed by the American Registry of Practical Technologists in Radiography
3. Discuss continuing education and reciprocity in all states that recognize Practical/Limited Scope Technology

committed to their mission. More information about membership in ASPTR can be obtained by contacting their national headquarters.*

WORK ENVIRONMENT

As a limited radiographer, your direct supervisor may be a physician, a nurse, a registered technologist, an office manager, or a radiology administrator. Work environments vary greatly, depending on the type of organization, its size, and the organizational structure. You will often work directly with one or more physicians (Fig. 1-4). These physicians may be primary care physicians whose specialty is general practice and who see both adults and children for a wide variety of complaints. On the other hand, you may work for a specialist, a physician who has completed extensive additional training to qualify as an expert in a particular aspect of medical or chiropractic care. Box 1-3 lists health care specialists and describes their

*The American Society for Practical Technologists in Radiography, 15232 North 49th Street, Scottsdale, AZ 85254-2279, (602) 996-0018.

areas of clinical interest. It will be helpful to understand these terms, since you will encounter them frequently in various aspects of your work.

The work of outpatient clinic facilities is divided into two general areas, often referred to as the front

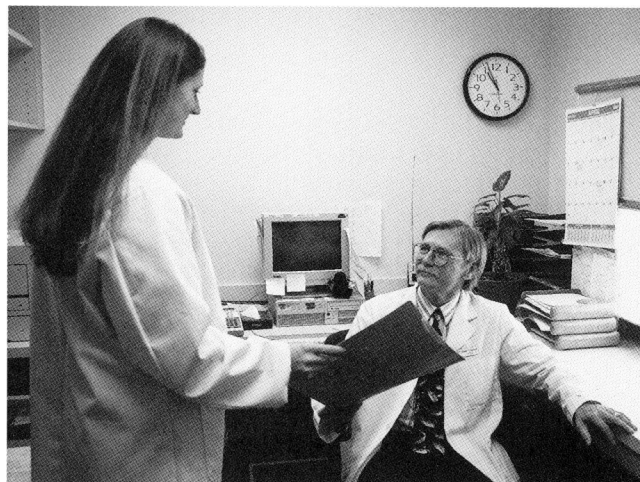

Fig. 1-4 Radiographers work directly with physicians.

BOX 1-3

Abbreviated Listing of Healthcare Specialties

Anesthesiologist: administers anesthetics and monitors patient during surgery

Dermatologist: diagnoses and treats conditions and diseases of the skin

Emergency department physician: specializes in trauma and emergency situations; a triage expert in disaster situations

Family practice physician: treats individuals and families in the context of daily life

Gastroenterologist: diagnoses and treats diseases of the gastrointestinal tract

Geriatrician: specializes in problems and diseases of elderly persons

Gynecologist: treats problems and diseases of the female reproductive system

Internist: specializes in diseases of the internal organs

Neurologist: specializes in functions and disorders of the nervous system

Obstetrician: specializes in pregnancy, labor, delivery, and immediate postpartum care

Oncologist: specializes in tumor identification and treatment

Ophthalmologist: diagnoses and treats problems and diseases of the eye

Pathologist: specializes in the scientific study of body alterations caused by disease and death

Pediatrician: treats and diagnoses disorders and diseases in children

Psychiatrist: specializes in diagnosis, treatment, and prevention of mental illness

Radiologist: specializes in diagnosis by means of medical imaging

Surgeons

Abdominal: specializes in surgery of the abdominal cavity

Neurologic: specializes in surgery of the brain, spinal cord, and peripheral nervous system

Orthopedic: diagnoses and treats problems of the musculoskeletal system

Plastic: restores or improves the appearance and function of body parts

Thoracic: specializes in problems of the chest

Urologic: diagnoses and treats problems of the urinary tract and the male reproductive system

Chiropractic specialties: specialty certification is available to chiropractic physicians in the fields of radiology, orthopedics (nonsurgical), neurology, nutrition, sports medicine, and other fields.

Both medical and chiropractic physicians may limit their practices to specific areas of interest with or without certification. Those with certification may have a general practice outside the scope of their specialty.

Modified from Ehrlich RA, McCloskey ED, and Daly JA: *Patient care in radiography,* ed 5, St Louis, 1999, Mosby.

office and the back office. The **front office** (Fig. 1-5) is a public area and includes the reception desk, the patient waiting area, and the desks or offices of those who deal with medical records, billing, and insurance claims. The **back office** (Fig. 1-6) is the area where patients are examined and treated. The back office includes consulting rooms, examination rooms, treatment rooms, and laboratory facilities, as well as the x-ray department. Utility and storage areas will also be found here.

The radiography suite will include one or more x-ray rooms, a darkroom for film processing, a desk or countertop for paperwork (including, perhaps, a computer), a film viewing and sorting area, and a film file area. Dressing rooms and restrooms are usually convenient to x-ray rooms. The limited radiographer is a "back office employee" and may be assigned other back office duties in addition to radiography.

Fig. 1-5 The front office includes the waiting room and reception area.

Radiographers employed by hospitals will work in a much different environment. The radiology department will be supervised by a director, usually a registered radiologic technologist. Radiographers work directly with **radiologists,** the physician specialists who interpret the radiographs and perform special imaging procedures. There will be less contact with the patients' primary care physicians, although some will visit the radiology department. The radiographer's duties are likely to be limited to radiography, with other personnel handling paperwork, patient transportation, and much of the communication with physicians' offices and other hospital departments. On the other hand, limited radiographers who also have medical assisting skills may be employed in ways that utilize these skills as well.

The hospital radiology staff may also include a number of radiographers with assignments involving specific procedures or work areas and some with responsibility for supervision or quality control. The staff may be scheduled in three shifts around the clock. In small institutions, the department may be closed during late night hours and on Sundays and holidays. When the department is closed, radiographers may take turns being "on-call," that is, available by telephone or pager to come to the hospital when needed.

Hospitals are complex, highly structured institutions. Each has many rules and procedures that must be mastered for the safety of patients and the efficient performance of the health care team. It is beyond the scope of this text to prepare radiographers to cope with all the situations and judgments they might face in an acute hospital setting. A thorough orientation to the institution is needed, and the use of additional texts and references is highly recommended.

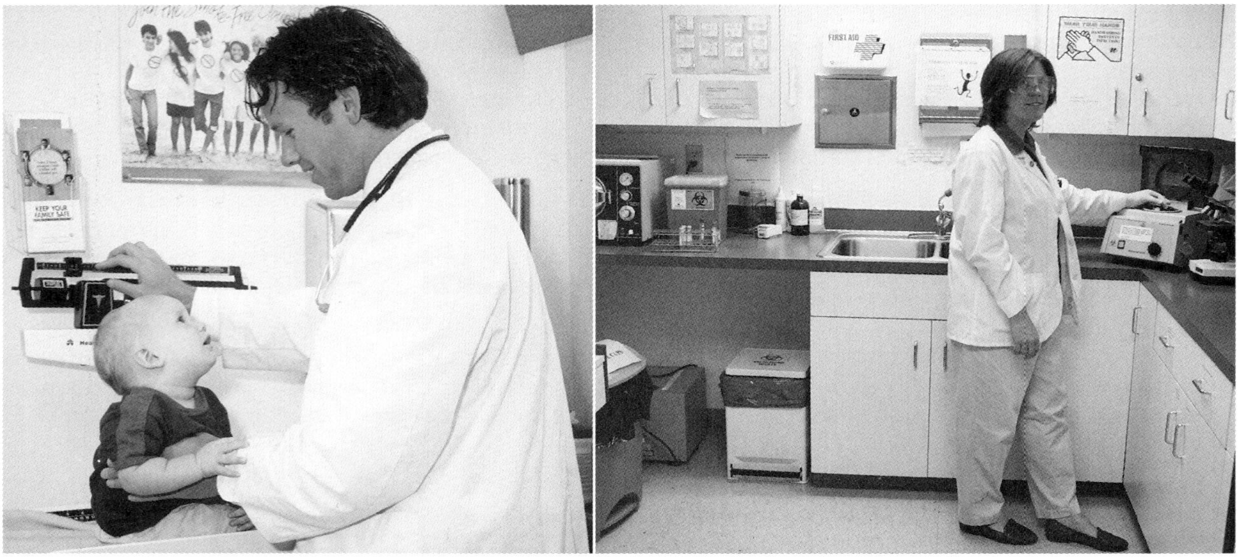

Fig. 1-6 The back office includes the consulting and treatment areas.

TYPICAL DUTIES OF A LIMITED RADIOGRAPHER

The radiographer encounters the patient after he or she has been admitted to the clinic or radiology department. A physician will have examined the patient, and one or more specific x-ray procedures will have been ordered. The physician may give the order directly to the radiographer or may instruct a nurse or medical assistant to communicate the order. The order may be verbal or in the form of a written requisition. The necessary x-ray paperwork may be completed by a clerical employee but is often the radiographer's responsibility.

Once the paperwork is completed, the radiographer greets the patient and determines whether the patient will need to undress and don a gown before radi-

ography. A dressing room or examining room is usually used for this purpose. The exact clothing to be removed is determined according to the examination. Generally, patients must remove outer clothing from the body area to be examined. Specific instructions for patient preparation are included in the appropriate sections of the text (Chapters 13 through 17).

The patient is then taken into the x-ray room. At this point, the radiographer provides a brief explanation and answers any questions about the procedure. When you have completed this text, you should be prepared to respond appropriately to most patient questions and concerns.

The next step is to assist the patient into the general position required for the x-ray examination (Fig. 1-7). For example, if a hand is to be x-rayed,

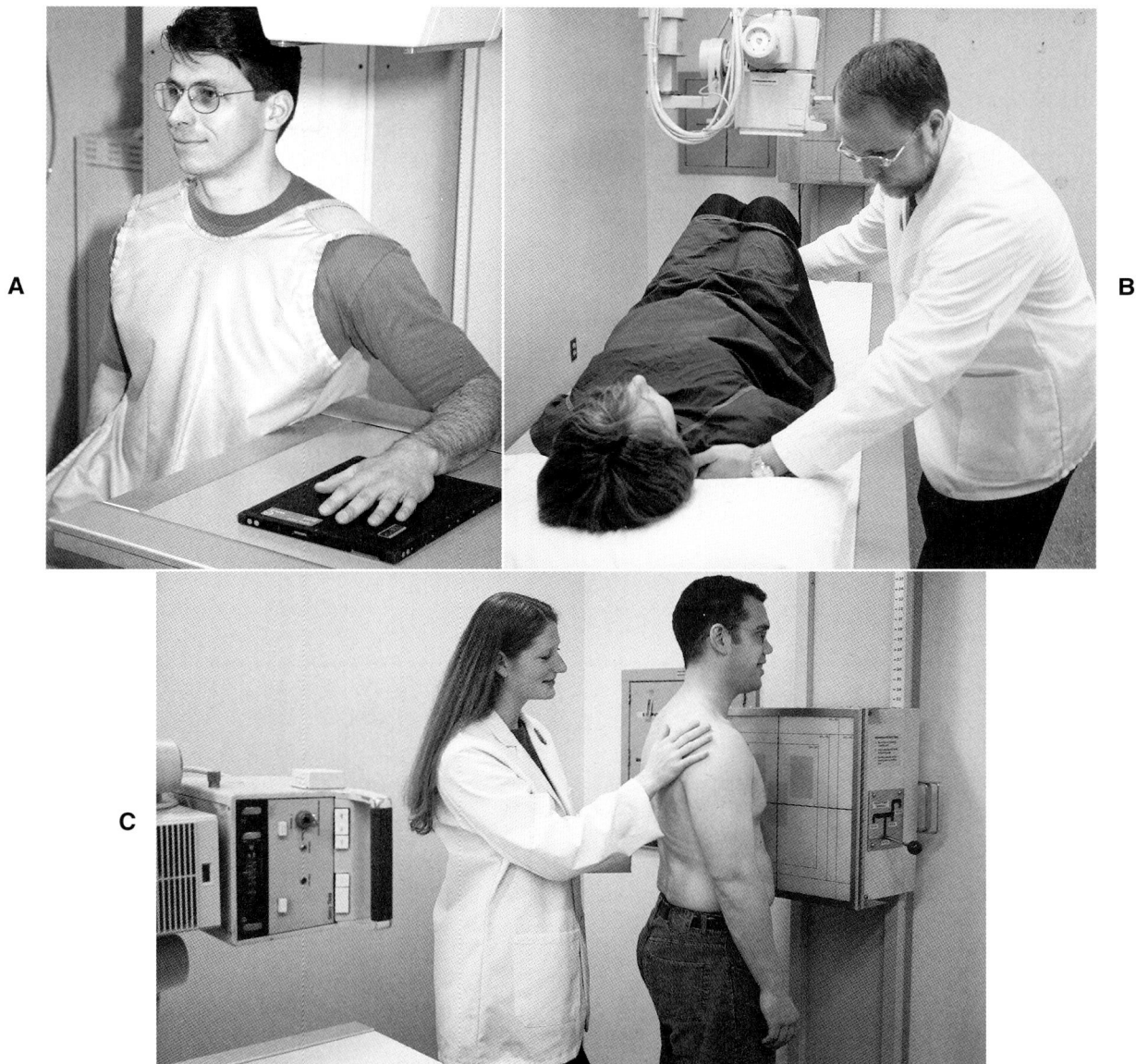

Fig. 1-7 A, Patient seated at table for hand radiograph. **B,** Radiographer assisting patient to lie on radiographic table. **C,** Radiographer assisting patient into position at upright cassette holder.

the patient can be seated at the end of the x-ray table. For a spine examination, the patient may need to lie on the table. If a chest examination is ordered, the patient will stand at an upright cassette holder. The radiographer then selects the correct film and places it in position. Next, the patient is positioned precisely and the x-ray tube is aligned to the body part and the film at a specific distance. The body part must be measured to determine the proper exposure factors from a technique chart. At this point, lead shields are positioned for radiation protection. The radiographer then goes to the control booth, consults the technique chart, and sets the x-ray control panel to the desired exposure. Final instructions are given to the patient, and the exposure is made. If more than one exposure is needed, the film is changed, the patient repositioned, and the steps repeated until the examination is complete.

After ensuring that the patient is safe and comfortable, the radiographer takes the film to the darkroom and processes it. Films may be processed in an automatic processing machine or hung on hangers and moved by hand through the processing solutions. Both types of processing are explained in Chapter 8. Film processing usually requires less than 10 minutes before the film can be evaluated. If the film is satisfactory and no further exposures are needed, the patient is returned to an examination room or dressing room. The radiographer then readies the x-ray room for the next examination and prepares the films for reading.

The exact nature of a radiographer's duties will vary with the place of employment, the size of the staff, and the equipment available.

SUMMARY

Limited radiography is a relatively new professional role in the field of health care. Radiographers work closely with physicians, and their duties involve direct patient contact. Most limited radiographers are employed in outpatient facilities such as clinics, but some are also employed by hospitals, and their work may vary considerably, depending on their place of employment.

Requirements and credentials for limited radiographers differ greatly from state to state, and limited radiographers are responsible for knowing and following the regulations that apply to them. The American Society of Practical Technologists in Radiography is working to establish national standards for the certification and practice of limited radiographers.

■ Review Questions ■

1. When, where, and by whom were x-rays discovered?
2. What is the purpose of the ARRT? Why might this organization be important to a limited radiographer?
3. List possible consequences of practicing radiography outside the limitations imposed by local regulations.
4. Describe the present status of the ASPTR. What are its primary goals?
5. Explain what is meant by "reciprocity."
6. List three activities that might take place in the "front office" of a clinic and four that typically occur in the "back office."
7. List five typical duties of a limited radiographer.

Introduction to Radiographic Equipment

Learning Objectives

At the conclusion of this chapter the student will be able to:

- Use correct terminology when discussing x-ray equipment and its parts
- Demonstrate the radiation field and define the central ray
- Explain the differences between primary radiation, scatter radiation, and remnant radiation
- List two effects of scatter radiation
- List the components of the image receptor system
- List the essential features of a typical x-ray room
- Explain the purposes of the control booth and the transformer cabinet
- Safely change the positions of the radiographic table and the x-ray tube
- Demonstrate a detent and explain its function
- Explain the purpose of a collimator
- Describe precautions to be taken to ensure personnel safety from radiation exposure

Key Terms

attenuation
bucky
cassette
central ray
collimator
computed radiography (CR)
control booth
control console
detent
image receptor system
latent image

radiation field
remnant radiation
scatter radiation
scatter radiation fog
tissue density
Trendelenburg position
tube housing
tube port
upright cassette holder
x-ray beam
x-ray tube

This chapter introduces the useful x-ray beam, discusses the equipment found in a typical x-ray room, and provides some fundamentals of radiation safety. Many of these topics are covered in greater detail later in the text, but it will be helpful for you at this point to have an orientation to the equipment and safety considerations that are central to your work as a radiographer.

PRIMARY X-RAY BEAM

The source of x-rays is the **x-ray tube.** The internal structure and function of the tube are discussed in Chapter 4. X-rays are formed within a very small area inside the tube. From this point, the x-rays diverge into space. The x-ray tube is surrounded by a lead-lined **tube housing.** Some of the x-rays are absorbed by the tube housing. X-rays that are traveling in a useful direction exit the housing through an opening called the **tube port.** These x-rays form the cone-shaped **x-ray beam** (Fig. 2-1). The cross section of the x-ray beam at the point where it is utilized is called the **radiation field.** An imaginary line in the center of the x-ray beam and perpendicular to the long axis of the x-ray tube is called the **central ray.**

During a radiographic exposure, x-rays from the tube are directed through the patient to the film (Fig. 2-2). As the x-rays pass through the patient, some of them are absorbed by the patient and others are not. Anatomic structures that have greater **tissue density** (mass), such as bone, will absorb more radiation than less dense tissues, such as fat. This results in a pattern of varying intensity in the x-ray beam that

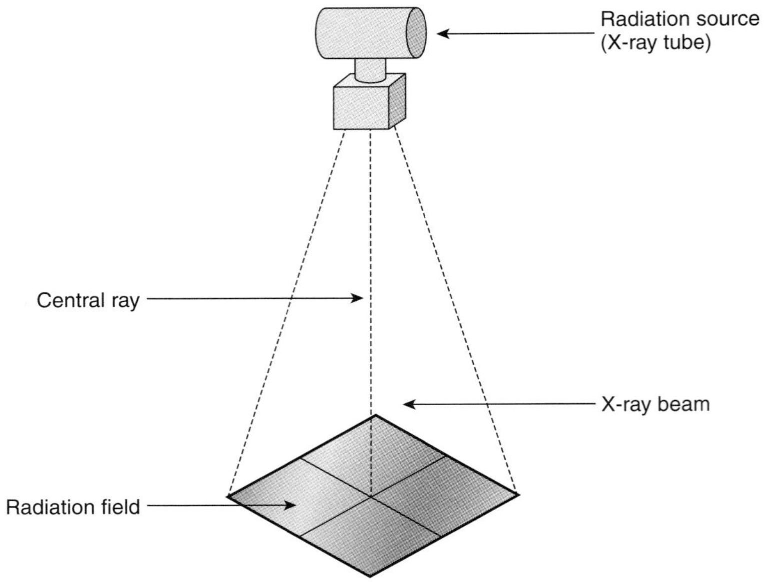

Fig. 2-1 Terms for basic concepts in radiography.

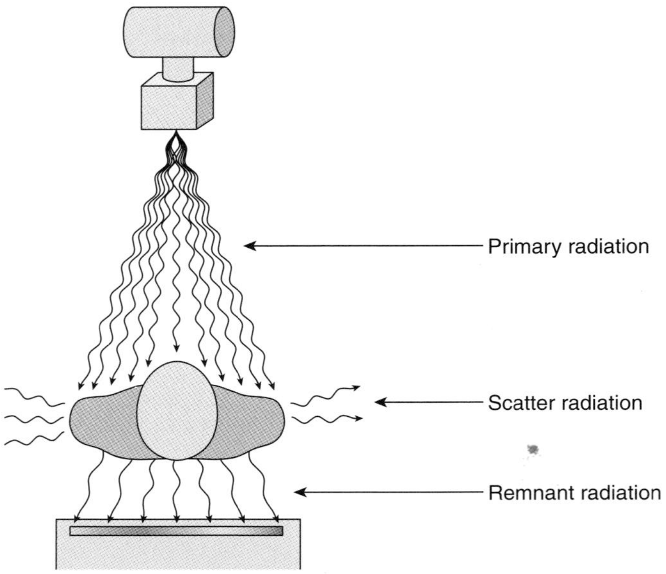

Radiation source (X-ray tube)

Central ray

X-ray beam

Radiation field

Fig. 2-2 X-ray beam.

Primary radiation

Scatter radiation

Remnant radiation

exits on the opposite side of the patient. This radiation, called **remnant radiation** or exit radiation, then passes through the film holder and exposes the film. The film now has a pattern of exposure called a **latent image.** The purpose of processing the film is to convert the latent image into a visible one.

SCATTER RADIATION

When the primary x-ray beam encounters matter, such as the patient, the x-ray table, or the film holder, a portion of its energy is absorbed within the matter. Absorption of the x-ray beam is called **attenuation.** Attenuation results in the production of **scatter radiation.** This radiation generally has less energy than the primary x-ray beam, but it is not as easily controlled. It travels out from the absorbing matter in all directions, causing unwanted exposure to the film and to anyone who is in the room. This is an important reason why the study of radiation safety is so essential to the radiographer. Radiation safety is discussed briefly at the end of this chapter and more extensively in Chapter 11. The unwanted film exposure caused by scatter radiation is called **scatter radiation fog.** The production of scatter radiation and control of the fog it produces are addressed in Chapter 9.

See Box 2-1 for a summary of primary, remnant, and scatter radiation.

IMAGE RECEPTOR SYSTEM

In most radiography departments, the **image receptor system** consists of the x-ray film and the film holder, which is called a **cassette** (Fig. 2-3). The cassette protects the film from damage and exposure to light during radiography. It performs other functions as well. Films and cassettes come in standard sizes. You will work more effectively when you have learned to recognize them at a glance. The most common sizes are listed in Box 2-2. Films and cassettes are discussed in detail in Chapter 7.

Some large x-ray departments now use **computed radiography (CR),** sometimes called a "filmless system." These imaging systems use special image receptors instead of film. Images are stored electronically in a computer and viewed on a computer monitor. The basic x-ray equipment and radiographic procedures are essentially the same for computed radiography systems as for those using film.

BOX 2-1
Synopsis of Primary, Remnant, and Scatter Radiation

Primary Radiation
- Definition: the x-ray beam that leaves the tube and is unattenuated, except by air.
- Its direction and location are predictable and controllable.

Remnant (Exit) Radiation
- Definition: what remains of the primary beam after it has been attenuated by matter.
- Since the pattern of densities in the matter results in differential absorption of the radiation, this pattern will be inherent in the remnant radiation.
- The pattern of the remnant radiation creates the film image.

Scatter Radiation
- Definition: radiation that is scattered or created as a result of the attenuation of the primary x-ray beam by matter.
- Scatter radiation travels in all directions from the scattering medium and is very difficult to control.
- Generally, it has less energy than the primary beam.

BOX 2-2
Common Sizes of X-ray Films and Cassettes

8 × 10 inches (20 × 25 cm)
9.5 × 9.5 inches (24 × 24 cm)
10 × 12 inches (24 × 30 cm)
11 × 14 inches (30 × 35 cm)
7 × 17 inches (18 × 43 cm)
14 × 17 inches (35 × 43 cm)
14 × 36 inches (36 × 91 cm)

Fig. 2-3 Cassettes.

X-RAY ROOM

The x-ray room (Fig. 2-4) includes the x-ray equipment itself, a counter area, and a protective **control booth.** The x-ray machine consists of the x-ray tube, the tube support, the **control console** (located in the control booth), and the transformer cabinet. There is usually also a radiographic table and a wall-mounted cassette holder for upright studies. In chiropractic offices, there may not be an x-ray table, since chiropractic radiography is often done upright to provide weight-bearing information (Fig. 2-5).

Radiography involves the positioning of the patient, the film, and the x-ray tube, as well as setting the control panel and making the exposure. Before you learn the specifics of radiographic positioning, it is important to understand how the equipment works so that you can position it safely and efficiently.

X-ray equipment may vary considerably, depending on age, manufacturer, and the complexity of procedures for which it was designed. The equipment descriptions provided here highlight the important

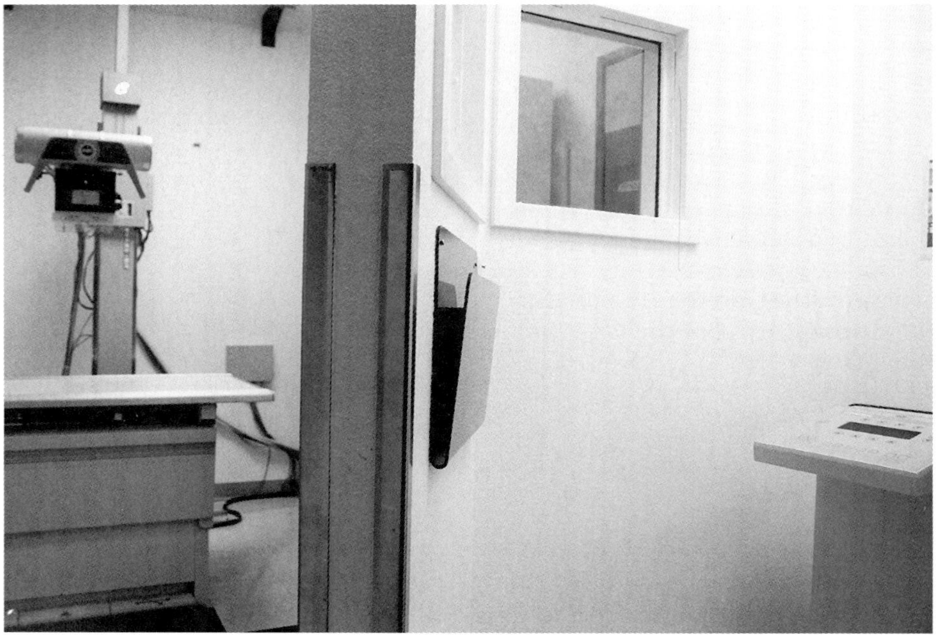

Fig. 2-4 X-ray room and control booth.

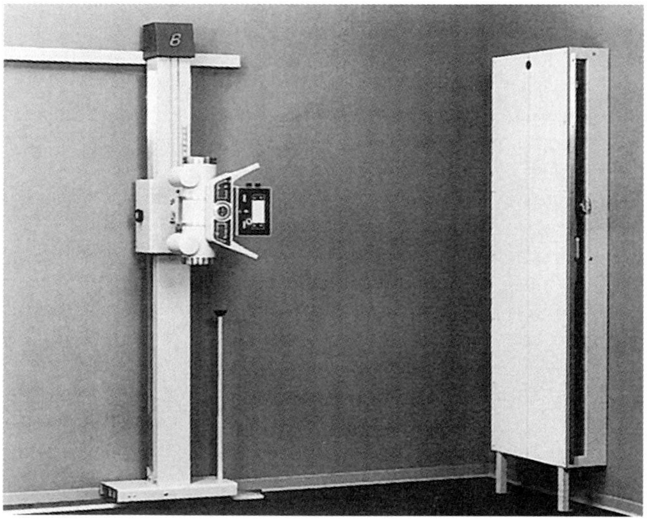

Fig. 2-5 Upright x-ray machine.

features of most general-purpose x-ray machines and some of the common variations. If you are currently associated with a clinical facility that has x-ray equipment, it will be helpful to consider the information that follows in relation to the equipment you will be using. When you have learned to use several different x-ray machines, it will be relatively easy to orient yourself to new equipment.

POSITIONING THE X-RAY TUBE

As stated before, the x-ray tube is encased in a barrel-shaped tube housing (Fig. 2-6). This housing is lead-lined for radiation control. It protects and insulates the tube and provides a mounting for attachments that the radiographer uses to position the x-ray tube and to control the size of the radiation field.

The tube housing may be attached to a ceiling mount or to a tube stand. Both types of mountings provide support and mobility for the tube. These tube supports allow the radiographer to move the tube, aligning it to the patient and the film, and adjusting its height. A ceiling-mounted tube support (Fig. 2-7), sometimes called a "ceiling crane" or "tube hanger," suspends the x-ray tube from a system of tracks, allowing the radiographer to position the tube at locations throughout the room. A tube stand (Fig. 2-8) is a vertical support with a horizontal arm that suspends the tube over the radiographic table. The tube stand rolls along a track that is secured to the floor (and sometimes also the ceiling or wall), parallel to

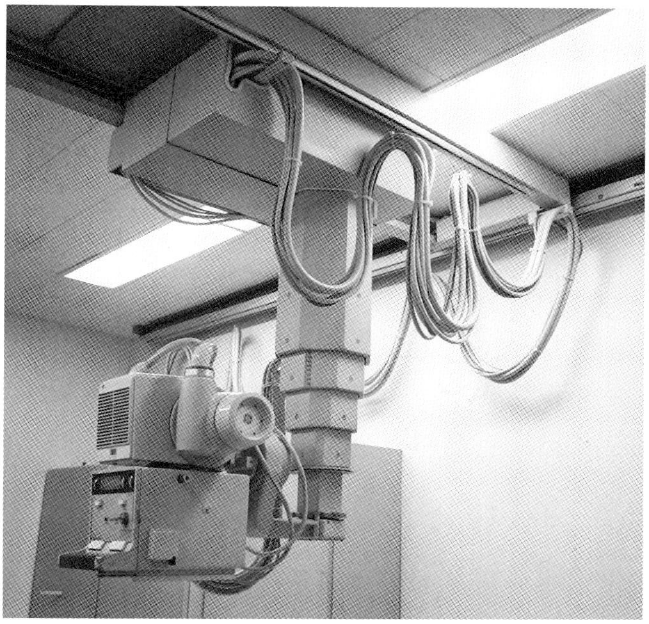

Fig. 2-7 Ceiling-mounted tube support.

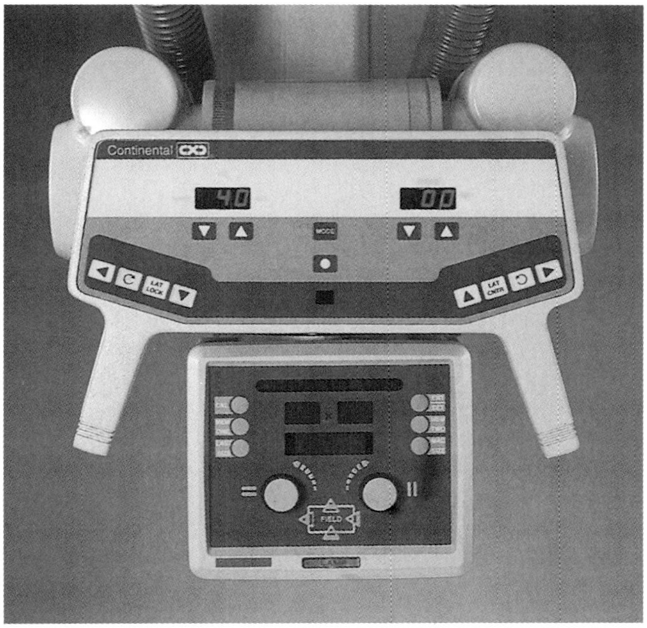

Fig. 2-6 X-ray tube housing and attachments.

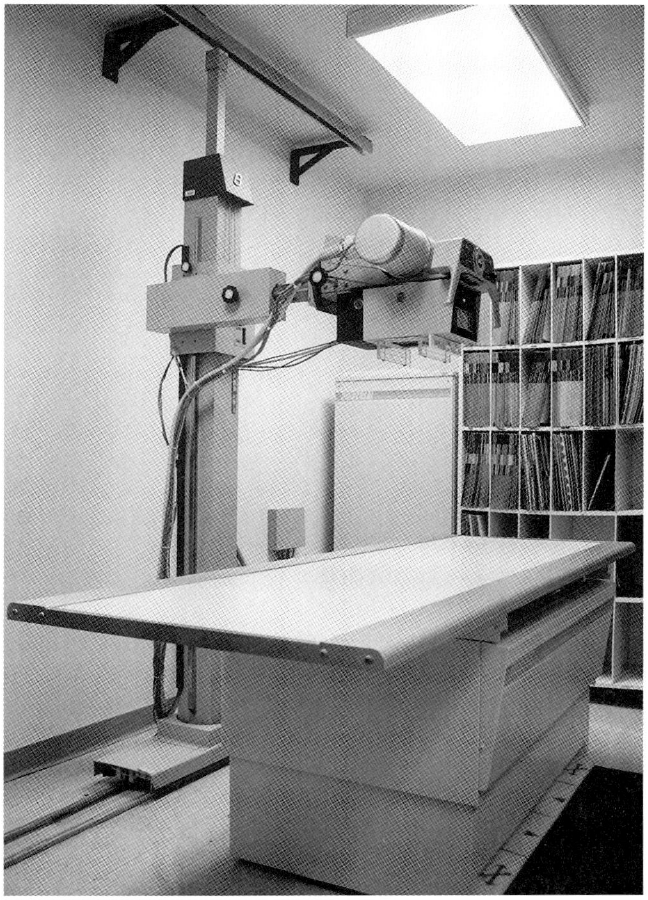

Fig. 2-8 Tube stand.

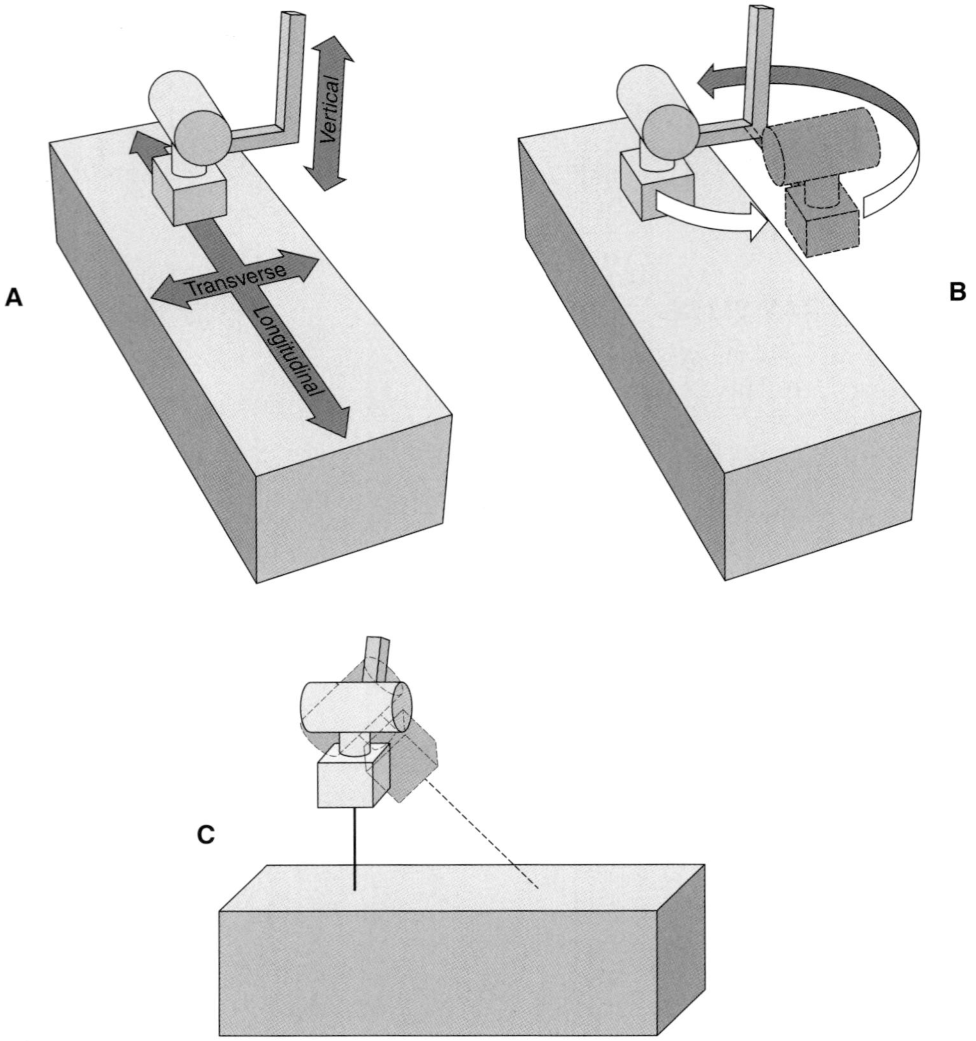

Fig. 2-9 Typical tube motions. **A,** Longitudinal, transverse, and vertical; **B,** rotation; **C,** angulations.

the x-ray table. This enables the tube to move longitudinally along the length of the table.

Typical tube motions (Fig. 2-9) include:

- Longitudinal—along the long axis of the table
- Transverse—across the table, at right angles to longitudinal
- Vertical—up and down, increasing or decreasing the distance between the tube and the table
- Angle (tilt, roll)—permits angulation of the tube along the longitudinal axis of the table and allows the tube to be aimed at the wall, rather than at the table
- Rotation—allows the entire tube stand to turn on its axis, changing the angle at which the tube arm is extended

A system of electric and/or mechanical locks holds the tube in position. The control system for all, or most, of these locks is usually an attachment on the front of the tube housing (see Fig. 2-6). To move the tube in any direction, a locking device must be released. It is sometimes possible to force tube movement without first releasing the lock, but this practice will damage the lock, making it impossible to secure the tube in position. *Do not attempt to move the tube without first releasing the appropriate lock.*

The tube stand and/or the movable portion of the upright cassette holder may move unexpectedly when the electrical supply to electric locks is turned off. To avoid damaging the equipment, it is important to be certain that these units are safely positioned before turning off the power to the locks. For example, it is usual to position the x-ray tube immediately above a pillow on the tabletop before shutting down the power supply to the tube support locks.

A **detent** is a special mechanism that tends to stop a moving part in a specific location. Detents are built into tube supports to provide ease in attaining placement at standard locations. For example, a vertical detent may indicate when the distance from tube to

Fig. 2-10 Collimator.

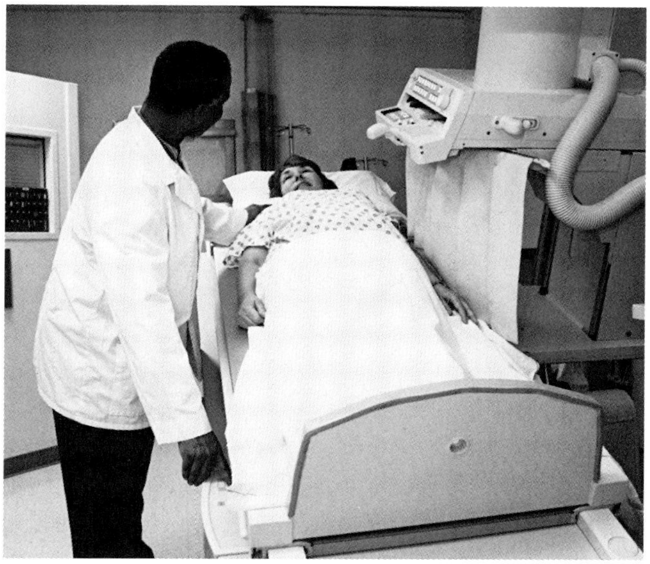

Fig. 2-11 Tilting table.

film is 40 inches, a common standard distance. Other detents provide "stops" when the transverse tube position is centered to the table and when the tube tilt position is such that the central ray is perpendicular to the table or to the wall.

COLLIMATOR

The **collimator** is a box-like device attached under the tube housing (see Fig. 2-6). It allows the radiographer to vary the size of the radiation field. The collimator includes a light that indicates the beam size and location and the center of the field. There is usually also a centering light that aids in aligning the cassette tray (Fig. 2-10). Controls on the front of the collimator allow the radiographer to adjust the size of each dimension of the radiation field. These dimensions are indicated on a scale on the front of the collimator. A timer controls the collimator light, turning it off after a certain length of time, usually 30 seconds. This helps to avoid accidental overheating of the unit by prolonged use of its high-intensity light.

RADIOGRAPHIC TABLE

The radiographic table is a specialized unit that is more than just a support for the patient. While the table is usually secured to the floor, it may be capable of several types of motion: vertical, tilt, and "floating tabletop."

For vertical table motion, a hydraulic motor, activated by a hand, foot, or knee switch, raises or lowers the height of the table. The table may be lowered, so that the patient can sit on it easily, and then elevated to a comfortable working height for the radiographer.

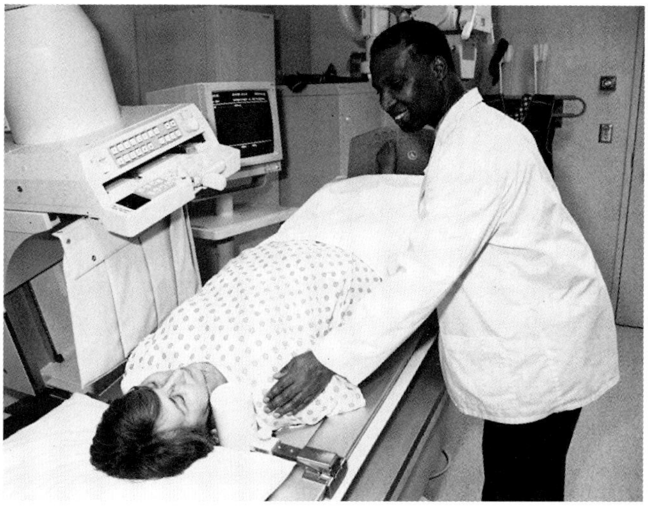

Fig. 2-12 Trendelenburg position.

There will be a detent that stops the table in the standard position for routine radiography. This standard table height corresponds to indicated distances from the x-ray tube. It is important that standard tube/film distances be used, so it is necessary to return the table to the detent position after lowering it for patient access. Not all tables are capable of vertical motion.

A tilting table (Fig. 2-11) also uses a hydraulic motor to change position. In this case, the table turns on a central axis to attain a vertical position. This allows the patient to be placed in a horizontal position, in a vertical position, or at any angle in between. The table may also tilt in the opposite direction, allowing the head end to be lowered at least 15° into the **Trendelenburg position** (Fig. 2-12). A detent stops the table in the horizontal (level) position.

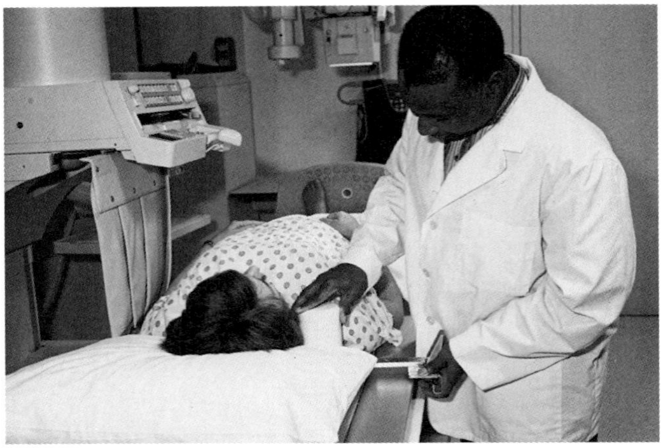

Fig. 2-13 Attach shoulder guards and footboards securely for patient safety.

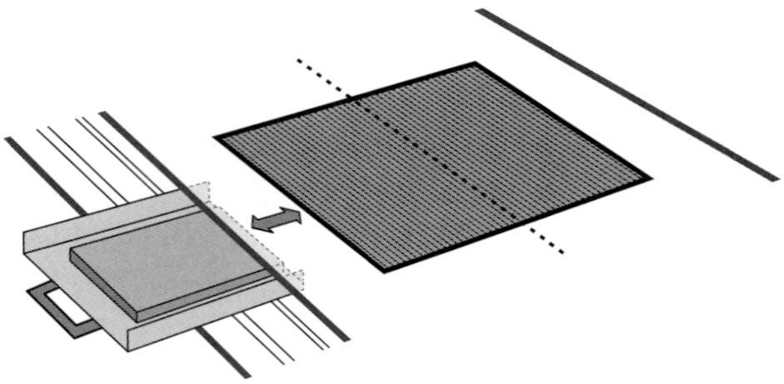

Fig. 2-14 The bucky with its film tray is mounted under the tabletop for scatter radiation control.

> ███ BOX 2-3
> ███
>
> *Safety Precautions When Moving X-ray Equipment*
>
> ---
>
> - Be sure that footboard and shoulder guard are secure before tilting a table with a patient.
> - Check that no equipment is under the table before tilting it.
> - Release locks before attempting to move the x-ray tube.
> - Move the x-ray tube out of the way before assisting a patient to or from the table, to avoid injuring the patient.
> - Be sure that equipment is in a safe position before shutting off power to the locks.

Special attachments for the tilting table include footboards and shoulder guards for patient safety when the table is tilted (Fig. 2-13). You should pay particular attention to the locking mechanisms on these attachments so that you will be able to apply them correctly when needed. *Always test the footboard and shoulder guards to be certain they are securely attached before tilting the table with a patient.*

The motor that tilts the table is quite powerful and can overcome the resistance of obstacles placed in the way. Many step stools and other pieces of movable equipment have been crushed because they were under the end of the table and out of view when the table motor was activated. Such a collision can also damage the table. *Be certain that the spaces under the table are clear before tilting the table.* Box 2-3 lists important safety precautions for moving x-ray equipment.

A "floating tabletop" allows the top of the table to move independently of the remainder of the table,

making it easy to align the patient to the x-ray beam. This is a common feature of modern tables. This motion may involve a mechanical release, allowing the radiographer to shift the position of the tabletop manually, or the movement may be power-driven, activated by a small control pad with directional switches.

Under the table surface is a grid device to protect the film from being fogged by scatter radiation. Most of these devices include a small motor that moves the grid during the exposure. When this is the case, the grid device is called a **bucky** (Fig. 2-14). Some tables have a stationary grid that does not move during the exposure. These are sometimes also referred to as buckys but are more accurately called grid cabinets. The bucky or grid cabinet incorporates a tray to hold the cassette. The entire unit can be moved along the length of the table and locked into position where desired. When the table has a floating tabletop, the grid device and film tray do not move with the tabletop. Grids are generally used only for radiography of body parts that measure 10 to 12 cm or more in thickness. This is about the size of the average adult neck or knee. When the grid is not needed, the cassette is placed on the tabletop or in an upright cassette holder that does not incorporate a grid. Grids and buckys are discussed in Chapter 9.

UPRIGHT CASSETTE HOLDER

The **upright cassette holder,** as its name implies, is a device to hold the film in the upright position for radiography (Fig. 2-15). It is usually placed against a wall and is adjustable in height. It may incorporate a bucky or grid. When a grid is included, it may be referred to as a grid cabinet. If the grid is a moving

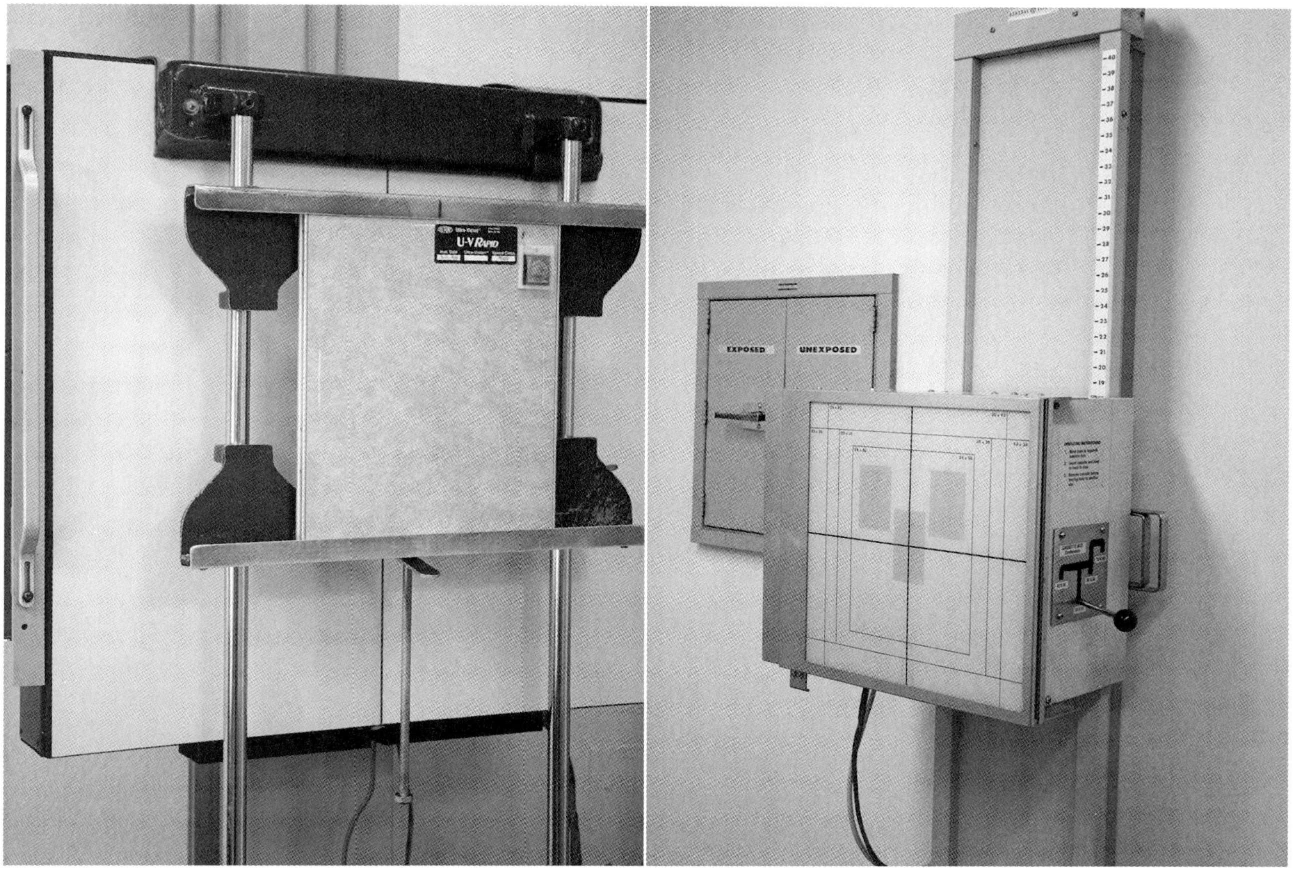

Fig. 2-15 Upright cassette holders. **A,** Nongrid cassette holder with cassette in place. **B,** Upright bucky.

grid, the device is called an upright bucky. Even when the table tilts for upright radiography, it is usual to have a separate upright cassette holder for some examinations. When the patient is to be sitting or standing at the upright cassette holder for radiography, the tube is angled to face the wall and cassette holder. The distance may be adjusted to 40 inches or to 72 inches, depending on the requirements of the procedure.

CONTROL CONSOLE

The control console is located in the control booth. This area is separated from the x-ray room by a lead barrier to protect the radiographer from scatter radiation during exposures. There is usually a lead glass window so that the radiographer can observe the patient from the control booth. The control console, or "control panel," is the access point for the radiographer to set the exposure factors and to initiate the exposure (Fig. 2-16). A typical radiographic control console will have buttons or switches for controlling the exposure and dials or digital readouts that indi-

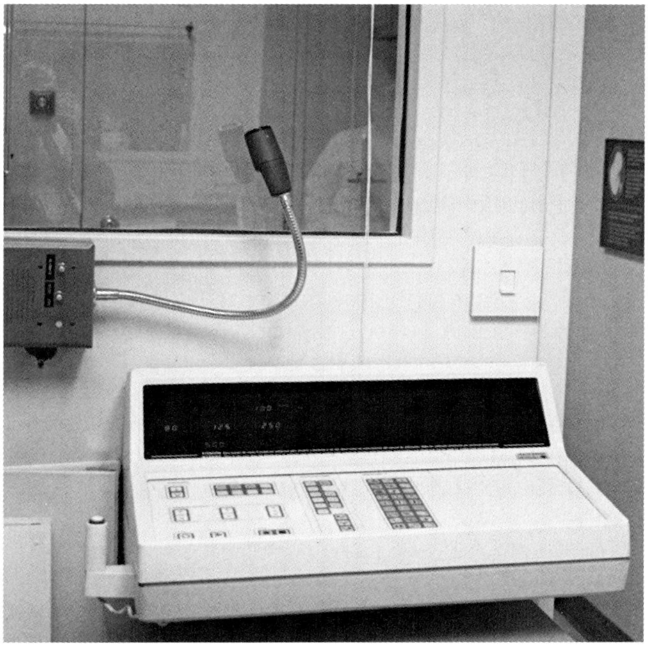

Fig. 2-16 Control console.

cate the settings. Details of the control panel are discussed in Chapter 5.

TRANSFORMER CABINET

The transformer is an essential part of the x-ray machine. Its function is to produce the high voltage required for x-ray production, as discussed in Chapter 4. The radiographer's work does not involve contact with the transformer. It is simply a large box, usually found in the x-ray room, that is connected by cables to both the control panel and the x-ray tube.

FUNDAMENTAL RADIATION SAFETY

Radiation exposure may pose a health hazard to radiographers if proper safety precautions are not observed. This subject is treated in greater depth in Chapter 11. The potential hazard is greater for the radiographer than for the patient because the radiographer is in frequent daily contact with the possibility of exposure. At this point, it is important for you to feel confident that you are not endangering yourself or others as you become acquainted with the radiography department.

As stated earlier in this chapter, scatter radiation is present throughout the x-ray room during an exposure. It is important to be aware that x-rays travel at the speed of light. They do not linger in the room after the exposure, and they are not capable of making the objects in the room radioactive. The only time that a radiation hazard exists is during the x-ray exposure itself.

The sources of radiation are the x-ray tube and any matter that is in the path of the primary x-ray beam. The principal source of scatter radiation is the patient. When a safety barrier, such as a lead wall, is placed between the sources of radiation and the radiographer, the radiographer is safe from exposure. X-rays travel in straight lines and do not turn corners. Scatter radiation is not powerful enough to generate additional radiation of concern when it interacts with matter, so it is not necessary for the control booth to be sealed.

Radiographers should always be completely behind the lead barrier of the control booth during exposures. They should *never* hold patients or films during exposures. Before making an exposure, the radiographer performs a safety check (Box 2-4) to ensure that only the required persons are in the x-ray room (usually this means only the patient), that everyone in the control booth is safely behind the lead barrier, and that the x-ray room door is

---

BOX 2-4
Preexposure Safety Check

Before making an exposure, be *certain* that:
- The x-ray room door is closed.
- No nonessential persons are in the x-ray room.
- All persons in the control booth are completely behind the lead barrier.
- No cassettes are in the room except the one in use.

---

closed. It is also wise to make sure that no cassettes have been left lying about. Only the cassette that is in immediate use should be in the x-ray room, since scatter radiation fog will damage the film.

SUMMARY

The primary x-ray beam originates at a tiny point within the x-ray tube. It exits in one general direction through the tube port and diverges into space. Objects in its path attenuate the beam, forming scatter radiation. This scatter radiation is present throughout the x-ray room during the exposure, creating a potential radiation hazard that requires proper precautions for safety.

The x-ray table, tube support, and cassette holder are capable of many possible motions, allowing alignment of the tube and film for radiography of all body parts in many positions. Special care is needed to move equipment correctly so that patients will not be injured and the equipment will not be damaged.

▪ Review Questions ▪

1. How can you determine the location of the central ray?
2. What is the location of remnant radiation?
3. What is meant by "attenuation"?
4. What component of the x-ray machine is located in the control booth?
5. What should you do before attempting to move the x-ray tube?
6. Where would you look to find a collimator?
7. How might you determine the size of the radiation field without actually measuring it?
8. List the steps in a pre-exposure safety check.
9. How soon is it safe to re-enter the x-ray room after an exposure?

PART 11

X-ray Science

3

Basic Physics for Radiography

Learning Objectives

At the conclusion of this chapter the student will be able to:

- Define matter and list its three forms
- Name the fundamental particles of the atom and list characteristics of each
- Draw or describe a conceptual model of atomic structure
- Given the chemical symbol of an element and a periodic table of elements, state the number of protons, neutrons, and electrons in the most abundant neutral atom of the element
- List and describe five forms of energy
- Draw a sine wave and measure its amplitude and its wavelength
- Relate the wavelength of a sine wave to its velocity and frequency
- Compare and contrast the characteristics of x-rays with the characteristics of visible light
- Correctly draw a diagram of a simple electric circuit containing a battery, an ammeter, a voltmeter, and a resistor, using conventional circuit diagram symbols
- Explain the relationship between electromotive force, current, and resistance in an electric circuit and state the units used to measure each
- Draw or describe the difference in voltage waveform between a direct current circuit and an alternating current circuit
- State the frequency of alternating current in the United States and Canada using the correct units
- Describe the process of electromagnetic induction
- Draw simple diagrams of a step-up transformer and of a step-down transformer

Key Terms

ampere (A)	kilovolt (kV)
atom	mass
chemical compound	matter
circuit	milliampere (mA)
conductor	neutron
current	nucleus
electromotive force (emf)	photon
electron	proton
element	resistance (R)
frequency	sine wave
intensifying screen	transformer
ionization	wavelength

Radiographers do not require an extensive background in physics, but some basic principles of physical science are essential to an understanding of x-rays and their use. This chapter covers the basic concepts of matter, energy, and electricity and relates these principles to radiography. It also discusses the nature of radiation.

If your educational background includes course work in physics or chemistry, this chapter will provide a comprehensive review of the pertinent material. If you are unfamiliar with these subjects, it will be important for you to master them so that you can relate well to the material that follows.

Everything of a physical nature in the universe can be classified as either **matter** or energy. Both matter and energy can exist in several forms.

MATTER

Matter is defined as anything that occupies space and has shape or form. The three basic forms of matter are solids, liquids, and gases. The quantity of matter that makes up any physical object is called its **mass.** Although the scientific definitions differ somewhat, mass is essentially the same thing we think of as "weight." An object may change in form, but its mass is unchangeable. For example, a 20-pound bucket of water may freeze into a 20-pound bucket of ice, or it may evaporate, resulting in 20 pounds of water vapor. The form changes, but the mass remains the same.

Atoms

All matter is composed of "building blocks" called **atoms.** Scientists have determined that atoms may be made up of nearly 100 different subatomic particles, but only three basic particles concern us here. The fundamental particles that compose atoms are **neutrons, protons,** and **electrons.** All neutrons are identical, as are all protons and all electrons. It is the number and arrangement of these particles in the atom that account for the differences in matter.

The neutrons and protons together form the **nucleus** of the atom, its center. The electrons circle the nucleus in orbits called shells. A useful model for visualizing atomic structure is that of the solar system, with the nucleus as the sun and the electrons as planets in orbit around the sun (Fig. 3-1). This model was first described by Neils Bohr in 1913 and is referred to as "Bohr's Atom."

The mass of each proton and neutron is *approximately* 1 atomic mass unit (amu), and the mass of an electron is $\frac{1}{1836}$ of an amu. That is, it would take 1836 electrons to equal the weight of one proton or one neutron. Atoms are tiny beyond our normal percep-

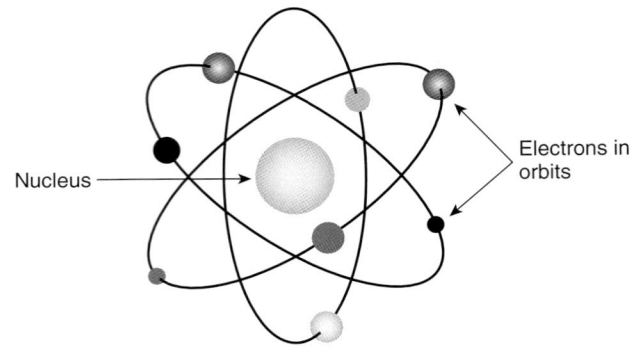

Fig. 3-1 Bohr's concept of the atom.

Laws of Conservation

> Matter can be neither created nor destroyed, but it can change form.
>   Energy can be neither created nor destroyed, but it can change form.

tions, so their actual size is difficult to comprehend. A row of 10 million atoms would be less than half an inch long! The mass of the atom is concentrated in the nucleus, and its form is largely made up of empty space. If the nucleus of an atom were expanded to the size of a football and placed on the 50-yard line of a huge stadium, the nearest electron would be outside the back row of the bleachers.

Atomic particles differ from one another with respect to electrical charge. Neutrons are electrically neutral (0); that is, they have no electrical charge. Protons have a positive charge (+). Electrons have a negative charge (−) that is equal to, but opposite, the charge of a proton. A particle's charge is important because it results in a magnetic effect. Opposite charges attract one another, seeking a neutral state. Like charges repel one another. Neutral particles neither attract nor repel and are not attracted or repelled by charged particles. Table 3-1 contains a summary of the characteristics of the fundamental atomic particles.

In its "normal" or neutral state, an atom has an equal number of protons and electrons, so the electrical charges are balanced and the atom as a whole has no charge. The electrons are arranged in their orbits with a specific number of electrons allotted to each shell. The shell nearest the nucleus may contain one or two electrons. Each additional shell is greater in size and can accommodate a larger number of electrons than the previous shell. The shells are lettered alphabetically, beginning with the letter K nearest the nucleus. Table 3-2 lists atomic shells with their letter symbols and the maximum number of electrons in each.

TABLE 3-1

Fundamental Atomic Particles

Particle	Location	Mass Number	Elemental Mass (amu)	Charge
Proton	Nucleus	1	1.00728	+1
Neutron	Nucleus	1	1.00867	0
Electron	Orbital shells	0	0.000549	−1

TABLE 3-2

Electron Shells

Shell Number	Shell Symbol	Maximum Number of Electrons
1	K	2
2	L	8
3	M	18
4	N	32
5	O	50
6	P	72
7	Q	98

NOTE: The number of electrons in each shell is equal to the shell number squared then multiplied by 2. For example, the L shell is #2: $2^2 = 4 \times 2 = 8$, the maximum number of electrons in the L shell.

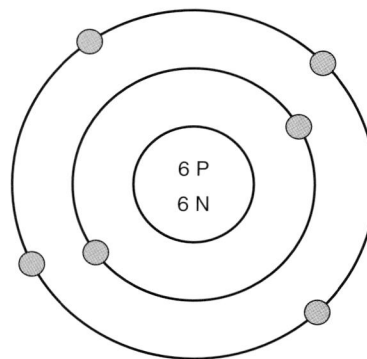

Fig. 3-2 Carbon atom.

The orbital shells are sometimes called energy levels because different quantities of binding energy are required to hold electrons in each shell. Electrons near the nucleus are attached with greater binding energy than those in outer shells. The binding energy of each shell varies with different atoms; larger atoms have greater binding energy than small ones.

Regardless of the maximum number of electrons permitted in a given shell, the maximum number of electrons in an *outer* shell is limited to 8. Thus the outer shell may contain any number of electrons from 1 to 8. The number of electrons in the outer shell determines the atom's physical properties and chemical behavior.

Elements

The essential characteristic of an atom that determines its type is the number of protons in the nucleus. An **element** is a substance made up of only one type of atom; that is, all atoms of an element have the same atomic number. Scientists have identified 108 different elements. Many of these are rare, and some of them are man-made. Each element has a name and a chemical symbol, consisting of one or two letters. Each also has an atomic number that represents the number of protons in the nucleus. The mass number of the element is the combined total of the protons and neutrons in the nucleus. For example, the element carbon (Fig. 3-2) is represented by the symbol C. Its atomic number is 6 and its mass number is 12, indicating that the nucleus contains 6 protons and 6 neutrons. The number of neutrons is determined by subtracting the atomic number from the mass number.

The periodic table of elements (Fig. 3-3) displays the elements in order of their atomic number and arranges them so that each column contains atoms with the same number of electrons in their outer shells. For this reason, each column in the table represents a group of elements with similar characteristics. For example, all of the elements in Group I, with the exception of hydrogen, are soft metals that readily combine with oxygen and cause a violent reaction when placed in water. Group VIII is called the "noble gases" because all of these elements occur normally in a gaseous state and are resistant to chemical reactions with other elements.

The number above each chemical symbol in the periodic table is the element's atomic number, and the number below each represents its elemental mass. The elemental mass is similar to the mass number but is very precise and is measured in atomic mass units. There is a slight difference between the elemental mass of a proton and that of a neutron, and this is not accounted for in the mass number. In

Periodic table of elements

Alkali metals — Group I
Alkaline earth metals — Group II
Halogens — Group VII
Noble gases — Group VIII
Transitional Elements

Group →	I	II	III	IV	V	VI	VII	VIII
Period 1	H 1, Hydrogen 1.09							He 2, Helium 4.00
Period 2	Li 3, Lithium 6.94	Be 4, Beryllium 9.02	B 5, Boron 10.82	C 6, Carbon 12.01	N 7, Nitrogen 14.09	O 8, Oxygen 16.00	F 9, Fluorine 19.00	Ne 10, Neon 20.18
Period 3	Na 11, Sodium 23.0	Mg 12, Magnesium 24.32	Al 13, Aluminum 26.97	Si 14, Silicon 28.06	P 15, Phosphorus 30.98	S 16, Sulfur 32.06	Cl 17, Chlorine 35.46	Ar 18, Argon 39.99
Period 4	K 19, Potassium 39.096	Ca 20, Calcium 40.08	Ga 31, Gallium 69.72	Ge 32, Germanium 72.60	As 33, Arsenic 74.91	Se 34, Selenium 79.00	Br 35, Bromine 79.92	Kr 36, Krypton 83.7
Period 5	Rb 37, Rubidium 85.48	Sr 38, Strontium 87.63	In 49, Indium 118.70	Sn 50, Tin 121.77	Sb 51, Antimony 127.6	Te 52, Tellurium 126.93	I 53, Iodine 126.92	Xe 54, Xenon 131.3
Period 6	Cs 55, Cesium 132.9	Ba 56, Barium 137.4	Tl 81, Thallium 204.4	Pb 82, Lead 207.2	Bi 83, Bismuth 209.0	Po 84, Polonium 210	At 85, Astatine 211	Rn 86, Radon 222
Period 7	Fr 87, Francium 223	Ra 88, Radium 226						

Transitional Elements

Sc 21, Scandium 45.10	Ti 22, Titanium 47.90	V 23, Vanadium 50.95	Cr 24, Chromium 52.01	Mn 25, Manganese 54.93	Fe 26, Iron 55.85	Co 27, Cobalt 58.94	Ni 28, Nickel 58.69	Cu 29, Copper 63.57	Zn 30, Zinc 65.38
Y 39, Yttrium 88.92	Zr 40, Zirconium 91.22	Nb 41, Niobium 92.91	Mo 42, Molybdenum 96.0	Tc 43, Technetium 99	Ru 44, Ruthenium 101.7	Rh 45, Rhodium 102.9	Pd 46, Palladium 106.7	Ag 47, Silver 107.88	Cd 48, Cadmium 112.41
Rare Earths 57-71	Hf 72, Hafnium 178.6	Ta 73, Tantalum 180.9	W 74, Tungsten 183.9	Re 75, Rhenium 186.3	Os 76, Osmium 190.2	Ir 77, Iridium 193.1	Pt 78, Platinum 195.2	Au 79, Gold 197.2	Hg 80, Mercury 200.6
Actinide Series 89-103	Rf 104, Rutherfordium 261	Db 105, Dubnium 262	Sg 106, Seaborgium 266	Bh 107, Bohrium 264	Hs 108, Hassium 265	Mt 109, Meitnerium 266			

Rare Earths

La 57, Lanthanum 138.91	Ce 58, Cerium 140.12	Pr 59, Praseodymium 140.91	Nd 60, Neodymium 144.24	Pm 61, Promethium 147	Sm 62, Samarium 150.35	Eu 63, Europium 151.96	Gd 64, Gadolinium 157.25	Tb 65, Terbium 158.92	Dy 66, Dysprosium 162.50	Ho 67, Holmium 164.93	Er 68, Erbium 167.26	Tm 69, Thulium 168.93	Yb 70, Ytterbium 173.04	Lu 71, Lutetium 174.97

Actinide Series

Ac 89, Actinium 227	Th 90, Thorium 232.04	Pa 91, Protactinium 231	U 92, Uranium 238.03	Np 93, Neptunium 237	Pu 94, Plutonium 242	Am 95, Americium 243	Cm 96, Curium 245	Bk 97, Berkelium 249	Cf 98, Californium 251	Es 99, Einsteinium 254	Fm 100, Fermium 255	Md 101, Mendelevium 256	No 102, Nobelium 254	Lr 103, Lawrencium 257

Fig. 3-3 Periodic table of elements.

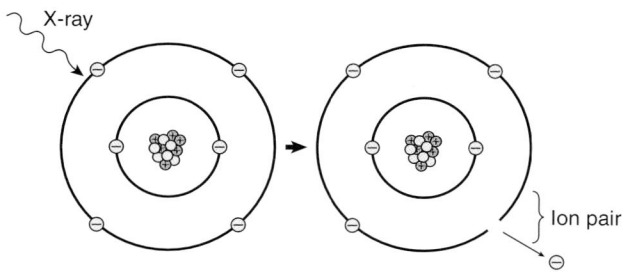

Fig. 3-4 Ionization.

Fig. 3-5 Radioactive decay.

addition, some elements occur in more than one form, having atoms with varying numbers of neutrons.

Atoms with more or fewer neutrons than the most abundant form of the element are called isotopes. Carbon, for example, occurs primarily as ^{12}C*; that is, most of its atoms have a mass number of 12. Some carbon atoms, however, have 2 additional neutrons and occur in the form of ^{14}C, which is an isotope of carbon. The elemental mass of carbon is 12.0111 amu, indicating the presence of a small number of carbon atoms with a mass number greater than 12 in any given sample.

Two or more atoms may combine chemically to form molecules. This combination occurs with the sharing of one or more outer shell electrons between atoms. A substance that consists of only one type of molecule is called a **chemical compound.** Water is an example of a chemical compound. Its chemical symbol is H_2O, indicating that it is made up of 2 atoms of hydrogen and 1 atom of oxygen. Substances that contain more than one type of molecule are called mixtures.

Ionization

An ion is defined as a charged particle. When a neutral atom gains or loses an electron, the electrical charges of its protons and electrons are no longer equal. This process, which is called **ionization,** produces an atom with an electrical charge. If an electron is added to a neutral atom, electrons will outnumber the protons and the atom will have a negative charge. If an electron is removed, there will be more protons than electrons, so the atom will have a positive charge. Because the outer orbital electrons are not tightly bound to the nucleus, the application of a small amount of energy can remove an outer orbital electron from the atom, resulting in two ions: a negative electron and the remainder of the atom, which now has a positive charge (Fig. 3-4). Ionization of a mole-

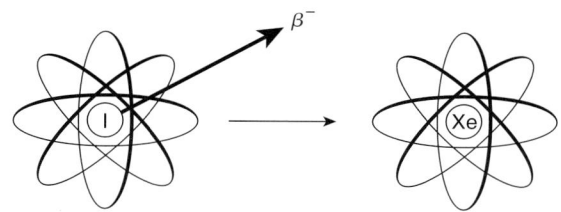

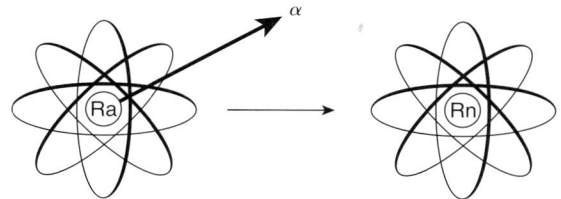

cule may disrupt the electron bond between atoms, also producing charged particles.

A familiar example of ionization is the "bad hair day" that occurs when the weather is cold and dry. The friction of a hairbrush removes electrons from atoms in the hair. In very dry air, the electrons cannot readily return to their orbits, and each hair is left with a positive charge. Since like charges repel each other, the hairs are repelled from one another and will not lie smoothly together.

X-rays cause ionization, which explains many of the effects of radiation that are discussed later in the text.

Radioactive Atoms Radionuclides are atoms with unstable nuclei that exist in an "excited" state. To reach stability, the nucleus emits particles and energy, transforming itself into a different atom. This process is called radioactive disintegration or radioactive decay (Fig. 3-5). The elements radium and uranium are examples of radionuclides. While many factors may influence nuclear stability, the most important factor is the number of neutrons. When an atom contains too few or too many neutrons, it may undergo radioactive disintegration until it reaches a more stable proportion of protons and neutrons.

Isotopes may be stable or unstable. Unstable radioactive isotopes are called radioisotopes. ^{14}C is an example of a naturally occurring radioisotope. Many radioisotopes can be created in the laboratory, and many such isotopes are used in health care for both diagnosis and treatment.

ENERGY

Energy is defined as the ability to do work. It occurs in several forms and can be changed from one form to another. Some familiar forms of energy include heat,

---

*Do not confuse this abbreviation with the format of the periodic table of elements. In atomic nomenclature, used here, the atomic number and the mass number precede the chemical symbol. The superscript (upper) number is the mass number and the subscript (lower) number is the atomic number. For carbon, the complete notation would be $^{12}_{6}C$. When discussing isotopes, as in this case, the atomic number is often omitted.

light, and electricity. Scientists have categorized energy in various ways. One method classifies energy into the following types: mechanical, chemical, thermal, nuclear, electrical, magnetic, and electromagnetic.

Mechanical energy can be further classified as either kinetic energy or potential energy. Kinetic energy is energy of motion, the ability of a moving object to do work. For example, a bowling ball in motion has energy to knock down the pins. Potential energy can be thought of as "stored" energy. When a bowling ball has been lifted, the work required to raise it is "stored" in the ball because of its position. When the ball is released, its potential energy is also released and is converted into kinetic energy.

Chemical energy is released through chemical changes in atoms or molecules. An example of chemical energy is fire. A gasoline engine converts the chemical energy of gasoline into mechanical energy. Chemical energy from the food we eat produces the energy needed for muscle movement and many other vital processes.

Thermal energy is commonly called heat. It is the result of atomic motion. As temperature rises, electrons move faster in their orbits and the orbits expand, causing the electrons to move farther from the nucleus. This phenomenon explains why matter expands in size when heated and contracts when cooled.

Nuclear energy is the energy released by radionuclides. This is the energy used to produce electricity in a nuclear power plant or the explosion of a nuclear bomb.

Electrical energy, or electricity, is the ability of electrical charges to do work. Although this process may seem mysterious, it is familiar to all of us. We use it to light our homes, run our computers, and make toast. Electrical energy also may exist in the form of potential energy. Potential electrical energy exists in a battery or at an unused wall socket. When we turn on a flashlight or plug in an appliance, this potential energy is converted into electricity.

Magnetic energy, or "magnetism," is defined as the ability of substances to attract iron. It is a property of certain materials that enables them to do work because of the arrangement of their atoms. The molecular arrangement of most materials is random, and these materials are nonmagnetic. But certain substances, particularly ferrous materials (those containing iron), consist mostly of atoms whose electrons orbit their nuclei in the same direction. When these atoms are also *aligned* in one direction, the substance has magnetic properties. Objects made of magnetic materials are surrounded by an invisible "field of force" called a magnetic field (Fig. 3-6). The magnetic field is polarized, that is, it has two ends that behave in opposite ways. The ends of the field are called the north and south poles of the magnet. They behave somewhat like electrical charges: unlike poles

Fig. 3-6 An invisible magnetic field surrounds a magnet.

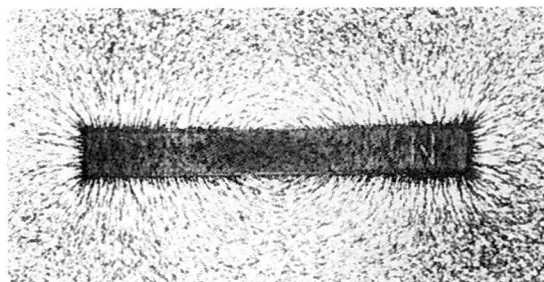

Fig. 3-7 Demonstration of the magnetic field.

Fig. 3-8 An x-ray photon can be visualized as two sine waves traveling in a straight line at the speed of light.

attract and like poles repel. When a piece of paper is placed over a magnet and iron filings are sprinkled on the paper, the iron will align itself with the lines of magnetic force, demonstrating the invisible magnetic field (Fig. 3-7).

Light, x-rays, radio waves, and microwaves are all forms of electromagnetic energy, or electromagnetic wave radiation. These energies have both electrical and magnetic properties, changing the field through which they pass both electrically and magnetically (Fig. 3-8). These changes in the field occur in the form of a repeating wave, a pattern that scientists call a "sinusoidal form" or **sine wave** (Fig. 3-9).

A more comprehensive understanding of magnetic energy, electrical energy, and electromagnetic

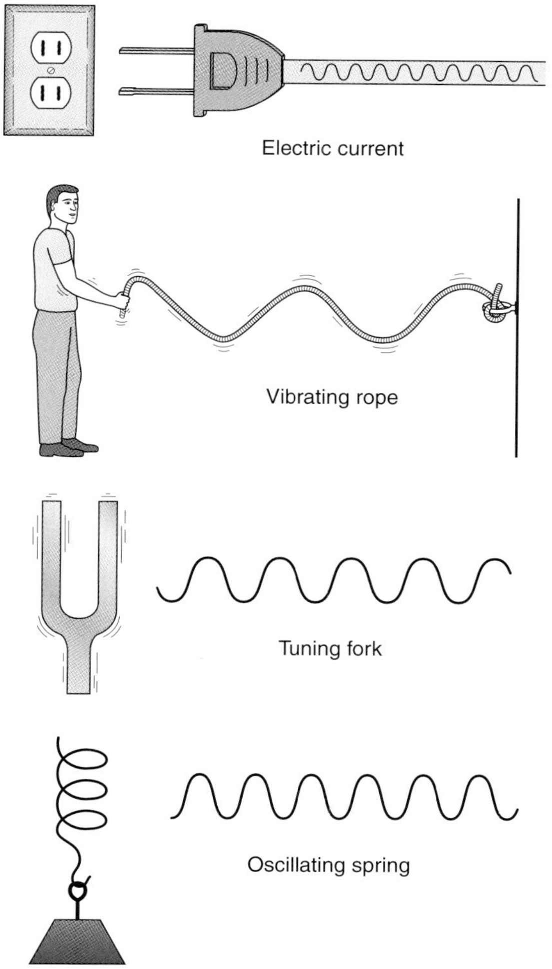

Electric current

Vibrating rope

Tuning fork

Oscillating spring

Fig. 3-9 Sine waves are energy expressed in a recurring wave form. Sine waves are associated with many naturally occurring phenomena, including electromagnetic radiation.

Amplitude a

Crest

Amplitude b

Trough

Amplitude c

Fig. 3-10 Sine wave amplitude, the distance from crest to valley. These three sine waves are identical except for their amplitude.

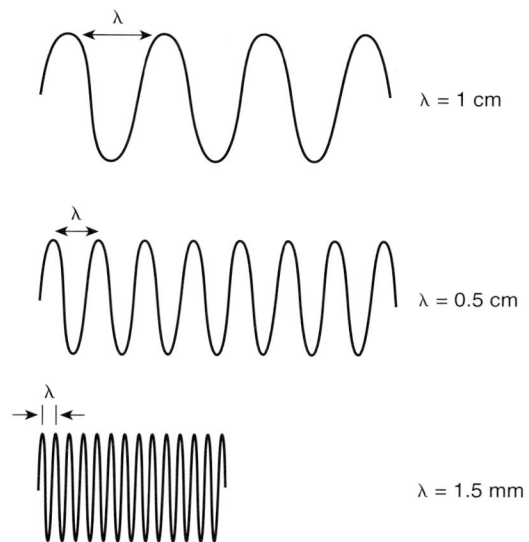

$\lambda = 1$ cm

$\lambda = 0.5$ cm

$\lambda = 1.5$ mm

Fig. 3-11 Sine wave wavelength, the distance from crest to crest. These three sine waves have different wavelengths. The shorter the wavelength, the higher the frequency.

energy is essential to the radiographer. These energy forms are discussed in greater detail in the sections that follow.

ELECTROMAGNETIC ENERGY

As stated above, electromagnetic energy occurs in the form of a sine wave. Several characteristics of this waveform are significant. The distance between the crest and the trough of the wave (its height) is called the amplitude (Fig. 3-10). More important to radiographers is the distance from one crest to the next, or **wavelength** (Fig. 3-11). The **frequency** of the wave is the number of times per second that a crest passes a given point (Fig. 3-12).

Electromagnetic energy moves through space at the velocity (speed) of approximately 186,000 miles per second (3×10^{10} centimeters per second). Since all electromagnetic energy moves at the same velocity, it is apparent that a relationship exists between wave-

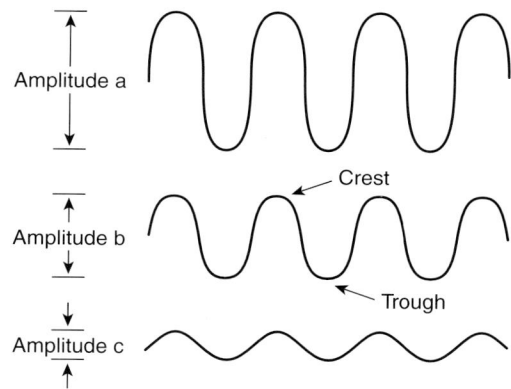

length and frequency. When the wavelength is short, the crests are closer together, so more of them will pass a given point in a second, resulting in a higher frequency. Longer wavelengths will have a lower frequency. This may be expressed mathematically:

$$\text{Velocity} = \text{Wavelength} \times \text{Frequency}$$

The more energy the wave has, the greater its frequency and the shorter its wavelength. We can therefore use either wavelength or frequency to describe the energy of the wave. In radiologic science, wavelength is more often used to describe the energy of the x-ray beam. The average wavelength of a diagnostic x-ray beam is about a billionth of an inch. *X-rays with greater energy have shorter wavelengths and are more penetrating.*

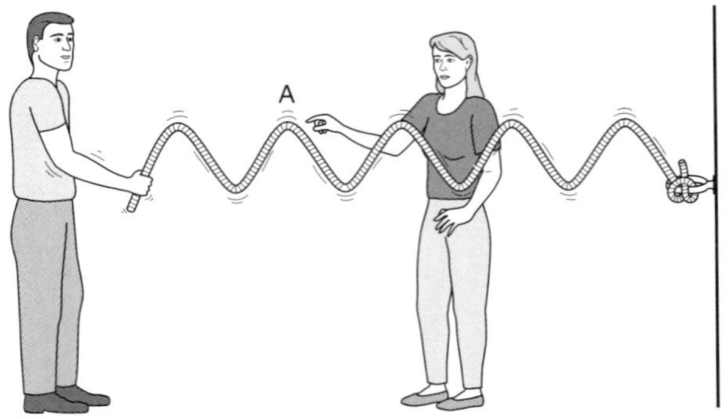

Fig. 3-12 Sine wave frequency, the number of crests or valleys that pass a fixed point per unit of time.

Sine Wave Velocity

Wavelength × Frequency = Velocity

The velocity of electromagnetic radiation is 186,000 miles/sec—3 × 10^{10} cm/sec. Note that ALL electromagnetic radiation has the same velocity.

Sine Wave Energy

- Sine waves with shorter wavelengths (higher frequency) have more energy.
- X-rays with shorter wavelengths are more penetrating.

Ionizing Radiation

- Sufficient energy to remove an electron from its orbit
- Wavelength of 1 nanometer or less
- X-rays are one form of ionizing radiation.

The wavelength of electromagnetic radiation varies from exceedingly short, even shorter than that of diagnostic x-rays, to very long, more than 5 miles. This range of energies is known as the electromagnetic spectrum. It includes x-rays, gamma rays, visible light, microwaves, and radio waves (Fig. 3-13). Radiation with a wavelength shorter than 1 nanometer (0.000000001 meter) is often called ionizing radiation, because it has sufficient energy to remove an electron from its atomic orbit. X-rays are one type of ionizing radiation.

The smallest possible unit of electromagnetic energy (analogous to the atom with respect to matter) is

Applications:	Wavelength:	
Therapeutic x-ray	1/100,000 nm	Ionizing
	1/10,000 nm	
Gamma rays	1/1000 nm	
	1/100 nm	
Diagnostic x-ray	1/10 nm	
	1 nm	
Ultraviolet rays	10 nm	
	100 nm	
Visible light	1000 nm	
Infrared rays	10,000 nm	
	100,000 nm	
	1/1000 m	
Radar	1/100 m	Nonionizing
	1/10 m	
	1 m	
Television	10 m	
Radio	100 m	
	1 nanometer = 10^{-9} meters	

Fig. 3-13 Electromagnetic spectrum.

the **photon,** which may be thought of as a tiny "bullet" of energy. Photons occur in groups or "bundles" called quanta (singular, quantum).

CHARACTERISTICS OF X-RAYS

Since x-rays and visible light are both forms of electromagnetic energy, they share some similar characteristics. Both travel in straight lines at the same velocity, and both have an effect on photographic emulsions. The photographic effect of x-rays is significant in the production of radiographs. It is also important to remember, because accidental exposure can occur when film is placed too near an x-ray source.

Both x-rays and light have a biologic effect, that is, they can cause changes in living organisms. For example, excessive exposure to either sunlight or x-rays may cause burns to the skin. X-rays are capable of producing more harmful effects than light because of their greater energy.

Unlike light, x-rays cannot be detected by the human senses. This fact may seem obvious, but it is important to consider. If x-rays could be seen, felt, or heard, we would have an increased awareness of their presence and radiation safety might be much simpler. Because x-rays are undetectable, safety requires that you learn to know when and where x-rays are present without being able to perceive them.

X-rays can penetrate matter that is opaque to light. This penetration is differential, depending on the mass density and thickness of the matter. For example, x-rays penetrate air very readily. There is

BOX 3-1

Characteristics of X-rays

- X-rays are a form of electromagnetic wave radiation that is capable of ionizing matter.
- They travel in straight lines at the speed of light, diverging into space from the point of origin.
- They are undetectable by the human senses.
- They can penetrate matter that is opaque to light.
- They cause certain crystals to fluoresce.
- They have a photographic effect, causing exposure to photographic film emulsions.
- They cannot be refracted by a lens.
- They have a biologic effect, causing alterations in living cells.

BOX 3-2

Electrical Supply Requirements: Household vs. X-ray Tube

Typical household circuit:	120 V	20 A
Typical x-ray tube circuit:	120,000 V (120 kV)	0.3 A (300 mA)

less penetration of fat or oil, even less of water, which has about the same mass as muscle tissue, and still less of bone.

X-rays cause certain crystals to fluoresce, that is, to give off light when exposed. Among crystals that respond in this way are barium platinocyanide, barium lead sulfate, calcium tungstate, and several salts of rare earth elements. These crystals are used to convert the energy pattern of remnant radiation into a visible image that can be viewed directly, as in fluoroscopy, or recorded on photographic film. Plates coated with these fluorescent crystals are called **intensifying screens** and are used to expose radiographs. Intensifying screens greatly reduce the quantity of radiation needed to produce an image. They are incorporated in cassettes and are discussed in Chapter 7.

Unlike light, x-rays cannot be refracted by a lens. The x-ray beam diverges into space from its source until it is absorbed by matter. Box 3-1 summarizes the characteristics of x-rays.

ELECTRICITY

X-ray energy is man-made and is produced electrically. To understand this process, it is helpful to know something about electric **current.** *Current is the flow of electrical charges.* Electrical charges will drift or flow in a vacuum, in certain gases (as in a neon light), in certain liquids (salt water, for example), and through certain metals called **conductors.** Copper wire is an excellent conductor and is commonly used for electrical wiring. It is connected to form a **circuit,** a continuous path. Current will flow in the circuit when there is a difference in electrical charge, a potential difference, between two points in the circuit. Current is produced when negatively charged electrons flow toward a positive charge. A positive potential can be maintained at one point in the circuit by means of electrical energy from a battery or a public utility.

Electrical Units

Three electrical factors are important in the function of an electric circuit: resistance, current, and electromotive force.

Resistance (R) is any property of the circuit that opposes or hinders the flow of current. The unit used to measure resistance is the ohm, represented by the Greek letter omega, Ω. Resistance depends on four factors: the material of the conductor, its length, its diameter, and its temperature. The longer the conductor, the more resistance it will provide. Resistance is decreased when the wire diameter is greater. Resistance increases as temperature increases.

The *total quantity of electrical charge* is measured in units of coulombs, abbreviated C. The **ampere,** abbreviated **A,** is used to measure the *rate of current flow* in the circuit. One ampere equals an electrical charge of 1 coulomb flowing through the conductor each second.

Electromotive force (emf) is another term for potential difference. Electromotive force is the positive electrical potential that attracts electrons, causing current to flow. The volt, abbreviated V, is the unit used to measure emf. One volt is the quantity of emf needed to cause a current of 1 A to flow in a circuit with a resistance of 1 Ω.

The relationship between emf, resistance, and current is expressed by Ohm's Law:

$$I = V/R$$

That is, the current (I) is equal to the voltage divided by the resistance. If the voltage is increased, more current will flow. If the resistance is increased, the current will be diminished.

It is significant that Ohm's Law does not apply to the x-ray tube circuit under normal conditions. This is because Ohm's Law assumes that there is always an unlimited number of electrons available in the circuit. As you will see in the following chapter, the electron supply in the x-ray tube is limited and controlled. This allows variations in voltage without significantly affecting the current. Radiographers can therefore *separately* control both the x-ray tube's voltage and its current.

The electrical requirements for x-ray tubes are much different from those needed by household appliances (Box 3-2). The voltage provided by public utilities for general household use is 120 V, and a common household circuit can carry a current of 15 to 30 A. X-ray tubes use much greater voltage and less amperage. A typical x-ray tube operates at between 40,000 and 150,000 V with a current of only 0.025 to 0.5 A. For this reason, it is convenient to use the **kilovolt (kV),** equal to 1000 volts, to measure the potential difference across an x-ray tube, and the **milliampere (mA),** equal to 1/1000 of an ampere or 0.001 A, to measure x-ray tube current. The significance of these units in x-ray production is discussed in Chapter 4. Table 3-3 lists units of electri-cal measurement that are important for radiogra-phers to remember.

Electric Circuits

An electric circuit is a continuous path for the flow of electrical charges from the power source through one or more electrical devices and back to the source (Fig. 3-14). Electric circuit diagrams are "maps" of circuits that show how current flows through the devices connected in the circuit. Table 3-4 contains some common symbols used in these diagrams. Circuit dia-

---

TABLE 3-3

Electrical Units

Measurement	Unit (Abbreviation)	Equivalent
Electrical charge	Coulomb (C)	A standard quantity
Current (I)	Ampere (A)	1 coulomb per second
	Milliampere (mA)	0.001 A (1/1000 A)
Electromotive force (emf)	Volt (V)	1 V causes current of 1 A to flow through 1 Ω of resistance
	Kilovolt (kV)	1 kV = 1000 V
Resistance (R)	Ohm (Ω)	A standard amount of hindrance to current flow

---

TABLE 3-4

Electric Circuit Elements: Their Symbol and Function

Circuit Element	Symbol	Function
Resistor		Inhibits flow of electrons
Battery		Provides electrical potential
Capacitor (condenser)		Momentarily stores electrical charge
Ammeter	Ⓐ	Measures electric current
Voltmeter	Ⓥ	Measures electrical potential
Switch		Turns circuit on or off by providing infinite resistance
Transformer	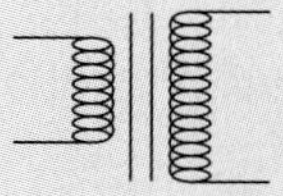	Increases or decreases voltage by fixed amount (AC only)
Rheostat		Variable resistor
Diode		Allows electrons to flow in only one direction
Transistor		Electronic switch that can also amplify signals
Electrical ground		Absorbs excess electrical charges

Modified from Bushong SC: *Radiologic science for technologists,* ed 5, St Louis, 2000, Mosby.

grams are used in this text to demonstrate electrical principles and to explain the function of the x-ray machine, so it will be helpful for you to become familiar with these symbols.

Devices may be connected in the circuit "in series" or "in parallel." In a series circuit, the wiring runs continuously from the source, through the device, and back to the source. An ammeter, for example, is a device for the measurement of current and is always connected in series (Fig. 3-15). Devices connected in parallel are wired *across* the circuit, creating a more complex circuit. A voltmeter measures the difference in electrical potential between two points in the circuit, so it must be connected in parallel, across the circuit between these two points (Fig. 3-16).

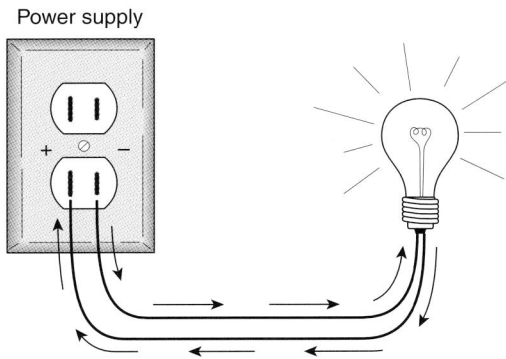

Direct Current and Alternating Current

A battery provides a constant positive charge at its positive pole, called the anode, and a negative charge at its negative pole, the cathode. Current flows in one direction, from the cathode to the anode (Fig. 3-17). This current is called direct current (DC). Because the battery's voltage is constant (Fig. 3-18), this direct current flows at a constant rate.

The electric service provided by a public utility is in the form of alternating current (AC). The polarity (positive or negative electrical potential) of the power source reverses at regular intervals, causing the current to flow first in one direction, then in the opposite direction (Fig. 3-19). The change is not instantaneous. In a household circuit, for example,

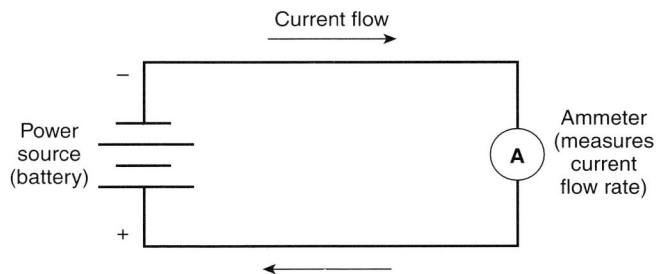

Fig. 3-14 Current flow through a circuit.

Fig. 3-15 Simple circuit wired in a series. An ammeter is always connected in a series.

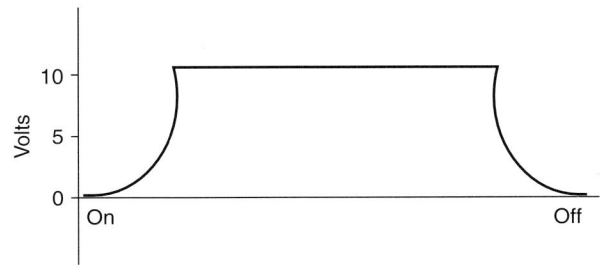

Fig. 3-16 Parallel circuit. Voltmeters are always connected in parallel.

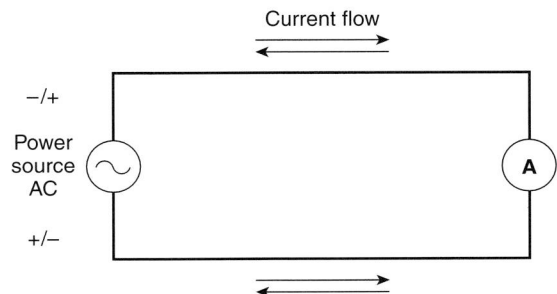

Fig. 3-17 Simple electric circuit, direct current. The power source is constant. Current flows from negative to positive.

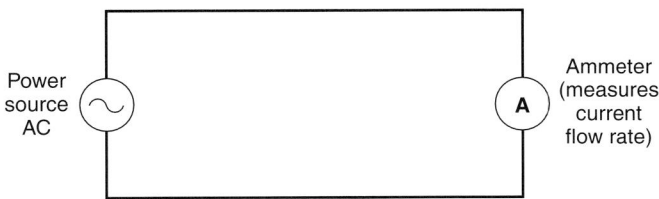

Fig. 3-18 Direct current. Voltage is constant when the circuit is "on."

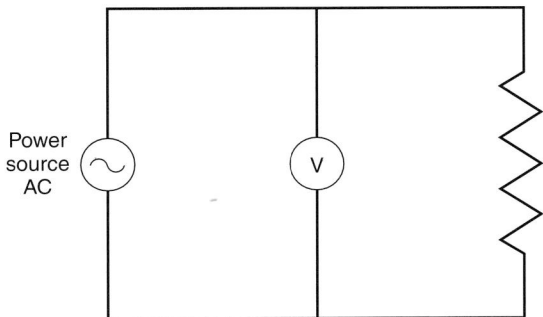

Fig. 3-19 Simple electric circuit, alternating current. The polarity of power source alternates between positive and negative at regular intervals.

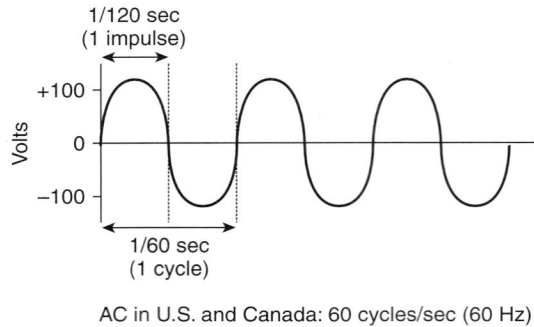

Fig. 3-20 Voltage wave form of alternating current.

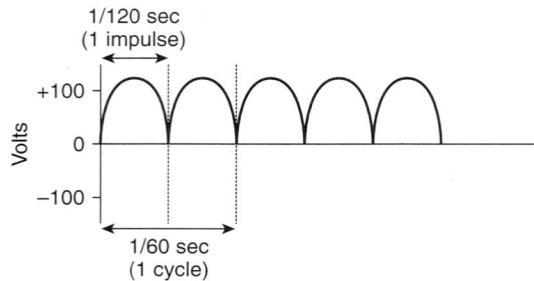

Fig. 3-21 Voltage wave form of rectified alternating current.

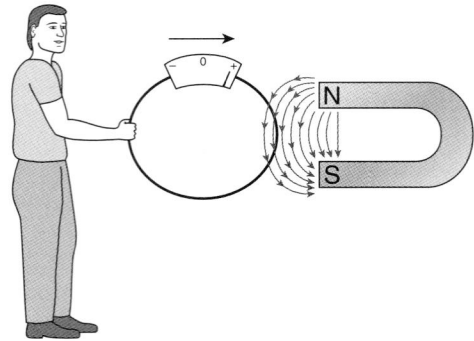

Fig. 3-22 When a conductor moves in and out of a magnetic field, alternating current will flow in the conductor.

the electrical cycle begins with the voltage at 0, increases to a *positive* 120 V, where it peaks, and then declines to 0 again. At this point the polarity changes and the voltage increases from 0 to a *negative* 120 V, peaks, and again returns to zero (Fig. 3-20). In the United States and Canada, public utilities deliver alternating current at 60 cycles per second, termed 60 Hertz (Hz). Thus the duration of each cycle is $\frac{1}{60}$ second. Half a cycle is called an impulse. There are 120 impulses per second with 60 Hz AC, so the duration of one impulse is $\frac{1}{120}$ second.

This alternating polarity produces electric current that is constantly changing. The current flow increases, peaks, and declines as the voltage changes. The current flow changes direction when the polarity changes.

Alternating current can be converted so that it flows in one direction only. This process, called rectification, is discussed in Chapter 5. Rectified alternating current is sometimes also referred to as direct current; it differs from DC produced by a battery in that it is pulsating, rather than constant (Fig. 3-21).

ELECTROMAGNETIC INDUCTION AND TRANSFORMERS

Magnetic fields and electrical energy are interrelated. Magnetic fields can be used to produce electricity, and conversely, electric currents create magnetic fields.

When a conductor is placed in a magnetic field and there is movement between the lines of magnetic force and the conductor, current will flow in the conductor. You could demonstrate this principle by moving a circuit in and out of the force field surrounding a magnet (Fig. 3-22). The same result is obtained by moving the magnet in relation to the conductor (Fig. 3-23). This process is called electromagnetic induction. When the direction of the movement changes, the direction of the current flow is reversed, creating alternating current. This effect also occurs with a change in the influencing pole of the magnet. For these reasons, *induced current is always alternating current*. It is the principle used to generate electric power. Public utilities use some other form of energy (steam generated by coal, gas or nuclear energy, or water flowing over a dam) in order to move either the magnet or the conductor.

When current is flowing through a circuit, it creates a magnetic field surrounding the conductor (Fig. 3-24). If this current is alternating, its magnetic field will be in constant motion. This moving magnetic field can be used to *induce* current to flow in another conductor (Fig. 3-25). The circuit that is connected to the power supply is called the primary circuit. The circuit that carries the induced current is called the secondary circuit. Note that the two circuits are not connected to each other.

Fig. 3-23 When a magnetic field moves in relation to a conductor, alternating current will flow in the conductor.

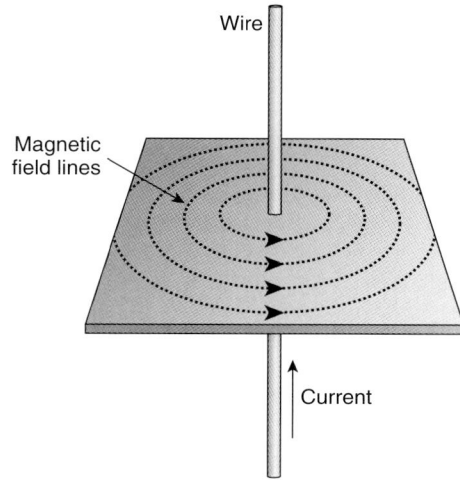

Fig. 3-24 A magnetic field is created around a conductor when current flows through the circuit. When the current is alternating, the magnetic field is in constant motion.

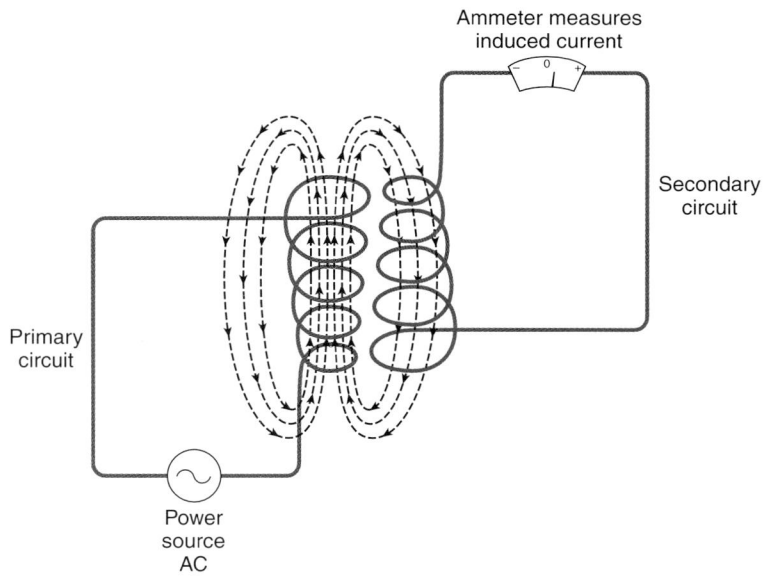

Fig. 3-25 When a conductor is placed in the magnetic field of an alternating current circuit, induced current will flow through the conductor.

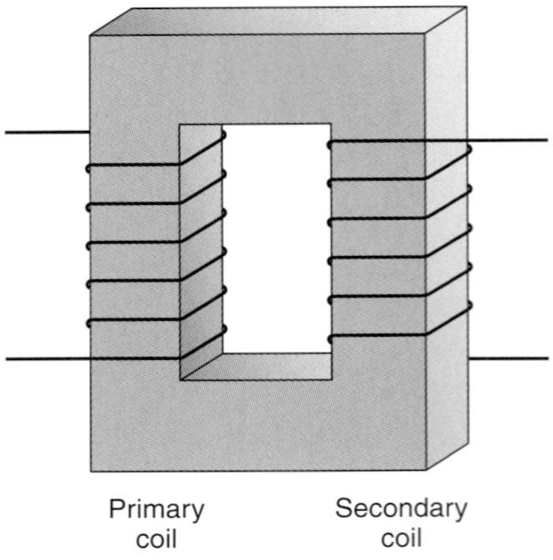

Fig. 3-26 A transformer consists of two circuits coiled around an iron core that enhances the magnetic fields. Since the circuits are insulated from the core, current cannot flow directly between the circuits.

Fig. 3-27 A step-up transformer has more windings on the secondary side.

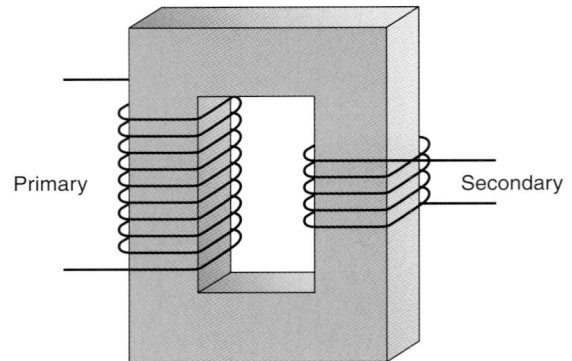

Fig. 3-28 A step-down transformer has more windings on the primary side.

Electromagnetic induction is the basis for the **transformer,** the device used to produce the high voltage needed for x-ray production. When the conductor is formed into a coil, the magnetic field around the coil is enhanced and polarized, creating a positive pole at one end of the coil and a negative pole at the other. This creates a potential difference (voltage) between the two poles. A transformer consists of the primary and the secondary coils, usually surrounding an iron core (Fig. 3-26). The iron core further enhances the magnetic fields of the coils.

When different numbers of turns, or "windings," are used in the coils of the primary and secondary circuits, the potential difference across the two coils will also be different. This makes it possible to change voltage by means of electromagnetic induction, which is the primary purpose of a transformer. When there are fewer turns in the primary coil than the secondary coil, the voltage on the secondary side is greater, and the transformer is called a step-up transformer (Fig. 3-27). On the other hand, if the secondary side has fewer turns, the secondary voltage will be less than the primary voltage and the transformer is called a step-down transformer (Fig. 3-28). The voltage increase or decrease produced by a transformer is proportional to the number of turns in each coil. For example, if the secondary side has twice as many turns as the primary side, the secondary voltage will be twice the primary voltage.

A transformer always increases or decreases the incoming voltage by a set multiple called the transformer ratio. For a step-up transformer, the ratio states the number of turns on the secondary side for each turn on the primary side. For example, if the voltage across the primary side were 200 V and the transformer

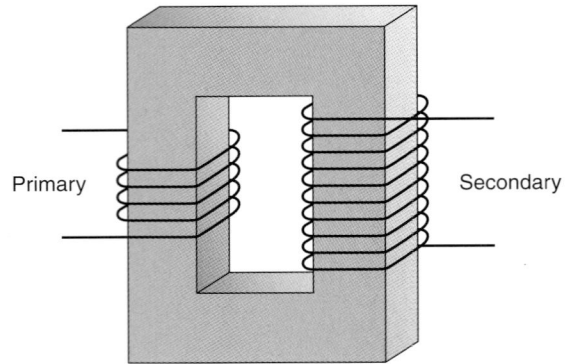

had 500 secondary turns for each primary turn, the ratio would be 1:500 and the secondary voltage would be 200×500, or 100,000 V (100 kV). For a step-down transformer, the ratio indicates the number of turns on the primary side for each turn on the secondary side. For example, a transformer with 90 turns on the primary side and 30 turns on the secondary side would have a 3:1 ratio. If the primary voltage were 150 V, the secondary voltage would be $150 \div 3$ or 50 V. Mathematical calculations involving transformer ratios and voltage are covered in Chapter 26.

SUMMARY

The atom is the basic building block of all matter, consisting of positively charged protons and uncharged neutrons in its nucleus and negatively charged electrons in orbits around the nucleus. Ionization takes place when a neutral atom or molecule gains or loses an electron, resulting in one or two charged particles. Unstable atoms called radionuclides emit energy and particles in the process of radioactive disintegration.

Energy exists in many forms and can be converted from one form to another. X-rays are one form of electromagnetic energy. Electromagnetic energy exists in the form of sine waves, and their energy is a function of their wavelength. X-rays cannot be perceived by humans and cannot be refracted by a lens. They travel in straight lines at the speed of light and can penetrate matter that is opaque to light.

Electric current is the flow of electrical changes in a circuit. Electromotive force (voltage) causes current to flow through resistance. A conductor in which alternating current flows is surrounded by a moving magnetic field. This field can induce current to flow in a circuit that is adjacent to it. This electromagnetic induction is the principle of the transformer, a device used to change voltage in the x-ray circuit.

▪ Review Questions ▪

1. State the Law of Conservation of Energy.
2. Name the electron orbit nearest the nucleus of an atom.
3. Name two forms of electromagnetic wave radiation that have a longer wavelength than diagnostic x-rays.
4. How does wavelength affect the usefulness of an x-ray beam?
5. What is meant by ionization? What determines the ionizing capability of electromagnetic radiation?
6. List six characteristics of x-rays.
7. What is the velocity of x-rays? Are x-rays faster or slower than visible light?
8. State the units used to measure current, emf, and electrical resistance.
9. What does an ammeter measure? A voltmeter? How is each connected in a circuit?
10. What is the duration of an electrical cycle in the United States? An electrical impulse?
11. What is meant by electromagnetic induction?
12. What is the primary purpose of a transformer?

4

X-ray Production

Learning Objectives

At the conclusion of this chapter the student will be able to:

- Draw a simple x-ray tube and label its parts
- Describe both the composition and the function of the basic parts of the x-ray tube
- Associate the terms "anode" and "cathode" with the appropriate parts of the x-ray tube
- Describe the production of both Bremsstrahlung and characteristic radiation, and explain what determines the wavelength of each
- Explain what is meant by a dual-focus tube and describe its advantages
- Explain the significance of the target angle with respect to the line focus principle and the maximum field size
- Define "effective focal spot" and state its significance with respect to the radiographic image
- Explain the function of a rotating anode and state its purpose
- State the effect of changes in mA and kVp on the resulting x-ray beam

Key Terms

actual focal spot
anode heel effect
Bremsstrahlung
characteristic radiation
concentric
dual focus
effective focal spot
electron stream
exposure time
filament
filtration
focal spot

focal track
focusing cup
heterogeneous
kilovolts peak (kVp)
line focus principle
milliampere-seconds (mAs)
millisecond (msec)
rotating anode
space charge
target
target angle
thermionic emission

This chapter is about x-ray tube structure and function and how these factors affect the primary x-ray beam. The electrical factors that control x-ray production are introduced in this chapter. Chapters 4 and 5 contain a tremendous amount of detail, and most of it is probably unfamiliar to you. Although it is all interrelated and is presented in a logical order, you may feel a bit overwhelmed if you try to comprehend it too quickly. Do not attempt to assimilate it all at once. When this material is taken in "small bites" and reviewed as needed, the entire process of creating and controlling x-rays will gradually come into focus.

Roentgen discovered x-rays while working with a Crookes tube (Fig. 4-1), a cathode ray tube that was the forerunner of the fluorescent tube and the neon light. These tubes were used in physics laboratories in the late nineteenth century for the investigation of electricity. In 1913, the General Electric Company introduced the Coolidge tube (Fig. 4-2), a "hot cathode tube" that was the prototype for modern x-ray tubes.

X-RAY TUBE

Fig. 4-3 illustrates a simple x-ray tube with its principal parts labeled. There are four essential requirements for the production of x-rays: (1) a vacuum, (2) a source of electrons, (3) a target, and (4) a high potential difference (voltage) between the electron source and the target.

A glass envelope forms the basic structure of the x-ray tube. It is made of strong, heat-resistant glass and contains both the source of electrons and the target. The air is removed from the glass envelope to form a near-perfect vacuum so that gas molecules will not interfere with the process of x-ray production. The tube is fitted on both ends with connections for the electrical supply.

The source of electrons is provided by a **filament** at one end of the tube. The filament consists of a small coil of tungsten wire. Tungsten, chemical symbol W, is a metal element; it is a large atom with 74 electrons in orbits around its nucleus. An electric current flows through the filament to heat it. As explained in Chapter 3, heat speeds up the movement of the electrons in their orbits and increases their distance from the nucleus. Electrons in the outermost orbital shells get so far from the nucleus that they are no longer held in orbit but are flung out of the atom, forming an "electron cloud" around the filament (Fig. 4-4). This process is called **thermionic emission.** The electron cloud is called a **space charge** and is the source of free electrons for x-ray production.

Fig. 4-1 Hittorf-Crookes tube used by Roentgen.

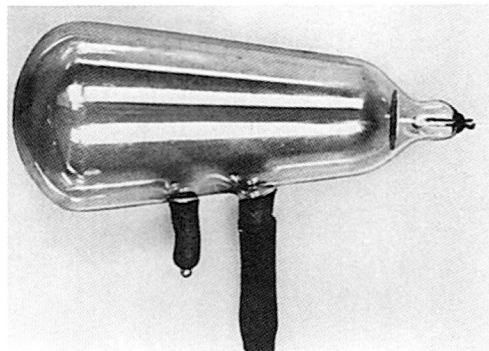

Fig. 4-2 Coolidge "hot cathode" tube.

At the opposite end of the tube is the **target,** a hard, smooth, slanted metal surface that is also made of tungsten. The electrons are directed toward the target, which is the place where x-rays are generated.

A high-voltage electrical source provides acceleration of the electrons. A large step-up transformer supplies the voltage required for x-ray production (40 to 150 kV). The two ends of the x-ray tube are connected in the transformer circuit so that during an exposure, the filament end is negative and the tar-get end is positive. The positive, target end of the tube is called the anode; the negative, filament end is called the cathode.

The high positive electrical potential at the target attracts the negatively charged electrons of the space charge, which move rapidly across the tube, forming an **electron stream.** When these fast-moving electrons collide with the target, the kinetic energy of their motion must be converted into a different form of energy. The great majority of this kinetic energy (99+%) is converted into heat, but a small amount is converted into the energy form that we know as x-rays (Fig. 4-5).

BREMSSTRAHLUNG AND CHARACTERISTIC RADIATION

X-rays are produced at the target as a result of the sudden deceleration of the electron stream. This deceleration may occur in one of two ways (Box 4-1). When an incoming electron nears the nucleus of a target atom, it slows abruptly and changes direction. The kinetic energy of the electron's motion is converted into an x-ray photon (Fig. 4-6). X-rays formed by this process are called **Bremsstrahlung,** German for "braking rays." Sometimes an incoming electron collides with an inner orbital electron of a

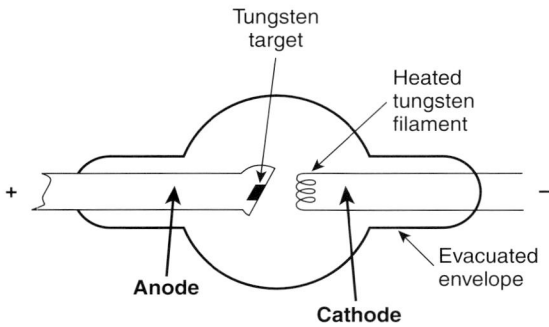

Fig. 4-3 Simple x-ray tube. The anode is the positive end of the tube; the target is part of the anode. The cathode is the negative end of the tube; the filament is part of the cathode.

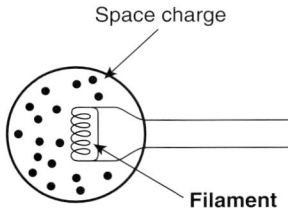

Fig. 4-4 Thermionic emission. As tungsten is heated, electrons in the tungsten atom orbits spin faster, moving farther from the nucleus. Electrons in outer orbits are flung out of the atom, forming an "electron cloud" or space charge. Space charge provides the electron source for x-ray production.

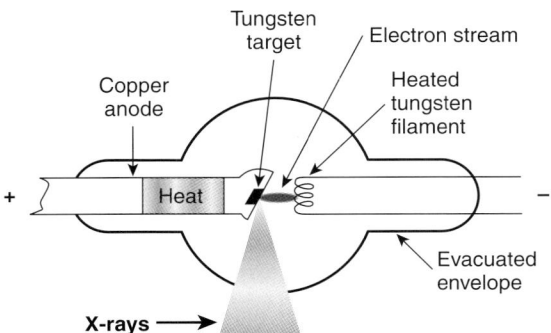

Fig. 4-5 The energy of the electron stream is converted at the anode into heat (99+%) and x-rays (1%).

▮ *BOX 4-1*

Target Interactions that Produce X-rays

Bremsstrahlung ("braking rays"): formed by sudden halting of electron stream as kinetic energy is converted to other forms

Characteristic radiation: formed within target atoms as result of interaction with incoming electron stream

The majority of the x-ray beam consists of Bremsstrahlung.

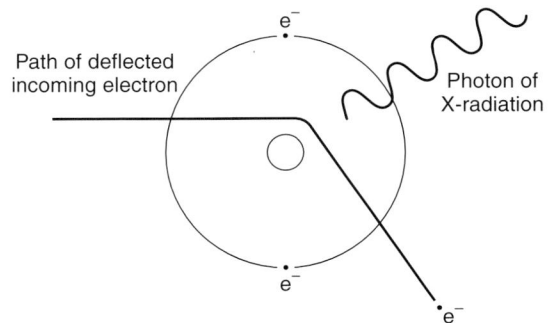

Fig. 4-6 Bremsstrahlung radiation is created when an incoming electron slows abruptly near the nucleus of a target atom.

Fig. 4-7 Distribution of wavelengths in the x-ray beam.

Fig. 4-8 Formation of characteristic radiation. An incoming electron removes an inner orbital electron from the target atom, creating a "hole" in the K shell. When an L shell electron fills the hole, a characteristic photon is emitted.

target atom, causing an interaction of a different type that also produces an x-ray photon. X-rays formed by this process are called **characteristic radiation.** There is no real difference between Bremsstrahlung photons and characteristic photons. The difference lies in the interactions that produce them. The primary x-ray beam is made up of both Bremsstrahlung and characteristic radiation.

The wavelength of Bremsstrahlung radiation varies because of the varying speed of the electron stream. The rectified AC power source provides constantly pulsating voltage, causing variations in the speed of the electron stream. When the electron speed is very rapid, at the peak of the voltage cycle, the resulting photons have more energy, expressed as short wavelengths. During the portions of the cycle when voltage is lower, the electrons' speed is slower, and the resulting photons have less energy and longer wavelengths. The resulting x-ray beam is said to be **heterogeneous,** made up of many different wavelengths.

Fig. 4-7 is a graph that shows the quantity of photons of various wavelengths in the x-ray beam. The bell-shaped curve represents the heterogeneous distribution of the wavelengths of Bremsstrahlung. The graph is high in the middle because there are many photons of average wavelength. It is low on both ends because there are fewer photons with very long or very short wavelengths.

As stated before, the creation of characteristic radiation involves a collision between an electron from the electron stream and an electron in the K shell of a tungsten atom. When this occurs, the K shell electron is removed from its orbit, leaving a vacant space or "hole" in the K shell. Because the proper number of atoms in the K shell is 2, an electron from the L shell will move into the K shell to fill the hole. Movement of this electron from the L shell to the K shell results in a release of energy in the form of an

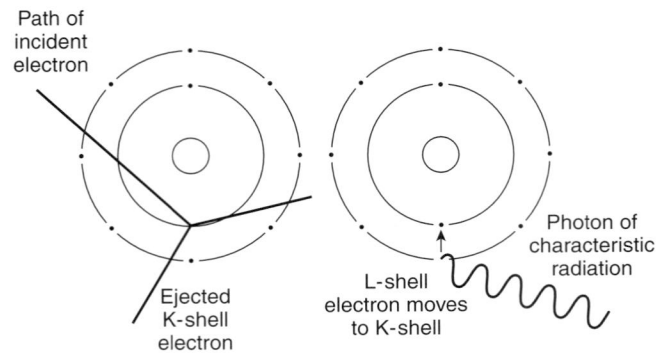

x-ray photon (Fig. 4-8). This photon's energy is equal to the difference in binding energy between the two shells. Because this binding energy difference is always exactly the same for the element tungsten, all *characteristic* photons produced in tungsten will have the same wavelength. This radiation is called characteristic radiation because its wavelength is typical or "characteristic" of the target element. Tungsten is a particularly good material for x-ray tube targets because the characteristic radiation it produces has a wavelength that is useful in radiography.

The vertical line above the bell-shaped curve in Fig. 4-7 shows a high number of photons with one specific wavelength. This line represents the portion of the x-ray beam that consists of characteristic radiation. Bremsstrahlung makes up the majority of the x-ray beam. The exact percentage of each type of radiation in the beam varies with the kilovoltage. At least 70 kV is required for the production of characteristic radiation. At 100 kV, characteristic radiation makes up about 15% of the useful x-ray beam, and the remainder is Bremsstrahlung.

CHARACTERISTICS OF THE ANODE AND THE CATHODE

Cathode

While it is essential to have at least one filament for x-ray production, modern multipurpose x-ray tubes are **dual-focus** tubes (Fig. 4-9). They contain two filaments, one large and one small. Only one filament at a time is used.

Each filament is situated in a hollow area in the cathode called a **focusing cup** (Fig. 4-10). The focusing cup has a slight negative charge. The shape of the focusing cup and its negative electrical charge cause the electrons to be repelled in the direction of a very precise area on the target called the **focal spot** (Fig. 4-11).

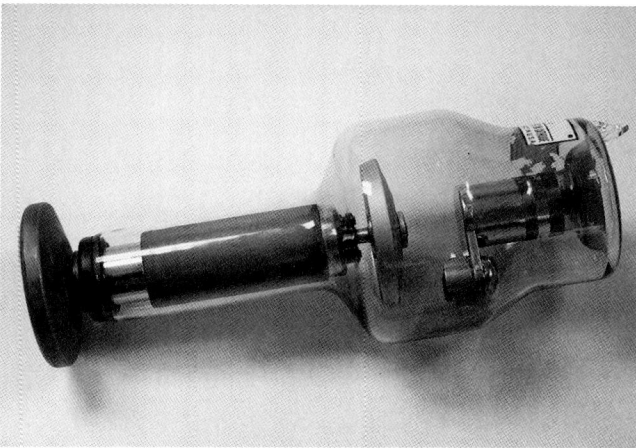

Fig. 4-9 Modern, dual-focus x-ray tube.

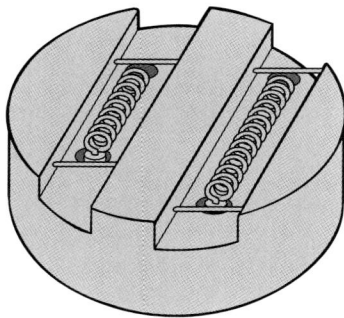

Fig. 4-10 Dual filaments in their focusing cups.

Fig. 4-11 A, Without a focusing cup, the electron stream spreads beyond the target area. **B,** Negatively charged focusing cup repels electrons, focusing them on a small target area, the focal spot.

When the small filament is activated, its electrons are directed to a tiny focal spot on the target. The small filament and focal spot provide finer image detail when a relatively small exposure is appropriate. The large filament provides more electrons and is aimed at a larger target area. The larger focal spot can better absorb the heat generated by the increased exposure. The combination of large filament and large focal spot is used when a large exposure is required. The two focal spots are **concentric,** that is, they have the same center (Fig. 4-12).

Anode

As stated earlier in this chapter, the vast majority of the energy of the electron stream is converted into heat. This energy conversion takes place at the target, *so the anode tends to get very hot.* Anodes are there-

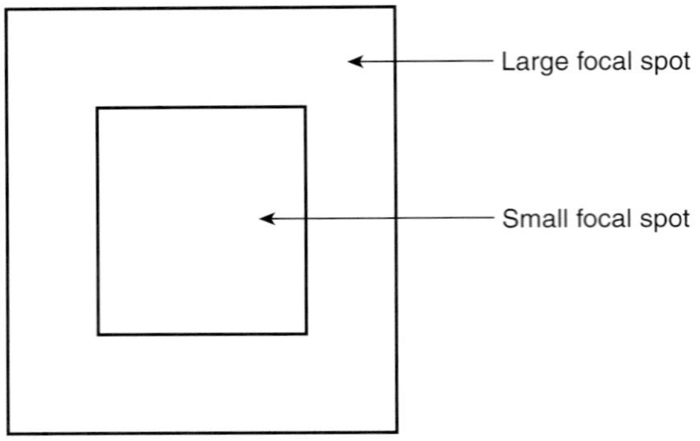

Fig. 4-12 Actual focal spot, the area on the target where the electron stream is focused. Dual-focus tubes have two focal spots, one large and one small, and they are concentric.

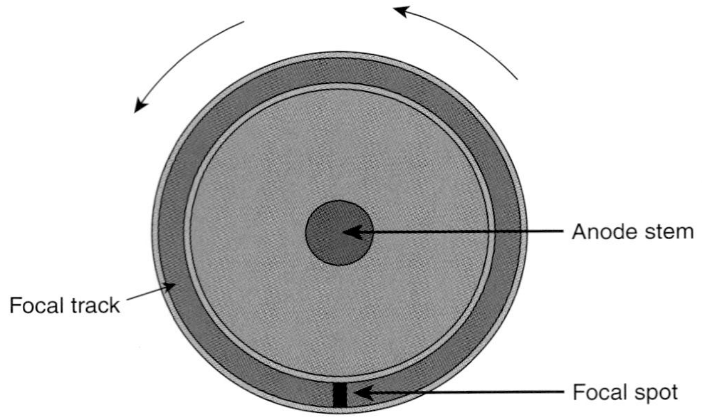

Fig. 4-15 Rotating anode face. The electrons stride the anode in the tiny focal spot area, but the heat is spread around the entire focal track of the spinning anode face.

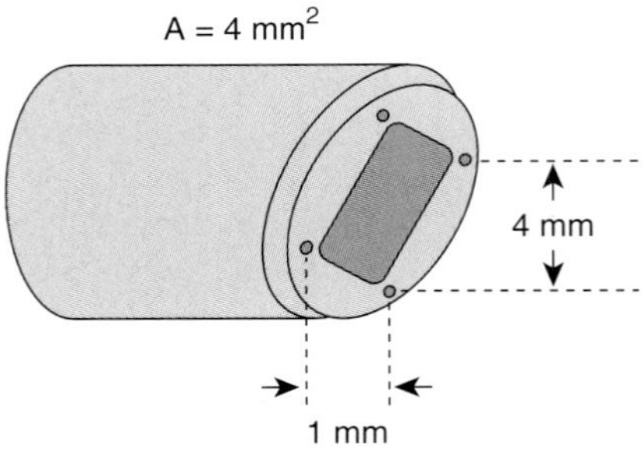

Fig. 4-13 Stationary anode.

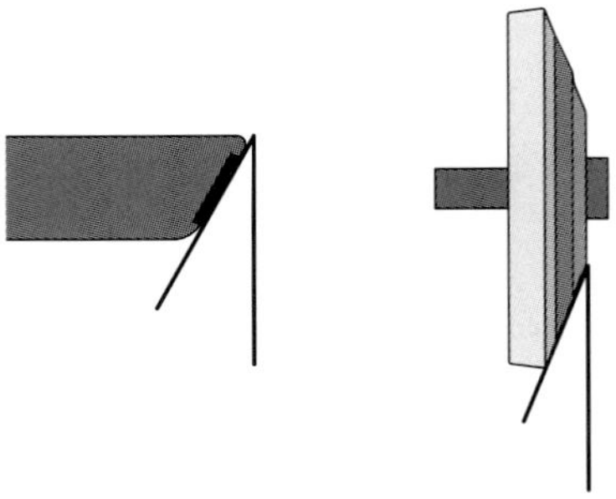

Fig. 4-16 Target angle.

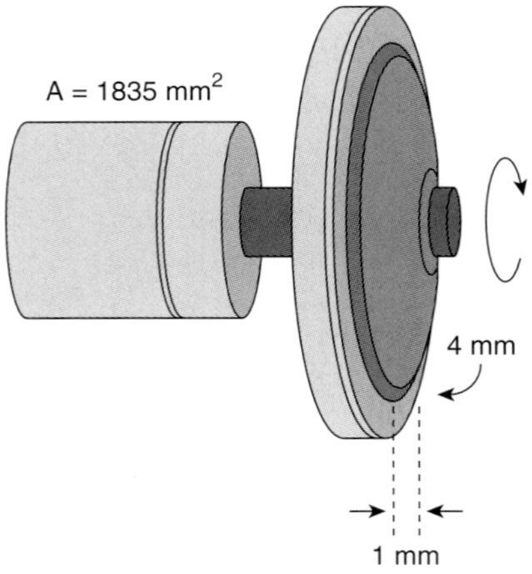

Fig. 4-14 Rotating anode.

fore constructed to dissipate heat. Another reason why tungsten is an excellent material for x-ray tube targets is because it has a very high melting point. A mass of copper is incorporated into the anode, which serves to conduct heat away from the target's focal area.

Early x-ray tubes, and some modern tubes with very low x-ray output, have a solid, stationary cop-

per anode with a slanted tungsten face (Fig. 4-13). Modern tubes for general radiography have a **rotating anode** (Fig. 4-14). The rotating anode is in the form of a disk with a beveled edge. This disk spins during the exposure, so the heat is distributed all around the circumference of the disk (Fig. 4-15). The tungsten focal area all around the beveled edge of the rotating anode is called a **focal track.** The focal spot remains in the same location in space, but the target metal is spinning. The metal struck by the electron stream is constantly changing during the exposure, distributing the anode heat over a larger area and increasing the heat capacity of the tube.

Both the slanted face of the stationary anode and the beveled edge of the rotating anode present an angled surface to the oncoming electron stream. The slant of the anode surface is called the **target angle** (Fig. 4-16). X-ray tube target angles are between 10° and 20°, with 12° to 15° being most common. The target angle is built into the x-ray tube and cannot be changed. The target angle affects the tube's heat capacity, the sharpness of the radiographic image, and the maximum size of the x-ray beam. These effects of the target angle are discussed in the following section.

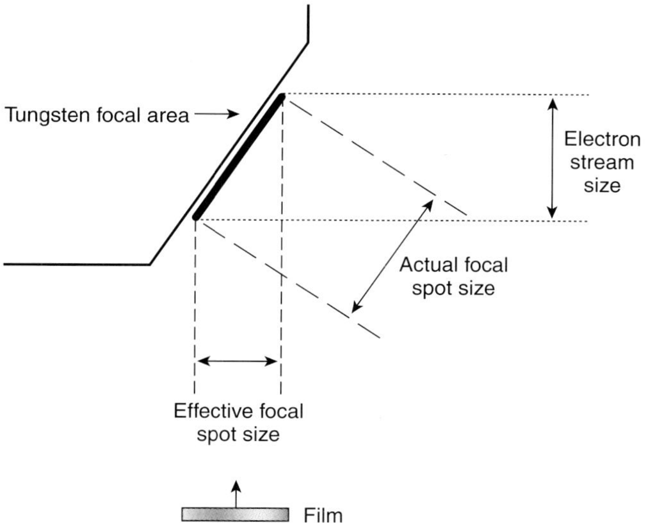

Fig. 4-17 The effective focal spot is the vertical projection of the actual or "true" focal spot.

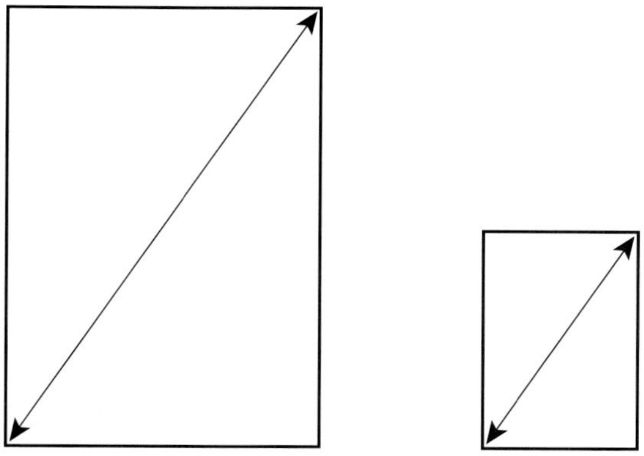

Fig. 4-18 Effective focal spot sizes.

Focal Spot Size

Actual
- Measurement of focal spot on target surface
- Affects tube heat capacity
- **BIGGER** is Better!

Effective
- Measurement of vertical projection of actual focal spot
- Affects image sharpness
- SMALLER is Better!

Line Focus Principle

- The size of the EFFECTIVE focal spot determines image sharpness.
- The relative size of the effective focal spot is determined by the target angle.
- The steeper the target angle, the greater difference between the actual and the effective focal spot sizes.

Line Focus Principle

The term **actual focal spot** refers to the area on the target surface that is struck by the electron stream. The **effective focal spot** is the *vertical projection* of the actual focal spot (Fig. 4-17). The size of the *effective* focal spot influences image sharpness. This fact is called the **line focus principle.** When vertical lines are drawn from each corner of the slanted actual focal spot, these lines define an "image" of the focal spot as viewed from the film. The effective focal spot is always smaller than the actual focal spot.

You can easily demonstrate this principle using your own hand to represent the focal spot. First, hold the palm of your hand in front of your face and note its size. This is the "actual size." Then flex your wrist, tipping your fingertips toward your nose. In this position, the distance between wrist and fingertips appears shorter due to the angle of view. This perspective demonstrates the "effective size" of your hand when seen from an angle. It corresponds to the appearance of the focal spot from the "film's eye view."

Focal spots are measured from corner to corner, on the diagonal (Fig. 4-18). Common effective focal spot sizes for general-purpose tubes are 0.6 to 1.0 millimeter (mm) for the small focal spot and 1.2 to 2.0 mm for the large focal spot.

Just how the effective focal spot influences image detail is explained in Chapter 6. At this point, simply note the fact that a smaller effective focal spot size will result in a sharper, more precise image and that a larger effective focal spot will have the opposite effect.

The angle of the target face determines the size difference between the actual focal spot and the effective focal spot. Fig. 4-19 shows two targets with the same-size actual focal spot but different target angles. It demonstrates that the "steeper" (more vertical) target has a smaller effective focal spot. *The smaller the target angle, the greater the size difference between the actual and effective focal spots.*

While a *small effective* focal spot is desirable for greater image sharpness, a large actual focal spot is desirable to dissipate the heat of large exposures. The best solution would seem to be to use the steepest possible target angle. There is a practical limit, however, on how steep the target angle can be. As you

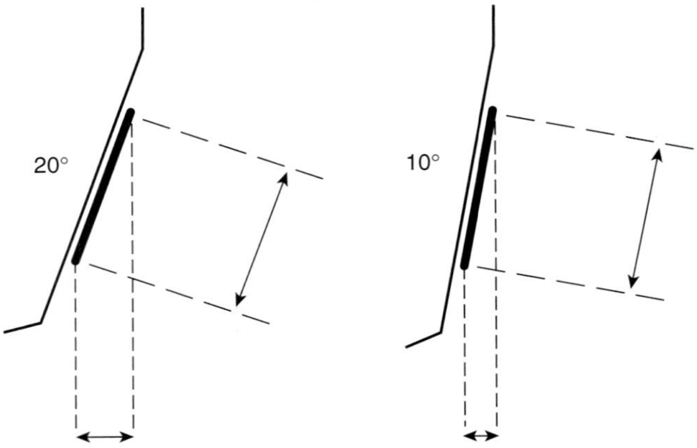

Fig. 4-19 A steeper target angle results in a smaller effective focal spot with a given actual focal spot size.

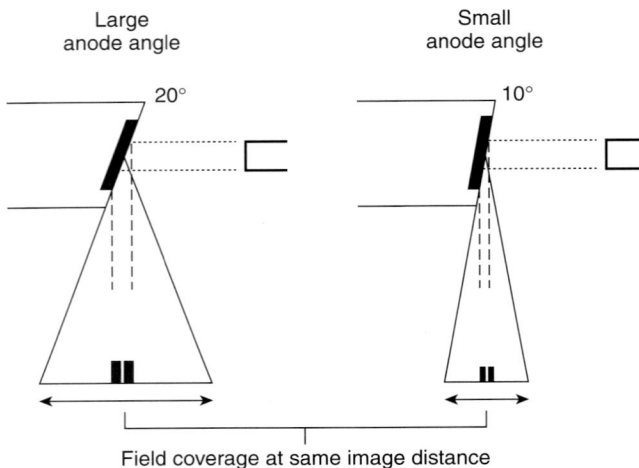

Fig. 4-20 The anode angle controls the maximum field size. The anode margin of field extends from the anode face at the same angle. The cathode side is the mirror image of the anode side. A 12° target angle is needed to cover a 14 × 17-inch film at a 40-inch distance from the source.

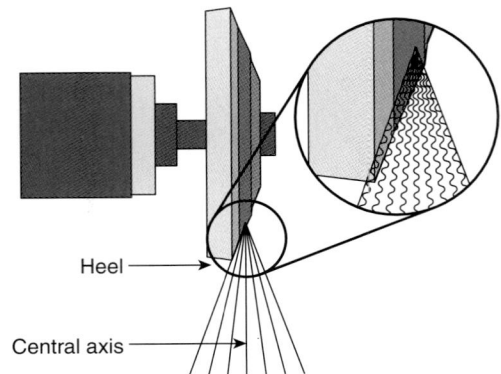

Fig. 4-21 Anode heel. X-rays are formed within the target material and are absorbed by the target as they exit. The sloping target face causes uneven absorption of the primary beam.

Anode Heel Effect

- Variation in radiation intensity across the length of the radiation field
- Greater radiation intensity toward the cathode end of the field
- Only significant when using the whole beam (14 × 17-inch film at 40 inches or full spine at 72 inches)
- Place thinner portion of body part toward anode end of tube

can see in Fig. 4-20, a straight line extended from the target face defines the margin of the x-ray beam on the anode side. The opposite margin of the beam is controlled by the tube port so that the beam is symmetrical with the perpendicular central ray in its center. Thus the target angle determines the maximum possible size of the x-ray beam and the radiation field. A target angle of at least 12° is needed to produce a radiation field that will cover a 14 × 17-inch film at a distance of 40 inches—the largest film in general use and a common, convenient working distance.

Anode Heel Effect

Most x-rays are not produced on the absolute surface of the target. Incoming electrons may penetrate the target to a depth of several layers of atoms before interacting with the target material. X-rays produced *within* the target must then pass through a portion of the target to get out (Fig. 4-21). Some of the x-rays will be absorbed by the target in this process. Because of the slanted face of the target, some x-rays will have to pass through more target material than others, depending on their direction. Those x-rays that are directed away from the cathode are more likely to be absorbed than those that are directed toward it. This results in uneven distribution of radiation intensity in the x-ray beam and is called the **anode heel effect.**

Fig. 4-22 illustrates the relative intensity of the x-ray beam from one end to the other. If the intensity of the beam is measured at the central ray and that intensity is designated as 100%, it is apparent that the majority of the x-ray beam has an intensity within ±10% of the central measurement. This 10% variation is not noticeable on the film. At the anode end of the beam, however, the intensity is only 75%, a noticeable difference.

The anode heel effect is only significant in radiography when the entire beam is in use. This is the case when a large film (14 × 17 inches) is used at a distance of 40 inches. Examples include examinations

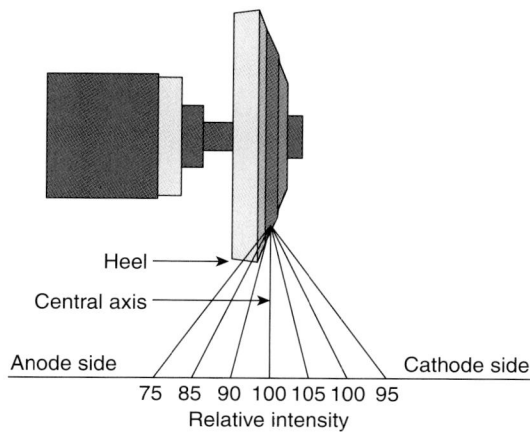

Fig. 4-22 Anode heel effect. The intensity of the x-ray beam is least toward the anode end of the field.

of the lower leg or the femur (thigh bone) and some views of the spine. The anode heel effect is also significant when an extra long (14 × 36-inch) film is used for radiography of the full spine or the entire leg. In these cases, it is advantageous to place the patient so that the thinnest portion of the anatomy to be radiographed is toward the anode end of the tube. If the anode heel effect is not used correctly, the result will be that the thinner portions of the anatomy appear too dark on the film and the thicker portions will be too light.

Obviously, to effectively use the anode heel effect, the radiographer must know which end of the tube is which. Sometimes this is easily determined by examining the tube housing. The cable connections to the housing are often labeled. If no labels are apparent on or near the tube housing, try following the cables to their connections at the other end, the transformer cabinet, and check for labels there. If these connections are not easily visible or are not labeled, try placing your hands on the two ends of the tube while the anode is spinning or listening closely to the tube from each end. The end with the most vibration and/or the most noise during anode rotation is the anode end. Tubes are usually installed so that the anode is up when the tube faces the upright cassette holder, but this is not always the case. A definitive determination can be made by taking a radiograph. Expose a 14 × 17-inch film at a 40-inch distance using approximately 1 mAs and 44 kVp. If these settings are not available, use the least mAs and kVp that are available. Open the collimator so that the radiation field covers the film. Place a lead marker at one end of the film so that you will be able to tell from the finished radiograph how it was placed during the exposure. The radiograph will demonstrate the anode heel effect because one end will be lighter than the other. The light end will signify the anode end of the tube.

ELECTRICAL CONTROL OF X-RAY PRODUCTION

Kilovoltage

Voltage is measured at the *peak* of the electrical cycle. When the voltage across the x-ray tube is measured, the units are often stated as **kilovolts peak,** abbreviated **kVp.** The terms kV and kVp are used interchangeably in radiography, with kVp being the preferred term.

The voltage applied to the x-ray tube controls the speed of the electrons in the electron stream. As stated previously, the electrons move faster when the voltage is increased, producing x-rays with shorter wavelengths and greater energy. For this reason, when the kVp setting is increased, the electron speed at the peak of the voltage cycle is increased, producing x-rays with shorter wavelengths. In other words, kilovoltage controls the energy (wavelength) of the x-ray beam. Since x-rays with shorter wavelengths are more penetrating, *the penetrating power of the x-ray beam is controlled by varying the kVp.*

Milliamperage

Milliamperage (mA) is a measure of the *rate of current flow* across the x-ray tube, that is, the number of electrons flowing from filament to target *each second* (see Box 4-2). Under normal conditions, electrons join the electron stream as soon as they are liberated from the filament. The number of available electrons is determined by the filament heat. When filament heat is increased, more electrons are available each second to cross the tube. Thus increasing the filament heat increases the mA in the x-ray tube circuit. When more electrons strike the target, more x-rays are produced, so *milliamperage controls the rate of x-ray production* and thus also the *rate of exposure.* In other words, high mA settings produce more x-rays per second, and low mA settings produce x-rays at a slower rate. Stated differently, mA controls the intensity of the x-ray beam, determining the number of photons that will strike the film during 1 second.

Exposure Time

Exposure time refers to the length of time that voltage is applied across the x-ray tube. It is the duration of the x-ray exposure. Exposure time is measured in units of seconds or fractions of seconds. Sometimes exposures are so short that it is convenient to measure them in **milliseconds (msec),** thousandths of seconds: 1 msec equals 0.001 sec.

A timer in the x-ray circuit terminates the exposure after a preset length of time. The quantity of x-rays produced is directly proportional to the exposure time. For example, it is apparent that if the exposure time is doubled, twice as many x-rays will be produced.

Milliampere-Seconds

In radiography it is often useful to know the *total quantity* of an exposure. The quantity of x-ray photons in an exposure cannot be determined by either the mA or the exposure time alone. While mA determines the *rate* of x-ray production, it does not indicate the total quantity because it does not indicate how long the exposure lasts. Exposure time does not indicate the total quantity either, because it does not measure the rate of x-ray production. To determine the quantity of radiation involved in an exposure, both mA and time must be considered.

The unit used to indicate the quantity of exposure is **milliampere-seconds,** abbreviated **mAs.** This unit is the product of mA and exposure time:

$$mA \times Time\ (seconds) = mAs$$

The mAs is actually a measure of the electrical charges (electrons) that cross the x-ray tube during an exposure. Because the number of charges determines the number of photons, mAs is an accurate and convenient indicator of the total quantity of an exposure.

To better understand the concept of mAs, imagine for a moment that the x-ray beam consists not of millions of photons but of only a few hundred and that each milliampere of current produces only one x-ray photon per second. If this were true, an exposure rate of 100 mA would produce 100 photons per second. If the exposure time were 2 sec, the mAs would equal 200 and the total number of photons in the exposure would be 200.

A desired quantity of exposure may be obtained by any combination of mA and time that, multiplied together, equals the desired mAs. In the example above, for instance, 200 photons could also be obtained using 200 mA and an exposure time of 1 sec.

A more familiar analogy might be to compare mAs to the distance driven in a car. In this case, the rate is not mA, but miles per hour, and the time is measured in hours rather than seconds. Suppose your destination is 50 miles away. If you drive your car at the rate of 100 mph, you will arrive in half an hour. The rate (100 mph) times the time (0.5 hour) equals the distance driven (50 miles). If you obeyed the traffic laws and drove at 50 mph, the trip would take 1 hour (50 mph × 1 hour = 50 miles). The mAs represents the final destination of an exposure; mA is the rate at which the goal is reached, and the exposure time tells how long it will take to reach the goal. This analogy is summarized in Box 4-2.

X-RAY BEAM FILTRATION

As explained earlier in this chapter, the x-ray beam is heterogeneous, consisting of photons with many

---

Milliampere-Seconds (mAs)

Measure of total quantity of electricity involved in exposure

$$mA \times Time\ (seconds) = mAs$$

Indicative of total quantity of radiation produced by an exposure

---

X-ray Beam Filtration

- Filter material placed between the tube housing port and the patient removes the long-wavelength radiation from the primary beam.
- Since this radiation does not have sufficient energy to penetrate the patient, the cassette, and the table, it does not contribute to the image.
- Filtration lowers patient dose significantly.
- Filtration decreases the average wavelength of the x-ray beam.

BOX 4-2

mAs Analogy

Driving a Car	X-ray Exposure
Rate of speed = 100 mph	Exposure rate = 100 mA
Duration of trip = 0.5 hour	duration of exposure = 0.5 second
Rate × Time = Distance	Rate × Time = Total exposure
100 × 0.5 = 50 miles	100 × 0.5 = 50 mAs

In both cases, the product of rate × time equals the destination or goal.
The goal can be achieved by any appropriate combination of rate and time.

different wavelengths. Those photons with long wavelengths are easily absorbed by matter and are unlikely to penetrate the subject and expose the film. They do not contribute to the x-ray image. If these photons are not eliminated from the x-ray beam, they will be absorbed by the patient. To avoid this unnecessary patient exposure, the primary x-ray beam is filtered. **Filtration** is the process of removing the long-wavelength photons from the x-ray beam. Filtration material placed between the x-ray tube and the patient absorbs these long-wavelength photons (Fig. 4-23).

The material commonly used to filter the x-ray beam is aluminum. One or more aluminum plates installed between the tube port and the collimator serve this purpose. Any other material through which the beam passes also provides filtration. This includes the glass of the tube itself, plus the mirror and the faceplate of the collimator. Because aluminum is the primary filtration material, all filtration is measured in units of millimeters of aluminum equivalents (mm Al equiv.), the amount of filtration provided by a millimeter of aluminum.

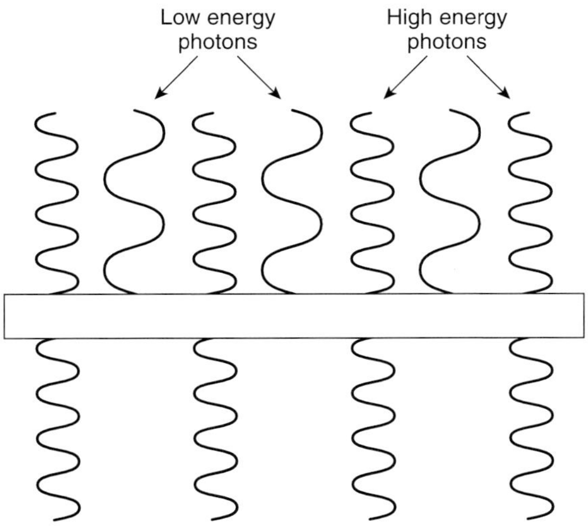

The filtration provided by the glass of the tube is called inherent (built-in) filtration and is approximately equal to 0.5 mm Al equiv. All other filtration is referred to as added filtration. The total filtration is equal to the inherent filtration plus the added filtration (Fig. 4-24). *X-ray equipment capable of producing 70 kVp or more (all general-purpose equipment) is required to have total filtration of at least 2.5 mm Al equiv. permanently installed.* Specialty equipment with a maximum kVp of less than 70, such as that used for mammography, requires less filtration.

The required added filtration is an important safety feature in radiography. It reduces patient dose to approximately 25% of the dose that would be received with only inherent filtration in place. Because the long wavelengths are removed from the x-ray beam, the filtered beam has a shorter average wavelength.

Fig. 4-23 Filtration absorbs long-wavelength photons.

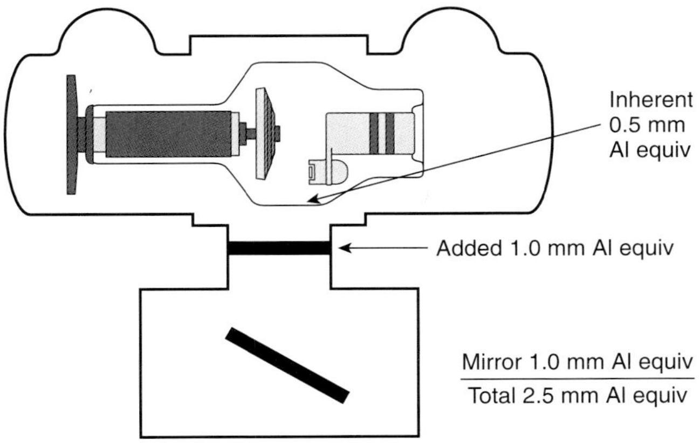

Fig. 4-24 Total filtration equals inherent filtration plus added filtration.

SUMMARY

X-rays are produced in a vacuum tube when high-speed electrons suddenly decelerate at the tube target. The electrons are liberated by heating the tungsten filament and are accelerated by high voltage from a step-up transformer.

Modern x-ray tubes for general radiography are dual-focus tubes with rotating anodes. Focusing cups that are part of the cathode assembly direct the electrons to the focal area on the target. The small filament and focal spot are used to provide fine detail with small exposures. The large filament and focal spot provide more electrons and greater heat capacity for large exposures. According to the line focus principle, the effective focal spot is always smaller than the actual focal spot, and its size affects image sharpness. The rotating anode increases tube heat capacity.

The intensity of the radiation field varies from one end to the other, because of the anode heel effect. The radiographer must know which end of the x-ray tube is the anode end and position the patient so that the anode heel effect is used correctly when using large films at a 40-inch distance.

The *quality* of the x-ray beam refers to its wavelength and penetrating power and is controlled by the kilovoltage. The *quantity* of the exposure is indicated by the mAs, the product of the mA and exposure time.

Filtration of the x-ray beam significantly reduces patient dose and decreases the average wavelength of the x-ray beam. A total of 2.5 mm Al equiv. filtration is required to be permanently installed on all equipment capable of operating above 70 kVp.

■ Review Questions ■

1. Describe the element tungsten. List two reasons why it is a good material for x-ray tube targets and at least one reason why it is used for x-ray tube filaments.
2. What is meant by "thermionic emission"? What is the purpose of thermionic emission in the x-ray tube?
3. What is meant by the term "heterogeneous"? What type of target interaction produces heterogeneous radiation?
4. What does a dual-focus tube have that a single-focus tube does not?
5. How much target angulation is needed in a general-purpose tube? Why?
6. What is the effect on the x-ray beam of an increase in kVp?
7. Why might it be desirable to increase mA? Why might it be undesirable?
8. An exposure is made using 100 mA and ¼ sec. What is the value of the mAs? State another combination of mA and time that will produce the same quantity of exposure.

5

X-ray Circuit and Tube Heat Management

Learning Objectives

At the conclusion of this chapter the student will be able to:

- Given an unlabeled x-ray circuit diagram, label the parts and state the function of each
- Explain what is meant by rectification and compare the three basic types
- Draw the current waveform for each of the following types: unrectified, half-wave rectified, full-wave rectified, three-phase rectified, and high-frequency
- List the primary features of all x-ray control panels and discuss the principal differences between conventional and computerized control consoles
- List five possible causes of x-ray tube failure and describe methods to prevent each

Key Terms

autotransformer
diode
electrical ground
exposure switch
exposure timer
full-wave rectification
half-wave rectification
heat units (HU)
high-frequency generator
ionization chamber
kVp meter
line meter

line voltage compensator
mA selector
milliammeter (mA meter)
phototimer
rectification
rectifier
rheostat
rotor switch
self-rectification
synchronous timer
three-phase current
tube rating chart

The radiographer is not required to build, alter, or repair an x-ray circuit. This chapter centers on a greatly simplified diagram of an x-ray circuit and is intended to aid your understanding of the various components of the circuit and how they work together to produce and control x-rays. It may be helpful to review the circuit diagram symbols introduced in Chapter 3 (see Table 3-4). The various features of the x-ray circuit and the x-ray control panel are discussed here in detail. Because x-ray tubes may be damaged by improper use and are expensive to replace, this chapter provides guidelines for the safe operation of tubes and suggestions for prolonging tube life.

X-RAY CIRCUIT

As indicated in Fig. 5-1, the x-ray circuit is divided into three sections or subcircuits: the *control circuit*, the *filament circuit*, and the *high-voltage circuit*. The various components of each section are numbered so that you can easily refer to them in the discussion that follows.

Control Circuit

The control circuit is illustrated at the upper left portion of Fig. 5-1 and is expanded in Fig. 5-2. It is the subcircuit between the AC power supply (1) and the primary (input) side of the high voltage (step-up) transformer (11). If you trace this circuit, beginning at the AC power supply, you will note that current flows through several devices before reaching the primary side of the step-up transformer. From the transformer, it returns to the power source, forming an enclosed loop. With the exception of the step-up transformer, all of the devices in this subcircuit are actually located within the control console. They include the main switch (2), line meter (3), line voltage compensator (4), autotransformer (5), kVp selector (6), kVp meter (8), exposure switch (9), and exposure timer (10).

The AC power supply (1) is wired into the building, providing electric power from the local power company. Most outpatient facilities have a 240 V power supply to the x-ray machine. Hospitals with more powerful equipment may have a larger supply. The main switch (2) controls the power to the control console.

The **line meter** (3) is a voltmeter, wired in parallel across the circuit to measure the incoming voltage. While the power supply may be rated at 240 V, the actual voltage varies somewhat, depending on the demand for power in the building or the neighborhood. Since this voltage will be multiplied many times by the step-up transformer, small variations in the incoming line may cause large variations in the kilo-voltage to the x-ray tube. For this reason, the incoming voltage must be monitored.

The **autotransformer** (5) is a single-coil transformer that works on the principle of self-induction (magnetic influence between the windings). It serves three functions: it provides the means for kVp selection, it provides compensation for fluctuations in the incoming line voltage, and it supplies power to other parts of the x-ray circuit.

The autotransformer's primary purpose is to vary the voltage to the primary side of the step-up transformer. This is accomplished by the kVp selector (6), which is a variable contact on the secondary (output) side of the autotransformer. It can be adjusted to control the number of autotransformer windings that are included in the outgoing portion of the circuit. Since the step-up transformer simply multiplies whatever voltage it receives, the autotransformer varies the kilovoltage to the tube by controlling the input to the step-up transformer.

The **line voltage compensator** (4) is a variable contact on the primary (input) side of the autotransformer. Its purpose is to adjust the number of autotransformer windings included in the incoming portion of the circuit. These adjustments are necessary to ensure that the outgoing voltage is accurate, regardless of fluctuations in the incoming line.

On the circuit diagram, connections from the autotransformer (7) form a part of the filament subcircuit, indicating that the filament circuit voltage is supplied by the autotransformer. The x-ray machine has other needs for voltage that are not shown in the diagram, and these may be supplied by the autotransformer as well. For example, the autotransformer may provide power for target rotation, lights on the control console, the collimator light, and the locks that secure the tube support.

The **kVp meter** (8) is connected in parallel across the circuit, secondary to the autotransformer. It measures the voltage output from the autotransformer. By measuring this voltage and taking into account the increase provided by the step-up transformer further on in the circuit, this meter predicts the kilovoltage across the x-ray tube. For this reason, it is sometimes called a "prereading voltmeter."

The **exposure switch** (9) closes the circuit, allowing current to flow through the primary side of the step-up transformer. When this occurs, current is induced to flow through the secondary side of the transformer, creating voltage across the x-ray tube. As previously discussed, this voltage causes the electron stream to flow across the tube, producing x-rays. The **exposure timer** (10) is a device that terminates the exposure by opening a switch after a preset time has elapsed. Exposure switches and timers are discussed in detail later in this chapter.

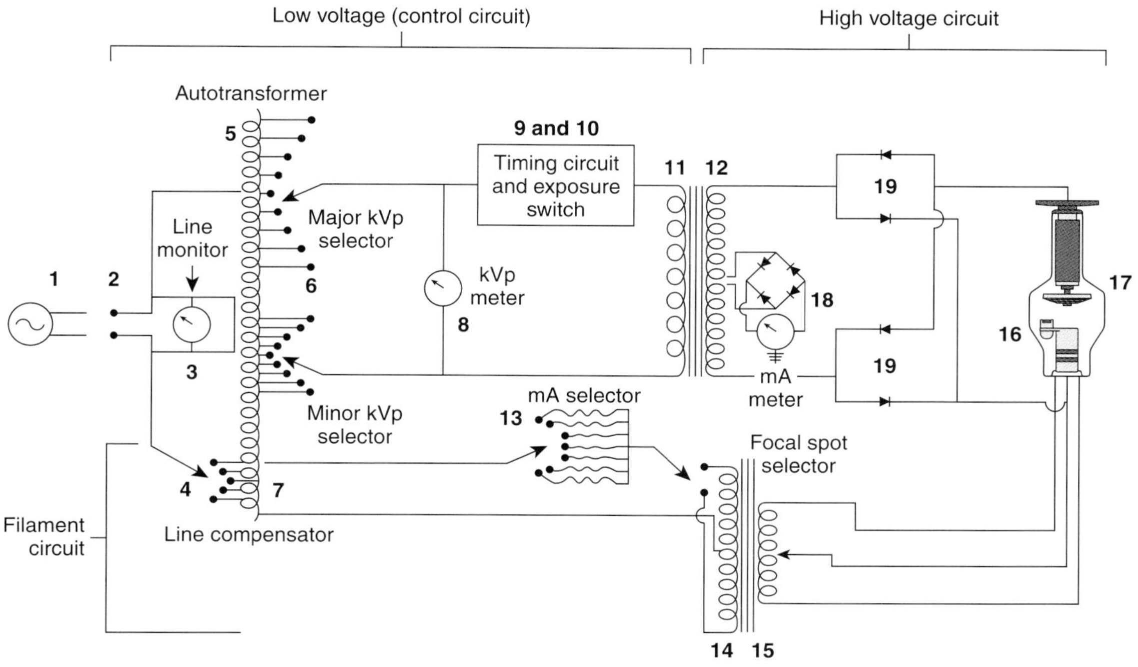

Fig. 5-1 Simplified diagram of an x-ray circuit.

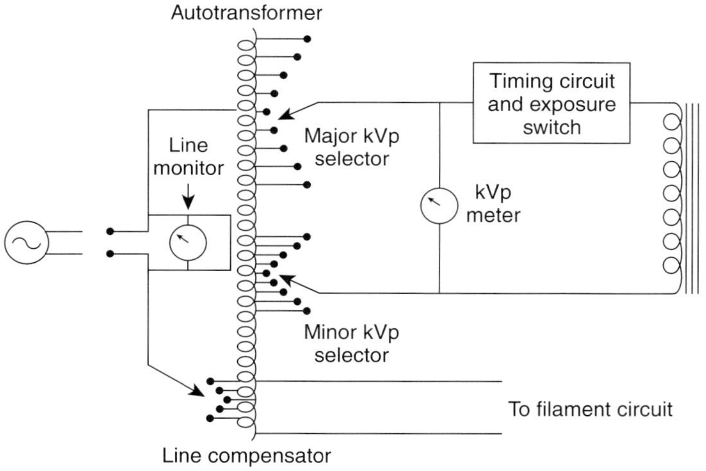

Fig. 5-2 Control circuit.

Filament Circuit

The filament circuit is the subcircuit of the x-ray circuit shown as the lower portion of Fig. 5-1. It is expanded in Fig. 5-3. It is divided into two parts by the step-down transformer *(14/15)*. The purpose of the filament circuit is to supply and control the heat required by the x-ray tube filament for thermionic emission. Adjustments made in this circuit will determine the milliamperage through the x-ray tube.

The primary side of this circuit begins and ends with the contacts on the autotransformer *(7)*. Current in this circuit flows from the autotransformer, through the mA selector *(13)* and the primary side of the step-down transformer *(14)*, and back to the autotransformer. The secondary side begins and ends with the secondary side of the step-down transformer *(15)*, conducting current through the x-ray tube filament *(16)*. The step-down transformer reduces the voltage on the secondary side, providing an appropriate current to heat the filament.

The **mA selector** *(13)* is a device called a **rheostat.** It is a variable resistance, providing several choices in the amount of resistance in the circuit. This control of resistance permits control of the amperage

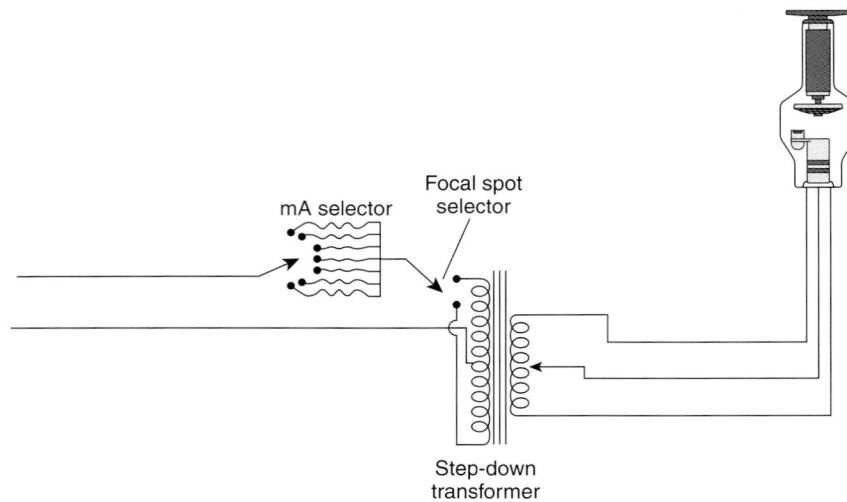

Fig. 5-3 Filament circuit.

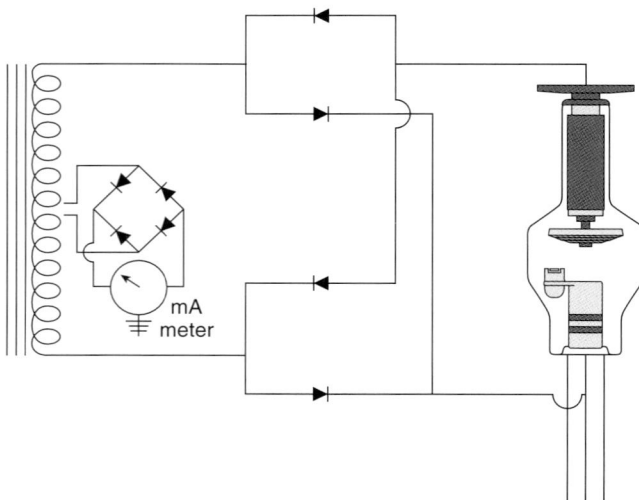

Fig. 5-4 High-voltage circuit.

in this circuit. Since the current through this circuit controls filament heat, this setting determines the number of available electrons in the x-ray tube and thus determines the mA in the high-voltage circuit that includes the x-ray tube.

High-Voltage Circuit

The high-voltage circuit is the subcircuit shown in the upper right portion of Fig. 5-1. It is expanded in Fig. 5-4. This circuit begins and ends with the secondary side of the step-up transformer *(12)*. It includes the x-ray tube *(17)*, the milliammeter *(18)*, and the rectifier unit *(19)*. *Current flows in this circuit only during an exposure.*

The step-up transformer is also referred to as the "high-voltage" or "high-tension" transformer. As explained in Chapter 3, it increases the incoming volt-

age by a set multiple called the transformer ratio. This transformer has a very high ratio. The primary voltage cannot be more than the voltage provided by the incoming line. It must be increased to provide the kilovolts needed to operate the x-ray tube.

The **milliammeter (mA meter)** is connected in series at the midpoint of the secondary side of the step-up transformer. This meter measures the current flowing in the high-voltage circuit during an exposure. It may serve as the exposure indicator on the control panel.

Note that the mA meter is "grounded," that is, connected to **electrical ground.** An electrical ground connection can absorb or "drain off" any number of excess electrical charges, resulting in zero electrical potential. This ground connection provides safety from high-voltage shock when the meter is mounted in the control panel. Because the voltage between this point and each end of the transformer is only half the voltage across the entire transformer, the ground reduces the voltage that must be carried by each cable that connects the transformer to the x-ray tube. This is advantageous because the cables would otherwise be much heavier and much more expensive.

As you have already learned, alternating current flows in two directions in a circuit, but it is desirable that the current flow in only one direction through the x-ray tube, from filament to target. The purpose of the **rectifier** unit *(19)* is to change the alternating current so that it flows in only one direction. As stated in Chapter 3, this process is called rectification.

RECTIFICATION

To *rectify* is "to make right," and the process of electrical **rectification** makes the current "right" by ensuring that it flows only in the right direction, in this case from filament to target.

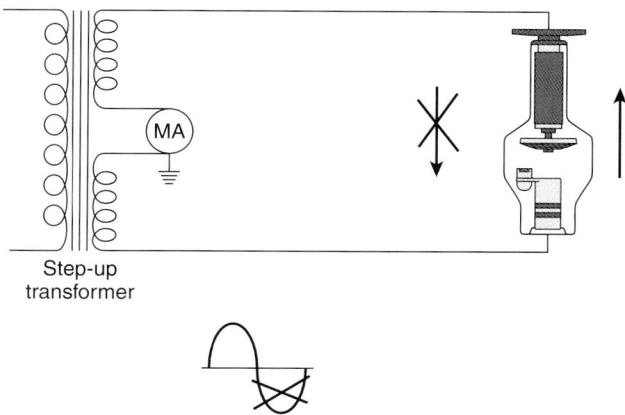

Fig. 5-5 Self-rectification. The nature of the x-ray tube prevents current from flowing from anode to cathode during half of the electrical cycle.

Fig. 5-6 Voltage waveform for both self-rectification and half-wave rectification.

The simplest high-voltage circuit employs **self-rectification** and is represented in Fig. 5-5. This is a primitive and inefficient form of rectification. It has no rectifier unit. Current flows in this circuit only during the portion of the electrical cycle when the anode is positive and the cathode is negative. During the other half of the cycle, there is a positive attraction at the cathode, but there are no electrons at the target to carry current back across the tube, so no current flows. The current waveform is illustrated in Fig. 5-6. This circuit design is called self-rectification because the x-ray tube itself controls the direction of current flow.

Self-rectification is not efficient because x-rays are produced during only half of the electrical cycle. This results in prolonged exposure times and wasted energy. Self-rectification is not totally effective either. As the anode heats during an exposure, its tungsten atoms may tend to liberate electrons through the process of thermionic emission, just like the tungsten atoms of the filament. Although the target is not constructed to facilitate thermionic emission, this can occur to some degree when the tube gets very hot. When this occurs, electrons flow "backward" from target to filament during the second half of the electrical cycle, damaging the tube.

Half-Wave Rectification

To prevent the damage that might occur if the tube failed to provide adequate self-rectification, one or

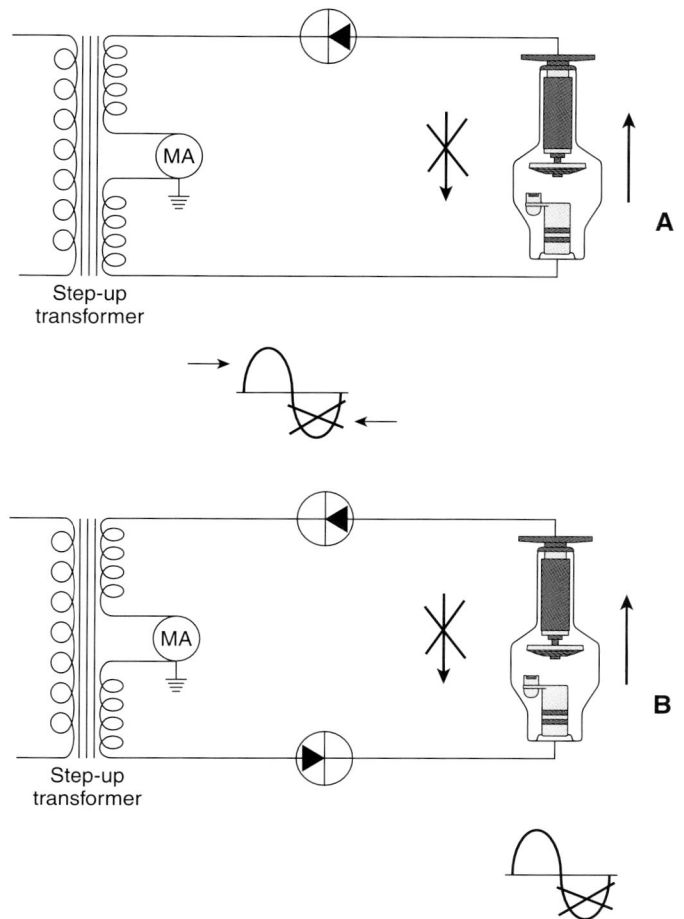

Fig. 5-7 Half-wave rectification. The diode prevents current from flowing from anode to cathode during half of the electrical cycle.

☀ *Rectification*

"To make right"
   The process of causing alternating current to flow in one direction only

---

two diodes may be included in the circuit. This circuit design (Fig. 5-7) provides **half-wave rectification.** A **diode** is an electronic device that permits current to flow in one direction only. If one diode is used, it is placed between the anode and the transformer. If two are used, the other is placed between the cathode and the transformer. Diodes prevent "backflow" of current during the second half of the electrical cycle. Note that the arrow direction of the diode symbol indicates the direction of current flow permitted by the diode. While half-wave rectification prevents tube damage, it is no more efficient than self-rectification. The current waveforms for the two circuit types are identical (see Fig. 5-6).

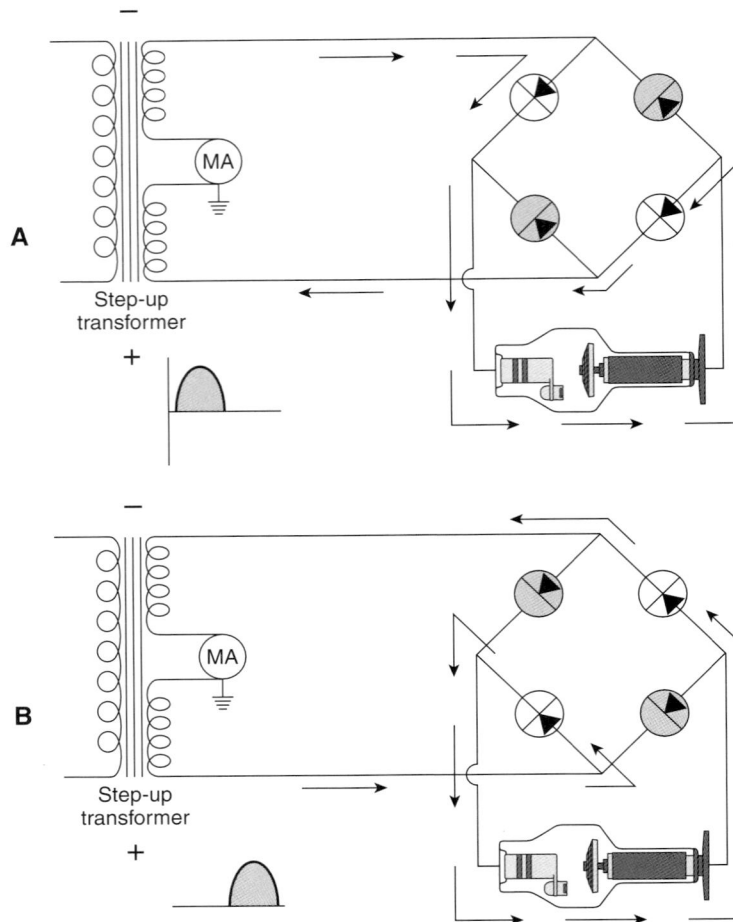

Fig. 5-8 Full-wave rectification. **A,** First half of cycle; **B,** second half of cycle.

Fig. 5-9 Full-wave rectification voltage waveform.

Full-Wave Rectification

By employing four diodes in the circuit, the current can be "redirected" during the second half of the electrical cycle so that current will flow in the same direction during both halves of the cycle. This process is called **full-wave rectification** because it utilizes the entire electrical cycle for x-ray production. It is best understood by tracing the current through the rectifier unit during each half of the electrical cycle, as illustrated in Fig. 5-8. Most modern general-purpose x-ray machines are full-wave rectified, employing the available energy more efficiently and shortening exposure times to half that required with either self-rectified or half-wave rectified units. The waveform of full-wave rectified current is shown in Fig. 5-9.

While this current is sometimes referred to as "direct current," it is very different from the direct current provided by a battery. True direct current is constant and is produced by a constant voltage. Rectified alternating current is pulsating. The lowest points in the cycle provide too little voltage to produce x-rays.

THREE-PHASE AND HIGH-FREQUENCY GENERATORS

A more constant and efficient voltage source is provided by a three-phase power supply that is produced by a special kind of transformer. Alternating current is generated in three overlapping cycles that produce the waveform illustrated in Fig. 5-10. When this current is rectified, its waveform has the appearance of a "ripple" with no real low points. The resulting waveform is shown in Fig. 5-11. **Three-phase current** generates x-rays more efficiently than single-phase current, decreasing exposure times by 40% to 50% and also decreasing patient dose. The purchase and installation of three-phase x-ray equipment are very

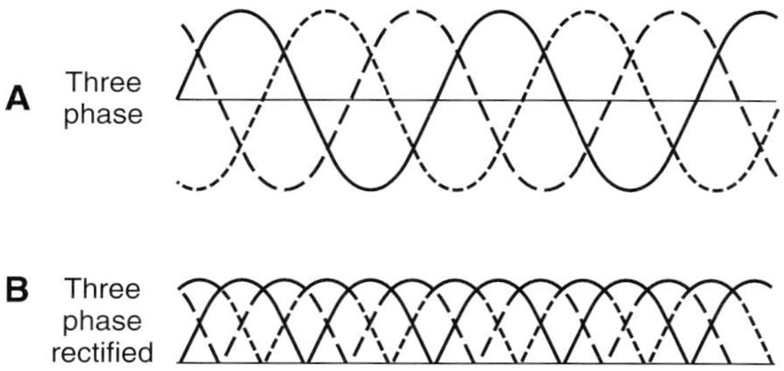

Fig. 5-10 Three-phase current. **A,** Unrectified; **B,** rectified.

Fig. 5-11 Rectified three-phase voltage waveform.

Fig. 5-12 High-frequency voltage waveform.

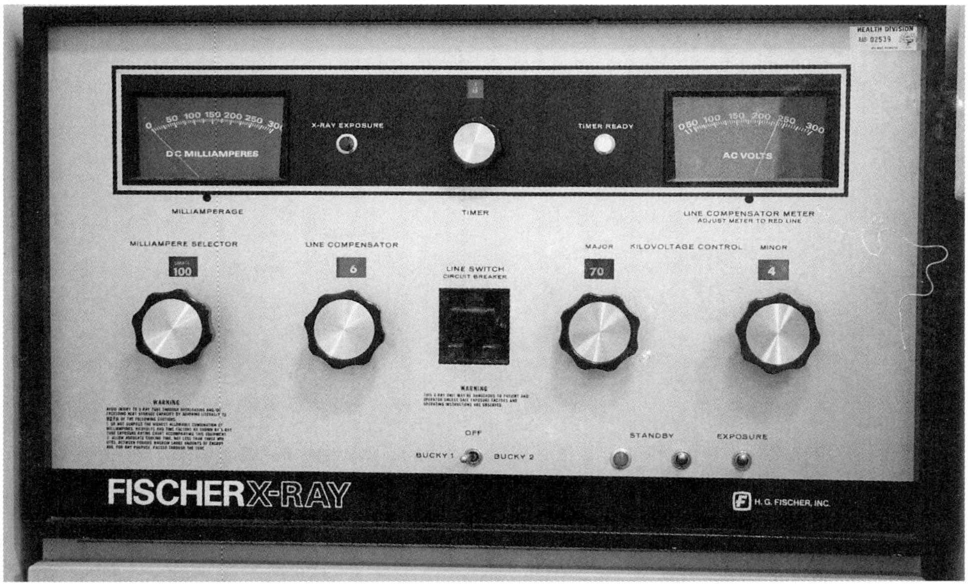

Fig. 5-13 Conventional control panel.

expensive compared to single-phase. Three-phase units are usually found only in hospitals.

A less expensive means of increasing the efficiency of x-ray production is provided by **high-frequency x-ray generators.** These units employ special "inverter circuits" that convert rectified AC into a series of square pulses. The latest high-frequency generators produce a near-constant voltage waveform (Fig. 5-12). Some of these units are as efficient as three-phase units, and all provide the same advantages to some degree.

X-RAY CONTROL PANEL

The devices of the x-ray control panel were introduced in the section on the control circuit. At this point, we consider how these devices appear on the control console and how the radiographer uses them.

X-ray control panels and the labels of their components vary, depending on age and manufacturer, but there are only two basic types. The older, "conventional" type is operated by means of knobs and switches. Dials or meters indicate the settings (Fig. 5-13). Newer, computerized models have button-like

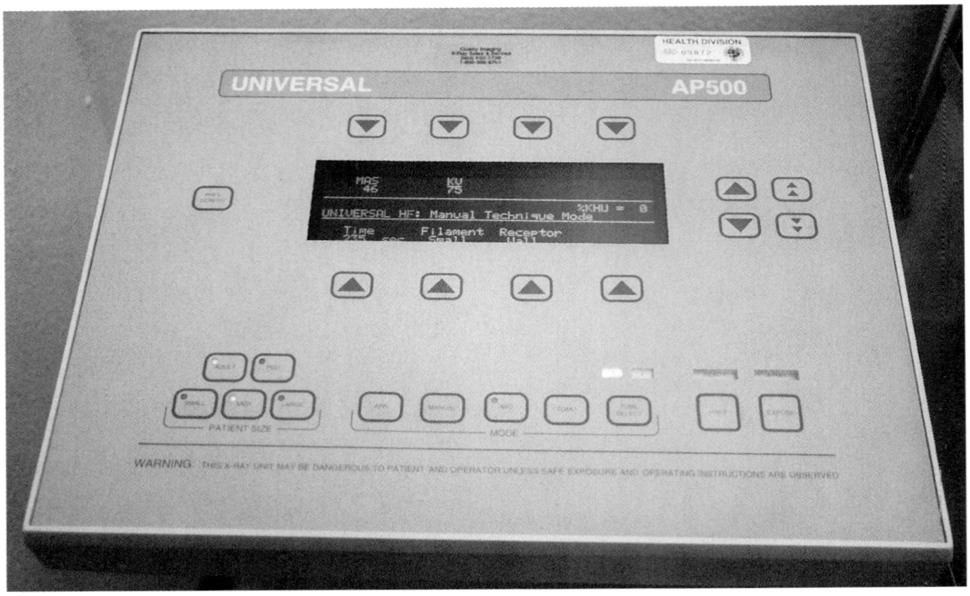

Fig. 5-14 Computerized control panel.

controls and digital readouts (Fig. 5-14). These computerized consoles automatically perform some functions that are done manually on conventional units.

All control panels provide some means for selecting kVp, mAs, and filament/focal spot size, and all indicate the current settings. There will also be switches to control power to the console and the bucky, plus rotor and exposure switches. Additional controls may also be present, depending on specific features of the equipment.

Power Control to Console

The on/off switch on the console corresponds to the main power switch in the circuit diagram (see Fig. 5-1, 2). This switch controls the power to the control panel and the entire x-ray generator. There is usually an electrical panel in the wall of the control booth that contains one or more circuit breakers. The appropriate circuit breaker must be turned on for the machine to receive power. Either a circuit breaker or the console power switch will activate the collimator and the tube support locks.

Some control consoles are designed to control either of two x-ray machines, so that two x-ray rooms may share a control booth and control console. When this is the case, there will be a selector on the control panel to activate the equipment that is in immediate use. These controls may be labeled simply "Tube 1" and "Tube 2," so the radiographer must know the significance of these designations.

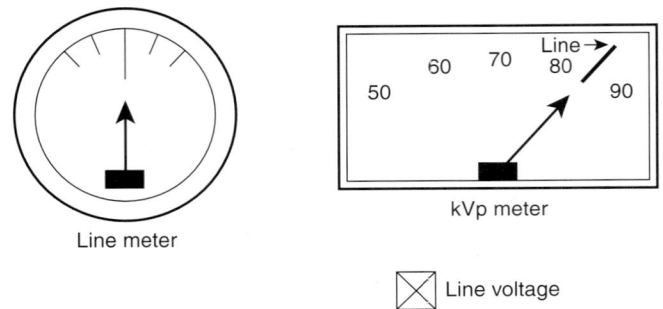

Fig. 5-15 Two types of line meters.

Line Voltage Compensation

Conventional control panels include a meter to monitor fluctuations in line voltage (see Fig. 5-1, 3) and a control knob to adjust for line voltage compensation (see Fig. 5-1, 4). When the meter indicates that the line voltage is above or below the optimum level, the voltage compensator must be adjusted until the meter reading is correct. Fig. 5-15 illustrates two types of line meter. The second type is actually a kVp meter, accompanied by a line meter button. When the button is pressed, it changes the contacts on the kVp meter, turning it temporarily into a line meter. The numbers on the kVp meter do not indicate line voltage. When serving as a line meter, this meter simply indicates whether or not the line voltage is correct.

Computerized control units monitor line voltage and compensate for fluctuations automatically, so there is usually no means for it to be monitored or controlled by the radiographer.

Milliamperage Control

Conventional control panels have an mA selector that provides several choices of milliamperage, often called "mA stations." The selector may be in the form of a knob or a series of buttons. Each setting corresponds to a specific resistance in the rheostat of the filament circuit (see Fig. 5-1, *13*). The number of possible mA settings is limited, and each is usually a whole number that is divisible by 50 or 100. For example, a typical radiographic unit may have the following mA stations: 50, 100, 150, 200, 300, 400, and 500. Some x-ray machines are capable of producing as much as 1500 mA, but equipment in outpatient facilities often has a maximum mA of 300.

Each filament in a dual-focus tube is connected to specific mA stations. The filament and its associated focal spot are chosen automatically when the mA is selected. The control panel will indicate which stations utilize which filament. For example, a setting marked "200 L" or "200 Large" signifies 200 mA with the large filament and large focal spot. Usually mA settings of 150 or less employ the small filament; the large filament supplies mA requirements of 200 or more. The rationale for selection of mA stations is discussed in Chapter 10.

Most computerized control consoles do not have mA selectors. These units provide variable mA that is computer-controlled according to the radiographer's selection of the desired mAs and focal spot size. Some are called "falling-load" generators. These units begin the exposure at the highest mA compatible with the desired filament/focal spot selection. The mA decreases during the exposure, if necessary, to avoid overheating the anode. The exact quantities of mA and time are determined by the computer to deliver the desired mAs as quickly as possible. Digital readouts will state the mAs setting and the exposure time. There may be an optional exposure time control that enables the radiographer to extend the time for an exposure, effectively lowering the mA without altering the mAs.

Some computerized controls are programmable so that exposure factors from a conventional technique chart can be made available from within the computer. These units may feature both a manual mode and an automatic mode. In manual mode, the radiographer sets the mAs and the kVp. In the automatic mode, the radiographer enters the examination and the body part measurement, and the computer selects the correct exposure factors.

Exposure Timers

All control consoles contain an exposure timer, whether it is set directly by the radiographer or not. Two types of manually set timers are used for x-ray exposure control: **synchronous timers** and electronic timers (Box 5-1). Computerized controls may

---

Types of Timers

Type	Minimum exposure time
Synchronous (impulse)	$\frac{1}{120}$ sec
Electronic	1 msec or less
Automated	Depends on back-up timer (may be synchronous or electronic)

---

select the exposure time on an electronic timer when the radiographer sets the mAs. In addition, some units have automated exposure controls, which are discussed in the following section.

A synchronous timer is controlled by a small electric motor, and its function is similar to that of an electric clock. Each impulse in the electrical cycle advances the timer, so that essentially it functions by "counting impulses." For this reason, it is sometimes called an "impulse timer." The time settings available will be stated in fractions, and all will be multiples of $\frac{1}{120}$ sec, the duration of an electrical impulse. The range of time settings is between $\frac{1}{120}$ sec and several seconds.

Electronic timers are more sophisticated devices designed for use with three-phase and high-frequency generators that have a current waveform with no discrete impulses. Electronic timers are capable of shorter exposure times than synchronous timers (1 msec or less), and exposure times are expressed in decimals.

Automated Exposure Control

Some x-ray machines provide automated exposure control (AEC), terminating the exposure when a certain quantity of radiation has been detected in the remnant portion of the x-ray beam. These systems limit the exposure time to the correct amount for proper exposure of the film, eliminating the need to know in advance just how much mAs a radiograph will require. There are two types of automated exposure controls: **phototimers** and **ionization chambers** (Fig. 5-16).

Phototimers are photomultiplier tubes that are sensitive to light. A fluorescent screen under the bucky tray gives off light when exposed by radiation that has passed through the patient and the film. The phototimer terminates the exposure after a specific quantity of light has been detected.

An ionization chamber is an enclosed volume of dry air connected into a circuit. When the air is ionized by the remnant radiation, a minute amount of current flows through the circuit. When a specific quantity of current has been measured in this circuit,

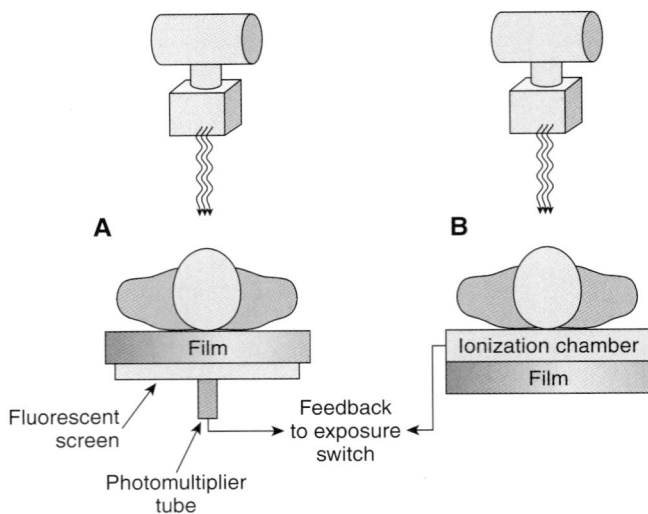

Fig. 5-16 Automated exposure systems. **A,** The photomultiplier tube responds to light from a fluorescent screen. **B,** The ionization chamber is charged by exposure from remnant radiation.

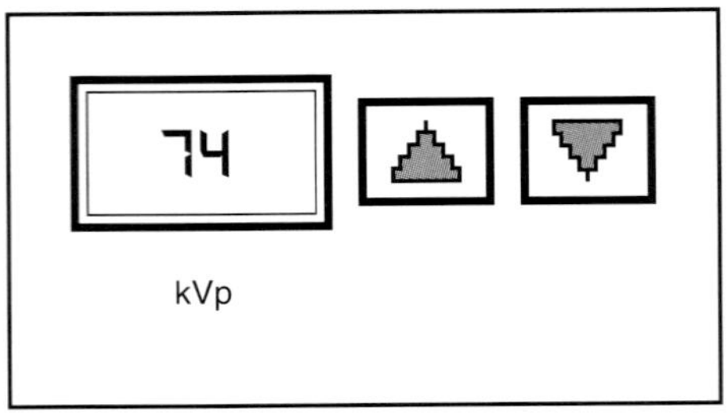

Fig. 5-17 kVp selector and display on a computerized control console.

the exposure is terminated. The ionization chamber is located between the patient and the film.

Both phototimers and ionization chambers usually have multiple detectors so that radiographers can select the location or locations within the radiation field where the radiation quantity will be measured. The selection of active detectors is determined by the film size and the specific radiographic examination. For example, a knee radiograph would utilize a central detector, while a chest film requires activation of two detectors with positions that correspond to the lungs.

Control panels for units with automated exposure systems will have a switch to engage the automated system and a means for activating the desired detector(s). The radiographer must set the kilovoltage and also set a "back-up time." This means that the exposure timer is set to a time that is greater than the anticipated requirement of the exposure. If the automated system fails for any reason, the back-up timer will terminate the exposure. If the back-up time setting is not long enough, the back-up timer will terminate the exposure prematurely.

Kilovoltage Control

Selection of kVp involves the contacts on the secondary side of the autotransformer (see Fig. 5-1, 6). Conventional control panels have two kVp selector dials: a major selector that changes the kilovoltage by 10 kVp at a time and a minor selector that has increments of 1 or 2 kVp. For example, if the major kVp selector is set at 70 and the minor selector is set at 4, the total kilovoltage will be 74 kVp. This dual selector system facilitates large changes in kVp. It would be cumbersome to adjust the kVp from 55 to 120 by turning a single dial "one click" for each kilovolt of

the change. The kVp setting on a conventional control panel may be indicated on the selector dials or may be read from a kVp meter.

Computerized controls will have a digital read-out for the current kVp setting and arrow buttons that can be pressed to increase or decrease the setting (Fig. 5-17). The rationale for changes in kVp is discussed in Chapter 10.

Bucky Switch

As mentioned in Chapter 2, the bucky is a moving grid that is used for radiography of the larger parts of the body. The bucky device incorporates a motor to move the grid. A switch (button or toggle type) on the control panel activates the motor circuit so that the grid will move during the exposure. The control panel may have a three-position switch so that it can control two buckys, one in the table and one in the upright cassette holder. The three positions are usually labeled "Bucky 1," "Bucky 2," and "Off."

Exposure Controls

Until about 25 years ago, exposure switches were incorporated into a "hand switch" that was attached to the control panel by a cable. Some of this equipment is still in use, but regulatory agencies usually require that it be permanently fastened to the control console so that exposures cannot be made from a position outside the control booth. More modern equipment has exposure switches in the form of buttons or toggle switches that are mounted on the control panel.

Two separate switches are needed to make an exposure (Table 5-1). The first is the **rotor switch,** which may be labeled "Rotor," "Prep," "Ready," or "Standby." This switch has two functions. When it is activated, the rotating anode begins to spin, and heat is applied to the filament to create the space charge. When this switch has been held in the "on" position for several seconds, a signal will indicate that the tube is ready for an exposure. The signal may be a particular sound or a light on the control panel or both. Making an exposure before the tube is ready may damage the tube. Nearly all x-ray machines in current use have a lockout feature that prevents the initiation of an ex-

TABLE 5-1

Synchronization of Exposure Switch and Timer Switch

	Exposure Switch	Timer Switch
Before exposure	_____ / _____	_____
During exposure	_____	_____
When preset time has elapsed	_____	_____ / _____
When exposure switch is released	_____ / _____	_____

Key to symbols: ____/ __ = off; _____ = on.

BOX 5-2

Setting Control Panel and Making an Exposure (Traditional X-ray Control)

- Check line voltage; adjust if needed
- Select mA
- Select exposure time
- Select kVp—major first, then minor
- Set bucky switch
- Activate and hold rotor switch
- On signal, activate and hold exposure switch
- Observe exposure indicator to validate exposure and to determine when it is complete
- Release rotor and exposure switches

posure before the rotor has reached operating speed and the filament has reached operating temperature.

When the tube is ready, *the exposure is initiated by continuing to hold the rotor switch and also pressing the second switch, the exposure switch.* An exposure indicator on the control panel will indicate when the timer has terminated the exposure. Only then are the two switches released. *Premature release of either the rotor switch or the exposure switch will abort the exposure before it is complete.*

On older equipment, the exposure indicator will be an mA meter on the control panel. It will indicate the mA during the exposure and return to zero as soon as the exposure is complete. Newer models will have an "exposure light," usually red in color, that is on during the exposure. When the light goes off, the exposure is complete. Sometimes there is also a sound signal.

The process of setting a *modern* control panel may take place in any order. It is wise to form a habit of setting the controls in the same order each time so that nothing is overlooked.

When setting an *older,* conventional control panel, the order may be very important. It is wise to check the line voltage and make any necessary adjustments before proceeding with the other settings. A change of mA station may result in an unintentional change in kilovoltage, so it is necessary to set the mA before setting the kVp. Box 5-2 lists in order the steps for setting the controls and making an exposure with a conventional control console.

PROLONGING X-RAY TUBE LIFE

The anode of the tube accumulates heat during exposures, and this is a primary cause of problems that may shorten tube life. The design of modern tubes incorporates several features for the purpose of rapid heat dissipation. The rotating anode prevents excess heat in any one area of the target, spreading it around the focal track. The anode disk consists of several layers of material (tungsten, molybdenum, and graphite), chosen for their heat-management characteristics. The stem of the anode conducts heat to a copper mass surrounding the rotor mechanism that serves as a "heat sink." Also, the space between the tube and the tube housing is filled with oil. This feature provides electrical insulation and also disperses heat from the glass envelope.

X-ray tubes that receive good care provide many years of service. Careless use, however, can significantly shorten the life of a tube and may result in sudden tube failure. Tube replacement is expensive and may cause the x-ray equipment to be out of service for some time. *The factors that affect tube life are controlled by the radiographer.* Responsible radiographers take care to ensure that x-ray tubes are not abused.

An excessive exposure on a cold tube will cause the anode to expand too rapidly and may cause it to crack and to fail. For this reason, a cool tube should be warmed up prior to any large exposure. The manufacturer may specify the warm-up procedure. Warm-up settings are preprogrammed into some computerized controls. In the absence of an established warm-up procedure, three exposures, 30 seconds apart, at a setting of 200 mA, 0.5 sec, and 80 kVp, will safely distribute heat throughout the anode. The warm-up procedure must be repeated if the tube has been idle for more than an hour. Exposures smaller than prescribed warm-up settings may be made without warming the tube. Warm-up exposures must be done before the patient enters the x-ray room. *Do not forget to do a pre-exposure safety check before doing warm-up exposures.*

A rapid series of large exposures or a single excessive exposure may damage the tube by melting the tungsten surface of the focal track. This melted tungsten boils and then cools to an irregular, pitted surface (Fig.

Fig. 5-18 Comparison of smooth, shiny appearance of rotating anodes when new *(a)* and their appearance after failure *(b-d)*. Examples of anode separation and surface melting shown were caused by *b*, slow rotation caused by bearing damage; *c*, repeated overload; and *d*, exceeding maximum heat storage capacity.

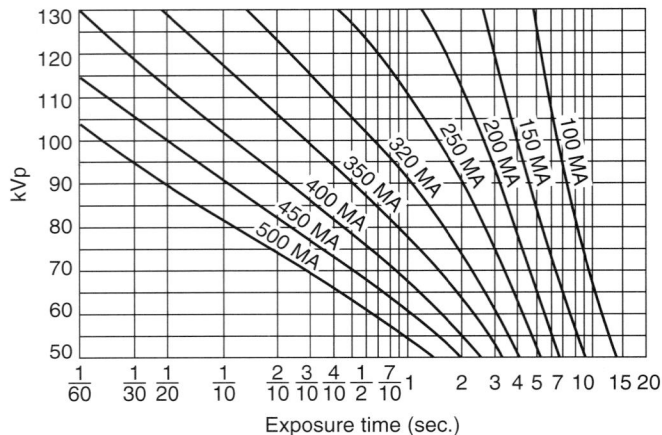

Fig. 5-19 Tube rating chart.

Tube Rating Charts

- Indicate maximum safe settings
- Specific to each tube design
- Used when creating a technique chart
- Check when using unusually large exposure

Calculating Percent Tube Load

1. At a given setting, determine the mAs.
2. Determine the maximum safe exposure time using the same mA and kVp.
3. Multiply maximum time by mA to determine maximum mAs.
4. Divide mAs determined in step 1 by mAs calculated in step 3.
   To prolong tube life, try not to exceed 80% tube load for routine work.

5-18). A tube that has been damaged in this way provides inconsistent radiation output and does not produce sharp images because of changes in the focal spot.

Modern x-ray generators have lockout circuits that prevent single exposures beyond the tube's maximum heat capacity, but this feature should not be relied upon as the only measure for tube heat protection. Tubes that are frequently used at or near their capacity will not last as long as tubes that are operated consistently at 80% of capacity or less.

Maximum tube capacity for a single exposure is determined by consulting the **tube rating chart** (Fig. 5-19). The chart is supplied by the tube manufacturer and is specific for each tube model. It should be displayed near the control console and consulted when a large exposure is contemplated. Tube rating charts are

used when creating technique charts or programming a programmable control to ensure that routine exposures are kept well below the limits of tube capacity.

The chart is read by noting the point on the graph where the horizontal line representing the kVp setting intersects with the vertical line that represents the exposure time. If this point is below the curved line that represents the mA setting, the exposure is safe. If it is above the mA line, the exposure exceeds tube capacity. From the chart it is apparent that much greater exposures can be obtained safely by using low mA settings and longer exposure times. Fig. 5-20 is a

Fig. 5-20 Reading a tube rating chart.

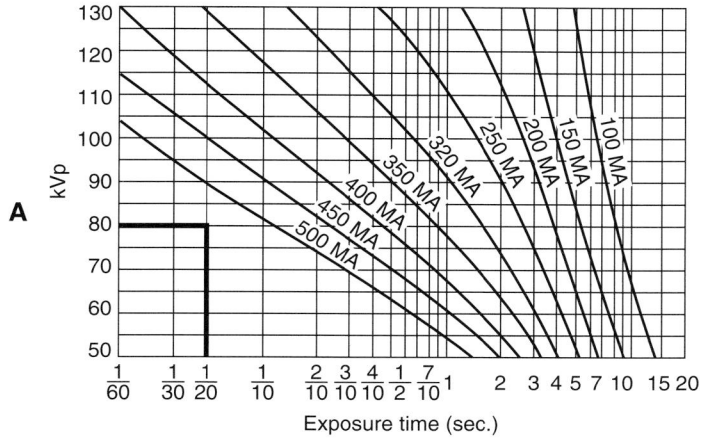

A, Safe exposure: 500 mA, ¹⁄₂₀ sec, 80 kVp.

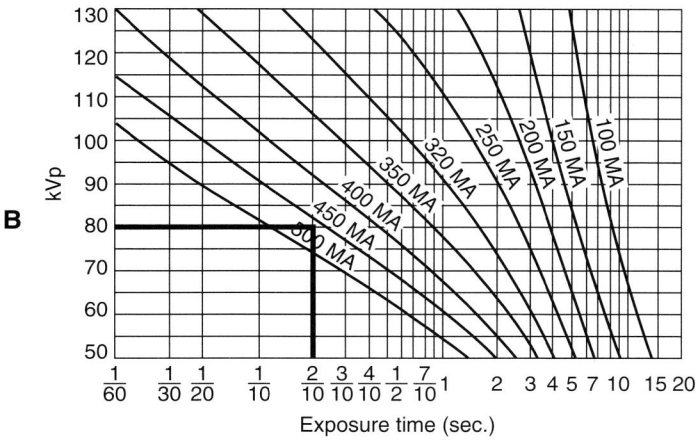

B, Unsafe exposure: 500 mA, ³⁄₁₀ sec, 80 kVp.

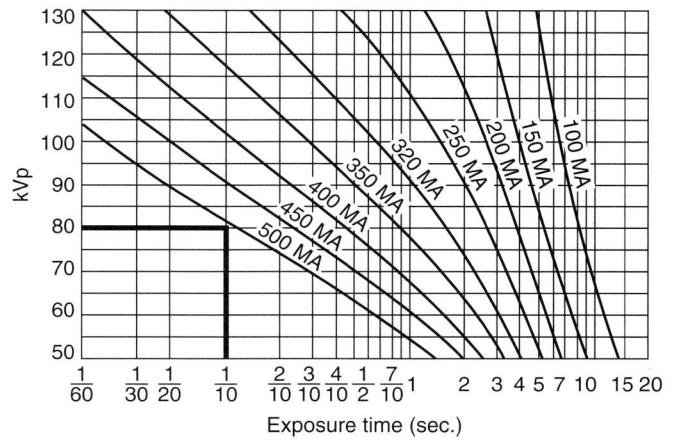

C, Maximum mAs at 80 kVp and 500 mA = 50 mAs (500 mA × ¹⁄₁₀ sec).

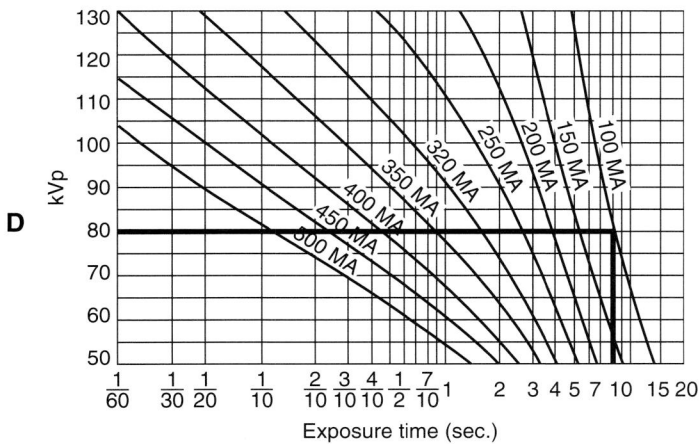

D, Maximum mAs at 80 kVp and 100 mA = 900 mAs.

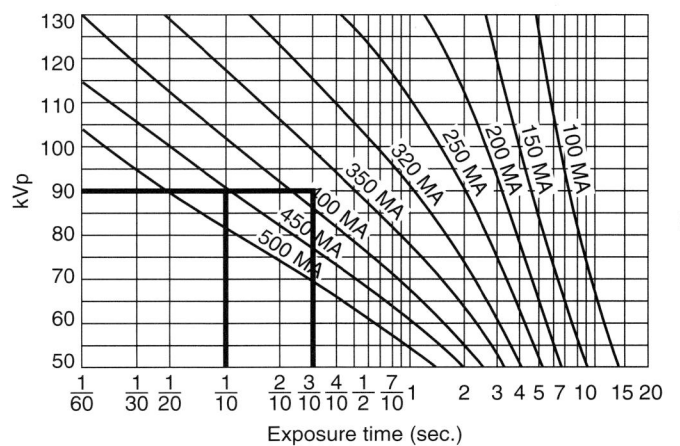

E, Calculating percent tube load. Exposure: 400 mA, ¹⁄₁₀ sec, 90 kVp. mAs = 40. Maximum exposure at 400 mA, 90 kVp = ³⁄₁₀ sec. mAs = 120. Divide desired exposure by maximum exposure: 40 mAs ÷ 120 mAs = 33.3% tube load.

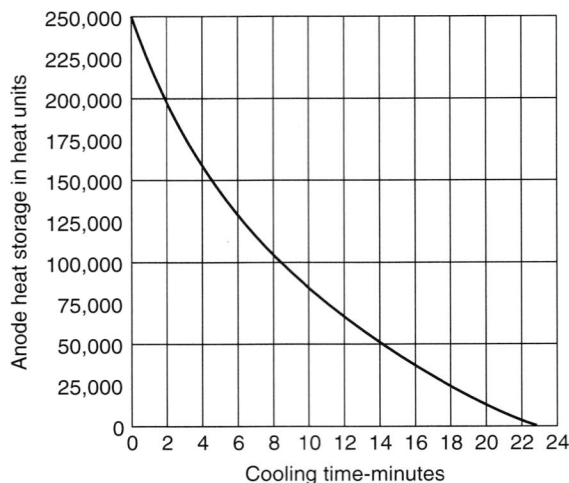

Fig. 5-21 Anode cooling chart.

series of exercises in reading a tube rating chart. A study of these illustrations will increase your confidence in reading tube rating charts accurately.

The maximum heat capacity of the anode is rated in **heat units (HU).** The heat units produced by an exposure are determined by multiplying the mAs by the kVp. The formula for heat unit differs depending on the design of the generator:

Single-phase: HU = mA × Time (sec) × kVp

Three-phase, six-pulse:
    HU = mA × Time (sec) × kVp × 1.35

Three-phase, 12-pulse:
    HU = mA × Time (sec) × kVp × 1.41

Heat units are used to calculate tube cooling time using a cooling chart (Fig. 5-21). Cooling charts also are provided by the tube manufacturer and are specific for each tube model. You will note that this chart was made for a tube with an anode heat capacity of 250,000 HU. If this tube were heated to its maximum capacity, half of the heat units would be dissipated within a period of 6 minutes. Cooling charts are used by radiographers involved in special procedures who need to calculate the effects of a rapid series of exposures, but they are seldom needed by limited radiographers. Tube heat dissipates rapidly following an exposure. The time needed to obtain a new film and reposition the patient between exposures usually allows for the tube to recover sufficiently to be ready for the next exposure. A rapid succession of large exposures must be avoided to prevent excessive anode heat accumulation. When a large number of films on a very large patient is contemplated for a machine with very limited capacity, it may be necessary to estimate the

heat units that will be produced and allow time for the tube to cool during the examination.

Even tubes that are never overheated eventually fail. The reason is almost always related to the events that occur when the rotor switch is activated. The anode rotor turns on precision ball bearings that are essential to its operation. These bearings wear out with use, and when this occurs, the tube must be replaced. During rotor activation, the filament is maintained at a high heat, resulting in tiny amounts of vaporization of the filament material. Over time, the diameter of the filament decreases until it is so thin that it breaks, similar to the failure of a light bulb filament that "burns out." The filaments are under greatest stress when used at the highest mA station to which they are connected. Using these settings only when needed will reduce filament wear and prolong tube life.

The life of rotor bearings and filaments can be greatly extended by minimizing the time that the rotor switch is activated. *Do not activate the rotor switch before you are completely ready to make the exposure.* Continuing to hold this switch, rather than proceeding with the exposure as soon as the tube is ready, increases wear on both bearings and filament, shortening tube life. Many radiographers develop the bad habit of starting the rotor switch before they begin to give breathing instructions to the patient. These instructions and their implementation take more time that is needed by the rotor, so this practice results in unnecessarily prolonged rotor time for every exposure. Most patients can easily hold their breath for the few seconds of preliminary rotor time, so it is best to start the rotor *after* the patient is ready. An exception to this rule is made when the patient is unable to cooperate and the radiographer must have instantaneous control over the timing of the exposure. This is often the case when performing radiography on infants and children.

SUMMARY

A simplified x-ray circuit diagram provides a means for studying the devices that make up the x-ray generator and how they work together to meet the requirements of radiography. The radiographer controls the function of these devices by means of a conventional or computerized control console. The control console provides the means for selecting the kVp and mAs and for making the x-ray exposure.

Rectification of alternating current is necessary for x-ray production. Although x-ray tubes tend to be self-rectifying, diodes in the high-voltage circuit provide more reliable rectification, preventing tube damage and increasing the efficiency of x-ray production. X-ray production is most efficient when the voltage waveform is nearly constant, as provided by three-phase and high-frequency generators.

The radiographer is responsible for implementing good practices to prolong x-ray tube life. These practices involve protecting the tube from sudden or excessive heating and minimizing the time that the rotor switch is activated.

■ *Review Questions* ■

1. What causes fluctuations in line voltage? What should the radiographer do about it?
2. What is the primary purpose of the autotransformer?
3. What is the name of the device in the circuit that is represented on the control panel as the mA selector? Where in the x-ray circuit is it located?
4. What is the name of the device used to provide rectification in the x-ray circuit? How many are needed to provide full-wave rectification?
5. List two advantages of a high-frequency x-ray generator compared to a full-wave rectified unit.
6. Explain in order the steps for making an x-ray exposure after the control panel has been set.
7. Name two things that occur when the rotor switch is activated.
8. Name three features of an x-ray tube that are designed for the purpose of handling heat.
9. Using the tube rating chart in Fig. 5-19, what is the maximum amount of mAs that can be obtained at 80 kVp? What combination of mA and time will produce this amount of mAs?
10. If an exposure is made using 300 mA, 1 sec, and 90 kVp, how many heat units are generated at the anode?

6

Principles of Exposure and Image Quality

Learning Objectives

At the conclusion of this chapter the student will be able to:

- List the prime factors of exposure
- State the formula for determining mAs and explain how this unit is useful to the radiographer
- Explain the radiographic effect caused by changes in each of the four prime factors of exposure
- Recognize changes in radiographic density and state the exposure factors used to control radiographic density
- Identify high, low, and optimum contrast on a radiograph and state the exposure factor that primarily controls radiographic contrast
- Define radiographic distortion and explain the difference between magnification and shape distortion
- Define radiographic definition and list factors that influence definition
- List and explain the geometric factors that affect radiographic definition and explain why magnification affects definition
- List and discuss methods for minimizing motion blur on radiographs

Key Terms

contrast
definition
distortion
film/screen contact
fog
inverse square law
latitude
long scale contrast
magnification
mottle
object-image distance (OID)
overexposure

parallax
penumbra
quantum mottle
radiographic density
resolution
short scale contrast
source-image distance (SID)
step-wedge penetrometer
subject contrast
tissue density
umbra
underexposure

This chapter explains the prime factors of radiographic exposure and their radiographic effects. You have already been introduced to some of them. In addition, it introduces the four primary factors of radiographic quality and the principal methods for controlling them. You will begin to observe the effects of exposure on radiographs and to understand how the various factors controlled by the radiographer affect the final image.

PRIME FACTORS OF RADIOGRAPHIC EXPOSURE

The prime factors of exposure are milliamperage (mA), exposure time (T), kilovoltage (kVp), and distance (SID). Each affects the quantity of exposure to the film in a different way.

Milliamperage

As explained in Chapter 4, changes in milliamperage (mA) affect the rate of exposure, that is, the number of photons produced per second during an exposure. For this reason, a change in mA will alter the quantity

---

Prime Factors of Radiographic Exposure

kVp (kilovolts peak)
- Controlled by manipulation of power source (adjustment of potential difference across tube)
- Increased kVp results in increased speed of electrons across tube
- Faster electrons have more kinetic energy
- Resulting photons will have more energy, expressed as shorter wavelength

Kilovoltage controls x-ray penetration

mA (Milliamperage)
- Controlled by filament heat
- Controls **rate of flow** of electrons across tube
- Controls **rate** of x-ray production
- **Quantity** of radiation produced is proportional to mA

Radiation exposure is proportional to mA

Exposure Time (Seconds)
- Controlled by timer in x-ray circuit
- Controls **duration** of exposure
- Controls **quantity** of x-rays produced
- Quantity of radiation produced is directly proportional to exposure time

Radiation exposure is proportional to exposure time

Source-Image Distance (SID)
Each dimension of the radiation field is proportional to the SID. Therefore the field area is proportional to the square of the SID and the radiation intensity is inversely proportional to the square of the SID.

---

of exposure to the film. An increase in mA will increase the quantity of exposure; decreased mA will reduce the quantity of exposure. Exposure is directly proportional to mA; that is, if the mA is doubled, the quantity of exposure will also be doubled.

Exposure Time

Exposure time also controls the exposure to the film. This factor affects the exposure by determining how long the tube operates and, therefore, how long the exposure will last. Obviously, a longer exposure time will increase the exposure to the film, and a decrease in exposure time will reduce the film exposure. The quantity of exposure is also directly proportional to the exposure time.

Milliampere-Seconds

As stated in Chapter 4, the unit used to indicate the total quantity of x-rays in an exposure is milliampere-seconds, abbreviated mAs. This unit is the product of mA and exposure time (mA × time = mAs). For example, if the control panel were set at 200 mA and 0.2 sec, the mAs would equal 200 times 0.2:

$$200 \text{ mA} \times 0.2 \text{ sec} = 40 \text{ mAs}$$

See Chapter 26 for instruction on multiplication with decimals and fractions, together with practice problems for determining mAs.

A desired quantity of exposure may be obtained by any combination of mA and time that, multiplied together, equals the desired mAs. For example, 40 mAs could also be obtained using 50 mA and 0.8 sec, 100 mA and 0.4 sec, or 400 mA and 0.1 sec.

The quantity of exposure and the patient dose are directly proportional to the mAs.

Kilovoltage

By controlling the wavelength of the x-ray beam, the kilovoltage determines the degree to which the x-ray beam penetrates the subject. This penetration determines how many of the total photons of the primary beam will reach the film. Thus kVp also has an effect on the quantity of exposure to the film. Unlike the effects of mA and time, changes in exposure are *not* directly proportional to kVp.

Source-Image Distance

The distance between the tube target and the film or imaging plane is called the **source-image distance,** abbreviated **SID.** At one time this distance was referred to as focal-film distance (FFD) or target-film distance (TFD), and you may still encounter these terms. Because the x-ray beam diverges into space,

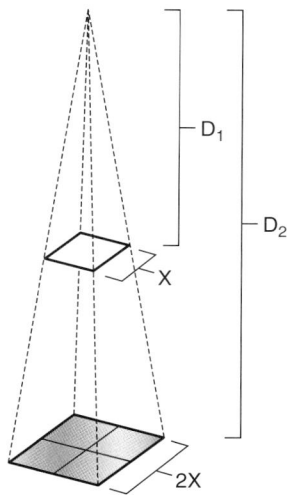

Fig. 6-1 Source-image distance affects both maximum field size and radiation intensity.

forming the shape of a cone, the photons get farther and farther apart as they get farther from the target (Fig. 6-1). Thus the SID affects the *intensity* of the x-ray beam, that is, the quantity of exposure in any given area of the radiation field. As noted in Fig. 6-1, each dimension of the radiation field is proportional to the distance. For this reason, the area of the field is proportional to the *square* of the distance. The greater the field area, the more the beam is spread out; as the area increases, the intensity of the beam and the exposure at any spot within the field are decreased.

The relationship between the SID and the intensity of the beam is expressed in the **inverse square law,** which states that the intensity is inversely proportional to the square of the distance. The inverse square law is expressed mathematically as a formula:

$$\frac{I_1}{I_2} = \frac{D_2{}^2}{D_1{}^2}$$

In this formula, I represents radiation intensity and D represents source-image distance. As the distance increases, the intensity is decreased. For example, if the distance were doubled, the intensity would be decreased to ¼ of the original intensity. In the illustration, suppose that D_1 is 40 inches and D_2 is 80 inches, twice as great. If the original intensity at D_1 had a value of 100, the intensity at D_2 would be 25. This value is determined by using the inverse square law formula.

Substitute the known values in the formula, using x for the unknown:

$$\frac{100}{X} = \frac{80^2}{40^2}$$ Reduce the fraction: $\frac{100}{X} = \frac{2^2}{1^2}$

Calculate the squares: $\frac{100}{X} = \frac{4}{1}$ Cross multiply: $4X = 100 \times 1 = 100$

To solve the equation, divide both sides by 4 to determine the value of X:

$$\frac{4X}{4} = \frac{100}{4}$$ Therefore: $X = 25$

Chapter 26 contains instructions and practice problems for working with squared numbers and for solving equations of this type.

RADIOGRAPHIC QUALITY FACTORS

There are four primary aspects of radiographic quality: density, contrast, distortion, and definition. An understanding of these factors is essential to the discussion and evaluation of film quality. Each quality factor is influenced and controlled differently. Knowledge about these concepts enables the radiographer to identify the nature of problems that relate to film quality and to solve these problems effectively.

Radiographic Density

Radiographic density refers to the overall blackness or darkness of the radiograph. A film that is neither too dark nor too light when seen on the viewbox is said to have the correct radiographic density. Fig. 6-2 provides examples of varying radiographic density. Note that density affects the *visibility of detail.* Some detail in the image is lost when the film is either too dark or too light.

Radiographic density is the result of exposure. Therefore, the quantity of exposure to the film determines the radiographic density. The greater the quantity of exposure, the darker the film will be. A film that is too dark is said to be **overexposed,** and one that is too light is **underexposed.**

Because exposure determines density, each of the four prime factors of exposure affects radiographic density. *Density is primarily controlled by varying the mAs,* usually by increasing or decreasing the exposure time.

Although kVp and SID also affect radiographic density, they are not used to control it. SID is usually kept constant, and kVp is used to control contrast, as explained in the section that follows.

Do not confuse tissue density with radiographic density. **Tissue density** refers to the mass density of the radiographic subject. Increased tissue density—bone, for example—causes a lighter area on the radiograph because it absorbs more of the primary radiation, leaving less exposure on the film. Fat, on the other hand, has less tissue density than muscle or bone. It will absorb less primary radiation and will

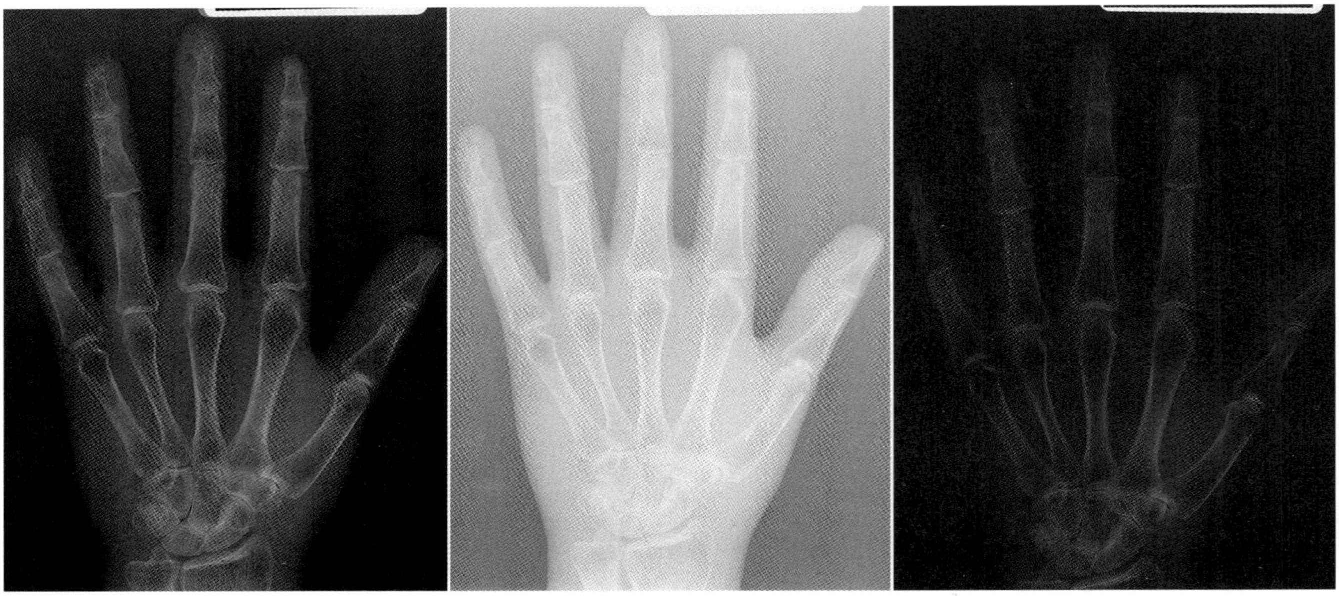

Fig. 6-2 Variations in radiographic density. **A,** Normal exposure; **B,** underexposed; **C,** overexposed.

produce a darker area on the film. Increased radiographic density means that the film is darker, while increased tissue density results in a lighter area on the film. In other words, radiographic density and tissue density are inversely related to each other. This can be quite confusing if the term density is used without qualification.

Radiographic Contrast

Radiographic **contrast** is defined as the difference in radiographic density between adjacent portions of the image. Fig. 6-3 illustrates differences in radiographic contrast. Adequate contrast is also a key factor in the *visibility of detail.* Films with a low level of contrast have little density difference within the subject. They have a flat, gray appearance, and details may be so similar in radiographic density that they are difficult to differentiate from each other. Films with too much contrast have a "black-and-white" appearance. They contain some areas that are very dark and others that are very light. It is difficult to see detail in areas with extremes of density. Optimum contrast provides sufficient difference in density to easily make out details in all portions of the image.

Contrast is primarily controlled by kilovoltage. A decrease in kVp produces increased contrast; increased kVp reduces contrast.

Fig. 6-4 illustrates a tool called a **step-wedge penetrometer.** It is a solid piece of aluminum with steps of varying thickness. A radiographic image of a penetrometer is a gray scale that shows the amount of penetration of each step. Fig. 6-5 illustrates the gray scale produced by radiography of a step-wedge penetrometer at different kVp settings. At 40 kVp, the number of gray tones between black and white is 5 (Steps 7 to 11). Note that there is considerable difference in radiographic density between each of these steps. This is a high degree of contrast. High contrast produced by low kilovoltage is called **short scale contrast.** At 100 kVp, there are more than 15 gray tones between black and white, but the difference in radiographic density between these steps is slight. This low level of contrast produced by high kilovoltage is called **long scale contrast.** A long scale of contrast is also called **latitude.** Latitude refers to the range of tissue densities that can be radiographed satisfactorily with a single exposure. Thus it can be seen that latitude increases as contrast decreases, and vice versa.

Radiographic contrast is significantly influenced by the tissue densities within the subject, referred to as **subject contrast.** For example, the abdomen has many structures with similar tissue density and therefore displays low subject contrast. Abdominal structures, such as the liver and the kidneys, will appear similar, so abdominal radiographs tend to have a gray appearance, even when taken at relatively low kVp settings (Fig. 6-6). The chest, on the other hand, has a high degree of subject contrast. The tissues are very dense in the center, where the x-ray beam must penetrate the sternum, the spine, and the heart, but the lungs are air-filled and easily penetrated. The contrast in tissue density between these structures

A
B
C

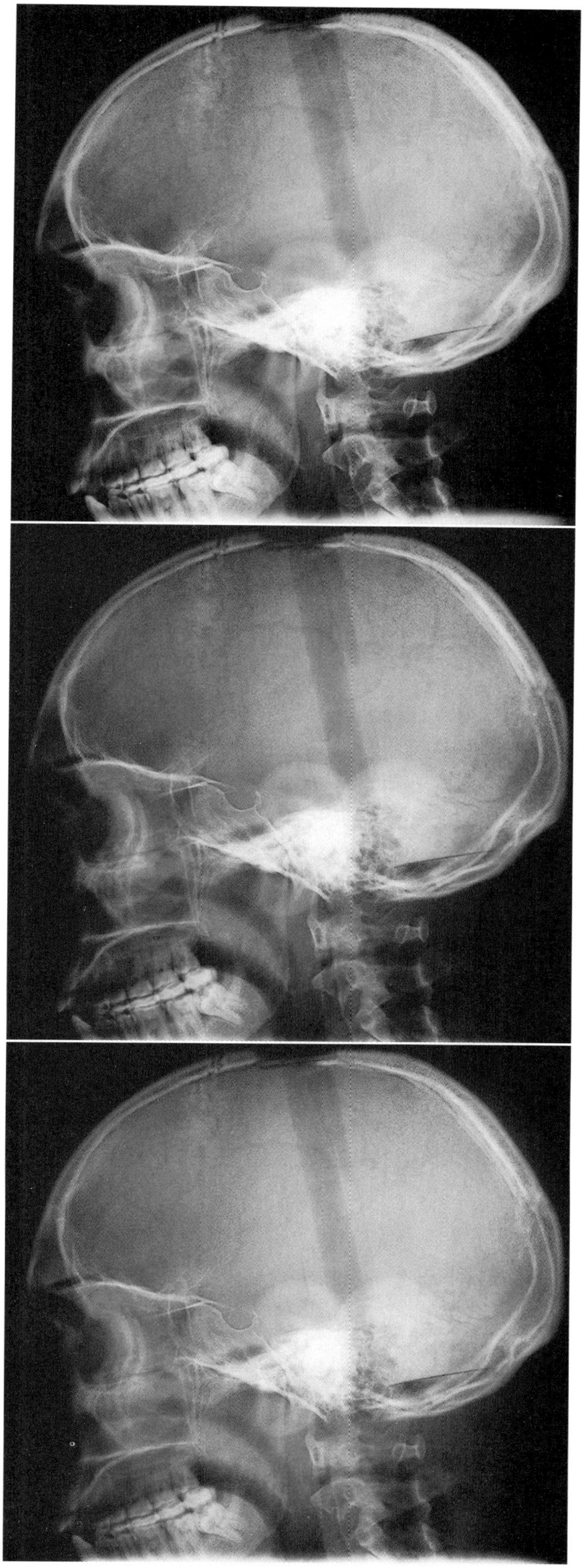

Fig. 6-3 Variations in radiographic contrast. **A,** Optimal contrast; **B,** high contrast (short scale); **C,** low contrast (long scale).

Fig. 6-4 Aluminum step-wedge penetrometer. Its radiographic image is a gray scale, as seen in Fig. 6-5.

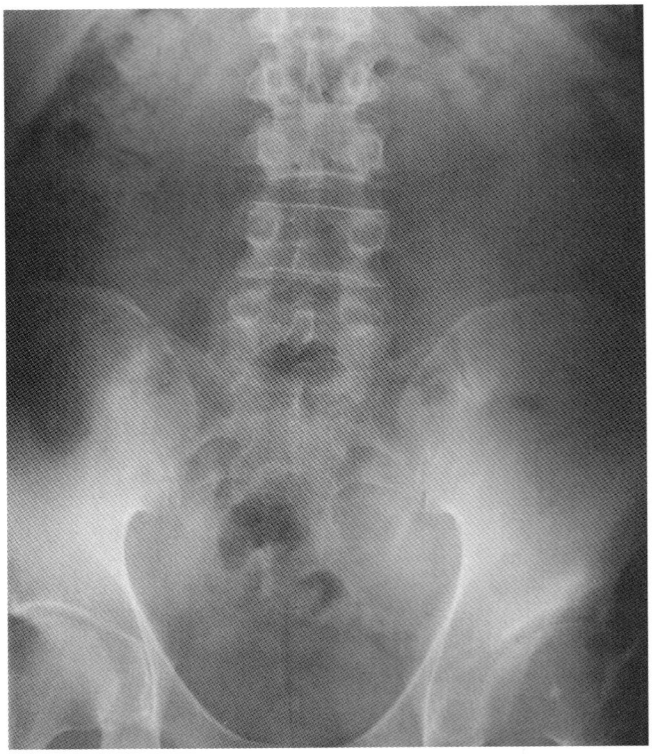

Fig. 6-6 Low subject contrast of the abdomen produces relatively low radiographic contrast at 75 kVp.

Fig. 6-5 Radiographs of step-wedge penetrometer demonstrate changes in contrast with varying kVp.

produces a black-and-white appearance, even at relatively high kVp settings (Fig. 6-7). A long scale of contrast is desirable for structures that have a high degree of subject contrast; conversely, when the subject contrast is low, a short scale of radiographic contrast produces a better image.

Contrast is also influenced by the presence of fog. **Fog** is a generalized, unwanted exposure to the film. Fog produces an overall increase in density that causes all parts of the image to appear as though seen

through a gray veil. It causes areas that would otherwise be bright or white to appear gray. Fog may be caused by scatter radiation or by other factors related to processing and film storage. These factors are discussed in Chapters 8 and 9. Fog decreases radiographic contrast, and measures taken to reduce fog will result in increased contrast. Although fog decreases contrast, it does not increase latitude. Latitude is affected only by kilovoltage and film characteristics.

Radiographic Distortion

Radiographic **distortion** refers to differences between the actual subject and its radiographic image. Because the subject is three-dimensional and the image is flat (two-dimensional), all radiographic images have some degree of distortion. Radiographic distortion may be categorized by whether it affects primarily the size of the object or its size as well as its shape. Size distortion is always in the form of magnification, enlargement. Radiographic images cannot be made to appear smaller than the objects they represent. Shape distortion is sometimes referred to as "true distortion." It is the result of unequal magnification.

Magnification occurs as a result of the geometry of the imaging setup. It is a function of the relationship between the source-image distance (SID) and

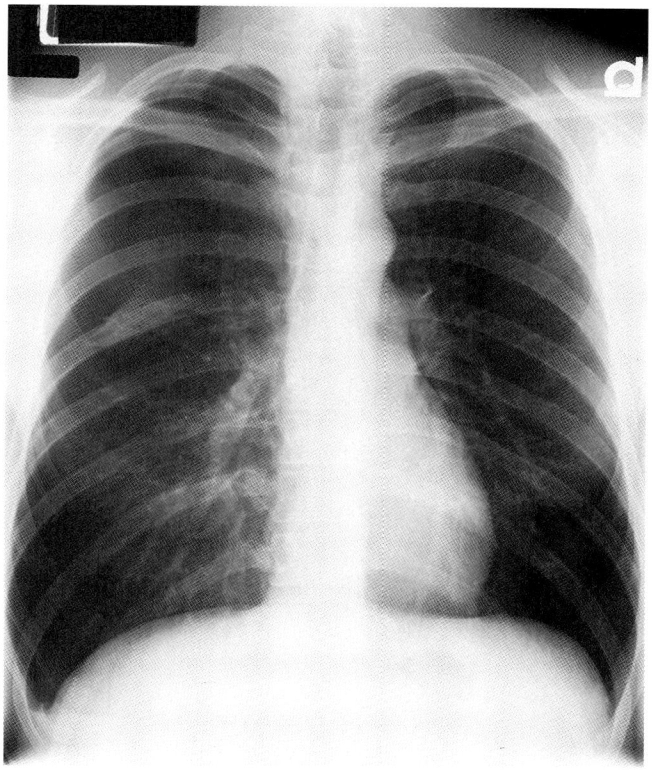

Fig. 6-7 High subject contrast of the chest produces relatively high radiographic contrast at 110 kVp.

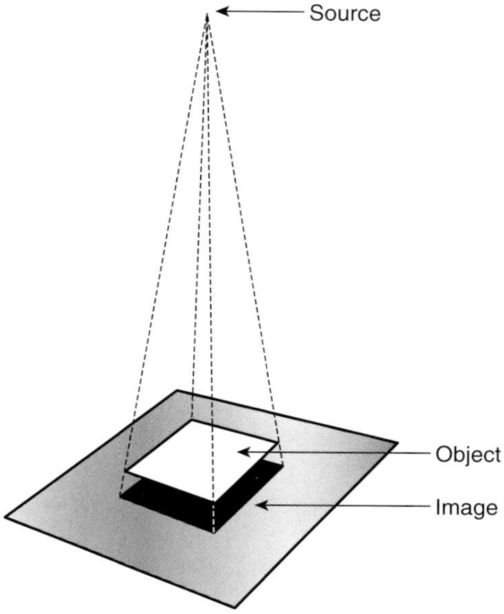

Fig. 6-8 When the object is near the film, the object and its image are nearly the same size.

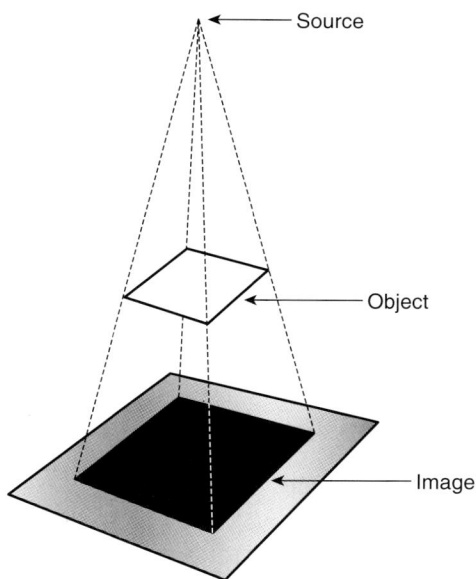

Fig. 6-9 With increased distance between the object and the film (OID), the image is magnified.

the distance between the subject and the film. This distance is called the **object-image distance (OID).** As you can see in Fig. 6-8, when the SID is great and the OID is minimal, there is little magnification. The object and its image are almost the same size.

As the OID is increased, the magnification increases (Fig. 6-9). You can demonstrate this principle by using a flashlight to project a shadow of your hand on a flat surface. The farther your hand is from the surface, the greater is the size of its shadow. Now hold your hand about 2 inches from the surface while you bring the flashlight closer. What is the result of decreasing the distance between the flashlight and the surface? Fig. 6-10 illustrates the increased magnification caused by a decrease in SID.

Shape distortion, as stated before, is the result of unequal magnification. The least shape distortion occurs when the plane of the subject is parallel to the plane of the film and the central ray is perpendicular to both (Fig. 6-11). Angulation of the part in relation to the film, or angulation of the x-ray beam, produces shape distortion (Figs. 6-12 and 6-13). For these reasons, effort is made to position the patient so that the object of clinical interest is as parallel to the film as possible and to minimize the need for tube angulation. Even when the x-ray beam is directed perpendicular to the film, only the central ray is truly per-

pendicular. Therefore the least distortion occurs at the center of the image. Structures at the outer edges of the field will exhibit some degree of distortion, especially when the film is large. For this reason, the object of primary clinical interest is usually placed in the center of the field.

Fig. 6-14 shows three wrist radiographs with different degrees of distortion. The first image shows foreshortening of the navicular bone, caused by its

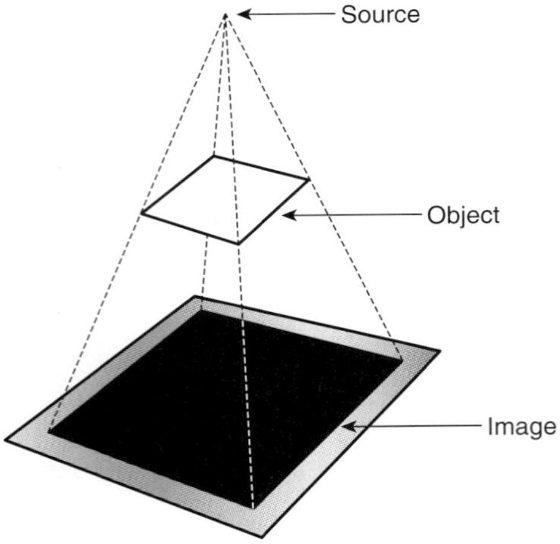

Fig. 6-10 With decreased distance between the radiation source and the film (SID), magnification is increased.

Fig. 6-11 Least distortion occurs when the object is parallel to the film and the central ray is perpendicular to both.

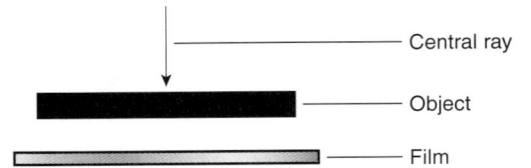

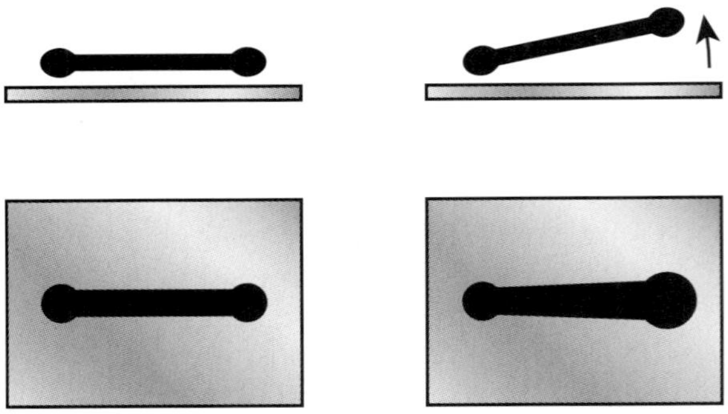

Fig. 6-12 When the object is not parallel to the film, unequal magnification creates distortion of shape.

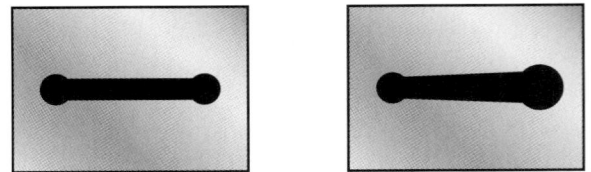

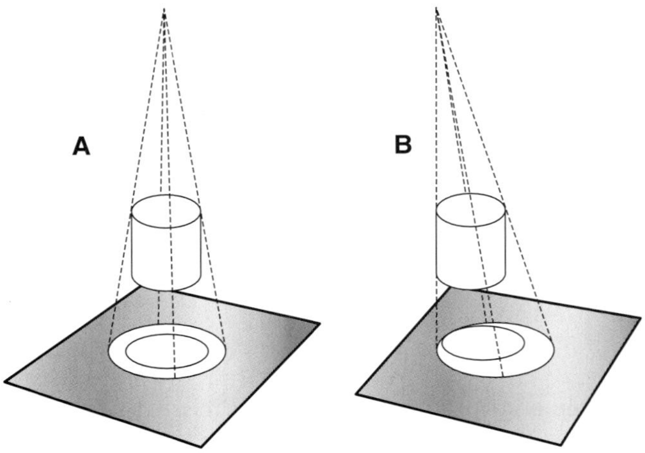

Fig. 6-13 A, When a can is radiographed with the central ray aligned to its center, the magnified image of the top of the can is seen as the outer ring, and its less magnified bottom is represented by the inner ring. **B,** When the can is radiographed with the central ray aligned to its margin, its top and bottom are superimposed under the central ray and separated by distortion in the remainder of the image.

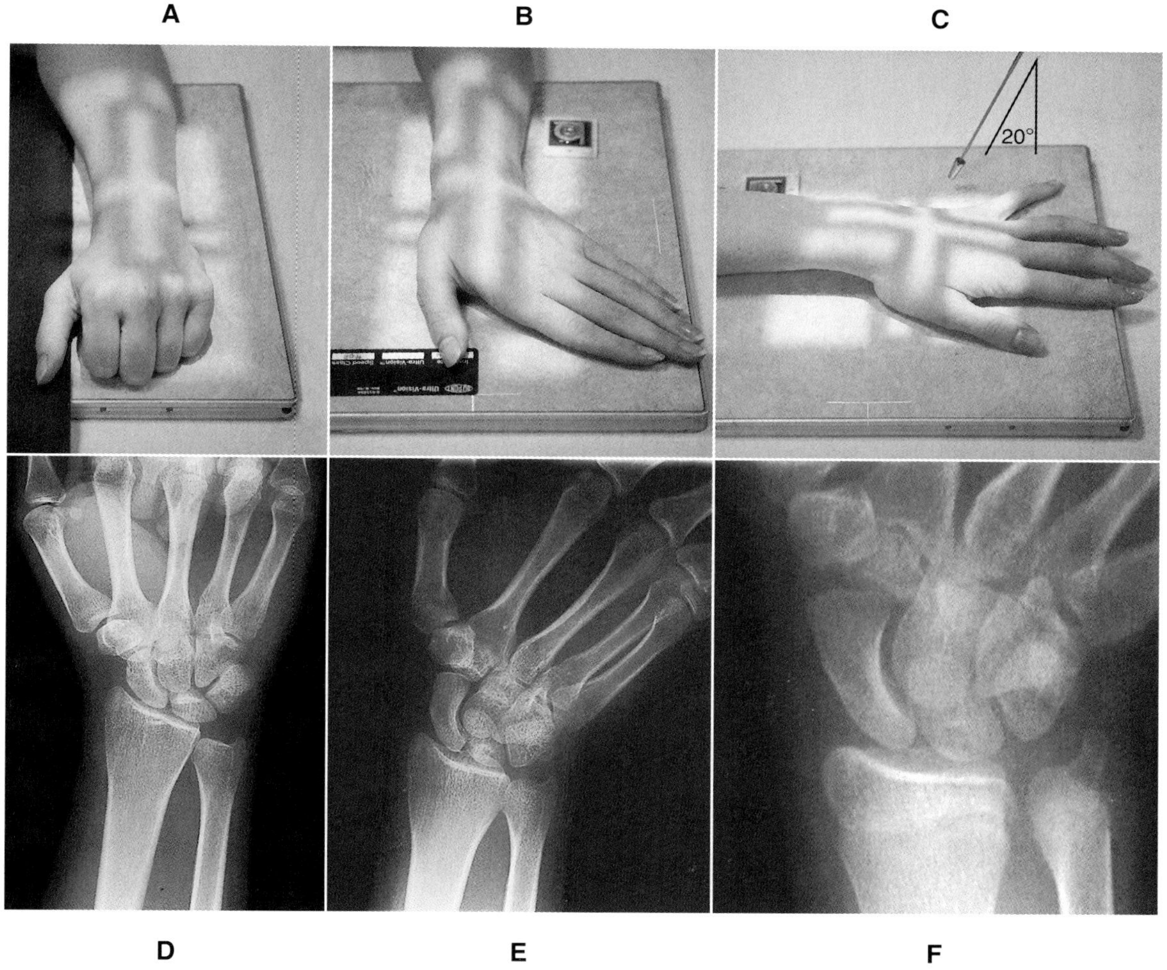

Fig. 6-14 With normal wrist position **(A)**, the scaphoid bone is somewhat foreshortened **(B)** because it does not lie parallel to the film. A change in position **(C)** provides less distortion of the scaphoid **(D)**. Angulation of the x-ray beam **(E)** elongates the wrist bones and also improves visualization of the scaphoid **(F)**.

angle in relation to the film. The second image is a special position that minimizes distortion. The third image shows elongation, a distortion caused by angulation of the x-ray tube.

Tube angulation, or positions that cause distortion, are sometimes used to prevent structures from being superimposed, that is, projected on top of one another. Because superimposition may obscure details in the primary subject, distortion is sometimes tolerated.

Radiographic Definition

Radiographic **definition** refers to the sharpness of the image. It is sometimes referred to as **resolution** or "sharpness of detail." It is the edge sharpness of all portions of the image that determines whether the image appears clear or blurred. When definition is optimum, the image has a crisp, clean appearance; it tends to appear "fuzzy" or unclear when definition is diminished. The factors that affect definition include geometric factors, motion, mottle, parallax, and film/screen contact.

Geometric Factors The geometry of image formation involves three factors: source-image distance (SID), object-image distance (OID), and focal spot size. In Chapter 4 it was stated that the effective focal spot size affects image sharpness. This section explains how and why this statement is true.

If x-rays arose from a point source on the target, the image would be very sharp and clear (Fig. 6-15). Because x-rays arise from many points on the focal area, x-ray beams from each point produce a slightly different image (Fig. 6-16). The portion of the image that is common to all points on the focal spot is the

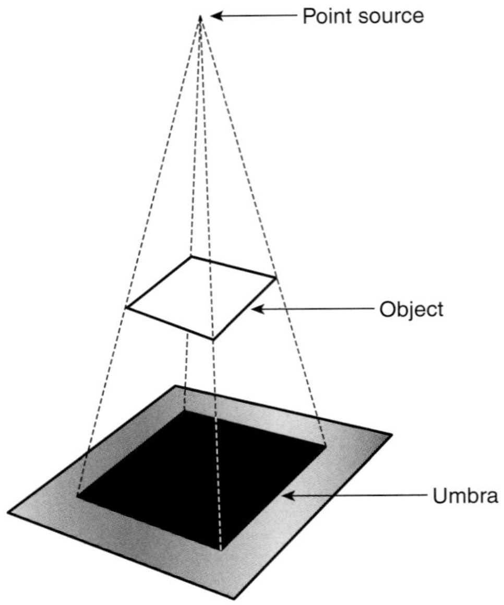

Fig. 6-15 When the radiation source is a *point*, the image has sharply defined borders.

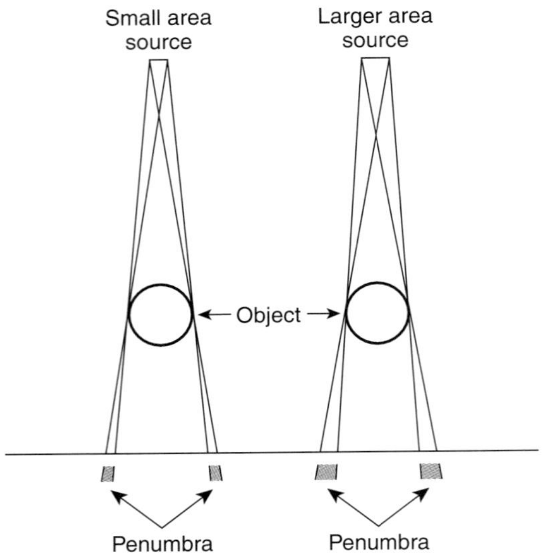

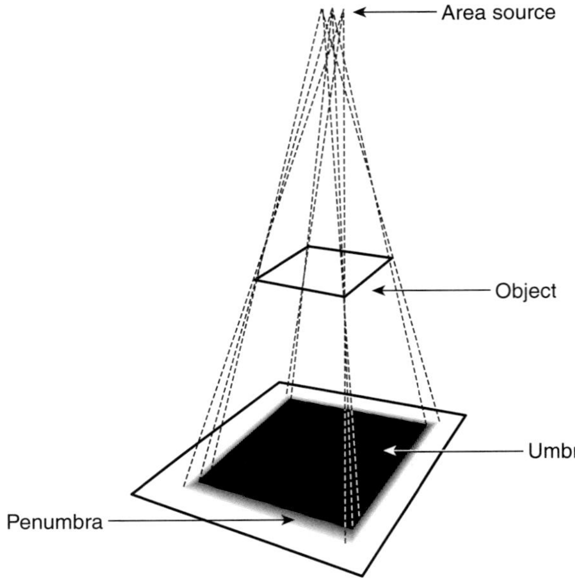

Fig. 6-16 When the radiation source is an *area*, the image has un-sharp (blurred) borders. This unsharpness is called penumbra.

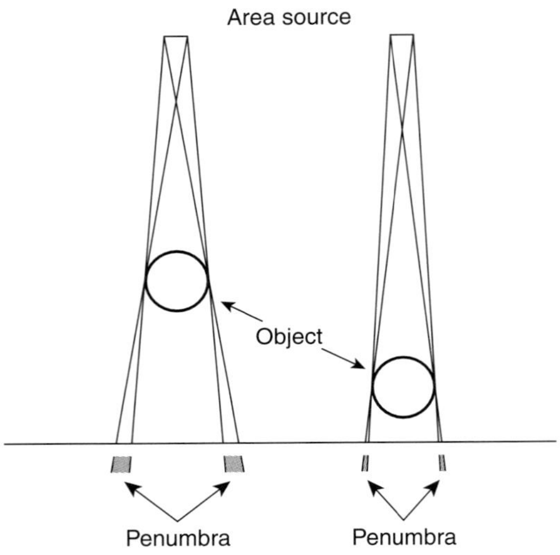

Fig. 6-17 A small area source produces less penumbra than a large area source. The size of the area source is determined by the size of the focal spot.

Fig. 6-18 Reducing the OID reduces penumbra.

true image or **umbra,** the Greek word for shadow. The blurred edge is called the **penumbra,** or "near-shadow," and represents slight variations in the image due to focal spot size.

Fig. 6-17 demonstrates that focal spot size affects the size of the penumbra. The larger the effective focal spot, the greater the penumbra and the less the definition. When the OID is increased, the penumbra is magnified (Fig. 6-18). When the SID is increased, the magnification and penumbra are decreased (Fig.

6-19). Thus whenever there is magnification of the image, there is also magnification of the penumbra. For this reason, *magnification results in image unsharpness.* These relationships are summarized in Table 6-1.

The quantity of magnification and unsharpness is proportional to the ratio between the SID and the OID. When the SID is divided by the OID, the resulting number will indicate the degree of unsharpness. The greater the number, the clearer the image; the smaller the number, the greater the magnification,

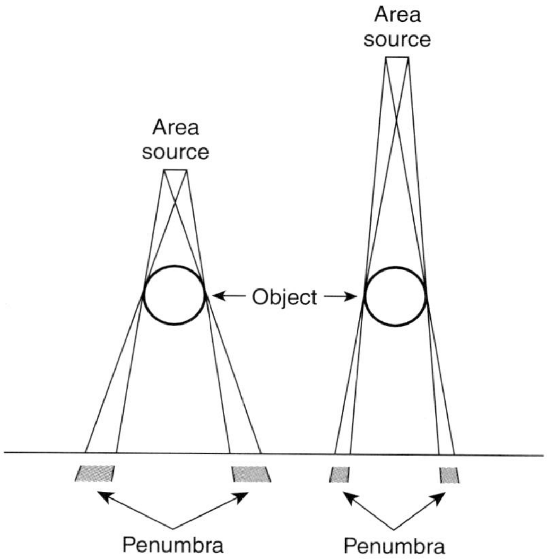

Fig. 6-19 Increasing the SID reduces penumbra.

TABLE 6-1

Effects of Image Geometry on Radiographic Definition

Factor	Direction of Change	Effect on Definition
Focal spot size	Increase	Decrease
Object-image distance	Increase	Decrease
Source-image distance	Increase	Increase

the penumbra, and the unsharpness. For example, if the OID were 4 inches and the SID 40 inches, the ratio would be 1:10. If the SID were increased to 80 inches, the ratio would be 1:20. An image taken with a geometry ratio of 1:20 will have only half as much magnification as when the ratio is 1:10.

When there is a significant object-image distance that cannot be minimized by positioning, increasing the SID and using the small focal spot will improve image quality. A good example is radiography of the lateral (side) projection of the cervical spine. Here the location of the shoulder prevents placement of the neck close to the film. To avoid undue loss of definition, the study is done upright at a 72-inch SID using the small focal spot.

It is important to remember that the distance between the patient's skin and the film does not necessarily represent the OID. It is the distance between the film and the object of clinical interest within the patient that matters. For this reason, an effort is made to perform radiography with the object of clinical interest as close to the film as possible.

Motion   Any movement during radiography will cause blurring of the radiographic image (Fig. 6-20),

reducing definition. This applies to patient motion, of course, but also to movement of the film or the x-ray tube. To avoid motion, the film is placed in a firm, stable location and the tube is locked in position. The discussion that follows is intended to assist you in avoiding patient motion.

Patient motion may be categorized as either voluntary or involuntary. Involuntary motion involves movements over which the patient has no control, such as tremors and heartbeats. Voluntary motion is normally controllable, but some patients may not be able to control motions that are in the voluntary category. For example, while most patients can hold their breath for a few seconds, an unconscious patient or a small baby will not be able to do this. In addition, patients who are in severe pain or who are unable to cooperate may present problems of motion during radiography.

The first step in avoiding motion is to make every effort to ensure that the patient understands what is expected and is willing to cooperate. Try to place the patient in a position that is as stable and comfortable as possible. When patients are standing for radiography, have them space their feet shoulder-width apart for a broad base of support and position them firmly

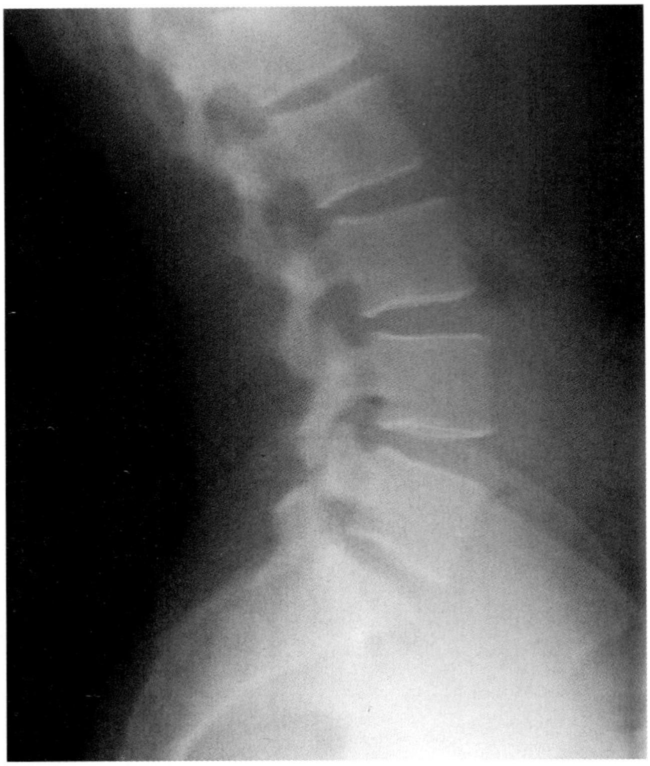

Fig. 6-20 Movement of the patient, tube, or film during exposure causes image unsharpness called motion blur.

against the film holder for stability. Minimize the time that a patient must maintain an awkward or uncomfortable position. Instructions should be clear and complete, with time allowed for the patient to comply.

Immobilization devices are used to minimize motion in certain cases. A sandbag placed strategically over the arm may aid in immobilization of the hand, for example. Devices and methods for pediatric immobilization are discussed in Chapter 18.

The principal means of controlling involuntary motion is a short exposure time. This is especially important for chest radiography, where heart motion tends to blur the lung image, and for radiography of children.

Mottle, Parallax, and Film/Screen Contact   **Mottle** is defined as a density variation that occurs in film areas that have received a uniform exposure. Mottle compromises definition. Several factors may be the cause. **Quantum mottle** is an uneven exposure caused by the random distribution of photons within the x-ray beam. Quantum mottle is only a problem when very fast image receptor systems are used with very short exposure times. Two other types of mottle, structural mottle and film graininess, are discussed in Chapter 7.

Parallax refers to slight variations in the image between the two sides of the film. This phenomenon results in decreased definition. Parallax is caused by the thickness of the imaging system and is greatest at the margins of the film or when the x-ray tube is angled. This will be more easily understood following a more complete explanation of the image receptor system and is therefore explained in greater detail in Chapter 7.

Film/screen contact is also essential to radiographic definition. The integrity of the cassette and its proper loading should keep the film in uniform tight contact with the two intensifying screens that line the cassette. When contact is poor, definition is decreased. Film/screen contact is discussed further in Chapter 7.

SUMMARY

The four prime factors of exposure are mA, time, kVp, and SID. The quantity of exposure is proportional to both the mA and the time. The mAs is the product of mA and time and indicates the total quantity of exposure. Kilovoltage affects the quantity of exposure by determining how much of the primary beam will penetrate the subject and expose the film. The quantity of exposure in a given area of the film is influenced by the SID according to the inverse square law, which states that radiation intensity is inversely proportional to the square of the distance.

Radiographic density refers to the overall blackness of a radiograph. It is influenced by all factors that affect exposure and is primarily controlled by mAs. An increase in exposure produces a darker film.

Kilovoltage is used to control the penetration of the x-ray beam and the contrast on the radiograph. High kilovoltage produces a long scale of contrast, providing the latitude needed to make radiographs of subjects with a wide range of tissue density. A short scale of contrast, produced by low kVp, results in greater density differences between portions of the subject that are similar in tissue density.

Distortion refers to both magnification and changes in the shape of the image as compared to the object. Magnification is enlargement of the image as a result of the relationship between the OID and the SID. Shape distortion is caused by unequal magnification. Shape distortion is controlled by alignment of the object to the film and by the alignment of the x-ray beam.

Definition refers to the sharpness of the radiographic image. It is affected by geometric factors (SID, OID, and focal spot size), as well as motion, mottle, parallax, and film/screen contact.

■ Review Questions ■

1. Which of the prime factors of exposure are directly proportional to the quantity of exposure?
2. What units are used to indicate the total quantity of exposure?
3. If an exposure is made using 300 mA, 0.3 sec, 85 kVp, and 40-inch SID, what is the value of the mAs?
4. If the radiographic image is too dark, which exposure factor(s) would you change to solve the problem?
5. When it is desired to differentiate tissues that have very similar tissue densities, is a long or short scale of contrast most desirable?
6. What type of subject requires a high latitude exposure? Would you use high or low kVp?
7. A radiograph appears gray and "flat"; list two possible causes.
8. If a large OID produced unacceptable loss of definition, what other factor could be changed to improve the image?
9. When a radiographic image appears blurred, what aspect of image quality is affected? Which exposure factor might be changed to solve this problem?
10. List measures that should be taken to avoid voluntary motion during radiography.

7

Image Receptor System

Learning Objectives

At the conclusion of this chapter the student will be able to:

- Identify the components of a typical radiographic cassette and explain the purpose of each
- State the purpose of intensifying screens and explain how this purpose is accomplished
- Identify screen characteristics that affect screen speed and resolution
- State the type of phosphors commonly used for intensifying screens and describe their spectral emission characteristics
- Demonstrate correct methods for the handling and cleaning of cassettes and screens
- Define parallax, state the two factors responsible for this problem, and list two methods of minimizing its effect
- Explain the importance of good film/screen contact and list three causes of poor film/screen contact; demonstrate an assessment of film/screen contact
- List or draw the layers of general-purpose film construction and explain the content and purpose of each
- List factors that may cause exposure to radiographic film
- State optimum conditions for film storage
- List three inherent characteristics of film response and identify each on a sensitometric curve

Key Terms

artifact
backscatter
D-max
film/screen contact
fluoresce
gross fog density
inherent
matrix
optical density (OD)

phosphor
radiolucent
rare earth phosphor
screen speed
sensitometric curve
silver halide
spectral emission
spectral sensitivity
structural mottle

This chapter is about imaging systems: cassettes, intensifying screens, and film. Although filmless imaging systems are being installed in large institutions at an expanding rate, it will be some time before these systems will be available to the vast majority of limited radiographers. When this time comes, there will be technological advances in these systems that will make any detailed discussion in this writing obsolete. For these reasons, this text focuses on imaging systems that involve the use of film.

Proper care and handling of cassettes and film are essential to radiographic quality and are skills needed by radiographers on a continuing basis. An understanding of the image receptor system will aid in the formation of good work habits for the production of quality radiographic images.

CASSETTES

Radiographic film holders, called cassettes, were first mentioned in Chapter 2. Common sizes are listed in Table 2-1. Cassettes serve three important functions:

1. They protect the film from exposure to light during use.
2. Their rigid structure protects the delicate film from bending and scratching during use.
3. They contain intensifying screens, keeping them in close contact with the film during exposure.

The exposure side of the cassette (Fig. 7-1) is considered to be its "front." A label on the front indicates the position of an area on the film that is protected from exposure. This area is often called the "ID blocker" because radiation is blocked from exposing the film in this location. The ID blocker reserves this area for the printing of patient identification at the time of processing. Cassettes may be purchased with the ID blocker at any corner and on either margin, but by far the most common location is the one illustrated.

The cassette "back" is the access side of the cassette (Fig. 7-2). When you are unloading and reloading the cassette, this side faces up. Note that this position usually places the ID blocker in the upper left corner.

The structure of the cassette is diagrammed in Fig. 7-3. The front is made of a **radiolucent** material, that is, a substance that is easily penetrated by the x-ray beam. It may be a very lightweight metal alloy or a plastic material made of a durable resin. Mounted to the inside of the cassette front is the front intensifying screen.

The back of the cassette may also be made of metal or plastic. Inside the back is a layer of lead foil. The purpose of this layer is to prevent **backscatter.** Backscatter is radiation from the cassette back and the

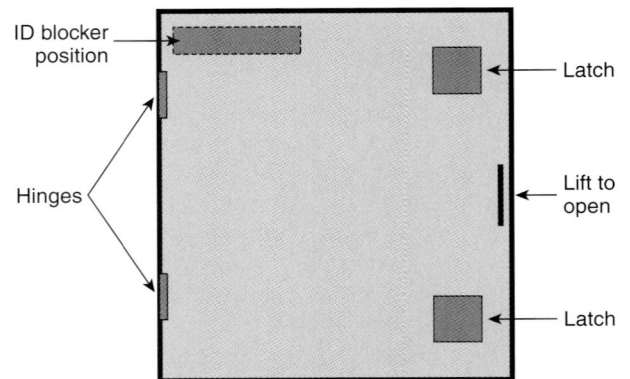

Fig. 7-2 Cassette back, access side. Properly positioned, the cassette opens like a book.

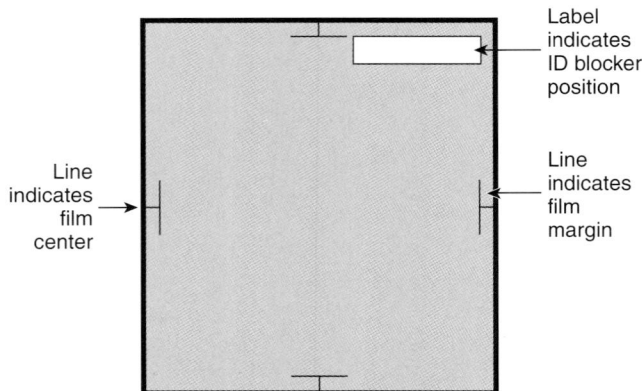

Fig. 7-1 Cassette front, exposure side.

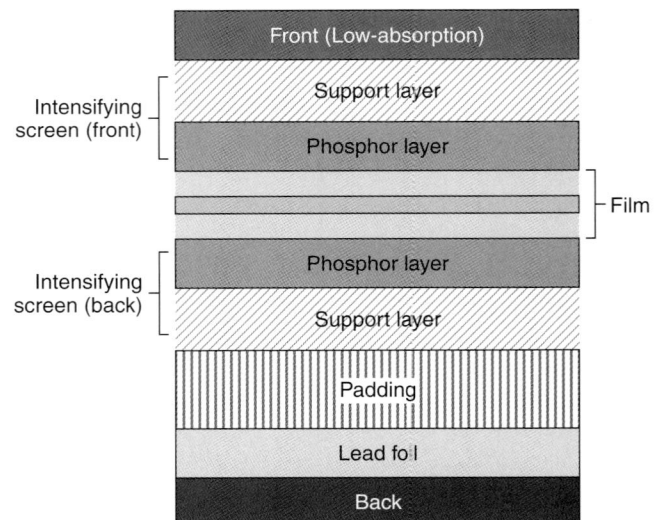

Fig. 7-3 Cross section of cassette.

cassette tray that is directed back toward the film. Backscatter would cause fog on the film if not absorbed by the lead foil layer.

Inside the foil layer is a layer of padding, usually a plastic foam. The purpose of this layer is to keep the intensifying screens pressed tightly against the film when the cassette is closed, so as to maintain good **film/screen contact.** The importance of good film/screen contact is discussed later in this chapter.

The back intensifying screen is mounted on the padding layer. The intensifying screens cover the inside surfaces of the cassette. When the film is placed in the cassette, it is sandwiched between the intensifying screens.

Special cassettes are used with phototiming systems. These cassettes must allow remnant radiation to pass readily through the cassette back to the detector. For this reason, they have a radiolucent back and no lead foil layer. Each is usually clearly marked "phototiming cassette" and should be used only for this purpose.

INTENSIFYING SCREENS

An intensifying screen is a flat surface coated with fluorescent crystals called **phosphors.** These phosphors glow, or **fluoresce,** giving off light when exposed to x-rays. The phosphors absorb x-ray energy and emit the energy in the form of light. While x-rays expose the film to some degree, by far the greatest amount of exposure on radiographs is from the light of the intensifying screens.

The purpose of intensifying screens is to reduce the amount of exposure required to produce an image. They enhance the efficiency of the exposure process. Direct x-ray exposures would require 25 to 400 times more exposure than the same images produced with intensifying screens. The use of intensifying screens greatly reduces both exposure time and patient dose. Screen use also reduces the generator size and x-ray tube capacity required for radiography.

Screen Construction

The structure of an intensifying screen is illustrated in Fig. 7-4. It consists of a plastic base, or support

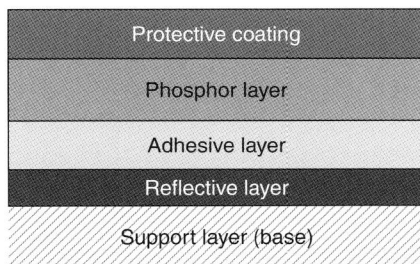

Fig. 7-4 Screen composition.

layer, with an adhesive coat that holds the phosphors in a thin, smooth layer. A clear protective coating helps to prevent stains and wear to the crystal layer.

Screen Speed

The efficiency of a screen in converting x-rays to light is called **screen speed.** A screen with greater efficiency requires less exposure, so the screen is said to be "faster." The response of a standard screen has been designated as a speed class of 100. This is a standard against which the speeds of screens are measured. A screen with a speed class of 200 would be twice as fast as a 100-speed screen and would require only half as much exposure. Typical speed classes are 50, 100, 200, 400, 600, and 800.

Screen Speed vs. Image Resolution

Intensifying screen crystals glow on an "all-or-none" basis. When a phosphor absorbs sufficient energy, the entire crystal will glow, not just the portion that was exposed. This results in an area of exposure on the film that is greater in size than the original area of exposure to the crystal. For this reason, the use of intensifying screens produces images with less definition than images produced by direct exposure to x-rays.

Large crystals and a thick crystal layer produce more film exposure from a given amount of x-ray exposure (greater speed) but also provide less radiographic resolution (Fig. 7-5). Loss of resolution due to crystal size, crystal layer thickness, and/or uneven crystal distribution is called **structural mottle.** You will recall from Chapter 6 that structural mottle is one of the factors that influences radiographic definition. Larger crystals produce a larger image of a "point" of exposure. Light rays diverge from within the screen until they reach the film, creating an exposure area on the film that is larger than the crystal size. When the crystal layer is thick, there is more light divergence from deep within the screen, enlarging the film area exposed and decreasing image sharpness.

Screen speeds for routine work range from 200 to 800; 400 or 600 is typical. Screens in these classes may be called "rapid" or "regular" screens. They may be used either on the tabletop or in the bucky. Some fast screens also have a reflective layer between the base and the phosphors. This layer increases the efficiency of the screen by reflecting the screen light, directing it toward the film.

Fine crystals in a thin layer provide greater radiographic definition but require more exposure. Speed classes of 50 or 100 are slower than regular screens and provide high-detail images for small anatomical parts. They may be referred to as "extremity" or

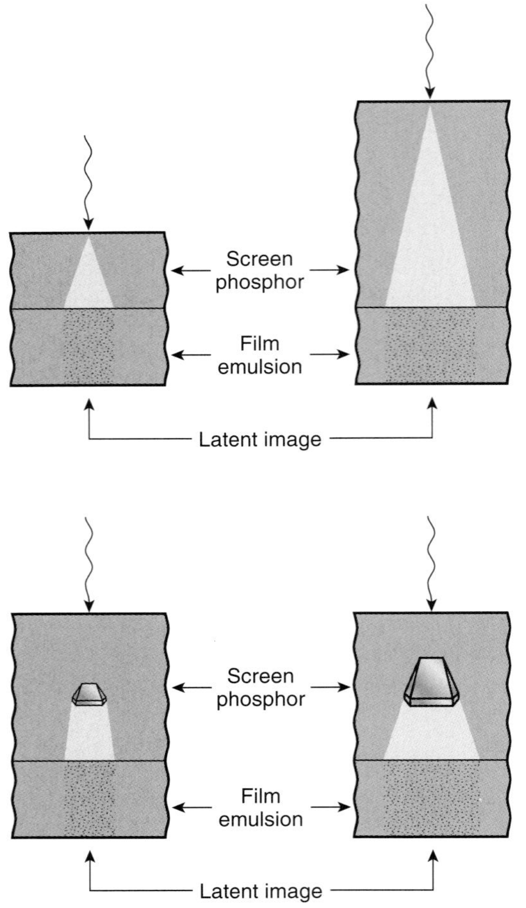

Fig. 7-5 Crystal size and phosphor layer thickness affect image resolution.

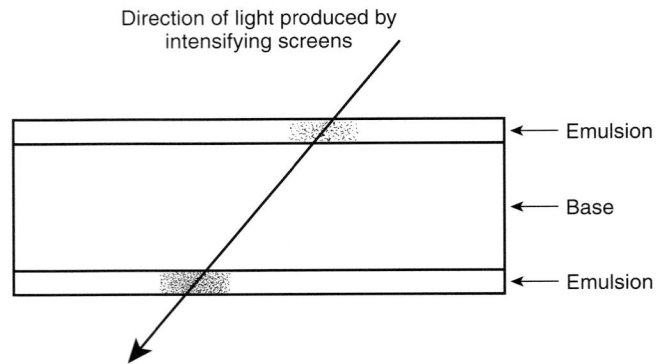

Fig. 7-6 Parallax decreases radiographic definition. Parallax is affected by the thickness of the image receptor system and the angulation of the x-ray beam.

BOX 7-1

Typical Modern Screen Characteristics

Regular (Rapid)	Detail (Extremity)
Rare earth phosphors	Rare earth phosphors
Moderate crystal size	Small crystal size
Medium layer thickness with tabular (flat) crystal grains	Thin crystal layer with tabular (flat) crystal grains
Reflective layer	No reflective layer
No dye in screen matrix	Dye in screen matrix
High crystal concentration	High crystal concentration
Speed class: 200-800	Speed class: 50-100

"detail" screens and are usually used only on the tabletop. Detail screens may have a dye added to the **matrix,** the substance that holds the phosphors together. This dye absorbs light that is not directed at the film, further enhancing resolution.

It is common for most x-ray departments to have some of both types of cassettes: rapid and extremity classes. Box 7-1 lists screen design characteristics that enhance either speed or detail.

Extremely fast screens in the 1000- to 1200-speed class are available, but these screens do not provide sufficient image quality for general work and are used only when very high speed is required. Hospitals may have extra-fast cassettes for very large patients or for special applications.

Parallax

This term was introduced in Chapter 6 as one of the phenomena that cause a decrease in radiographic definition. Parallax refers to slight differences in the image on the two sides of the film. This difference occurs as a result of the x-ray beam passing at an angle through the image receptor system (Fig. 7-6). The image is the same on both sides of the film when the x-ray beam is perpendicular to the film. The degree of parallax depends on the thickness of the screen/film combination and the angulation of the x-ray beam. For this reason, parallax is most pronounced at the margins of the film, where the divergent x-ray beam comes in at an angle, and when the x-ray tube is angled for specific procedures. Parallax is another reason why thicker screens produce less sharp images.

To reduce parallax for high-detail procedures such as mammography, single-screen cassettes are used with single-emulsion film. Because screen pairs are used with double-emulsion film for routine radiography, parallax is an important factor to consider for best image detail. To reduce the effect of parallax, the area of clinical interest is placed in the center of the film and tube angulation is minimized as much as possible.

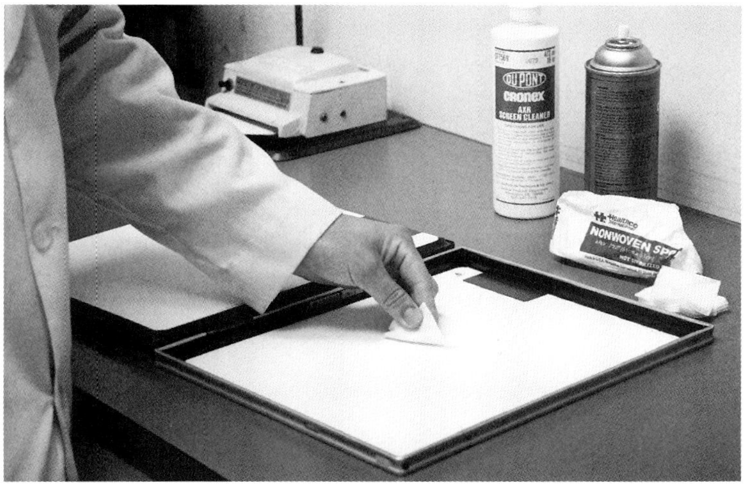

Fig. 7-7 Radiographer cleaning a screen with liquid cleaner.

Screen Phosphors

For several decades, the "conventional" screen phosphor was calcium tungstate, which is now obsolete. During the 1970s, new screen technology was introduced using **rare earth phosphors.** These are crystals containing one of the rare earth elements—gadolinium, lanthanum, or yttrium. Screens using rare earth phosphors are about four times more efficient than the old calcium tungstate screens, reducing exposure to approximately one fourth of the amount previously required, with no negative consequences in terms of film quality.

Spectral emission refers to the color of light emitted by a phosphor. Typical colors of light emitted by rare earth phosphors are green or yellow-green and blue or blue-violet.

Screen Care and Cleaning

Intensifying screens tend to build up a static electrical charge. The static charge on the screens attracts dust and dirt. A discharge of this static electricity may expose the film, creating black static **artifacts** (unwanted marks or images) on the radiograph. These marks and their prevention are discussed in Chapter 8.

Dirt on screens prevents the screen light from reaching the film, causing unexposed (white) areas on the film image. To prevent these artifacts, intensifying screens must be kept clean. The frequency of cleaning is determined by the frequency of use and by the amount of dust in the environment. All screens should be inspected and cleaned at least every 3 months. Screens may be marked in one corner so that an identifying number or letter is evident on all the films exposed in the cassette. The outside

Cleaning Intensifying Screens

1. Remove dust with pressured air or a very soft brush.
2. Using a recommended commercial screen cleaner that contains an antistatic ingredient and a very soft, lint-free applicator, apply screen cleaner sparingly (never pour liquid on screen).
3. Stand open cassette on edge to dry before reloading.

of the cassette carries the same identifying mark. When a film exhibits evidence of screen dirt, it is easy to locate the offending cassette for cleaning without inspecting all cassettes of that size.

Dust and dry dirt may be removed with a *soft* brush (camel hair or sable) or with a spray of pressurized air from an aerosol can. Pressurized air and suitable brushes are available at camera supply stores. If screen dirt is not removed with a soft brush or pressurized air, a liquid cleaner is needed. In the past, soap and water or grain alcohol was used for cleaning screens. Since modern rare earth screens may be damaged by these products, a commercial screen cleaner recommended by the screen manufacturer should be used. Commercial screen cleaners have an additional advantage in that they contain an antistatic ingredient. This reduces the static charge build-up on the screen, minimizing the accumulation of screen dirt and helping to prevent artifacts on the film from either screen dirt or static.

When cleaning screens with a liquid cleaner, use a soft, lint-free applicator (Fig. 7-7). Pressed cotton squares or nonwoven gauze sponges work well. *Never pour liquid directly onto the screen,* since the liquid will be absorbed in the foam pad layer of the cassette and/or

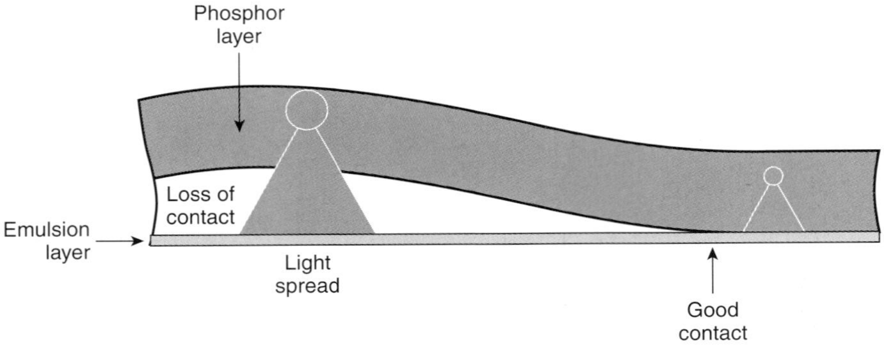

Fig. 7-8 When there is space between the screen and the film, divergence of light from the screens causes blurring of the film image.

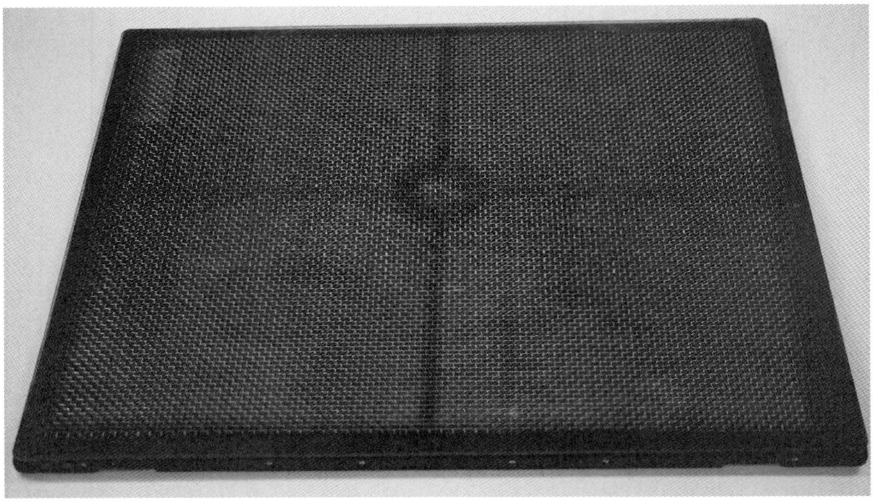

Fig. 7-9 Screen contact test mesh positioned for test.

the support layer of the screen. This may damage the padding or warp the screen. Moisten the applicator sparingly and use a light touch to wipe the screen clean. Do not pick, scratch, or scrape at screen dirt; this practice damages the screen and worsens the problem.

White or light-colored "dirt" may be difficult to detect on screens in ordinary light, but it can be easily seen when the screen is illuminated with an ultraviolet light in a darkened room. Scanning with an ultraviolet light will also reveal areas where the screen crystals have been worn away or lost due to damage. Loss of screen crystals will produce artifacts similar to those caused by specks of dirt.

When the screens are clean, stand the cassette on edge to dry completely before reloading it with film.

Cassettes and screens are expensive and easily damaged. Proper care and cleaning are essential to ensure that they provide good service for a reasonable period (3 to 5 years). Develop the habit of keeping your hands clean and dry. Remove film from the cassette without completely opening it. Never leave the cassette lying open on the loading bench, since it is easy to drop or scrape something on the screens

when working in the dark. Additionally, this practice increases wear on the hinges of the cassette. Take care that cassettes are not dropped or bumped, since damage to the cassette frame may cause the screens to fit improperly, leading to poor film/screen contact. Keeping the loading bench clean will reduce screen dirt, minimizing the need to clean the screens.

Film/Screen Contact

Poor film/screen contact is a condition that exists when the two screens are not in firm contact with both surfaces of the film. The usual causes are warping of the cassette frame or the screens, dents in screens, deterioration of the padding layer, damaged hinges, or damaged latches. Poor film/screen contact results in poor image quality because the light rays diverge in the space between the film and the screen (Fig. 7-8), creating a film image that differs from the precise screen image, appearing blurred.

Film/screen contact is tested using a special tool called a screen contact test mesh. This coarse wire mesh is laminated in plastic to keep it flat (Fig. 7-9).

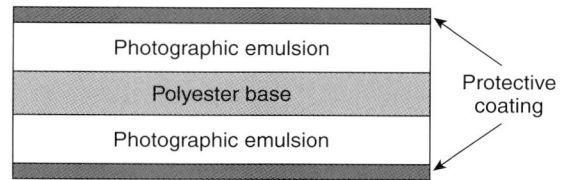

Fig. 7-10 Film/screen contact test showing area of poor contact.

Fig. 7-11 Film construction. The thin outer layers are an "overcoat" that protects the emulsion.

Composition of Radiographic Film Emulsion

- Gelatin
- Silver halide microcrystals
  - Silver bromide—90%
  - Silver iodide—10%
- Catalyst (sulfur contaminate, usually silver sulfide)

Causes of Film Exposure

Film emulsion readily assumes the "exposed" state. Exposure may be caused by:
- Light
- X-rays
- Heat
- Certain chemicals or chemical fumes
- Pressure
- Static electricity discharge
- Age

It is placed on top of the cassette and an exposure is made. A suggested exposure will be provided with the tool. If none is available, try approximately 5 mAs and 50 kVp when using a 400-speed cassette. At a viewing distance of 9 feet, the radiograph should show a consistent image of the mesh. Dark, unsharp areas on the test film indicate areas of poor film/screen contact (Fig. 7-10).

If your facility does not have a test mesh, it may be possible to borrow one from your x-ray supply dealer or the technical representative of the company that sold the cassettes. If no test tool is available, the test can be performed using paper clips. Use a new box of paper clips and spread them evenly and close together over the surface of the cassette. Expose and evaluate as with the test tool.

This test should be performed quarterly on all cassettes and on the offending cassette when unsharp areas are noted on a radiograph.

RADIOGRAPHIC FILM

A processed x-ray film is similar to a photographic negative. The exposed areas appear dark and the unexposed areas are light. Film for general use differs from photographic film, however, in that it is "duplitized," that is, it has a photographic emulsion on both sides. Because x-rays can penetrate film and because each side is in contact with an intensifying screen, the exposure to the film is doubled, decreasing the exposure time and patient dose by half.

Film Construction

The structure of film is illustrated in Fig. 7-11. The film base is made of a polyester plastic. It must be optically clear, strong, and of consistent thickness. It is tinted a pale blue or blue-gray color. This tint reduces eye strain for the physician when reading the films. The photographic emulsion is coated on both sides of the base. A thin, soluble "overcoat" on both sides protects the emulsion to some degree during film handling.

Film emulsion consists of a mixture of gelatin and **silver halide** crystals. Silver halides are salts of the halogen elements: fluorine, chlorine, bromine, and iodine (Class VII in the periodic table of elements). Most x-ray film emulsions consist mainly of silver bromide, with a small percentage of silver iodide. A tiny amount of sulfur, usually in the form of silver sulfide, is added to the emulsion as a catalyst to enhance the exposure and development process. The sulfur compound forms "sensitivity specks" within the emulsion that attract electrons, providing a focus for chemical changes within the emulsion crystals. Exposure causes the silver halide to assume an ex-

Fig. 7-12 Film is placed on edge on the shelf. **A,** Right way; **B,** wrong way.

BOX 7-2

Film Storage

- Clean, dry location
- 40° to 65° F
- Away from chemical fumes
- Safe from radiation exposure
- Standing on edge
- Expiration date clearly visible

---

posed state by changing its chemical arrangement, thus forming the latent image.

Film Storage

Unexposed film is quite vulnerable to accidental exposure. The unexposed crystal state is unstable and may be changed to an exposed state by exposure to light, x-rays, chemicals or chemical fumes, pressure, heat, static electricity, and age. For this reason, it is important that unexposed film be properly stored and carefully handled. Improper storage results in a generalized exposure response, fogging the film. Optimum storage requires a cool, dry place that is protected from x-rays and chemical fumes. Conditions for optimum film storage are listed in Box 7-2.

Film boxes should be stored on edge (Fig. 7-12) so that the labels and expiration dates are visible. Stacking of boxes should be avoided because fog may be caused by the pressure of the weight of the boxes. Film stock should be rotated so that the oldest boxes are used first to avoid film becoming fogged from age. The expiration date on new film should be at least 1 year after the purchase date. It is false economy to stock more film than can be used within a few months.

Boxes that are opened and in use are kept in a light-proof film bin in the x-ray darkroom for convenience in reloading cassettes. The darkroom is not a good place for longer-term film storage, however, because it tends to be warm, damp, and contaminated with the fumes of processing chemicals.

Film Characteristics and Types

Film is manufactured in a variety of different types, providing different responses to exposure. **Inherent** film characteristics (those "built into" the film by the manufacturer) include sensitivity (speed), contrast, and **spectral sensitivity,** the portion of the electromagnetic spectrum to which the film is most sensitive.

The size of the silver halide crystals and the thickness of the emulsion determine both the speed of the film and the degree of resolution in the finished radiograph. As with intensifying screens, larger crystals and a thicker crystal layer produce a faster response. Thinner layers of finer crystals are used to produce greater detail. Film response is also tailored by the manufacturer to vary the amount of radiographic contrast.

The characteristics of film response may be plotted on a graph called a **sensitometric curve.** This curve is sometimes also referred to as an H & D curve, after its originators, Hirter and Driffield. This S-shaped curve (Fig. 7-13) demonstrates the response of a film to a full range of exposure. The horizontal axis of the graph

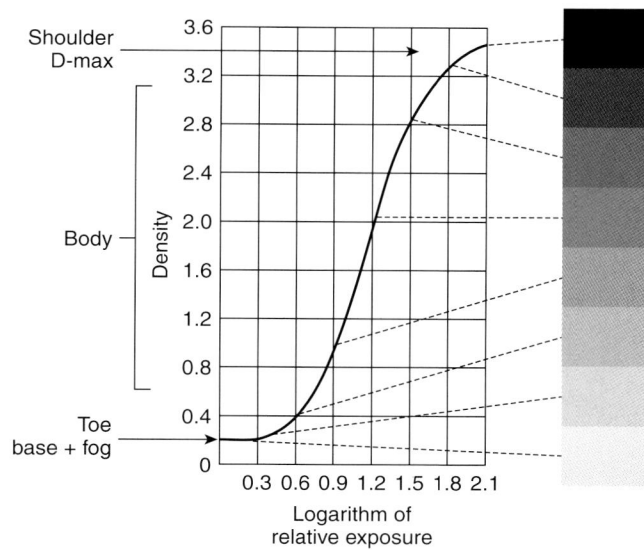

Fig. 7-13 A, Sensitometric (H & D) curve.

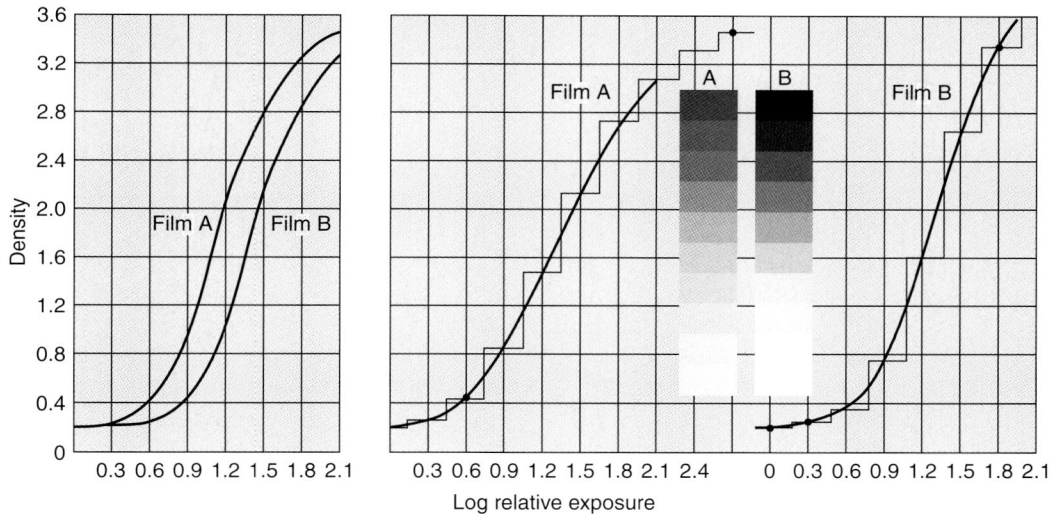

B, Film A is faster than Film B. Both have high contrast.

C, Film A has more latitude (low contrast). Film B has high contrast.

represents the logarithm of the exposure and the vertical axis represents the **optical density (OD)** of the film. Optical density is a numerical representation of the film's ability to transmit light. Specifically, it is the logarithm of the percentage of light transmitted through the film. The gray tones on a film have an OD of 0.5 to 2.0. An OD of 1.0 is a medium gray tone, about halfway between black and white. An OD greater than 2.5 appears black.

Note in Fig. 7-13 that each film has an OD of 0.15 to 0.20 when it is unexposed (at the extreme left edge of the graph). This density is called the base + fog density, or **gross fog density.** It is the measure of

light absorbed by the film base plus any fog exposure to the film prior to processing. The toe of the curve is the flat portion at the lower left of the graph. This demonstrates that film does not respond to exposure until a certain amount of exposure has been received. The length of the toe indicates the sensitivity of the film to fog.

The body of the curve is the relatively straight, vertical portion. *The location of the body on the graph indicates film speed.* The closer the body is to the left side of the graph, the faster the film. *The slope of the curve indicates the film's inherent contrast.* A steep, vertical curve indicates a short scale of contrast; a less vertical slope indicates a film with a longer scale of contrast, that is, more latitude.

The horizontal upper portion of the curve is called the shoulder. The location of the shoulder on the graph represents the film's maximum density, or **D-max.** Further exposure does not produce more density because the film is as dark as it can get.

Film can be manufactured with particular sensitivity to a certain portion of the electromagnetic spectrum, called spectral sensitivity. In the past, certain films were made to be especially sensitive to x-rays and were used for direct exposure without intensifying screens. To differentiate between x-ray-sensitive and light-sensitive film, modern film is often referred to as "screen film." Since screens may emit either green or blue light, screen film is made to be sensitive to one of these colors. *It is important that the spectral sensitivity of the film be matched to the spectral emission of the screens with which it is used.* Inappropriate matching of film and screens increases the exposure required to produce an image, thus increasing patient dose unnecessarily.

General-purpose film has double emulsion and is usually quite fast with moderate to high contrast. Latitude film provides less contrast for examinations of body parts that have high subject contrast, such as the chest. High-detail films are slower and may be employed when high resolution is required, but more commonly this is accomplished with regular film and detail screens. Single-emulsion films are used in special cassettes with single intensifying screens for mammography (breast examinations) and certain high-detail studies, but they are not commonly used by limited radiographers. Outpatient facilities typically stock only one type of general-purpose film.

Special single-emulsion film is available for making duplicates of radiographs. Unexposed duplicating film will be black when processed. Duplicating film responds to light or ultraviolet exposure by getting lighter. If your facility has equipment for duplicating radiographs, there may be a supply of duplicating film in your film bin. It is easily recognized in the dark by feeling the top edge, which is notched for easy identification.

Procedures for correct handling and processing of film are described in Chapter 8.

SUMMARY

Cassettes are an integral part of the image receptor system. They prevent light exposure to the film and keep the intensifying screens in close contact with the film surface. The purpose of the intensifying screens is to reduce the exposure required to produce an image. Rare earth phosphors in the screens emit blue or green light when exposed to x-rays, and this light is primarily responsible for the exposure to the film. Screens are available in a number of speed classes. Fast screens reduce exposure but produce less radiographic resolution; slower screens are used when fine detail is desired. Screens require good care, proper cleaning, periodic inspection, and testing for film/screen contact.

General-purpose radiographic film has a double emulsion coated on a clear, blue polyester base. The spectral sensitivity of the film should be matched to the spectral emission of the screens. The silver halide crystals of the emulsion are unstable in their unexposed state and easily become exposed when subjected to light, x-rays, heat, fumes, pressure, and age. They must therefore be stored properly and handled with care. The inherent characteristics of film include speed, contrast, and spectral sensitivity. Speed and contrast can be demonstrated on a sensitometric curve by plotting the response of film to a full range of exposures.

▪ Review Questions ▪

1. What is the purpose of the lead foil layer in the cassette? Where in the cassette is it located?
2. List three types of damage that may occur to a cassette and describe the problems created by each.
3. Compare the design characteristics and performance of fast intensifying screens to those of detail screens.
4. What colors are typical of the spectral emission of rare earth intensifying screens?
5. You note a white image in the shape of a hair on one of your radiographs. What should you do?
6. If you were going shopping for screen cleaning supplies, what would you put on your shopping list?
7. Why is it a poor practice to stack film boxes on top of each other?
8. What is the optimum temperature for long-term film storage?
9. State the film characteristic represented by each of the aspects of a sensitometric curve: toe, body placement, body slope, shoulder.
10. Why is it important to match the spectral sensitivity of the film to the spectral emission of the screens?

8

X-ray Darkroom and Film Processing

Learning Objectives

At the conclusion of this chapter the student will be able to:

- List the essential features and equipment found in an x-ray darkroom and state the purpose of each
- List possible causes of darkroom fog and demonstrate the procedure for testing the safety of safelights
- Demonstrate correct methods for film handling
- Name the chemical components of processing solutions for both manual and automatic processing and state the function of each
- Demonstrate or describe, in order, the steps involved in manual film processing, giving the purpose and time required for each
- Demonstrate or describe the process of cleaning manual processing tanks and mixing new solutions
- State the two purposes of replenishment in any film processing system
- Demonstrate appropriate safety precautions for mixing or pouring processing solutions and for cleaning up spills
- List the systems that comprise an automatic film processor and state the function of each
- Demonstrate the procedure for processing a film with an automatic processor
- List steps to be taken daily at processor start-up and processor shut-down
- Demonstrate correct use of equipment for processor quality control evaluations and plot the results of an evaluation on a typical graph
- Identify common radiographic artifacts, state their cause, and explain how to avoid them

Key Terms

automatic film processor
crescent mark
densitometer
film bin
film dryer
film identification printer

gross fog
manual film processing system
passbox
safelight
sensitometer

This chapter introduces the x-ray darkroom and covers the specific information needed for correct film handling and film processing. Both manual and automatic processing methods are presented in detail, as is information about processor maintenance and quality control. This chapter can only be understood when you have mastered the information on film quality in Chapter 6 and have a clear understanding of cassettes and film from Chapter 7.

Many problems with radiographic quality are traceable not to the x-ray room but to the darkroom. An understanding of film handling and film processing will aid in the formation of good work habits for the production of quality radiographic images. This knowledge enables the radiographer to identify and correct problems when they occur.

DARKROOM

An x-ray darkroom is illustrated in Fig. 8-1. The essential features include the film processor (automatic) or processing tanks (manual), a loading bench, a film bin, a film identification printer, and one or more safelights. There may be a passbox in the wall, which allows transfer of films to and from the darkroom while ensuring that no light is admitted to the darkroom.

The walls of the darkroom are painted a light color so that the safelight is reflected throughout the room, increasing visibility in the dark. There should be ample ventilation, because of the presence of chemical fumes. The doors and any seams in the walls must be effectively sealed so that no white light is admitted during film processing. Foam weather stripping around the door jamb is effective for this purpose and must be replaced as soon as it begins to fail.

Safelights

Safelights are special light fixtures fitted with filters (Fig. 8-2). The safelight filter should permit only the passage of light frequencies to which the film is least sensitive. For this reason, the filter must match the spectral sensitivity of the film. Kodak 6B filters are dark orange in color and are used for blue-sensitive film only. GBX filters are a deep cherry-red and are safe for use with blue-sensitive, green-sensitive, and x-ray sensitive films. Fluorescent safelights are also available for use with both blue- and green-sensitive films.

The safelight is not safe under all conditions. It must be mounted far enough from the loading bench (at least 3 feet) and should be located so that it is never too close to the film. The wattage of safelight bulbs is also important; 7 to 15 watts is usual, depending on location. Never replace a safelight bulb with a bulb of higher wattage than the one removed. Safelights that are safe for normal use will fog film when the exposure is prolonged. The safety of the safelight should

Fig. 8-1 Loading bench and work area in typical x-ray darkroom.

Fig. 8-2 X-ray darkroom safelight.

be determined for a period of time that exceeds typical safelight exposure time, usually 3 minutes. Care should be taken that exposure to safelight illumination does not exceed this period of time.

Safelight Testing   Safelights are tested using film that has been "sensitized" by preexposing it to a very light x-ray exposure (e.g., 1 mAs at 40 kVp). This exposure results in a film density beyond the toe of the sensitometric curve, where the film responds more readily to additional exposure. In total darkness, the preexposed film is placed in an envelope or folder so that one edge extends a distance of about 1 inch (Fig.

8-3). The film is placed on the loading bench (or other area to be tested) and the safelight is turned on. After 30 seconds, the film is extended from its cover by an additional inch. This is repeated every 30 seconds until a total of at least 3 minutes has elapsed. The last inch or so of film is never exposed to the safelight. The safelight is then turned off and the film is processed in darkness. An even density on the processed test film indicates that the safelight is safe for the period and location tested. Strips of varying density on the processed film (Fig. 8-4) indicate safelight fog. The location of the first visible strip of fog indicates the time when fog began to occur.

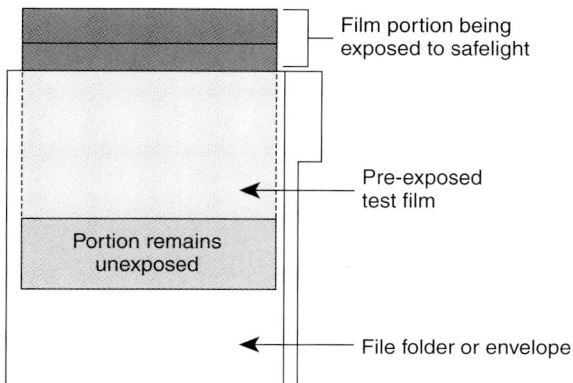

Film portion being exposed to safelight

Pre-exposed test film

Portion remains unexposed

File folder or envelope

Fig. 8-3 Pre-exposed film is placed in folder or envelope, as shown, for safelight test.

A

B

Fig. 8-4 Safelight test results. **A,** Film density is even throughout the film, indicating that the safelight is safe for the location and period of time tested. **B,** Strips of varying density on the test film indicate safelight fog.

Darkroom Fog

May be caused by:
- Unsafe safelights
- White light leak
- Excessive developer time, strength, or temperature

May cause:
- Decreased radiographic contrast
- Increased radiographic density

If this test indicates that the safelight is not safe for normal use, check the integrity of the filter (filters tend to crack and peel with age), the wattage of the bulb, and the distance from the safelight to the testing site. Decreasing the bulb wattage or increasing the distance may solve the problem. White light leaks may also result in a positive test. If the test indicates fog but the safelight appears to be satisfactory, check the darkroom carefully to ensure that no light is leaking around the door, the vents, or the processor. This is done most effectively after your eyes are well adapted to the dark. When you believe you have solved the problem, be sure to repeat the safelight test to confirm safelight safety.

Film Bin

The **film bin** is a special container in the darkroom to hold film for reloading cassettes. It is usually located under the loading bench. It opens from the front by pulling its handle toward you. It is light-tight and is divided to hold open boxes of each size of film. *The film bin is opened under safelight illumination only.*

Film Identification

A **film identification printer** (Fig. 8-5), sometimes referred to as a "flasher" or "stamper," is used to print patient information on each film prior to processing. A card with the necessary information is inserted into the printer. The corner of the film that was protected by the ID blocker is then inserted into the printer and the printer is activated. A light in the printer exposes the film through the card, leaving an image of the card data on the film. These devices vary in design and are very simple to operate. Become familiar with the film identification printer in your facility before trying to use it in the dark.

Some facilities have a daylight identification system that uses special cassettes. The corner of the cassette is inserted into the printer in the x-ray room so that film can be identified immediately before or after exposure.

Information to be imprinted on the film should include at least the patient's name, an identifying number (file number, chart number, or birth date), the

Fig. 8-5 Film identification printer is used to identify radiographs prior to processing.

date of the examination, and the name and location of the x-ray facility. Preprinted forms are usually used for this purpose. The patient data may be typed or handwritten. Some x-ray department computer systems generate an identification card along with other patient records when patient data are entered. If cards are typed or handwritten, do not use correction fluid to fix errors. Since the data are exposed by light passing through the card, the correction fluid will obscure any information on that portion of the card. The identification information will be most legible on the film when the card stock has a smooth, even texture and the printing is as dark as possible.

Passbox

A **passbox** is a convenient darkroom feature when there is a frequent need to transfer film to and from the darkroom while it is in use. Passboxes have two compartments, one for exposed film entering the darkroom and one for unexposed film leaving the darkroom after the cassette has been reloaded. Each compartment has two doors, one on the outside of the darkroom and one on the inside. An interlock between the inside and outside doors makes it impossible for both doors to be opened at the same time. Most passboxes are designed to protect film from exposure to x-rays. When the passbox location is convenient to the x-ray room, the compartment for unexposed film may be used as a storage place for cassettes that are ready for use.

FILM HANDLING

Radiographic film is handled when loading and unloading cassettes and when feeding it into the processing machine or preparing for manual processing.

Fig. 8-6 When holding film with one hand, let it hang vertically.

Fig. 8-7 To hold film horizontally, use both hands and grasp at opposite corners.

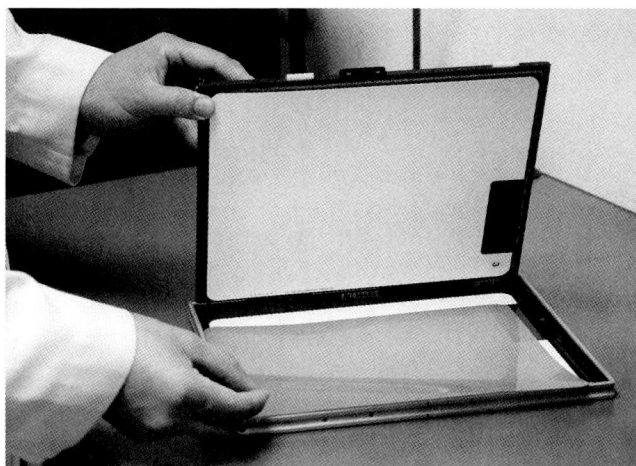

Fig. 8-8 To reload cassette, open it partially with your left hand and place the film with your right hand. When film is properly situated, tightly close all latches.

These activities take place in the darkroom under "safelight conditions," that is, a very low level of red or orange light with very limited visibility. It will be helpful for you to practice film handling with waste film in the daylight before you need to handle patient films in the darkroom.

Film must be handled with clean, dry hands and touched only on the corners. Dirt or chemical residue on your fingers will cause marks on the film and may also stain the screens. When holding a film with one hand, let it hang vertically (Fig. 8-6). To place the film horizontal, use both hands, holding it at opposite corners (Fig. 8-7). Attempting to hold the film horizontal with one hand causes it to bend, creating pressure exposure in the form of "crinkle" or **crescent marks.** Attempting to avoid this by extending your fingers to balance the film will result in finger marks. Crescent marks and finger marks are discussed and illustrated later in this chapter in the section on artifacts.

To unload a cassette, place the cassette on the loading bench with the back side up and the ID blocker positioned in the upper left corner. Release the latches, partially open the cassette, and grasp the film in the lower right corner with your right hand (Fig. 8-8). Remove the film, holding it vertically. Grasp the opposite corner with your left hand, holding with both hands to position the film in the identification printer.

Cassettes are reloaded as soon as possible after unloading. Since it is impossible to tell by looking at a closed cassette whether it is loaded or not and whether the film is exposed or not, it is important to follow standard procedures to avoid errors. For this reason, empty cassettes are never completely closed; they are almost closed but the latches are not fastened. They are reloaded promptly so that they are ready for use when needed. There should be separate standard locations for exposed and unexposed cassettes to avoid the possibility of accidentally reusing an exposed cassette, creating a double exposure.

To reload, open the film bin and select a film of the proper size, grasping it by the upper right corner with your right hand. With your left hand, close the film bin and open the cassette. Place the film in the cassette so that it is properly situated. Since double-emulsion films are identical on both sides, there is no right or wrong side to the film and no top or bottom. Simply place the film so that it fits properly into the cassette. Close the cassette, ensuring that all latches are tightly closed.

It is a good habit to double check that the film bin is closed and that there are no films out in the darkroom before opening the door or turning on the lights.

MANUAL FILM PROCESSING

As stated before, the purpose of film processing is to change the latent image on the film into a visible image. While most x-ray facilities today use automatic film processing machines, there are two good reasons why the study of manual processing is helpful to the radiographer. First, there are still some facilities that use manual processing systems, and many of these employ limited radiographers. In addition, an understanding of the manual system provides a good basis for understanding all film processing and the operation of automatic processors.

Manual Processing Systems

The **manual film processing system** consists of a large tank plumbed for constant water flow. Water flows in at the bottom of the tank and out at the top. The flow rate should be sufficient to provide eight complete changes per hour. The water temperature

Fig. 8-9 Thermal mixing valve for manual processing system water supply.

is controlled manually by a thermal valve (Fig. 8-9). Two or more smaller tanks (usually 5-gallon capacity) are placed inside the water tank (Fig. 8-10). These contain the chemical solutions, and their temperature is controlled by the temperature of the water that surrounds them. The principal chemical solutions are the developer and the fixer. Additional tanks may contain a "stop bath," between the developer and the fixer, and a final rinse. The largest part of the water tank does not have chemical tanks in it. It serves as a wash tank.

Manual Processing Chemistry

The chemical composition of manual processing solutions is summarized in Table 8-1.

The first step in film processing is development. The purpose of the developer is to change the exposed silver halide crystals of the film into black, metallic silver, creating the black or dark portions of the image. This chemical change is accomplished by the reducing agents. The gelatin of the emulsion is softened by the activator, allowing the reducing agents to penetrate the emulsion. The activator also maintains the alkaline chemical balance (pH) of the solution for optimal performance of the reducing agents. The function of the preservative is to prevent the reducing agents from losing their strength through interaction with air. The restrainer prevents chemical fog on the film that might occur as a result of overactivity of the developer when it is fresh and strong. The solvent for all of these chemicals is water.

Following development, the film must be "fixed." The fixer solution has two purposes: to "clear" the film and to "tan" it. The clearing process dissolves the unexposed silver halide crystals that remain in the emulsion, creating the clear areas that will appear light or white when the film is illuminated. The tanning process shrinks and hardens the emulsion to preserve the film image. The purpose of the activator in the fixer is to neutralize the developer that remains in the emulsion, stopping the development process. The activator also helps to maintain the acidic pH of the fixer solution, enhancing its function. A preservative is also included. The solvent for the fixing chemicals is also water.

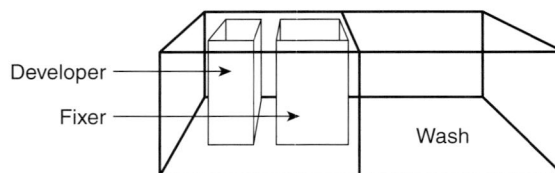

Fig. 8-10 Manual processing system consists of a large tank filled with running water. Smaller chemical tanks for developer and fixer are placed inside the water tank.

Some manual processing systems include a stop bath between the developer and the fixer. This is a dilute solution of acetic acid. Its purpose is to neutralize the developer on the film before it is placed in the fixer. The stop bath adds precision to the timing of the development process and prolongs the life of the fixer solution. In the absence of a stop bath, a fresh water rinse serves this purpose.

The wash tank contains clear, running water. Its purpose is to remove the processing chemicals from the film emulsion after fixing and before drying.

A final rinse bath is an optional part of the manual processing system. This is a small tank containing a water bath to which a small amount of detergent has been added. The detergent serves as a wetting agent, breaking the surface tension of water droplets and creating a "sheeting action." Use of a final rinse speeds drying and avoids the formation of water droplets on the surface that may run down the film, causing streaks on the finished radiograph.

Preparation for Manual Processing

The active chemical ingredients of the processing solutions tend to settle to the bottoms of the tanks when not in use. For this reason, the solutions are stirred prior to processing. Separate paddles or stirring rods are used for each solution to avoid cross-contamination of the chemicals. Chemicals are stirred gently to avoid splashing and to minimize mixing air into the solutions.

The next step is to check the temperature of the developer solution. Optimum temperature for manual development is 68° F. Minor variations in temperature are tolerable with appropriate adjustments in developing time.

A special darkroom timer is used to time development. The timer can be preset to the desired time. Since developing time may vary somewhat with temperature, the timer is preset after the temperature has been measured.

Under safelight illumination only, the cassette is unloaded, the film identified, and the film hung on a hanger. Film hangers are usually stored on racks on the darkroom wall or under the loading bench and are arranged according to size. Each corner of the hanger has a clip to hold the film. The clips are opened by pinching. It is most effective to attach the film to the bottom of the hanger first, since these bottom clips are stationary. The top clips are attached to wire springs that must be stretched to reach the film

---

TABLE 8-1

Chemical Components of Processing Solutions

Developer Solution

Reducing Agents	
Hydroquinone	Reduce EXPOSED silver halide to black metallic silver
Metol or Elon	
Phenidone	
Activator	
Sodium carbonate	Softens gelatin, maintains alkaline pH ($\uparrow$pH)
Hardener	
Glutaraldehyde	Prevents overswelling of gelatin in automatic developer
Preservative	
Sodium sulfite	Antioxidant—prevents oxidation of developer
Restrainer	
Potassium bromide	Prevents chemical fog in new developer
Solvent	
Water	

Fixer Solution

Clearing Agent	
Ammonium thiosulfate	Dissolves undeveloped silver halide
Tanning Agent	
Chrome alum or other aluminum salt	Shrinks, hardens, preserves emulsion
Activator	
Acetic acid	Neutralizes developer, maintains acid pH ($\downarrow$pH)
Preservative	
Sodium sulfite	Prevents oxidation, prolongs solution life
Solvent	
Water	

Stir Solutions

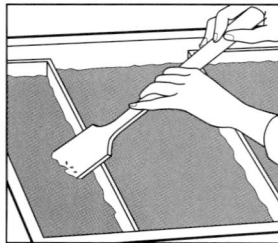

A, Stir chemicals using a separate paddle for each tank. Rinse paddles after use.

Check Temperature

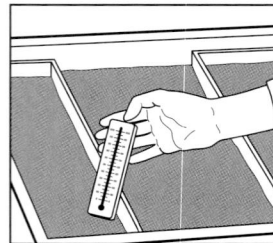

B, Check developer temperature. Temperature should be 68° F.

Pre-Set Timer

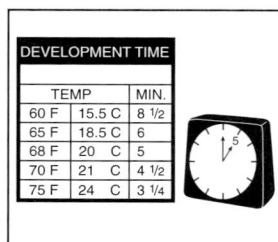

C, Preset developing timer, making any needed adjustments for temperature variations. Then turn off the white lights, remove film from cassette, and identify it.

Load Film on Hanger

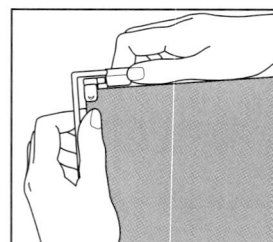

D, Load film on hanger, attaching bottom edge first. Film should be taut in hanger. Handle with care.

Immerse Film in Developer

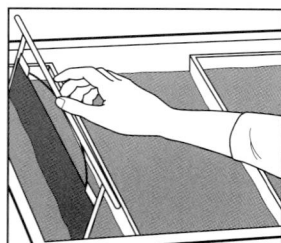

E, Immerse film in developer tank. Plunge it in smoothly, without pause, to avoid streaks. Activate developer timer.

Agitate in Developer

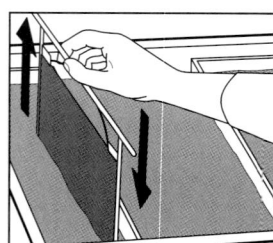

F, Agitate film in developer to avoid bubbles clinging to film surface and to bring fresh chemicals in contact with film.

When timer rings, remove film from developer

G, When timer rings, remove film from developer tank immediately. Do not allow excess developer to drain back into developer tank.

Rinse Thoroughly

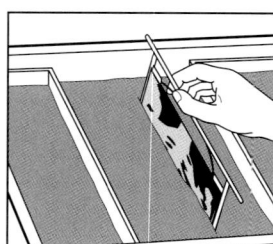

H, Rinse thoroughly by agitating continuously in fresh, running water for 30 seconds. Alternatively, use acid rinse bath.

Fig. 8-11 Manual processing procedure.

Fix Adequately

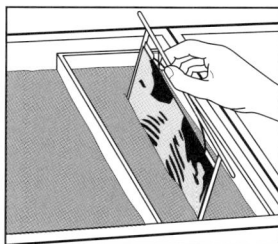

I, Place film in fixer tank and agitate vigorously at first. Allow film to fix for a period at least twice the clearing time.

Wash Completely

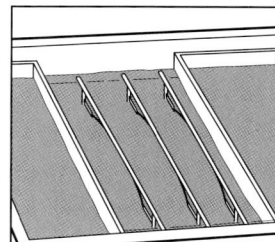

J, Wash thoroughly by placing in running water of wash tank for 20 to 30 minutes. Keep space between films.

Final Rinse

K, If facilities permit, dip film in final rinse solution containing wetting agent. This rinse speeds drying and prevents streaks and watermarks.

Allow Film to Dry

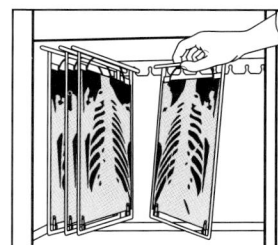

L, Allow film to dry. Hang on rack over drip tray or sink or use film dryer. Take care that films do not touch each other during drying process. When dry, remove from hanger and trim corners.

Fig. 8-11, cont'd Manual processing procedure.

edge. These help to keep the film taut and straight in the hanger.

Several films may be prepared and processed at one time. The number will depend on the size of the solution tanks. Take care that films do not touch one another or scrape against other hangers when processed together.

Manual Processing Procedure

Manual film processing is not difficult when you are familiar with the procedure and the equipment. It is helpful to practice this procedure in daylight before attempting it under safelight illumination. The steps for manual film processing are illustrated in Fig. 8-11.

The safelights are turned on before the cassette is opened, and the procedure takes place under safelight illumination until the film has been in the fixer for a least 1 minute. At this point, the developer has been neutralized and exposure to white light will not fog the film.

At several points in the processing procedure, the film is "agitated." This refers to the practice of moving the film up and down briskly in the tank to remove bubbles and to bring fresh solution or water into contact with the film surface. Be careful when agitating

TABLE 8-2

Manual Developing Time

Temperature	Minutes
63° F	7½
65° F	6
68° F	5
70° F	4½
73° F	4
75° F	3½

so that the film does not rub against the sides of the tank or come into contact with other films or film hangers, since the wet emulsion is soft and easily damaged. Contact with other films results in incomplete processing.

The timing of development is critical. *Optimum developing time is 5 minutes at 68° F.* Small variations in temperature may be compensated by varying the developing time (Table 8-2). Too little developing time will result in incomplete development, causing the film to be too light. Too much time in the developer may cause chemical fog, reducing film contrast.

When the developing time has elapsed, the film is immediately removed from the developer tank and placed in either the stop bath or the water between the developer and fixer tanks. The film must not be allowed to drain back into the developer tank as it is removed. If a stop bath is used, the time in this tank is about 5 seconds. If there is no stop bath, the film is agitated in the water for 30 seconds and the excess water is allowed to drain off.

The film is then placed in the fixer. Fixing requires less precise timing than development. The strength of the fixer solution varies greatly, depending on age, how much water is carried in with each film, and whether or not a stop bath is used. The strength of the fixer can be gauged by the time required to clear the film. Early in the fixing process, the film has a "milky" appearance due to the presence of unexposed silver halide crystals in the emulsion. As these crystals are dissolved by the fixer, the film takes on its typical black-and-white appearance and the film is said to be "cleared." Clearing should occur within 1 to 3 minutes of fixing time. The tanning process takes approximately twice as long as the clearing process and cannot be assessed visually. From 5 to 8 minutes of fixing time is usually adequate for both clearing and tanning, but 10 to 15 minutes will not affect the film adversely. Excessive fixing time results in saturation of the film base by fixer and requires extra washing time.

After fixing, the film is washed. The purpose of washing the film is to remove processing chemicals before the film is dried. Failure to remove fixer from the film causes deterioration of the film during storage. When this occurs, the film turns yellow and eventually the emulsion separates from the base. Washing time of 20 to 30 minutes is usually adequate. The time should be extended if the water flow rate is diminished, if the wash tank is crowded, or if the film has been in the fixer for an excessive length of time. If there is a final rinse, the film is dipped in this tank briefly at the conclusion of the wash.

Film may be dried in a **film dryer** or by hanging it on a rack over a draining tray or sink. A film dryer is a metal cabinet with an electric heating coil and a fan to circulate the air. Take care that films are not touching each other while drying. Optimum drying temperature is 110° F. Too high a dryer temperature tends to melt the gelatin of the emulsion, causing blistering of the film surface.

When the film is dry, it is removed from the hanger. The corners of the film within the hanger clips tend to retain processing chemicals and may not dry thoroughly. There may also be sharp bits of film at the corners due to piercing by the hanger clips. For these reasons, the corners of the film are cut off before the films are sorted and prepared for reading. Commercial corner cutters are available for remov-

---

Mixing Fresh Solutions for Manual Processing

Developer	Fixer
1 Gal concentrate + 4 Gal water + 4 oz restrainer ——————— 5 Gal developer solution	1 Gal concentrate + 4 Gal water ——————— 5 Gal fixer solution

---

ing the corners from several sheets of film at a time. Corners can also be removed with utility shears.

Mixing New Processing Solutions

A chemical service company may be contracted to clean the processing tanks and supply fresh solutions on a monthly schedule. In some facilities, however, this may be the duty of the radiographer.

The first step in cleaning tanks is to empty the water tank. Then, with the bottom drain open and the water flowing, empty the developer tank. Developer may be allowed to flow down the drain of the water tank. Because used fixer contains silver, the fixer must be dipped or siphoned into holding containers for recycling unless the silver has been removed from it. Some facilities have silver salvage units that remove the silver from used fixer. If this is the case, the fixer also may be allowed to flow down the drain. *Never empty both the developer and the fixer tanks at once.* This results in the mixing of developer and fixer in the bottom of the wash tank, producing toxic ammonia gas.

When the tanks are empty, attach a hose to a nearby faucet. Using a stiff, nonmetal brush, scrub each tank and rinse thoroughly using the hose. Do not use steel wool or metal cleaners to clean processing tanks. Abrasive cleaners will scratch and damage the tanks; chemical agents are likely to contaminate the processing chemicals. When the tanks have been cleaned and thoroughly rinsed, replace the stoppers in the chemical tanks.

Both developer and fixer are supplied in 1-gallon containers of concentrate. Each gallon of concentrate makes 5 gallons of solution. The developer restrainer solution is supplied separately in a small bottle. Partially fill each tank with water from the hose and then add the chemical concentrate to each. The accompanying box summarizes the proportions of concentrate and water for mixing fresh solutions. Add the restrainer to the developer. Finish filling each tank with water, stirring thoroughly. When each chemical tank is partially filled, close the water tank drain and allow it to refill. When all solutions have reached the desired level and temperature, the system is ready to use.

Mixing Replenisher Solutions for Manual Processing

Developer	Fixer
I Gal concentrate + 2 Gal water ——————— 3 Gal replenisher	Fixer is replenished with undiluted fixer concentrate
Do not add restrainer	

Replenishing Processing Solutions

As films are removed from each chemical tank during processing, each carries a certain amount of solution with it. As a result, the volume of solution in the tanks gradually decreases. In addition, the solution that remains in the tanks has lost some of its strength due to chemical interaction with the film. Without replenishment, the solutions would soon be too weak to perform their functions, and there would be too little solution in the tank to cover the film. To replace both the lost volume and the lost strength, the tanks must be replenished with a solution that is stronger than the original.

Developer replenisher is made by mixing 1 part of concentrate to 2 parts of water. *Restrainer is not added,* since its purpose applies only to fresh developer. Bromide ions from the film mix with the developer during use and serve as a restrainer as the developer ages. As noted in the processing procedure, it is important to lift the film out of the developer quickly, allowing as little solution as possible to drain back into the tank. This ensures that a consistent volume of solution is removed from the tank for each film size and so ensures that the replenishment process maintains consistent chemical strength.

Fixer volume decreases more slowly than developer volume, since liquid is carried into this tank as well as out of it. For this reason, fixer is replenished with undiluted fixer concentrate. If the fixer becomes sluggish and there is no room to add replenisher, a quart of used fixer may be dipped out of the tank and replaced with fixer concentrate.

Be sure to stir chemicals thoroughly following replenishment.

Processing Chemical Safety

While it is quite safe to place one's hands in developer or fixer during film processing, these chemicals may be irritating to the skin with prolonged contact. Hands should be washed thoroughly after contact with processing solutions. Avoid skin contact with concentrated chemicals.

Processing solutions are especially irritating to the eyes and may cause permanent injury with prolonged contact. Chemical eye splashes should be flushed immediately with running water. An eyewash station should be used if one is convenient. If an eyewash station is not available, place the eye under a gentle stream of cool, running water for 2 minutes. If pain persists after eyewash treatment, consult a physician without delay.

The Occupational Safety and Health Administration (OSHA) requires the use of splash-proof goggles and nitrile* gloves when mixing or pouring chemicals and when cleaning up spills. A plastic apron is also a wise idea. Material safety data sheets (MSDS) must be available to radiographers for all chemicals used in the department. It is your duty to be familiar with these documents and to know where they are kept.

As stated before, used fixer contains silver unless the silver has been reclaimed from it. Silver is an environmental hazard when released into rivers and streams, causing death to plants and fish. Since public sewer facilities do not precipitate for silver when processing waste water, there are regulations against disposing of silver into the sewer system. For this reason, used fixer is placed in plastic containers and recycled. The companies that provide processing chemicals usually also provide a recycling service for used fixer, crediting the customer's account for the value of the reclaimed silver.

AUTOMATIC FILM PROCESSORS

Automatic film processors vary in design according to age and manufacturer. There are large models that stand on the floor and small units that rest on a countertop (Fig. 8-12). The essential features of all processors are similar, however, and the specifics of each model are fully explained in the processor operation manual supplied with the unit. This manual is a helpful reference for understanding the operation of and proper settings for the processor.

As compared to a manual processing system, which takes about an hour to process a film "dry-to-dry," modern automatic processors require 3 minutes or less. This is made possible principally by the increased strength and temperature of solutions, but all aspects of the system are designed to support and enhance this speed.

The operation of an automatic processor may be divided into six parts: transport system, chemistry, replenisher system, recirculation system, water system, and dryer. Processors contain three processing tanks: developer, fixer, and wash (water) tank.

---

*Nitrile is a coating that makes the gloves impervious to chemicals. These gloves are available at hardware stores and at sources where industrial clothing is sold.

Fig. 8-12 Automatic film processors. **A,** Floor model; **B,** countertop model.

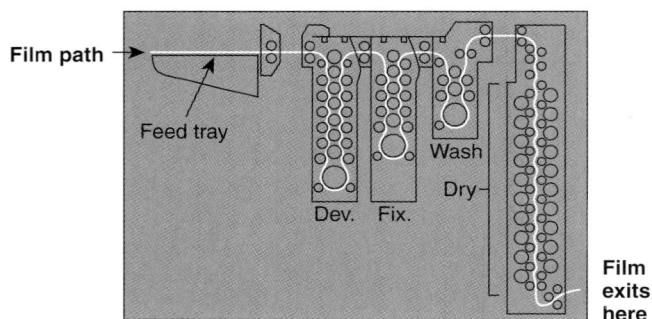

Fig. 8-13 Diagram of film path through automatic processor.

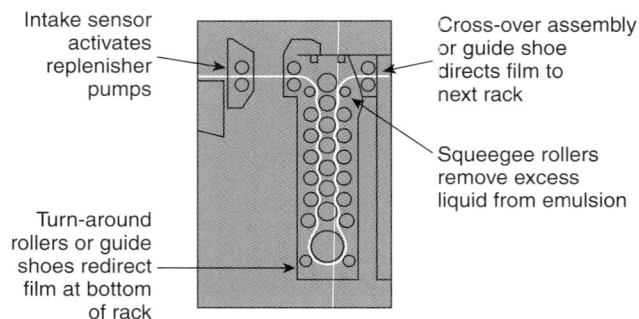

Fig. 8-14 Close-up view of developer rack in automatic processor.

Automatic Film Processor Systems

- Transport system
- Chemistry
- Replenisher system
- Recirculation system
- Water system
- Dryer system

Transport System

The transport system (Fig. 8-13) consists of the drive motor, gear assembly, and a roller assembly or "rack" for each tank. The roller action moves the film through the processor. Roller action also helps to agitate the solutions, keeping fresh solution in contact with the film surface. A cross-over roller assembly or curved metal "guideshoe" transports the film between the tanks and from the wash tank to the dryer. Turn-around rollers or guideshoes at the bottom of each rack change the direction of film travel when it has reached the bottom of the tank. The exit rollers on each rack provide a "squeegee" action to remove excess solution from the film before it enters the next rack. A close-up diagram of the developer rack (Fig. 8-14) illustrates these features.

The transport mechanism moves at a constant rate and governs the time the film is subjected to each portion of the processing cycle. Broken gear teeth or cogwheels in the transport mechanism may occur as a result of wear or film jams, causing slippage of the film in the area where the damage has occurred. A stopwatch test of cycle time, from the instant the leading film edge enters the processor until it emerges from the dryer, provides a measure of the timing ac-

BOX 8-1

Preservation of Developer Strength

- Do not run processor for prolonged periods when not in use.
- Install a stand-by system that maintains temperature but stops water flow and recirculation when not in use (if not included with processor).
- Use floating lid on developer replenisher tank to reduce deterioration of developer caused by exposure to air. (Observe color: developer should not be darker than light beer.)
- Do not store excess quantities of mixed solutions.

BOX 8-2

Correct Replenishment Is Essential to Proper Processing

Underreplenishment	Overreplenishment
• Loss of contrast	• Loss of contrast (↑ fog)
• Loss of density	• Decreased D-max
• Inadequate clearing	• Loss of density
• Inadequate hardening	• Waste
• Failure to dry completely	
• Transport failure	

Chemistry Modifications for Automatic Processing

- Stronger preservative decreases deterioration of developer at high temperatures.
- Greater developer strength permits more rapid development.
- Glutaraldehyde, a tanning agent in the developer, prevents overswelling of gelatin.
- Stronger restrainer called "starter" prevents chemical fog in new developer.

curacy of the transport mechanism. Actual cycle time varies with the make and model of the processor and is stated in the processor operation manual.

Automatic Processing Chemistry

The chemicals used in automatic processors are essentially the same as those for manual processing as listed in Table 8-2, with some modification to accommodate the requirements of higher temperatures and the critical tolerances of the transport system. Greater developer strength promotes rapid development, and stronger preservatives decrease deterioration of the solutions at high temperatures. A stronger restrainer called "starter" prevents chemical fog in fresh developer.

Glutaraldehyde, a tanning agent, is included in the *developer* to prevent overswelling of the gelatin. The correct amount of this chemical is essential. Too little causes the emulsion to become too soft, absorbing too much liquid. This condition results in "roller marks" on the film from the transport rollers and may also prevent the film from drying completely. In extreme cases, the emulsion becomes so thick and sticky that it cannot pass through the rollers, causing a jam. Too much glutaraldehyde, on the other hand, shrinks the emulsion before sufficient developer has been absorbed, causing underdevelopment.

See Box 8-1 for suggestions to preserve developer strength.

Replenishment of Automatic Processing Chemicals

Replenisher solutions for the processor are stored in separate holding tanks, connected to the processor by hoses. Replenisher pumps in the processor bring enough fresh solution into the processor tanks for each film as it is processed. As fresh solution is added, the tanks overflow. Developer is allowed to overflow into the drain. Fixer may be pumped through a silver recovery unit before flowing into the drain, or it may flow into a holding container for recycling.

An intake sensor on the processor feed tray, seen in Fig. 8-14, activates the replenisher pumps as the film passes over the sensor. The size of the film thus determines the duration of the pumping action and the amount of replenisher added for each film.

The replenishment rates are adjustable and are measured using a 14 × 17-inch film. To measure replenishment rates, the tubes that carry replenisher to the processing tanks are diverted into measuring cups while the test film is fed into the processor. Approximately 100 to 120 ml. of replenisher is usually added to both the developer and fixer for each 14 × 17-inch film. The replenishment rate must compensate not only for the chemical strength used to process the film but also for strength lost to deterioration when the processor is idle. For this reason, ideal replenisher rates vary according to the volume of film processed. The processor operation manual will state the desired rates for fixer and developer replenishment based on the average number of films processed per week. When film volume changes significantly, replenishment rates must be adjusted accordingly. Higher rates are required when film volume is low, and rates may be reduced when volume is high. Box 8-2 lists consequences of incorrect replenishment rates.

Recirculation

When the processor is in operation, recirculation pumps constantly move developer and fixer from the tanks, through a heat exchanger, and back into the tanks. Recirculation through the heat exchanger maintains solution temperature at the desired level, usually between 92° and 96° F. The ideal temperature is stated in the processor operation manual. Recirculation also keeps the solutions in motion, continuously placing fresh solution in contact with the film surfaces to increase the speed of the chemical action.

Water System

Fresh water runs into the wash tank and overflows into the drain continuously when the processor is in operation. This fresh water flow efficiently washes the film. The water system may also provide cooling when needed in the heat exchanger portion of the recirculation system. Depending on the incoming water supply, a filtration system may be necessary to maintain adequate water quality. Filters may be installed on the water line before the water enters the processor. Unlike manual processing systems, modern automatic processors do not require a warm water supply. Cold water only is supplied to the processor and is heated, as needed, within the processor.

Dryer

The dryer section of the processor consists of air tubes with long slits. The air tubes are mounted on both sides of the dryer rack, directing warm air at both sides of the film as it passes through this portion of the roller system. The air is warmed to approximately 110° F and is filtered to prevent dirt from clogging the tubes or sticking to the film.

The dryer temperature is adjustable but is usually quite stable. When films are damp as they exit the processor, there are several possible causes. Inadequate air supply may be caused by too little space around the processor or clogging of the air filters with dirt. When a small processor is used to process too many films in a short time period, moisture may build up within the dryer, decreasing its effectiveness. The most common cause of damp films, however, is inadequate developer replenishment. Insufficient replenishment causes a decrease in the delicate glutaraldehyde balance, permitting the emulsion to swell with liquid to the point where it cannot dry at standard drying time and temperature.

Processor Operation and Maintenance

Automatic processors are very simple to use. After the film has been identified, it is placed on the processor feed tray and moved into the processor until it is picked up and moved forward by the intake rollers.

The width of the feed tray and the recommendations of the manufacturer (see the processor operation manual) will determine whether the film is fed with its long or short dimension parallel to the rollers. Usually, the long dimension is parallel to the rollers unless it cannot be accommodated by the feed tray. When the entire film has passed beyond the intake rollers, a tone or a red light, or both, will signal that it is safe to feed the next film or to turn on the lights. Feeding in another film before this signal is given will result in overlap of the two films during processing. This causes incomplete processing of both films and may also cause a jam. White light in the darkroom prior to the signal will fog the trailing edge of the film. The finished film will drop into a receiving bin 1½ to 3 minutes after the film enters the processor.

When the processor is first turned on, it takes some time for it to warm to operating temperature. Film processed before the processor is ready will be too light and will lack contrast. Most processors take about 30 minutes to warm up, and most provide an indicator that the desired temperature has been reached. There may be a sound signal or a light indicator. Some processors have digital thermometers that indicate the developer temperature.

The upper portions of the top rollers on each rack are above the level of the solution. When the processor is idle, solutions tend to dry on the exposed rollers. This dried chemical residue is deposited on the first film that is processed after the idle period, leaving undesirable marks. To avoid this problem when the processor has been idle for a long period, a "clean-up" film is fed through the processor before processing patient films. A clean-up film is a 14 × 17-inch film that has not been processed. Film that has aged beyond the expiration date or has been accidentally fogged or exposed may be kept for this purpose. The use of a previously processed film as a clean-up film is not recommended because any fixer residue in the emulsion may contaminate the developer.

At the end of the day, the processor and its water supply are turned off. Some manufacturers recommend that the lid be partially opened when the processor is turned off. This practice allows chemical fumes to escape, preventing fixer vapor from condensing in the developer and contaminating it. Gelatin residue and warm temperatures in the wash tank may promote the growth of algae. When this occurs, the algae form a scum in the water tank that may cling to the film. Growth of algae can usually be controlled by emptying the wash tank at the end of the day, allowing it to dry out during the night.

Box 8-3 contains a checklist for daily and monthly processor maintenance. Many facilities have a contract for monthly maintenance provided by a processor service company. Every 6 months, a service company should perform a thorough processor cleaning using special systems cleaners to clean the insides of

BOX 8-3

Darkroom and Processor Maintenance Checklist

Daily Start-Up
- Close wash tank drain.
- Turn on water supply.
- Close processor cover.
- Turn on processor.
- Allow processor to warm to operating temperature.
- Wipe down loading bench.
- Check solution levels in replenisher holding tanks and used fixer overflow container.
- Run a "clean-up" film.
- Expose and process sensitometric test film.
- Evaluate film and graph result. Troubleshoot any problems indicated.

Daily Shut-Down
- Turn off processor.
- Turn off water supply.
- Open wash tank drain.
- Leave processor cover partially open.

Monthly Processor Maintenance
- Completely drain processing tanks.
- Remove, clean, and inspect roller racks.
- Clean tanks and rinse thoroughly.
- Fill developer and fixer tanks with new solutions; add starter to developer.
- Refill wash tank.
- Turn on processor and warm to operating temperature.
- Check developer temperature with separate, calibrated thermometer.
- Check replenishment rates and adjust, if necessary.
- Measure cycle time with stopwatch.
- Expose and process sensitometric test film and compare to standard.

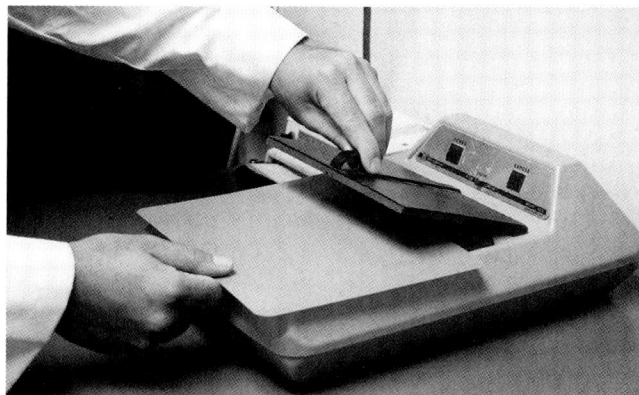

Fig. 8-15 Sensitometer uses precisely controlled light source to expose gray scale on film.

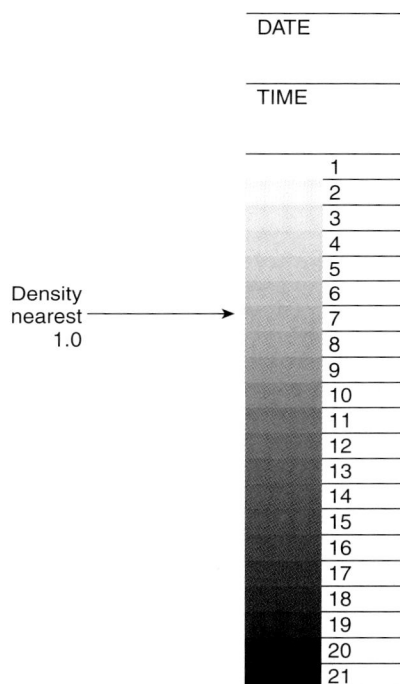

Fig. 8-16 Processed film that has been exposed by sensitometer on both sides.

the pumps and the recirculation system. This service should include a complete inspection of the processor with replacement of any worn or damaged parts.

PROCESSOR QUALITY CONTROL

Considering the complexity of the automatic processor, it is understandable that variations may occur in temperature, film volume, replenisher rates, and cycle time and that these variations may have a significant impact on film quality. The purpose of processor quality control is to monitor these variations and solve any problems before they become serious. Some state agencies have regulations requiring processor quality control monitoring. These agencies may review quality control records when conducting radiation safety inspections.

Two special pieces of equipment are necessary for monitoring processor performance: a **sensitometer** and a **densitometer.** A sensitometer (Fig. 8-15) is a

device that prints a standard gray scale on film (Fig. 8-16). Each tone on the gray scale is called a "step," and each step is numbered. The exposure is made by a precisely controlled light source, so that the amount of exposure is always exactly the same. A fresh sheet of film is exposed in the darkroom with the sensitometer and is processed. It is a good practice to expose two edges of the film with the sensitometer, one with each side of the emulsion down. The different positions of the two sides of the emulsion with respect to the developer rack sometimes result in slightly different responses to processing. These variations may cause erratic test results unless they are taken into account. The resulting film is evaluated using the densitometer.

Fig. 8-17 Densitometer is used to measure optical density of the gray scale steps on sensitometric film.

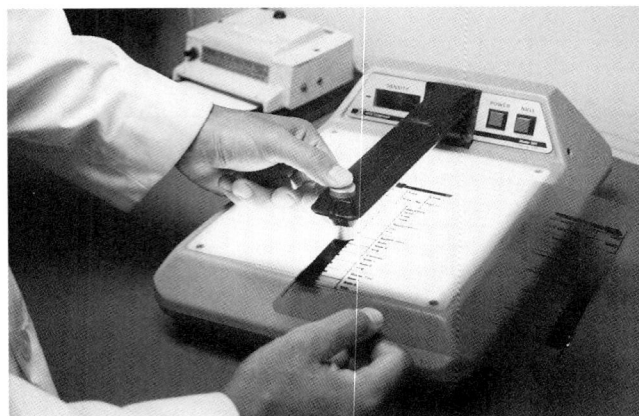

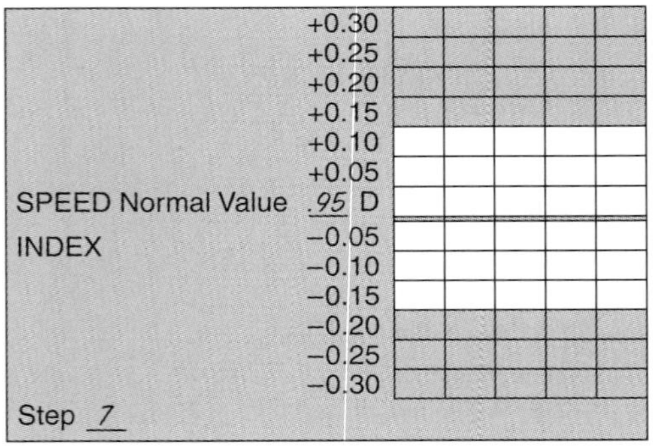

Fig. 8-18 Speed index portion of quality control record. The number of the step on the baseline film with an OD nearest 1.0 is entered at the bottom and its optical density is entered as the normal value. This establishes the value of the center line of the graph.

A densitometer (Fig. 8-17) is used to measure the transmission of light through a tiny area of the film. It is a digital device that measures the optical density of the film. The densitometer is first calibrated to a null (zero) value by engaging it with no film. The test film is then measured and the optical density value is recorded. When both sides of the emulsion have been exposed, both sides are measured and the results are averaged.

Setting Up a Quality Control System

Processor performance is monitored by comparing the densities of standardized exposures on the test film to those of a similar film (the control standard or baseline film) that was processed when the processor was performing at an optimal level. The best time to establish this standard is soon after the processor has been cleaned but when sufficient film has been processed to indicate consistent satisfactory performance. It is desirable to do preliminary tests with the sensitometer over a period of several days to be certain that processor performance is stable.

Three or four measurements are compared for each processor test: film density (speed index), contrast, **gross fog** (base + fog), and sometimes D-max. Each of these factors is measured on the control standard film to establish baseline measurements for comparison with future tests. These measurements

are used to set up a graph on which the results of future tests will be plotted.

To begin, the control film is exposed with the sensitometer and processed when the processor is performing optimally. The density of each step in the gray scale is measured with the densitometer. Comparable steps on the two sides of the film are averaged and the average density recorded.

To establish the speed index, the baseline for film density, the step on the gray scale is chosen that measures closest to an optical density of 1.0. This density represents a middle-gray tone, about halfway between the densities that appear black and white on the viewbox. The number of this step and its average measurement are recorded on the processor record graph (Fig. 8-18).

Since radiographic contrast is defined as a *difference* in radiographic density, the contrast index is a measure of the difference between two densities on the gray scale. A good choice is to measure the difference between the speed index and the average density of the step that is two steps darker than the speed index. For example, in our illustration the speed index is measured at step 7 and the contrast index is equal to step 9 minus step 7 (Fig. 8-19). The actual value of the contrast index will vary, depending on the inherent contrast of the film and the developer temperature. A value between 1.4 and 1.7 is usual for medium- to high-contrast films at recommended processor temperatures.

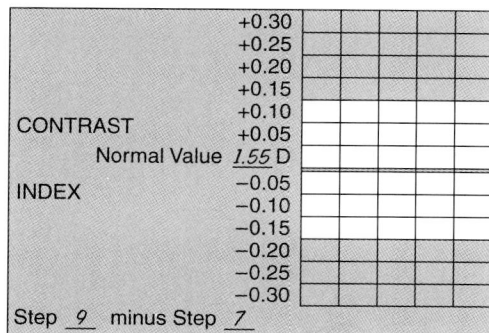

Fig. 8-19 Contrast index portion of quality control record. The numbers of the baseline film speed index step and the step that is two above it (speed step + 2) are entered at the bottom. The difference in optical density between the two steps is entered as the normal value. This establishes the value of the center line of the graph.

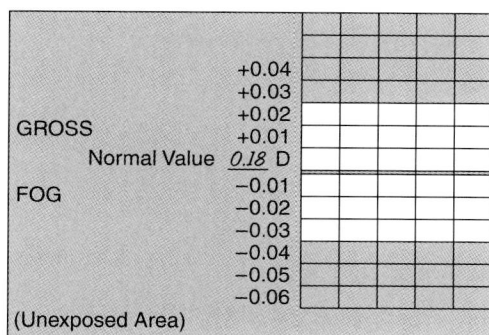

Fig. 8-20 Gross fog index portion of quality control record. The optical density of the unexposed portion of the baseline film is entered as the normal value, establishing the value of the center line of the graph.

The gross fog index is the average measurement of step 1 on the test film. It can also be measured in any unexposed portion of the film. The value of this step is also entered on the processor record form (Fig. 8-20). The usual value for the gross fog index is an optical density between 0.15 and 0.20. This value varies with film speed, faster films having a higher gross fog index.

If a record is to be kept of D-max, this index is measured on the darkest step of the gray scale. The average density of this step is measured and entered on the processor record in the same manner as for the other indices.

The control baseline sensitometric film is kept for reference and is marked to indicate the date and time it was made, the developer temperature, and the replenishment rates for both developer and fixer. If there is a change in the brand or type of film used, a new baseline film must be made to establish a new standard.

An additional way to assess the consistency of processing solutions is to test the pH, that is, the degree of alkalinity of the developer and the acidity of the fixer. Test strips are available for this purpose. These paper strips are dipped into the solutions and their color compared to a standard color strip to obtain a reading. Directions are provided with the strips and should be followed precisely. If pH testing is part of your quality control procedure, the pH of both developer and fixer should be tested when the quality control record is established. The readings are recorded on the baseline film and the processor record form.

BOX 8-4

Sensitometric Processor Evaluation

- Expose both sides of a fresh 8 × 10-inch film with sensitometer.
- Process film (be sure processor has reached operating temperature).
- Measure key steps with densitometer and average readings from the two sides.
  - Speed step (Step 7 in Fig. 8-21, *A*)
  - Speed step + 2 (Step 9 in Fig. 8-21, *B*)
  - Gross fog level (unexposed area; Fig. 8-21, *C*)
- Calculate average density readings for each step measured.
- Calculate contrast index (Step 9 minus Step 7 in Fig. 8-21, *B*).
- Plot readings on graph.
- Troubleshoot any readings outside acceptable range.

Monitoring Processor Performance

Once the baseline indices and the processor performance record have been established, it is easy to maintain a record of processor performance. Box 8-4 lists the steps for sensitometric processor evaluation. A sensitometric film is exposed and processed daily after the processor has reached operating temperature. The densities of the key steps are measured and averaged and their values plotted on the graph (Fig. 8-21). Each

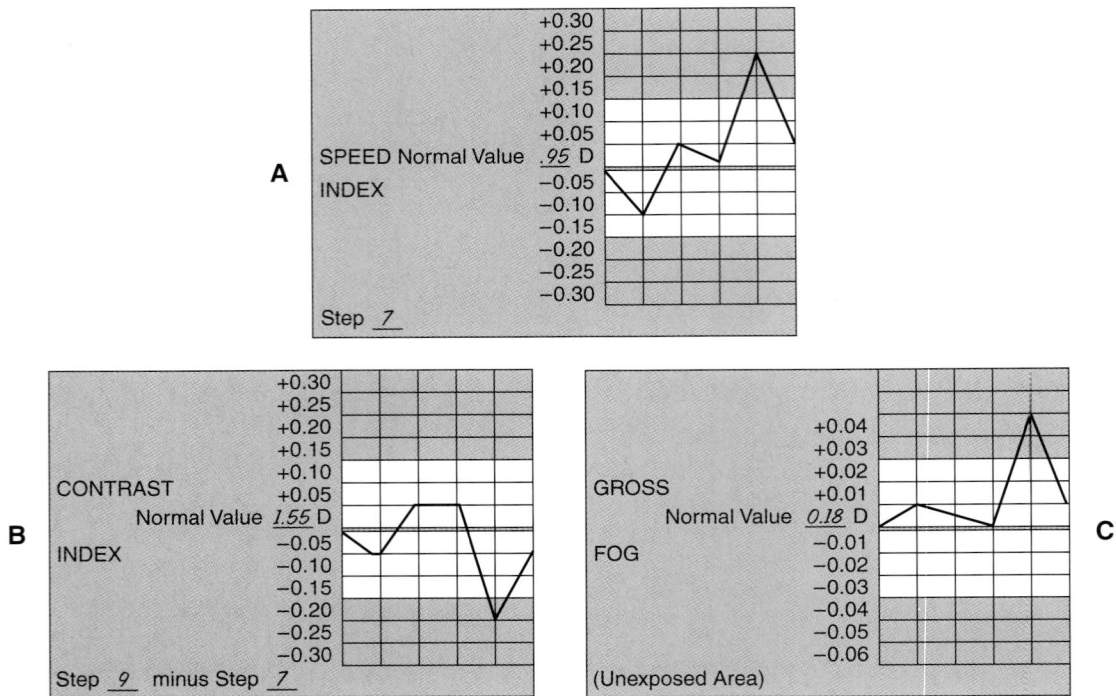

Fig. 8-21 In this example, daily sensitometric films are measured with the densitometer at Step 7, Step 9, and the unexposed area. Each vertical line represents a separate day and a separate test film. **A,** The graph of speed index is a plotting of the readings of Step 7. **B,** Graph of contrast index. **C,** Graph of gross fog index. Note that on day 5, something happened that caused all three indices to be outside the normal range. Check Table 8-3 to determine possible causes.

TABLE 8-3

Troubleshooting Chart

	Developer Temperature	Developer Depleted	Developer Contaminated	Developer Overdiluted	No Starter in Fresh Developer	Incorrectly Mixed Developer	Fixer Depleted	Replenishment Rates Incorrect	Water Problems	Dryer Problems	Loss of Circulation	Dirty Rollers	Dirty Water	Filter Size Too Small	Misaligned Guideshoes	Improper Film Handling	Safelights	Storage
Increased base fog	1	5	4		2	4	4					3					2	2
Reduced contrast	1	1	2	3	1	1	2	2			4	3					2	2
Increased contrast	1					1		1										
Reduced film speed	1	2	3	3		2	2				5							
Increased film speed	1		1		1	2	2	3			5	3				4	4	3
Wet or damp films	3	2	2	2		2	1	2		3	2	4						
Improper clearing	2	2	2	2	5	3	1				2							
Dirty films	2	2	2	2		2	3		1	3	2	2	2	2		3		
Scratches on films	2	2	3	2		3	2		2			1			1	2		

From Dupont Diagnostic Imaging and Information Systems.
1, Check first; *2,* check second; *3,* check third; *4,* check fourth; *5,* check last.

vertical line on the graph represents one test. The date, time, developer temperature, and pH values are also noted. On dates when replenisher rates are measured, these data are recorded on the processor record also. Replenisher rates should be checked at least monthly and whenever the test film indicates a problem that may be related to replenishment.

There will be normal fluctuations in all of the indices, but they should stay within a tolerable range. This range is ±0.15 for the speed and contrast indices and is ±0.03 for the gross fog index. Processor record forms are usually marked to indicate clearly when measurements are beyond the acceptable limits. When this occurs, it is important to identify and rectify the problem immediately. Table 8-3 is a guide to causes of common processor problems that may be identified in sensitometric quality control tests.

ARTIFACTS

Film artifacts are marks, exposures, or images on the radiograph that are not part of the intended image. Artifacts appear dark if they result from unintended exposure and light when caused by interference with exposure or processing.

Cassette and Screen Artifacts

Light leaks in the cassette cause unintended exposure to the film (Fig. 8-22). This problem may be caused by wear or damage to the cassette. Leaky cassettes usually cannot be repaired. The cassette must be replaced. A light leak is also caused when the film is not loaded properly and overlaps the cassette edge (Fig. 8-23). This results in a black edge on the film and a blurred image adjacent to it that is caused by poor film/screen contact. This problem is avoided by checking to be certain that the film is properly situated in the cassette during the loading process.

Screen artifacts result when dirt on the screen prevents screen light from exposing the film (Fig. 8-24). These artifacts can be minimized by keeping the loading area clean and by cleaning screens with a product that contains an antistatic ingredient. Similar artifacts may occur when the screen is stained or damaged, in which case the screens must be replaced. Even though only one screen is damaged, the pair must be replaced. It is best to have new screens installed by the dealer's local technical representative.

Film-Handling Artifacts

Common film-handling artifacts include creases, crescent marks, finger marks, and abrasions. Creases and crescent marks are caused by bending or crimping the film (Fig. 8-25). This can be avoided by remembering to let the film hang vertically when holding it with only one hand.

Finger marks (Fig. 8-26) are caused whenever the film is touched and are worst when the hands are not clean and dry. These marks are avoided by using clean, dry hands and handling film only at the corners. Finger marks may appear as either dark or light artifacts. Oily skin causes light finger marks by preventing development, while developer on fingers causes dark marks.

Abrasions occur when film rubs against any surface, especially if pressure is applied at the same time.

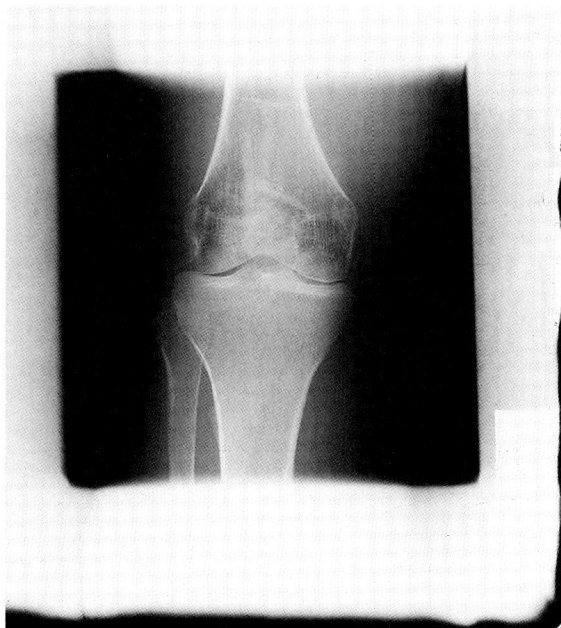

Fig. 8-22 Dark areas of fog on the margins of this film indicate that cassette has a light leak. Leaky cassettes should be replaced.

Fig. 8-23 A, Cassette is not properly loaded and film extends at the latch edge. **B,** Film taken in improperly loaded cassette shows light fog on the latch edge. **C,** Film/screen contact test with cassette improperly loaded shows fog and loss of image sharpness on film margin.

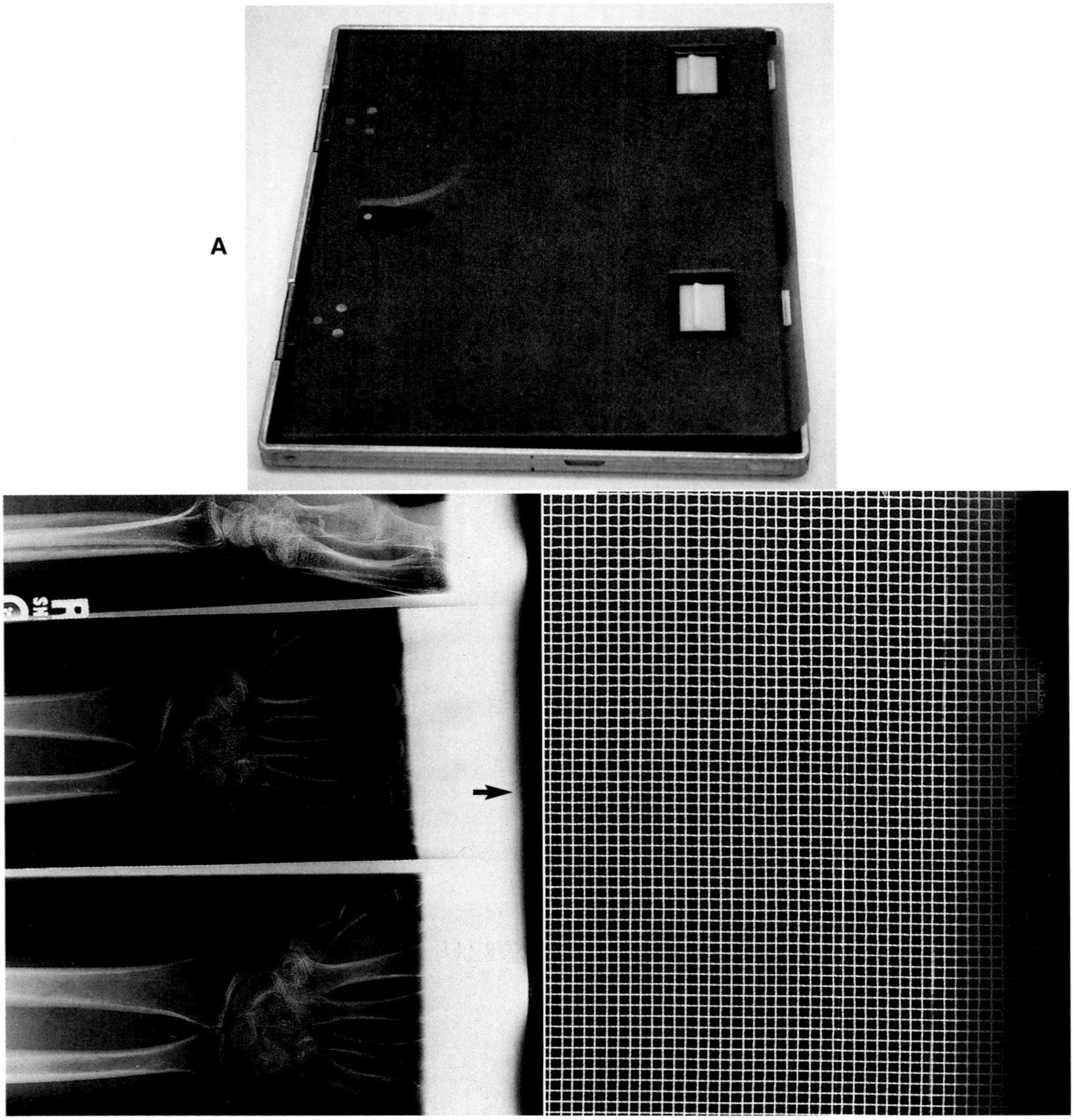

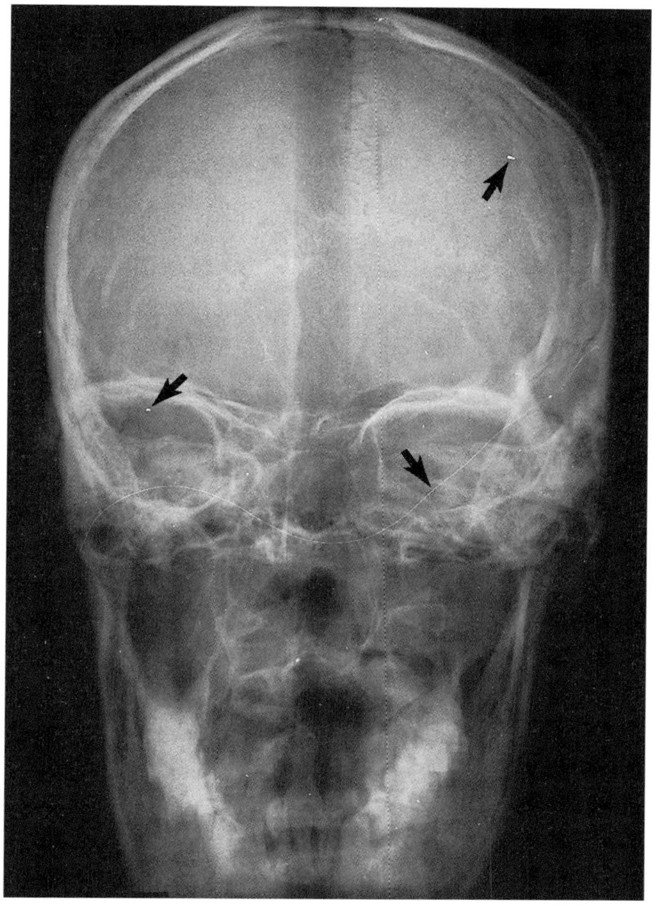

Fig. 8-24 Light-colored artifacts on this film *(arrows)* are unexposed areas resulting from dirt on intensifying screens.

Fig. 8-27 shows abrasions caused by excess finger pressure on the film against the film tray when feeding the film into the processor. Dropping a film on the floor or placing a cassette on top of it will cause abrasions as well.

Static artifacts result when the discharge of static electricity exposes the film prior to processing (Fig. 8-28). These artifacts may occur as clusters of tiny black dots that have the appearance of smudges. Larger discharges may create the shapes of asterisks ("star static"), clumps of grass ("crown static"), or trees. This can be a particularly difficult problem when the atmosphere is very dry. Static discharge may occur when you are simply touching the film, removing a fresh film from the box, loading or unloading a cassette, or when the film makes contact with the loading bench or processor feed tray. The most likely event is discharge between the intensifying screens and the film; this is greatly diminished by cleaning the screens with a cleaner that contains an antistatic ingredient. Antistatic solutions can also be purchased for treating the loading bench and the processor feed tray. If these measures do not solve the problem, it may be necessary to install a humidifier to decrease the build-up of static in the environment.

Manual Processing Artifacts

"Air bells" are small, round, light-colored artifacts caused by bubbles clinging to the film in the developer. These areas are not developed because the developer solution cannot reach them. Air bells are prevented by agitating the film when it is placed in the developer.

Film emulsion is soft and delicate during processing and is subject to scratches or abrasions. This is especially true while the film is in the developer. Abrasion artifacts may occur if the film is rubbed against the side of a tank. Scratches occur when film is rubbed against the hanger clips of other films in the tank.

"Kiss marks" are the result of films in contact with one another during processing. Kissing in the developer causes undeveloped areas on the films. These areas will clear completely in the fixer, resulting in matching light areas on the films. Avoid these artifacts by agitating films individually in the developer to ensure that they are not sticking together. Kissing

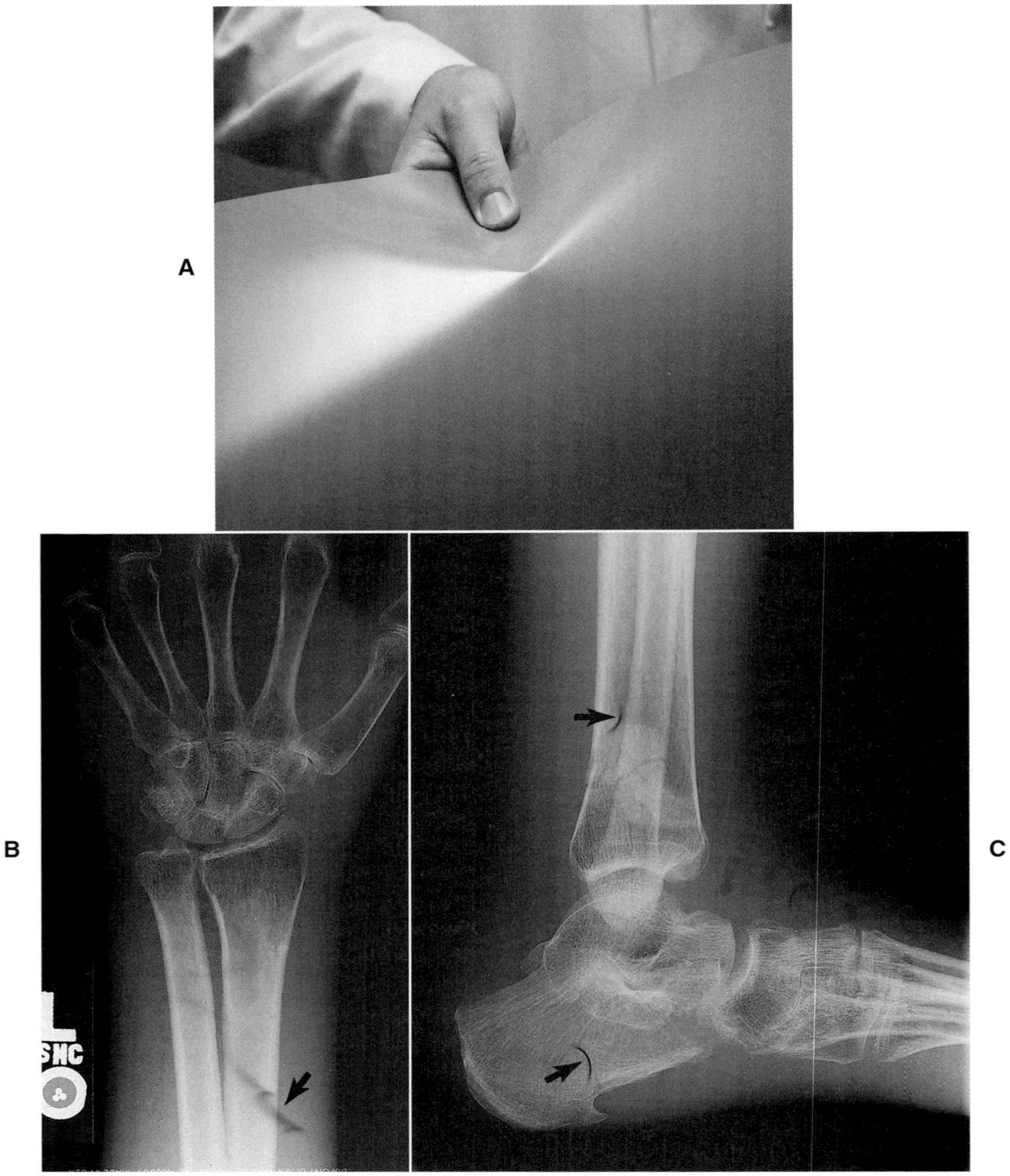

Fig. 8-25 A, Attempting to hold film horizontal with one hand causes film-handling artifacts. **B,** Crease mark *(arrow)* caused by gripping film too tightly. **C,** Crescent marks *(arrow)* on this film are caused by film bending at a crease while operator attempted to hold it horizontal with one hand.

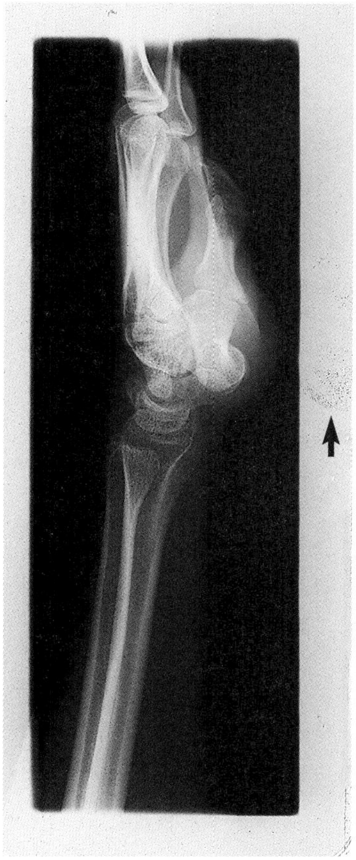

Fig. 8-26 Fingermarks seen here can be prevented by handling films by corners with clean, dry hands.

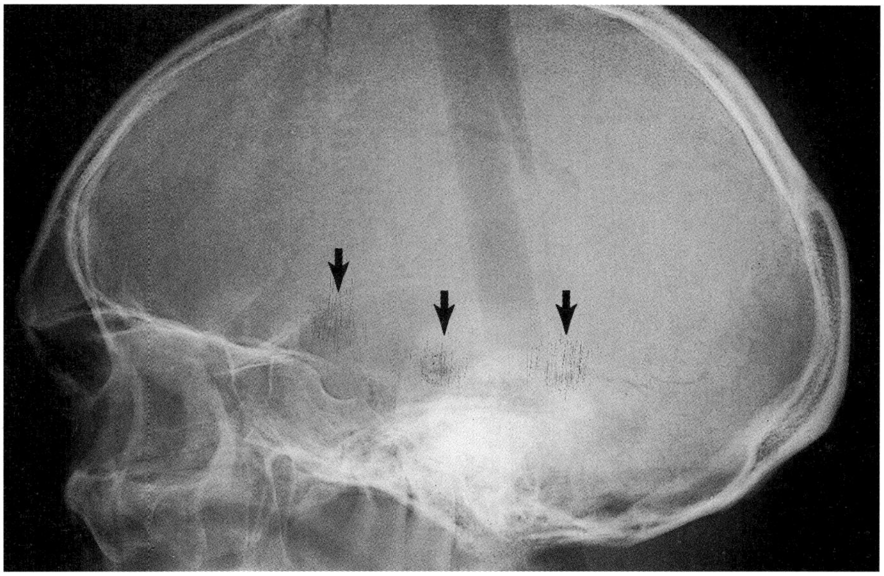

Fig. 8-27 Film abrasions *(arrows)* occur when film rubs across a surface. These marks occurred with finger pressure on the film as it moved across the processor feed tray.

Fig. 8-28 Examples of static marks due to discharge of static electricity.

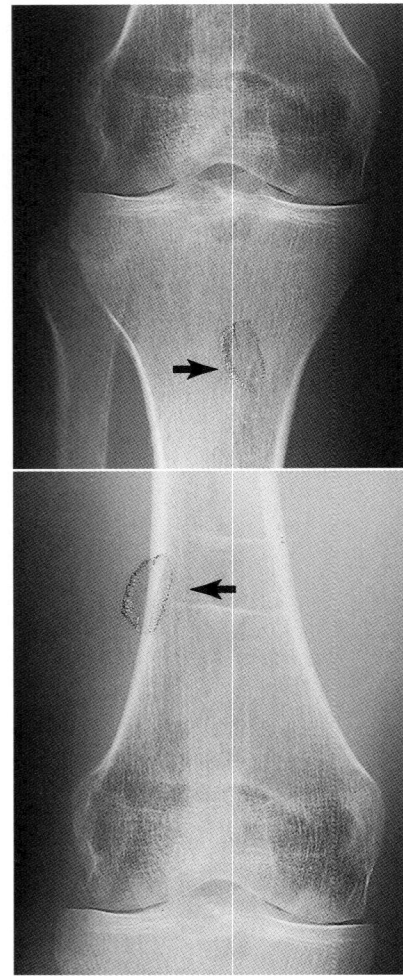

Fig. 8-29 These films were "kissing" in the dryer. Although these films were separated, the matching oval artifacts *(arrows)* define the area of contact.

in the fixer results in failure of the films to clear properly, leaving a milky or streaky appearance. This can be corrected by separating the films and extending the fixing time. When films kiss in the dryer, they may be married for life! When films dry thoroughly while stuck together, the two emulsions become one, and separation removes a portion of the emulsion from one of the films. If the films are still damp in the contact area, it may be possible to separate them, but a kissing artifact will result (Fig. 8-29).

Dark streaks extending from the corners of the films are the result of chemical residue accumulation in the hanger clips. This can be corrected by cleaning the hangers with a stiff brush and a solution of trisodium phosphate in water. Trisodium phosphate is abbreviated TSP and is available at hardware stores. Dark streaks over the entire film result from light exposure while developing (Fig. 8-30). This occurs when the developer lid is removed under white light illumination while processing is taking place.

"Pin holes" appear as tiny bright spots on the film and are caused by too high a temperature in the dryer. The emulsion actually blisters. The blisters break when the film cools, leaving tiny holes in the emulsion. This problem is solved by maintaining a consistent dryer temperature of 110° F.

Automatic Processing Artifacts

Chemical marks from uneven development, especially on the leading edge of the film, are usually due to concentrated chemical residue on the rollers. This can be eliminated by feeding a clean-up film, as discussed above.

Roller marks or "pi lines" are artifacts that occur repeatedly at right angles to the direction of film travel. These may be caused by a damaged roller or by an imbalance in glutaraldehyde that permits over-softening of the emulsion. If these artifacts occur when the solutions are fresh, the processor racks should be inspected for roller damage and the offending roller(s) replaced. Fig. 8-31 shows prominent roller marks that occurred in combination with chemical contamination.

Scratches along the length of film travel (Fig. 8-32) may be caused by a sharp point on a guideshoe or by bits of foreign matter lodged in a roller or extending from a dryer tube. These problems are best handled by service personnel.

A brown, "lacy" artifact is the result of algae scum on the film from the wash tank (Fig. 8-33). As

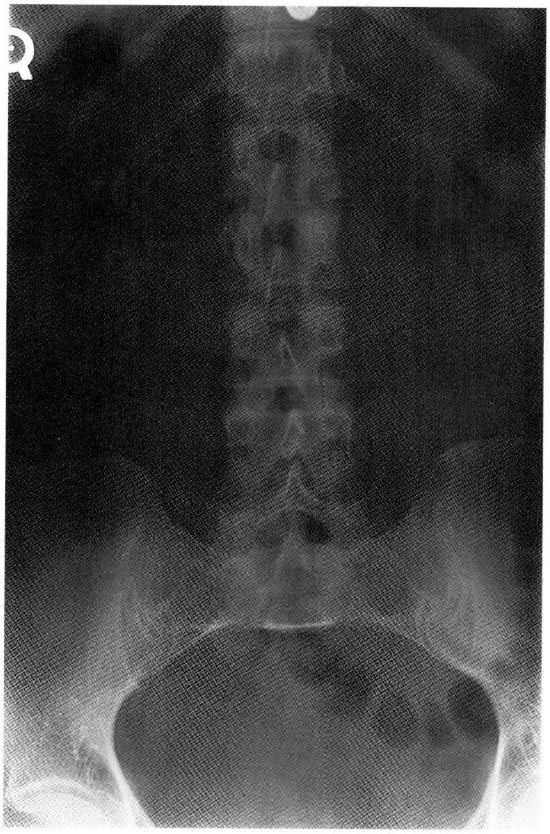

Fig. 8-30 Dark streaks resulting from light exposure during manual development.

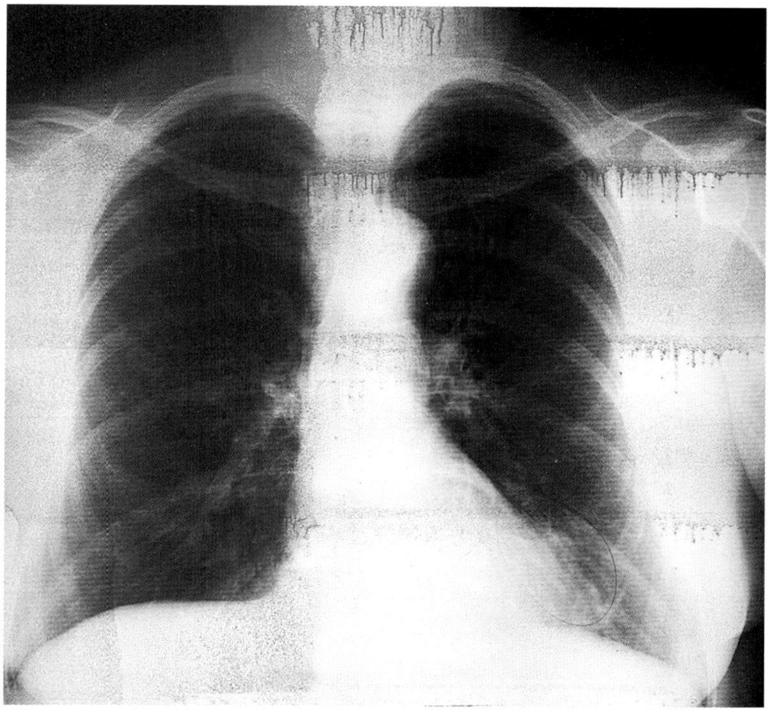

Fig. 8-31 Roller marks and chemical contamination.

Fig. 8-32 Scratches caused by rough projections from corroded guideshoes.

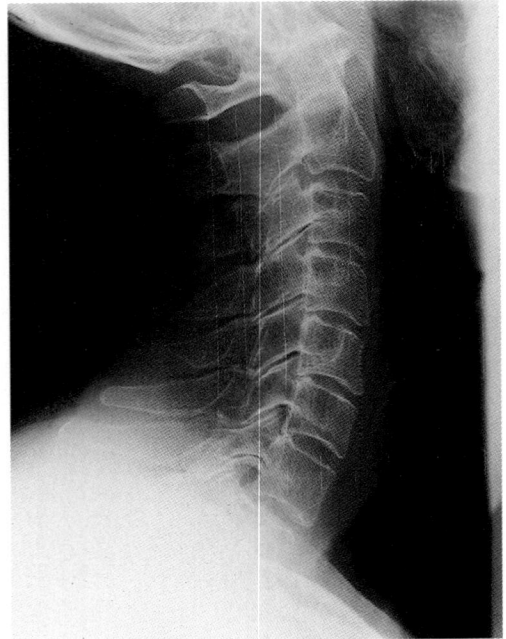

Fig. 8-33 The dark, lacy pattern on this film represents algal scum from the automatic processor wash tank. The light artifact is a fly caught in the cassette, actually a "screen dirt" artifact.

mentioned earlier, this problem is usually cured by draining the wash tank at night. If the problem persists, it may be necessary to clean the wash tank with a bleach solution and rinse thoroughly, to kill the algae spores. Algaecide tablets are available for use in the wash tanks of both processors and manual systems, but these tend to leave a white residue on the films and may create almost as many problems as they solve. They should be used only as a last resort.

SUMMARY

The x-ray darkroom is the place where cassettes are loaded and unloaded and where films are identified and processed. Safelights permit limited visibility in the darkroom during these activities. They may cause fog if they are too bright, too close, or improperly filtered. Periodic testing with presensitized film ensures the safety of safelights.

Careless film handling causes film artifacts. Film should be handled with clean, dry hands and only at the corners. Care must be taken that the film is not creased, bent, scraped, or dropped.

For manual film processing, films are placed on hangers and moved progressively through chemical baths of developer, rinse, fixer, and running water wash and then dried. Optimum manual development requires 5 minutes at 68° F, and the entire process takes about an hour.

Automatic processors shorten the dry-to-dry processing time to 3 minutes or less by using stronger, warmer solutions in constant motion. These systems require a precise and consistent chemical strength and temperature. Manufacturers' recommendations for operation and maintenance must be followed. To assure that all conditions are optimal for best radiographic quality, the processor is monitored for quality control using standard exposures from a sensitometer that are measured on the film with a densitometer. A graphic record of these tests is maintained.

Film artifacts may be caused by cassette damage, dirty screens, careless handling, static electricity, and problems within the automatic processor. Improper cassette loading and safelight fog also compromise radiographic quality. Radiographers must be able to recognize the signs of these problems, identify their causes, and take appropriate steps to rectify them.

▪ Review Questions ▪

1. What might cause you to suspect that a safelight was not safe? How would you find out whether it was safe? What would you do if it was not?
2. What information must be included in the film identification?
3. State the primary function of the developer solution and two functions of fixer.
4. How would you determine the minimum fixing time for a manual processing system?
5. List three factors that would make it desirable to extend the wash time in a manual processing system beyond the usual 20 minutes.
6. Explain why replenisher solution for manual developer needs to be stronger than the original solution. Why does fixer replenisher need to be stronger than developer replenisher?
7. Two physicians from your clinic have gone on vacation and you are now processing half as many films as usual in your automatic processor. This requires a change in replenishment rates. Should the rates be increased or decreased? How would you determine what the actual rates should be?
8. Your quality control film indicates an unacceptable increase in the gross fog index and a decrease in the contrast index. What are the possible causes? What could you do to identify the actual cause?
9. There has been an accidental spill of the used fixer holding container. What equipment will you need to clean it up? If you get some in your eye, what should you do?

9

Scatter Radiation and Its Control

Learning Objectives

At the conclusion of this chapter the student will be able to:

- List and explain three types of interactions between radiation and matter that produce scatter radiation
- Explain the problems caused by scatter radiation in radiography
- List factors that affect the quantity of scatter radiation fog on a radiograph
- Identify scatter radiation fog on a radiograph
- List four measures that can be taken to reduce the quantity of scatter radiation fog on radiographs
- Define grid ratio, grid frequency, and grid radius
- List common grid ratios and state the appropriate application of each
- Define what is meant by "grid cut-off" and list four causes of this phenomenon
- Explain the difference between a bucky and a stationary grid
- State the criteria for determining whether grid use is appropriate

Key Terms

air gap
backscatter
bucky
classical scatter
Compton effect
crosshatch grid
focal range
focused grid
grid
grid cassette
grid cut-off

grid frequency
grid lines
grid radius
grid ratio
parallel grid
photoelectric effect
scatter radiation
secondary radiation
spot film
stationary grid

Scatter radiation, introduced in Chapter 2, is produced as a result of the attenuation of the x-ray beam by matter. This chapter explores the production of scatter radiation and the factors that influence its formation. In addition, this chapter covers methods used to minimize the fog that this radiation causes on radiographs.

RADIATION INTERACTIONS WITH MATTER

When x-rays *penetrate* matter, there is no interaction with the matter and no scatter or scattered radiation is formed as a result. When x-rays are *absorbed,* however, their energy is "scattered," or converted into new, scatter x-rays. Five types of interactions occur when radiation is absorbed by matter: classical scatter, Compton effect, photoelectric effect, pair production, and photodisintegration. Only the first three occur with radiation energies in the diagnostic x-ray range. Pair production and photodisintegration occur only with very high–energy radiation and are beyond the scope of this text.

The result of either classical scattering or the Compton effect is properly called **scatter radiation** or simply "scatter." Radiation produced by the photoelectric effect is correctly referred to as **secondary radiation.** Since more than one type of interaction takes place during radiography and the resulting radiation is so similar, the terms are often used interchangeably. When referring to both scatter and secondary radiation, this text uses the term "scatter radiation."

The interactions that produce scatter radiation in radiography occur primarily within the patient. Some scattering also occurs as a result of interactions between the x-ray beam and the tabletop or film holder, the cassette, and any other matter that happens to be within the radiation field.

Classical Scatter

Classical scatter is also known as "coherent scatter" and "Thompson scatter." This type of interaction takes place throughout the energy spectrum but is most prevalent at relatively low energy levels (low kVp). When the incoming x-ray photon is absorbed by an atom, the atom becomes "excited" (Fig. 9-1). The excited atom produces a new x-ray photon that has the same wavelength as the incoming photon but exits the atom in a slightly new direction. The effect is similar to a ricochet; the photon is redirected because of its interaction with the atom. The direction of classically scattered x-rays is forward, that is, in the same general direction as the incoming x-ray beam.

Compton Effect

The **Compton effect** occurs at energy levels throughout the diagnostic x-ray range. The incoming x-ray

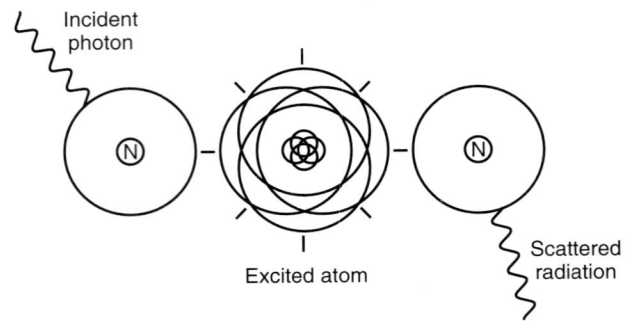

Fig. 9-1 Classical scatter occurs in three phases. *1,* The x-ray photon enters the atom. *2,* The photon's energy is momentarily transferred to the atom, causing an excited state. *3,* The energy is given up by the atom as a photon of the same energy but with an altered direction.

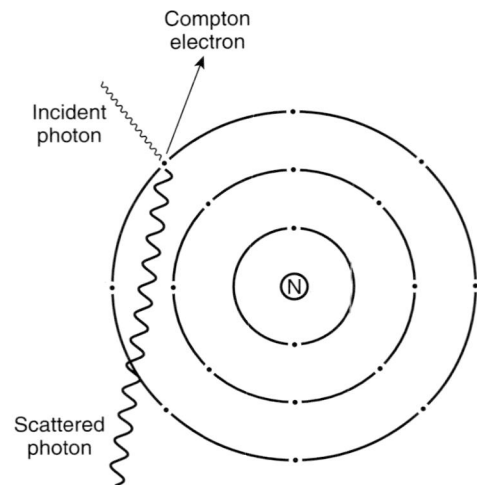

Fig. 9-2 Compton scatter occurs when an x-ray photon collides with an outer orbital electron of an atom. The electron is ejected from its orbit. The photon is deflected from its path and continues with decreased energy.

photon collides with an outer orbital electron of an atom, removing it from the atom (ionization). Part of the energy of the photon dislodges the electron from its orbit and part of the energy is imparted to this "Compton electron" as energy of motion. The majority of the photon's energy, however, is converted into a new photon of scatter radiation (Fig. 9-2). This new photon has a longer wavelength than the incoming photon, because it has less energy. It also travels in a new direction. Compton scatter travels in all directions. If it is directed back toward the x-ray tube, it is termed **backscatter.**

Photoelectric Effect

The **photoelectric effect** is similar to that which forms characteristic radiation in the x-ray tube. In this case, however, the incoming energy is an x-ray photon

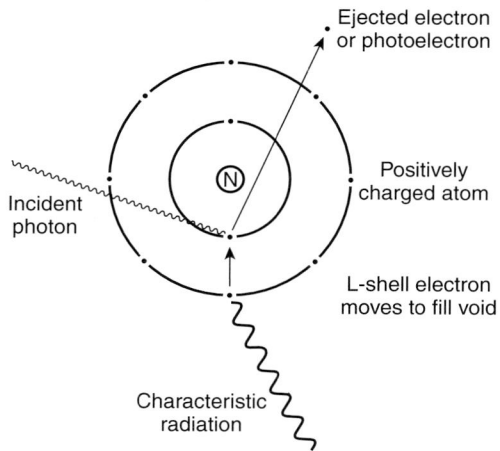

Fig. 9-3 Photoelectric effect occurs when an x-ray photon collides with an inner orbital electron of an atom. The photon's energy is absorbed in the process of ejecting the electron. When an outer orbital electron moves to the inner orbit to fill the space vacated by the departing electron, the difference in binding energy between the two electron shells is emitted in the form of a new photon of characteristic radiation.

rather than an electron. In a photoelectric interaction, the incoming photon collides with an inner orbital electron of an atom. Part of the photon's energy removes the electron from its orbit; the remainder is imparted to the electron as energy of motion. The electron's departure leaves a "hole" in the orbit, which is filled by an electron from an outer shell. The difference in binding energy between the two shells is emitted as a scatter photon (Fig. 9-3). This photon will have a new direction. Its energy will be less than that of the primary photon and will be characteristic of the atom in which the interaction occurred. Photoelectric interactions are less prevalent in the diagnostic energy range than Compton interactions. The likelihood of a photoelectric interaction is determined by both the kVp level and the electron-binding energy of the atom in which the interaction occurs.

Because no part of the energy of the incoming photon exits the atom, photoelectric interactions are sometimes referred to as "true absorption." The radiation produced by these interactions is most properly termed secondary radiation. In this text, references to scatter also apply to secondary radiation formed by the photoelectric effect.

RADIOGRAPHIC EFFECT OF SCATTER RADIATION

The production of scatter radiation during an exposure results in fog on the radiograph. Fog is unwanted exposure to the film. It does not strike the film in a pattern that represents the subject, and it contributes nothing of value to the image. This fog

---

BOX 9-1

Factors Affecting Scatter Radiation Fog

↑ Tissue thickness = ↑ interactions = ↑ scatter = ↑ fog
↑ Field size = ↑ photons = ↑ interactions = ↑ scatter = ↑ fog
↑ mAs = ↑ photons = ↑ interactions = ↑ scatter = ↑ fog
↑ kVp = ↓ scatter reabsorption = ↑ fog

---

produces an overall increase in radiographic density. The result is a reduction in radiographic contrast, as stated in Chapter 6. While increased density in the darker areas of the image is scarcely noticeable, areas that would otherwise be bright or white will instead be gray because of the fog. The intermediate gray tones will appear more similar to each other, making it difficult to distinguish details within those portions of the image that have similar tissue densities. In other words, *scatter radiation fog reduces visibility of detail by decreasing contrast* (Fig. 9-4).

FACTORS AFFECTING QUANTITY OF SCATTER RADIATION FOG

Several factors affect the quantity of scatter radiation fog on the radiograph (Box 9-1). The primary consideration is the volume of tissue irradiated. When there is a greater quantity of matter in the path of the x-ray beam, there will be greater absorption of the x-ray beam and more interactions that produce scatter radiation. The volume of tissue irradiated is determined by the thickness of the subject and the size of the radiation field. When the subject is more than 10 cm in thickness, the amount of fog becomes objectionable unless the field size is very small.

Because a thicker subject requires a greater quantity of exposure, there will be more primary x-ray photons and more interactions. This is another reason why there will be more scatter radiation when the thickness of the subject is increased.

Kilovoltage, too, affects the quantity of scatter radiation that reaches the film. It is tempting to leap to the conclusion that higher kVp will result in less scatter radiation. With greater penetration by shorter wavelengths, it seems logical that higher kVp would produce less absorption, fewer interactions, and therefore less scatter radiation. In fact, *higher kVp results in more scatter radiation fog*. When high-energy photons interact with matter, the scatter radiation that is produced is also of a higher energy. This high-energy radiation is better able to escape the subject without being reabsorbed and so is more likely to cause fog on the radiograph.

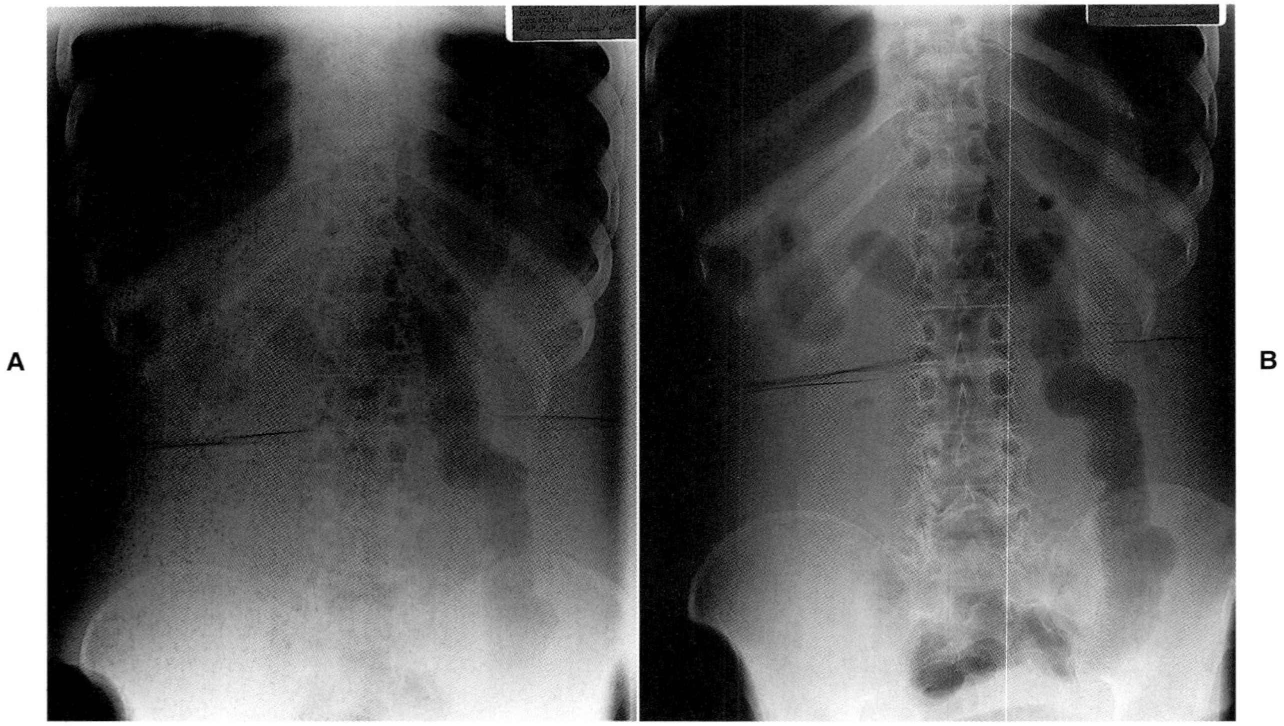

Fig. 9-4 A, Uncontrolled scatter radiation causes decreased contrast on this radiograph of a phantom abdomen. **B,** When the bucky is used to control scatter radiation, contrast improves significantly.

The density of the absorbing matter also influences the quantity of scatter radiation fog. A very dense subject, such as a lead brick, will absorb a large quantity of primary radiation, but the scatter will be readily absorbed and very little will exit the subject. On the other hand, little scatter radiation would be produced in an air-filled balloon because very little absorption of primary radiation would occur. The prevalence of scatter radiation in radiography indicates that there must be some matter that produces scatter radiation and permits it to escape. In fact, this is the case with all matter that is similar in density to water. Since the human body is largely made up of water, *the patient is the principal source of scatter radiation in radiography.*

CONTROLLING SCATTER RADIATION FOG

From the previous discussion it is clear that scatter radiation fog becomes increasingly objectionable as the thickness of the subject increases. To obtain diagnostic quality radiographs of the trunk of the body, it is absolutely essential to employ some means of limiting the fog effect on the film. The principal method for reducing scatter radiation fog is the use of a radiographic grid. Additional strategies include the use of an air gap, limitation of field size, and reduction of kilovoltage.

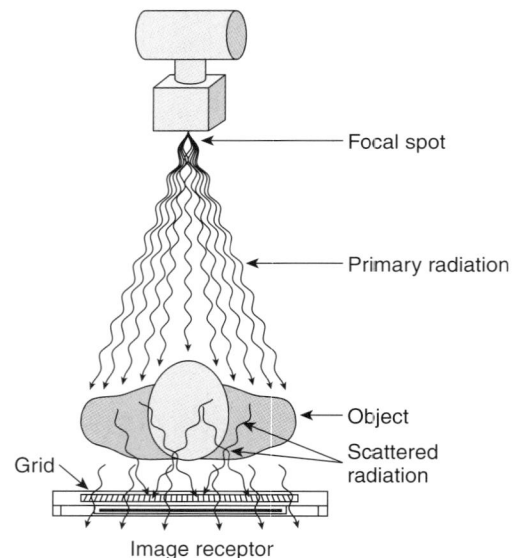

Fig. 9-5 A grid is placed between the patient and the film to absorb scatter radiation. Scattered x-rays not moving in the same direction as the primary x-ray beam are absorbed by the grid.

Grids

A **grid** is a device placed between the patient and the film (Fig. 9-5). It has the appearance of a thin metal plate (Fig. 9-6). It is constructed of tiny, tissue-thin lead strips, placed on edge. The lead strips are held in

Fig. 9-6 A grid has the form of a thin plate and is covered by a protective aluminum coating.

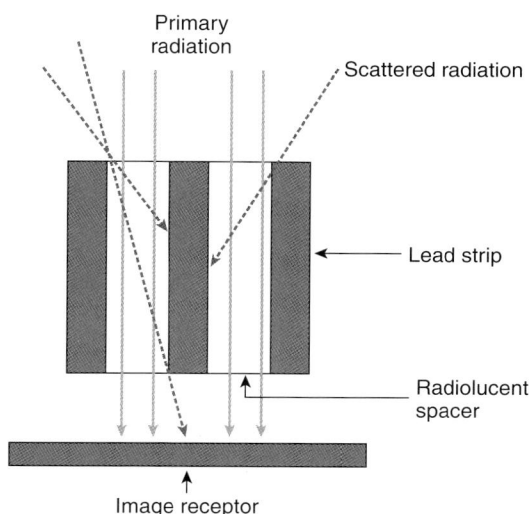

Fig. 9-7 Cross section of grid demonstrating its function. Grid consists of lead strips on edge, separated by radiolucent interspacing material.

Both grids have an 8:1 ratio.
B has a greater frequency.

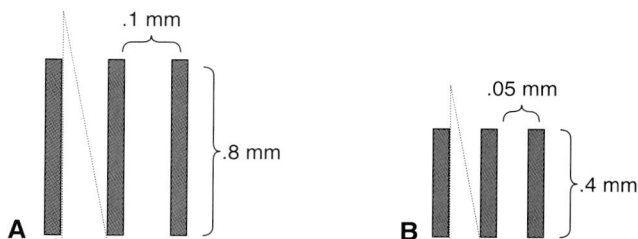

Fig. 9-8 Grid ratio is relationship between the height of the lead strips and the width of the interspacing material. These two grids are different but have the same ratio. Note that the same degree of variation in direction of the x-ray is permitted by both.

TABLE 9-1

Grid Applications

Ratio	Application
5:1 and 6:1	Portable use, mobile radiography
8:1	Head units, chest radiography
10:1	Chest radiography, general purpose
12:1	General purpose
16:1	High kVp radiography only

Grid Focal Range

The range of source-image distances at which the grid will not absorb significant amounts of primary radiation
EXAMPLE: A grid with a 40-inch radius may have a focal range of 36 to 48 inches.

place by a light metal interspacing material, usually aluminum or an alloy than contains aluminum. Because they are aligned to the x-ray beam, the lead strips tend to absorb scatter radiation while permitting remnant radiation to pass through (Fig. 9-7).

The effectiveness of a grid is determined by the **grid ratio,** that is, the relationship between the height of the lead strips and the width of the spaces between them (Fig. 9-8). This ratio determines how much variation in the direction of the incoming photon is allowed without the photon being absorbed by the grid. The higher the ratio, the less variation is permitted and the greater is the efficiency of the grid in absorbing the unwanted photons. Typical grid ratios range from 5:1 to 16:1. Table 9-1 lists common grid ratios and their usual applications.

As seen in Fig. 9-8, grids with the same ratio may have many strips close together or fewer strips farther apart. The number of lead strips per inch is called the **grid frequency.** Grid frequencies range between 60 and 170 lines per inch.

Grids for general-purpose use are called **focused grids** because the lead strips are aligned to the direction of the diverging primary x-ray beam (Fig. 9-9). The lead strips of a focused grid are precisely aligned to the x-ray beam at a specific source-image distance, which is called the **grid radius.** Since the alignment does not need to be exact for the useful photons to pass through the grid, there is a range of distance within which the grid will not absorb an undue amount of useful radiation. This is referred to as the **focal range** of the grid. The SID employed with a

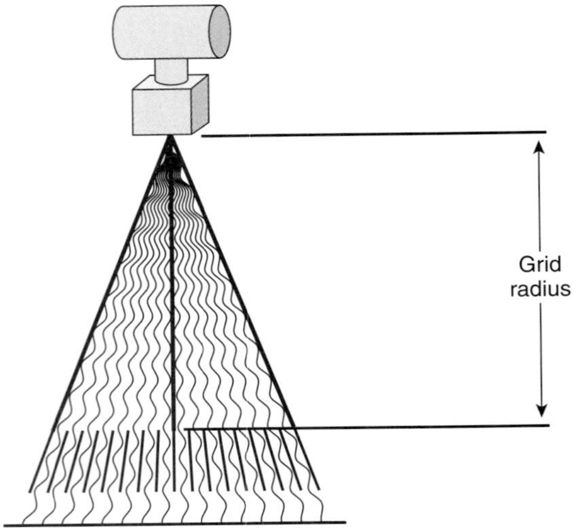

Fig. 9-9 Grid radius is the distance at which the primary x-ray beam is parallel to the focused lead strips of the grid.

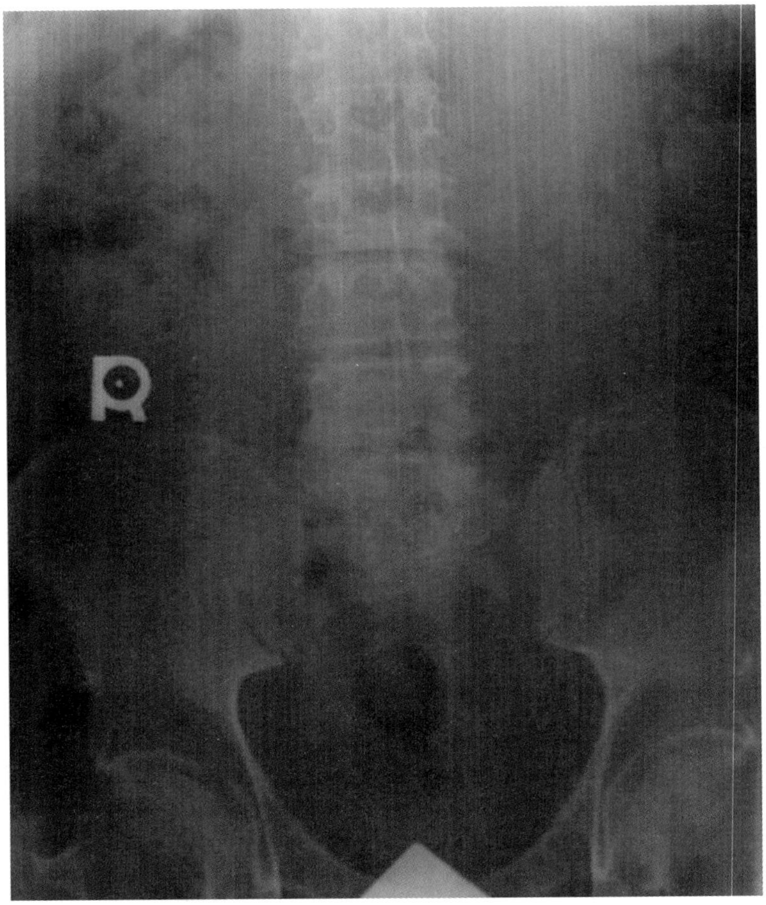

Grid radius

Fig. 9-10 Grid lines are the radiographic image of the grid itself. They are most obvious when the grid is stationary and the grid frequency is low.

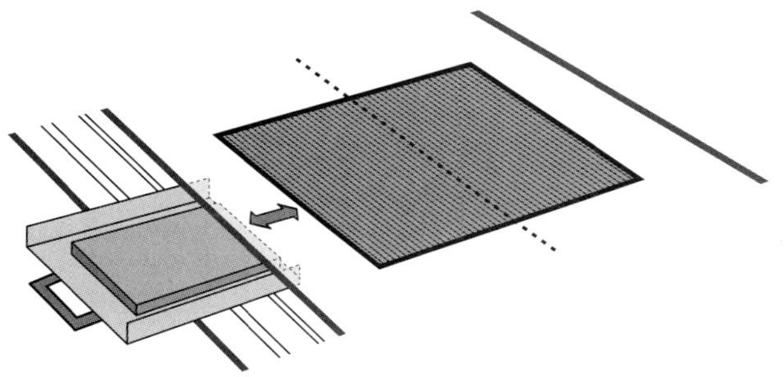

Fig. 9-11 A bucky is a moving grid installed under the tabletop or in an upright cabinet.

A moving grid is called a **bucky.**

Moving the grid during the exposure blurs the image of the grid lines so that the grid image is not visible on the film.

Stationary Grids

- Do not move during the exposure
- Should have many very fine lines (high frequency) to avoid objectionable grid lines on images
- Commonly used today in upright cassette holders

grid should always be within the grid's focal range. The most commonly used source-image distances are 40 inches and 72 inches. Each usually requires a different grid with a suitable focal range. Grids with extra long focal ranges are now available, however. Some have a focal range of 40 to 72 inches.

Each grid has a label indicating its ratio, its frequency, and its radius or focal range. This label is placed on the side of the grid that faces the x-ray tube when the grid is in use. There is usually also a line down the middle that indicates the long axis of the lead strips and the grid's focal center.

Since the grid absorbs some useful radiation, the radiographic image includes an image of the grid itself. This grid image is called **grid lines** or "grid striping" (Fig. 9-10). Two methods are employed to avoid objectionable grid lines. The grid may be moved during the exposure, or the grid may have a very high frequency, that is, many fine lines very close together.

A moving grid is called a **bucky.** A bucky may be part of a radiographic table (Fig. 9-11) or an upright cassette holder. The bucky grid is mounted in a frame that incorporates a small motor. The motor causes the grid to oscillate back and forth rapidly during an exposure. This movement blurs the image of the grid lines, making them invisible on the radiograph. The bucky device also includes the film tray that holds the cassette in place when making bucky exposures. A table bucky has rollers that allow it to be moved along a track under the tabletop. This permits placement of the bucky and the film at any location along the length of the table. Bucky grids typically have a ratio of 10:1 or 12:1 and a frequency of 60 to 80 lines per inch.

High-frequency grids that do not move during the exposure are called **stationary grids.** They produce grid lines, but the lines are very tiny and close together and are not readily seen at a normal viewing distance. Their appearance is acceptable because they are almost invisible and do not adversely affect the diagnostic quality of the image. Stationary grids may be part of a bucky-like device in the radiographic table, and they are often used in upright cassette holders called grid cabinets. Stationary grids for per-

manent installations typically have a ratio of 10:1 or 12:1 and a frequency of at least 103 lines per inch.

Portable stationary grids come in various sizes and may be attached to a cassette when needed in locations other than the permanent installations. **Grid cassettes** are special cassettes with a grid built into the front side. Both portable grids and grid cassettes are used for bedside radiography and for special applications when the patient cannot be positioned on the table or at the upright grid cabinet. Grid cassettes typically have lower ratios than the grids used in permanent installations. Ratios of 6:1 and 8:1 are common. Grid cassettes should be clearly marked and kept in a separate place from the regular cassettes. They must never be used in the bucky.

Grids are precision instruments and are very expensive to replace. Special care must be taken that they are not damaged by dropping, striking, or bending. Such damage causes the lead strips to become misaligned, making the grid useless.

Grid Cut-Off Focused grids are designed to allow the passage of radiation that is aligned with the lead strips. Any misalignment of the primary x-ray beam will result in undesirable absorption of useful radiation by the grid. Excessive absorption of useful radiation by the grid is called **grid cut-off.** Grid cut-off appears as decreased radiographic density on the radiograph, frequently accompanied by the appearance of obvious grid lines (Fig. 9-12).

No grid cut-off occurs when the x-ray beam is correctly aligned to the grid (Fig. 9-13). Grid cut-off occurs when the x-ray tube is centered to one side of the grid rather than to the focal center line (Fig. 9-14). Cut-off also occurs when the x-ray tube is angled toward one side of the grid, rather than perpendicular to its center. The same effect is encountered if the grid is tipped side-to-side in relation to the primary beam (Fig. 9-15).

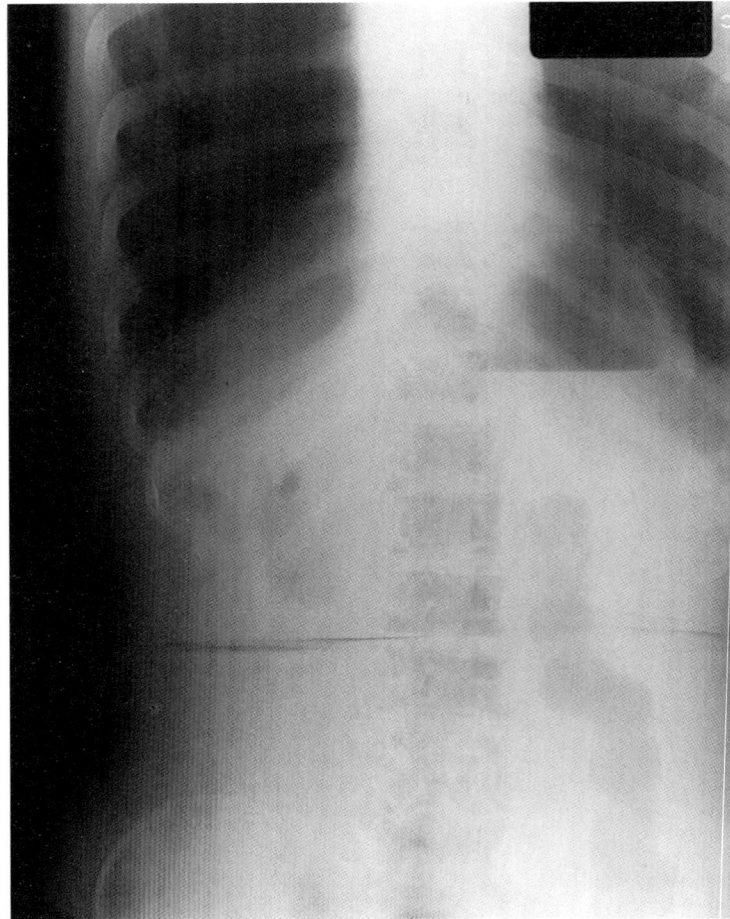

Fig. 9-12 Grid cut-off appears as areas of decreased radiographic density with obvious grid lines. It occurs when the x-ray tube and the grid are not correctly aligned with each other.

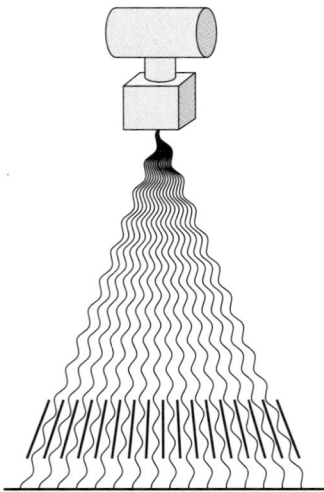

Fig. 9-13 When the x-ray beam is correctly aligned to a focused grid, no cut-off occurs.

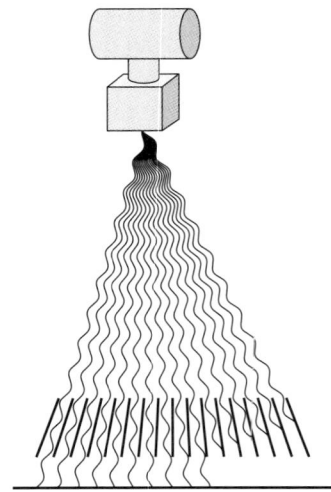

Fig. 9-14 Grid cut-off occurs when the x-ray beam is off-center to one side of the grid.

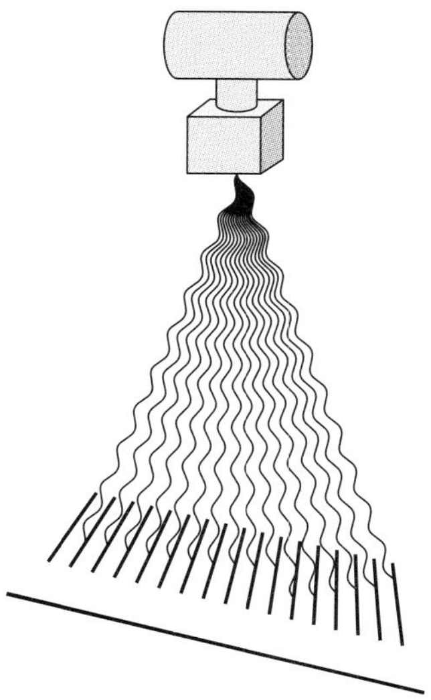

Fig. 9-15 Cut-off occurs when the grid is tilted, or when the x-ray beam is angled across the grid.

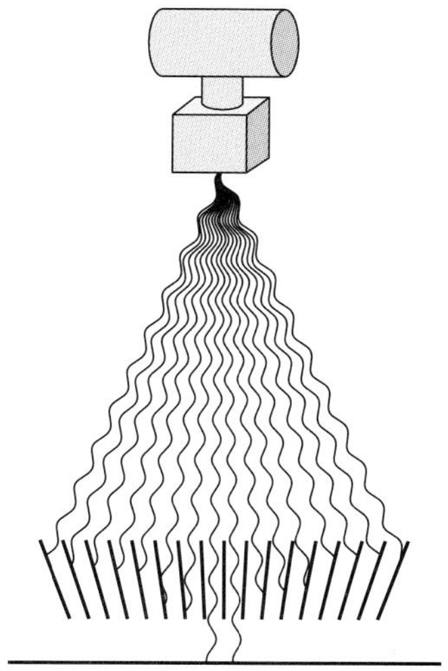

Fig. 9-17 Severe cut-off occurs when the grid is reversed. The image is visible only in the center, and grid lines are apparent.

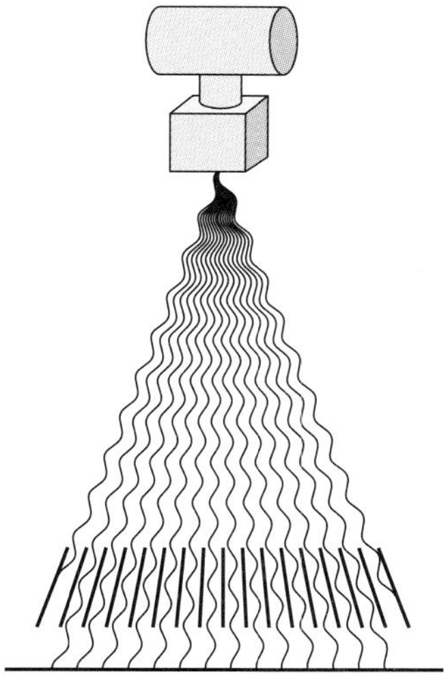

Fig. 9-16 Cut-off occurs on the sides of the film when the SID is outside the focal range of the grid.

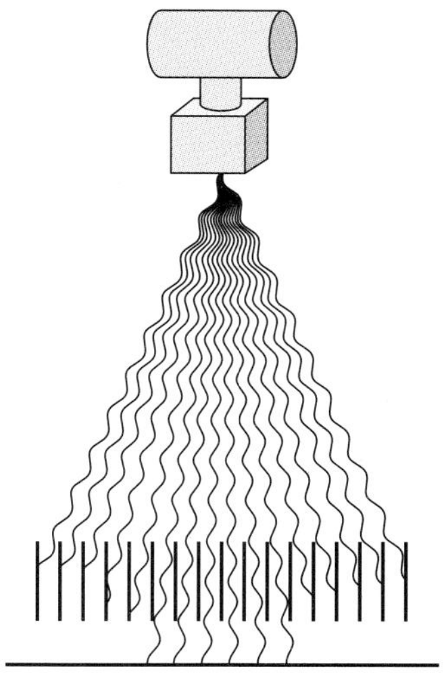

Fig. 9-18 Lead strips of a parallel grid are not focused. It is practical only for very small fields. The x-ray beam may be centered to the grid at any point.

Both of these problems involve only centering and angulation with respect to the *sides* of the grid. Cut-off does *not* result when the x-ray tube is off-center lengthwise or angled along the length of the lead strips.

When a grid is used at a distance outside its focal range, cut-off will occur on both sides of the film (Fig. 9-16).

When the grid is reversed, that is, the wrong side is facing the x-ray tube, grid cut-off prevents most of the primary radiation from reaching the film (Fig. 9-17). A film taken with the grid reversed will show exposure only in a line down the center. The remainder of the film will be very light and streaked with grid lines.

Grid cut-off is prevented by ensuring that the x-ray beam is always properly aligned to the center of the grid at the appropriate distance. The precision required when aligning the x-ray beam to the grid is determined by the grid ratio. The higher the grid ratio, the more precise the alignment must be. This is why relatively high-ratio grids are used only in permanent installations where alignment is easily maintained. Lower-ratio grids are used for bedside radiography and other portable applications where it is more difficult to ensure precise alignment.

Specialty Grids Specialty grids are available for special situations but are not suitable for general-purpose use. A grid with strips that are parallel to each other, rather than focused, is called a **parallel grid** (Fig. 9-18). The radius of a parallel grid is infinity. Parallel

Grid Cut-Off

Undesirable attenuation of the primary x-ray beam by the grid
  Caused by misalignment between grid and x-ray beam
  • Lateral decentering
  • SID outside focal range
  • Lateral angulation or grid off-level
  • Grid reversed

Fig. 9-19 A crosshatch grid consists of two grids laminated together with their lead strips perpendicular to each other. The x-ray tube cannot be angled when using a crosshatch grid.

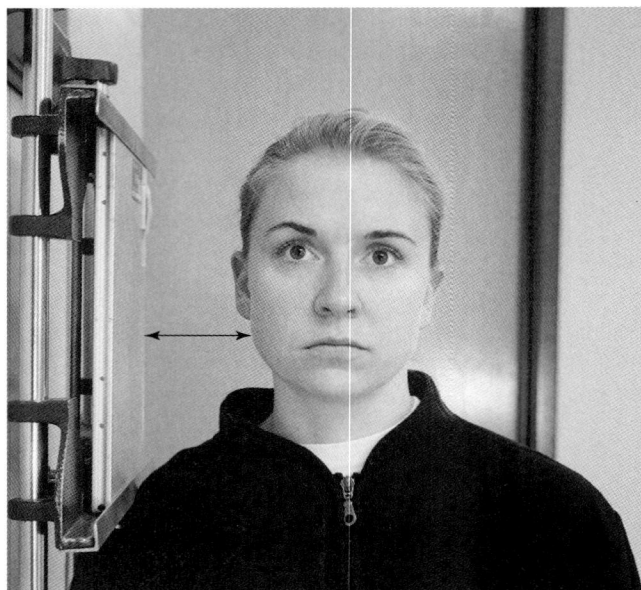

Fig. 9-20 With lateral cervical spine radiography, space between the patient and the film creates an air gap that reduces scatter radiation fog.

grids can be used for radiography at very long source-image distances because the useful portion of the x-ray beam is nearly perpendicular to the grid when the distance is great. Parallel grids are also used in fluoroscopic spot film devices. These devices make multiple small areas of exposure, each centered over a different portion of the film. Parallel grids are ideal for this application because they have no focal center and the x-ray beam can be aligned to them at any point.

Another type of specialty grid is the **crosshatch grid,** sometimes called a "crossed grid." This is actually a composite of two grids with the lead strips at right angles to each other (Fig. 9-19). A crosshatch grid is desirable because it has an effective ratio that is greater than the ratios of the two grids combined. For example, a crosshatch grid made of two 8:1 grids would be more efficient at preventing scatter radiation fog than a regular 16:1 grid but would have the alignment flexibility of an 8:1 grid. The limitation of the crosshatch grid is that it produces unacceptable grid cut-off when the x-ray tube is angled. It is useful only for procedures done with the central ray perpendicular to the grid and the film. This consideration makes it unsuitable for general use.

Air Gap

Another method for reducing scatter radiation fog on radiographs is the use of an **air gap.** An air gap, as its name implies, is an air space between the patient and the film—in other words, a large object-image distance (OID). Since the patient is the source of most scatter radiation and all radiation obeys the inverse square law, increasing the OID will decrease the intensity of scatter radiation at the film.

As stated in Chapter 6, an increase in OID causes increased magnification and decreased definition.

These are significant problems that must be overcome if an air gap is to be utilized successfully. For this reason, air gap technique uses a long source-image distance and, where possible, the small focal spot.

In the past, air gaps were used in place of grids for chest radiography and some other applications. Today, with improved grid technology, the air gap is seldom utilized. There is one procedure, however, where the air gap is often employed: the lateral projection of the cervical spine (Fig. 9-20). A bucky or stationary grid is used for most views of the cervical spine. In the lateral position, however, the patient's shoulder must be accommodated in a space between the neck and the film. This creates an air gap that cannot be prevented. Because of the air gap, a grid is not needed in this application. Lateral cervical spine radiographs are often taken without a grid at a 72-inch source-image distance (SID) using the small focal spot.

Field Size

Because field size significantly affects the volume of tissue irradiated, a reduction in field size will decrease the scatter radiation fog on the radiograph. For this reason, the radiation field is collimated to expose only the area of clinical interest. Decreasing field size improves contrast even when a grid or bucky is used, decreasing the quantity of scatter that penetrates the grid. When fog compromises the ability to see specific details in a large body part, it is common to take a **spot film** (Fig. 9-21). This term refers to a radiograph of a very small area of the sub-

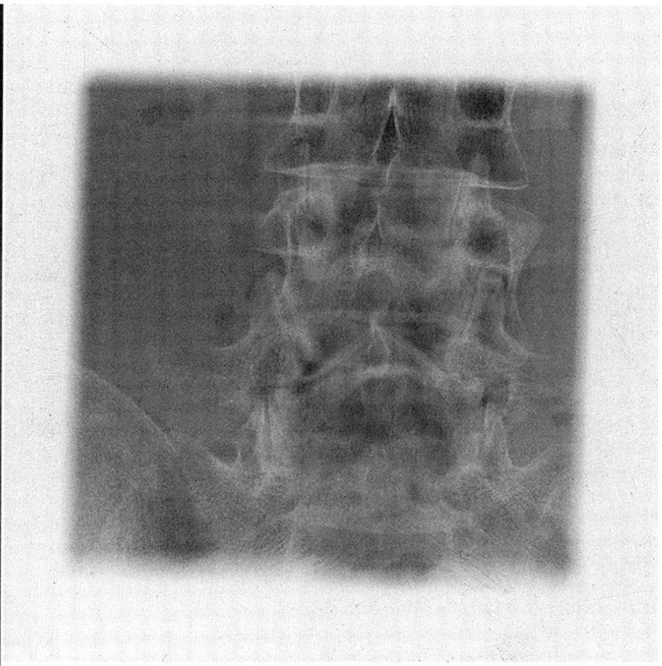

Fig. 9-21 Both of these radiographs were taken without a grid using the same exposure factors. **A,** A 14 × 17-inch radiation field results in excessive scatter radiation fog. **B,** A 5 × 5-inch radiation field significantly reduces scatter, improving visibility of detail on the film.

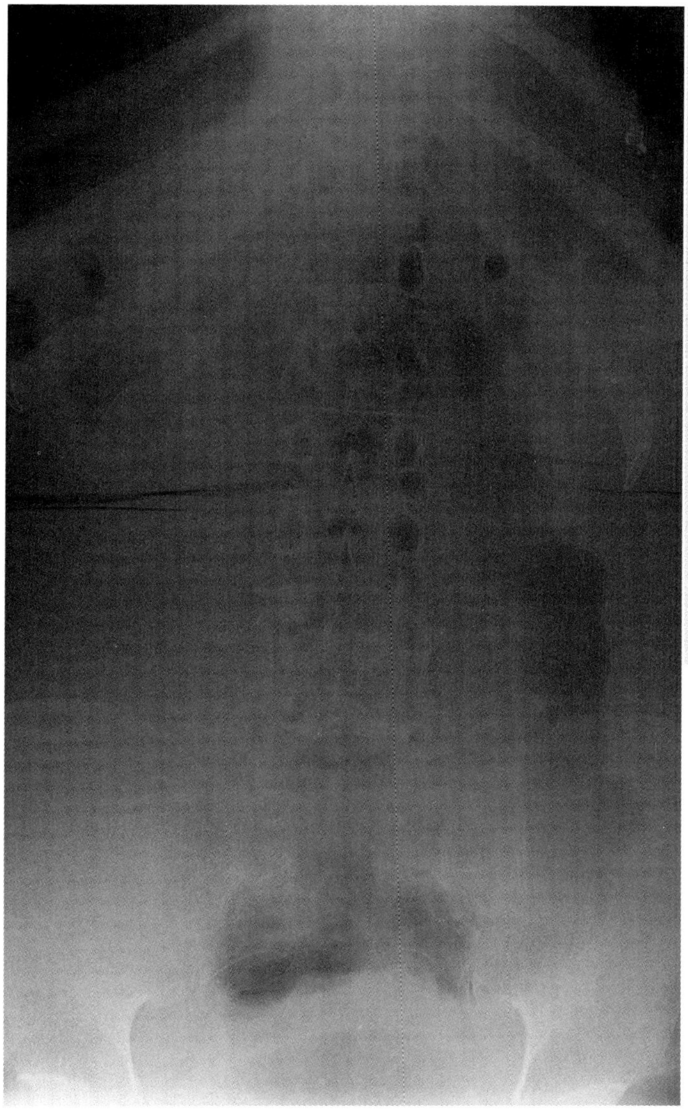

ject. Most spot films are taken using an 8 × 10-inch (20 × 25-cm) film and a 5- or 6-inch square field. They are centered precisely to the area of interest. Spot films demonstrate increased contrast compared to the same anatomy seen within a larger field. When a specific area of interest is centered to the field, as in a spot film, the decreased distortion and parallax also improve image quality.

Decreasing kVp

When other methods fail to provide sufficient contrast, a decrease in kVp should be considered. Decreasing kVp increases contrast in two ways. First, this shortens the scale of contrast, as explained in Chapter 6. Second, lower kVp levels decrease the energy of the scatter radiation and decrease the fog, as explained earlier in this chapter. Of course, when kVp is decreased, radiographic density will also de-

Methods of Scatter Radiation Fog Control

- Grid device—placed between patient and film to absorb scatter
- Air gap—increased OID decreases scatter intensity at film
- Minimize field size—decreases volume of scattering tissue
- Decrease kVp—decreases energy of scatter and increases contrast

crease. This requires an increase in mAs to compensate. The mathematical calculations for maintaining radiographic density when changing kilovoltage are discussed in Chapter 10. This change in exposure factors results in increased radiation dose to the patient, so it should be used only when necessary.

SUMMARY

When diagnostic x-rays are absorbed by matter, they may be "scattered," forming scatter and secondary radiation by means of classical scatter, the Compton effect, or the photoelectric effect. The greatest quantity of scatter formed during routine radiography is due to the Compton effect, which scatters photons in all directions. This scattered radiation causes fog on the radiograph, decreasing radiographic contrast. The fog level becomes objectionable when the thickness of the subject is greater than 10 to 12 cm.

Scatter radiation fog is controlled primarily by using either a stationary grid or a moving grid device called a bucky. The grid ratio indicates the efficiency of the grid in cleaning up scatter radiation, as well as the precision with which the x-ray beam must be aligned to the grid. When the x-ray beam and the grid are not properly aligned to each other, grid cut-off results. Other means for reducing scatter radiation fog include using an air gap, reducing the field size, and decreasing kilovoltage.

▪ Review Questions ▪

1. Which type of radiation interaction produces scatter radiation that is characteristic of the subject irradiated?
2. List two factors that affect the volume of tissue irradiated.
3. When the kVp is increased, will the quantity of scatter radiation fog be increased or decreased?
4. What is the principal source of scatter radiation that causes fog in radiography?
5. State a grid ratio that is typical of a table bucky.
6. What is the usual minimum frequency for a stationary grid in an upright grid cabinet?
7. What is the typical radiographic appearance of grid cut-off caused by using an SID that is outside the grid's focal range?
8. For what radiographic examination is an air gap commonly used instead of a grid?
9. How might the image of a vertebra on a spot film differ from the image of the same vertebra as part of 14 × 17-inch (35 × 43-cm) radiograph of the spine?
10. How would you determine whether or not to use a grid?

10

Formulating X-ray Techniques

Learning Objectives

At the conclusion of this chapter the student will be able to:

- Read and use an x-ray technique chart
- List methods for obtaining and/or creating an x-ray technique chart
- Accurately measure a body part using an x-ray caliper
- Compare fixed kVp technique charts with variable kVp technique charts and state which is preferable
- Explain what is meant by "optimum kVp" and how this value is determined
- Select an appropriate mA station for a given set of circumstances
- Take appropriate steps when the technique chart fails to provide an appropriate exposure
- Calculate exposure adjustments for changes in patient/part size
- Estimate the technique change required when radiographs are too dark or too light
- Suggest appropriate technique changes for lengthening or shortening the scale of contrast
- Calculate technique changes for variations in SID
- Calculate technique changes required when using a grid or changing grid ratio
- Calculate technique changes required for changes in the speed of the image receptor system

Key Terms

caliper
fixed kVp chart
optimum kilovoltage

radiographic phantom
technique chart
variable kVp chart

Most radiography departments have a functional technique chart that provides appropriate exposures for most circumstances. This chapter explores the components of technique charts and explains their limitations. When there is no chart or when the existing chart is inadequate, the radiographer must obtain or create a suitable chart. It is important to remember that no chart meets the requirements for every circumstance. When the planned procedure differs from the chart, the radiographer must make appropriate changes in the exposure settings. These radiographic examinations might include patients who are larger or smaller than the measurements provided on the chart or patients whose conditions affect the amount of exposure required. Changes in the type of cassette, film, grid, or source-image distance (SID) will require a change in technique.

There is not one perfect set of factors that must be used for each exposure. A number of possible combinations may produce a diagnostic radiograph. The radiographer is responsible for ensuring that the technique chart is available, complete, and consistent. It must meet the requirements of the equipment and the preferences of the physician who will interpret the radiographs.

This chapter contains a number of formulas and mathematical calculations that all radiographers must master. The basic mathematical skills needed to perform these calculations are reviewed in Chapter 26, where you will also find additional examples and practice problems to increase your proficiency.

TECHNIQUE CHARTS

A **technique chart** is a listing of the various radiographic examinations performed in the facility. It provides exposure factors for each body part according to its thickness. Fig. 10-1 is an example of a portion of a technique chart. It includes the following information: type of examination, view, screen type, SID, patient/part measurement (in centimeters), kVp, mA, exposure time, and a grid (bucky) notation.

Some radiographers become so familiar with the operation of their equipment over time that they tend to memorize or estimate exposures and do not feel the need for a technique chart. This practice may result in charts being outdated or unavailable and may cause unnecessary exposure errors. Radiation control regulations may require posting of a current technique chart and may also specify the information to be included. In Oregon, for example, regulations require that technique charts include a notation that gonad shielding is required for specific examinations. The Joint Commission for the Accreditation of Healthcare Organizations (JCAHO) establishes standards for institutions that receive Medicare payments, and these standards include requirements for

Lumbar Spine						
AP and Oblique			Lateral			
40" SID Bucky			40" SID Bucky			
XYZ Film			XYZ Film			
Rapid Screens			Rapid Screens			
Cm	mA	Sec	kVp	mA	Sec	kVp
18-19	200	.04	86			
20-21	200	.05	86			
22-23	200	.06	86			
24-25	200	.08	86	200		96
26-27	200	.1	86	200	.15	96
28-29	200	.15	86	200	.2	96
30-31	200	.2	86	200	.25	96
32-33	200	.25	86	200	.37	96
34-35	200	.37	86	200	.5	96
36-37	200	.5	86	200	.65	96
38-39	200	.65	86	200	.85	96
40-41	200	.85	86	200	1.2	96

Notes: Measure at level of ASIS
Use gonad shielding

Fig. 10-1 Example of portion of radiographic technique chart.

x-ray technique charts in radiography departments. Radiographers must be aware of requirements for technique charts and ensure that their charts conform to the regulatory standards.

Technique charts are unique to each x-ray machine and each facility. The x-ray machine manufacturer cannot supply a definitive chart with the machine because the exposures will vary considerably depending on the types of film, screens, grids, and processing that are used. When a new chart is needed there are several possible sources.

Some of the major film companies will supply computer-generated charts for their customers. The local technical representative of your film company may come to your department, on request, and do some testing to obtain the necessary data. These data are submitted to the company and the chart is sent to your facility.

In some communities experienced technologists will prepare a technique chart for a fee. This option is often advantageous because the chart can be made specifically for your facility and equipment, using only settings available on your control panel. Exposures may be provided for any procedures unique to your facility. Such a chart is also more likely to conform to local radiation control regulations.

If an existing chart needs to be changed because *all* of the exposures are too light or too dark, you can

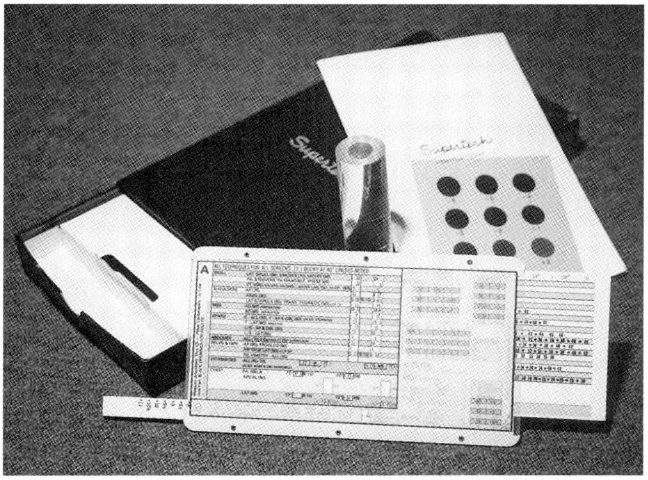

Fig. 10-2 Supertech® calculator kit.

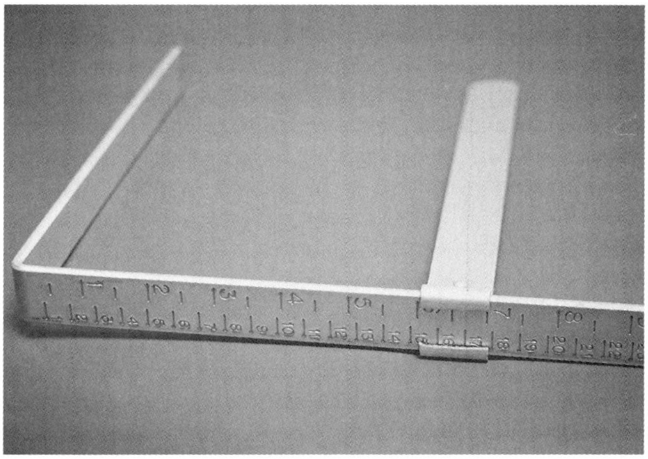

Fig. 10-3 X-ray caliper for measurement of body part size.

probably modify it easily. When the chart is consistent throughout—that is, all of the exposures are too light or too dark to about the same degree—it is a simple matter to increase or decrease all of the exposure times by a specific percentage to correct the radiographic density for all exposures. If your existing chart is not consistent, the use of a consistent chart borrowed from another facility or from Appendix B may provide a starting point.

Finally, if you decide to prepare a chart yourself, products are available to assist you. Supertech®, Inc.* offers both computer software and a handheld slide rule to calculate exposures. Both products are supplied with a small penetrometer and a master film (Fig. 10-2). These are used to test your system and gather the basic data needed to tailor the tool to your x-ray department. Complete instructions are included. The Supertech computer software generates technique charts that conform to JCAHO standards.

Regardless of whether you prepare the chart yourself or arrange for its preparation by someone else, some testing is needed to establish baseline data for your system. This testing should be done when the processor is functioning at optimum levels. The same film and cassettes that will be used with the chart should be used for testing. It is helpful to have a record of exposures kept over a period of time, listing the examination, measurement, exposure factors, and an assessment of the film produced.

Before using your technique chart to take radiographs of patients, it is helpful to test some of your exposures using a **radiographic phantom.** A radiographic phantom is a human skeleton, or portion of a skeleton, encased in a plastic material that is similar in density to human tissue. You may already be fa-

miliar with phantoms through experience in your radiography education program. A good phantom provides an excellent simulation of radiography of a human patient. It may be possible to borrow one from your film company's technical representative, from your x-ray supply dealer, or from a radiography education program.

Computerized control units with automated exposure settings or "anatomic programming" will automatically select the kVp and mAs for an examination when the radiographer selects the body part and enters the measurement. To function accurately, these units must first be programmed with exposures that meet the requirements of the facility. In other words, a technique chart must be created and entered into the control's computer. Even with this type of equipment, radiation control agencies may require posting of a printed technique chart.

PATIENT MEASUREMENT

Technique charts are based on the measurement of the body part to be radiographed. The radiographer must measure the body part accurately in order to select the correct exposure from the technique chart or to obtain the correct exposure with a programmable computerized control.

The tool for body part measurement is called a **caliper** (Fig. 10-3). The main shaft of the caliper is a flat strip of metal, calibrated in both inches and centimeters. There are two perpendicular extensions from the shaft: one is permanently affixed to one end of the shaft, the other slides up and down the shaft. These two extensions form "jaws" between which the body part is measured.

It is usual for technique charts and computerized controls to specify the part measurement in centimeters. The dimension to be measured is the thickness through which the x-ray beam will pass. Measure-

*Supertech®, Inc., P.O. Box 186, Elkhart, IN 48615; www.supertechx-ray.com

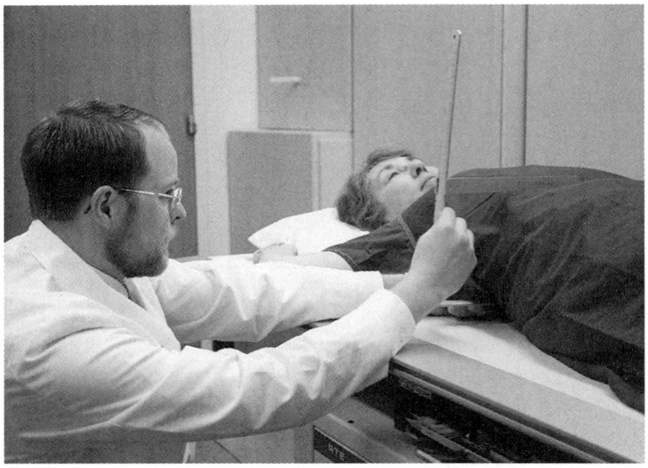

Fig. 10-4 When measuring, the fixed jaw of the caliper is placed under or against the part and the movable jaw is brought snugly and firmly against the patient on the opposite side. Take care that the jaws of the caliper remain parallel to each other.

ments of the trunk of the body should always be made in the same general position as for radiography because measurements may change significantly when the patient changes position. For example, upright measurements of the abdomen are often 4 to 6 cm greater than measurements of the same region when the patient is lying down.

When measuring, the fixed jaw of the caliper is placed under or against the part and the movable jaw is brought snugly and firmly against the patient on the opposite side (Fig. 10-4). You must take care that the jaws of the caliper remain parallel to each other. Pressing the jaws too tightly against the patient may cause them to spread apart at the open end, resulting in an inaccurate measurement. You must also take care that you do not measure air space. For instance, if the patient is lying on the table and you are measuring the thickness of the patient at the waist, the arch of the patient's back may leave a space between the back and the tabletop. Measuring from the tabletop to the surface of the patient's abdomen will give an inaccurate measurement. *Both jaws of the caliper must be firmly in contact with the body part.*

Body parts are usually measured through the path of the central ray. Some parts, however, may be measured through their thickest portion, or another method may be used. For example, exposures for the AP open-mouth projection of the upper cervical spine are usually based on the cervical spine measurement taken in the midcervical region. When there is variation in measurement method, it should be stated in the technique chart. The technique chart is designed for a specific measurement method and will not produce accurate results unless the measurement is consistent with the method intended by the chart.

FIXED kVp VS. VARIABLE kVp

Years ago, it was common to construct technique charts based on a specific mAs value for each view and to vary the kilovoltage by 2 to 3 kVp per centimeter for changes in patient/part size. This type of chart is called a **variable kVp chart.** While a variable kVp chart may be adequate for patient/part sizes within a moderate "normal" range, it tends to fail when taken to extremes. For parts that may vary greatly, such as examinations of the trunk of the body, the kVp will also vary greatly. This will produce inappropriate contrast levels for patients who are larger or smaller than average.

A more desirable type of chart is called a **fixed kVp chart.** For this method, an optimum kVp value is established for each view and the mAs is varied according to the patient/part thickness. When the kVp levels are kept to the high end of the optimum range, exposures will have more latitude, or leeway for exposure error. Exposures may be designated for small, medium, large, and extra large patients, rather than having a separate listing for each centimeter measurement. When this is the case, each size category should state the size range in centimeters.

Some control panels do not have a sufficiently wide range of possible mAs combinations to provide ideal exposures at fixed kVp for all patient/part sizes. Small kVp changes can be used to "fine tune" exposures between mAs settings to obtain a proper exposure for each measurement. This results in a modified fixed kVp chart where mAs is the primary variable but kVp fluctuates within a range of ±4 kVp from the optimum level.

OPTIMUM KILOVOLTAGE

The kVp setting for any examination must first provide sufficient penetration to create an image. Kilovoltage settings that are too low will result in loss of detail in the dense portions of the image. The best policy is to use the highest kVp setting that will produce sufficient contrast for acceptable film quality. This setting is referred to as **optimum kilovoltage.** This approach will result in the least patient exposure and the greatest exposure latitude.

The optimum kVp for a specific examination may be determined by taking a series of films with a radiographic phantom. Fig. 10-5 illustrates such a series, taken at 10-kVp increments. In each case, the mAs was adjusted to maintain approximately the same radiographic density on all of the radiographs. It is not possible to evaluate the contrast of a radiograph that is too dark or too light. The radiographic density must be in an acceptable range in order for contrast to be apparent. The mAs must be adjusted to maintain the desired radiographic density when changing kVp. These adjustments are explained later in this chapter.

A
56 kVp

B
66 kVp

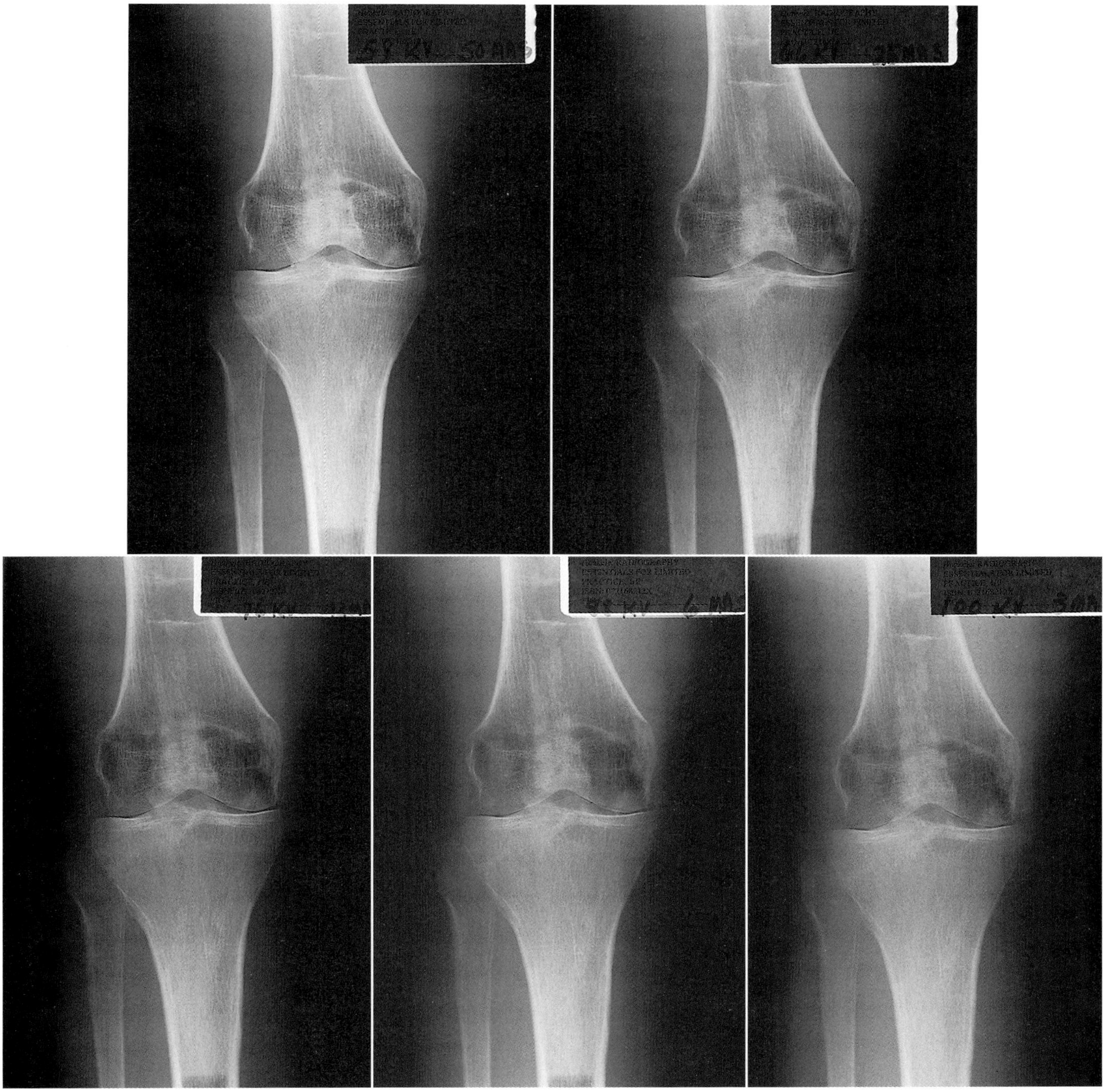

C
76 kVp

D
86 kVp

E
96 kVp

Fig. 10-5 Typical film series using radiographic phantom to determine kVp. In this series, **C** or **D** would probably be considered optimum, or perhaps a compromise between the two.

The ideal kVp levels for your facility will depend on the film type and grid ratio and should also correspond with the preferences of the physician who will read the radiographs. Useful ranges have been well established by the experience of others. Appendix C provides a listing of suggested optimum kVp ranges that will assist you by providing a starting point.

CRITERIA FOR mA SELECTION

Many of the exposure changes in this chapter involve changes in the value of the mAs. Most computerized control panels provide the option of setting the mAs directly. Older units, however, will require you to decide how a given quantity of mAs will be obtained. As explained in Chapters 4 and 6, mAs is the product of mA and time, and several possible combinations of mA and time may be used to achieve the desired quantity of exposure. For example, 10 mAs may be obtained using any of the following combinations: 50 mA and 0.2 sec, 100 mA and 0.1 sec, or 200 mA and 0.05 sec. When a technique chart is being created or mAs values are being adjusted, the question then arises as to which of several possible combinations is best. In choosing an mA setting, the radiographer should consider the tube rating, focal spot size, exposure time requirements, and the available mA and time settings. Table 10-1 provides general guidelines for selecting mA stations on an x-ray control panel with a maximum mA setting of 300.

First, the mA must be obtainable at the desired kVp and mAs values without exceeding the capacity of the x-ray tube. This is determined by consulting the tube rating chart or by attempting to set the desired exposure on a control panel that includes a tube load indicator. For very large exposures, there is greater tube capacity at relatively low mA settings with longer exposure times.

Since the focal spot size is determined by the mA selection, the choice of focal spot size will limit the mA choices available. As explained in Chapter 4, most x-ray generators are wired so that the small filament and focal spot are used with settings of 150 mA or less; settings of 200 mA and above will activate the large filament and focal spot. The small focal spot

provides better image sharpness for small body parts, such as extremities. The small focal spot is also desirable when the OID is great, for example, on the lateral projection of the cervical spine. For examinations of the trunk of the body, higher mA stations associated with the large focal spot are used to shorten exposure time.

When there is likelihood of motion, it is wise to select a high mA setting to minimize exposure time. This is especially important for chest radiography, for small children, and for patients with tremors. On the other hand, the highest setting should not be used routinely when it is not needed, since tube life is extended when the highest mA setting is used sparingly.

Finally, the choice of mA station may be determined by the value of the mAs. When there is a limitation on the possible combinations of mA and time that will produce the desired mAs, this consideration may determine the choice of mA station.

CALCULATING EXPOSURE TIME

When the desired mAs is known and you have selected an mA setting according to the criteria in the previous section, the next step is to determine the exposure time. To determine the exposure time, divide the desired mAs by the selected mA. This is a variation of the mAs formula:

$$\frac{mAs}{mA} = \text{Time (sec)}$$

Example: Suppose the desired mAs is 50, and you have decided to use 200 mA.

Using the formula:

$$\frac{50 \text{ mAs}}{200 \text{ mA}} = 0.25 \text{ sec } (¼ \text{ sec})$$

It is helpful to make an mAs chart for your machine. Create a column for each mA station, and label the rows with the exposure time settings available on your control panel. Then calculate the mAs for all possible combinations and enter them on the chart. An example of such a chart is included in Appendix D. An mAs chart is a helpful reference when creating a technique chart or when departing from

TABLE 10-1			
Guidelines for Selecting Milliamperage			
50 mA	**100 mA**	**200 mA**	**300 mA**
Extremities	Extremities	Thoracic spine	Chest
Breathing techniques	Cervical spine	Lumbar spine	Patient unable to hold still
	Facial bones	Abdomen	
		Skull	

the usual exposure for a particular case, such as a patient who is unable to hold still.

TECHNIQUE CHART "FAILURE"

When a technique chart produces inconsistent results, it is often because the kVp levels are not optimal. Too low a kVp range will tend to produce too much contrast for smaller part measurements. For larger part measurements, insufficient penetration will result in films that are too light. When the fixed kVp is too high, radiographs will lack contrast, especially when the part measurement is at the upper end of the size range. These problems are solved by adjusting the level of the fixed kVp for those categories that are causing problems. Changes in kVp will require adjustments in mAs. The calculation of these adjustments is discussed later in this chapter.

Even the best technique charts do not always produce ideal radiographs. There are several possible causes. One common cause is variation in processor performance. When films are too dark or too light, it is wise to perform a processor quality control check (see Chapter 8) before modifying the technique chart.

Inconsistency in patient/part measurement will produce inconsistent results. It is important that the measurement method conform to the technique chart. Some radiographers always measure body parts through the path of the central ray. Others may measure at the thickest portion of the part for some or all examinations. It is helpful to include measurement instructions on the technique chart. Accurate body part measurement is explained earlier in this chapter.

Technique charts are constructed to meet the requirements of average or normal tissue densities for each body part. Tissue density may vary significantly because of disease, age, or muscle tone. An athlete or laborer may have muscle tissue that is much greater in density than average, requiring more exposure. Elderly patients usually have diminished bone density and muscle tone, requiring less exposure. As you gain experience you will learn to recognize patients whose tissue density requires adjustments to the exposures provided in the technique chart. Box 10-1 lists conditions that require an increase in exposure; Box 10-2 lists conditions that require an exposure decrease. Some lists of this type suggest a specific amount of exposure change for each condition. The proper adjustment will depend on the severity of the condition. An increase or decrease of 2 to 4 cm from the patient measurement is a convenient method for arriving at a suitable technique.

BOX 10-1

Conditions Requiring an Exposure INCREASE (Hard to Penetrate)

Chest Conditions
Atelectasis
Bronchiectasis
Carcinoma, advanced
Edema, pulmonary
Empyema
Hemothorax
Hydropneumothorax
Metastases (blastic)
Pleural effusion
Pneumoconiosis diseases
Pneumonia
Thoracoplasty
Tuberculosis (calcific, military)

Conditions of Bone
Acromegaly
Arthritis, rheumatoid

Conditions of Bone—cont'd
Charcot joint
Osteochondroma
Osteomyelitis, healed
Osteopetrosis (marble bone)
Paget's disease

Abdomen
Ascites
Cirrhosis of liver

Soft Tissue
Edema

Generalized Conditions
Heavy musculature
Large bones

For each of the conditions listed, increase body part measurement by 2 to 4 cm, depending on severity.

Casts and Splints
Wet plaster cast	Multiply mAs × 3 <u>or</u> add 16 kVp
Dry plaster cast	Multiply mAs × 2 <u>or</u> add 10 kVp
Fiberglass cast	Add 4-6 kVp
Aluminum splint	Add 4 kVp
Pneumatic splint	No change required

Conditions Requiring an Exposure DECREASE (Easy to Penetrate)

Chest Conditions
Chronic obstructive pulmonary disease (COPD) (emphysema)
Pneumothorax
Tuberculosis, active

Conditions of Bone
Arthritis, degenerative
Gout
Hyperparathyroidism
Metastasis, lytic
Multiple myeloma
Necrosis
Osteomyelitis, active
Osteoporosis
Sarcoma
Syphilis, advanced

Abdomen
Bowel obstruction
Pneumoperitoneum

Generalized Conditions
Advanced age
Atrophy
Emaciation

For each of the conditions listed, decrease body part measurement by 2 to 4 cm, depending on severity of condition.

ADJUSTING TECHNIQUES

Variations in Patient/Part Size

As the thickness of the subject increases, radiographic density will decrease unless adjustments are made in exposure. This adjustment is usually in the form of a change in mAs. A 30% mAs increase will compensate for a 2-cm increase in part size. This change compounds, like compound interest, and may require multiple steps. When decreasing technique to compensate for a smaller part size, the mAs is reduced by 20% for each 2 cm of part size reduction.

Example: Your routine technique for a lumbar spine measuring 20 cm is 200 mA, 0.15 sec, 80 kVp, 40-inch SID, 12:1 grid. What technique would you use for a patient who measured 24 cm?

Adjust for part size by increasing mAs 30% for each 2 cm of additional size.

First, determine the mAs:

$$200 \text{ mA} \times 0.15 \text{ sec} = 30 \text{ mAs}$$

To increase to 22 cm, multiply the mAs by 1.3 (100% + 30%):

$$30 \text{ mAs} \times 1.3 = 39 \text{ mAs}$$

To increase to 24 cm, multiply the new mAs (39) by 1.3:

$$39 \text{ mAs} \times 1.3 = 50.7 \text{ mAs (round off to 50 mAs)}$$

In this case, as with others in this chapter, the "ideal" mAs value may be one that is not available on the control panel. Since mAs variations of up to 20% are scarcely noticeable on the radiograph, the "ideal" mAs can usually be rounded up or down to the nearest available mAs value without any significant effect on radiographic density. Now you must select appropriate mA and time settings to produce the calculated mAs, as explained earlier in this chapter.

It is also possible to compensate for part size variations using kVp, but this method should be used only when the size variation is relatively small, since a large change in kVp will also alter contrast and may have negative effects on radiographic quality. A change of 2 kVp/cm is sufficient below 85 kVp; above 85 kVp, a 3 kVp change is needed.

Example: A satisfactory lateral cervical spine radiograph is made on a patient measuring 12 cm using 100 mA, 0.05 sec, and 76 kVp at 72-inch SID. Vary the kVp to alter this exposure for a patient measuring 10 cm.

Since the size variation is small, kVp may be used in this instance. Since the original kVp is less than 85, the compensation will be a reduction of 2 kVp/cm. The size difference is 2 cm (12 cm − 10 cm).

$$2 \text{ cm} \times 2 \text{ kVp/cm} = 4 \text{ kVp}$$

$$76 \text{ kVp} - 4 \text{ kVp} = 72 \text{ kVp}$$

Therefore, the new exposure will be 100 mA, 0.05 sec, 72 kVp, 72-inch SID.

Chapter 26 contains further discussion of these methods of technique adjustment with additional examples and practice problems.

Altering Radiographic Density

When a film is too light, the best solution is usually an increase in mAs. Likewise, when the film is too dark, mAs may be increased. Changing mAs by a specific quantity does not always produce the same result. It is the *percentage change* that is significant. Fig. 10-6, illustrates changes in radiographic density produced by various percentages of mAs increase. Note that a 20% increase in mAs does not produce a significant change in density. A 50% increase is apparent, but not great. If the film is so light that it must be repeated to obtain diagnostic quality, at least a doubling (100% increase) is usually needed. When

A
Baseline: 5 mAs

B
20% increase: 6 mAs

C
50% increase: 7.5 mAs

D
100% increase: 10 mAs

E
300% increase: 15 mAs

F
400% increase: 20 mAs

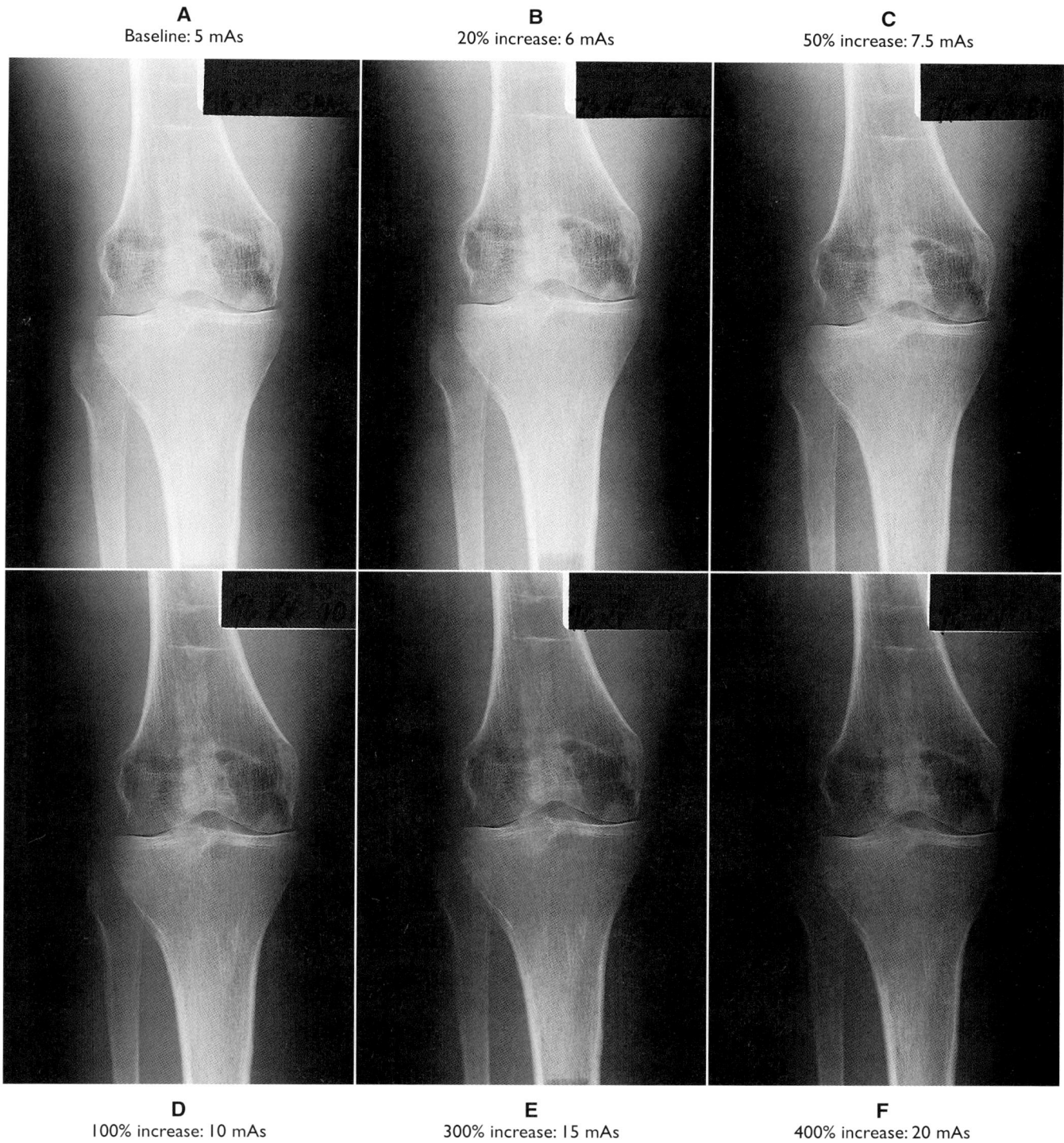

Fig. 10-6 Increasing mAs to correct a film that is too light.

A
Baseline:
5 mAs

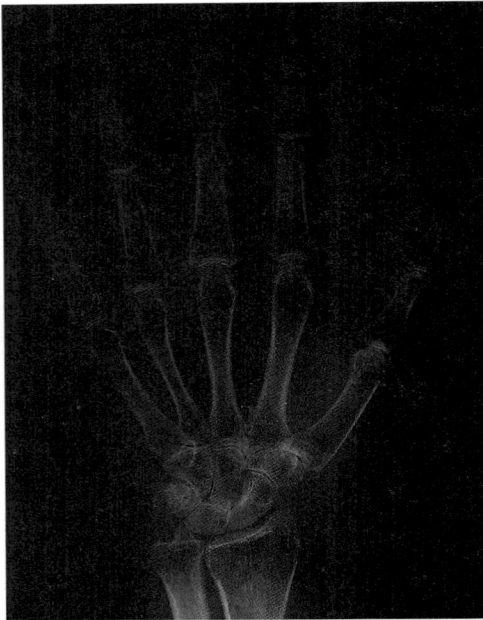

B
10% decrease:
4.5 mAs

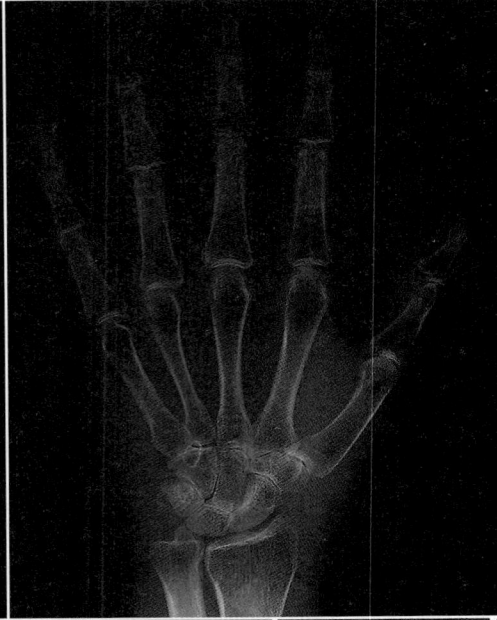

C
25% decrease:
3.75 mAs

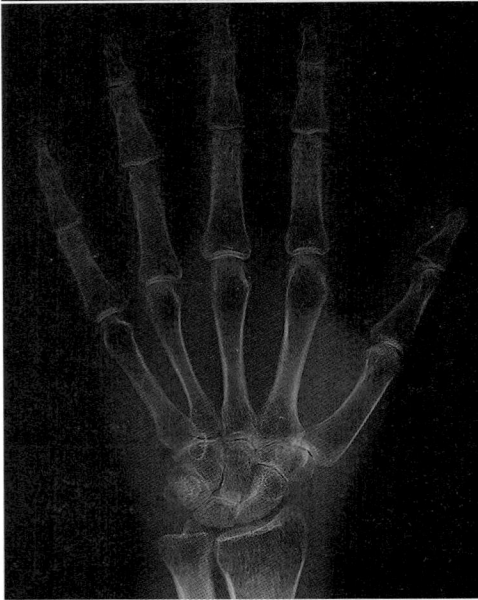

D
50% decrease:
2.5 mAs

E
65% decrease:
1.75 mAs

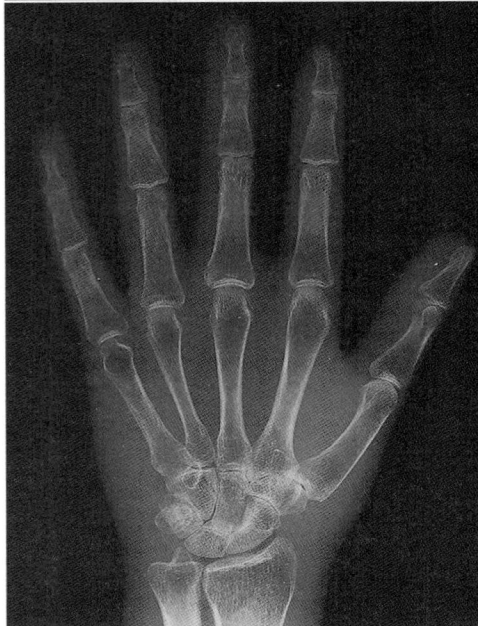

F
80% decrease:
1 mAs

Fig. 10-7 Decreasing mAs to correct a film that is too dark.

Fig. 10-8 Application of the 15% rule. **A,** 10 mAs and 70 kVp. **B,** 5 mAs and 80 kVp.

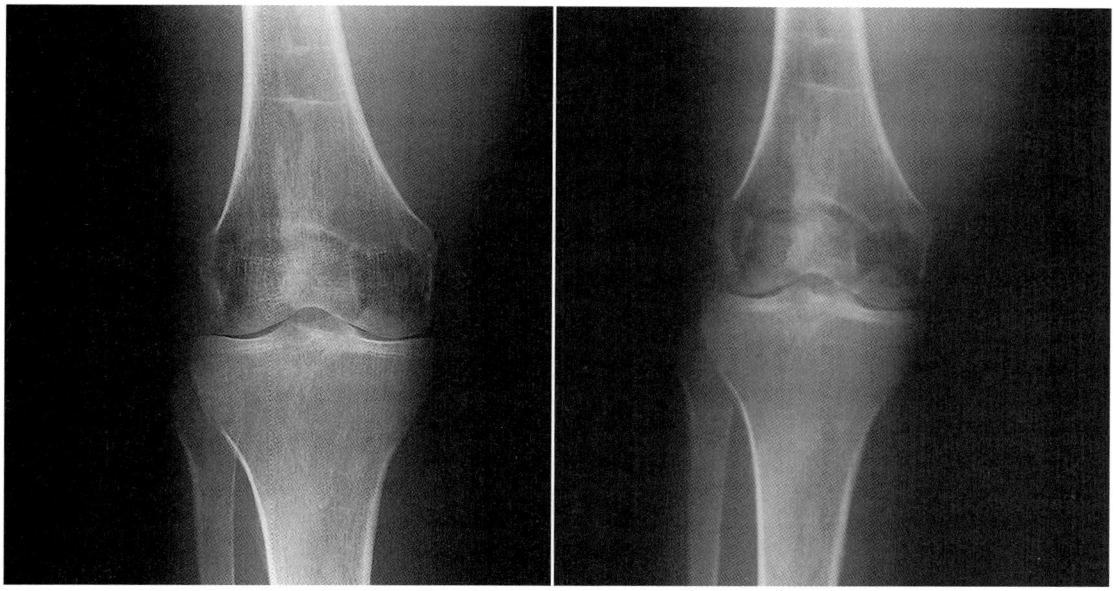

decreasing density, a smaller percentage change is required (Fig. 10-7). That is, a 50% mAs reduction produces the same amount of change as a 100% increase. While these changes will become much easier for you to estimate with experience, the main lesson here is that you must be bold when making changes. Trying to correct improper density with tiny increments of change is ineffective.

Altering Contrast Levels

A change in kilovoltage is used to alter contrast. Since kVp affects the quantity of exposure to the film, changing kVp will also affect radiographic density. For this reason, when a film of appropriate density requires a change in contrast, it is necessary to change both the kVp and the mAs. Whether working with an entire technique chart or a single exposure, the 15% rule can be used to change the level of contrast while keeping the density constant.

The 15% Rule The 15% rule is based on the fact that a 15% change in kVp will produce approximately the same change in radiographic density as a doubling or halving of the mAs.

> To lengthen the scale contrast, increase kVp by 15% and divide the mAs by 2. This application of the 15% rule increases latitude and decreases patient dose by approximately 30%.

Example: A radiograph is made using 20 mAs and 68 kVp. The radiographic density is acceptable, but a longer scale of contrast is desired. To calculate the new exposure, first increase kVp by 15% (100% + 15% = 115% or 1.15):

$$68 \text{ kVp} \times 1.15 = 78.2 \text{ kVp (round off to 78 kVp)}$$

Next, divide the mAs by 2:

$$20 \text{ mAs} \div 2 = 10 \text{ mAs}$$

Therefore, the new exposure is 10 mAs at 78 kVp.

> To shorten the scale contrast, decrease kVp by 15% and multiply the mAs by 2. This application of the 15% rule decreases latitude and increases patient dose.

Example: A radiograph is made using 10 mAs and 78 kVp. The radiographic density is acceptable, but a shorter scale of contrast is desired. To calculate the new exposure, first decrease kVp by 15% (100% − 12% = 88% or 0.88)

$$78 \text{ kVp} \times 0.85 = 66.3 \text{ kVp (round off to 66 kVp)}$$

Next, multiply the mAs by 2:

$$10 \text{ mAs} \times 2 = 20 \text{ mAs}$$

Therefore, the new exposure is 20 mAs at 66 kVp.

Fig. 10-8 illustrates application of the 15% rule. See also Chapter 26.

Variations in SID

Changes in SID are not routine. A standard distance is established for each procedure, and the technique

TABLE 10-2

Approximate Calculations for Changes in SID

From 40-inch SID to	60 inches	72 inches	80 inches
Multiply mAs by	2	3	4

chart provides the correct exposure for the standard distance. Only on rare occasions will it be necessary to modify exposures for changes in distance. For example, bedside radiography with mobile equipment may not permit the usual distance.

As explained in Chapter 6, variations in SID result in changes in radiation intensity. If the SID is to be changed without altering radiographic density, the mAs must be modified accordingly. The formula for this change is:

$$\frac{mAs_1}{mAs_2} = \frac{D_1^2}{D_2^2}$$

Example: Suppose that the usual technique for chest radiography of a patient measuring 20 cm is 4 mAs, 110 kVp, at 72-inch SID. In this case it is necessary to perform the examination at a 60-inch SID. How should the technique factors be changed? The solution to this distance problem involves a change in the mAs according to the formula given above. To solve it, substitute known values in the equation:

$$\frac{4 \; mAs}{X} = \frac{72^2}{60^2}$$

Reduce the fraction:

$$\frac{4 \; mAs}{X} = \frac{72^2}{60^2} = \frac{6^2}{5^2}$$

Calculate the squares:

$$\frac{4 \; mAs}{X} = \frac{36}{25}$$

Cross multiply:

$$36X = 100$$

To solve for X, divide both sides of the equation by 36:

$$X = 2.77 \; mAs$$

Table 10-2 provides guidelines for approximate exposure changes for common variations in SID. See also Chapter 26.

Grid Changes

Certain body parts, such as the lumbar spine, are *always* radiographed using a grid or bucky. Small extremities, such as the hand, are always radiographed "tabletop," that is, without a grid. Technique charts indicate the exposure for grid or nongrid use according to the usual procedure. Occasionally, however, it may be desirable to use a grid for procedures where grids are not usually employed or to make an exposure without a grid when grids are generally used. These changes can be implemented successfully when the body part thickness falls within the 8- to 16-cm range.

The use of a grid causes a significant decrease in radiographic density when compared to the same exposure made without a grid. This occurs because the grid absorbs both primary and secondary radiation that would otherwise expose the film. Since the quantity of radiation absorbed by the grid is determined by the grid ratio, the exposure change required by the grid is also a function of grid ratio.

Either kVp or mAs may be used to adapt a nongrid technique for use with a grid or vice versa. Because contrast is affected by both kVp and grid use, a kVp reduction for the absence of a grid aids in maintaining the desired contrast level. The use of a grid results in increased contrast, even when the kVp is increased to compensate for the grid. When an air gap is substituted for a grid, the more desirable change is obtained by altering the mAs.

Table 10-3 provides the conversion factors needed to change from nongrid to grid exposures and vice versa, according to grid ratio, using either kVp or mAs. This table provides useful information within a limited part size range. Fig. 10-9 illustrates the use of these grid conversion factors. When the part size is larger, more scatter radiation is involved and the change must be greater. It would be very unusual, however, to radiograph a larger body part without a grid.

The quantity of kVp change required is equal to the value of the grid ratio plus 4. When changing from one grid ratio to another, the quantity of kVp change is equal to the difference in the two grid ratios.

When mAs is used to compensate for a change in grid ratio, divide the original mAs by the grid factor to obtain a nongrid technique. Then multiply the result by the factor for the new grid.

Example: Suppose your usual technique for a hip measuring 18 cm is 15 mAs and 76 kVp at 40-inch SID using the table bucky with a 12:1 grid ratio. You wish instead to do this examination with the patient on a stretcher using a grid cassette with a ratio of 8:1.

According to the table, when going from a 12:1 grid to no grid, the mAs is divided by 3:

$$15 \; mAs \div 3 = 5 \; mAs \; for \; no \; grid$$

TABLE 10-3

Grid Conversion Table

Grid ratio	From nongrid to grid, add kVp / From grid to nongrid, subtract kVp		From nongrid to grid, multiply mAs by grid factor / From grid to nongrid, divide mAs by grid factor
6:1	10		1.5
8:1	12	OR	2
10:1	14		2.5
12:1	16		3
16:1	20		4

This table is useful for body part sizes in the 8- to 18-cm range.

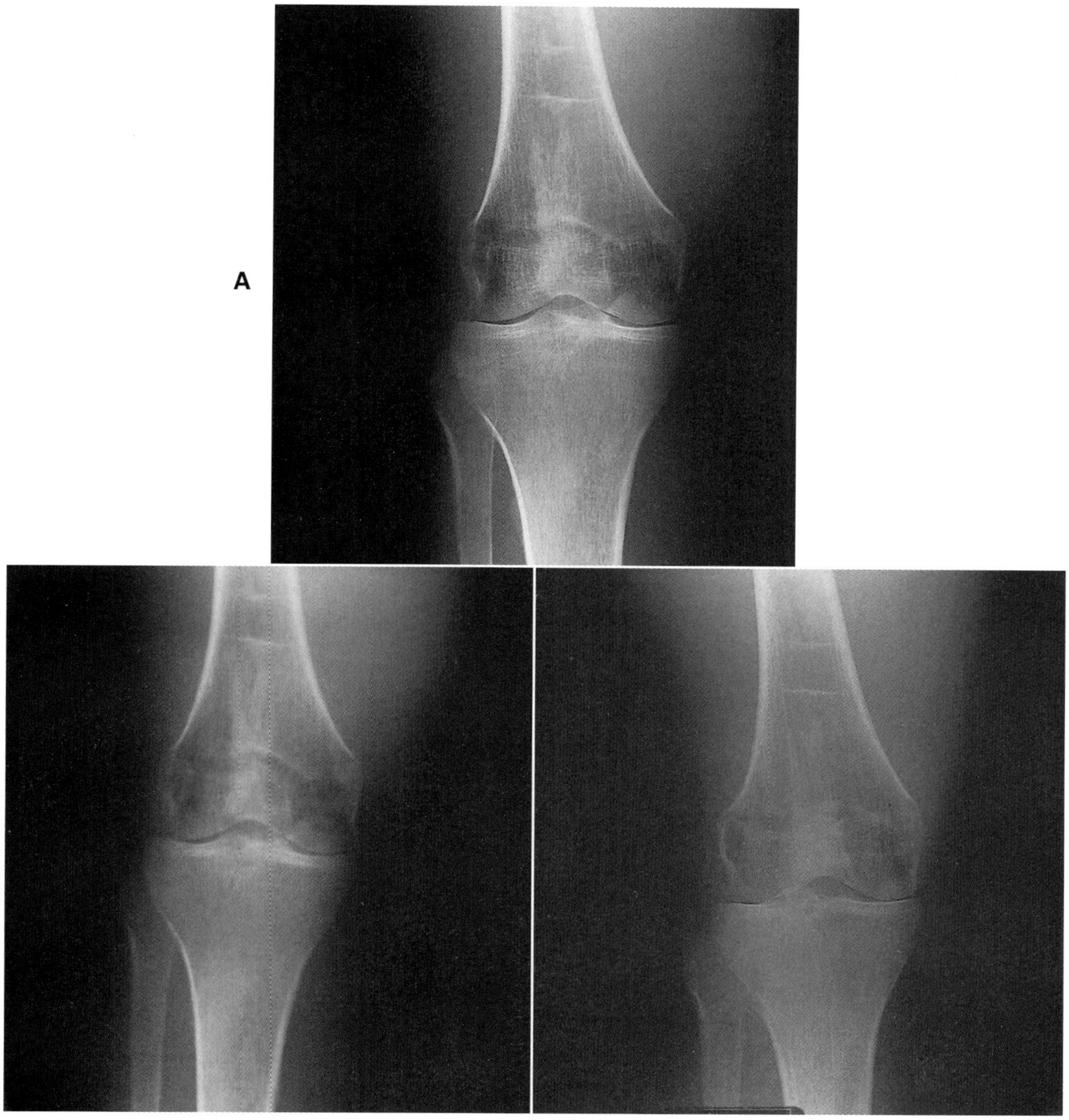

Fig. 10-9 A, Phantom knee exposed using 3 mAs, 78 kVp, and 12:1 ratio bucky. **B,** Same subject exposed at 3 mAs and 62 kVp nonbucky. **C,** Same subject exposed using 1 mAs and 76 kVp nonbucky.

According to the table, when going from no grid to an 8:1 grid, the mAs is multiplied by 2:

$$5 \text{ mAs} \times 2 = 10 \text{ mAs for an 8:1 grid}$$

Changes in the Image Receptor System

Changes in the speed class of either the intensifying screens or the film will necessitate a change in exposure. The formula for calculating this change is:

$$\frac{\text{Relative speed}_1}{\text{Relative speed}_2} \times \text{mAs}_1 = \text{mAs}_2$$

Example: Suppose you usually use 10 mAs for radiography of the ankle with extremity screens that have a relative speed of 100. You wish instead to use rapid screens with a relative speed of 400. The faster speed would require less exposure. Using the formula, your calculation would look like this after substituting known values in the equation:

$$\frac{100}{400} \times 10 \text{ mAs} = \text{mAs}_2$$

Reduce the fraction:

$$\frac{\cancel{100}}{\cancel{400}} - 10 \text{ mAs} = \text{mAs}_2$$

Multiply to solve:

$$\frac{1}{4} \times 10 \text{ mAs} = 2.5 \text{ mAs}$$

Other examples and practice problems are included in Chapter 26.

SUMMARY

A good fixed kVp technique chart is required to produce consistent exposure results. Technique charts are based on body part measurements in centimeters, taken using a caliper. Proper measurement according to the method specified by the technique chart is essential. Optimum kVp ranges are established for each body part, and the mAs is varied to adjust the exposure for changes in patient/part thickness. The mA is selected to conform to the requirements of the tube rating chart and to provide the desired focal spot size. The highest mA settings are used when short exposure times are required. Exposure times are determined by calculation, based on the required mAs and the desired mA.

Technique charts are needed to program computerized controls and may be required by radiation control regulations and JCAHO standards. They should be updated when there is a change in the system that affects exposure requirements. Technique charts are designed for normal tissue density and usual radiographic procedures. They must be adjusted for patients whose tissue density is outside the normal range and for procedures that depart from the usual SID, grid, or image receptor system speed.

The mAs is used to adjust radiographic density. Bold changes are needed when films must be repeated because they are too dark or too light. Contrast levels may be adjusted without affecting density by using the 15% rules.

▪ *Review Questions* ▪

1. Using the technique chart from your facility or the one provided in Appendix B, state the exposure factors for a lateral chest radiograph on a patient measuring 32 cm.
2. What tool and what units are used to measure body part thickness for radiography?
3. List steps to take in preparation for making a new technique chart.
4. Using the table of optimum kVp ranges in Appendix C, state the optimum kVp ranges for AP views of the cervical spine, thoracic spine, and lumbar spine.
5. Assume your x-ray control panel has the following mA settings: 50, 100, 200, and 300. Which might you use for radiography of the elbow? The lumbar spine? The chest?
6. You are about to take a radiograph that requires 10 mAs and you have decided to use 100 mA. What should the exposure time setting be?
7. List two conditions that require an exposure increase and two that require a decrease.
8. An acceptable radiograph is made using 200 mA, 0.3 sec, and 70 kVp. Calculate a new exposure that will provide more latitude and less patient dose for the same examination on the same patient.
9. If a satisfactory radiograph is made using 20 mAs at 40-inch SID, how much mAs is needed to produce a similar radiograph at a 72-inch SID?
10. A satisfactory radiograph is made using 10 mAs and 76 kVp with a 10:1 grid. How would you change the technique to perform the same examinations without a grid?

11

Radiobiology and Radiation Safety

Learning Objectives

At the conclusion of this chapter, the student will be able to:

- State the units used to measure radiation intensity, radiation dose, and dose equivalents in both the conventional and the SI systems
- Given a set of x-ray exposure factors, calculate the entrance skin exposure using a dose graph
- Discuss the potential effects of radiation injury to cells
- Define and compare radiation risks according to type: somatic vs. genetic, stochastic vs. nonstochastic, short-term vs. long-term
- Discuss the risks of exposure to low doses of ionizing radiation and compare these to other familiar health risks
- Explain the significance of the ALARA principle
- List and explain methods for minimizing patient dose during radiography
- Explain what is meant by "low-dose techniques"
- List and explain precautions for the safety of radiographers
- List potential risks of radiation exposure during pregnancy and explain ways to reduce these risks

Key Terms

ALARA principle
chromosome
coulombs per kilogram (C/kg)
DNA
dominant gene
dose equivalent
entrance skin exposure (ESE)
enzyme
erythema
focus-skin distance (FSD)
free radical
gene
gonad
gonad shield

Gray (Gy)
mutation
nonstochastic
optically stimulated
 luminescence (OSL)
quality factor (QF)
rad
recessive gene
rem
roentgen (R)
Sievert (Sv)
stochastic
thermoluminescent
 dosimeter (TLD)

The health risks involved in radiation use are not well understood by the general public. Diagnostic radiography involves very low doses of radiation exposure, and the risks to both patients and radiographers are extremely small. It is important for radiographers to understand the risks associated with radiography and to commit themselves to the practice of radiation safety in all aspects of their work.

This chapter is about the measurement of radiation, the effects of radiation exposure, and the ways in which radiographers can minimize the potential hazards to their patients, their coworkers, themselves, and future generations.

RADIATION MEASUREMENT

Two systems are used to measure radiation and radiation dose: the conventional (British) system and the SI (Système Internationale) units established by the International Commission on Radiation Units in 1981. The conventional system is still the most commonly used in the United States. Table 11-1 lists the units used in both systems.

Units of Exposure

The **roentgen,** abbreviated R, is the conventional unit of radiation exposure. It represents a measurement of the radiation intensity in the air. This is determined by the ionization of air resulting from interaction with the x-ray beam. The roentgen is defined as the quantity of radiation that will produce 2.08×10^9 ion pairs in a cubic centimeter of air. It is measured by placing a container of dry air in the x-ray beam. The container is connected to an electric circuit with a meter. The ionization of the air causes a tiny current to flow in the circuit. The total quantity of electrical charges that flows in the circuit during the exposure indicates the quantity of the exposure.

The corresponding SI unit for measuring radiation exposure is **coulombs per kilogram (C/kg),** specifying the quantity of electrical charge in coulombs produced by the exposure of 1 kilogram of dry air.

$$1\ R = 2.58 \times 10^{-4}\ C/kg$$

Units of Dose

While the roentgen is useful for measuring the quantity of radiation present, it is not a useful dose measurement. Dose varies with the depth of measurement and the quantity of radiation energy absorbed in the tissue that is exposed. To measure both therapeutic radiation doses and specific tissue doses received in diagnostic applications, the conventional unit is the **rad.** Rad stands for radiation absorbed dose and is equal to 100 ergs (an energy unit) ab-

TABLE 11-1

Units of Exposure and Dose

Conventional Units	SI Units
Roentgen (R)	Coulombs per kilogram (C/kg)
rad	Gray (Gy)
rem	Sievert (Sv)

TABLE 11-2

Quality Factors for Different Types of Ionizing Radiations

Type of Radiation	Quality Factor
X-ray photons	1
Beta particles	1
Gamma photons	1
Thermal neutrons	5
Fast neutrons	20
Alpha particles	20

sorbed per gram of tissue. One roentgen of exposure will result in approximately 1 rad of absorbed dose in muscle tissue. The SI unit for dose measurement is the **Gray (Gy).**

$$1\ Gy = 100\ rad$$

and conversely,

$$1\ rad = 1\ centigray\ (cGy, 0.01\ Gy)$$

Dose Equivalents

The biologic effect of radiation exposure varies according to the type of radiation involved and its energy. Equal doses of various types of radiation will not necessarily result in equal biologic effects. Some radiation workers, such as engineers in nuclear power plants, nuclear submarine construction workers, or technologists in nuclear medicine laboratories, may be exposed to several types of radiation with unequal levels of biologic effect. Neither the roentgen nor the rad is a useful unit for measuring the occupational dose of combined radiations with different levels of effects.

To simplify the process of measuring occupational dose, a **quality factor (QF)** number is assigned to each type of radiation, based upon its relative biologic effect as compared to x-rays. Table 11-2 lists the quality factors for different types of ionizing radiations. For example, since 1 rad of thermal neutrons has been shown to cause biologic effects equal to those produced by 5 rads of x-ray energy, thermal neutrons have a QF of 5. The absorbed dose is multiplied by the QF to obtain the **dose equivalent.** The resulting unit in the conventional system is called the **rem,** which stands for <u>r</u>oentgen <u>e</u>quivalent in <u>man</u>. Thus the worker exposed to 1 rad of fast neutrons would receive 5 rem of occupational dose.

$$rad \times QF = rem$$

The SI unit used to measure dose equivalents is the **Sievert (Sv).** The Sievert is determined by multiplying the dose in Gy times the QF; thus,

$$Gy \times QF = Sv$$

and

$$1 \, Sv = 100 \, rem$$

Since the radiation quantities involved in diagnostic radiology are so small, radiographers commonly use units that represent 1/1000 of the common units: milliroentgen (mR), millirad (mrad), and millirem (mrem).

It may be confusing to determine which dose units should be used in a given situation. This is made more difficult by the tendency of many radiographers to use the traditional roentgen, rad, and rem units interchangeably. This practice does not cause serious inaccuracy when speaking only of diagnostic x-rays, because exposure to 1 roentgen of x-ray energy will result in approximately 1 rad of absorbed dose. Since the QF of diagnostic x-rays is 1, 1 rad is also equal to 1 rem.

In general, the reason for the measurement determines which unit is most appropriate. The roentgen is used to measure the presence of x-radiation without any reference to its absorption, that is, the quantity of radiation present in air. For example, an x-ray machine might produce 10 mR during the exposure for a chest radiograph.

The rad is used to measure dose. It is the unit commonly used to prescribe radiation therapy. The amount absorbed by a specific tissue is what is being measured, so a statement indicating the part of the body involved usually modifies the rad dosage. For example, a radiation oncologist may prescribe a treatment involving 250 rad *to the liver.* During the 10-mR chest exposure, the patient may have received a dose of 4 mrad *to the thyroid.*

Dose equivalents are usually assumed to represent whole body dose or dose to unspecified tissues. The report sent monthly by the laboratory that processes the personnel radiation monitor badges will report occupational dose in rem or mrem. Dose equivalents are not used only for occupational dose measurements, however. Studies that measure population dose and exposure to naturally occurring radiation, such as radon, are usually reported in rems. The chest radio-

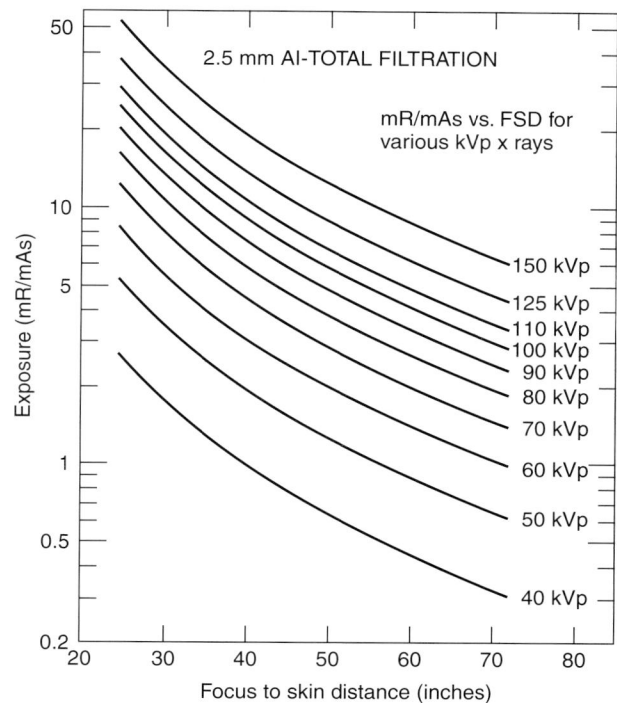

Fig. 11-1 Graph used to calculate estimated patient skin dose from x-ray exposure factors. To calculate dose, use the full size version of this graph in Appendix E.

graph mentioned above may have added 10 mrem to the total population dose.

Estimating Dose from X-ray Exposure Factors

Radiographers usually think of exposure in terms of mAs, kVp, and SID. A special graph (Fig. 11-1) is needed to convert these factors into units of exposure and dose. A larger version of this graph is included in Appendix E.

The data needed to calculate exposure with the graph include the mAs, the kVp, and the **focus-skin distance (FSD),** that is, the distance from the radiation source to the patient. This distance is determined by measuring patient thickness *in inches* and subtracting this measurement from the SID. If a bucky or grid cabinet is used, an additional 2 or 3 inches must be added to the patient measurement to allow for the distance between the patient and the film:

Example:

SID:	40 inches
Patient measurement (+ 2 inches if bucky):	−10 inches
Focus-skin distance:	30 inches

To practice using the graph, suppose you wished to determine the dose for an exposure of 40 mAs and 90 kVp at a focus-skin distance of 30 inches. Use a sheet of paper to assist in reading the graph. Place the paper over the lower left corner of the graph so that

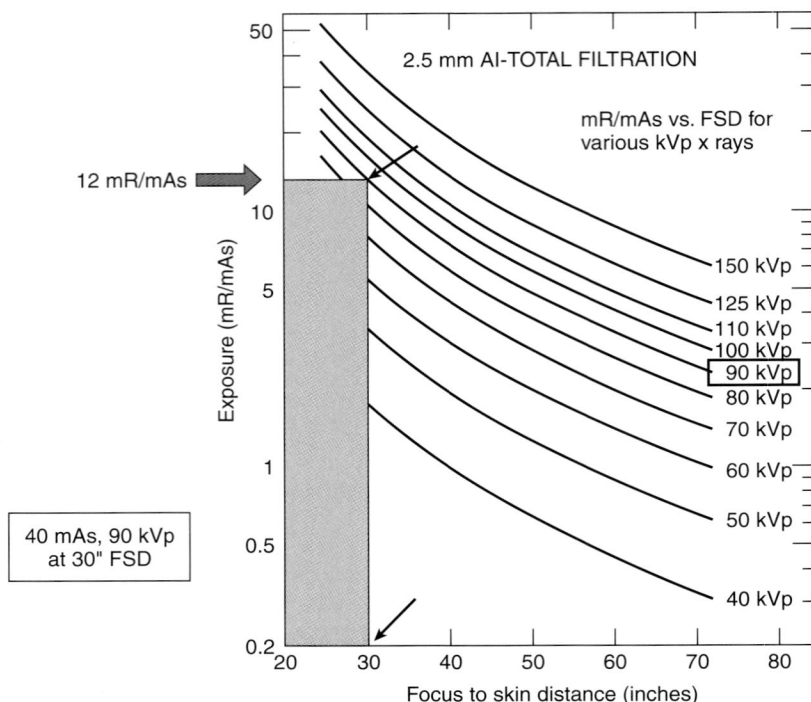

Fig. 11-2 To calculate dose, place a piece of paper over the lower left corner, aligning its margins with the correct focus-to-skin distance and the appropriate kVp curve. Read the dose rate on the left margin of the graph. Multiply this number by the mAs to determine dose.

TABLE 11-3

Typical Doses for Radiographic Examinations

Examination	Entrance Skin Exposure (mrad)	Mean Marrow Dose (mrad)	Gonad Dose (mrad)
Skull	200	10	<1
Chest	10	2	<1
Cervical spine	150	10	<1
Lumbar spine	300	60	225
Abdomen	400	30	125
Pelvis	150	20	150
Extremity	50	2	<1

its right edge is aligned to 30-inch FSD at the bottom of the graph. Then adjust the paper so that the upper right corner is aligned to the 90 kVp curve (Fig. 11-2). Read the *exposure rate* on the left side of the graph—12 mR/mAs, that is, 12 milliroentgens for each milliampere-second. To determine the dose, the mAs must be multiplied by the exposure rate:

Example: 40 mAs × 12 mR/mAs = 480 mR

Patient dose in radiography is usually calculated according to the exposure level at the skin. This is called the **entrance skin exposure (ESE).** Because 1 R of exposure produces approximately 1 rad of

dose, the readings obtained from the graph may be interpreted as mrads ESE.

Some typical doses for radiographic exposures are listed in Table 11-3. Note the differences in dose for various examinations and compare the ESE with the dose to bone marrow and gonads (reproductive organs). It is apparent that the highest ESE doses are received with examinations of the skull, abdomen, and lumbar spine, while the greatest bone marrow and gonad doses are associated with examinations of the abdomen, lumbar spine, and pelvis.

Some radiation control agencies require that the ESE for an average patient be posted for each examination commonly done at the facility. This may be specified as a required part of the technique chart.

BIOLOGIC EFFECTS OF RADIATION EXPOSURE

Cellular Response

To understand how cells are affected by radiation exposure, it is helpful to understand something of the composition of a typical cell. Fig. 11-3 is a simplified diagram of a cell. The cell is surrounded by the plasma membrane. At its center is the nucleus, which contains the nucleoli. Inside the nucleoli are 23 pairs of **chromosomes,** microscopic bodies that contain the **genes.** Genes are the determiners of heredity and are made of a unique protein called **DNA** (deoxyribonucleic acid). The chromosomes contain the coded "information" that the cell needs in order to function.

As discussed in Chapter 3, x-rays can ionize substances, removing electrons from their orbits. This process results in a free, negatively charged electron and leaves the remainder of the atom with a positive charge. When cells are irradiated, ionization may occur to any part of the cell, such as the material that makes up its membrane, the water within the membrane, or the DNA. The initial ionization may produce a "domino effect," causing ionization in the surrounding area.

Two types of interactions, called "direct hit" and "indirect hit," may cause injury to living cells. A direct hit refers to the breakage of a DNA molecule as a result of being struck by an x-ray photon. An indirect hit causes a chemical reaction in the water that makes up most of the cell's substance. This process causes the formation of **free radicals,** temporary molecules and parts of molecules that occur as the result of ionization. Free radicals may interact directly with the DNA or may combine to produce toxic substances that are injurious to DNA.

Most of these effects are extremely short-lived. Electrons quickly find new homes in the orbits of other atoms, and the balance of charges returns to normal. Free radicals combine to form more stable compounds. Over 90% of radiation injuries to cells have been repaired or corrected within the first hour after exposure. Some cells may sustain damage that requires several days for the body to make repairs. Biologic chemicals called **enzymes** repair damage to cell membranes and DNA, correcting an additional 5% to 9% of the damage within a period of about 3 days. Occasionally, however, the damage is not resolved. A cell may be so injured that it cannot sustain itself and will die. Cell death is not serious unless it involves large numbers of cells. A cell may be damaged in such a way that its DNA "programming" is changed and the cell no longer behaves normally.

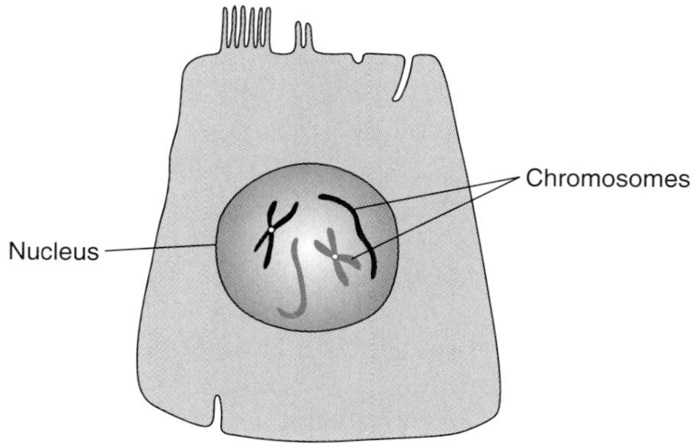

Fig. 11-3 Simplified diagram of a typical human cell.

This type of injury may cause malfunction of the cell or may affect its ability to divide and multiply. Another possible result is the runaway production of new, abnormal cells, causing cancer or a malignant blood disease such as leukemia.

The relative sensitivity of different types of cells is summarized in the Laws of Bergonié and Tribondeau, which state that cell sensitivity to radiation exposure depends upon four characteristics of the cell:

1. *Age.* Younger cells are more sensitive than older ones.
2. *Differentiation.* Simple cells are more sensitive than highly complex ones.
3. *Metabolic rate.* Cells that use energy rapidly are more sensitive than those that have a slower metabolism.
4. *Mitotic rate.* Cells that divide and multiply rapidly are more sensitive than those that replicate slowly.

According to these laws, blood cells and blood-producing cells have characteristics that cause them to be very sensitive. Cells that are in contact with the environment are quite simple, have relatively short lives, and are quite sensitive. These include the cells of the skin and the mucous membranes that line the mouth, nose, stomach, and bowel. Some glandular tissue is also particularly sensitive, especially that of the thyroid gland and the female breast. The tissues of embryos, fetuses, infants, children, and adolescents tend to be more sensitive than those of adults, both because of their age and because of their higher metabolic and mitotic rates. Nerve cells, which have a long life and are quite complex, are much less vulnerable to radiation injury. Cortical bone cells are also relatively insensitive.

Classification of Radiation Effects

Radiation effects are classified in various ways. Short-term effects are those observed within 3 months of the exposure. They are associated with high radiation doses—greater than 50 rad. Short-term effects may be further categorized according to the body system affected: *hematologic* (affecting the blood), *gastrointestinal* (affecting the digestive tract), and *central nervous system* or *CNS* (affecting the brain and spinal cord). Long-term effects, sometimes referred to as latent effects, are not observed until several years after exposure; in fact, they may not be apparent for as long as 30 years. Somatic effects are those that affect the body of the individual who is irradiated. Genetic effects occur as a result of damage to the reproductive cells of the irradiated person and are observed as defects in the children or grandchildren of the irradiated individual.

Radiation effects may also be classified as **nonstochastic** or **stochastic.** Nonstochastic effects are typical of relatively high doses. They are predictable, and the severity of the effect is proportional to the dose. Stochastic effects, on the other hand, are random and unpredictable. The *likelihood* of these effects is proportional to dose, but the severity is unrelated to dose. The effects of the low doses of radiation exposure from diagnostic procedures are stochastic effects.

Short-Term Somatic Effects Short-term somatic effects occur with high doses of radiation and are nonstochastic. One observable short-term effect is reddening of the skin, called **erythema.** This phenomenon is sometimes called a "radiation burn." In the very early days of radiation use, the amount of radiation necessary to produce reddening of the skin was called the "erythema dose" and was the first unit used to measure radiation.

Other short-term effects have been observed and studied in radiation therapy patients and in the victims of nuclear accidents and atomic bomb blasts. This is vastly more exposure than is delivered by diagnostic x-ray machines. Extremely high doses produce CNS effects, causing seizures and coma and resulting in death in a short period of time. Lesser doses will result in "radiation sickness," a gastrointestinal effect in which the mucosal lining of the digestive tract is damaged, breaks down, and becomes infected by the bacteria that normally inhabit the bowel. These victims also have a compromised immune system, due to the death of white blood cells, and are unable to fight the infection. Radiation sickness is usually fatal, but suffering may be prolonged. A lesser dose, affecting primarily the blood and blood-forming cells of the bone marrow, results in hematologic effects: anemia and compromise of the immune system. These victims are prone to infectious diseases that may or may not be fatal, depending on the radiation dose and the severity of the disease process. The whole body radiation dose that is fatal to 50% of the irradiated human population within 30 days (LD 50/30) is approximately 300 rad.

Long-Term Somatic Effects The time required for long-term effects to become apparent is generally considered to be 5 to 30 years, with the greatest percentage occurring between 10 and 15 years.

Short-term radiation effects are predictable and the quantity of exposure required to produce them is well documented. Long-term effects, on the other hand, are stochastic and apparently random. They may involve repeated small doses, such as those used in radiography.

Long-term radiation effects are not easily identified as such because they occur so long after exposure and because these same effects also occur in the absence of radiation exposure. Only extensive research with large populations using computer analysis can demonstrate the role of radiation in causing these effects. The incidence of certain conditions is shown to be increased when results for irradiated groups are compared to those of nonirradiated control groups. The documented latent effects of low doses of ionizing radiation include the following:

- *Cataractogenesis:* the formation of *cataracts,* clouding of the lens of the eye. This effect is of concern to radiologists and radiographers who work extensively in fluoroscopy and who perform other work that involves repeated exposure to the eyes.
- *Carcinogenesis:* increased risk of malignant disease, particularly cancer of the skin, thyroid, and breast, and leukemia, a malignant disease of the blood that has been clearly demonstrated to be associated with radiation exposure.
- *Life span shortening.* A study of the life span of radiologists who died during a 3-year period prior to 1945 showed that they had shorter life spans than physicians who did not use radiation in their practices. This group of radiologists involved those who had been using radiation since the early days of x-ray science. More recent studies show that the decreased occupational exposure typical today has no measurable effect on the life span of radiologists. Radiation exposure is still definitely linked to life span shortening, however. This is a public health concern and another reason to practice a high level of radiation safety.

Genetic Effects

Genetic effects are changes or **mutations** to the genes of the reproductive cells. They occur as a result of radiation exposure to the reproductive organs called **gonads,** the female ovaries or the male testes. In the female, all the ova (egg) cells that the individual will ever produce are present in the ovaries in an immature state at birth. Since no new egg cells are produced as the individual ages, the effect of radiation exposure to the ovaries is cumulative. Radiation to the testes also has longer-term genetic effects than might at first be presumed, because damage to the stem cells that produce the sperm may result in the continued production of sperm that carry the genetic mutation. The vast majority of genetic mutations are considered to be negative, or less well suited to survival than nonmutated cells.

Reproductive cells have only half the number of chromosomes of other cells. Each parent contributes one chromosome to each pair in the new individual, and nature makes the choice as to which gene of each pair will determine the characteristics of the offspring. Those genes that are "chosen" are said to be **dominant,** and those that are not selected are called **recessive.** Genes that have mutated are usually recessive and so do not affect the characteristics of the child. Both dominant and recessive genes, however, occur in the reproductive cells of the child and may be passed on to future generations.

As the population is exposed to radiation from natural, occupational, and health care sources, there is likelihood that individuals will be conceived with mutation of both genes in a strategic pair, resulting in some type of deformity, defect, or characteristic that is less well suited to survival. Public health officials and governments are very concerned about preserving the integrity of the population's gene pool by minimizing radiation that may cause defects in future generations. This concern should motivate those who use ionizing radiation to minimize gonad doses in every way possible. Gonad shielding for this purpose is addressed later in this chapter.

Genetic effects from mutations caused by x-ray exposure have long been demonstrated in animal research. Interestingly, very little genetic effect has so far been confirmed by continuing research involving the Japanese populations exposed to radiation when the atomic bombs were dropped on Hiroshima and Nagasaki during World War II. Individuals who were children at that time are now becoming grandparents. Studies of this new generation and those that follow will be needed before the genetic effects to bomb survivors can be completely evaluated.

Comparative Risks

The average American is exposed to an annual dose of 360 mrem of radiation from all sources—less than 1 mrem per day. Naturally occurring radiation from space, from the earth, and from radon gas accounts for 82% of this exposure. Only 18% of this dose is from manmade sources, and only 11% is due to x-ray use in health care, including radiation therapy. In other words, a very small percentage of the average radiation dose is attributable to diagnostic x-rays.

Certainly, the percentage of observable effects from the radiation involved in typical x-ray examinations is extremely low, and the risk to any one patient is minimal. Most of us take greater risks daily when we drive a car or cross a busy street. The occurrence of these stochastic effects might be compared to being struck by lightning. Many people may be outdoors during a thunderstorm. Few, if any, will be struck by lightning. People struck by lightning may be killed or only slightly injured. The chance of being struck by lightning is extremely remote, but it is greater if you make it a habit to stand in high places during thunderstorms. Scientists can predict fairly accurately the annual rate at which lightning will strike human beings, but it is impossible to predict who will be struck and who will not.

Similarly, radiation causes increased *risk* of these effects, but the effects cannot be predicted with respect to any one individual. Table 11-4 provides some

TABLE 11-4

Decrease in Life Expectancy from Various Causes

Cause	Days
Unmarried male	3500
Cigarette-smoking male (20 cigarettes/day)	2250
Heart disease	2100
Unmarried female	1600
Overweight 30%	1300
Coal miner	1100
Overweight 20%	900
Less than eighth-grade education	850
Cigarette-smoking female (20 cigarettes/day)	800
Low socioeconomic status	700
Stroke	520
Pipe smoking	220
Increasing food intake 100 cal/day	210
Job with radiation exposure (1 rem/year for 40 years)	**40**
Natural radiation (BEIR)	8
Medical x-ray films	6
Coffee	6
Oral contraceptives	5
5 rem/year (occupational exposure)	5
Diet drinks	2
Reactor accidents (Rasmussen)	0.02*
Radiation from nuclear industry	0.02*
Papanicolaou test	−4
Smoke alarm in home	−10
Air bags in car	−50
Mobile coronary care unit	−125

*These items assume that all U.S. power is nuclear.

interesting comparisons between the risks involved in radiography and other more familiar risks.

Scientists agree that any one individual's risk from radiography is *extremely small,* but exposure to the entire population does pose public health risks. Even when the chance of serious effects is one in a million, that adds up to 250 serious problems in a nation of 250 million people. While the risk from a chest x-ray is frequently quoted as typical, the dose for a lumbar spine examination may be 50 times greater, increasing the risk. All who are involved in applying ionizing radiation to human beings share the responsibility for ensuring that everything possible is done to keep these risks as low as possible.

RADIATION SAFETY

Clearly, exposure to x-rays creates some risk for both patients and radiographers. It is therefore an essential part of your education and your ethical responsibility to be knowledgeable about radiation safety and to use this knowledge to avoid all unnecessary radiation exposure to your patients, your coworkers, and yourself.

The federal government issues regulations and recommendations to ensure the safety of patients and radiation workers. State agencies incorporate federal guidelines into the state regulations. Additional laws and regulations may apply in individual states. The state radiation control agencies are responsible for the administration of both state and federal regulations. Your state may have a regulation that requires the posting of a summary of the regulations that apply to your facility. Radiographers are responsible for knowing and following the laws and regulations that apply to their work.

Patient Protection

There is no arbitrary limit on the amount of radiation exposure a patient may receive. The guiding philosophy is called the **ALARA principle.** The ALARA principle states that all radiation exposure to humans should be limited to levels that are As Low As Reasonably Achievable. Radiation control agencies use this guideline to compare the quantities of radiation used for specific procedures within the community. If the average ESE for a specific examination in your community is 20 mrad, this provides evidence of what is reasonably achievable. If the dose in your facility for the same examination is much greater than the average, the level is unacceptable and must be reduced to meet regulatory requirements.

The greatest cause of unnecessary radiation to patients that can be controlled by radiographers is repeat exposures. Repeat exposures are undesirable for many reasons. They require extra time and materials that increase health care costs, in addition to increasing patient dose, so it is important to avoid the need for repeats. On the other hand, exposures *must* be repeated when radiographs are inadequate. Reduction of patient dose is not a valid reason for failing to repeat a film that is not diagnostic.

To minimize the need for repeat exposures, radiographers must take care to avoid mistakes. Double check requisitions and patient identification so that the right patient gets the right examination. Establish good routine procedures and follow them strictly so that careless errors do not necessitate repeat exposures. Provide clear instructions to patients so that they will cooperate in obtaining a successful examination.

Another cause of unnecessary exposure that radiographers control is the size of the radiation field. Radiation exposure can be significantly controlled by the proper use of collimation. *Use the smallest radiation field that will cover the area of clinical interest.* In no case should the size of the radiation field be greater than the size of the film. Whenever possible, collimate to exclude sensitive tissues such as the eyes, the thyroid gland, the female breasts, and the gonads. As you practice positioning and centering with precision, you will gain confidence that the essential anatomy can be visualized successfully without using an excessive field size.

In addition to developing good habits of procedure and collimation, the radiographer can reduce patient exposure by using "low-dose techniques." Low-dose techniques involve using optimum kVp, fast screens and film, a minimum SID of 40 inches, and nongrid techniques when appropriate.

The highest kVp that will produce acceptable contrast results in less exposure to the patient than a low kVp technique. An increase in kVp with no other change will increase the patient dose rate slightly. However, when the mAs is adjusted to compensate for the kVp increase, the net result is a reduction in dose. The use of the 15% rule to increase kVp and decrease mAs (see Chapter 10) results in a dose reduction of approximately 30%. You can demonstrate this principle using the dose graph in Appendix F. Determine the dose for a sample exposure. Apply the 15% rule to your original technique and recalculate the dose.

Dose is also significantly affected by the speed of the image receptor system. Radiographers should use the fastest films and screens that are consistent with the desired film quality. While you may have little choice in the types of film and screens available in your facility, there may be more than one type. You should select the slower system only when a high degree of definition is needed.

Routine radiography should never be performed at less than a 40-inch SID. The tube housing permits leakage of some radiation that increases patient dose without being useful in image formation. In addition, interaction between the primary x-ray beam and the parts of the collimator produces scatter radiation. Patient dose from these sources is relatively insignificant at a 40-inch SID but increases dramatically, according to the inverse square law, when the tube is closer to the patient.

As stated in Chapter 9, the use of a grid requires a significant increase in exposure compared to the same examination performed without a grid. For this reason, grids and buckys should be used only when needed to control scatter radiation. Small body parts that do not generate large quantities of scatter should be radiographed on the tabletop.

Gonad Shielding Lead shields that prevent unnecessary radiation to the reproductive organs are required by regulation in most jurisdictions. **Gonad shields** are used to reduce the likelihood of genetic radiation effects. Gonad shields must be used when the patient is of reproductive age or younger, whenever the gonads are within the primary x-ray beam, and when the shield will not interfere with the purpose of the examination. Generally, this applies to most patients under the age of 55. A shield device consisting of at least 0.5 mm lead equivalent is placed between the x-ray tube and the patient. Shields attached to the collimator are called shadow shields. They are positioned by viewing their shadows within the collimator light field (Fig. 11-4). Shields placed on or near the patient's body are called contact shields and are somewhat more effective than shadow shields (Fig. 11-5). Both types meet the legal requirements for gonad shielding. Fig. 11-6 demonstrates shield placement for both males and females. It is helpful to note that the pubic symphysis (the center of the pubic bone) is at the same level as the greater trochanter of the femur, avoiding the necessity of palpating the pubic bone for proper shield placement.

Gonad shields should also be used when the primary radiation field is *near* the gonads, even though this may not be required by regulation. Whenever the gonads are within 6 to 8 inches of the margin of the radiation field, gonad dose will be significantly reduced by shielding. When the field is more than 8 inches from the gonads, shielding has little or no effect with respect to protection from primary radiation. On the other hand, scatter radiation may provide some level of gonad dose for *any* examination when the gonads are not shielded. Little extra effort is required to provide a lead apron or lead shield. Most radiographers feel better about their work when they shield conscientiously, and patients also appreciate this level of concern.

Shields may be purchased that provide precise shielding of the gonads when doing radiography of adjacent structures. It is almost always possible to shield the male gonads, regardless of the examination. A female gonad shield may sometimes interfere with the purpose of the examination. The abdomen, sacrum (pelvic portion of the spine), and coccyx (tail bone), for example, cannot be well visualized with an ovary shield in place. When the ovaries cannot be shielded, the ovarian dose is greatly reduced if the patient can be radiographed prone (face down) or facing away from the x-ray tube. In this position, the tissues of the buttocks and the bones of the pelvis absorb a significant quantity of radiation that would otherwise increase the gonad dose. Variations from

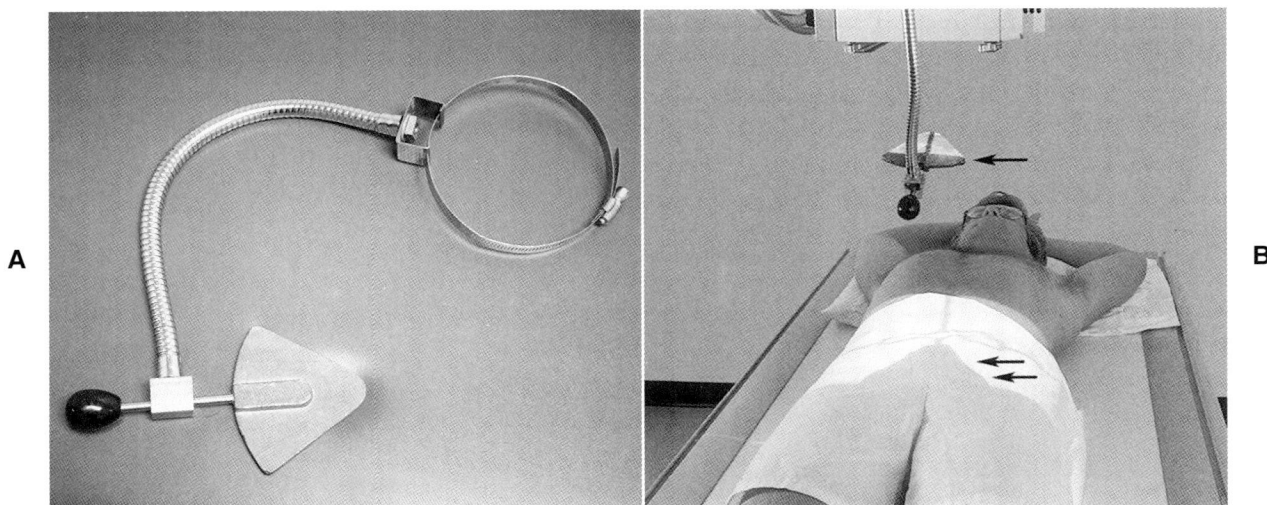

A

B

Fig. 11-4 Shadow shield is attached to the collimator or tube housing and placed by observing location of its shadow in the collimator light field.

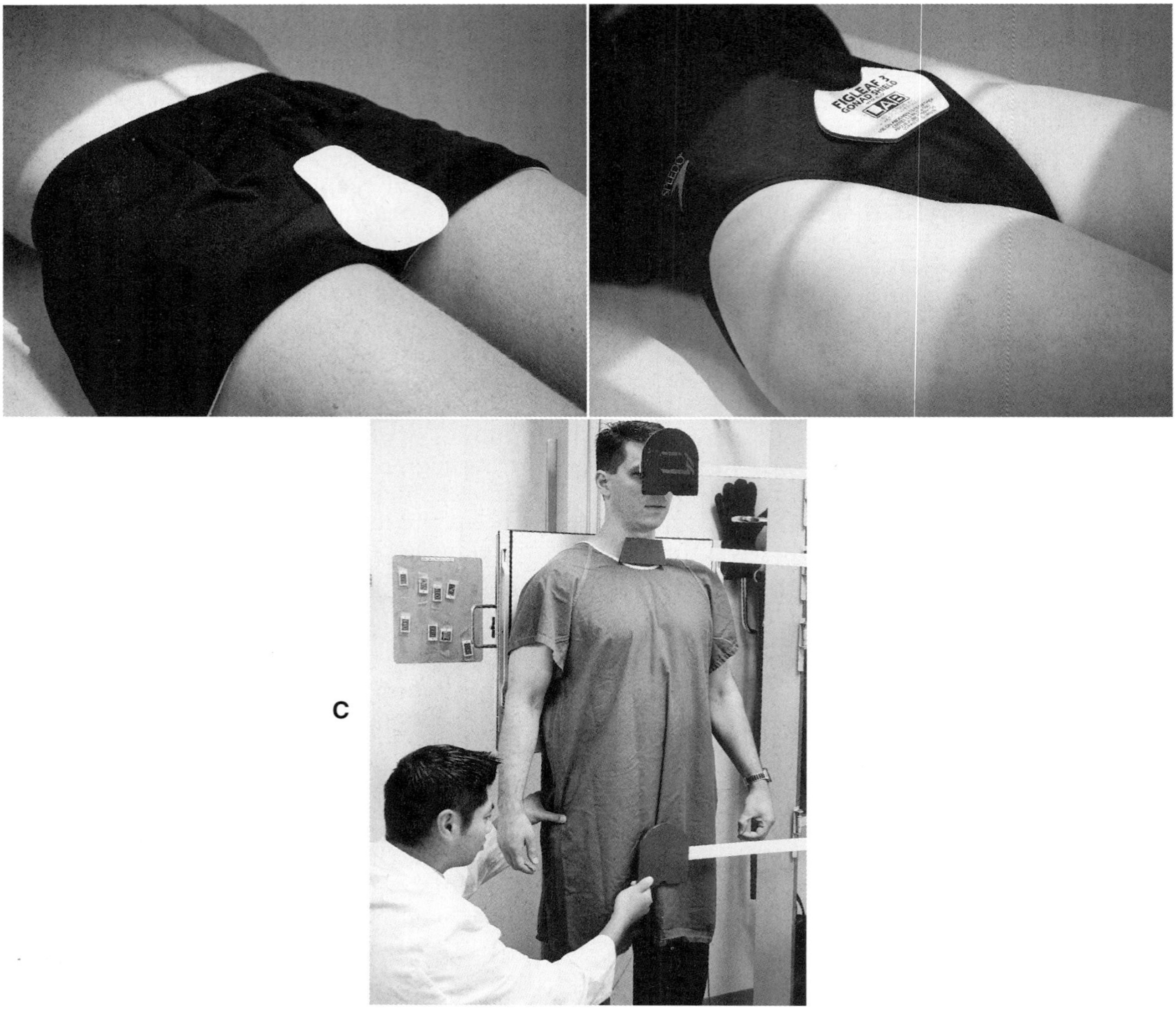

Fig. 11-5 Contact shields are placed on the patient's body during exposure. **A,** Male shielding; **B,** female shielding. **C,** Shield stand provides contact shielding for upright radiography. This device provides shielding for gonads, eyes, thyroid, and breasts (not shown).

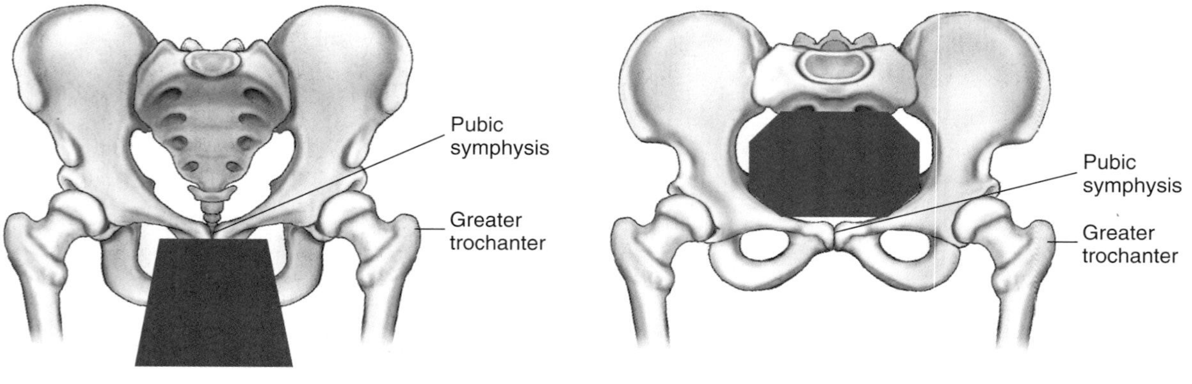

Fig. 11-6 Shield placement. **A,** Top of male shield is 1 inch inferior to top of pubic symphysis, which is at the level of the greater trochanter. **B,** Female shield is placed in the midline, midway between the level of the anterior superior iliac spine and symphysis pubis.

standard positioning should first be approved by the physician.

Personnel Safety

Radiographers may potentially be exposed to radiation from either the primary x-ray beam or from scatter radiation. Because radiographers are considered to be "occupationally exposed individuals," they are prohibited from activities that would result in direct exposure to the primary x-ray beam. This means that *radiographers are not allowed to hold patients or cassettes during x-ray exposures.*

The radiographers at greatest risk from occupational exposure are those involved in fluoroscopy and mobile radiography. These procedures are not commonly performed by limited radiographers. Fluoroscopy involves direct observation of the x-ray image in motion during procedures commonly used to visualize the digestive tract or the circulatory system. Special fluoroscopic x-ray equipment is required. Radiographers assisting with fluoroscopy may have to be in the room to change cassettes and to assist with patient positioning and the administration of contrast media during these procedures. Mobile radiographic examinations are sometimes referred to as "portable radiography." These examinations are performed in the surgical suite or at the patient's bedside where there is no protective control booth.

The three principal methods used to protect radiographers from unnecessary radiation exposure are *time, distance,* and *shielding.* Time and distance apply principally to radiographers who are involved in fluoroscopy and mobile radiography. Shielding is employed to protect all radiographers.

The amount of exposure received is directly proportional to the time spent in a radiation field, so occupational dose is decreased when this time is minimized. For example, a radiographer might shorten the time of exposure by stepping into the control booth during fluoroscopic procedures when not required to be near the patient.

The second method involves using distance. Increasing the distance between yourself and a radiation source decreases your exposure in proportion to the square of the distance, so small increases in distance have a relatively large effect. Mobile x-ray units have long cords on the exposure switches, enabling the radiographer to get as far from the machine and the patient as possible while making an exposure.

The third method is shielding, and this is by far the most common method of protection used by radiographers. The lead wall of the control booth provides protective shielding and is the radiographer's primary defense. Radiographers are unlikely to be exposed to any significant amount of radiation when standing well within the protection of the control booth. Other types of shielding include lead aprons, gloves, and thyroid shields (Fig. 11-7). These types of shielding are worn during fluoroscopic procedures and mobile radiographic examinations.

The pre-exposure safety check introduced in Chapter 2 is an essential safety practice. It is the principal method used to ensure that you do not accidentally expose your coworkers. When you do the safety check, you make certain that no one is in the x-ray room unnecessarily, that everyone in the control booth is completely behind the lead barrier, and that the x-ray room door is closed. Doors are usually open or ajar except during exposures. A closed door indicates that it is not safe to enter.

Personnel Monitoring Devices for monitoring radiation exposure to radiation workers are called personnel dosimeters. Film badges were once the principal type of dosimeter. They consist of a piece of dental film enclosed in a badge-type holder. Several filters are incorporated in the badge so that, if the unfiltered exposure exceeds the capacity of the film, additional exposure can be measured in the filtered area. The disadvantage of this type of personnel monitor is that the dental film is subject to fog when exposed to heat or fumes, and this exposure could result in a false

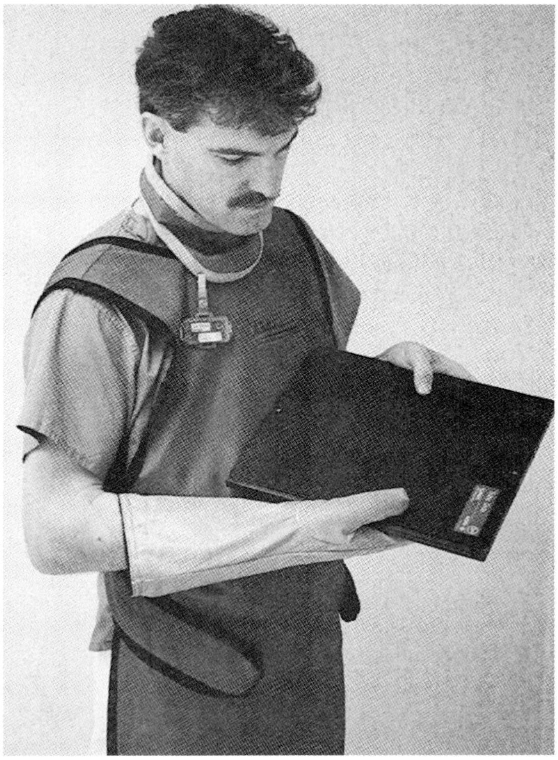

Fig. 11-7 Radiographer wears protective apparel during fluoroscopic and mobile radiographic examinations. Protective eyewear is also available.

reading. The film is also ruined if it is laundered! After a period of use, the film is returned to a laboratory that processes it and measures the film density. Based on this measurement, the exposure is calculated and reported. Many radiographers still refer to their dosimeters as "film badges," but today they are more likely to be TLDs or OSLs.

TLD stands for **thermoluminescent dosimeter.** The roots of this word mean, "dose-measuring device that gives off light when heated." The TLD is the principal type of personnel monitor used by radiographers. It consists of a plastic badge containing one or more lithium fluoride crystals (Fig 11-8). These crystals (and several others with similar characteristics) absorb x-ray energy and give off the energy again in the form of light when heated. The TLD is more durable than the dental film of the film badge and responds only to ionizing radiation exposure. At the end of the measurement period, the badge is sent to a laboratory where the crystals are placed in a special tray and inserted into the TLD analyzer. This instrument heats the crystals to the required temperature, measures the light emitted, and transmits the data to a computer.

OSL stands for *optically stimulated luminescence* and refers to a recently developed monitoring dosimeter (Fig. 11-9) that uses aluminum oxide as a radiation detector. This dosimeter is processed using laser light rather than heat, as with TLDs. OSL dosimeters are similar to TLDs and have several additional advantages. They can measure small doses more precisely and can be reanalyzed to confirm results. They are accurate over a wide dose range and have excellent long-term stability.

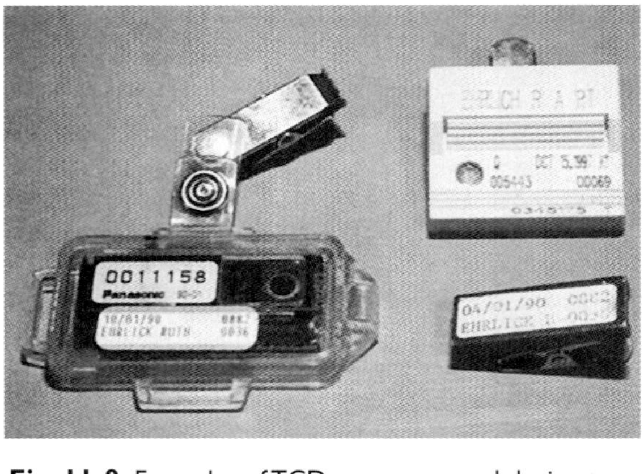

Personnel dosimeter laboratories provide badges, processing service, and reports, and they keep permanent records of the radiation exposure of each person monitored. Service may be arranged on a weekly, monthly, or quarterly basis. Personnel who receive relatively high doses of occupational exposure change their badges most frequently. Since these dosimeters cannot accurately measure total exposures of less than 5 mrem, personnel who receive very small amounts of exposure will get more accurate measurements with less frequent badge changes. Personnel involved in diagnostic radiography who are always, or nearly always, in a control booth during exposures get the most accurate reports with quarterly service. Those who work in fluoroscopy and do mobile radiography are usually best monitored with monthly service.

Service companies provide an extra badge in every batch that is marked CONTROL. The purpose of this badge is to measure any radiation exposure that might occur to the entire batch while in transit. Any amount of exposure measured from the control badge will be subtracted from the amounts measured from the other badges in the batch. The control badge should be kept in a safe place, away from any possibility of x-ray exposure. *It should never be used to measure occupational dose or for any other purpose.*

Radiation badge service companies will want to know the name, birth date, and Social Security number of all persons to be monitored so that all records will be accurately identified. If there has been a history of previous occupational radiation exposure and the dose is known, this information should also be provided so that the record will be complete and accurate. Dosimeter reports are sent to the subscriber for each batch, and an annual summary of personnel exposure is also provided. Personnel should be advised of the radiation exposure reported from their badges and should be provided with copies of the annual reports for their own records. Occupationally exposed personnel should not leave their employment without a complete record of their radiation exposure history. Employers are required to provide this information.

Badges should be worn in the region of the collar and should be outside the lead apron when a lead apron is worn. Workers who receive exposure to their hands while their bodies are protected may wear ring or wrist dosimeters. A larger dose is permitted for this limited exposure.

Fig. 11-8 Examples of TCD-type personnel dosimeters.

Fig. 11-9 Example of OSL.

Effective Dose Equivalent (EDE) Limits

The effective dose equivalent–limiting (EDE) system is used to calculate the upper limits of occupational exposure that is permitted. For occupationally exposed personnel, the upper EDE limit is 5 rem (50 mSv) per year. This applies to workers over the age of 18 who are not pregnant, and it is assumed to be a whole body dose. These limits apply to occupational exposure only, not to exposure that workers may receive as a result of imaging or tests related to their own health care.

The EDE system also states a retrospective or cumulative dose limit that is equal to 1 rem (10 mSv) times the worker's age. For example, a 30-year-old worker with no previous occupational exposure would have a cumulative EDE limit of 30 rem (300 mSv). This is referred to as the *rem bank*. The worker is permitted to exceed the annual limits by small amounts as long as the rem bank is not depleted. The EDE system specifies dose equivalents for specific body organs and tissues. These limits are less than for the whole body and are used to calculate the limits of EDEs for whole body or unspecified exposures.

The established EDE limits ensure that the safety of radiation workers is comparable to that of workers in other occupations. The risk from the allowable exposure is considered to be so low as to pose an insignificant risk. The occupational dose received by radiographers is usually well below the established limits.

The ALARA principle is the guiding philosophy associated with the use of EDE limits. It is important that radiographers not be complacent simply because their dose is below the limit. It is required by radiation control agencies that employers and employees make every effort to ensure that occupational dose is kept to the lowest levels that are reasonably achievable.

The upper boundaries of occupational dose were formerly referred to as "maximum permissible dose" or MPD. This term is now out of favor because it implies that exposure in excess of the lowest achievable dose is permissible.

Radiation and Pregnancy

It has long been recognized that radiation exposure poses risks to the developing embryo or fetus. In general, we now know that radiation during pregnancy may result in spontaneous abortion, congenital defects in the child, increased risk of malignant disease in childhood, and an increase in significant genetic abnormalities in the children of parents who were exposed prior to birth.

Animal studies first alerted scientists to the fact that radiation could cause spontaneous abortion of the developing embryo and could increase the rate of congenital abnormalities seen in those that survived to birth. These findings have been confirmed in humans by studying the pregnancies of women who survived the atomic bomb blasts of Hiroshima and Nagasaki and the nuclear accident at Chernobyl. In the 1950s, Alice Stewart, an English researcher, demonstrated a 14-fold increase in the incidence of childhood leukemia among children who had been exposed to radiation in utero as a result of x-ray pelvimetry examinations in the third trimester of pregnancy.

Studies of smaller groups of women exposed to radiation as a result of diagnostic and therapeutic procedures confirm that radiation *in excess of 5 rad to the uterus* is cause for concern. Table 11-5 lists the average fetal dose factors associated with various radiographic examinations. To use the table, determine the ESE to the patient and multiply it by the fetal dose factor. When determining ESE from the graph in Appendix E, you will need to convert the dose from mR to R by dividing it by 1000.

The greatest risks for spontaneous abortion, fetal death, and birth defects exist when significant levels of exposure occur during the first trimester of pregnancy, that is, the first 3 months. The embryo is most vulnerable to radiation insult while tissues are in the process of differentiation. Unfortunately, this creates the greatest hazard at a time when a woman may not yet be aware that she is pregnant.

Radiation control agencies address the issue of radiation exposure to pregnant radiation workers. Regulations regarding pregnant workers use the term "declared pregnant woman." When a worker voluntarily notifies her employer, in writing, that she is pregnant, the employer is responsible for ensuring that her EDE remains below the established limit for pregnancy. The EDE limit for a pregnant worker is 0.5 rem over the 9-month course of the pregnancy. The work assignment should be evaluated to minimize exposure. For a pregnant radiographer, the safest work

TABLE 11-5

Estimating Fetal Dose of Radiation (Fetal Dose Factor)

X-ray Examination	mR/R ESE
Skull	<0.01
Cervical spine	<0.01
Full-mouth dental	<0.01
Chest	2
Lumbar spine	250
Pelvis	295
Extremity	<0.01

Modified from Bushong, S: *Radiologic Science for Technologists*, ed 6, St. Louis, 1997, Mosby.

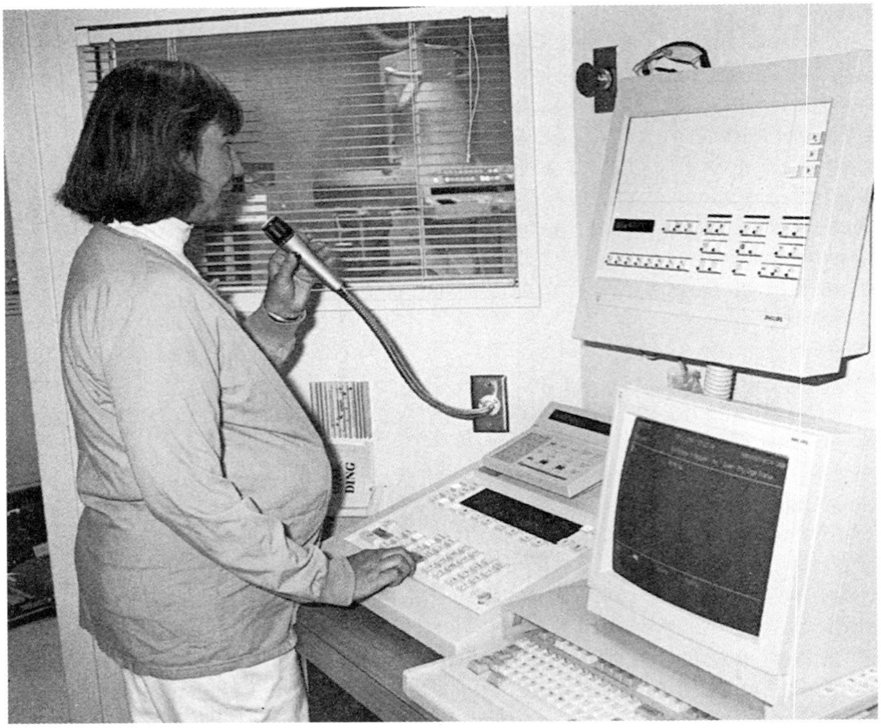

Fig. 11-10 Lead barrier of control booth protects pregnant radiographer and her unborn child.

assignment would be one where a permanent lead barrier (control booth) always shields the worker during exposures (Fig. 11-10). Here again, the ALARA principle is important. Every effort should be made to minimize exposure, keeping the dose received as far below the established limit as possible. Pregnant radiographers, or those of childbearing age who *may* be pregnant, should pay particular attention to personal safety measures when assisting with fluoroscopy or using mobile x-ray equipment.

The public is generally aware that x-radiation is to be avoided during pregnancy, and this may lead to irrational fears on the part of pregnant women or their families. The chance is extremely remote that a routine x-ray examination of the chest or an extremity would result in harm to the developing child. The risk of abnormality in the child as a result of a diagnostic x-ray examination in the first trimester of pregnancy is considered to be 1 in 1000 or less, depending on the examination. On the other hand, examinations requiring direct radiation to the pelvis, especially relatively high-dose fluoroscopic studies or computed tomography (CT) scans of the abdomen or lumbar spine, may be cause for concern.

Radiation control regulations require that women of childbearing age be advised of potential radiation hazards prior to x-ray examination. This requirement is usually met by posting signs in the radiology department advising women to tell the radiographer before the examination if there is any possibility that they may be pregnant (Fig. 11-11). Such signs should be written in all languages commonly used in the community.

The patient's physician is in the best position to be aware of an early pregnancy. The patient's history may indicate the possibility of pregnancy, and specific questions to rule out pregnancy should be a part of any medical history that precedes the ordering of pelvic or abdominal x-ray examinations. If pregnancy is a possibility, an early pregnancy test, easily and quickly performed in the physician's office, may clarify the situation. If the patient is pregnant and the proposed x-ray examination involves direct pelvic radiation, the physician must weigh the potential risks and benefits of the examination and discuss them with the patient before proceeding with the study. In the case of minor or chronic complaints, it is common to delay the examination until after the birth of the child.

In practice, however, the possibility of pregnancy may not even be considered. This is especially true in the case of accident or injury, where the emergency room or office visit is brief and the history is limited to the injury complaint. For this reason, *it is essential that the radiographer be mindful of the possibility of pregnancy whenever the patient is a female of childbearing age.* Specific questions should be asked to determine that the patient's physician has addressed the issue of preg-

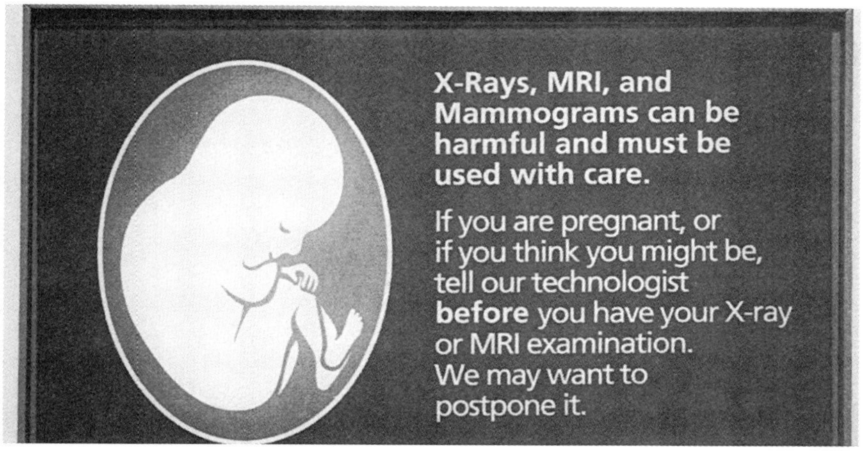

Fig. 11-11 Signs in dressing rooms, waiting areas, and imaging suites alert patients to the potential hazard of examination when pregnant.

nancy prior to ordering the examination. When the radiographic examination does not involve the abdomen and pelvis, it is good practice to provide a lead apron for women of childbearing age, whether pregnant or not (usually age 10 to 55).

Some radiology departments use the "10-day rule" to screen potentially pregnant patients prior to radiography. This rule states that nonemergency x-ray examinations of the female abdomen and pelvis are done only during the first 10 days of the menstrual cycle, the onset of menstruation being considered as day 1. Because menstruation usually indicates that the patient is not pregnant and pregnancy cannot commence until 13 to 15 days later when the patient ovulates, the first 10 days of the cycle is the period when the patient is least likely to be pregnant. Where this rule is used, exceptions are usually made only after the radiologist has considered the case and spoken with the referring physician.

At one time, the American College of Radiology recommended the 10-day rule for all radiography practice. It is no longer recommended, however, because of the decreased dose used with modern intensifying screens and the low risk now shown to be associated with routine radiography. There is still some risk, however, and there is also a *perceived* risk. Since there is some risk of problems with any pregnancy, limiting the chance that a birth problem will be attributed to radiography may be an important aspect of risk management for the physician or the institution. The 10-day rule is still used in many facilities to avoid the potential liability that may fall to the facility when radiography is associated with an undesirable outcome of pregnancy, regardless of the cause.

Neither the 10-day rule nor an early pregnancy test can guarantee that the patient is not pregnant,

but these methods can greatly decrease the likelihood of irradiating an otherwise unsuspected embryo.

If x-ray examinations of a pregnant patient must be done, modifications in procedure can help to minimize dose to the embryo or fetus. If the part to be examined is *not* the abdomen or pelvis, this area can be shielded with a lead apron. If the abdomen or pelvis is to be evaluated, the number of views and/or the size of the radiation field may be minimized, resulting in less radiation exposure than that required for a routine procedure. The decision to do a limited study, and the determination of the exact limitations to be imposed, are up to the patient's physician or the radiologist.

SUMMARY

Either conventional or SI units may be used to measure radiation exposure and dose. Conventional units are most commonly used in the United States. They include the roentgen for measuring exposure, the rad for measuring absorbed dose, and the rem, the dose equivalent unit for occupational and population dose measurement. For diagnostic x-rays, 1 roentgen of exposure produces approximately 1 rad of dose and is equal to a dose equivalent of 1 rem. ESE stands for entrance skin exposure, the most common dose measurement in diagnostic radiography.

Cellular response to radiation exposure is the result of ionization that may involve a direct hit to the DNA of a cell's chromosome or may damage the cell indirectly as a result of the ionization of water and the formation of free radicals. Most cellular damage is repaired within a very short period of time. Cellular sensitivity is greatest for cells that are young, simple, rapidly dividing and multiplying, and have a rapid metabolism. Blood cells and blood-forming cells are very sensitive, as are skin cells, mucosal cells, and the cells of the thyroid gland and female breast; brain cells and cortical bone cells are relatively insensitive.

Low doses of radiation, typical of those received in diagnostic radiography, produce effects that are long-term and stochastic. They include the formation of cataracts, cancer, and leukemia and the possibility of birth defects in children irradiated during early gestation. Genetic effects are the result of mutations to genes caused when the reproductive organs are exposed. They may result in malformations or other defects in future generations.

Radiographers can reduce radiation risk to their patients by minimizing the need for repeat exposures, collimating well, and using low-dose techniques. Gonad shielding is required in some cases to reduce the likelihood of genetic effects.

Personnel safety is ensured by the proper use of time, distance, and shielding. Pre-exposure safety checks help prevent accidental exposure to co-workers. Radiographers' occupational dose is monitored using personnel dosimeters, usually the TLD type. They are worn on the collar and evaluated monthly or quarterly by a qualified laboratory. Reports should be available to radiographers. The occupational dose for radiographers is typically well below the established EDE limit of 5 rem (50 mSv) per year.

Warning signs, early pregnancy tests, interviews, and double-checks by the radiographer are used to avoid the possibility of inadvertent exposure to a developing embryo or fetus that would increase the risk of spontaneous abortion, birth defects, or childhood cancer.

▪ Review Questions ▪

1. State the SI equivalents for the conventional units R, rad, and rem.
2. If a radiation worker received 1 rad of x-ray exposure, 1 rad of thermal neutron exposure, and 1 rad of fast neutron exposure, state the total dose in both rads and rems.
3. Using the dose graph in Appendix E, calculate the ESE for an exposure of 20 mAs and 80 kVp at a 60-inch FSD.
4. Explain the differences between long-term and short-term effects. Which are nonstochastic? How would you classify erythema?
5. Explain the risks involved in an x-ray examination of the knee as you would explain it to a patient.
6. Explain the difference between a dominant characteristic and a recessive one. Which is typical of a mutation?
7. List the common regulatory requirements for gonad shielding. What is the purpose of gonad shielding?
8. What should a radiographer do to minimize the need for repeat examinations?
9. What is meant by "low-dose technique"?
10. What is the radiographer's responsibility for ensuring that an embryo is not inadvertently exposed to x-rays?
11. What is the primary method used to provide radiation safety for limited radiographers?
12. What is the EDE limit for nonpregnant workers over age 18? How does this compare to the limit for pregnancy? How is it affected by the ALARA principle?

PART III

Radiographic Anatomy, Positioning, and Pathology

12

Introduction to Anatomy, Positioning, and Pathology

Learning Objectives

At the conclusion of this chapter, the student will be able to:

- Explain the differences between cells, tissues, organs, and systems
- List the systems of the human body and state the basic components and function of each
- Describe the structure of bone
- List the three types of joints and give an example of each
- Use correct terminology to describe joint motions
- Demonstrate anatomic position
- List and define the planes of the body
- Use correct terminology when referring to radiographic positions and projections
- Given a position/projection description from one of the following chapters, select, mark, and place the cassette correctly
- Define common terms used to describe or classify disease processes
- Explain the differences between acute and chronic conditions and between benign and malignant conditions
- Define inflammation and describe its possible consequences

Key Terms

abrasion
acute
anatomic position
anomaly
articulation
atrophy
benign
cartilage
central nervous system (CNS)
chronic
congenital
contusion
degeneration
diagnosis
dislocation

edema
fracture
gastrointestinal (GI) tract
hormone
iatrogenic
idiopathic
infection
inflammation
ischemia
laceration
lesion
ligament
lymph
malignant
metastasis

microorganism
neoplasm
nosocomial
pathology
peripheral
prognosis
sign
sprain
strain
symptom
syndrome
tendon
trauma
ulcer
vascular insufficiency

This chapter introduces the subjects of anatomy, positioning, and pathology in a general way. At first it may seem strange to find these three different subjects in one place. The purpose is to provide concepts and terminology to prepare you for the more specific material in the remainder of this section of the text. Each part of the body will then be considered with respect to these three aspects.

Many of the concepts in this chapter will be familiar. It is largely the terminology and the organization of the concepts that may be new to you. The terms presented are essential vocabulary for radiographers. There are many new terms in this chapter, and not all of them could be included in the key terms list. In many cases there is more than one term with the same meaning. It would obviously be easier to use only one term, but this practice would handicap you when you encounter alternative terminology in examinations and when reading other resources. While the number of new terms in this chapter may seem intimidating, time taken to memorize them now will save time later and will greatly increase your ability to learn the details of anatomy, pathology, and radiography in the chapters that follow.

The term anatomy refers to the *structure* of the body. A comprehensive knowledge of the anatomy to be radiographed is essential for accurate radiographic positioning and for the correct evaluation of finished radiographs. Physiology, on the other hand, refers to the function of the body. While the primary emphasis in this text is on anatomy rather than physiology, function will be explained to a limited degree.

As preparation for the study of radiographic positioning, this chapter includes the terms used to describe body positions and radiographic projections. Also included are details of procedures that are connected with radiographic positioning and preparation for radiography, such as cassette selection, tube angulation, and alignment.

Pathology is defined as the study of abnormal conditions. This is a vast subject that can only be treated quite superficially here. A general knowledge of pathology will help you understand the reasons for many of the procedures you perform. This awareness will alert you to physical signs that may be helpful to the physician. Basic knowledge of disease processes will form a foundation on which to build a more comprehensive understanding as you gain experience. Your work will be more interesting and more rewarding when you can see how it fits into the overall picture of health care.

ANATOMY

There are six levels of structural organization of the human body (Fig. 12-1). The body, like all matter, is made up of atoms and molecules. This is referred to as the *chemical level* of organization. The next level is the *cellular level*. Cells are the smallest units of living things. Groups of similar cells that work together to perform a common function are called *tissues*. An *organ* is more complex than a tissue. It is a group of tissues that act together to perform a special function. A *system* is a group of organs that work together to perform complex functions. The sixth and highest level of structural organization is the body as a whole.

Cells

Cells are the smallest units of all living things. The human body is made up of many trillions of living cells. Cells are too small to be seen with the naked eye but can be examined using a microscope. There is great variation among cells with respect to size, shape, and function. For example, the ovum (female sex cell) is more than 100 times the size of a red blood cell. Some cells are flat, some are brick shaped, some are thread-like, some have irregular shapes, and some are capable of changing shape. Cells function as parts of tissues, so their function is explained under that heading.

The three main parts of a cell are the plasma membrane, the cytoplasm, and the nucleus. The plasma membrane encloses the cytoplasm and forms the outer boundary of the cell. The membrane also serves to provide communication between the cell and the rest of the body. Hormones or other chemical compounds can attach to the cell membrane and affect the activities within the cell. The cytoplasm is a highly specialized living material between the plasma membrane and the nucleus. The cytoplasm consists of fluid and a variety of tiny structures called *organelles* that perform the work of the cell. The nucleus of the cell contains the chromosomes, the hereditary structures that contain the "blueprint" for cell structure and function. They are made up of the complex protein, DNA.

Tissues

Four main types of tissues compose the body's many organs:

- Epithelial tissue
- Connective tissue
- Muscle tissue
- Nervous tissue

There are a number of types of epithelial tissue, but all perform the basic function of protecting underlying tissues. The skin is made up of epithelial tissue, as are the linings of the stomach and the air passages of the lungs. Some epithelial tissues also absorb

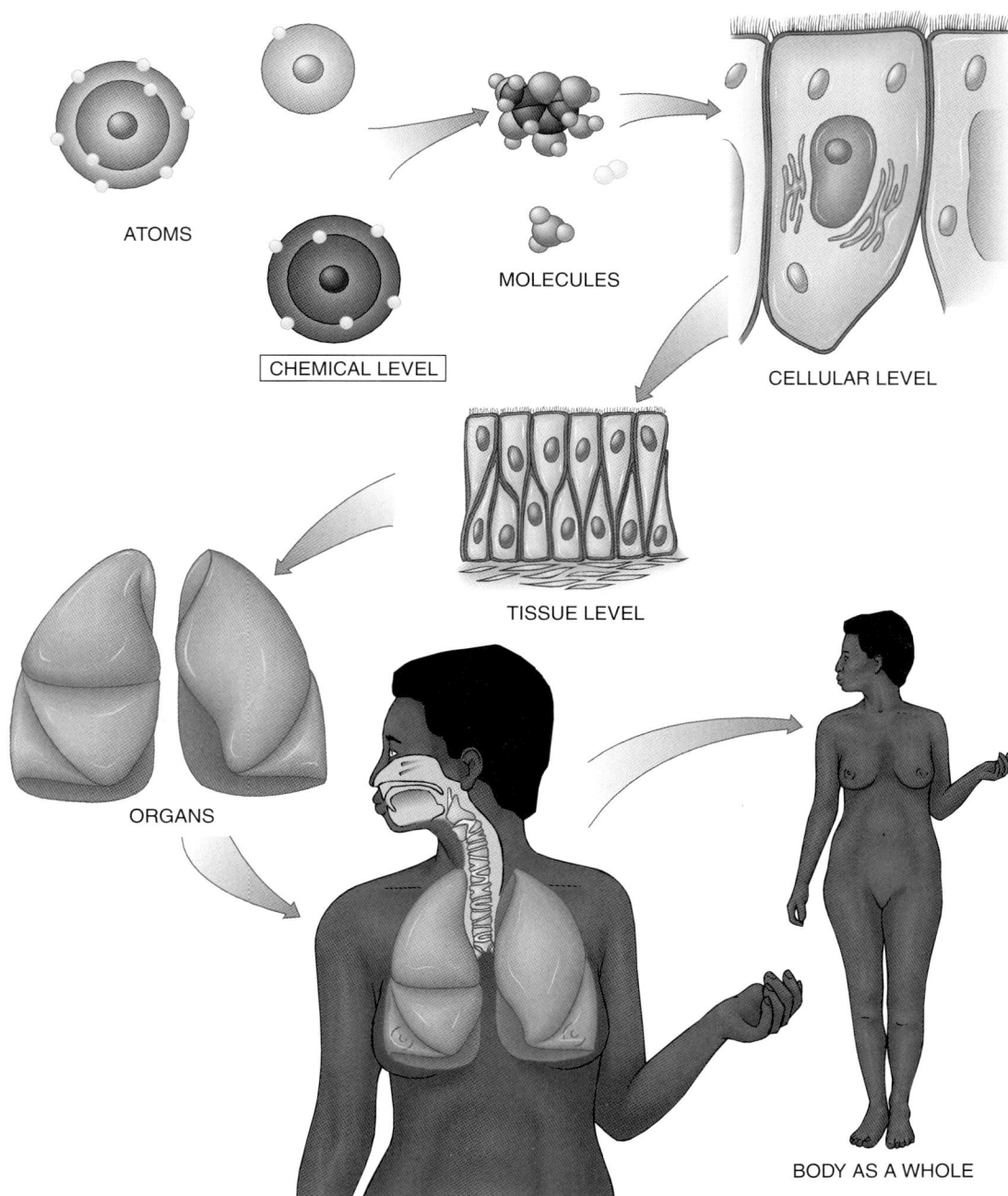

Fig. 12-1 Levels of structural organization.

and/or secrete substances. For example, the lining of the stomach absorbs nutrients from food and secretes chemicals that aid in digestion.

Connective tissue is the most widely distributed of all tissues and has the greatest variety of form and function. Connective tissue is found between tissues and between organs and serves to hold them together. It also makes up the structure of bone, cartilage, and adipose tissue (fat).

Muscle tissue is capable of stretching and contracting. Its function is to produce movement. Different types of muscle tissue serve to move the bones, cause the heart to beat, and provide the movements required by other body organs.

Nervous tissue consists of the nerve cells, called *neurons,* and support cells. Neurons conduct electrical impulses, providing rapid communication between body structures and control of body functions.

Body Systems

Organ systems are the largest and most complex structural units of the body. The 11 major organ systems that compose the human body are as follows:

- Integumentary
- Skeletal
- Muscular
- Nervous
- Endocrine
- Circulatory
- Lymphatic
- Respiratory
- Digestive
- Urinary
- Reproductive

This section of the text focuses heavily on the skeletal system because this system is the principal subject of x-ray examinations by limited radiographers. Section III also covers radiography of the chest, which contains the major portion of the respiratory system, and of the abdomen, which contains the major parts of the digestive and urinary systems.

Integumentary The integumentary system (Fig. 12-2) consists principally of the skin. It also includes the hair and the nails. The glands within the skin that secrete oil and sweat are parts of the integumentary system as well. Special microscopic organs in the skin sense contacts with the outside world, enabling the body to respond to pain, pressure, touch, and changes in temperature.

Muscular The muscular system (Fig. 12-3) consists of the voluntary muscles, which control the movements of the skeleton and are under conscious control, and the involuntary muscles, which function to produce the movements of organs. Muscles produce heat, maintaining a constant body temperature. Some specific muscles will be introduced in Chapter 24 when discussing intramuscular injections.

Nervous The nervous system (Fig. 12-4) consists of the brain, the spinal cord, and the nerves. The brain and the spinal cord are referred to as the **central nervous system (CNS).** The nerves that carry information between the central nervous system and all parts of the body are called the *peripheral nervous system.* Nervous system functions include communication, integration and control of body functions, and recognition of stimuli. Stimuli are perceptions or sensations that "stimulate," such as light, heat, pressure, and sound. The functions of the nervous system are accomplished by the transmission of tiny electrical impulses along the nerve pathways.

Endocrine The endocrine system (Fig. 12-5) is composed of glands that secrete special chemicals called *hormones.* Endocrine glands are sometimes referred to as "ductless glands" because, unlike most other glands, their secretions are released directly into the bloodstream. They serve some of the same functions as the nervous system in that they communicate, integrate, and regulate body functions. The hormones of the endocrine system provide slower, longer-lasting control of body function than the rapid electrical impulses of the nervous system.

Circulatory The circulatory system (Fig. 12-6) is also referred to as the *cardiovascular system.* It consists of the heart and the blood vessels. There are three types of blood vessels: *arteries,* which carry blood away from the heart; *veins,* which carry blood back to the heart; and *capillaries,* tiny vessels between the arteries and the veins that provide oxygen and nutrients to the cells. The heart provides a pumping action to keep blood flowing throughout the circulatory system. The major portions of the circulatory system will be addressed in more detail with the anatomy of the chest and abdomen in Chapter 16. Specific arteries and veins will be introduced when learning the procedures for taking a pulse in Chapter 22 and for drawing blood samples in Chapter 24.

Lymphatic The lymphatic system (Fig. 12-7) consists of the lymph nodes and lymph vessels, plus the spleen, tonsils, and thymus gland. The lymphatic system provides the fluid called *lymph* that surrounds the cells and serves to move fluid and certain large molecules from the cells to the circulatory system. The lymphatic system communicates with the circulatory system by means of the thoracic duct in the chest. An important function of the lymphatic system is its role in the immune system, which is discussed later in this chapter.

Fig. 12-2 Integumentary system.

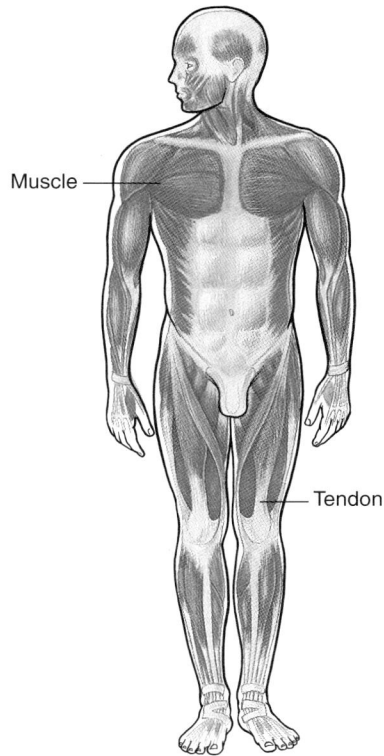

Fig. 12-3 Muscular system.

Fig. 12-4 Nervous system.

Fig. 12-5 Endocrine system.

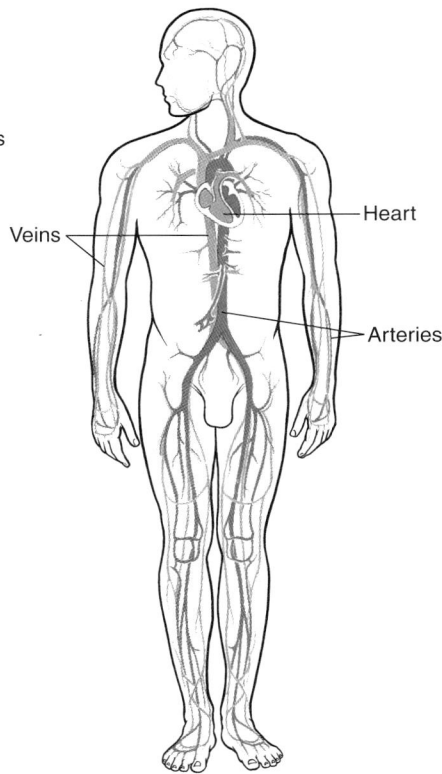

Fig. 12-6 Circulatory system.

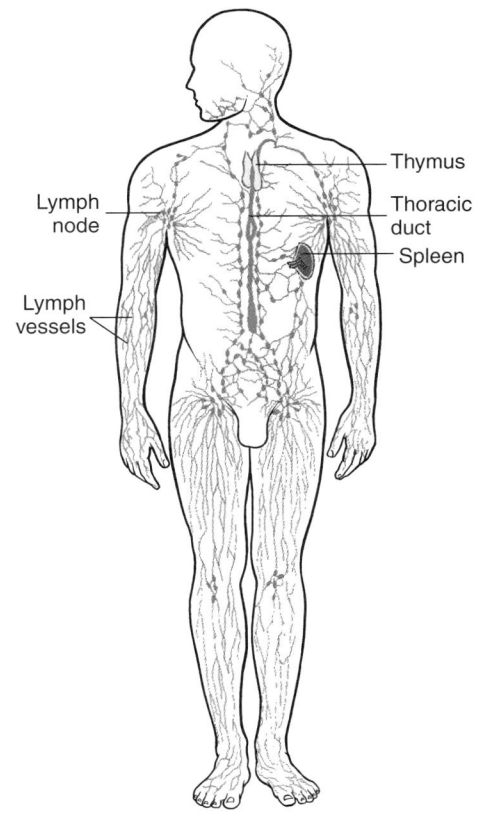

Fig. 12-7 Lymphatic system.

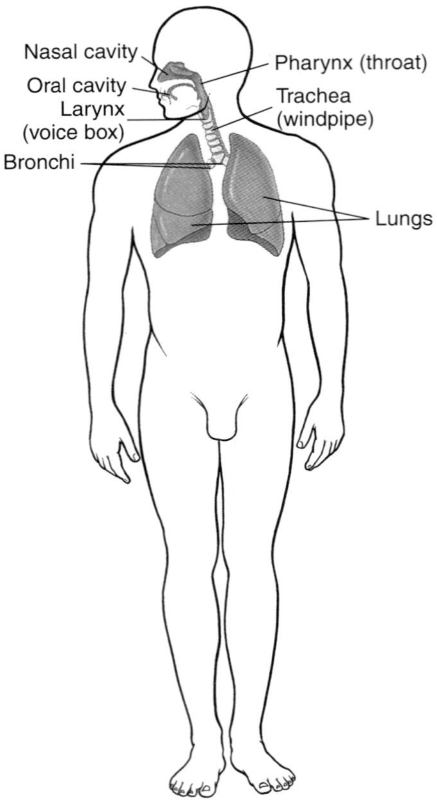

Fig. 12-8 Respiratory system.

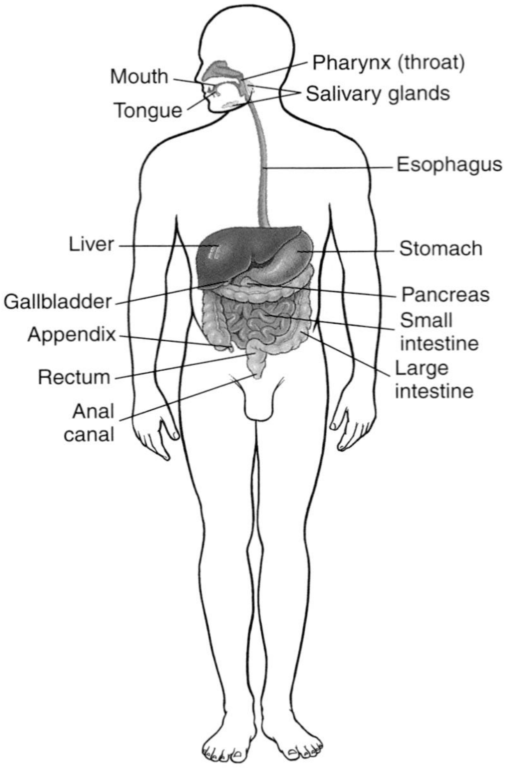

Fig. 12-9 Digestive system.

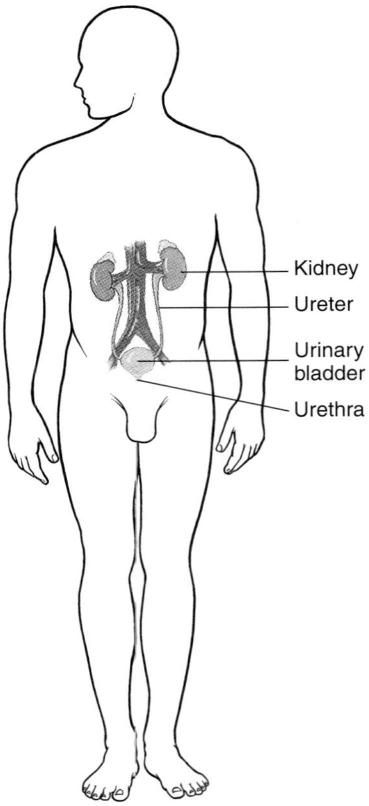

Fig. 12-10 Urinary system.

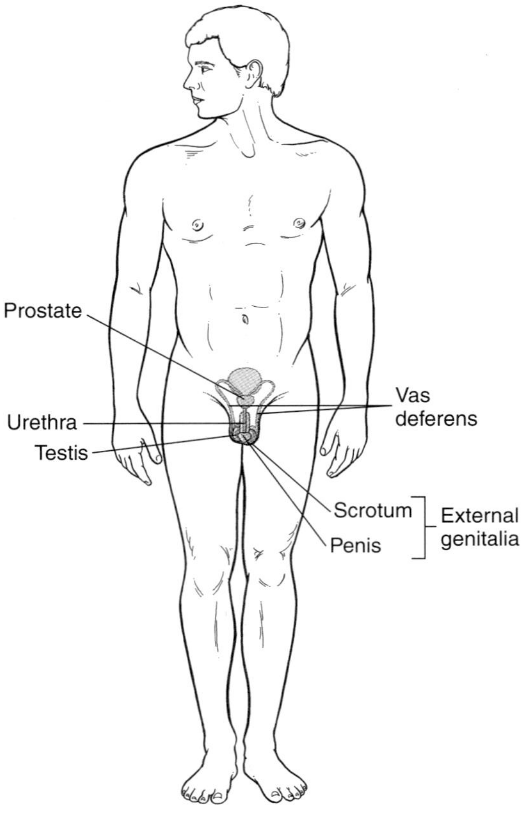

Fig. 12-11 Male reproductive system.

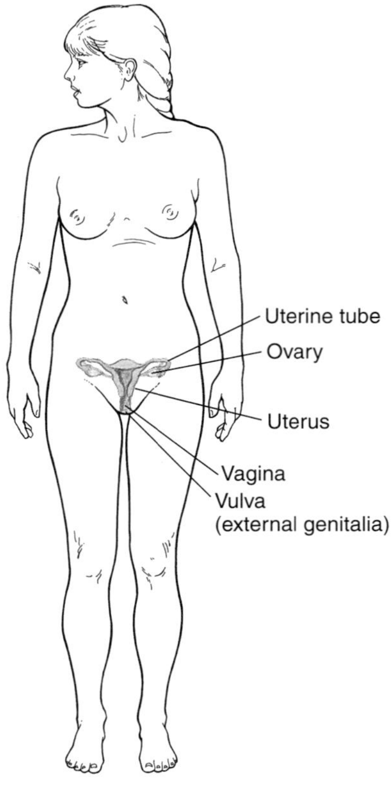

Fig. 12-12 Female reproductive system.

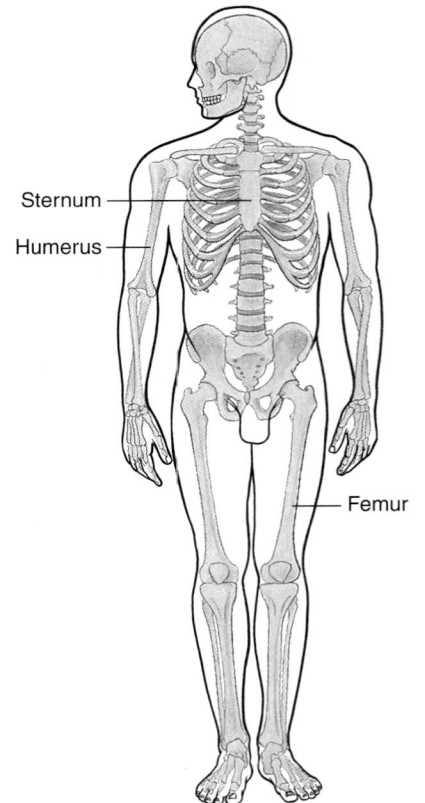

Fig. 12-13 Skeletal system.

Respiratory The respiratory system (Fig. 12-8) consists of the breathing passages of the body. Included are the nose and mouth, the *pharynx* (throat), the *trachea* (windpipe), the *bronchi* (singular, bronchus), and the lungs. The bronchi are branching tubes from the trachea to the tissues of the lungs. The small, peripheral branches of the bronchi are called the *bronchioles.* The bronchioles terminate in tiny sacs called *alveoli* that are surrounded by blood vessels. Oxygen from the air is transferred to the blood from the alveoli. Carbon dioxide, a gaseous waste produced when oxygen is utilized, is transferred from the blood to the alveoli so that it can be exhaled from the lungs. More detailed anatomy of the respiratory system is covered in Chapter 16.

Digestive The primary organs of the digestive system (Fig. 12-9) are the mouth, pharynx, esophagus, stomach, small intestine, large intestine, rectum, and anal canal. These organs constitute the **gastrointestinal (GI) tract,** also called the *alimentary canal,* a hollow tube that is open at both ends. The GI tract is the path of food from the time it enters the mouth until it is excreted as waste. It is lined with epithelial tissue called *mucous membrane.* The digestive system also includes accessory organs that aid in digestion: the tongue, teeth, salivary glands, liver, gallbladder, pancreas, and appendix. The major portion of the digestive system is contained within the abdominal cavity and is further discussed in Chapter 16.

Urinary The urinary system (Fig. 12-10) consists of the kidneys, ureters, bladder, and urethra. The function of the urinary system is to eliminate excess fluid and the waste products of cellular activity from the body. The fluid and chemical waste are removed from the blood by the kidneys, forming *urine.* The urine flows through long tubes called *ureters* and into the bladder, where it is stored. Urine is empties from the bladder through a tube called the *urethra.*

Reproductive The reproductive system, unlike other organ systems, does not function for the survival of the individual. Its purpose is the survival of the human race. Hormones produced by the reproductive organs promote the development of sexual characteristics. External organs or structures of the reproductive system are called *genitalia.* The organs that produce reproductive cells are called *gonads;* they were introduced in Chapter 11.

In the male reproductive system (Fig. 12-11) the external genitalia consist of the scrotum and the penis. The gonads are called testes (singular, testis). The testes are located within the scrotum. Sperm (male reproductive cells) from the testes travel through the vas deferens (a small, tubular structure) to the urethra, which opens to the outside. In the male, the urethra is a part of both the urinary system and the reproductive system. The prostate gland is an accessory organ of the male reproductive system. It surrounds the urethra at its junction with the bladder.

In the female reproductive system (Fig. 12-12) the genitalia consist only of the vulva, the soft tissues that surround the vaginal opening. The gonads are the ovaries, which are located within the pelvic portion of the abdomen. Fallopian tubes, which are also called uterine tubes, capture the ova when they are released by the ovaries and transmit them to the uterus. If the ovum is fertilized by a sperm, it attaches to the wall of the uterus where the new life is nourished and protected as it grows until it is mature enough to be born.

Immune The immune system is *not* one of the body's organ systems. It is a complex of organs and tissues of several body systems that protects the body from disease organisms and other foreign substances. The immune system includes organs and tissues of the bone marrow, the spleen, the thymus gland, and the lymphatic system. The immune system is responsible for the manifestation of inflammation, which is discussed later in this chapter.

Skeletal The skeletal system (Fig. 12-13) provides a rigid framework for the body. It consists of 206 bones with other associated tissues, such as cartilage and ligaments. **Cartilage** is a tough, fibrous connective tissue that is both stiff and flexible. The lay term for cartilage is "gristle." **Ligaments** are flexible bands of fibrous tissue that bind joints together and provide connections between bones and cartilage. **Tendons** are also bands of fibrous tissue; they attach muscles to bones.

The skeletal system may be divided into two basic parts: the axial skeleton and the appendicular skeleton. The axial skeleton consists of the skull, spine, sternum (breast bone), and ribs. The appendicular skeleton includes the bones of extremities (arms and legs) and those of the pelvis and shoulders.

TABLE 12-1

Bone Types

Bone Type	Description	Examples	Bone Type	Description	Examples
Long	Long shaft with thick cortex and medullary canal; the two ends form joints	Humerus (upper arm), femur (thigh bone)	Flat	Two layers of compact bone with a thin cancellous layer (diploë) between them	Cranium (outer skull), scapula (shoulder blade)
Short	Small bones made primarily of cancellous bone with a thin cortex	Bones of wrist and ankle	Irregular	Wide variety of shapes and structures	Vertebrae (spine), bones of face

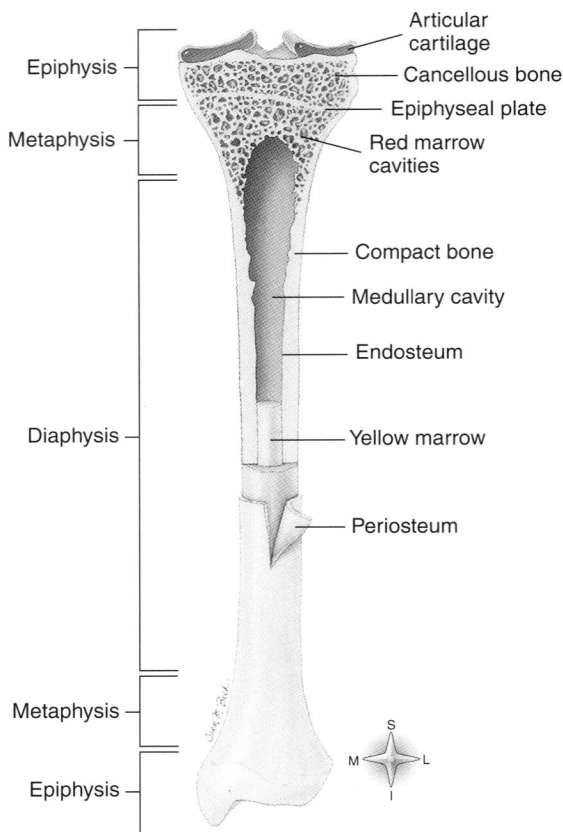

Articular cartilage
Cancellous bone
Epiphyseal plate
Red marrow cavities
Compact bone
Medullary cavity
Endosteum
Yellow marrow
Periosteum

Epiphysis
Metaphysis
Diaphysis
Metaphysis
Epiphysis

Fig. 12-14 Structure of long bones.

Structure of Bone

The structure of bone varies considerably, depending on the specific type of bone. The four basic types are listed and described in Table 12-1.

The outer portion of most bones is a layer of hard, compact bone called the *cortex*. Inside the cortex is bone tissue that has a "honeycomb," or *trabecular*, structure and is called *cancellous* bone. Long bones have a long cavity in the center called the *medullary canal*. The medullary canal and the spaces within cancellous bone contain *marrow*, a fatty substance containing blood vessels and immature blood cells. The surfaces of bones that form moving joints are covered with *joint cartilage*. All other bone surfaces are covered by a tough, fibrous membrane called the *periosteum*.

The long bones are all found in the extremities. They include the major bones of the arms and legs and those that make up the fingers and toes. The parts of a long bone are illustrated in Fig. 12-14. The long shaft of the bone is called the *diaphysis*. At each end of the diaphysis is a flared portion called the metaphysis. The rounded ends that form joints are called the *epiphyses (sing. epiphysis)*. Between the metaphysis and the epiphysis is the epiphyseal plate, or "growth plate." Early in life, this plate is made of cartilage and is the center for bone growth. When the bone is mature, the growth plate will ossify, that is, it will turn to bone. The ossified growth plate is often seen on radiographs and is referred to as the *epiphyseal line*.

The features of bone shapes may be characterized as either *projections*, processes that grow out from the bone surface, or *depressions*, which are indentations or hollows in the surface. Specific terms are used to describe these features according to their characteristics:

Projections

condyle   a rounded process that forms part of a joint
coracoid   a pointed projection
coronoid   a beak-like projection
crest   a bony ridge

epicondyle   a projection above a condyle
facet   a small, smooth process that forms part of a joint
head   the rounded, wide end of a long bone
malleolus   a club-shaped projection
process   a general term for a projection
protuberance   a general term for a projection
spine   a sharp process or a sharp ridge
styloid   a long, sharp process
trochanter   one of the large, rounded processes of the femur (thigh bone)
tubercle   a small, rounded process
tuberosity   a rounded process larger than a tubercle (the terms are sometimes used interchangeably)

Depressions

fissure   a linear depression, a groove
foramen (pl. foramina)   a hole in a bone for the passage of blood vessels and nerves
fossa (pl. fossae)   a pit or hollow
groove   a shallow linear depression
sinus   a cavity or hollow space
sulcus   a trench-like depression, a deep fissure

Joints

The places where bones are joined together are called joints or *articulations.* There are three classifications of joints, based on their ability to move:

- *Synarthrosis* refers to a joint that does not move. With the exception of the mandible (jaw bone), the joints of the skull are all synarthrodial joints and are called *sutures.*
- *Amphiarthrosis* refers to a joint that has very limited motion. The articular surfaces that form these joints are covered by fibrous cartilage or cushioned by disks of fibrous cartilage. The joints between the bodies of the spinal vertebrae and the sacroiliac joints (between the spine and the pelvis) are examples of amphiarthrodial joints.
- *Diarthrosis* refers to a joint that can move freely. The bones that form these joints are shaped to fit together to accomplish the required movement, and their articular surfaces are covered by articular cartilage. The joint is surrounded by a fibrous capsule that is lined with synovial membrane. This membrane secretes *synovial fluid,* providing moisture to lubricate the joint.

Some diarthrodial joints have sacs filled with synovial fluid. These are called *bursae (sing. bursa).* They serve to cushion the movements of tendons or muscles. Important bursae are located at the shoulder, elbow, hip, and knee.

Four types of movement are possible in diarthrodial joints:

- *Circumduction* is the ability to move in a circle. This motion is typical of the shoulder joint and the hip joint. These joints are often referred to as "ball-and-socket" joints because of the shapes of their articular surfaces.
- *Rotation* is the ability to turn on an axis. Rotation of the elbow joint allows turning the hand palm-up or palm-down. Rotation in the upper spine allows turning the head from side to side.
- *Angular motion* refers to the ability to move back and forth in one plane, that is, to "bend." Joints that move in this way are often called "hinge joints." The knee and the joints of the fingers move in this way.
- *Gliding motion* occurs when one bone slides over another. A good example of this type of motion is the movement of the kneecap when the knee joint is bent.

Some joints are capable of more than one type of movement. Combinations of joint movements provide a wide variety of possible body motions. In addition to the basic joint movements just defined, the following terms are used to describe the movements of joints:

- *abduct, abduction*   movement of an arm or leg away from the central part of the body
- *adduct, adduction*   movement of an arm or leg toward the central part of the body
- *evert, eversion*   to turn outward (typically used to describe ankle motion)
- *extend, extension*   straightening of a hinge joint; backward bending of the spine
- *flex, flexion*   bending of a hinge joint, decreasing the angle between the bones that make up the joint; forward bending of the spine; the opposite of extension
- *invert, inversion*   to turn inward (typically used to describe ankle motion); the opposite of eversion
- *pronate, pronation*   to turn the arm so that the palm of the hand faces backward
- *supinate, supination*   to turn the arm so that the palm of the hand faces forward; the opposite of pronation

RADIOGRAPHIC POSITIONING

The description of body positions requires an understanding of commonly used terms. Terms that indicate the surfaces, directions, and planes of various body locations are based upon *anatomic position.* In anatomic position, the subject is standing, facing the observer, with the palms of the hands forward (Fig. 12-15). The terms in the list that follows are used to accurately describe locations on and within the body:

- *anterior*   forward or front portion of the body or body part
- *caudal, caudad*   away from the head
- *central*   the middle area or main part of an organ or body part
- *cephalic, cephalad*   pertaining to the head; toward the head; the opposite of caudal
- *distal*   away from the source or point of origin. For example, the wrist is *distal* to the elbow, being farther from the point of origin of the arm, which is at the shoulder.

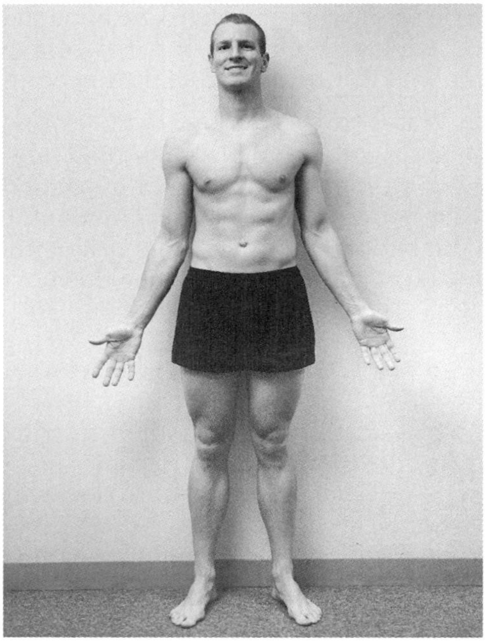

- *dorsal*   back part or surface of the body or part; the top surface of the foot or the back of the hand
- *external*   to the outside, at or near the surface of the body or a body part
- *inferior*   below, farther from the head
- *internal*   deep, near the center of the body or a part; the opposite of external
- *lateral*   referring to the side, away from the center to the left or right
- *medial, mesial*   toward the center of the body or the center of a part; the opposite of lateral
- *palmar*   referring to the palm (anterior surface) of the hand
- *parietal*   referring to the walls of a cavity
- *peripheral*   parts away from the central mass of an organ, toward its outer limits; the opposite of central
- *plantar*   referring to the sole of the foot
- *posterior*   backward or back portion of the body or body part, the opposite of anterior
- *proximal*   toward the source or point of origin, the opposite of distal
- *superior*   above, toward the head, opposite of inferior
- *ventral*   forward, front part; the opposite of dorsal
- *visceral*   pertaining to organs

Procedures for radiographic positioning are described using body planes (Fig. 12-16). The *sagittal plane* divides the body into right and left parts; the *midsagittal* or *median plane* divides the body into equal right and left parts. The *coronal plane* divides the body into anterior and posterior parts. The midcoronal or *midfrontal plane* divides the body into relatively equal parts; it passes through the external auditory meatus (the opening of the ear), the center of the shoulder, the greater trochanter (the bony promi-

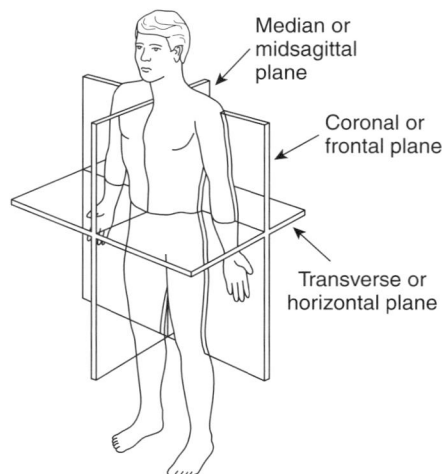

Fig. 12-15 Anatomic position.

Median or midsagittal plane

Coronal or frontal plane

Transverse or horizontal plane

Fig. 12-16 Sagittal, coronal, and transverse planes.

nence in the lateral hip area), and the lateral malleolus (the bony prominence on the lateral surface of the ankle). The *transverse* or *horizontal plane* divides the body into superior and inferior portions. It may be drawn at any level.

Body Positions

The following terms are used to describe body positions.

- *decubitus*   lying down
- *prone*   lying face down
- *recumbent*   lying down. The position may be further described by adding the name of the body surface on which the patient is lying:
    dorsal recumbent   lying on the back, supine
    lateral recumbent   lying on the side
    ventral recumbent   lying face down, prone
- *supine*   lying on the back
- *upright*   standing or seated

Projections

A radiographic **projection** indicates the relative positions of the patient, film, and x-ray tube for an exposure.

For a *frontal projection,* the coronal plane of the body or body part is parallel to the film plane and the central ray is perpendicular to both. If the patient is supine, or facing the x-ray tube, the projection is said to be *anteroposterior* or *AP* (Fig. 12-17). If the patient is prone, or facing the film, the projection is said to *posteroanterior* or *PA* (Fig. 12-18). These terms indicate the direction of the x-ray beam, from front to back or back to front.

Lateral projections are those in which the sagittal plane of the body or body part is parallel to the film. Lateral projections are always named for the side of the patient that is nearest the film (Fig. 12-19).

Oblique projections are those in which the body is rotated so that the projection is neither frontal nor lateral. Oblique projections also are named for the side of the body that is nearest the film. For example, in a right anterior oblique (RAO) projection, the patient's right anterior aspect is closest to the film. Fig. 12-20 illustrates all four oblique projections: RAO, right posterior oblique (RPO), left anterior oblique (LAO), and left posterior oblique (LPO).

Decubitus projections are those for which the patient is recumbent and the x-ray beam is parallel to the floor. Decubitus projections are named for the body surface on which the patient is lying. For example, if the patient is lying on the right side, the projection is called right lateral decubitus. A radiograph of this projection will be similar to a frontal projection. Dorsal or ventral decubitus projections will produce radiographs that are similar in appearance to lateral projections. Fig. 12-21 illustrates the four decubitus projections.

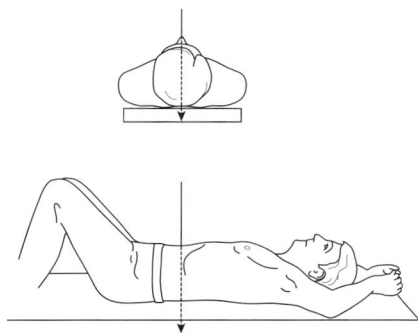

AP (anteroposterior)

Fig. 12-17 AP, anteroposterior projection.

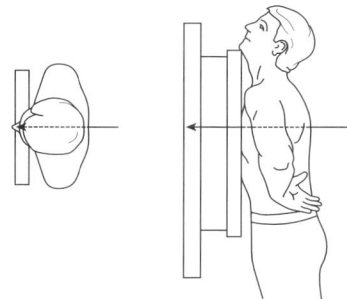

PA (posteroanterior)

Fig. 12-18 PA, posteroanterior projection.

Left lateral

Right lateral

Fig. 12-19 Lateral projection.

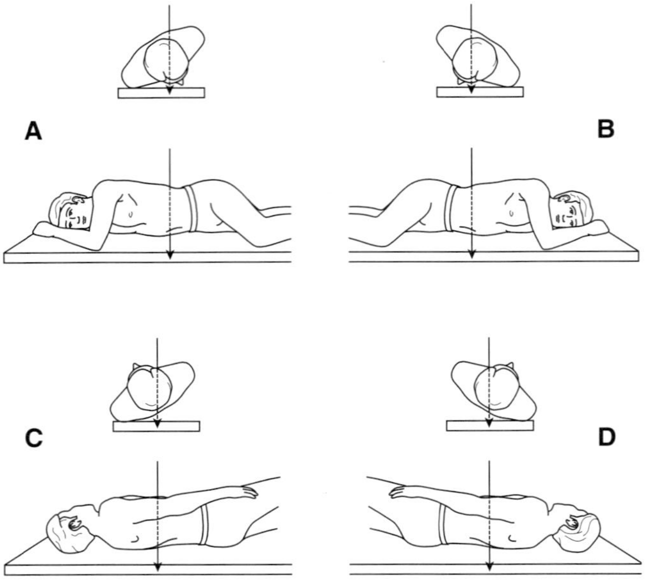

Fig. 12-20 Oblique projections. **A,** RAO, right anterior oblique. **B,** LAO, left anterior oblique. **C,** LPO, left posterior oblique. **D,** RPO, right posterior oblique.

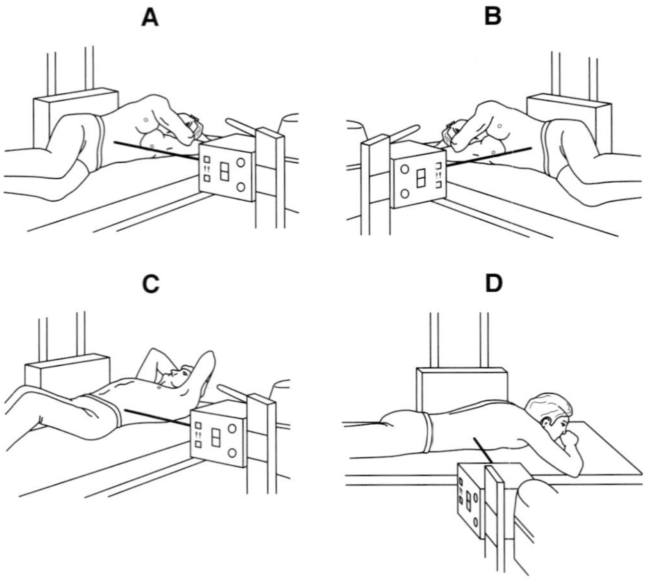

Fig. 12-21 Decubitus positions. Note horizontal orientation of the central ray. **A,** Left lateral decubitus position results in a frontal projection. **B,** Right lateral decubitus position results in a frontal projection. **C,** Dorsal decubitus position results in a lateral projection. **D,** Ventral decubitus position results in a lateral projection.

Fig. 12-22 Axial projections use angulation of the central ray along the long axis of the body or part. **A,** Cephalad angulation; **B,** caudad angulation.

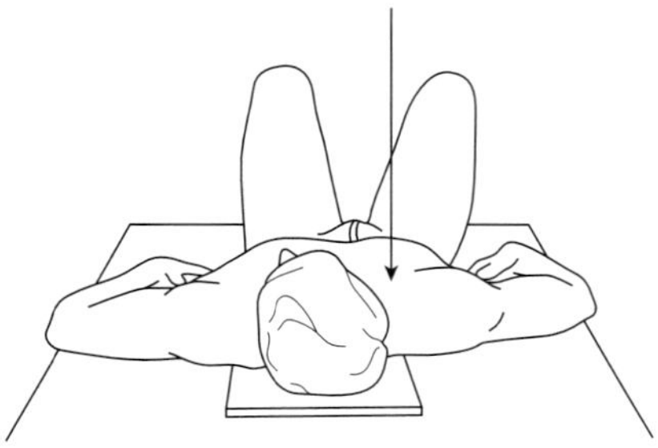

Fig. 12-23 Tangential projection. Central ray "skims" the profile of the subject.

Axial projections are radiographs taken with a longitudinal angulation of the x-ray beam. Sometimes they are referred to as semiaxial projections. The central ray is angled along the long axis of the body, either cephalad (toward the head) or caudad (away from the head) (Fig. 12-22).

Tangential projections are taken in such a way that the central ray tends to "skim" the profile of the subject (Fig. 12-23). Such a projection might be helpful, for example, to demonstrate the depth of the hollow area of a depressed skull fracture.

RADIOGRAPHIC PROCEDURES

Cassette Selection

The first step in the technical aspects of radiographic procedure is cassette selection. The radiographer will choose a cassette of the correct size and screen type.

Fig. 12-24 Lengthwise cassette placement. Long dimension of cassette is parallel to axis of the body.

Fig. 12-25 Crosswise cassette placement. Long dimension of cassette is perpendicular to axis of the body.

The screen type should be stated on the technique chart, and a suggested screen type is indicated in the position/projection descriptions that are provided in the following chapters. The selected cassette should be somewhat larger than the body part to be radiographed. You will recall from Chapter 6 that all radiographic images are magnified to some degree. A cassette that is the exact size of the body part will not be large enough to contain the entire image. Since the size necessary for each examination on an adult has been well established by the experience of others, suggested cassette sizes are provided with the position/projection descriptions in this text.

Cassette Placement

Most cassettes are not square. They have a long dimension and a short one. The correct placement of the cassette requires alignment of its long dimension parallel to the greatest dimension of the body part. The terms *lengthwise* and *crosswise* are commonly used to indicate cassette placement.

Lengthwise placement means that the long dimension of the cassette is aligned with the long axis of the body. This is sometimes called "portrait style." Lengthwise placement is appropriate for examinations of the spine (Fig. 12-24), for instance, because the spine is a long, narrow, vertical structure that is parallel to the long axis of the body.

Crosswise placement means that the long dimension of the cassette is perpendicular to the long axis of the body. This is sometimes called "landscape style." It is used for structures whose longest dimension is at right angles to the long axis of the body.

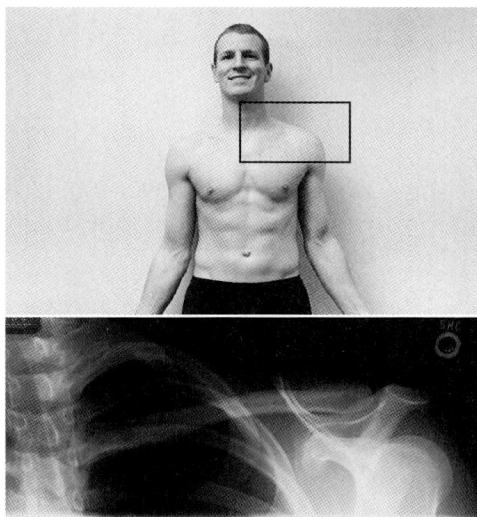

Crosswise placement would be appropriate for examinations of the clavicle (collar bone), for example, because its length is perpendicular to the long axis of the body (Fig. 12-25).

When a field length greater than the greatest cassette dimension is needed for tabletop (nongrid) studies, the body part is sometimes aligned to the cassette diagonally. This may be a useful approach for long bone examinations, such as the forearm or lower leg, on some patients. When diagonal placement is used, position the cassette so that the ID blocker is in a corner of the film that is not being used. Collimate the field to the body part rather than to the film.

Dividing Cassettes for Multiple Exposures

When extremity examinations are done tabletop, it is often possible to use a single film for more than one exposure. For example, two or more projections of the same body part may be radiographed side by side (Fig. 12-26). This practice is both convenient and economical.

The unused portion of the cassette is covered with a lead mask to protect it from scatter radiation exposure, and the remaining portion is used as if it were a small cassette. After the first exposure, the exposed portion is covered with a second lead mask and the original lead mask is removed to provide an unexposed area for the next exposure (Fig. 12-27). Some radiographers prefer not to use lead masks, using only collimation to limit exposure to a specific portion of the film. The use of lead masks is preferred because it avoids the possibility of fogging or overlapping exposures.

Some general rules apply when you are using a single cassette for more than one exposure. All exposures should be of the same body part. (An exception to this rule is sometimes made when similar projections of right and left extremities are radiographed on the same film for comparison.) All projections should be aligned in the same general direction, so that all of the anatomy is viewed "right side up" on the radiograph at the same time. The visual quality of the image is improved when care is taken to align the joints similarly, as seen in Fig. 12-26.

Alignment of Tube, Part, and Film

Radiographic procedure requires precise alignment of both the film and the x-ray tube to the body part. There is more than one possible sequence for accomplishing this alignment.

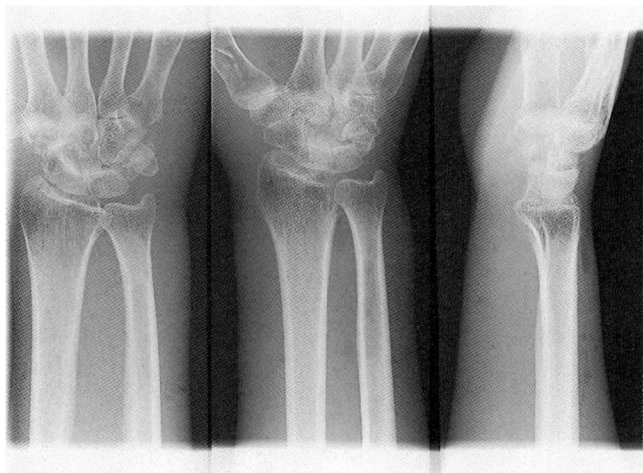

Fig. 12-26 Three separate exposures on one film.

For nongrid exposures, such as the extremities, the cassette is placed on the tabletop in a convenient location for the patient. The center of the body part is then placed in the center of the film. The x-ray tube is adjusted to the correct SID and the central ray is aligned to the center of both the body part and the film. The tube may be centered by aligning its crosshairs to the center marks on the cassette margins, or the center of the field may be aligned to the center of the body part. The result should be the same. If the tube does not align accurately to both the body part and the film, it is because the part was not aligned correctly to the film at the beginning.

When a bucky or grid cabinet is used, the cassette is placed in the bucky tray and latched firmly into the center of the tray.

When an upright bucky or grid cabinet or a radiographic table with a stationary tabletop is used, the body part is adjusted to the right or left so that it is aligned to the center of the long axis of the grid. The tube must also be aligned to the center of the grid and adjusted to the correct distance. For longitudinal alignment, the tube may be aligned to the center of the body part. The cassette tray is then aligned to the tube, using the centering light from the collimator. On the other hand, it is sometimes more convenient to align the bucky tray to the patient and then center the tube to the tray. This is often the case when alignment depends on the location of an outer portion of the body part rather than its center.

When the table bucky with a floating tabletop is used, the tube must be locked into the cross-table detent so that it is aligned to the longitudinal center of the grid. The bucky tray is placed in a convenient location along the length of the table and the x-ray tube is aligned to it at the correct SID. The tabletop is then released and moved so that the center of the body part is in the center of the radiation field.

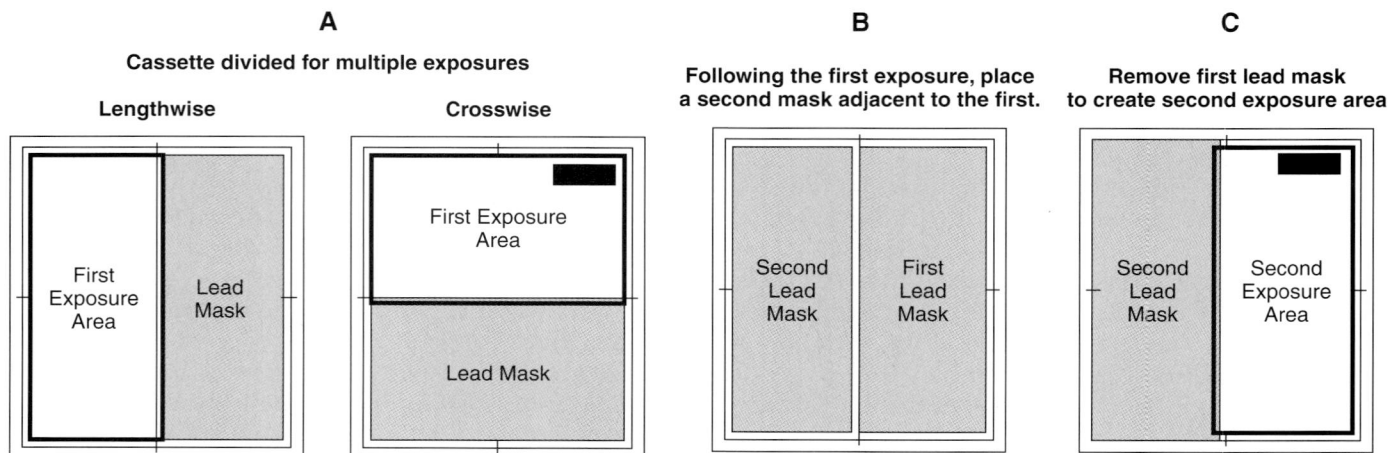

A

Cassette divided for multiple exposures

Lengthwise

First Exposure Area Lead Mask

Crosswise

First Exposure Area

Lead Mask

B

Following the first exposure, place a second mask adjacent to the first.

Second Lead Mask   First Lead Mask

C

Remove first lead mask to create second exposure area

Second Lead Mask   Second Exposure Area

Fig. 12-27 Multiple exposures on a cassette.

For some procedures, the x-ray tube must be angled. This means that it is not aligned with the central ray perpendicular to the film. When this is the case, the distance is adjusted first. The tube is then angled to the correct degree, and finally the central ray is aligned to the body part and/or the film. *When the tube is angled, the tube, part, and film must be aligned along the path of the central ray* (Fig. 12-28).

When the tube is angled, the actual SID is greater than the perpendicular distance from tube to film, so it may be necessary to adjust the perpendicular distance from the film so that the actual SID is correct. A handy rule for this adjustment is to decrease the perpendicular distance by 1 inch for every 5° of tube angulation. For example, if the tube is angled 30°, the perpendicular distance is reduced by 6 inches (30° ÷ 5° = 6 inches). The perpendicular distance in this case would be 40 inches minus 6 inches, or 34 inches

Fig. 12-28 When angling tube, take care that both body part and cassette are aligned to path of central ray.

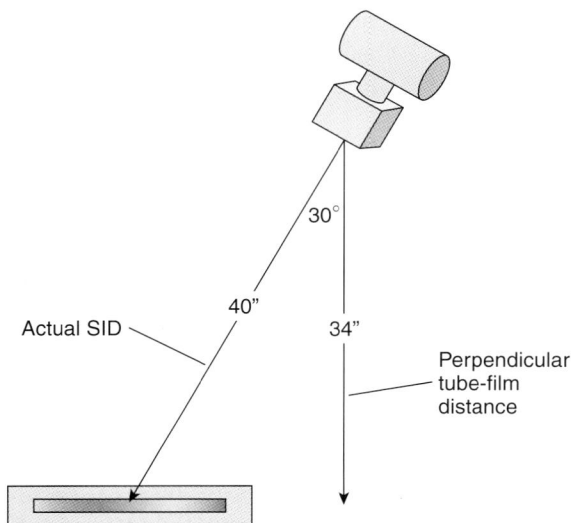

Fig. 12-29 When angling tube, maintain SID at 40 inches by reducing perpendicular distance between tube and film by 1 inch for every 5° of tube angulation.

(Fig. 12-29). If you are using an automatic or semi-automatic collimator, it may not be possible to adjust the SID for angles less than 15°. These collimators permit exposures only when the perpendicular distance is set at a standard location where there is a microswitch for the system. When tube angulation is greater than 15°, these collimators change to a manual mode of operation, allowing a change in distance.

Film Markers

Lead markers are placed on the cassette, cassette holder, or tabletop during exposures so that their image becomes a part of the radiographic image. They are usually held in place with tape. Some facilities have cassettes and markers with Velcro attachments.

A side marker is an indicator of the patient's right or left side on a radiograph. Side markers are pieces of lead in the shapes of the letters R and L. Some side markers contain the entire words, *right* and *left.*

The general rules for marker placement are:

1. Every radiograph must be marked with the correct side marker.
2. Place markers within the cassette margins and within the radiation field.
3. Markers must be placed so as not to interfere with the visualization of structures of clinical interest.
4. A right marker must *never* be placed to the left of the body's midline and vice versa.
5. While an extremity radiograph may be marked at any location to indicate whether it is the left or right extremity, shoulders and hips must have the marker placed on the *lateral* film margin. (Markers placed near the body's midline are meaningless.)
6. Markers should appear right side up when the radiograph is seen on the viewbox in normal viewing position.
7. When multiple exposures of the same body part are made on a single film, only one of the exposures is required to include a side marker.

Some side markers incorporate a plastic bubble that contains a droplet of mercury. The purpose of this feature is to indicate the position of the cassette during the exposure. If the cassette is horizontal, the mercury droplet will appear in the center of the bubble on the radiograph. If the cassette is upright or on edge, the mercury will be seen in the portion of the bubble that was nearest the floor. It is important for the physician who interprets the radiograph to know the position of the cassette during the exposure. If the side markers used in your facility do not have mercury indicators, it is a good practice to add an additional marker when making radiographs in upright or decubitus positions. Some x-ray departments have special lead markers that say *upright* and *decubitus.* Some facilities use a lead arrow for this purpose. When an arrow is used, it should be placed so that it is pointing toward the ceiling during the exposure.

Some facilities provide each radiographer with a set of side markers or require radiographers to provide their own. Since markers are easily lost or misplaced, this is one way of holding radiographers responsible for the care of markers. "Personal" side markers may also include the radiographer's initials or an identifying number, so that it is possible to determine by looking at the radiograph who took the film. Another means of identifying the radiographer who takes each film is for the radiographer to initial the film identification card.

Your facility may have other lead markers as well, depending on the procedures performed and the preferences of the physician who reads the radiographs. The radiographer is responsible for knowing the accepted procedure for marker use in the workplace and marking films accordingly.

Breathing Instructions

Normal patient breathing results in motion that may cause blurring of the radiographic image. For this reason, it is usual to instruct the patient to suspend breathing for examinations of structures in or near the trunk, including studies involving the head or shoulder.

For examinations of the trunk, the respiratory phase is also important. The positioning instructions in this text include patient breathing instructions for each examination where breathing instructions are needed. Since inspiration (inhalation) expands the lungs and lowers the diaphragm, the patient is usually instructed to hold a deep breath for examinations of the chest and for structures lying in the thorax above the diaphragm, such as the upper ribs and sternum (breastbone). When maximum lung expansion is essential, as in chest radiography, it is wise to instruct the patient to take two deep breaths and to hold on the second inspiration. This practice usually results in greater lung expansion and also provides a warning to the radiographer if deep inspiration is likely to cause a cough.

Elevation of the diaphragm is important for radiography of the supine abdomen, the lower ribs, and the lumbar spine. Expiration (exhalation) raises the diaphragm, avoiding superimposition of lung tissue over other structures of interest. When a patient is instructed to hold the breath on expiration, it is important to first tell the patient to take a deep breath. This ensures that the patient has taken in ample oxygen to hold the breath on expiration for the duration of the exposure. In this case, the instruction is, "Take a deep breath. Now blow your breath all out and hold it out."

It is important to have the patient's attention when giving breathing instructions. Give the instructions clearly, one at a time, and allow time for compliance with each instruction before giving the next one and before making the exposure. This practice ensures that you do not rush the patient and accidentally expose the film before the patient has had time to comply with the instruction. When patient hearing or comprehension may be a problem, it is wise to explain the breathing instructions in advance, to allow for demonstration or practice before proceeding to the exposure.

"Breathing technique" is a method that uses breathing motion to blur the rib and lung structures when they might otherwise interfere with the visualization of structures of interest. Breathing technique is used to enhance visualization of the thoracic spine in the lateral projection, the sternum in the oblique projection, and the transthoracic view of the upper humerus (upper arm bone) in cases of shoulder trauma. The patient is instructed to breathe in and out during the exposure while holding the body as still as possible. An exposure time of 1 to 2 seconds is needed for the breathing technique to be fully effective. This is achieved by using a low mA setting to obtain the desired mAs. Computerized controls that do not permit direct setting of the mA will usually allow lengthening of the exposure time, adjusting the mA automatically to provide the proper exposure.

PATHOLOGY

The term "pathology" refers to the study of disease processes. Disease may be defined as any abnormal change in the structure or function of the body. A localized area of destructive change in body tissue is called a **lesion.** Wounds, rashes, and tumors are all examples of lesions.

Disease Identification

Diagnosis is the process of identifying a disease. Diseases are identified by means of their manifestations, that is, the changes in the patient that are caused by the disease. There are three types of manifestations: symptoms, signs, and syndromes. **Symptoms** refer to the patient's reported perception of the condition. Symptoms are *subjective,* such as pain or dizziness, and can only be identified by the patient. **Signs,** on the other hand, are objective manifestations that can be observed by the examiner. Swelling, fever, and discoloration of the skin are examples of signs. A **syndrome** is a group of manifestations that, taken together, are typical of a specific condition. Marfan's syndrome, for example, is a disease characterized by elongation of the bones, weakness of the ligaments, and changes in the circulatory system.

The process of making a diagnosis begins with taking a history (Fig. 12-30), making a record of the patient's symptoms and other information about the patient's life and health, past and present, that may be relevant. The next step is physical examination (Fig. 12-31). This process may involve observing, palpating (feeling), listening with a stethoscope, measuring temperature, pulse, respirations, and blood pressure, and other specific procedures. If the history and physical examination do not provide a definitive diagnosis, the next step is the ordering of additional tests. A great variety of diagnostic tests is available to assist the physician in confirming or ruling out possible diagnoses. These tests may include laboratory analysis of blood or other body fluids and radiography or other imaging studies.

Fig. 12-30 Taking a medical history.

When the diagnosis has been made, the physician makes a treatment plan. The information gathered is also used to determine the *prognosis,* a prediction of the course of the disease and the prospects for the patient's recovery.

Disease Classification

Diseases may be classified in a number of different ways. Many classifications divide diseases into two groups, depending on whether or not they fit a certain criterion. The structure/function classification, discussed next, has to do with whether the disease has any structural manifestations. Diseases may be labeled either acute or chronic, depending on the course of the disease process. Contagious and noncontagious classifications are determined by whether or not the disease may be transmitted from one person to another. Other categories are based on the cause of the disease or on the nature of its manifestations. Depending on the classification system used, one disease may logically fall into several categories.

Structural vs. Functional Disease An organic disease involves changes to the cells of the body and may be referred to as a "structural disease." A functional disease, on the other hand, refers to an abnormal change in function with no structural change. A typical example of a functional disease is a migraine headache. Because radiographers are concerned with the identification of conditions that produce physical changes, the discussion of pathology in this text is limited to organic conditions.

Hereditary and Congenital Disease Hereditary diseases are those that are caused by abnormalities in the genetic make-up of the individual and are inher-

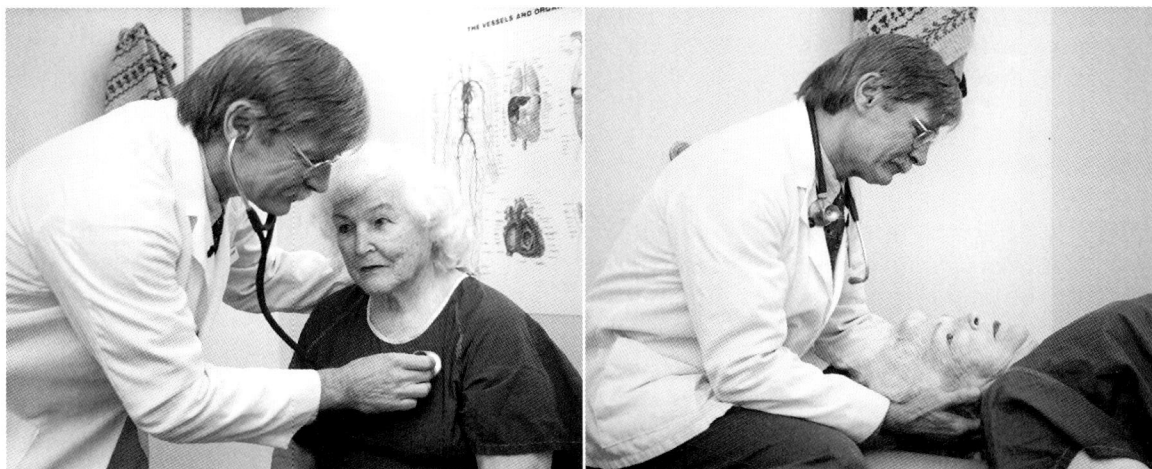

Fig. 12-31 Physical examination.

ited from a parent. Diseases that are not hereditary are classified as *acquired* diseases.

Conditions that are present at birth are said to be **congenital.** Congenital diseases are usually hereditary but may also be caused by events that occur prior to birth. Fetal alcohol syndrome is an example of an acquired congenital disease.

An example of hereditary congenital disease is trisomy 21, also known as Down syndrome, a condition marked by mental retardation and physical defects. Some hereditary diseases are not manifested until later in life. Diabetes mellitus is an example of a disease that may be inherited but is not present at birth. Some hereditary disorders are more accurately classified as "tendencies." Both hereditary and environmental factors may affect the manifestation of these diseases. Heart disease may fall into this category for some individuals. For example, an inherited tendency to heart disease is more likely to produce serious illness in an obese, sedentary individual with uncontrolled high blood pressure than in one who has normal weight and blood pressure and is physically fit.

Congenital conditions that cause abnormal variations in the shape or form of a body part are called **anomalies.** Anomalies in the circulatory system are not uncommon. Extra ribs, extra toes, or extra spinal vertebrae are anomalous structures.

Acute vs. Chronic Disease **Acute** conditions are characterized by a sudden onset. They are relatively severe and have a short duration. **Chronic** conditions are of long duration. They may come and go, in which case they are described as *recurrent,* or they may exist constantly at a low level and tend to "flare up" occasionally. Some diseases typically begin with an acute phase that resolves into a chronic condition.

Classification by Cause Diseases may be classified as *endogenous,* when the cause is an internal one, or *exogenous,* when caused by an external agent. Internal causes of disease may be further classified as circulatory, deficiency, or autoimmune.

The most common endogenous diseases are related to problems in the circulatory system. Problems in arteries may result in loss of blood supply to an organ or tissue caused by restriction or obstruction of the blood supply or by bleeding. Lack of adequate blood <u>flow</u> is called *vascular insufficiency.* This term describes the condition of the circulatory system itself. Lack of adequate blood <u>supply</u> is called *ischemia.* This term describes the condition of the tissues that are deprived of blood supply. Ischemia may cause the death of tissues, which is called *necrosis.* An area of tissue that has undergone necrosis is called an infarct. Strokes and heart attacks are typical diseases associated with vascular insufficiency.

Deficiency is any condition that compromises body function as a result of lack of required substances such as vitamins, minerals, or proteins. Scurvy, pellagra, and rickets are vitamin-deficiency diseases. Goiter (thyroid gland enlargement) may be caused by iodine deficiency. Some deficiencies compromise the function of the immune system. When immunity is low, resistance to infection is decreased and the patient is vulnerable to infectious disease.

Autoimmune conditions are a group of diseases that occur when the immune system attacks the body's own tissues. The cause is not known but may be related to previous viral infection. Rheumatoid arthritis (Fig. 12-32), lupus erythematosus, and ankylosing spondylitis are examples of autoimmune conditions, also called "collagen diseases." Ankylosing spondylitis is not common but may be seen by the radiographer because it causes fusion of the joints, particularly the joints of the spine. In extreme cases, the entire spinal column fuses together, creating the radiologic appearance referred to as "bamboo spine" (Fig. 12-33).

External causes of disease may be further classified as *physical, chemical,* or *microbiologic.*

Physical injury caused by an object is called *trauma.* Physical injury may also occur as a result of exposure to extreme temperatures, electricity, and radiation. Radiography is often used to evaluate the extent of trauma, and various manifestations of trauma will be addressed throughout this section of the text.

Trauma to the skeletal system may result in fracture, dislocation, or sprain. A **fracture** is a bone injury in which the tissue of the bone is broken. Fractures may be classified as simple or compound. A compound fracture is one in which the broken end of the bone penetrates the skin. If this does not occur, the fracture is said to be simple. Fig. 12-34 illustrates the basic types of simple fractures.

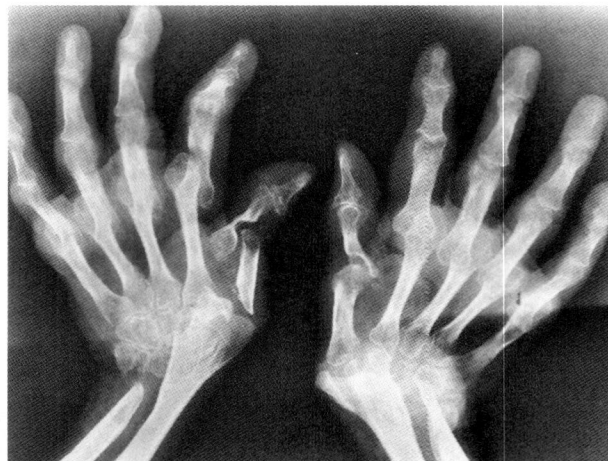

Fig. 12-32 Rheumatoid arthritis is a crippling autoimmune disease that affects joints.

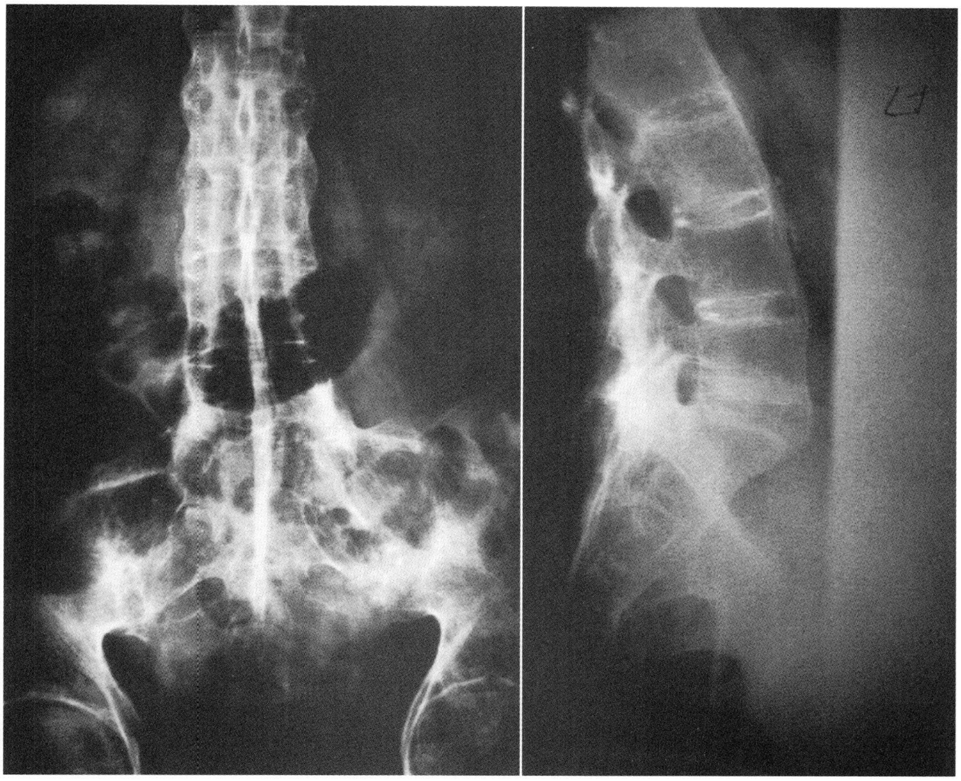

Fig. 12-33 Ankylosing spondylitis causes fusion of the spine, creating radiographic appearance known as "bamboo spine."

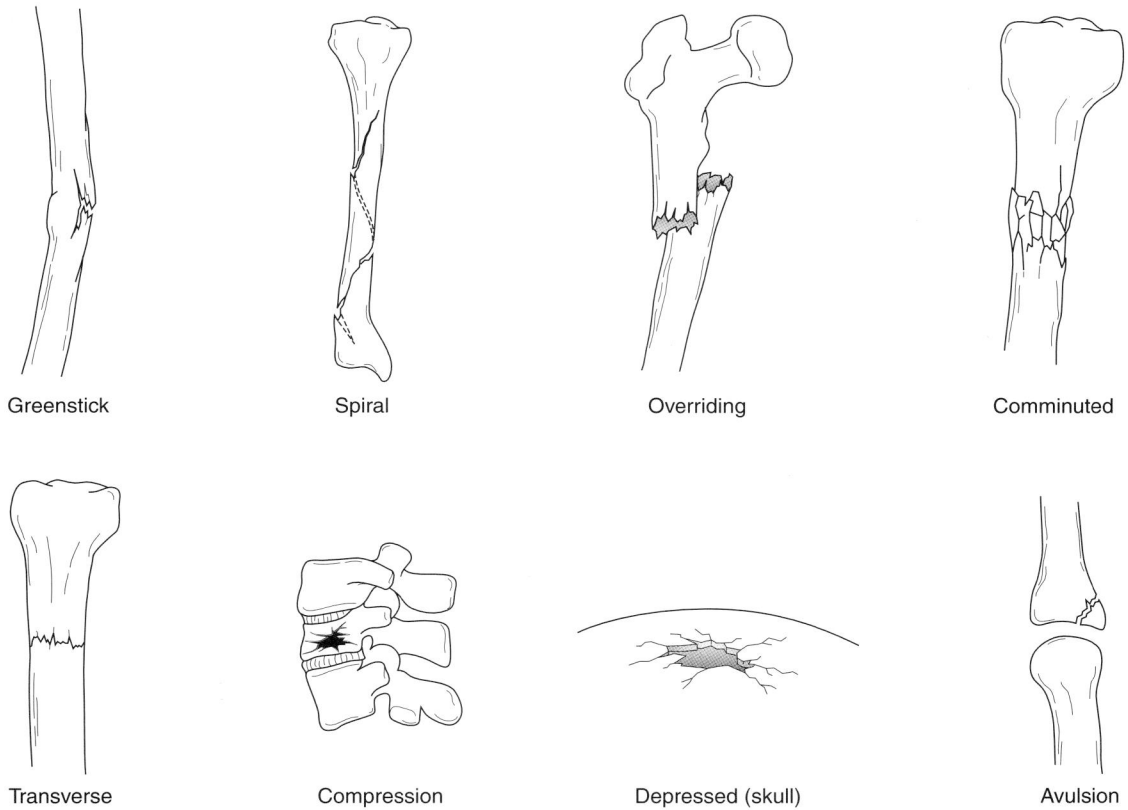

| Greenstick | Spiral | Overriding | Comminuted |

| Transverse | Compression | Depressed (skull) | Avulsion |

Fig. 12-34 Fracture types.

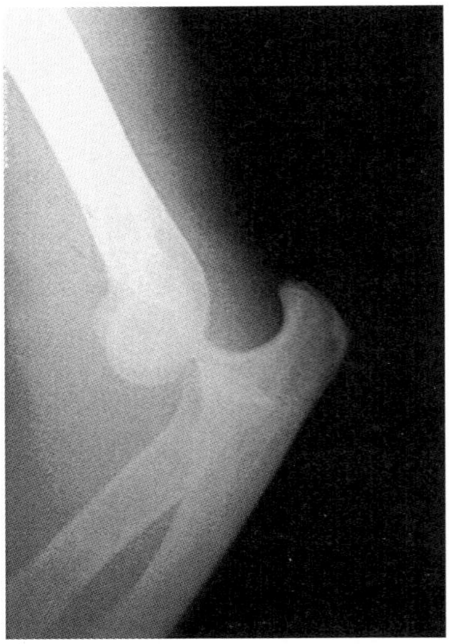

Fig. 12-35 Elbow joint dislocation.

Dislocation is the movement of a bone from its normal location within a joint (Fig. 12-35). A *sprain* is injury to the ligaments, tendons, and muscles that surround a joint. Muscle damage from excessive physical effort or force is called a *strain.* Other types of soft tissue trauma include *lacerations,* cuts or tears through the skin and underlying tissues; *abrasions,* scraping wounds to the skin; and *contusions,* closed wounds that cause bleeding under the skin and are commonly called bruises.

The principal chemical causes of disease are poisons and drug reactions. Inhalation of toxic fumes or the absorption of toxic substances through the skin may also cause chemical injury to the body.

Exogenous diseases caused by microbiologic agents are called **infections.** Infections may occur in almost any part of the body but are most likely to affect structures that are in some way in contact with the outside world. The most common infections are those that occur in wounds and those that involve the respiratory system. Microbiologic agents are living organisms too small to be seen with the naked eye and are termed *microorganisms.* Many of these agents cause contagious diseases. Microorganisms are discussed in Chapter 21, which deals with infection control.

A crater-like sore on the skin or on mucous membrane is called an *ulcer.* There are many possible causes. Ulcers in the lining of the stomach, for example, may be caused by chemical injury from the acids produced by the body to aid in digestion.

When the cause of a disease is unknown, it is said to be *idiopathic.* Diseases that occur as the result of treatment by health professionals are termed *iatrogenic.* The term *nosocomial* refers to diseases that are acquired in hospitals.

Classification by Disease Process Diseases are often classified according to the nature of the disease process itself. The disease process may involve inflammation, degeneration, or alterations in tissue growth.

Inflammation is the immune system's response to cellular injury. It is the initial part of the healing process. Inflammation is characterized by swelling, reddening, heat at the site, and pain. During this process, the blood supply to the injured area is increased. The term for swelling caused by vascular congestion is **edema.** White blood cells attack microorganisms and clean up dead tissue and other debris of injury. The terms used to define inflammatory conditions all end in "itis." For example, *arthritis* refers to inflammation of a joint and *sinusitis* is the term for inflammation of the sinuses.

Chronic inflammation causes **degeneration,** further injury to cells and tissues. Degeneration may lead to **atrophy,** a decrease in the size or number of cells. Atrophy causes tissue to waste away and also causes impairment or loss of function. There are four types of atrophy: senile, disuse, pressure, and endocrine. Senile atrophy occurs as a result of age. Decreased muscle mass and strength in the elderly are manifestations of senile atrophy. Disuse atrophy occurs to any body part that is not used. For example, when a cast is used to treat a fracture of an extremity, the muscles within the cast tend to shrink during the period that the cast is in place. An example of pressure atrophy is the deterioration of skin and underlying tissues that may occur in bedridden patients, causing bedsores, which are a form of ulcer. Endocrine atrophy is caused by a decrease in the supply of hormones. An example is the shrinkage of the ovaries and the uterus after menopause.

Hyperplasia and *hypertrophy* are both increases in the size of tissues or organs. Hyperplasia is defined as an increase in the number of cells, whereas hypertrophy refers to an increase in the size of the cells. These two conditions may exist together, and the two terms are often used interchangeably. These alterations in cell growth may be caused by inflammation or by an excess or deficiency of endocrine production.

Neoplasms are growths or tumors. They develop when changes in cells cause failure of the mechanisms that normally control cell growth. The names of neoplasms usually end with "oma." Some examples include *carcinoma, sarcoma, melanoma, lipoma, lymphoma,* and *adenoma.* Most neoplasms can be classified as either benign or malignant. **Benign** neoplasms are single masses of cells that remain at one location and are limited in their growth. **Malignant** neoplasms are cancers. They tend to invade surrounding tissues and are capable of **metastasis,** that is, they may be transplanted to other locations in the body. Metastasis may occur through "seeding," the migration of cells through the cavities of the body, or the tumor may be spread through either the lymphatic system or the circulatory system.

SUMMARY

The human body is a highly organized and complex structure. Its fundamental units are cells. Groups of similar cells form tissues. Organs are combinations of tissues with a special function. Groups of organs form each of the 11 organ systems of the body. Each system has a unique structure that is suitable for its special functions. The radiographer should be familiar with the general structure and function of all body systems.

The radiographer requires a deeper understanding of the skeletal system because of its significance in radiography. An understanding of the structure of bones and joints and the names of their parts prepares the radiographer for a deeper study of each part of the skeletal system.

The terminology of body positions and radiographic projections is the essential language of radiographic positioning. Frontal projections (AP and PA) are taken with the coronal body plane parallel to the film. For lateral projections, the sagittal plane is parallel to the film. Oblique projections refer to radiography taken when neither of these planes is parallel to the film. Lateral and oblique projections are named for the body surface nearest the film. Decubitus projections are taken with the patient recumbent and the central ray parallel to the floor. They are named according to the body surface on which the patient is supported.

Guidelines for cassette selection and placement are useful for making appropriate choices in the clinical setting. The film, the x-ray beam, and the body part must be correctly aligned to each other at the proper distance for all radiography. If the tube is angled, adjustments in alignment and distance are often necessary. All radiographs must be correctly marked with lead side markers. Markers indicating cassette position are needed for upright studies and decubitus projections.

Pathology is the study of disease. It includes many topics not usually thought of as diseases, such as injuries, birth defects, and anomalies. Diseases may be caused by circulatory system problems, by deficiencies or autoimmune conditions within the body, or by trauma, chemical injury, or infection from outside the body. Fractures, dislocations, and sprains are injuries caused by trauma to the skeletal system. Strains, lacerations, abrasions, and contusions are soft tissue injuries.

The response of the immune system to cellular injury is called inflammation. Chronic inflammation causes degeneration and atrophy of tissues, resulting in loss of function. Inflammation or endocrine imbalance may cause hyperplasia or hypertrophy, an increase in the size of an organ or structure. Neoplasms are tumors that may be benign or malignant. Benign neoplasms are self-limiting, but malignant neoplasms may invade surrounding tissues or spread to other parts of the body through a process called metastasis.

▪ *Review Questions* ▪

1. Name the three main parts of a cell.
2. Name a structure of the body that is composed of each of the following tissue types: epithelial, connective, muscle, nerve.
3. What is the difference between a tissue and an organ?
4. Name two organs that are part of each of the following systems: digestive, urinary, endocrine, and reproductive.
5. Describe the function of the lymphatic system.
6. List two endogenous conditions and two exogenous conditions.
7. List the four manifestations of inflammation.
8. Compare and contrast *degeneration* and *hyperplasia.*
9. What kinds of conditions are named with terms that end with "-itis"? With "-oma"?
10. Classify the disease "hay fever," an allergic response of the upper respiratory system to pollen. Is it inflammatory? Hereditary? Congenital? Endogenous? Autoimmune?
11. Name a structure that is superior to the knees, inferior to the neck, lateral to the spine, proximal to the hand, and distal to the shoulder.
12. What is the radiographic term for a body position in which the patient is lying on the left side? What is the name of the projection of this position when the central ray is horizontal? Vertical?
13. Since the width of the adult pelvis is greater than its height, what is the correct film placement for an AP projection of the pelvis?

13

Upper Extremity and Shoulder Girdle

Learning Objectives

At the conclusion of this chapter the student will be able to:

- Name the bones that comprise the upper extremity and shoulder girdle and identify each on an anatomic diagram and on a radiograph
- Name and identify the significant bony prominences and depressions of the upper extremity and shoulder girdle and identify significant positioning landmarks by palpation
- Demonstrate correct body and part positioning for routine projections and common special projections of the upper extremity and shoulder girdle
- Correctly evaluate radiographs of the upper extremity and shoulder girdle for positioning accuracy
- Describe and recognize on radiographs pathology common to the upper extremity and shoulder girdle

Key Terms

acromion process
axilla
bursitis
carpal digits
carpus, carpal
clavicle
fat pad sign
glenoid
humerus
joint effusion
metacarpal
olecranon process
osteoarthritis

osteoblastic
osteolytic
osteomyelitis
osteophyte
phalanx (pl. phalanges)
radial flexion
radius
scapula
sesamoid bone
tendinitis
ulna
ulnar flexion

While this chapter addresses mainly the upper extremity and shoulder girdle, many of the principles introduced in this chapter apply equally well to the lower extremity. For example, similar types of fractures and other pathologies occur in both the upper and the lower extremities. We begin with the distal portion of the extremity because the anatomy and positioning for these parts are somewhat simpler than that for the shoulder girdle.

ANATOMY

The upper extremity includes the fingers, hand, wrist, forearm, elbow, and arm (sometimes referred to as "upper arm") (Fig. 13-1).

Fingers and Thumb

The fingers are called the **carpal digits** and are considered to be part of the hand (Fig. 13-2). They are numbered from 1 to 5, beginning with the thumb. Digits 2 through 5 consist of three small long bones called **phalanges** (singular **phalanx**). The phalanges are distinguished from one another as proximal (nearest the hand), middle, and distal. For example, the bone at the tip of the "ring finger" is called the "fourth distal phalanx." The rounded tips of the distal phalanges are called ungual tufts. The hinge joints that connect the phalanges are called the interphalangeal (IP) joints and are distinguished as proximal and distal. The thumb has only two phalanges with one IP joint.

Hand

The bones of the hand are called **metacarpals.** They are numbered 1 through 5, starting on the lateral aspect. The numbers correspond to the digits with which they articulate. The distal end of a metacarpal is called its head, and the proximal end is referred to as the base. For example, the thumb is attached to the head of the first metacarpal. The hinge joints between the metacarpals and the proximal phalanges are called the metacarpophalangeal (MP) joints.

There are usually one or more small bones in the region of the first MP joint called **sesamoid bones.** These small, flat, oval bones within tendons are not counted among the bones of the body. They are called sesamoid bones because they resemble a sesame seed. They serve to protect the joint.

Wrist

The wrist consists of eight short bones called carpal bones. Together, they are referred to as the **carpus.** Most of them have two names that are in common use. They are arranged in two rows. Beginning with the proximal row on the lateral aspect (thumb side), they are named scaphoid (navicular), lunate (semilunar), triquetrum (cuneiform), pisiform, trapezium (greater multangular), trapezoid (lesser multangular), capitate, and hamate. There is a small curved projection on the anterior aspect of the hamate called

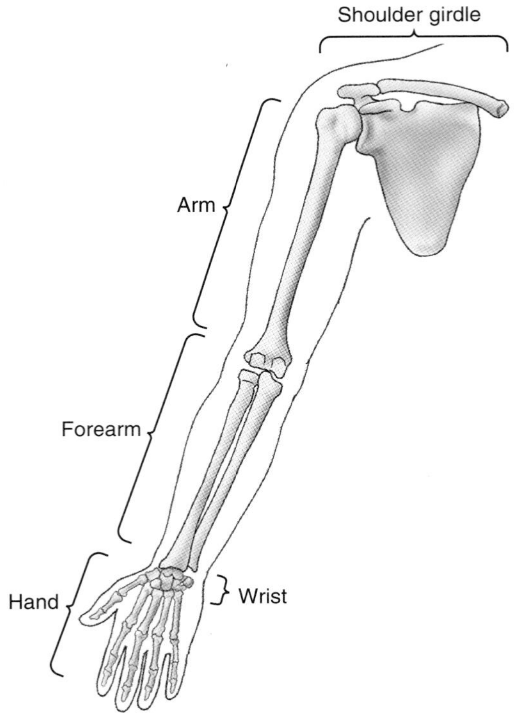

Fig. 13-1 Upper extremity.

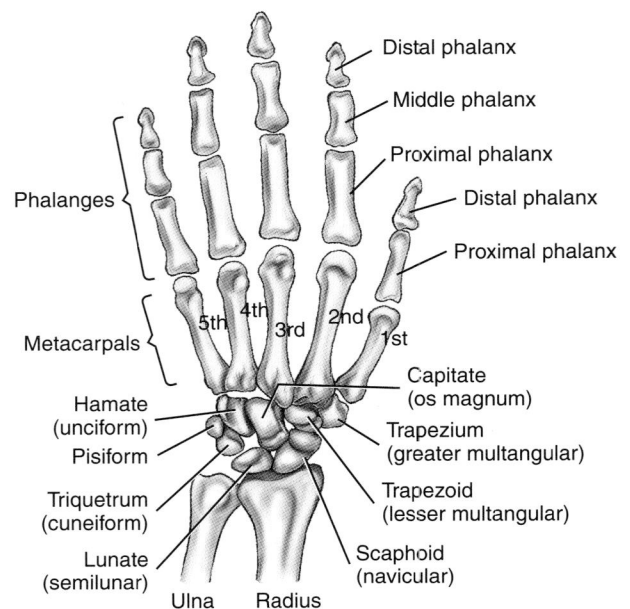

Fig. 13-2 Posterior aspect of hand and wrist.

the hook or hamulus. As a whole, the wrist is capable of all possible joint motions except rotation. It flexes in four directions: anterior, posterior, medial, and lateral. The greatest degree of movement occurs with anterior and posterior flexion and involves primarily the scaphoid and lunate as they articulate with the radius. Medial flexion (toward the body when in anatomic position) is called **ulnar flexion.** Lateral flexion (away from the body when in anatomic position) is called **radial flexion.**

Arm

The forearm (Fig. 13-3) consists of two long bones, the **radius** and the **ulna.** The ulna is the longer, thinner bone on the medial aspect. The radius is thicker and somewhat shorter and is located laterally. The styloid

process of the radius is a bony prominence that can be palpated on the lateral aspect of the wrist. The ulnar styloid can be felt on the medial aspect of the wrist when the hand is supinated; it protrudes on the dorsal surface when the hand is pronated. Pronation causes the radius to cross over the ulna (Fig. 13-4).

The proximal end of the radius is referred to as the radial head. The radial tuberosity, which is distal to the radial head, is a muscle attachment that is not normally palpable.

The proximal end of the ulna terminates in the olecranon process posteriorly. In lay terms, the **olecranon process** is sometimes called the "funny bone" or "crazy bone." Anterior to the olecranon process is the semilunar notch. The inferior lip of the semilunar notch is called the coronoid process.

The single bone of the upper arm is called the **humerus** (Fig. 13-5). The distal end of the humerus flares to form two palpable prominences, the medial and lateral epicondyles. Between the epicondyles posteriorly is the olecranon fossa, a depression where the olecranon process fits when the elbow joint is extended. There are two distal articular surfaces on humerus: the rounded capitulum (also called the capitellum), which articulates with the head of the radius, and the trochlea, a spool-shaped process that articulates within the semilunar notch of the ulna. The superior end of the humerus is called the head and the portion between the head and the shaft is the surgical neck. Two prominences on the head of the humerus are significant: the

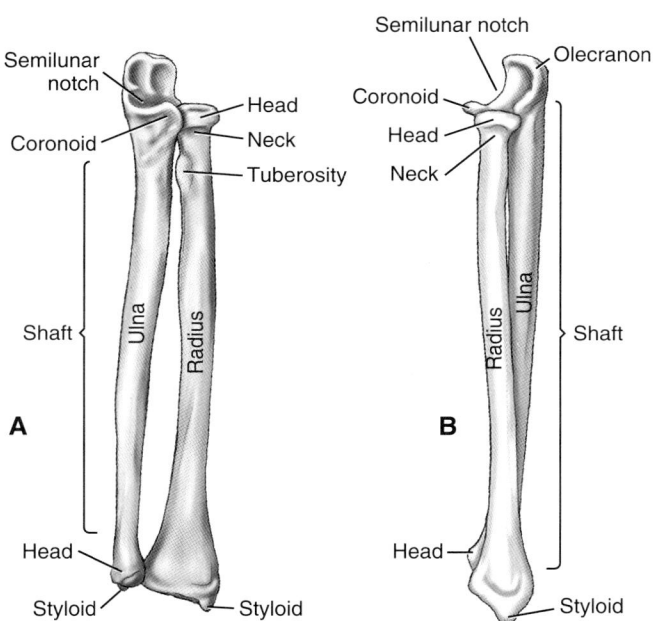

Fig. 13-3 Forearm. **A,** Anterior aspect; **B,** lateral aspect.

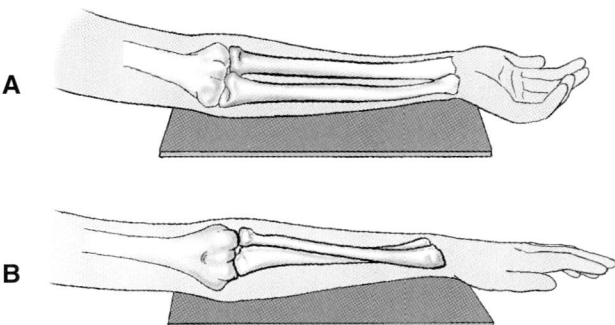

Fig. 13-4 Positions of forearm bones. **A,** Supination; **B,** pronation.

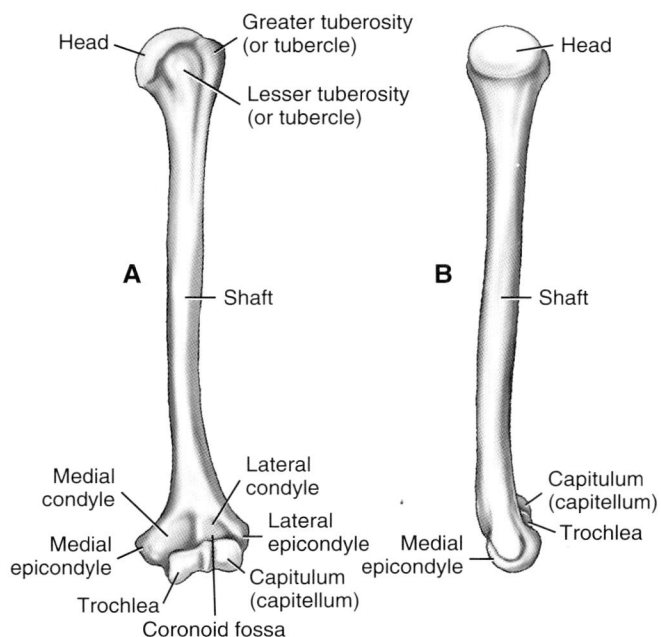

Fig. 13-5 Humerus. **A,** Anterior aspect; **B,** medial aspect.

Fig. 13-6 Shoulder girdle.

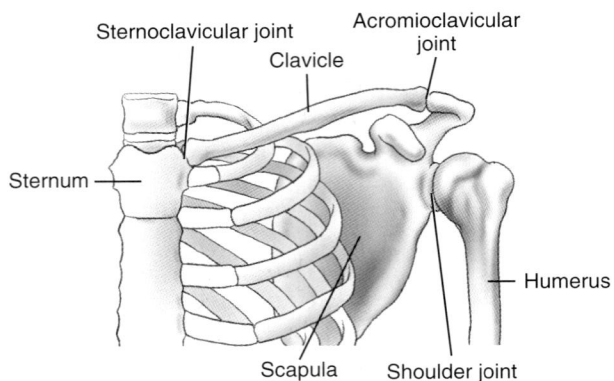

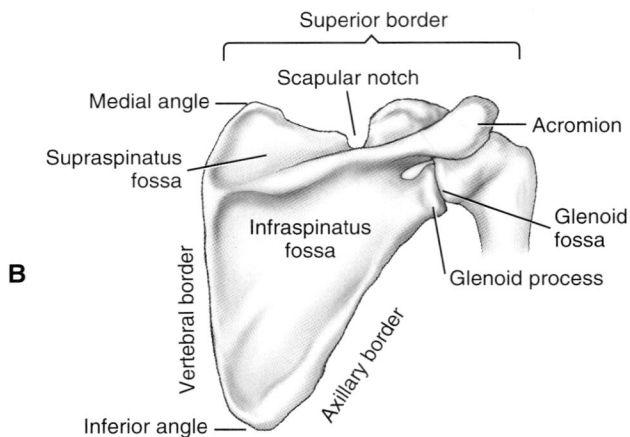

Sternoclavicular joint

Acromioclavicular joint

Clavicle

Sternum

Humerus

Scapula

Shoulder joint

Coracoid process

Acromion

Scapular notch

Medial angle

Glenoid fossa

Body

Vertebral border

Axillary border

Glenoid process

Inferior angle

A

Superior border

Scapular notch

Medial angle

Supraspinatus fossa

Acromion

Infraspinatus fossa

Glenoid fossa

Vertebral border

Glenoid process

Axillary border

Inferior angle

B

Acromion

Medial angle

Coracoid process

Glenoid fossa

Axillary border

Posterior surface

Anterior surface

Inferior angle

C

Fig. 13-7 Scapula. **A,** Anterior aspect; **B,** posterior aspect; **C,** lateral aspect.

Fig. 13-8 Anterior aspect of clavicle.

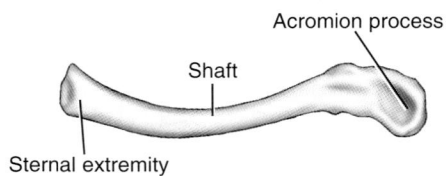

Acromion process

Shaft

Sternal extremity

greater tuberosity, which is superior and lateral, and the lesser tuberosity, which is medial and inferior. The greater tuberosity can be felt on the upper outer aspect of the shoulder when the patient is in anatomic position.

Shoulder Girdle

The bones of the shoulder are often referred to as the "shoulder girdle" (Fig. 13-6). They include the **scapula** (shoulder blade), the **clavicle** (collar bone), and the proximal portion of the humerus.

The scapula (Fig. 13-7) is a flat, triangular bone. Its three sides are called the superior border, the lateral border, and the vertebral (medial) border. The junction of the lateral and vertebral borders at the lower tip is called the inferior angle. The spine of the scapula is a bony ridge on the posterior surface that is inferior and somewhat parallel to the superior border. It serves as a muscle attachment. At the junction of the superior and lateral borders are three significant features: the acromion process, the coracoid process, and the glenoid process. The **acromion process** is a large, rounded projection that can be felt on the superior surface of the shoulder. The coracoid process is a muscle attachment on the anterior surface that is palpable just medial to the shoulder joint. The **glenoid** process is on the superior lateral aspect. It contains a cavity called the glenoid fossa that forms the socket of the shoulder joint. Its articulation with the humeral head is also called the glenohumeral joint.

The clavicle (Fig. 13-8) is a long, narrow bone located anterior to the upper portion of the rib cage. Its proximal end attaches to the sternum (breast bone), forming the sternoclavicular joint. Its distal end forms a gliding joint with the acromion process of the scapula that is called the acromioclavicular (A/C) joint.

Fig. 13-9 illustrates the palpable bony landmarks of the upper extremity.

POSITIONING AND RADIOGRAPHIC EXAMINATIONS

Examinations of the upper extremity from the fingertips through the elbow joint are usually done tabletop (nongrid) with the patient seated at the end of the radiographic table (Fig. 13-10).

For many of these examinations a single film is used for more than one exposure. The procedure for multiple exposures on a single cassette is explained in Chapter 12.

To prepare for these examinations, the patient removes any clothing or jewelry from the area that will be within the radiation field. This is usually a simple matter for examinations of the distal extremity. Patients may need to remove outer clothing from the upper torso and don a gown for examinations of the shoulder region.

While gonad shielding is not legally required for examinations of the upper extremity, it is wise to shield whenever possible. A full lead apron is shown here for these examinations.

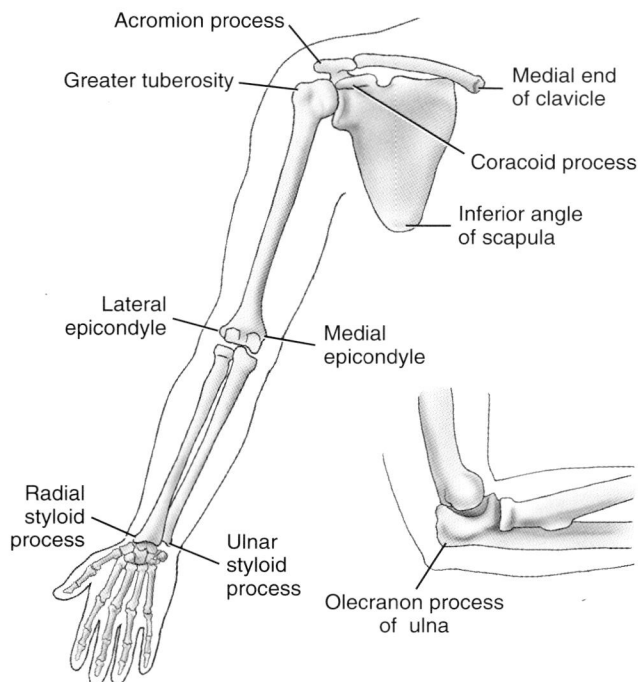

Fig. 13-9 Palpable bony landmarks of upper extremity.

Fig. 13-10 Body position for most views of distal upper extremity.

Basic Examination

The basic examination of the hand includes the PA, oblique (anteromedial), and lateral views.

Screens: Extremity (detail)

Grid: No

SID: 40 inches

Body position:
Seated at end of table with elbow flexed 90° and arm fully supported

Part position:

PA: Hand open, fingers extended, with palmar surface in contact with cassette, fingers moderately separated

> **TIP:** If injury or deformity prevents full extension of the fingers, the hand cannot be placed flat, palm-down on the cassette. In such a case, a better result will be obtained with the AP projection, placing the back of the hand on the cassette. All other aspects of the radiograph are the same.

Oblique: Anteromedial aspect of hand is in contact with cassette and coronal plane of hand forms 45° angle with cassette. Stair-step sponge is used to support and maintain position so that interphalangeal joints are clearly visualized. Alternatively, without stair-step sponge, "modified teacup" position (named for position of hand when holding a teacup) is used.

Lateral: Ulnar aspect of hand is in contact with cassette and hand is in true lateral position with coronal plane perpendicular to film. Thumb is in PA position and is supported on sponge or splinted against hand. Fingers may be separated so as not to be superimposed, if desired.

Central ray:

PA: Perpendicular to third metacarpophalangeal joint

Oblique and lateral: Perpendicular to second metacarpophalangeal joint

Patient instruction:
Do not move.

Structures seen:
Entire hand (including fingertips), carpus, and most distal aspects of radius and ulna. Oblique shows a second view of carpals and metacarpophalangeal joints with relatively little superimposition. Lateral demonstrates displacement of structures or fracture fragments and is essential for localization of foreign bodies.

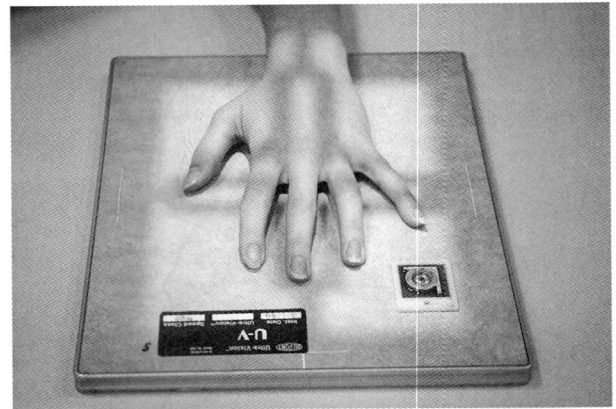

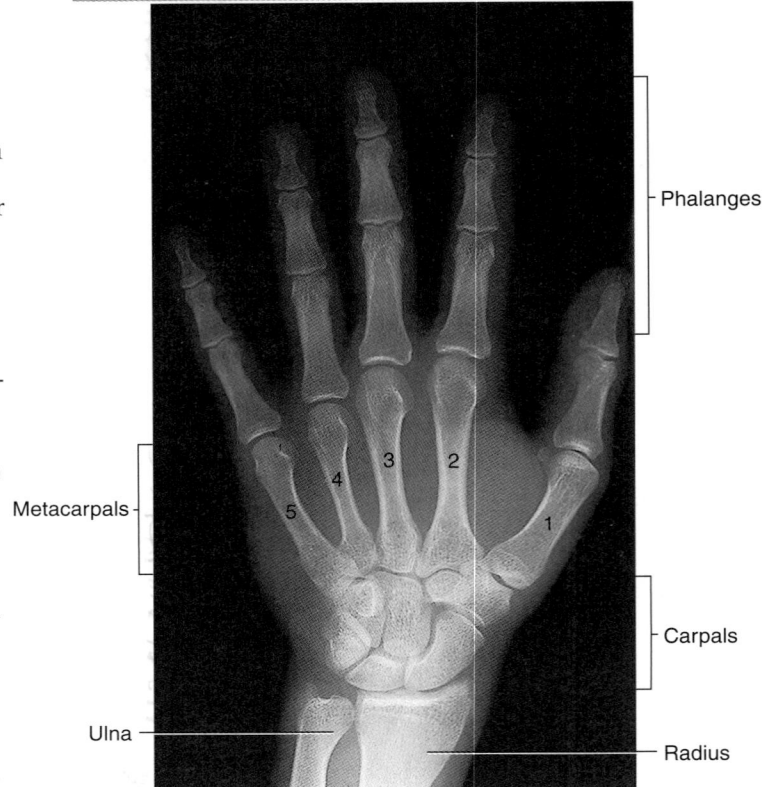

PA hand position and radiograph.

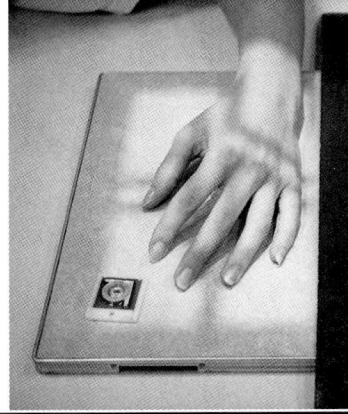

Oblique hand position and radiograph taken using stair-step sponge.

Oblique hand position and radiograph, "modified teacup" position.

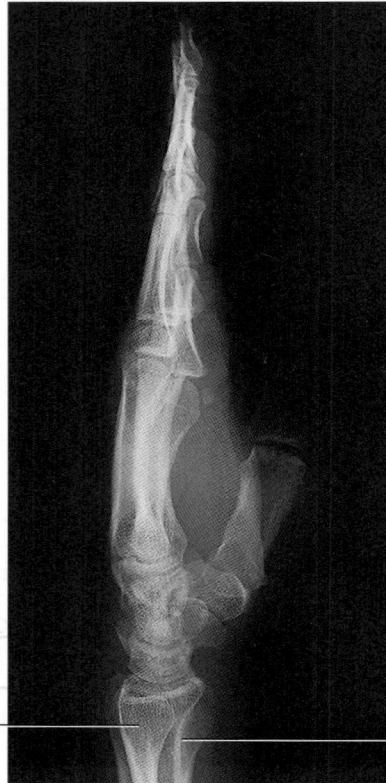

Phalanges

Metacarpals

Carpal bones

Ulna ——   —— Radius

Lateral hand position and radiograph.

While the fingers are included in the examination of the hand, separate finger studies are often performed when the area of clinical interest is limited to a specific finger. Depending on department protocol, the routine projections for the fingers may include a PA projection of the entire hand plus oblique and lateral projections of the affected finger, or the examination may be limited to the PA, oblique, and lateral projections of the affected finger only. Each projection of the finger should include at least the entire digit and the distal portion of the corresponding metacarpal.

The hand position for lateral projections of the fingers will vary, depending on which finger is involved and what motions are possible for the patient.

Keep the finger as close to the film as possible and maintain the finger in a position parallel to the film. When the finger is angled in relation to the film, visualization of the joint spaces is compromised (Fig. 13-11). A stair-step sponge is a desirable aid for oblique positioning, since it supports the finger parallel to the film. This position decreases distortion and improves visualization of the interphalangeal joints. A stair-step sponge or other radiolucent support also helps to avoid motion that is likely to occur if the finger is not supported during the exposure.

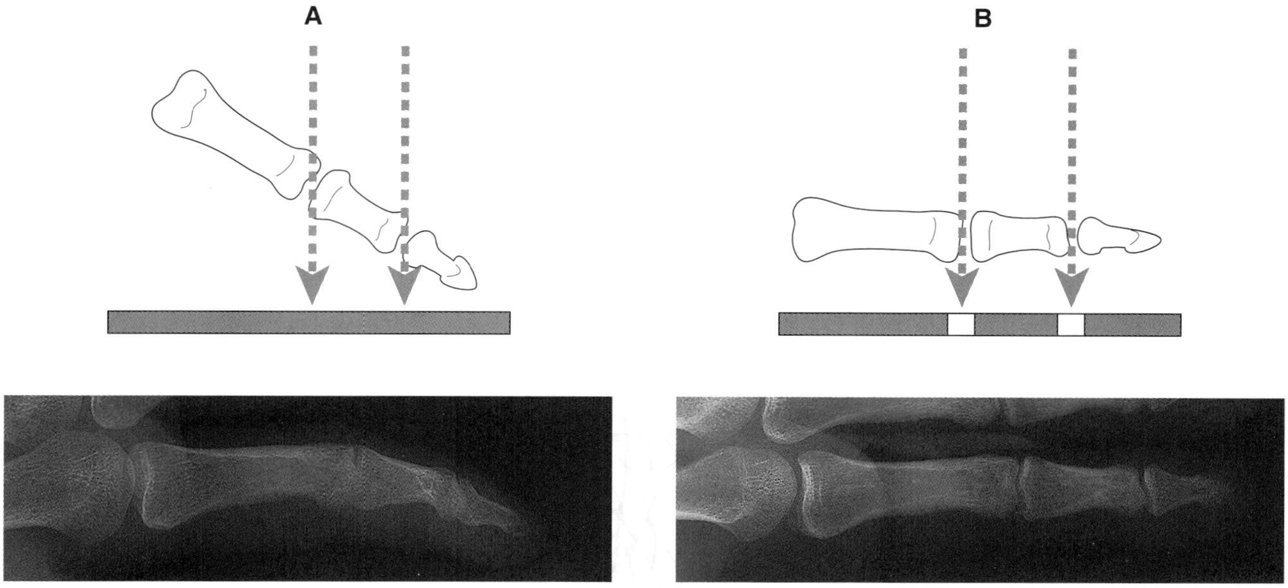

Fig. 13-11 A, Modified teacup position. Fingers are not parallel to film. X-ray beam causes superimposition of articular surfaces, obscuring joint spaces. **B,** Using a stair-step sponge, fingers are parallel to film. X-ray beam is parallel to articular surfaces, demonstrating joint spaces.

Basic Examination

The basic examination of the fingers includes the PA, oblique (anteromedial), and lateral views.

Screens: Extremity (detail)

Grid: No

SID: 40 inches

8″ × 10″

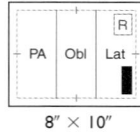

Body position:

Seated at end of table with elbow flexed and arm resting on table

Part position:

PA: Hand is open with palmar surface in contact with cassette. Fingers are moderately separated.

Oblique: Anteromedial (palmar/ulnar) surface of hand is in contact with cassette. Fingers are supported by stair-step sponge. (The conventional alternative to the use of a stair-step sponge is the "modified teacup" position, as shown for examination of the hand.)

Lateral: Medial or lateral surface of hand may be in contact with cassette, depending on which brings the finger of interest nearest to cassette. Other fingers are flexed or extended as needed to leave affected finger free of superimposition. Affected finger is supported parallel to cassette.

Central ray:

Perpendicular to midproximal phalanx

Patient instruction:

Do not move.

Structures seen:

Entire digit and distal half of metacarpal with interphalangeal and metacarpophalangeal joint spaces open and clearly visualized

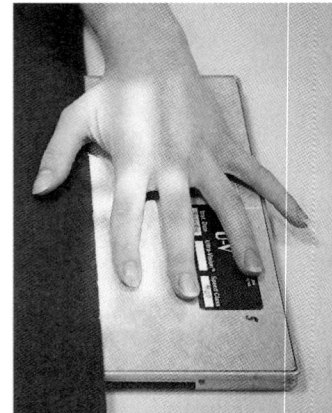

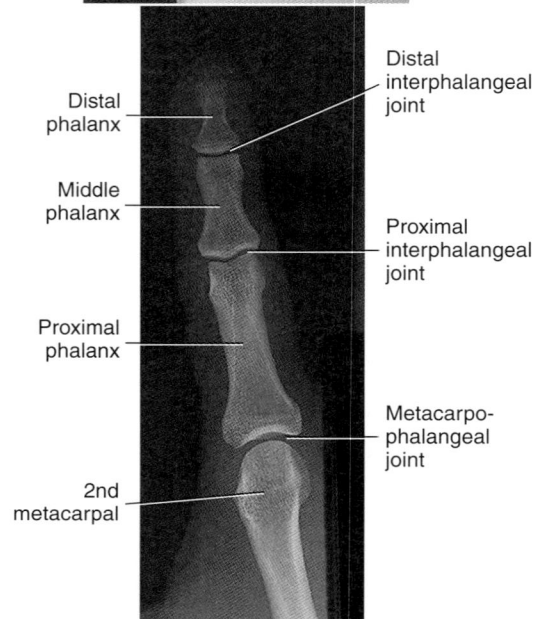

PA finger position and radiograph.

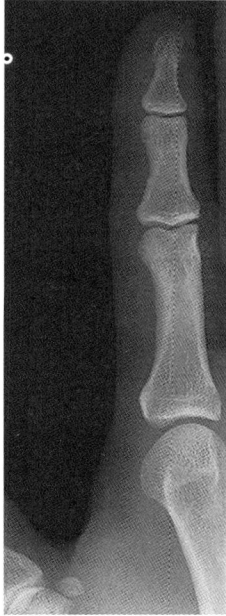

Oblique finger position and radiograph.

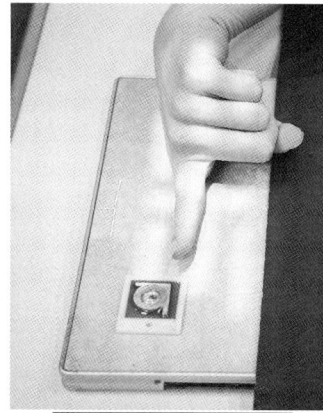

Lateral finger position and radiograph.

Examination of the thumb differs from examinations of the other fingers because the thumb attaches to the hand at a different angle. In addition, examinations of the thumb must include the entire first metacarpal rather than only a portion of it. As with the finger, the thumb examination may include a PA view of the entire hand. When this is the case, the PA hand provides the oblique projection of the thumb.

Basic Examination

The basic examination of the thumb includes the AP, oblique (anterolateral), and lateral views.

Screens: Extremity (detail)

Grid: No

SID: 40 inches

		R
AP or PA	Obl	Lat

8" × 10"

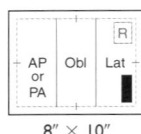

Body position:

AP: Seated at end of table with forearm rotated into exaggerated degree of pronation

> **TIP:** It is sometimes helpful to seat the patient with back to film and arm extended posteriorly.

PA, oblique, and lateral: Seated at end of table with elbow flexed 90° and arm fully supported

Part position:

AP: Dorsal surface of thumb is in contact with cassette. Coronal plane of thumb is parallel to film. Plane of palm of hand is perpendicular to film.

> **TIPS:**
> - Take care that medial aspect of hand does not superimpose the first metacarpal.
> - If patient is unable to assume a satisfactory position for this projection, substitute the PA projection.

Oblique: Palmar surface of hand is in contact with cassette as for PA fingers. Thumb naturally assumes a 45° oblique position.

Lateral: Beginning with hand positioned as for oblique thumb, patient flexes MP joints 2 through 5 with the fingers extended, "tenting" hand until thumb is in true lateral position.

Central ray:
Perpendicular to first MP joint

Patient instruction:
Do not move.

Structures seen:
Entire thumb and first metacarpal with all joint spaces open and clearly visualized

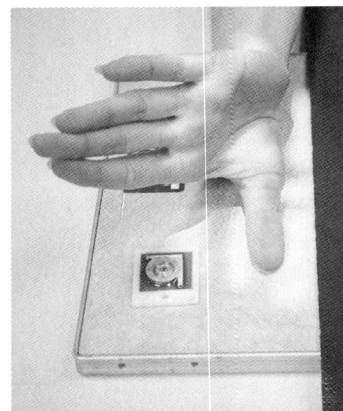

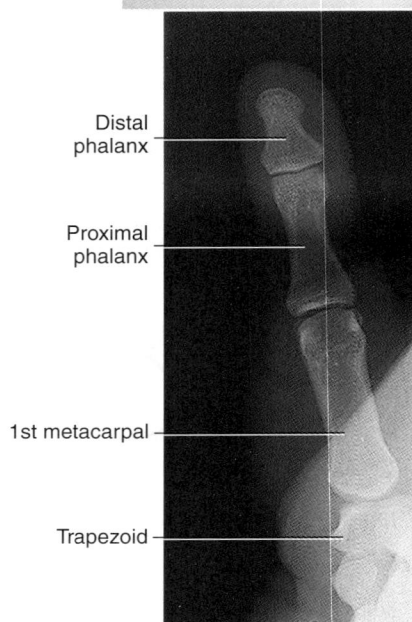

Distal phalanx

Proximal phalanx

1st metacarpal

Trapezoid

AP thumb position and radiograph.

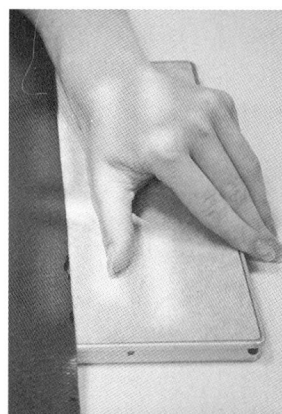

Oblique thumb position and radiograph.

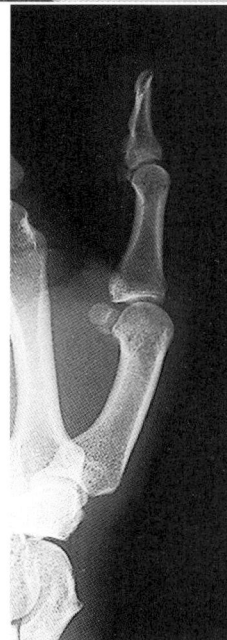

Lateral thumb position and radiograph.

Alternative Position

When patients are unable to assume a good position for the AP projection, the PA projection provides an alternative. The AP projection is preferred to the PA projection because the decreased OID produces better radiographic definition.

Body position:
Same as for oblique and lateral thumb projections.

Part position:
Medial aspect of hand is in contact with cassette and palm of hand is perpendicular to film. Thumb is supported parallel to cassette with coronal plane parallel to film.

Central ray:
Perpendicular to first metacarpophalangeal joint

> **TIP:** Be sure to use the small focal spot for the PA thumb for improved radiographic definition.

PA thumb position.

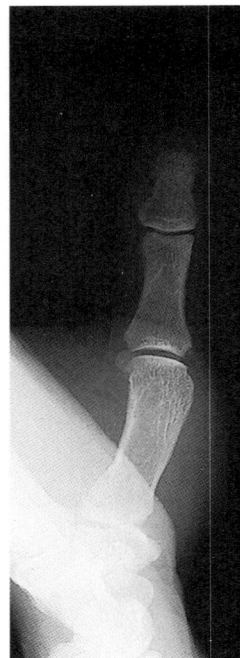

PA thumb radiograph. Image is magnified and definition is reduced compared to AP thumb radiograph.

The wrist is a complex structure with many small bones and joints, so there are a number of special views designed to visualize specific aspects of this region. The basic examination should include the metacarpal bones and at least the distal fourth of the forearm.

Basic Examination

The basic examination of the wrist includes the PA, oblique (anteromedial), and lateral views.

Screens: Extremity (detail)

Grid: No

SID: 40 inches

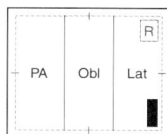

Body position:
Seated at end of table with elbow flexed 90° and forearm resting on table

Part position:

PA: Anterior surface of wrist is in contact with cassette. Fingers are flexed to form a loose fist, placing wrist in firmer contact with cassette and opening intercarpal joints.

Oblique: Anteromedial surface of wrist is in contact with cassette so that coronal plane of wrist forms a 45° angle with film. Position may be supported by wedge sponge, stair-step sponge, or patient's thumb.

Lateral: Medial surface of wrist is in contact with cassette. Coronal plane of wrist is perpendicular to film.

Central ray:

PA and oblique: Perpendicular to a point midway between styloid processes of radius and ulna

Lateral: Perpendicular to styloid process of radius

Patient instruction:
Do not move.

Structures seen:
Carpal bones, proximal halves of metacarpals, and distal portion of forearm

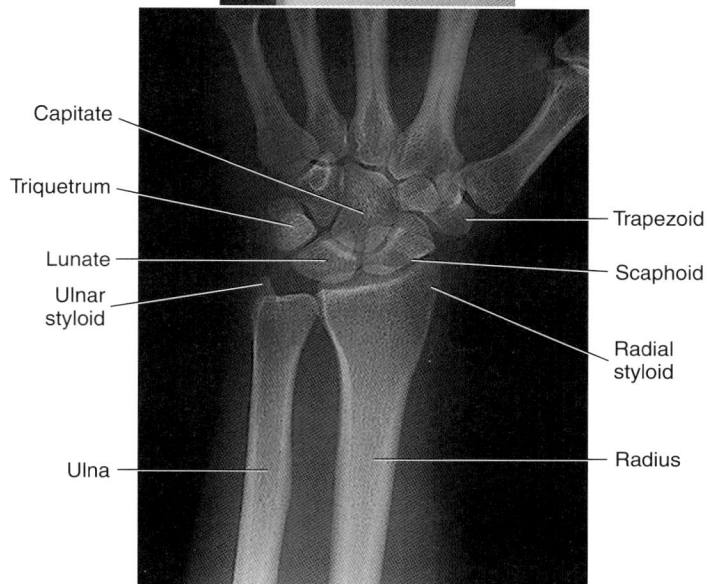

PA wrist position and radiograph.

Triquetrum

Lunate

Ulna

Trapezoid

Scaphoid

Radius

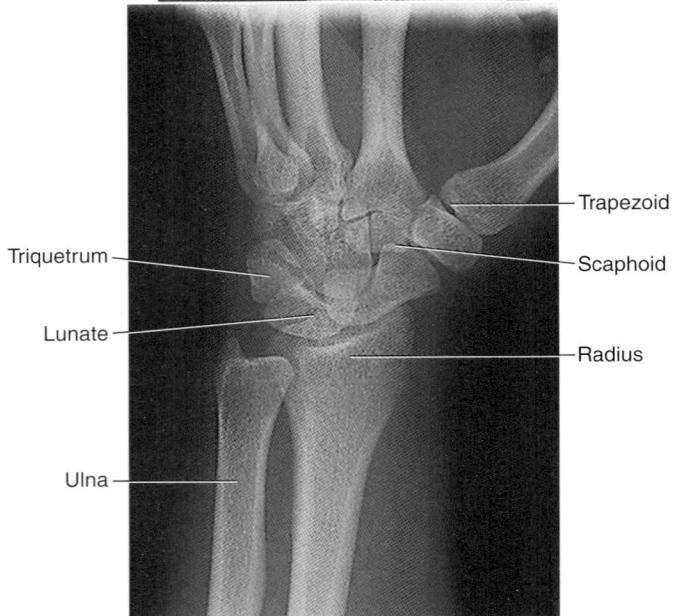

Anteromedial oblique wrist position and radiograph.

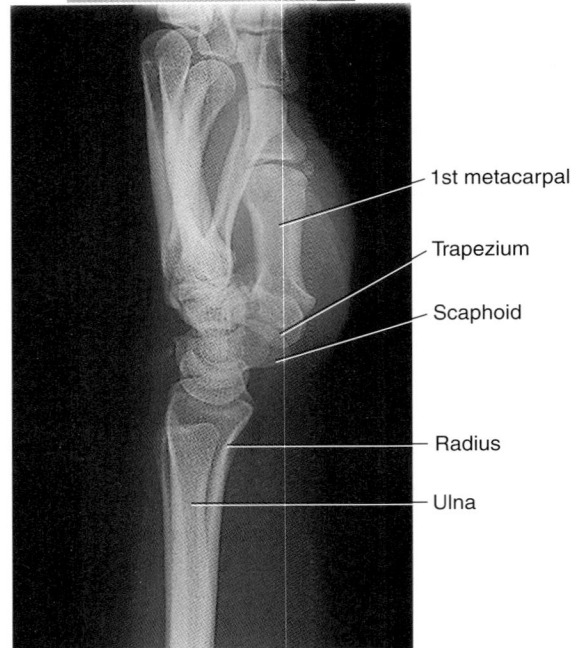

1st metacarpal

Trapezium

Scaphoid

Radius

Ulna

Lateral wrist position and radiograph.

Supplemental Projections

These views of the wrist may be added individually to the basic routine according to the area of clinical interest and the instructions of the physician.

POSTEROMEDIAL OBLIQUE

This oblique projection is the opposite of the usual oblique. This projection is useful for demonstration of the medial aspect of the carpus, particularly the lunate and the pisiform.

Body position:
Same as for other wrist projections

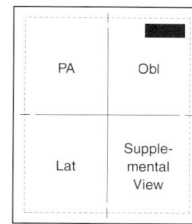

Part position:
Posteromedial surface of wrist is in contact with cassette.

Central ray:
Same as for other wrist projections

Cassette for four-view wrist examination.

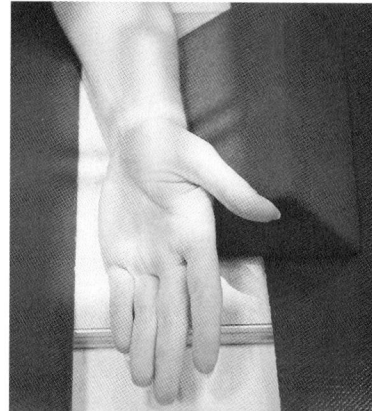

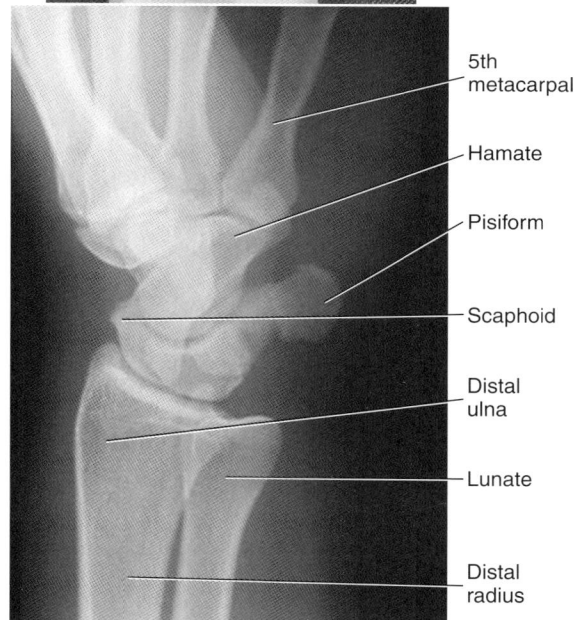

Posteromedial oblique wrist position and radiograph.

ULNAR FLEXION

This position is used specifically for visualization of the scaphoid (navicular).

Part position:
Same as for PA hand with fingers extended. Wrist is then flexed in direction of ulna to extent that patient can tolerate.

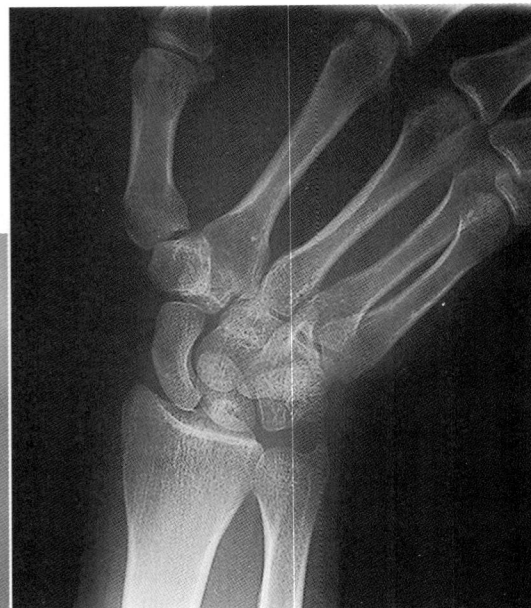

Ulnar flexion wrist position and radiograph. Position elongates scaphoid and opens parascaphoid joint spaces.

STECHER PROJECTION

The Stecher projection is another special view for the scaphoid (navicular).

Part position:
Arm is extended and parallel to long axis of table for correct alignment with angled x-ray beam. Wrist is positioned with anterior aspect on cassette as for PA hand.

Central ray:
Angled 20° in direction of elbow and centered to midpoint between styloid processes.

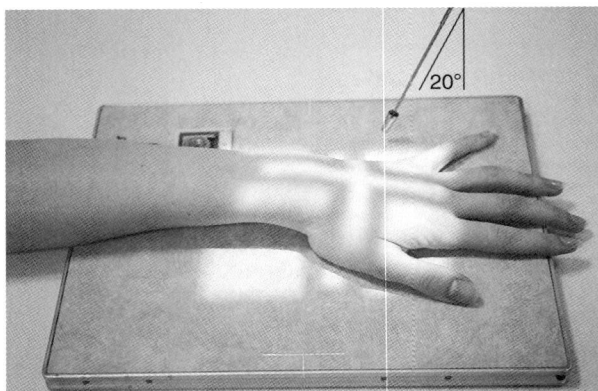

Stecher position for scaphoid demonstration.

Scaphoid

Stecher view of wrist.

CARPAL CANAL

Part position:
Arm is extended and parallel to long axis of table for correct alignment with angled x-ray beam. Wrist is extended as much as possible, placing palm of hand perpendicular to cassette.

Central ray:
Angled 25° to 30° in direction of elbow and centered to middle of palm of hand.

Structures seen:
Anterior arch of carpal bones, scaphoid, trapezium, pisiform, and hook of hamate

25°-30°

Position for tangential projection of carpal canal.

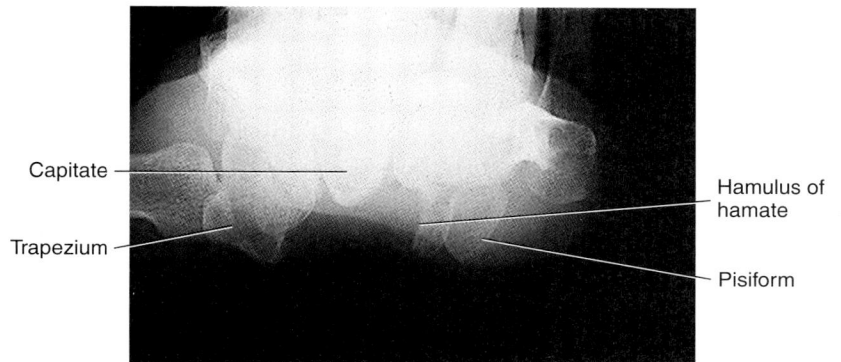

Capitate

Trapezium

Hamulus of hamate

Pisiform

Carpal canal radiograph.

Examination of the forearm is usually ordered when the area of clinical interest is in the shaft of the radius and/or the ulna. The examination should include the entirety of both bones and their articular surfaces.

Visualization of both joints is preferable. If only one joint is demonstrated, the *same* joint must be demonstrated on both projections.

Basic Examination

The basic examination of the forearm includes the AP and lateral views.

Screens: Extremity (detail)

Grid: No

SID: 40 inches

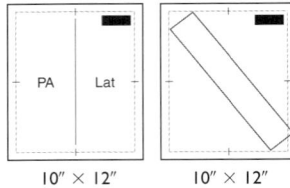

Body position:
Seated at end of table with **axilla** (armpit) at table level. This may be achieved by lowering the seat or by having patient lean toward table.

Part position:

AP: Arm is fully extended with hand supinated and posterior surface in contact with cassette. Both wrist and elbow are in true AP position with coronal plane of arm parallel to film. A small sandbag in palm of hand aids in maintaining position.

Lateral: Elbow is flexed 90° with medial surface in contact with cassette. Wrist is in lateral position with thumb up.

Central ray:
Perpendicular to center of film (midforearm)

Patient instruction:
Do not move.

Structures seen:
Entire forearm and at least one joint

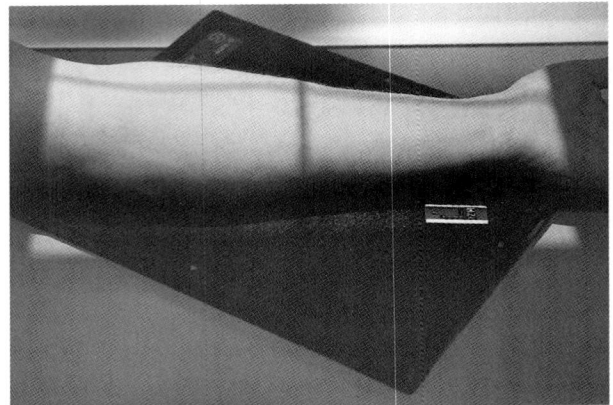

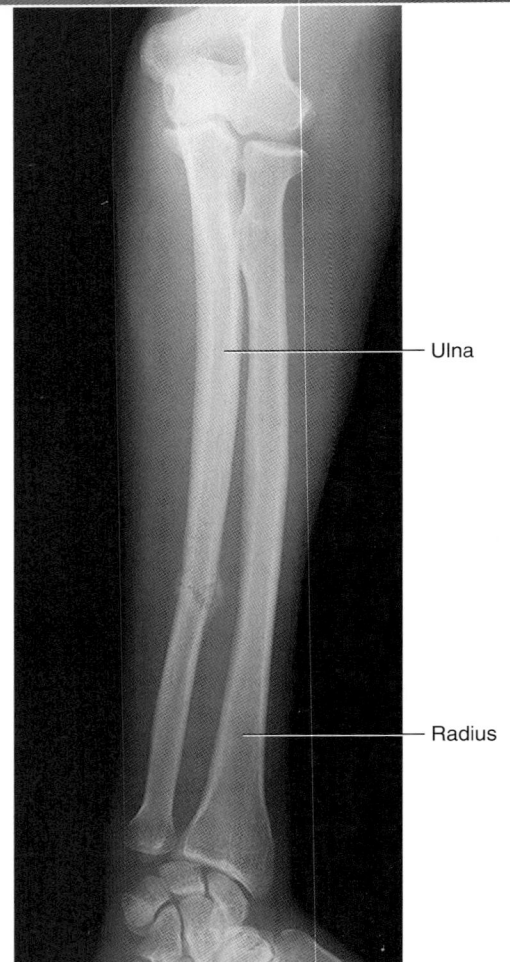

AP forearm position and radiograph.

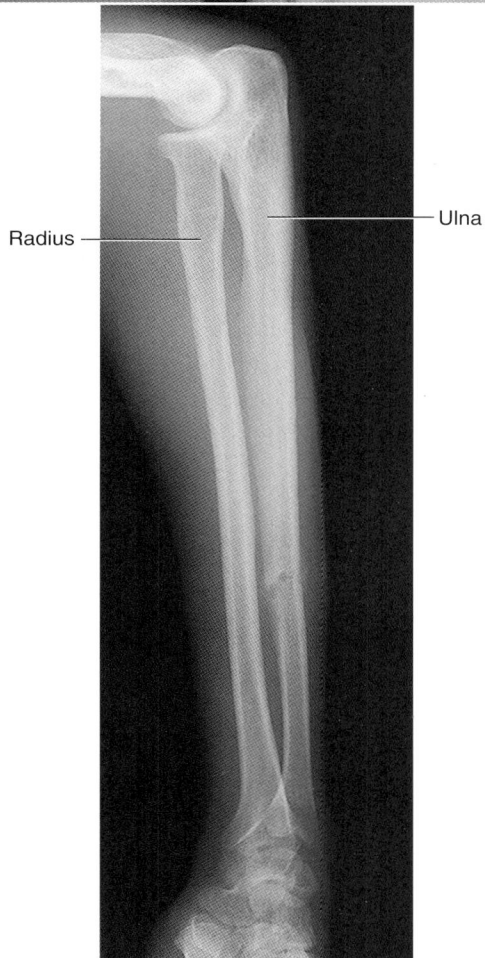

Lateral forearm position and radiograph.

The basic examination of the elbow is quite simple and the positions are the same as those for the forearm. Because it is often impossible for patients with injured elbows to fully extend the elbow joint, alternatives are presented for the AP projection with the elbow flexed. Supplemental views may be added to the basic examination for further visualization of specific aspects of the joint.

Basic Examination

The basic examination of the elbow includes the AP and lateral views.

Screens: Extremity (detail)

Grid: No

SID: 40 inches

10" × 12"

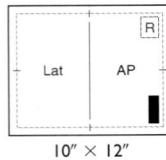

Body position:
Seated at end of table with axilla at level of table, as for AP forearm

Part position:

AP: Arm is fully extended with hand supinated and posterior surface in contact with cassette. If patient is unable to fully extend arm, substitute AP views for flexed elbow.

Lateral: Elbow is flexed 90° with medial surface in contact with cassette. Wrist is in lateral position with thumb up.

Central ray: Perpendicular to elbow joint. For AP projection, joint is midway between humeral epicondyles. For lateral projection, it is at lateral epicondyle.

Patient instruction:
Do not move.

Structures seen:
Elbow joint with portions of distal humerus and proximal forearm

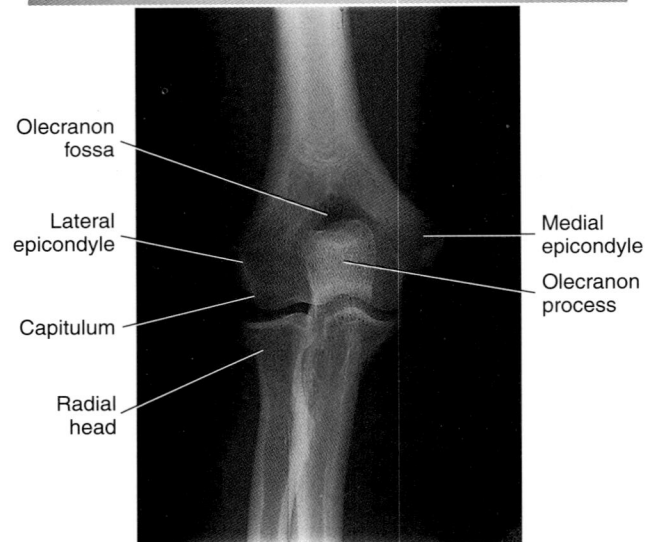

AP elbow position and radiograph.

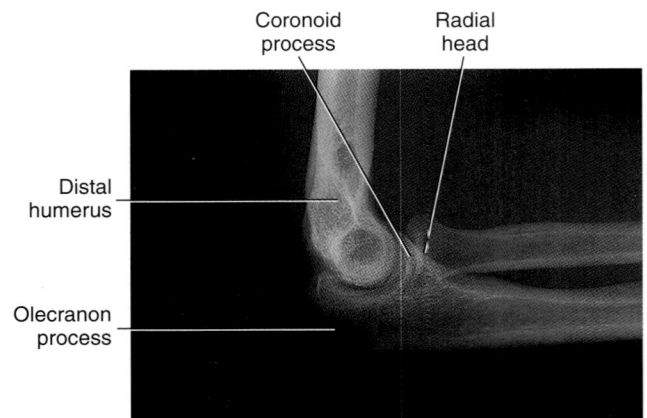

Lateral elbow position and radiograph.

Alternative Projections

When the patient is unable to fully extend the elbow joint for an AP projection, *both* of the following AP projections with flexed elbow are needed to substitute for the routine AP projection.

AP FLEXED ELBOW, FOREARM PORTION

Patient position:
Standing and leaning over table

Part position:
Posterior aspect of forearm rests on cassette and elbow is extended as much as possible.

Central ray:
Perpendicular to center of elbow. If the elbow is flexed 90° or more, it is necessary to angle central ray 5° to 15° in proximal direction.

> **TIP:** Take care that the patient's head is not in the path of the x-ray beam.

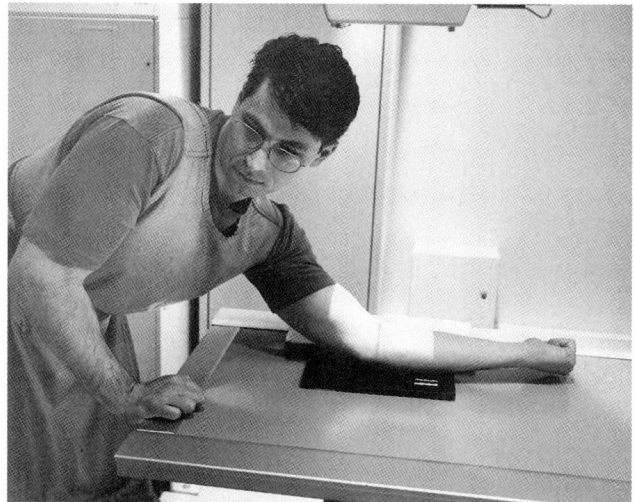

Position for AP flexed elbow, forearm portion.

AP FLEXED ELBOW, HUMERAL PORTION

Patient position:
With axilla at table level as for routine AP projection, elbow extended as much as possible and forearm supported

Central ray:
Directed to center of joint. If elbow is flexed 90° or more, it is necessary to angle central ray 5° to 15° in distal direction.

Position for AP flexed elbow, humeral portion.

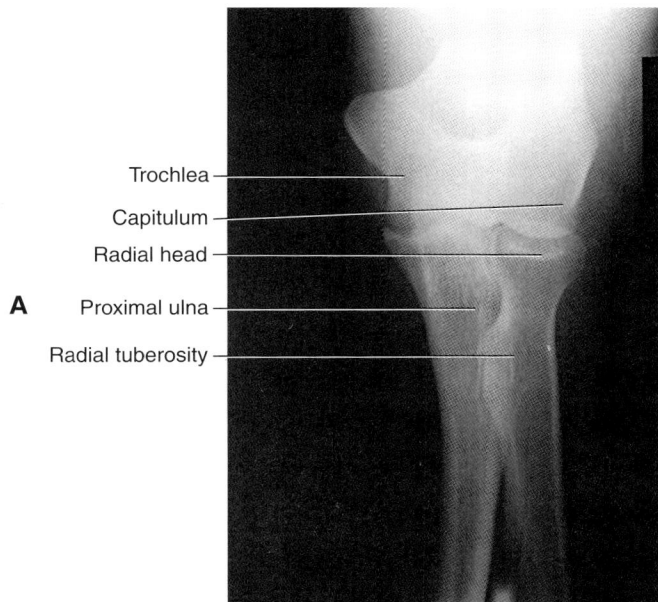

A

Trochlea
Capitulum
Radial head
Proximal ulna
Radial tuberosity

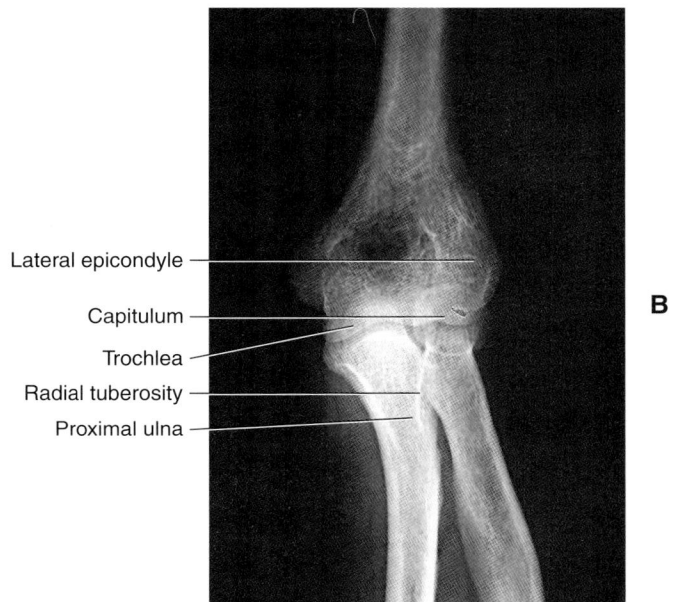

B

Lateral epicondyle
Capitulum
Trochlea
Radial tuberosity
Proximal ulna

AP flexed elbow radiographs. **A,** Forearm portion; **B,** humeral portion.

Supplemental Projections

LATERAL OBLIQUE

Body position:
From AP position, leaning back and rotating shoulder externally so that posterior lateral aspect of elbow is in contact with film

Part position:
Coronal plane of elbow forms angle of 35° to 40° with film plane.

Central ray:
Same as for AP projection

Structures seen:
Radial head and capitulum

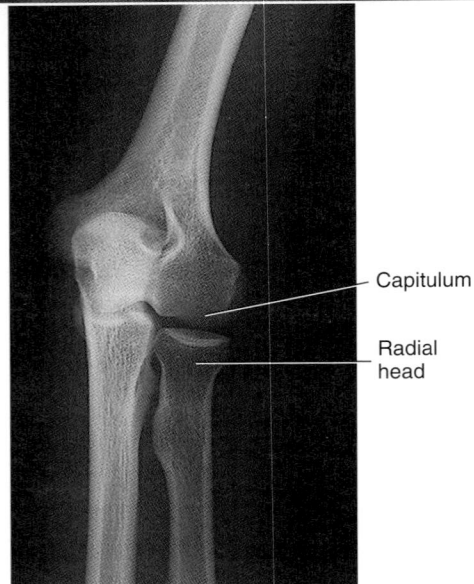

Lateral oblique elbow position and radiograph.

MEDIAL OBLIQUE

Body position:
Same as for AP projection

Part position:
Hand pronated, allowing coronal plane of elbow to assume 45° angle with film plane

Central ray:
Same as for AP projection

Structures seen:
Coronoid process and trochlea

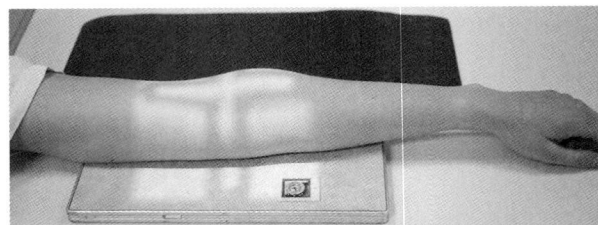

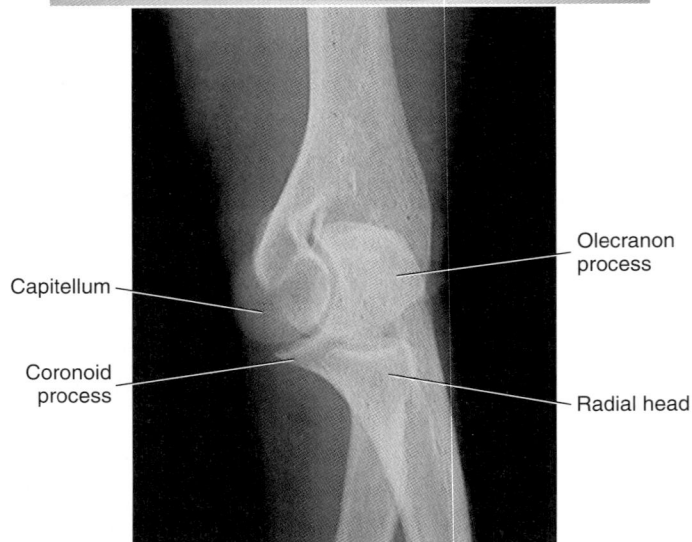

Medial oblique elbow position and radiograph.

FLEXED ELBOW, JONES VIEW

Part position:
Elbow is maximally flexed with posterior aspect of humerus in contact with cassette.

TIP: There is no rotation of elbow; palm of hand faces shoulder.

Central ray:
Perpendicular to point 2 inches distal to olecranon tip

Structures seen:
Frontal projection of olecranon process

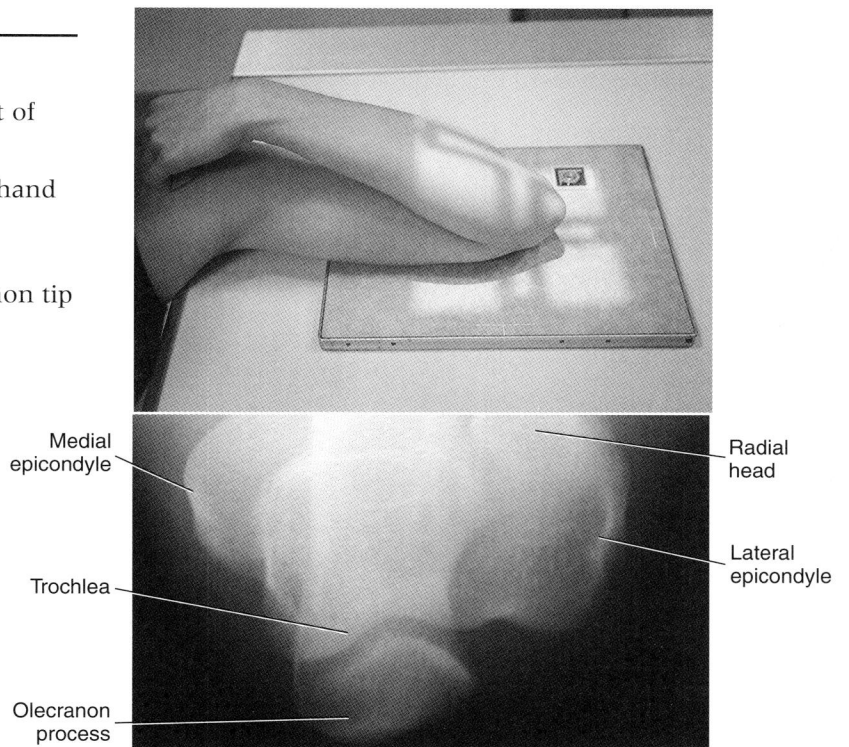

AP elbow position and radiograph, with acute flexion, Jones view.

The humerus may be radiographed with the patient either upright or supine. Both methods are presented here. The thickness of the shoulder joint usually dictates that this study be done using a bucky or grid. For relatively small patients, however, a grid is not necessarily required.

Basic Examination

UPRIGHT HUMERUS

The basic examination of the upright humerus includes AP and lateral views.

Screens: Regular (rapid)

Grid: Yes

SID: 40 inches

14" × 17" (need 2)

Body position:
Seated or standing with back to upright bucky or grid cabinet

Part position:

AP: True anatomic position with palm of hand anterior

Lateral: Elbow flexed approximately 45° and palm of hand against hip so that fingertips point down and elbow is lateral

Central ray:
Perpendicular to midpoint of film (midhumerus)

Patient instruction:
Stop breathing. Do not move.

Structures seen:
Entire humerus, shoulder, and elbow

> **TIP:** It is helpful for this examination to collimate first, align the light field to the part, and then center the film to the central ray. The top of the film should be about 1½ inches above the acromion process. Check to be sure shadows of acromion and olecranon processes are within the field.

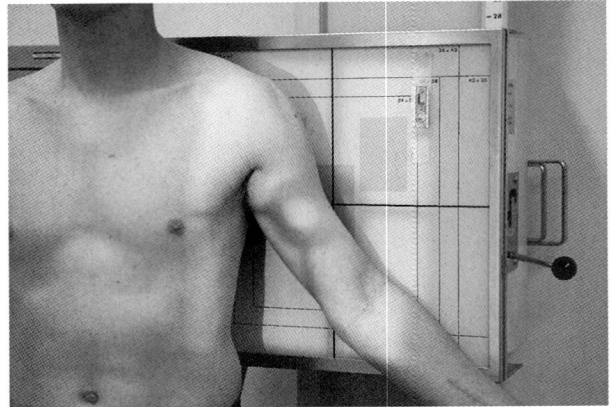

AP humerus position, upright.

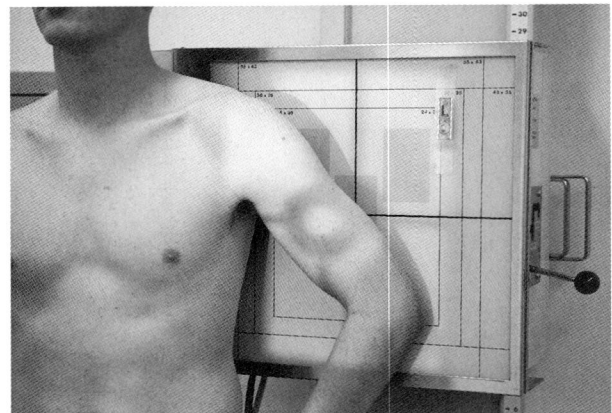

Lateral humerus position, upright.

RECUMBENT HUMERUS

The basic examination of the recumbent humerus includes AP and lateral views.

The procedure for radiography of the humerus in the recumbent position is similar to that used for upright studies. It varies somewhat because the width of the radiographic table may not allow room for the patient to lie safely when the arm is abducted and placed over the center of the grid. For this reason, the recumbent examination is done with the humerus nearer the trunk of the body, and the lateral projection may be done without flexing the elbow.

Screens: Regular (rapid)

Grid: Yes

SID: 40 inches

14″ × 17″ (need 2)

Body position:
Supine on radiographic table

Part position:

AP: Arm extended and supinated

Lateral: Arm rotated to exaggerated degree of pronation until coronal plane of elbow is perpendicular to table

Central ray:
Perpendicular to midpoint of film (midhumerus)

Patient instruction:
Stop breathing. Do not move.

Structures seen:
Entire humerus, shoulder, and elbow

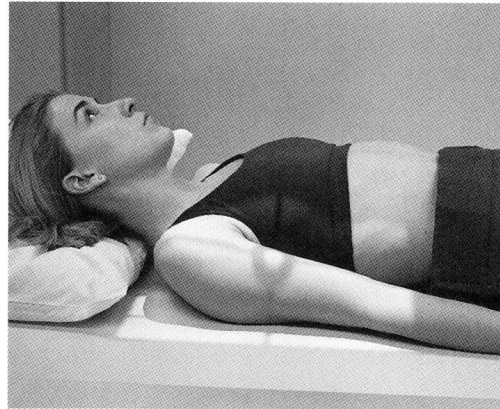

AP humerus position, recumbent.

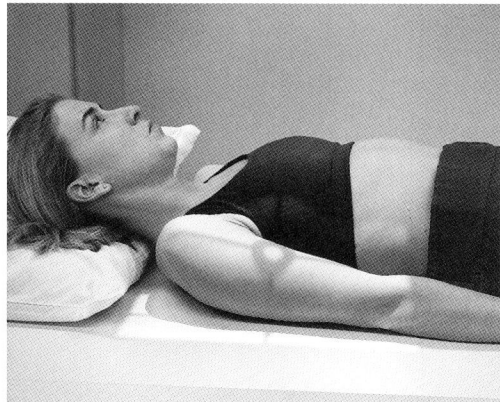

Lateral humerus position, recumbent.

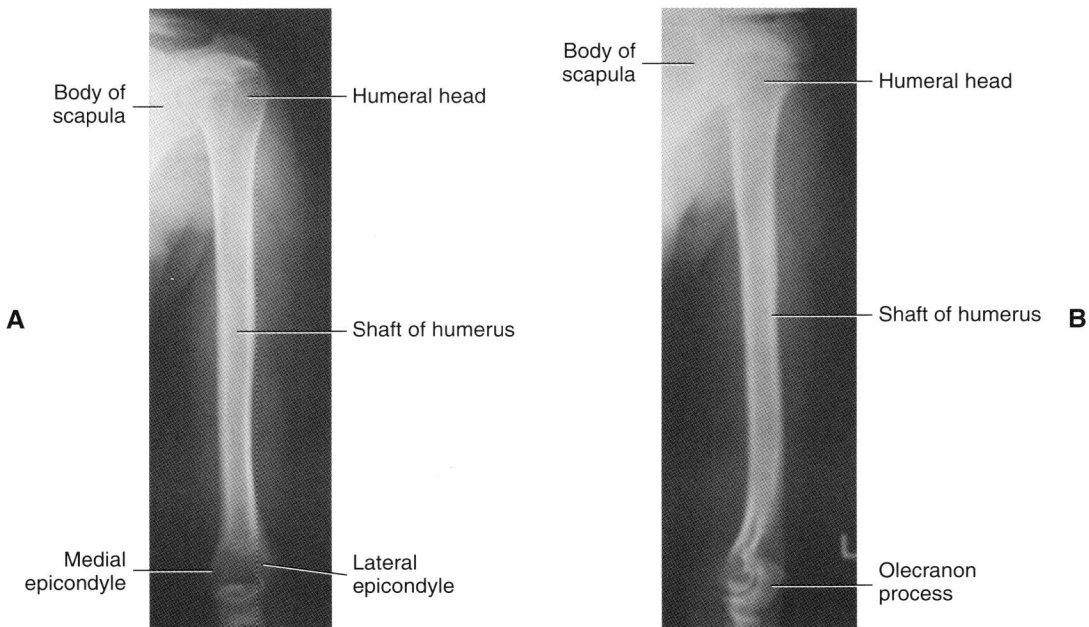

Recumbent humerus radiographs. **A,** AP; **B,** lateral.

Examinations of the shoulder girdle may be done with the patient recumbent on the radiographic table or upright using the upright bucky or grid cabinet. In hospitals it is usual for patients to be kept in a recumbent position until their injuries have been evaluated. In outpatient facilities, ambulatory patients are frequently more comfortable in the upright position. Because a full lead apron may interfere with the examination, gonad shielding is shown here using a half-apron for examinations of the shoulder region.

The routine examination of the shoulder requires that the patient be able to rotate the humerus. This is usually possible for relatively mild or chronic complaints. In such cases, it is desirable to examine the entire shoulder girdle, so the routine study should include the scapula, clavicle, and proximal humerus. The cassette is placed crosswise to accommodate the length of the clavicle.

Acute injuries to the shoulder may involve fractures of the proximal humerus or dislocation of the glenohumeral joint. In such cases, the patient cannot rotate the shoulder, and arm movement could cause additional injury. In cases of acute injury, the examination is performed without moving the arm. The cassette is placed lengthwise to include a greater portion of the humerus. Both the routine procedure and the acute injury procedure are presented.

Basic Examination

The basic examination of the shoulder includes AP projections in both internal and external rotation.

Screens: Regular (rapid)

Grid: Yes

SID: 40 inches

10″ × 12″
(need 2)

Body position:
Standing or seated with back to upright bucky or grid cabinet or supine on table. Coronal plane of body parallel to film.

Part position:

External rotation: True anatomic position with palm of hand anterior

Internal rotation: Humerus and arm rotated internally until back of hand is against thigh

Patient instruction:
Stop breathing. Do not move.

Central ray:
Perpendicular to a point 1 inch medial and inferior to coracoid process

Structures seen:
Entire clavicle and scapula and proximal third of humerus. External rotation demonstrates greater tuberosity in profile; internal rotation profiles lesser tuberosity.

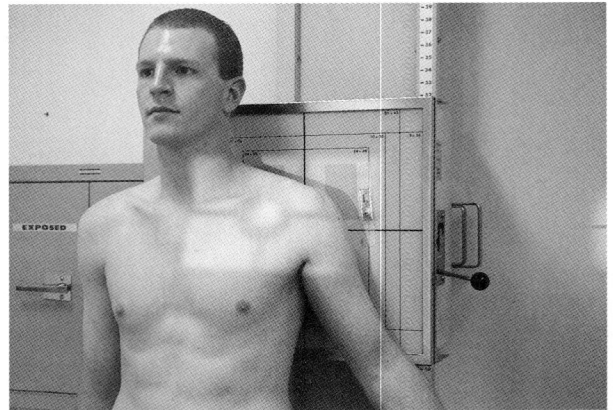

Upright AP shoulder position, external rotation.

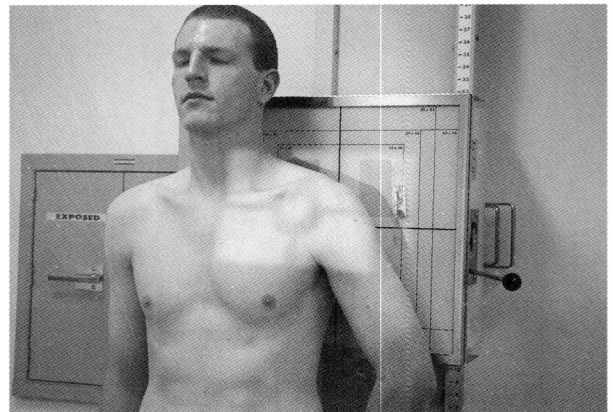

Upright AP shoulder position, internal rotation.

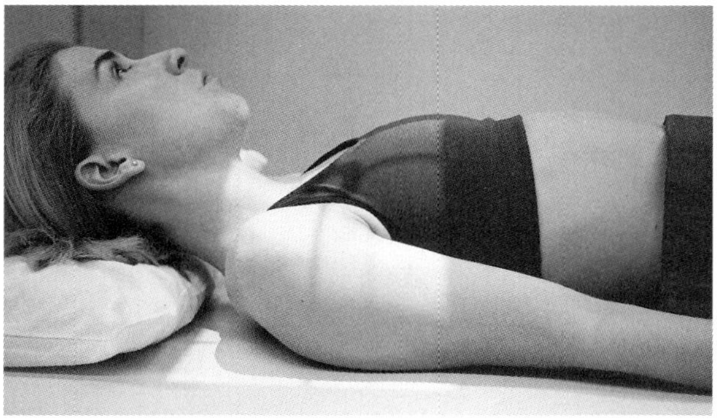

Recumbent AP shoulder position, external rotation.

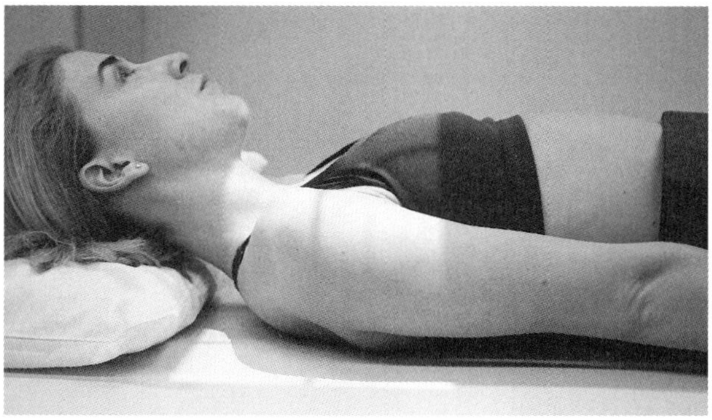

Recumbent AP shoulder position, internal rotation.

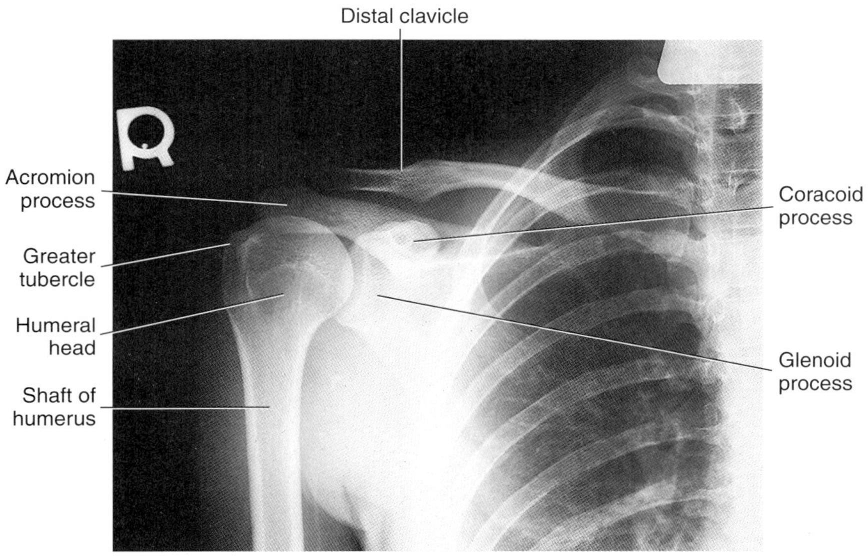

Distal clavicle

Acromion process

Greater tubercle

Humeral head

Shaft of humerus

Coracoid process

Glenoid process

AP shoulder radiograph, external rotation.

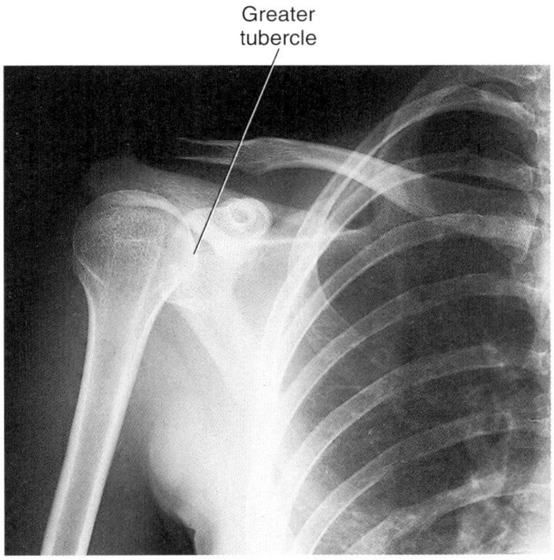

Greater tubercle

AP shoulder radiograph, internal rotation.

Alternative Projection

GRASHEY PROJECTION OF GLENOHUMERAL ARTICULATION

Film: 8 × 10-inch (20 × 25-cm) cassette, length-wise in bucky

Body position:
Coronal plane of body aligned to form 30° angle with film plane

Part position:
Arm in internal, external, or neutral (thumb-forward) rotation

Central ray:
Perpendicular to point midway between acromion process and axillary fold

Structures seen:
Glenohumeral articulation with open joint space and glenoid process in profile

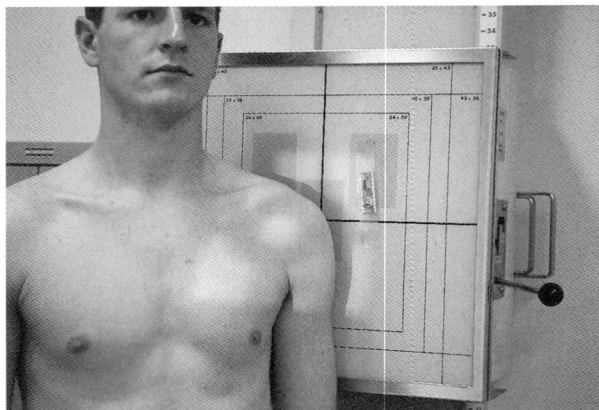

Grashey position for shoulder joint.

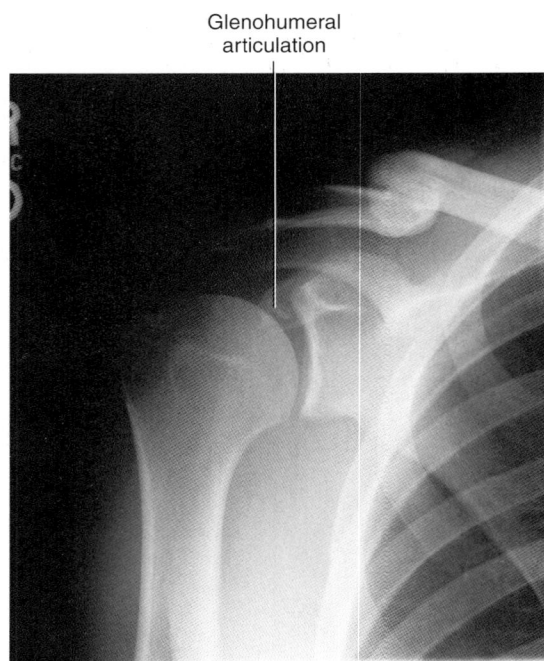

Glenohumeral articulation

Grashey view of shoulder joint.

Basic Examination for Acute Injury

The basic examination for acute injury to the shoulder includes AP and transthoracic lateral views.

Screens: Regular (rapid)

Grid: Yes

SID: 40 inches

10" × 12"
(need 2)

Body position:
Standing or seated at upright bucky or grid cabinet

Part position:

AP: With back to film, and without moving arm. Coronal plane of body is parallel to film.

Transthoracic lateral: Coronal plane of body is perpendicular to film with affected side nearest film. Unaffected arm is raised above head.

Central ray:
Perpendicular to center of film, with top of film 1½ to 2 inches above level of acromion process (approximately at midpoint of upper half of humerus)

Patient instruction:
"Breathing technique." Exposure is made during exhalation. This technique effectively blurs superimposing rib and lung structures, improving visualization of humerus.

Structures seen:
Proximal half of humerus and lateral half of scapula. Demonstrates fractures of upper humerus and aids in evaluation of glenohumeral dislocation.

TIPS:
- Measure for transthoracic lateral through body and affected arm (path of central ray).
- If chart does not provide exposure for this projection, use lateral thoracic spine chart.
- If patient is unable to raise contralateral arm high enough to avoid superimposition of shoulders, angle tube 5° to 10° cephalad.
- A low mA setting that provides the desired mAs with an exposure time of at least 1 second is needed to take full advantage of breathing technique.

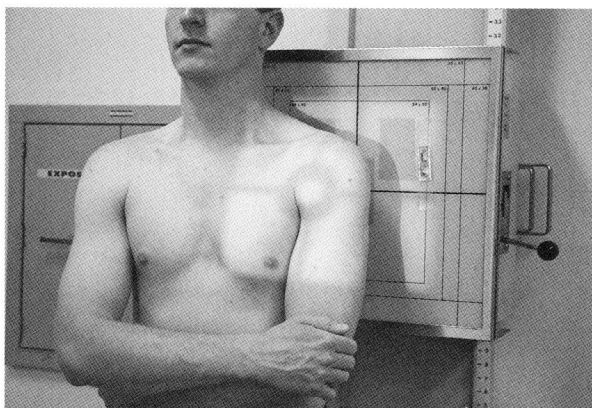

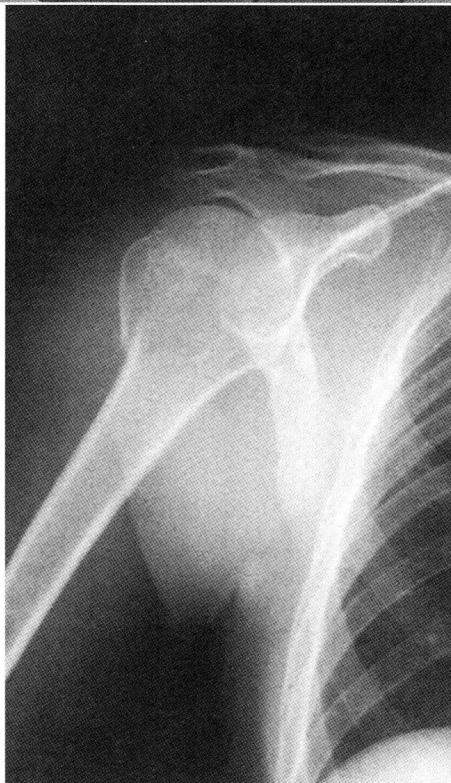

AP position for shoulder with acute injury and radiograph.

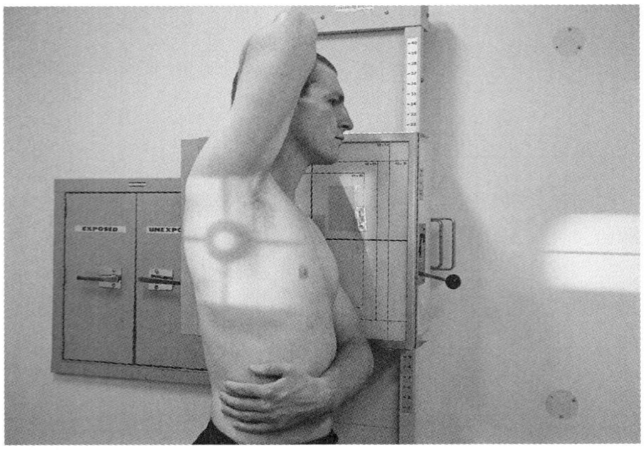

Position for transthoracic lateral position for shoulder and upper humerus with acute injury.

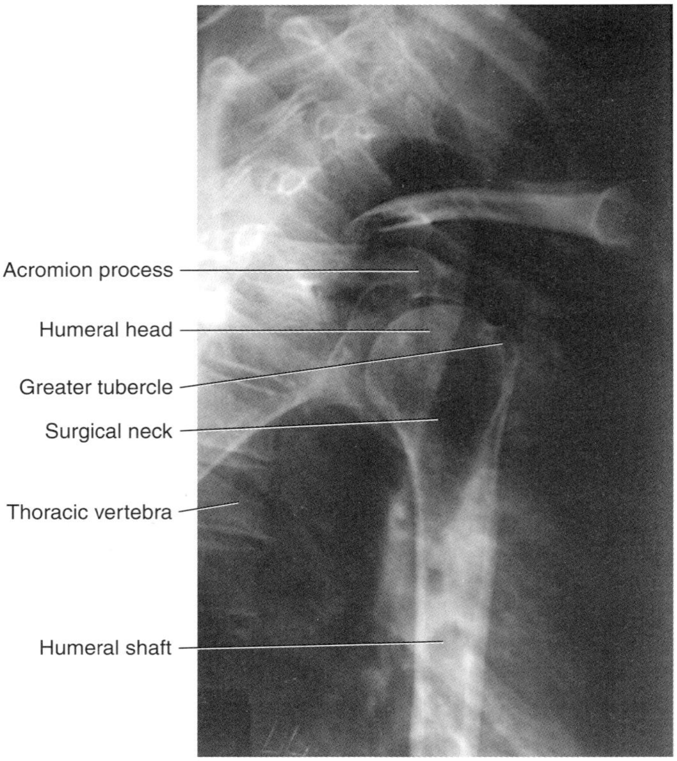

Acromion process ——————

Humeral head ——

Greater tubercle ——

Surgical neck ——

Thoracic vertebra ——

Humeral shaft ——

Transthoracic lateral view of shoulder and upper humerus.

Examination of the clavicle is routinely done in the PA and PA axial projections to keep the clavicle as close to the film as possible. If the patient is recumbent, however, a prone position may be very uncomfortable when the clavicle has been injured. If necessary, AP projections may be done in the supine position. The central ray must be angled *cephalad* for the AP axial projection.

Basic Examination

The basic examination of the clavicle includes the PA and PA axial views.

Screens: Regular (rapid)

Grid: Yes

SID: 40 inches

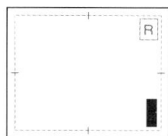
10″ × 12″ (need 2)

Body position:
Standing, seated, or prone facing bucky with coronal plane parallel to cassette. Turn head away from side of interest. Arm at side.

Part position:
See body position.

Central ray:

PA: Perpendicular to midclavicle

Axial: 30° caudad to midclavicle through trapezius ridge (large muscle above spine of scapula)

Patient instruction:
Stop breathing. Do not move.

Structures seen:
Entire clavicle and its articulations

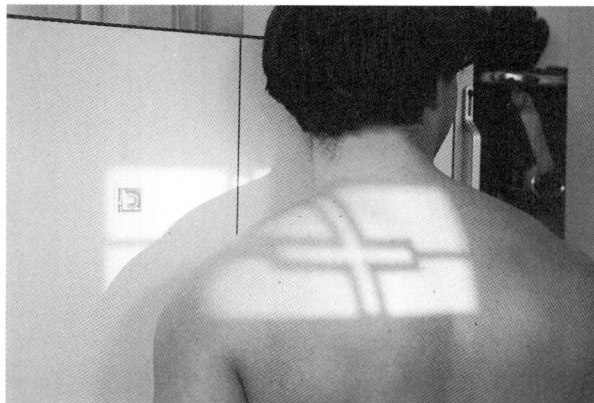
PA clavicle position, upright.

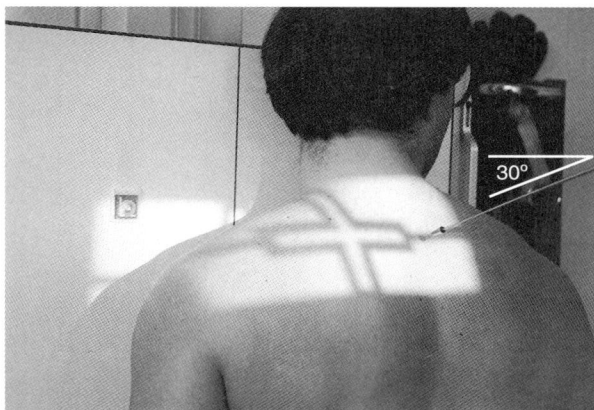
Axial clavicle position, upright.

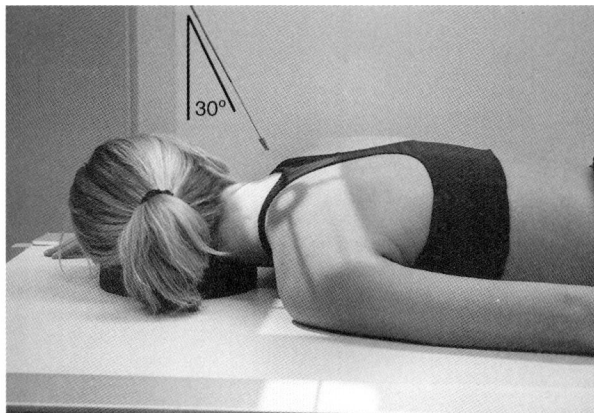
PA clavicle position, recumbent.

Axial clavicle position, recumbent.

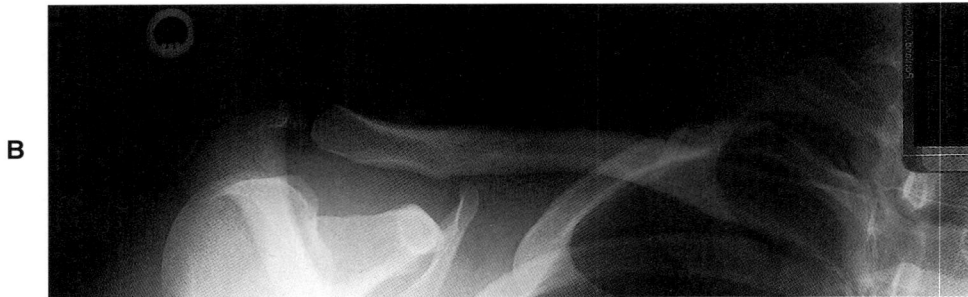

Clavicle radiographs. **A,** PA; **B,** axial.

Basic Examination

The basic examination of the scapula includes the AP and lateral views.

Screens: Regular (rapid)

Grid: Yes

SID: 40 inches

10″ × 12″
(need 2)

Body position:
Standing or seated at upright bucky or grid cabinet, or recumbent on table

Part position:

AP: Arm abducted so that humerus is perpendicular to long axis of body. Elbow flexed 90°. When patient is upright, patient may support position by grasping a pole.

Lateral: Anterior oblique body position with affected side nearest film. Adjust angle of body so that blade of scapula is perpendicular to film. Patient's forearm is positioned behind back with elbow flexed 90°. Alternatively arm may be positioned over head or across chest, depending on structures of interest and patient's ability to comply.

> **TIP:** To ensure that the blade of the scapula is perpendicular to the film, it may be helpful to grasp the scapula with your fingers on its lateral border and your thumb on the vertebral border. Rotate patient's body until a line between your fingers and your thumb is perpendicular to the film.

Central ray:
Perpendicular to midscapula, midway between acromion process and inferior angle

Patient instruction:
Stop breathing. Do not move.

Structures seen:
Entire scapula and its articulations with clavicle and humerus. For lateral projection, arm positioned behind back gives a superior view of acromion and coracoid processes but sometimes results in superimposition of humerus over blade of scapula. Arm placed over head provides unobstructed view of blade but humeral head superimposes superior structures. Arm positioned across chest also compromises visualization of superior structures but is often only position attainable by patient with scapular injury.

AP scapula position, upright.

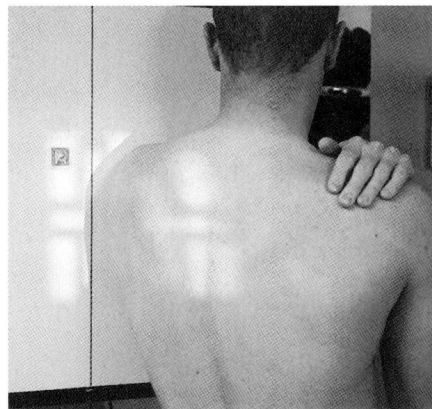

Lateral scapula position, upright.

AP scapula position, recumbent.

Lateral scapula position, recumbent.

Acromion process

Glenoid fossa

Lateral border

Inferior angle

Scapular spine

Vertebral border

Coracoid process

Body

AP scapula radiograph.

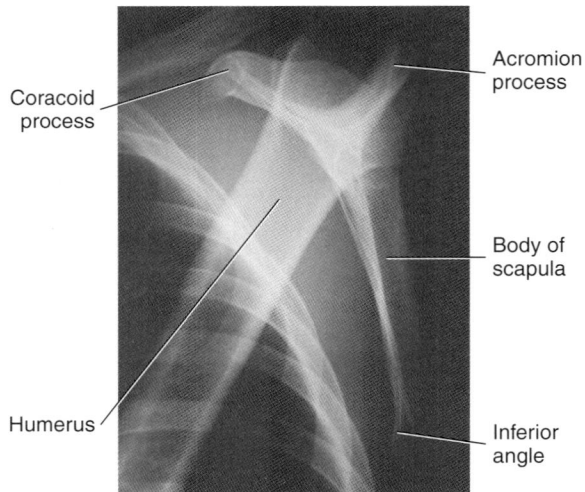

Coracoid process

Humerus

Acromion process

Body of scapula

Inferior angle

Lateral scapula radiograph taken with forearm behind waist.

The purpose of the acromioclavicular (A/C) joint study is to determine the status of the ligaments that bind the A/C joints together. While ligaments cannot be seen on radiographs, the relative positions of the bones of the joint provide information that is used for this evaluation. The study includes comparison views of the right and left A/C joints, taken both with and without weights. The weights are usually 20-pound sandbags. The patient must hold them in such a way that the weight bears on the A/C joints. This means that the elbows must be fully extended and the muscles of the arms relaxed.

The diagram below demonstrates why it is most desirable to radiograph both joints at the same time. The diverging rays at the periphery of the x-ray beam then pass through the joints at an angle that provides a direct view through the joint spaces.

A frequent problem with this examination is that the patient's shoulders are too broad for both joints to be radiographed at once, even when a 14 × 17-inch (35 × 43-cm) cassette is placed crosswise. In this case, two cassettes may be placed in a nongrid film holder (as shown below) as far apart as needed to accommodate patient size. It may be necessary to increase the SID beyond 40 inches to obtain a radiation field wide enough to cover the expanded area.

To perform this examination, position a 14 × 17-inch (36 × 43-cm) cassette crosswise in an upright bucky, grid cabinet, or nongrid film holder. If the pa-tients' shoulders are too broad for one film, use two 10 × 12-inch (25 × 30-cm) cassettes placed lengthwise in a nongrid film holder. Separate the cassettes to permit coverage of both A/C joints, allowing ample space for magnification. Center the tube to the patient and align the *lower half* of the film(s) to the central ray. With the patient's arms in neutral position, make the first exposure without weights. It is helpful to mark the film as to which view has weights and which does not.

Adjust the film height so that the central ray is now centered to the *upper half* of the film(s). Maintain patient position while providing a 20-pound sandbag for each hand, instructing the patient to allow the weight to pull from the shoulder. Quickly make a second exposure.

The thickness of the A/C joint, when measured carefully through this area only, is about the size of the elbow. It is therefore possible to get acceptable radiographs of this area without the use of a grid. The joints are very small, and the x-ray beam can be collimated to a thin horizontal strip, making it possible to get two exposures (with and without weights) on the same film. A lead mask is not needed, since the patient's body absorbs most of the scatter that might otherwise fog the film.

To further decrease scatter and to minimize radiation dose to the sensitive tissues of the thyroid gland and upper chest area, a lead strip may be taped across the center of the collimator.

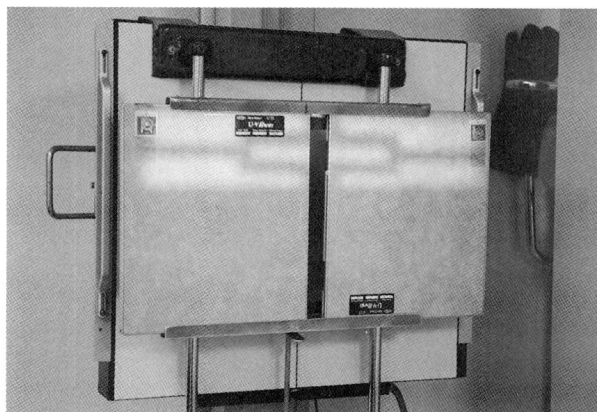

Seen from the top, it is apparent that the acromioclavicular joints lie at an angle. When the tube is centered to the body's midline, the diverging x-ray beam more accurately demonstrates the joint spaces.

Cassette placement for AP acromioclavicular joint study using two cassettes, nonbucky.

Screens: Extremity (detail) or regular (rapid)

Grid: Not required, but may be used

SID: 40 inches (or more as needed for wide shoulders)

Body position:
Standing with back to cassette(s)

Part position:
Same as body position

Central ray:
Perpendicular to midline at level of acromion processes

Patient instruction:
Stop breathing. Do not move.

Structures seen:
Bilateral AC joints for comparison to evaluate ligamentous integrity

TIPS:
- If the two sides are radiographed separately, which is acceptable, patient must hold sandbags in both hands for both weight-bearing views.
- All views must be taken using the same position, alignment, and technique.
- Measure through A/C joints only; do not use full shoulder measurement.
- If SID is increased beyond 40 inches to cover a wide field, adjust mAs accordingly.

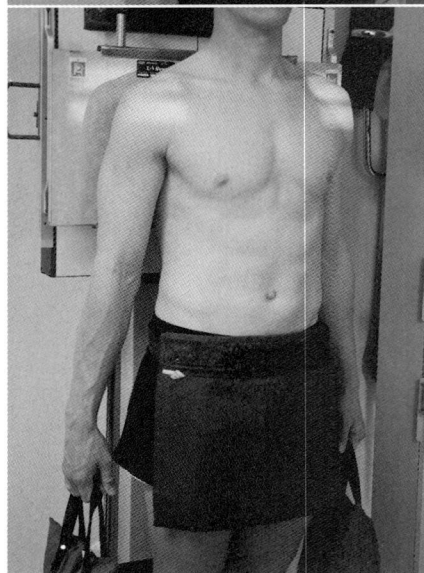

Positioning for acromioclavicular joint study using two cassettes and nongrid cassette holder. **A,** First exposure is made on lower third of cassettes without weight bearing. **B,** Second exposure is made on upper third of cassettes with patient holding 20-pound weights.

Bilateral study of acromioclavicular joints, with and without weight bearing, using two cassettes and nongrid cassette holder.

PATHOLOGY

Probably the most significant pathology affecting the upper extremity from the radiographer's viewpoint is trauma. Fractures and other injuries to this portion of the anatomy may vary greatly and only those most commonly seen in radiography are discussed here.

Common Fractures

The boxer's fracture is a common fracture of the fifth metacarpal, usually caused when the patient strikes a solid object with a closed fist (Fig. 13-12).

An important carpal bone injury is a fracture of the scaphoid (navicular), often caused by a fall on an outstretched arm. When they are new, scaphoid fractures may be occult, that is, very subtle or completely invisible on a radiograph. Special projections are often needed to demonstrate this fracture, or it may be necessary to have the patient return for another x-ray examination after about 10 days (Fig. 13-13). It

is especially important to identify these fractures because the scaphoid has a relatively inefficient blood supply, and there is a strong possibility of necrosis if the fracture is not identified early and treated correctly (Fig. 13-14). Necrosis of the scaphoid causes impairment of wrist function.

A Colles fracture (Fig. 13-15) is a common fracture of the distal radius, accompanied by posterior and medial displacement. The Monteggia fracture is a classic fracture of the forearm and is associated with dislocation of the radial head (Fig. 13-16).

The most common elbow fracture is a fracture of the radial head, which may occur as the result of a fall on an outstretched arm (Fig. 13-17). It is particularly important to demonstrate the soft tissues of the elbow joint in the lateral projection. When a fracture at the elbow causes **joint effusion** (increased fluid in the joint capsule), the fat pad in the joint region will be displaced. It moves upward from the joint area and can be seen radiographically as a dark shadow in

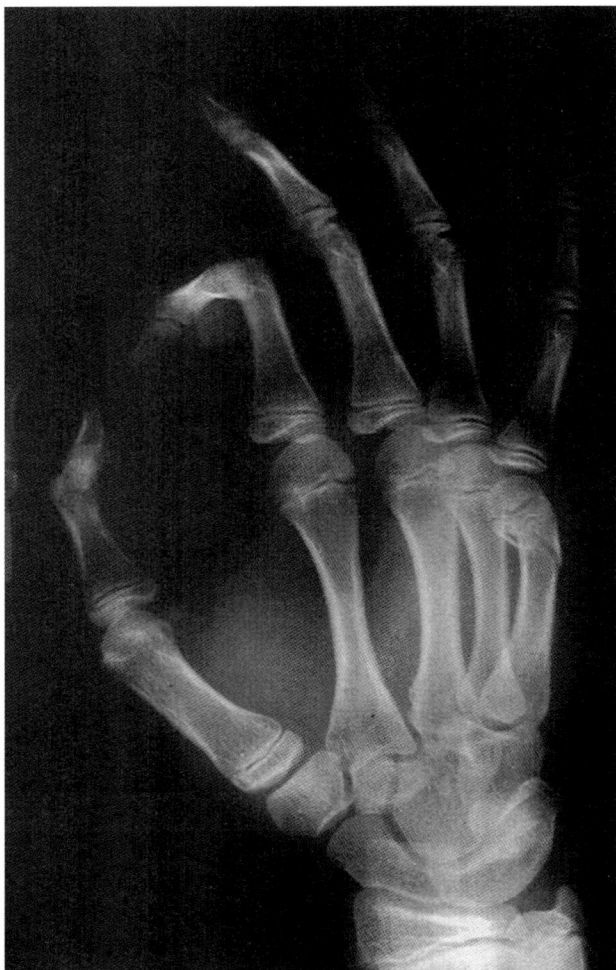

Fig. 13-12 Boxer's fracture.

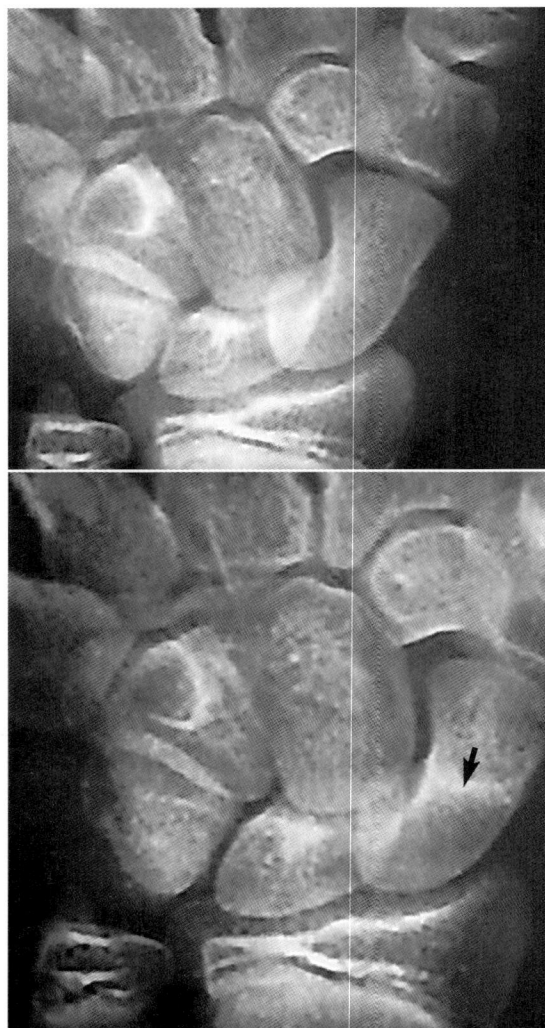

Fig. 13-13 Scaphoid fracture at injury and 3 weeks later.

Fig. 13-14 Avascular necrosis of scaphoid.

Fig. 13-15 Colles' fracture.

Fig. 13-16 Monteggia fracture.

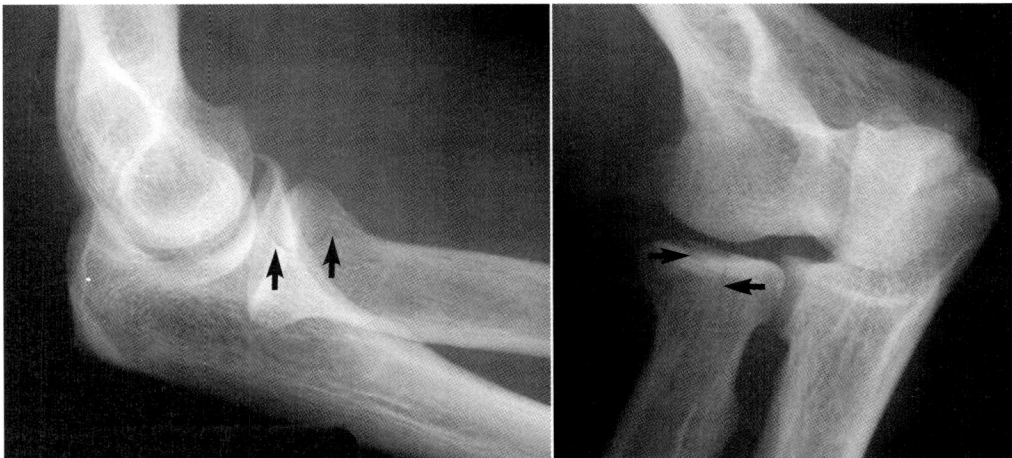

Fig. 13-17 Radial head fractures.

the soft tissues anterior or posterior to the humerus. This **fat pad sign** (Fig. 13-18) may be the only radiographic indication of a fracture involving the elbow joint.

The humerus is most commonly fractured at its weak point, the surgical neck. An example is seen in the positioning section of this chapter.

Fractures of the clavicle are frequently seen, especially in children (Fig. 13-19).

Other Trauma Conditions

Dislocation of the shoulder joint is a fairly common injury. The humeral head may be displaced from the glenoid fossa either anteriorly or posteriorly (Fig. 13-20).

Injuries that introduce foreign bodies into the soft tissues may also be evaluated radiographically. In the upper extremity, this is most commonly seen in the hand.

Nontraumatic Conditions

Chronic irritation to a bursa may lead to a condition called **bursitis,** inflammation of the bursa. Bursitis may cause calcific (calcium) deposits in the soft tissues of the joint region that are visible on radiographs. The shoulder is a very common site for calcific bursitis (Fig. 13-21). Inflammation of a tendon, called **tendinitis,** may occur at any tendon attachment in the body. It is common at the shoulder and in the wrist. Tendinitis may also produce calcific deposits in the soft tissues.

Arthritis (joint inflammation) may affect any part of the body, and there are a number of different types. Rheumatoid arthritis, illustrated in Fig. 12-35, is a crippling disease that often involves the hands and is a common reason for radiography. The most common type of arthritis is a degenerative joint disease called **osteoarthritis** (Fig. 13-22). It is a chronic condition that causes hypertrophy of the bone. The enlarged, deformed portions of the bone are called **osteophytes.**

Osteomyelitis (Fig. 13-23) is inflammation of bone, especially the marrow, caused by a pathogenic organism. Bone infection may be caused by a number of bacteria, including *Staphylococcus* and tuberculosis.

Fig. 13-24 illustrates a bone cyst of the humerus. The cyst is fluid-filled and has a wall of fibrous tissue. Cysts of this type have no symptoms and are usually discovered only when they have weakened the bone sufficiently to cause a fracture. Fractures caused by underlying disease are called *pathologic fractures.* The cyst is actually a type of benign neoplasm. Other types of neoplasm and metastatic bone diseases may also affect the bones of the upper extremity. They are called **osteoblastic** if they result in increased bone formation and **osteolytic** if they cause destruction of the bone.

Fig. 13-25 is an example of a bone infarct. Ischemia causes dense calcification in the medullary canal. It is sometimes necessary to take an additional radiograph with increased kVp to demonstrate the extent of necrosis.

Fig. 13-18 Fat pad sign *(arrow).*

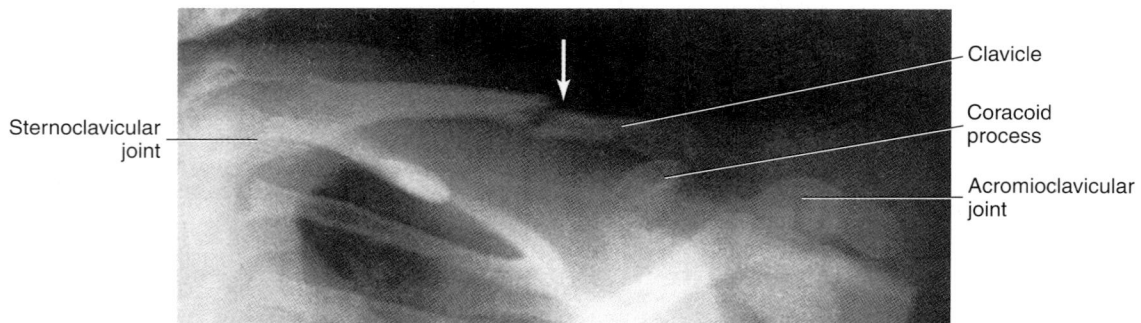

Fig. 13-19 Clavicle fracture.

A　　　　　　　　**B**

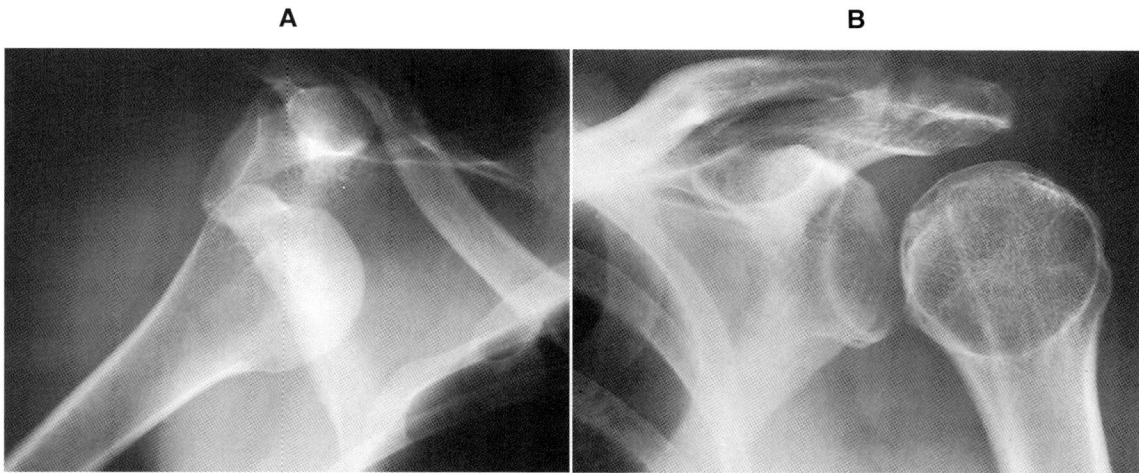

Fig. 13-20 Shoulder dislocations. **A,** Anterior; **B,** posterior.

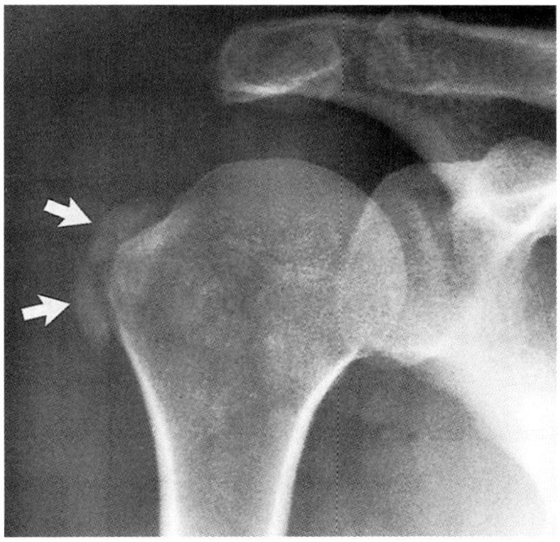

Fig. 13-21 Calcific bursitis.

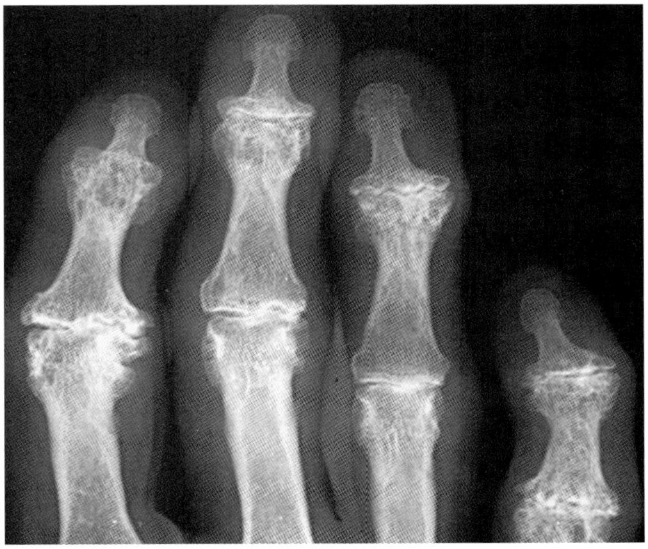

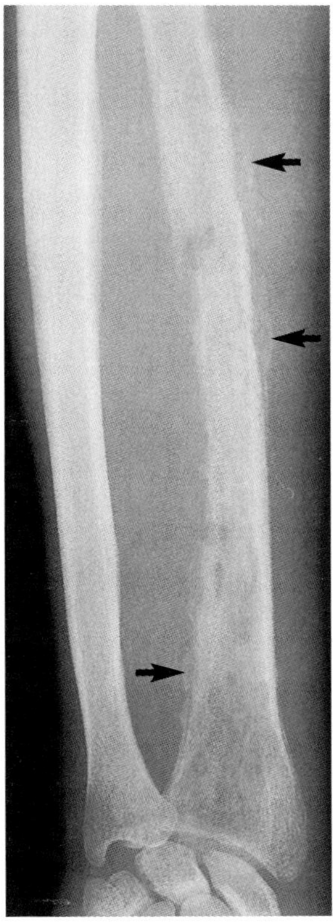

Fig. 13-23 Osteomyelitis. 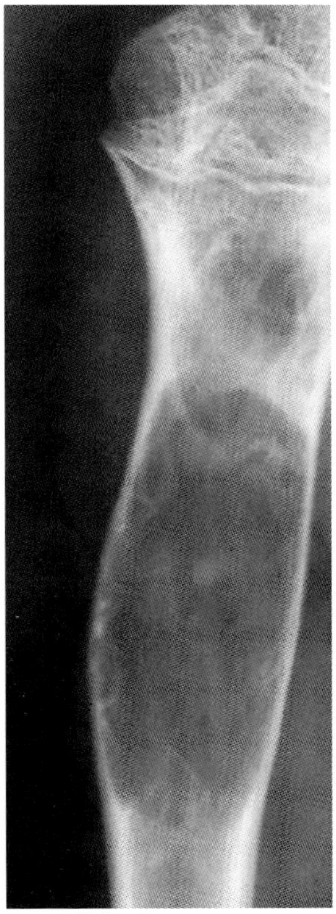　　**Fig. 13-24** Bone cyst in humerus.

Fig. 13-22 Osteoarthritis of hand.

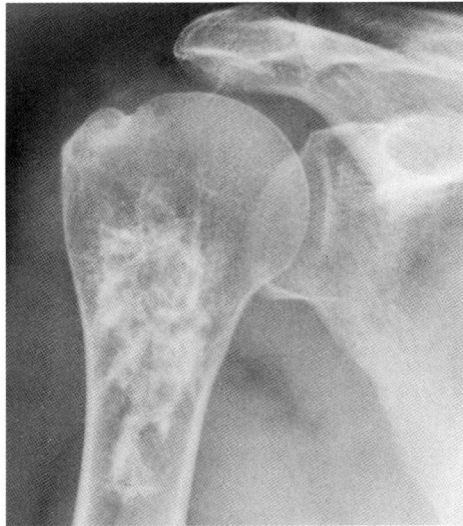

Fig. 13-25 Bone infarct.

SUMMARY

The bones of the hand and wrist include the phalanges, metacarpals, and carpals. The radius and ulna form the forearm, articulating at the wrist and the elbow. The humerus articulates with the forearm at the elbow and forms the shoulder joint where it articulates with the scapula. The scapula, clavicle, and upper humerus form the shoulder girdle. Important palpable bony prominences include the styloid processes of the distal radius and ulna, the olecranon process of the proximal ulna, the epicondyles and greater tuberosity of the humerus, and the acromion and coracoid processes of the scapula.

Radiography of the hand, wrist, forearm, and elbow is done tabletop using extremity cassettes. The patient is usually seated at the end of the table. Multiple views are often done on a single cassette. Examinations of the humerus and shoulder girdle, on the other hand, are done using rapid cassettes and grids or buckys. The patient may be recumbent or upright. Examination of the acromioclavicular joints requires a rather unusual combination of methods to produce bilateral views taken with and without weights.

The most common radiographic pathology occurring in this portion of the anatomy involves trauma, particularly fractures. Nontraumatic conditions, such as arthritis, bursitis, tendinitis, and neoplastic disease, are also seen.

▪ Review Questions ▪

1. Name the middle bone of the middle finger and a carpal bone that articulates with the first metacarpal.
2. Is the radius medial or lateral to the ulna?
3. Name the articular processes of the elbow joint.
4. Name and point to three bony prominences on your own scapula.
5. List two ways in which an examination of the thumb differs from an examination of a finger.
6. How does the position for a PA wrist differ from that for a PA hand?
7. Name two special projections used specifically to demonstrate the scaphoid.
8. If the patient is unable to fully extend the elbow for an AP elbow projection, what should you do?
9. How does a routine shoulder examination differ from a shoulder examination for acute injury? Why?
10. What views constitute a basic examination of the clavicle?
11. Demonstrate the possible arm positions for a lateral projection of the scapula and state a possible disadvantage of each.
12. List and describe four specific types of fractures of the upper extremity.
13. List three general types on nontraumatic pathology that may affect the bones of the upper extremity.

14

Lower Extremity and Pelvis

Learning Objectives

At the conclusion of this chapter, the student will be able to:

- Name the bones that make up the lower extremity and pelvis and identify each on an anatomic diagram and on a radiograph
- Name and identify the significant bony prominences and depressions of the lower extremity and pelvis and identify significant positioning landmarks by palpation
- Demonstrate correct body and part positioning for routine projections and common special projections of the lower extremity and pelvis
- Correctly evaluate radiographs of the lower extremity and pelvis for positioning accuracy
- Describe and recognize on radiographs pathology that is common to the lower extremity and pelvis

Key Terms

calcaneus
fabella
femur
fibula (pl. fibulae)
ilium (pl. ilia)
innominate bone
ischium (pl. ischia)
meniscus (pl. menisci)

metatarsal
patella (pl. patellae)
pedal digit
prosthesis
pubis (pl. pubes)
talus
tarsal bones
tibia

f you have studied the previous chapter on the upper extremity, you will find many parallels in this chapter. Many of the bones of the lower extremity correspond to similar structures with similar functions in the upper extremity. There are a number of significant differences, however, particularly in the ankle and knee joints. Your understanding of both upper and lower extremities will be enhanced if you compare and contrast the two extremities as you study.

ANATOMY

The lower extremity includes the foot, toes, ankle, lower leg, knee and femur (thigh bone) (Fig. 14-1).

Foot

The foot may be divided for discussion into three basic parts: the forefoot, the midfoot, and the heel (Fig. 14-2). The bones of the forefoot are called **metatarsals.** They correspond to the metacarpal bones of the hand. They are numbered 1 through 5, starting on the medial aspect. The numbers correspond to the digits with which they articulate. The distal end of a metatarsal is called its head, and the proximal end is referred to as the base. The hinge joints between the metatarsals and the proximal phalanges are called metatarsophalangeal (MP) joints.

There are usually two sesamoid bones in the region of the first MP joint. These small, flat, oval bones

were introduced in Chapter 13. They are located within tendons and are not counted among the bones of the body. They serve to protect the joint.

The midfoot consists of five short bones called **tarsal bones.** Some of them have two names that are in common use. The three cuneiform bones are distinguished as first, second, and third or internal, middle, and external. They articulate, respectively, with the first, second, and third metatarsals. Lateral to the third cuneiform is the cuboid, which articulates with the fourth and fifth metatarsals. Proximal to the cuneiforms is the tarsal navicular (scaphoid).

There are two additional tarsal bones. The heel bone is called the **calcaneus** (os calcis). Superior to the calcaneus is the **talus** (astragalus), or "ankle bone." The intertarsal (between the tarsal) joints are gliding joints with relatively small amounts of motion.

Toes

The toes are called the **pedal digits** and are considered to be part of the foot. They are numbered from 1 to 5, beginning with the great toe on the medial side. Digits 2 through 5 consist of three small long bones called phalanges, the same as the bones of the fingers. They are distinguished from one another

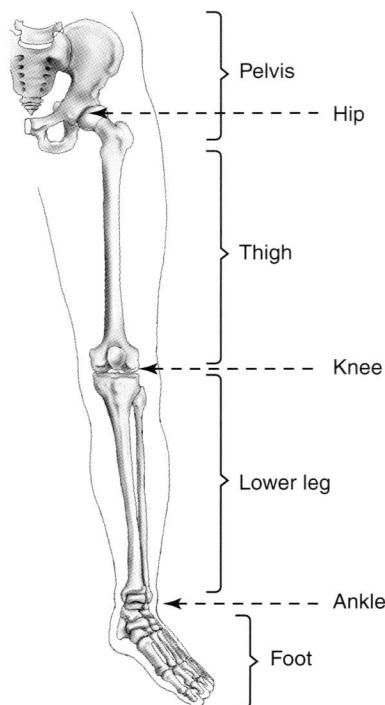

Fig. 14-1 Lower extremity.

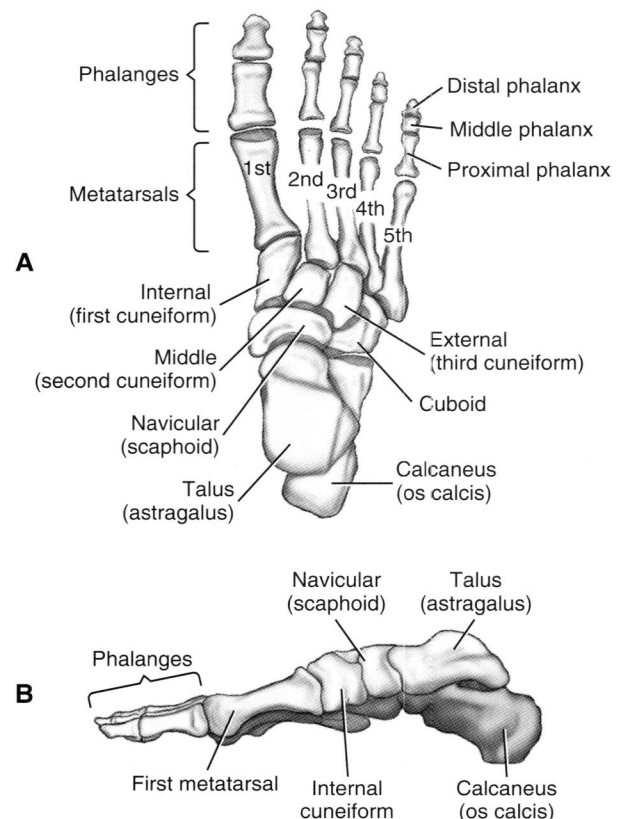

Fig. 14-2 Foot. **A,** Anterior (dorsal) aspect. **B,** Medial aspect.

as proximal (nearest the foot), middle, and distal. For example, the bone at the tip of the "little toe" is called the fifth distal phalanx. The rounded tips of the distal phalanges are called ungual tufts. The hinge joints that connect the phalanges are called the interphalangeal (IP) joints and are distinguished as proximal and distal. The great toe has only two phalanges with one IP joint.

Leg, Knee, and Thigh

The lower leg (Fig. 14-3) consists of two long bones, the **tibia** and the **fibula.** The tibia is the longer, thicker bone on the medial side. The fibula is much thinner and somewhat shorter and is located laterally. The medial malleolus is a bony prominence that can be palpated at the ankle on the medial aspect of the distal tibia. The lateral malleolus is the rounded prominence on the distal aspect of the fibula and can be felt on the lateral aspect of the ankle.

The talus articulates with the tibia and fibula to form the ankle joint. Weight is transferred from the shaft of the tibia through the talus, while the malleoli provide stability on either side. The entire joint is shaped like an inverted box (⊓) (Fig. 14-4). It is called a mortise joint because it resembles a carpenter's joint of the same name. The ankle is a hinge joint. When it flexes to raise the foot, the motion is called dorsiflexion; when it extends, pointing the toe downward, the motion is called plantar flexion. Lateral flexion of the ankle tends to roll the foot onto its medial aspect and is called eversion. Medial flexion causes the foot to roll onto its lateral aspect and is called inversion.

A knob-like protuberance on the anterior surface of the tibia near the proximal end of the shaft is called the tibial tuberosity. The articular surface of the prox-imal tibia is a large flat surface called the tibial plateau. The medial and lateral condyles are palpable projections on either side of the tibial plateau. Near the center of the tibial plateau are two superior projections called the intercondylar eminences or tibial spines.

The rounded proximal portion of the fibula is called the head. It terminates in a styloid process. The proximal fibula articulates with the metaphysis of the tibia at the inferior aspect of the lateral condyle. It is not a part of the knee joint.

The single long bone of the thigh is called the **femur** (Fig. 14-5). It is the largest, heaviest bone of the body. The distal end of the femur flares to form two palpable prominences, the medial and lateral condyles. Between the condyles posteriorly is the intercondylar fossa. The distal articular surfaces of the condyles articulate with the tibial plateau to form the knee, which is a hinge-type joint. The articular surface of each condyle is cushioned by a C-shaped cartilage called a **meniscus.**

Anterior to the distal femur is the **patella,** commonly called the kneecap. It is a flat bone in the shape of a rounded triangle with its apex on the inferior margin. The patella is actually a large sesamoid bone, the only sesamoid bone numbered among the bones of the body. It is not unusual for there to be an additional small sesamoid bone posterior to the knee. This normal variation is called a **fabella.**

The rounded superior end of the femur is called the head. A small indentation on its posterior superior surface is called the fovea capitis. The narrow portion between the head and the shaft is the neck. The neck extends from the shaft at an angle, projecting superiorly and medially. Just inferior to the neck, the proximal shaft of the femur flares to form two prominences, the greater and lesser trochanters. The greater trochanter is a large projection on the lateral aspect that is palpable on the side of the upper thigh. The lesser trochanter is inferior to the greater trochanter and projects medially. It is not normally palpable. Between the two trochanters posteriorly is a bony ridge called the intertrochanteric crest.

Fig. 14-6 illustrates the palpable bony landmarks of the lower extremity.

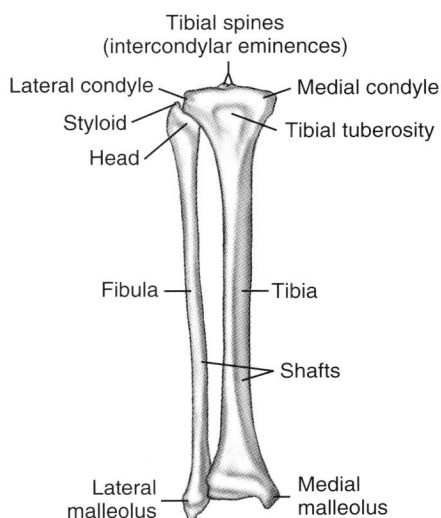

Fig. 14-3 Anterior aspect of tibia and fibula (lower leg).

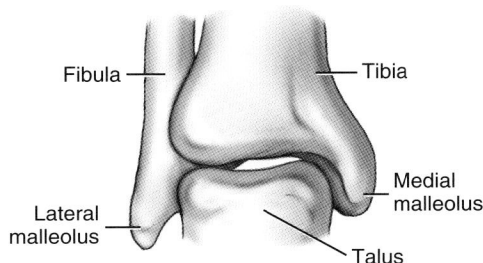

Fig. 14-4 Anterior aspect of ankle joint.

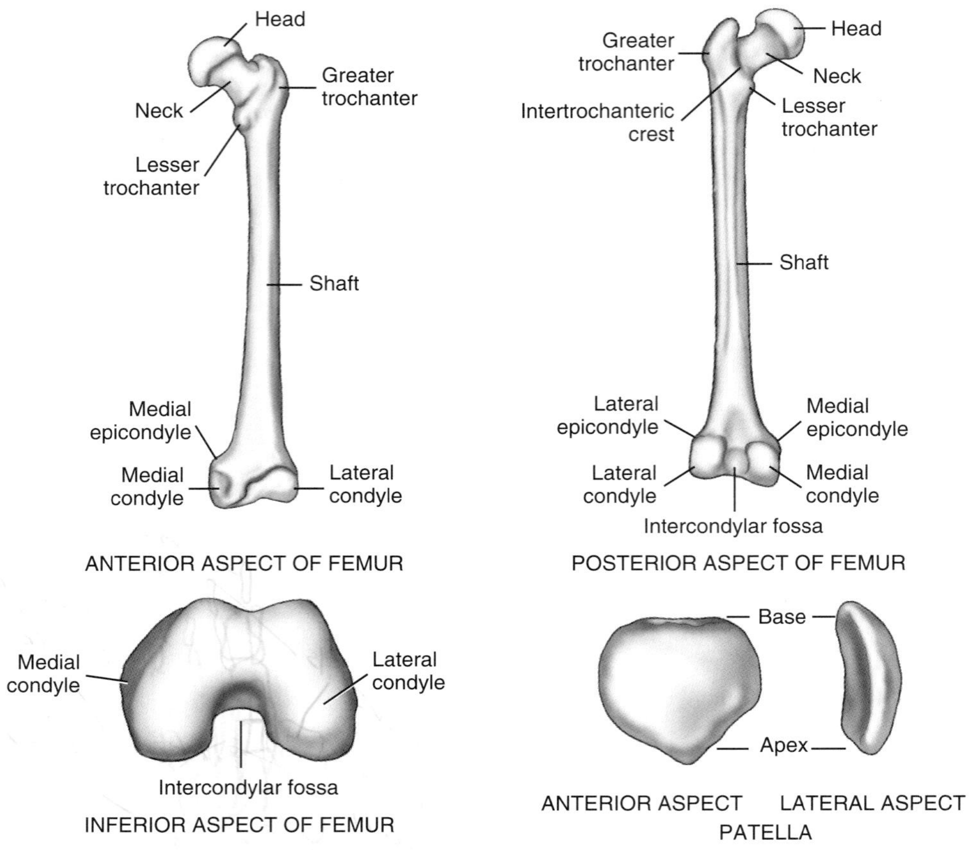

Fig. 14-5 Femur and patella.

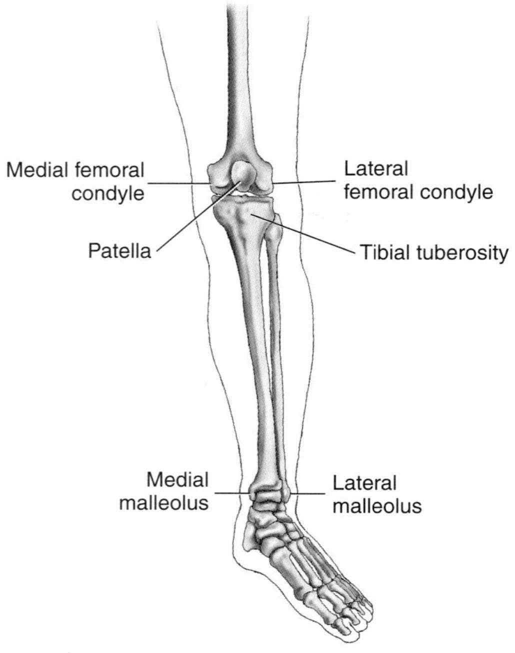

Fig. 14-6 Palpable bony landmarks of the lower extremity.

Pelvis and Hip

The two bones that make up the halves of the pelvis are called the **innominate bones** (Fig. 14-7). Each is a composite bone made up of three bones: the **ilium,** the **ischium,** and the **pubis.**

The ilium forms the upper portion of the pelvis. Its large, flat superior portion is called the ala, or wing. It articulates with the sacral portion of the spine medially, forming the sacroiliac joint. Its rounded superior margin is palpable and is a common positioning landmark called the iliac crest. On the lateral aspect of the ilium is an anterior projection called the anterior superior iliac spine (ASIS). The ASIS is palpable on the anterior surface of the body in the hip region and is also a common positioning landmark.

The inferior portion of the pelvis is formed by the ischium posteriorly and the pubis anteriorly. The most inferior portion of the ischium is a bony prominence called the ischial tuberosity. When sitting erect, the weight of the body is supported on the ischial tuberosities. They are palpable through the inferior portions of the buttocks. Together, the rami

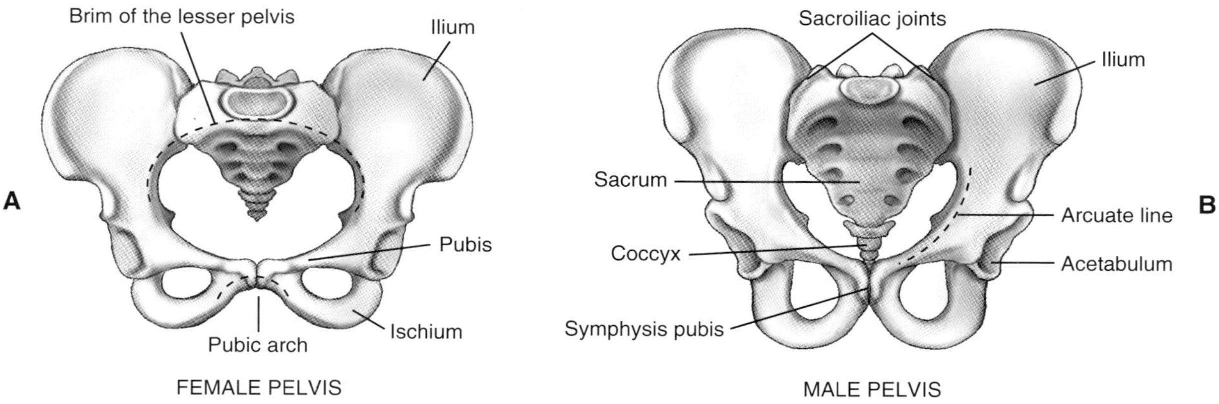

Fig. 14-7 Pelvis. **A,** Female. **B,** Male.

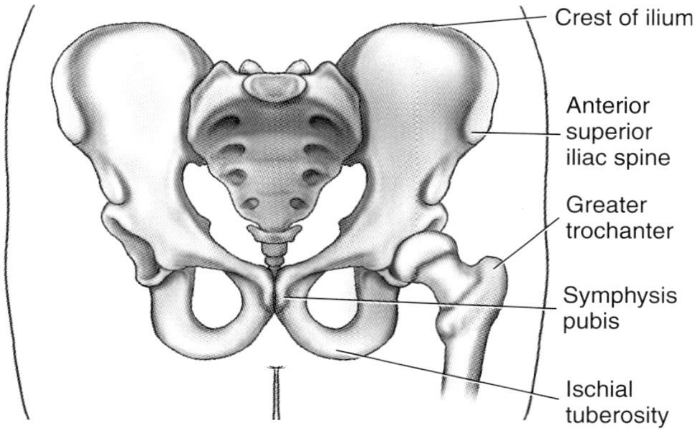

Fig. 14-8 Palpable bony landmarks of the pelvis and hip.

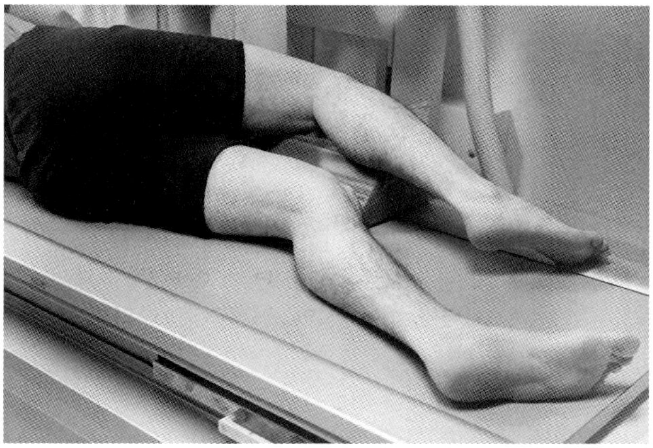

Fig. 14-9 Body position for lateral projections of lower extremity.

(branches) of the ischium and pubis form a bony ring. The hole within this ring is called the obturator foramen.

The ilium, ischium, and pubis join to form a synarthrodial joint at the acetabulum. The acetabulum is the rounded fossa that forms the socket of the hip joint. It articulates with the head of the femur. The right and left pubic bones join in the midline to form the symphysis pubis, an amphiarthrodial joint.

Fig. 14-8 illustrates the palpable bony landmarks of the pelvis and hip.

POSITIONING AND RADIOGRAPHIC EXAMINATIONS

Examinations of the lower extremity from the toes up to the knee joint are usually done tabletop (nongrid) with the patient sitting or lying on the radiographic table.

The shoe and stocking should be removed from the affected leg, as well as any jewelry in the region.

Trousers should be removed if they cannot be moved out of the radiation field, especially if the fabric is heavy, such as jeans. A rolled or bunched trouser leg must not be included in the field.

For many of these examinations a single film is used for more than one exposure. The procedure for multiple exposures on a single cassette is explained in Chapter 12.

While gonad shielding is legally required only for examinations of the femur, it is wise to shield whenever possible. A lead apron is shown here for these examinations.

The body position illustrated in Fig. 14-9 places the entire lower extremity in lateral position. This is an excellent body position for lateral lower extremity projections. Note that the coronal plane of the pelvis is perpendicular to the table, the knee is flexed approximately 45°, and the ankle is dorsiflexed so that the foot forms an angle of 90° with the leg. The unaffected leg is supported to prevent rotation of the pelvis.

Regardless of the fact that the top of the foot is considered to be its dorsal aspect, a dorsoplantar projection is also called an AP projection. The central ray is angled posteriorly for this projection to avoid foreshortening of the metatarsals and to better demonstrate the intertarsal articulations.

Basic Examination

The basic examination of the foot includes the AP (dorsoplantar), oblique (medial), and lateral views.

Screens: Extremity (detail)

Grid: No

SID: 40 inches

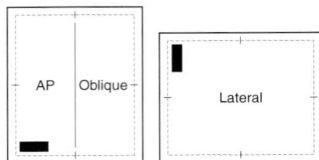

10″ × 12″ (need 2)

Body position:
Seated or recumbent on table with knee flexed

Part position:
For all views foot is centered to film or film portion so that toes, heel, and both malleoli are within field.

AP: Plantar surface of foot is in contact with cassette.

Oblique: Ankle is everted so that medial plantar aspect of foot is in contact with cassette. Coronal plane of foot forms 30° angle with cassette. Large 45°-angle sponge medial to distal leg may be used to support and maintain position.

Lateral: Lateral aspect of foot is in contact with cassette and foot is in true lateral position with coronal plane perpendicular to film. Ankle is dorsiflexed so that long axis of foot is perpendicular to tibia. Alternatively, medial aspect of foot may be placed in contact with cassette. This may be desirable when medial structures are of primary clinical interest. This position is often awkward for patient.

Central ray:

AP: Angled 10° posteriorly (toward heel) and centered to center of film

Oblique and lateral: Perpendicular to center of film

Patient instruction:
Do not move.

Structures seen:
Entire foot, including toes, metatarsals, and tarsal bones. On AP projection, calcaneus is obscured by superimposition of lower leg. Oblique projection should demonstrate the metatarsals without superimposition on one another. Too much superimposition of these structures indicates that angle between plantar surface of foot and film was too great; that is, foot was everted too much. Lateral projection shows superimposition of metatarsals. It should include ankle joint.

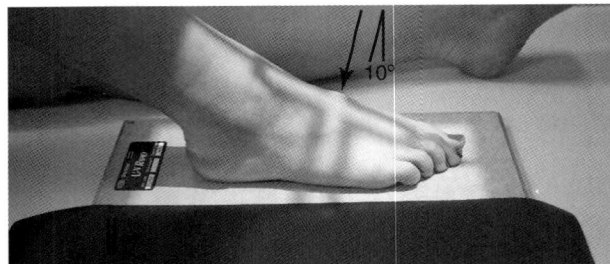

AP foot position.

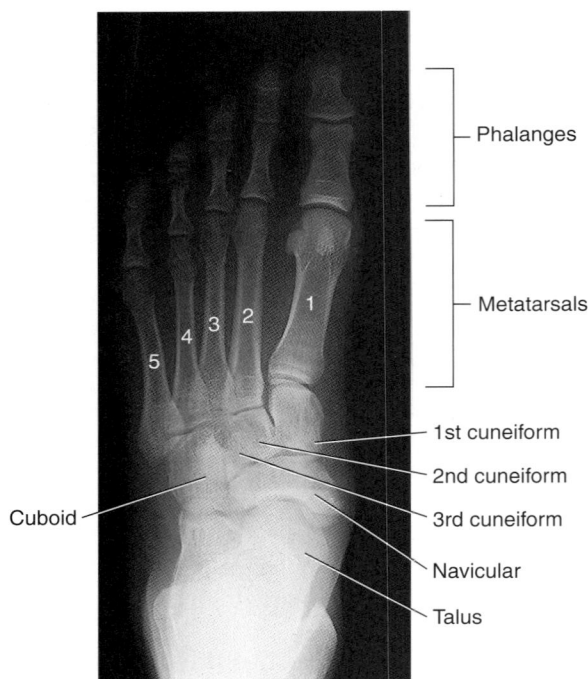

AP foot radiograph.

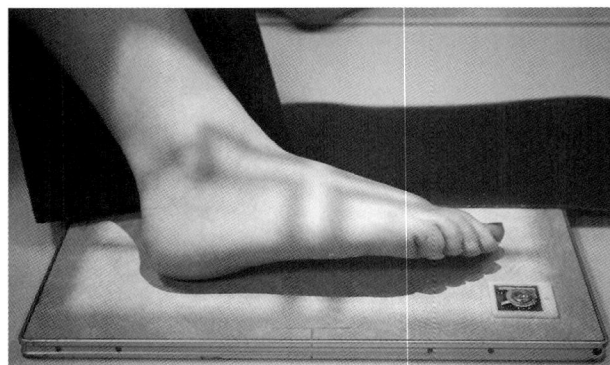

Medial oblique foot position.

TIP: A #2 wedge filter may be used to produce a more even radiographic density on the AP and oblique projections, avoiding overexposure of the toes and distal metatarsals. The filter is placed so that its thicker portion is projected over the toes and its thin edge is projected in the midmetatarsal region.

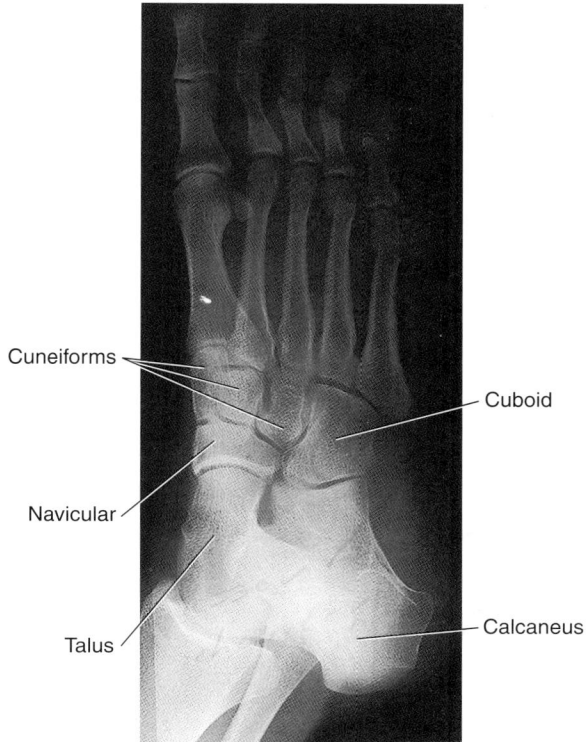

Medial oblique foot radiograph.

Lateral foot position.

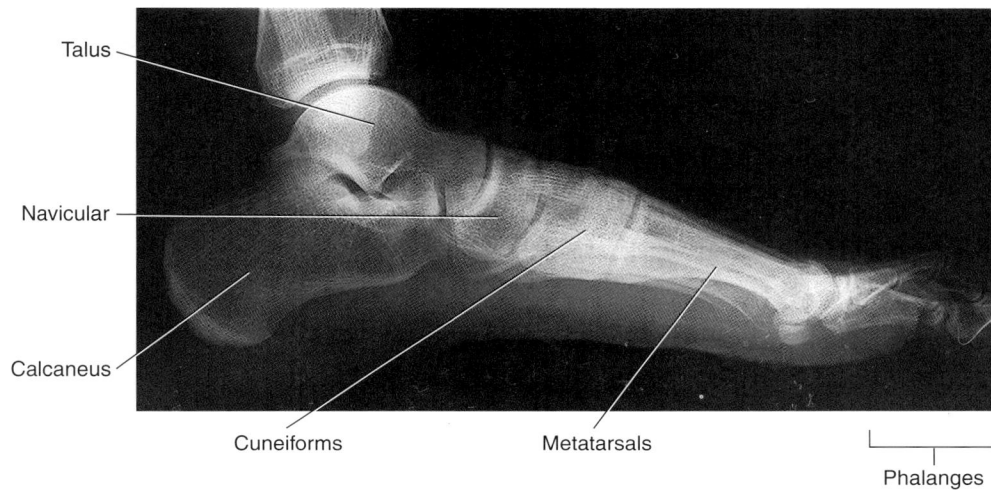

Lateral foot radiograph.

While the toes are included in the examination of the foot, separate studies of the toes may be performed when the area of clinical interest is limited to a specific toe. Each projection of the toe should include the entire digit and the distal portion of the corresponding metatarsal.

Toes 2 through 5 tend to curl downward. When this is the case, a small wedge sponge helps support the toes parallel to the film for the AP projection (Fig. 14-10). For lateral projections of the toes, the foot may be positioned with either its medial or lateral aspect in contact with the film, depending on which toe is involved. Keep the toe as close to the film as possible and maintain the toe in a position parallel to the film. When the toe is angled in relation to the film, visualization of the joint spaces is compromised. This concept was explained in Chapter 13 with respect of visualization of the IP joints of the fingers; it applies to examinations of the toes as well.

Fig. 14-10 Wedge sponge supports toes parallel to film.

Basic Examination

The basic examination of the toe includes the AP, oblique (plantomedial), and lateral views.

Screens: Extremity (detail)

Grid: No

SID: 40 inches

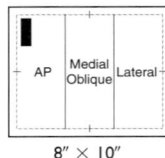

Body position:
Seated or recumbent on table with knee flexed

Part position:

AP: Plantar surface is in contact with cassette.

Oblique: Medial plantar surface of toe and forefoot is in contact with cassette. Coronal plane of forefoot and toes forms 30° angle with cassette. Distal lower leg may be supported medially with large 45°-angle sponge, as for oblique projection of foot.

Lateral: Medial or lateral surface of foot may be in contact with cassette, depending on which brings toe of interest nearest to cassette. Other toes are flexed or extended as needed to leave affected toe free of superimposition. Affected toe is supported parallel to cassette. Toes may be held in position using tape or a bandage. Positioning a single toe apart from the others often demands some creativity by radiographer. Variations may be required depending on which toe is involved, configuration of toe, and movements tolerable for patient.

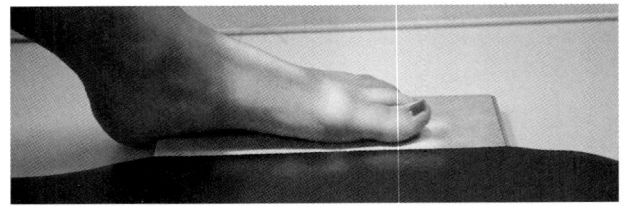

AP position for toe(s).

AP radiograph of great toe.

Central ray:

Perpendicular to MP joint

Patient instruction:

Do not move.

Structures seen

Entire digit and distal half of metatarsal with interphalangeal and metatarsophalangeal joint spaces open and clearly visualized

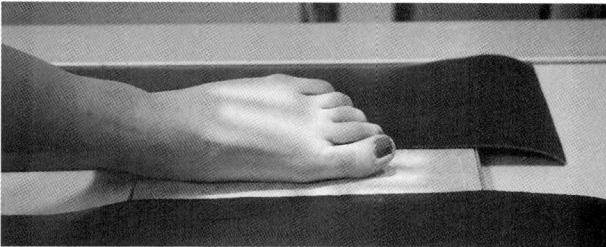

Medial oblique position for toe(s).

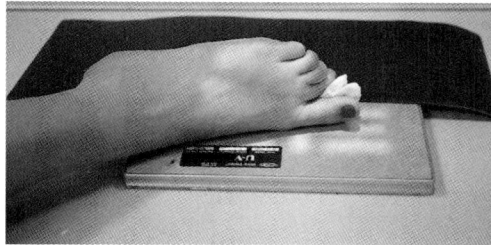

Lateral position for great toe.

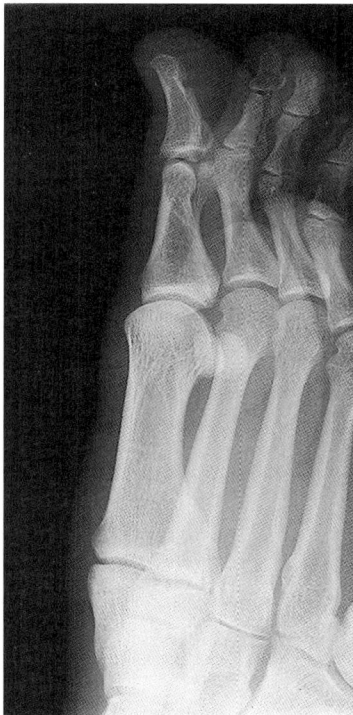

Medial oblique radiograph of great toe.

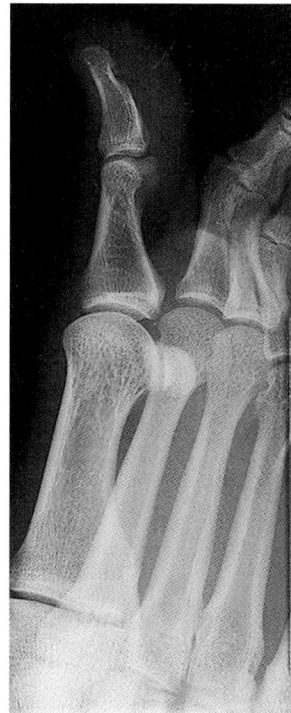

Lateral radiograph of great toe.

Basic Examination

The basic examination of the calcaneus includes the axial and lateral views.

Screens: Extremity (detail)

Grid: No

SID: 40 inches

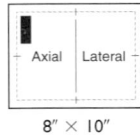

Body position:

Axial: Seated or recumbent on table with leg extended

Lateral: Seated or recumbent on table with knee flexed

Part position:

Axial: Posterior surface of ankle and heel is in contact with cassette. Place foot so that malleoli are centered to middle of cassette. Sagittal plane of foot is perpendicular to film. Ankle is dorsiflexed as much as possible and held in position by patient using a strap or bandage.

Lateral: Lateral surface of heel is in contact with cassette. Part is positioned as for lateral projection of foot, but with calcaneus centered to film.

Central ray:

Axial: Angled 40° cephalad to center of film

Lateral: Perpendicular to center of film

Patient instruction:
Do not move.

Structures seen:
Both projections demonstrate entire calcaneus and its articulation with talus. Lateral projection also shows calcaneal articulations with cuboid and navicular.

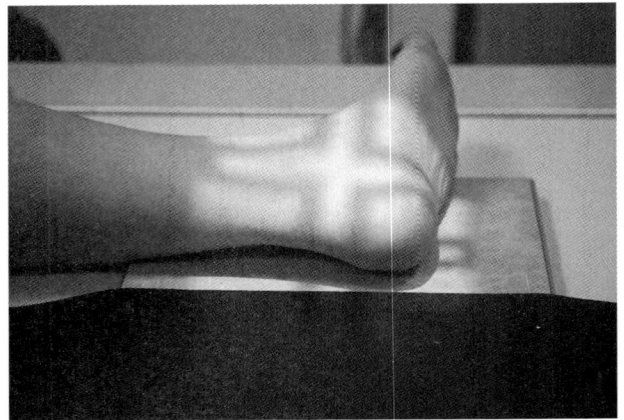

Position for axial calcaneus.

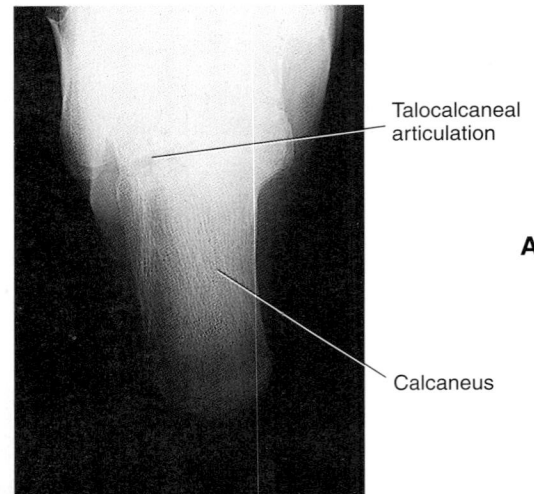

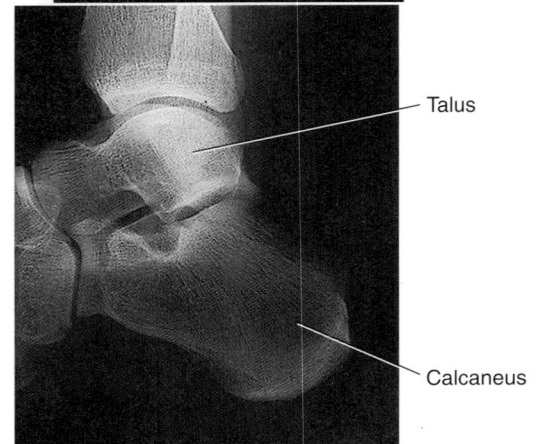

Position for axial calcaneus.

Calcaneus radiographs. **A,** Axial; **B,** lateral.

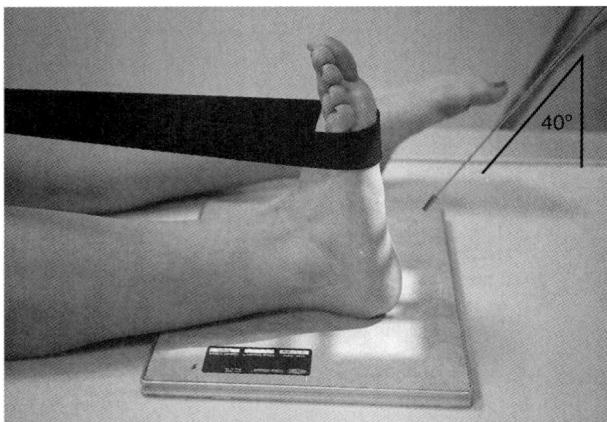

The medial oblique ankle projection described here is sometimes called the "mortise view," since it best demonstrates the entire ankle joint. Some authorities refer to it as the AP projection of the ankle and recommend an oblique projection with 45° medial rotation. The projections that constitute a basic examination must be determined by the physician who will interpret the films.

Basic Examination

The basic examination of the ankle includes the AP, medial oblique, and lateral views.

Screens: Extremity (detail)

Grid: No

SID: 40 inches

Body position:

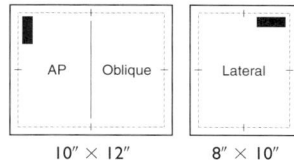

AP and oblique: Seated or recumbent on table with affected leg extended

Lateral: Recumbent or semirecumbent on affected side with knee flexed 30° to 45°

Part position:

AP: Posterior surface of heel and lower leg is in contact with cassette. Midpoint between malleoli is centered to film. Ankle is dorsiflexed so that axis of foot forms 90° angle with axis of lower leg. Sagittal planes of leg and foot are perpendicular to film. Foot may be held in position by patient using a strap or bandage.

Oblique: From AP position, entire leg is rotated medially 15° to 25°. Sagittal planes of foot and leg must remain aligned to each other. When position is correct, plane between malleoli should be parallel to film. Ankle is dorsiflexed so that axis of foot forms 90° angle with axis of lower leg.

Lateral: Lateral surface of ankle is in contact with cassette. Sagittal plane of foot and leg is parallel to film. Ankle is dorsiflexed so that axis of foot forms 90° angle with axis of lower leg.

> **TIP:** When pressure on the lateral malleolus is painful for the patient, a small level sponge under the distal portion of the leg helps to maintain this position without discomfort.

Central ray:

AP and oblique: Perpendicular to point midway between malleoli

Lateral: Perpendicular to medial malleolus

Patient instruction:
Do not move.

Structures seen:
Superior portion of talus and distal portions of tibia and fibula. Oblique projection demonstrates joint spaces surrounding talus without superimposition by fibula.

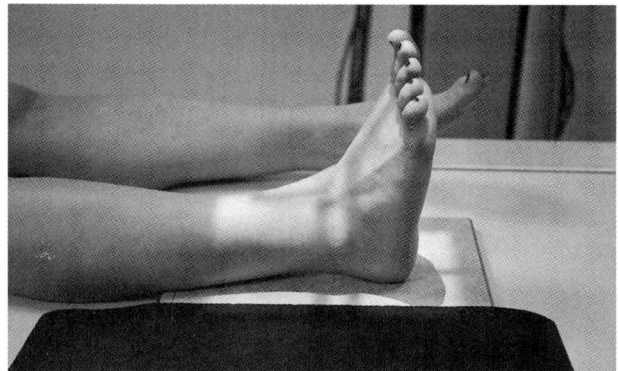

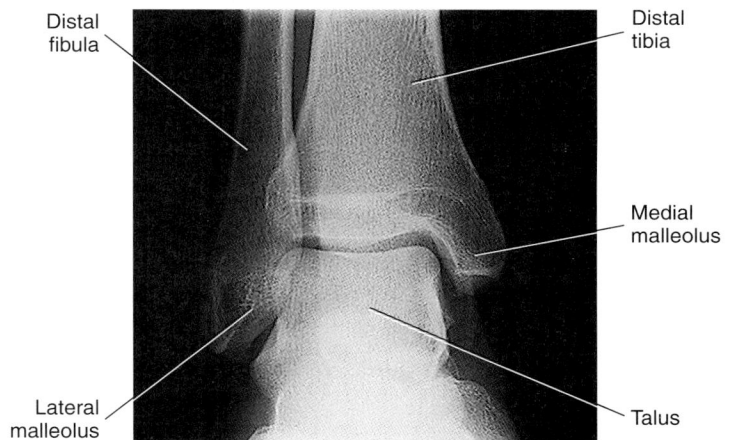

AP ankle position and radiograph.

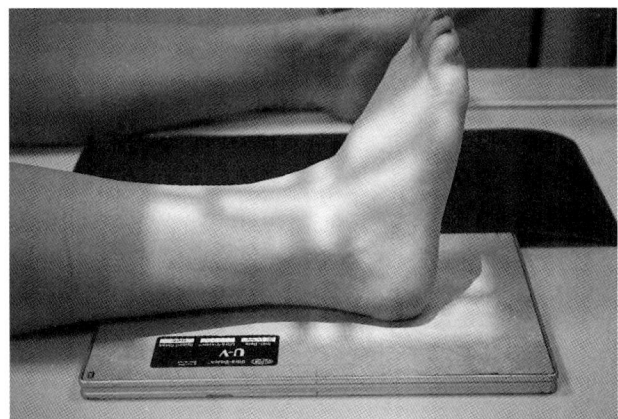

Medial oblique ankle position.

15-25°

Leg rotation for medial oblique ankle. Maintain normal alignment between leg and foot.

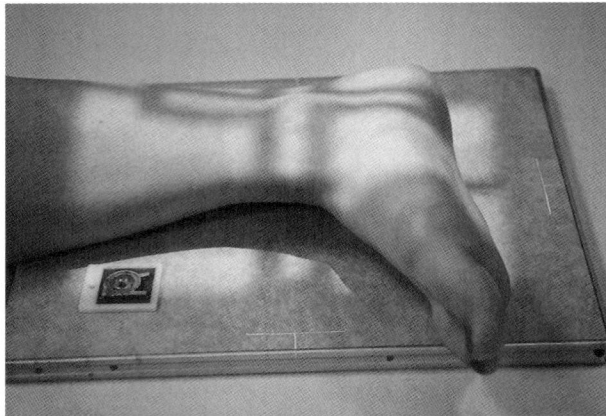

Talofibular articulation

Medial oblique ankle radiograph (mortise view).

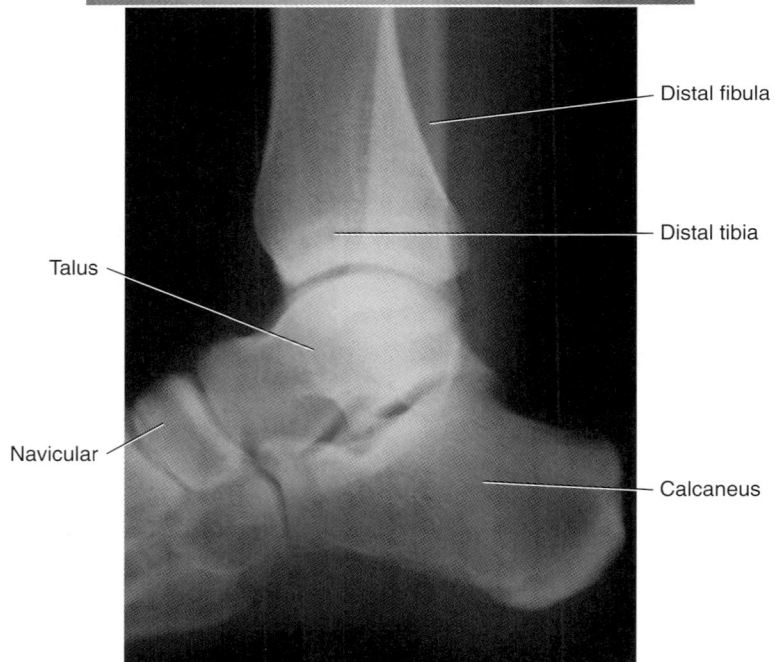

Distal fibula

Distal tibia

Talus

Navicular

Calcaneus

Lateral ankle position and radiograph.

The lower leg examination should include the entire tibia and the fibula and their articular surfaces. Visualization of both knee and ankle joints is preferable. If only one joint is demonstrated, the *same* joint must be demonstrated on both projections. It is often necessary to use two 14 × 17-inch (43 × 36-cm) cassettes diagonally to demonstrate both joints on adult patients.

Basic Examination

The basic examination of the lower leg includes the AP and lateral views.

Screens: Extremity (detail) or regular (rapid)

Grid: No

SID: 40 inches

Body position:

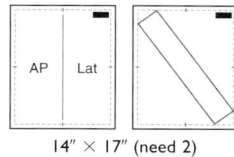

14″ × 17″ (need 2)

AP: Seated or recumbent on table

Lateral: Recumbent on affected side with contralateral leg supported anterior to affected leg

Part position:

AP: Leg is fully extended with posterior surface of lower leg in contact with cassette. Margin of cassette is placed 1 to 2 inches beyond joint of primary interest. Ankle is dorsiflexed so that axis of foot forms 90° angle with axis of lower leg. Sagittal planes of leg and foot are perpendicular to film. Foot may be held in position by patient using a strap or bandage.

Lateral: Knee is flexed 30° to 45° with lateral surface of lower leg in contact with cassette. Sagittal plane of leg is parallel to film. Margin of film is placed 1 to 2 inches beyond joint of primary interest.

Central ray:
Perpendicular to center of film

Patient instruction:
Do not move.

Structures seen:
Entire lower leg and at least one joint

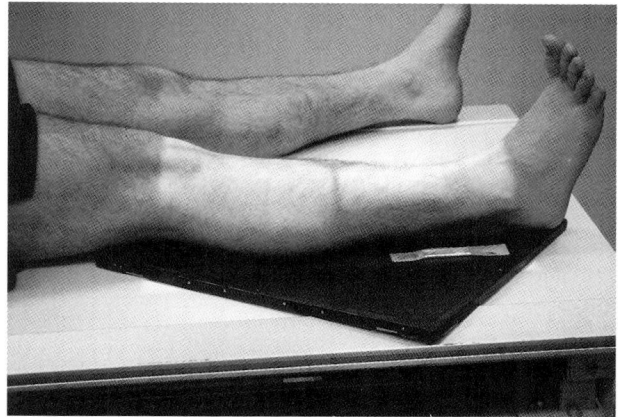

AP lower leg position.

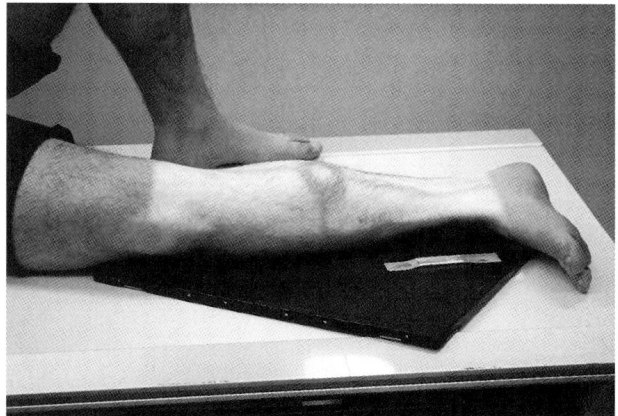

Lateral lower leg position.

Lower leg radiographs. **A,** AP; **B,** lateral.

While the basic examination of the knee consists of only AP and lateral projections, it is not uncommon to include also a view of the intercondylar fossa ("tunnel" or "notch" view) and a tangential projection of the patella ("sunrise" view). These additional views are frequently requested for the evaluation of chronic knee complaints.

When the area of clinical interest is the patella, the basic examination includes PA and lateral projections. A tangential view of the patella may be added for chronic conditions but should not be included when there is suspicion of patellar fracture. *Do not flex the knee more than 10° when there is suspicion of fracture of the patella.* When this is the case, the lateral projection is taken with the knee extended.

Basic Examination

The basic examination of the knee includes the AP and lateral views.

Screens: Extremity (detail) without grid *or* regular (rapid) with grid

Grid: With or without is acceptable. For large knees, radiographic contrast is superior with a grid.

SID: 40 inches

8″ × 10″ (need 2)

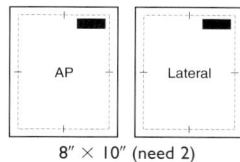

Body position:

AP: Seated or supine on table with leg extended

Lateral: Recumbent on affected side with femur aligned to center line of table. Unaffected leg is supported anterior to affected leg.

Part position:

AP: Leg is fully extended with sagittal plane of leg perpendicular to film.

Lateral: Knee is flexed 30° to 45°. Sagittal plane of femur and lower leg is parallel to film.

Central ray:

AP: Angled 5° cephalad and centered to inferior margin of patella

Lateral: Angled 5° cephalad and centered to inferior margin of medial condyle

Patient instruction:
Do not move.

Structures seen:
Knee joint with portions of distal femur and proximal lower leg. Lateral projection includes a profile of tibial tuberosity. It should demonstrate distal femur with condyles superimposed and joint space free of superimposition. Entire patella and retropatellar joint space should also be clearly visualized.

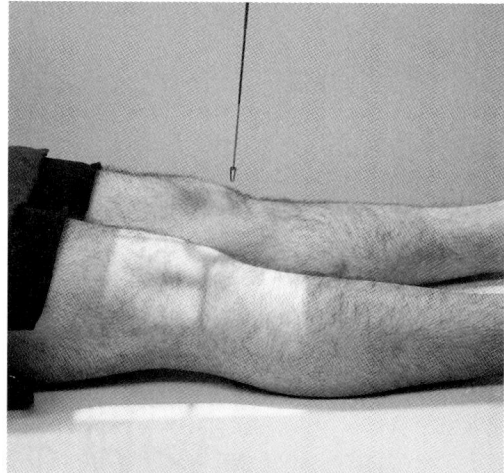

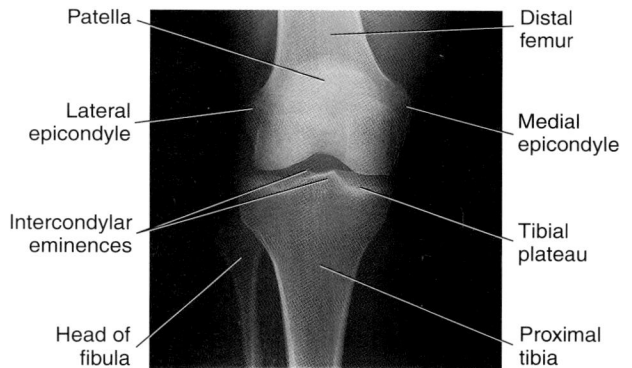

AP knee position and radiograph.

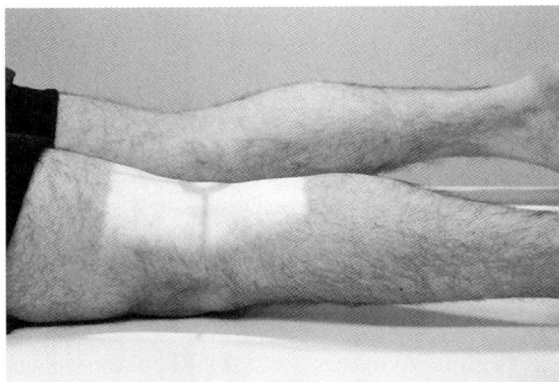

Lateral knee position and radiograph.

Alternative Projection

The PA projection of the knee is sometimes substituted for the AP projection. This is especially desirable when the patella is of particular clinical interest.

Body position:
Patient prone

Part position:
Affected leg extended and sagittal plane of leg perpendicular to film; foot on affected side is plantar flexed and rests on its dorsal aspect.

Central ray:
Perpendicular to center of film through knee joint

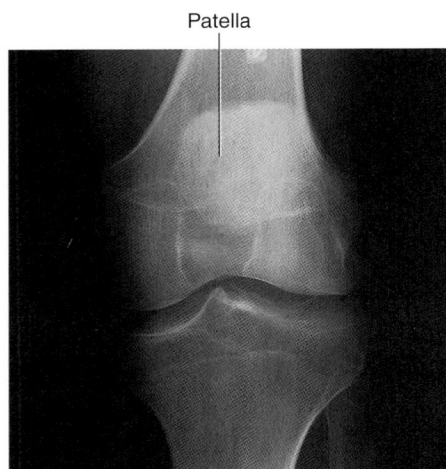

PA knee (and/or patella) position and radiograph.

Supplemental Projections

INTERCONDYLAR FOSSA ("NOTCH" OR "TUNNEL" VIEWS)

Two positions are presented for demonstration of the intercondylar fossa. The Holmblad method is often preferred because there is less distortion due to tube angulation. The Camp-Coventry method may be desirable if the patient is unable to assume the correct position for the Holmblad method. Both methods may be done either with or without the bucky.

HOLMBLAD METHOD

Body position:
Patient is on hands and knees on radiographic table with affected knee flexed so that angle between femur and table is 70°. Contralateral knee is flexed more and is advanced to provide support. Pelvis must remain level and sagittal plane of affected leg remain perpendicular to the film.

Central ray:
Perpendicular to center of film through center of knee joint

> **TIPS:**
> - Small level sponges under the knees help provide patient comfort in this position.
> - An increase in mAs of 50% from that used for the AP or PA projection is needed to compensate for the increased tissue thickness at the distal femur.

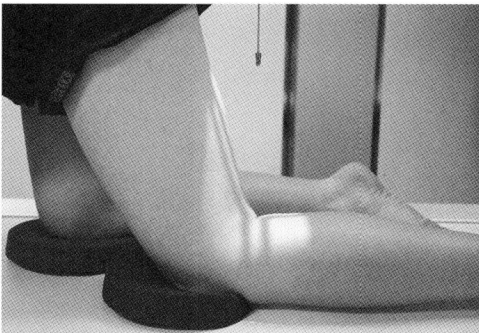

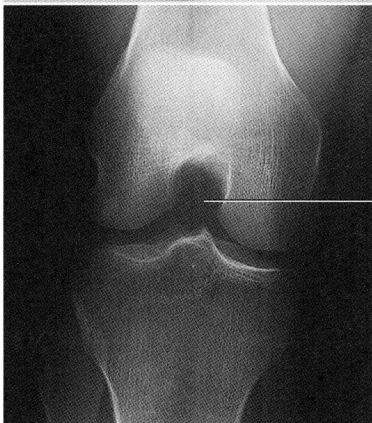

— Intercondylar fossa (notch)

Position and radiograph of intercondylar fossa ("tunnel view"), Holmblad method.

CAMP-COVENTRY METHOD

Position:
Prone with affected knee flexed to form angle of 40° between tibia and table

Central ray:
Angled 40° caudad through knee joint to center of film

> **TIP:** An increase in mAs of 50% from that used for the AP or PA projection is needed to compensate for increased tissue thickness caused by the angulation of the x-ray beam.

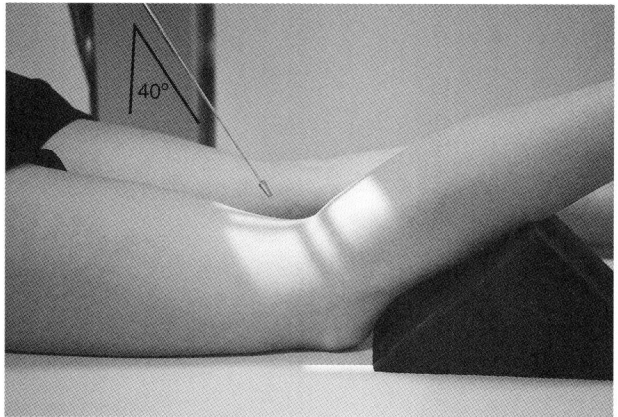

Position for intercondylar fossa, Camp-Coventry method.

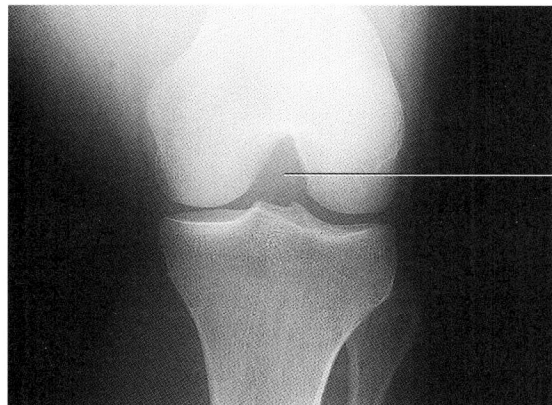

— Intercondylar fossa (notch)

Radiograph of intercondylar fossa ("tunnel view"), Camp-Coventry method.

TANGENTIAL (AXIAL) PATELLA ("SUNRISE VIEW"), SETTEGAST METHOD

This projection provides an axial view of the patella and the retropatellar space.

Position: Prone with affected knee flexed to form angle of 80° between femur and tibia. Sagittal plane of femur is perpendicular to film. Position may be supported by a strap around ankle that is extended over patient's shoulder and held by patient.

Central ray: Angled 10° to 20° cephalad and centered to inferior margin of patella. Angulation is adjusted so that central ray passes between patella and distal femur.

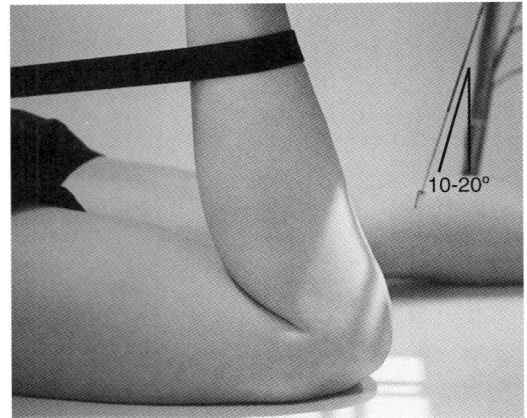

Position for tangential view of patella, Settegast method.

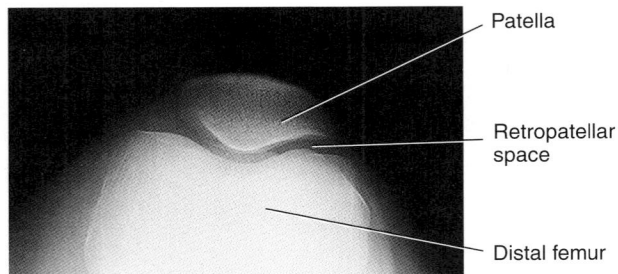

Tangential ("sunrise") view of patella.

For most adults, the image of the femur is too long to be included on a 14 × 17-inch (43 × 36-cm) film. The examination may be designated as a proximal femur, to include the hip joint, or a distal femur, to include the knee joint. If it is desired to examine the entire femur, a second examination is done to demonstrate the joint that was not included. For example, a distal femur examination might be supplemented with a hip study, or a knee examination might be done in addition to a proximal femur examination.

When the suspected lesion is in the midfemur, a distal femur examination is often performed. This is advantageous because the knee and midfemur are usually more similar in tissue density than are the midfemur and the hip, resulting in a more even radiographic density. In addition, a true lateral projection is more easily obtained of the distal femur than of the proximal femur. On the other hand, if the pathology of interest involves both the proximal femur and midfemur, it is preferable to do a proximal femur examination so that there is continuity to the image of the entire area of interest.

Basic Examination

DISTAL FEMUR

The basic examination of the distal femur includes the AP and lateral views.

Screens: Regular (rapid)

Grid: Yes

SID: 40 inches

AP	Lateral

14″ × 17″ (need 2)

Body position:

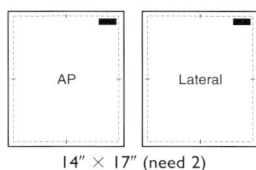

AP: Supine with affected femur aligned to center of table

Lateral: Recumbent on affected side with affected femur aligned to center of table. Knee and hip of unaffected extremity are flexed and leg is supported anterior to body.

Part position:

AP: Leg is extended with sagittal plane perpendicular to film. Film is placed with inferior margin 1 to 2 inches below knee joint.

Lateral: Knee of affected leg is flexed 30° to 45°. Sagittal plane of femur is parallel to film. Film is placed with its inferior margin 1 to 2 inches below knee joint.

Central ray:
Perpendicular to midpoint of film

Patient instruction:
Do not move.

Structures seen:
Knee joint and distal three fourths of femur

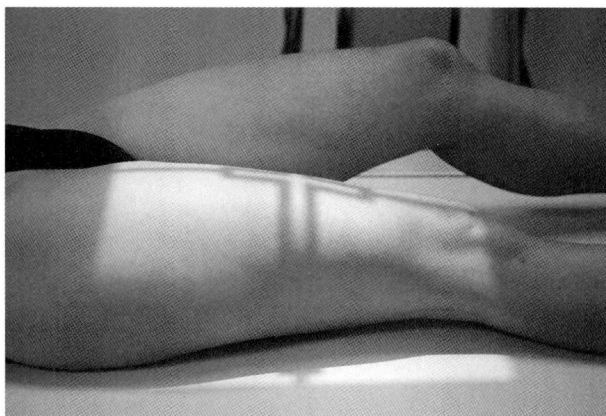

AP distal femur position.

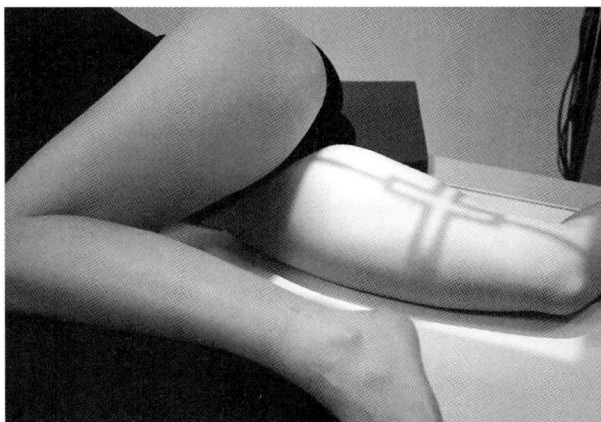

Lateral distal femur position.

Distal femur radiographs. **A,** AP; **B,** lateral.

PROXIMAL FEMUR

The basic examination of the proximal femur includes the AP and lateral views.

Screens: Regular (rapid)

Grid: Yes

SID: 40 inches

Body position:

AP: Supine with affected femur aligned to center of table

Lateral: Recumbent in oblique position on affected side with support under unaffected hip. Affected femur aligned to center of table. Knee and hip of unaffected extremity are flexed and leg is supported posterior to body.

Part position:

AP: Leg is extended with sagittal plane perpendicular to film. Film is placed with superior margin at level of ASIS.

Lateral: Knee of affected leg is flexed 30° to 45°. Sagittal plane of femur is parallel to film as much as possible. Film is placed with superior margin at level of ASIS.

Central ray:
Perpendicular to midpoint of film

Patient instruction:
Do not move.

Structures seen:
Hip joint and proximal three fourths of femur

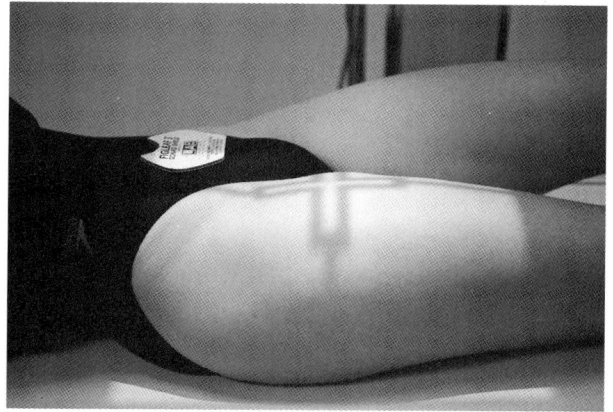

AP proximal femur position.

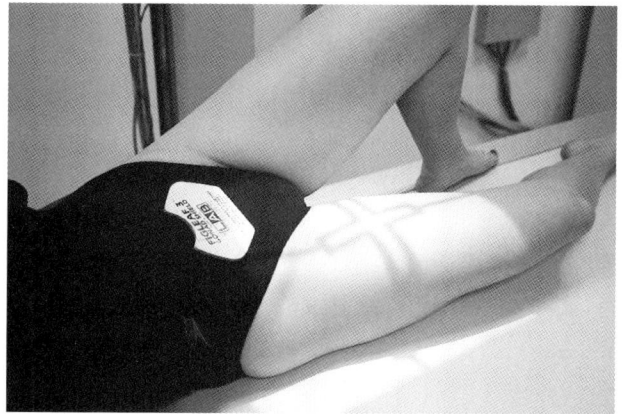

Lateral proximal femur position.

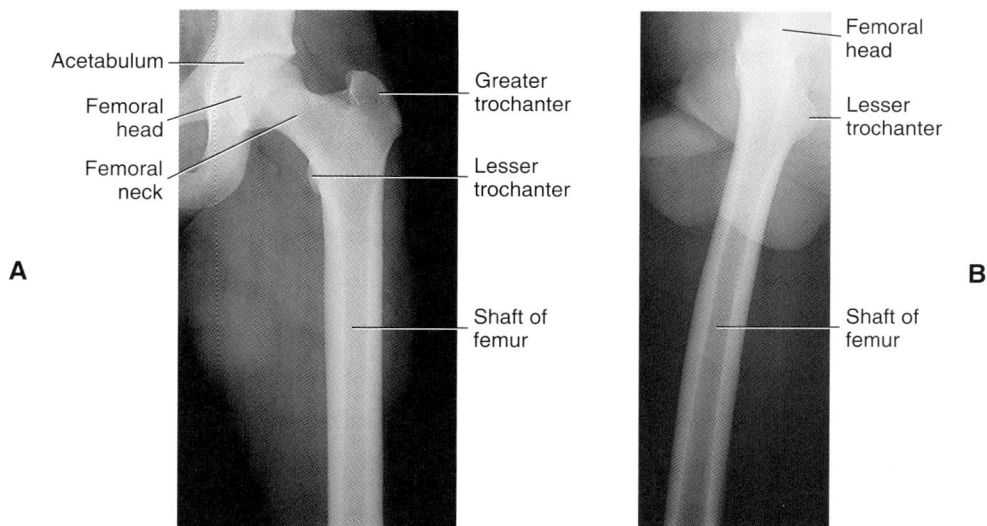

Proximal femur radiographs. **A,** AP; **B,** lateral.

Basic Examination

The basic examination of the pelvis includes the AP view.

Screens: Regular (rapid)

Grid: Yes

SID: 40 inches

AP

14″ × 17″

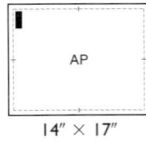

Body position:

Supine on table. Coronal plane of body is parallel to film. If there is no suspicion of recent fracture, femurs are rotated medially 15° to place femoral necks parallel to film. This involves medial rotation of feet approximately 45°.

Film placement:

Superior margin of film is aligned 1 to 2 inches superior to iliac crest. Check to be certain that greater trochanters are well within inferior margin of film.

Patient instruction:

Stop breathing. Do not move.

Central ray:

Perpendicular to midpoint of film

Structures seen:

Entire pelvis and proximal fourth of femurs

AP pelvis position.

Foot position for AP pelvis.

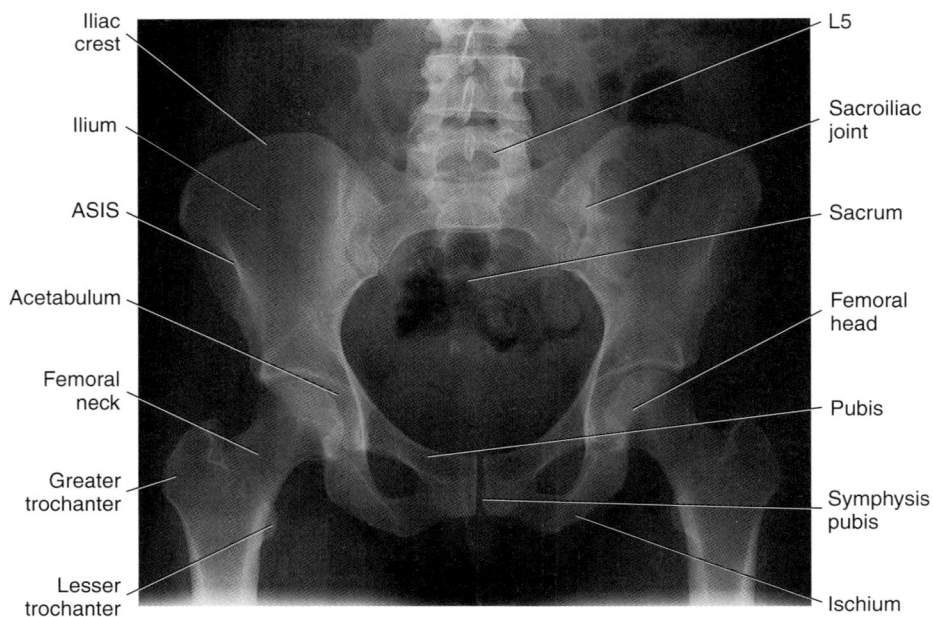

AP pelvis radiograph.

When there has been recent trauma to the hip with a possibility of hip fracture, it is usual for the examination to begin with an AP projection of the entire pelvis. *Do not move the affected leg in relation to the body, regardless of its position.* The pelvis film should be checked by a physician before attempting to move or rotate the femur into another position. If a lateral projection is needed in cases of recent hip fracture, a true lateral projection is done without moving the affected leg. The equipment available in outpatient facilities may not accommodate the true lateral hip procedure.

Routine hip examinations are most commonly performed as follow-up studies after treatment for hip fracture or to evaluate chronic hip complaints.

Fig. 14-11 is an aid to the localization of the hip joint, the axis of the femoral neck, and the center for hip radiographs. First, a line is drawn from the ASIS to the symphysis pubis. The midpoint of this line marks the superior margin of the acetabulum. This area must be included in examinations of the proximal femur as well as in all hip studies. When a perpendicular line is drawn inferior to the center of the first line, forming a T, this second line will indicate the long axis of the femoral neck. Approximately 1 inch inferior to the junction of the two lines is the location of the femoral head. The center of the femoral neck is approximately 2 inches inferior to the junction and is the center point for radiographic examinations of the hip.

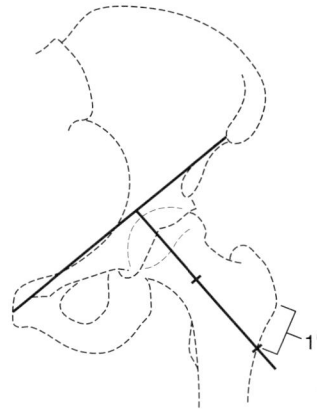

Fig. 14-11 Hip localization. Palpate the anterior superior iliac spine (ASIS) and the symphysis pubis. A line between these two points forms the crossbar of a T with the dome of the acetabulum at its center. The leg of the T is perpendicular to the crossbar and indicates the axis of the femoral neck. The midpoint of the femoral neck is the center point for hip radiographs. It is located along the leg of the T, 1 to 2 inches inferior to its junction with the cross bar.

Basic Examination

The basic examination of the hip includes the AP and frog-leg lateral views.

Screens: Regular (rapid)

Grid: Yes

SID: 40 inches

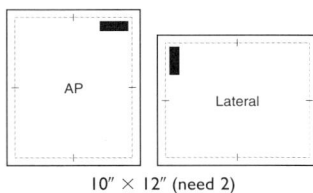

AP

Lateral

10″ × 12″ (need 2)

Body position:
Supine on table

Part position:

AP: Femur is medially rotated 15° as for pelvis.

Frog-leg lateral: Hip is flexed and femur abducted so that sagittal plane of femur is parallel or near-parallel to film. If patient cannot abduct hip sufficiently from supine position, pelvis may be rotated toward affected side and supported under unaffected hip.

Central ray:
Perpendicular to midfemoral neck

Patient instruction:
Do not move.

Structures seen:
Proximal fourth of femur, acetabulum, and portion of pelvis surrounding acetabulum

AP hip position.

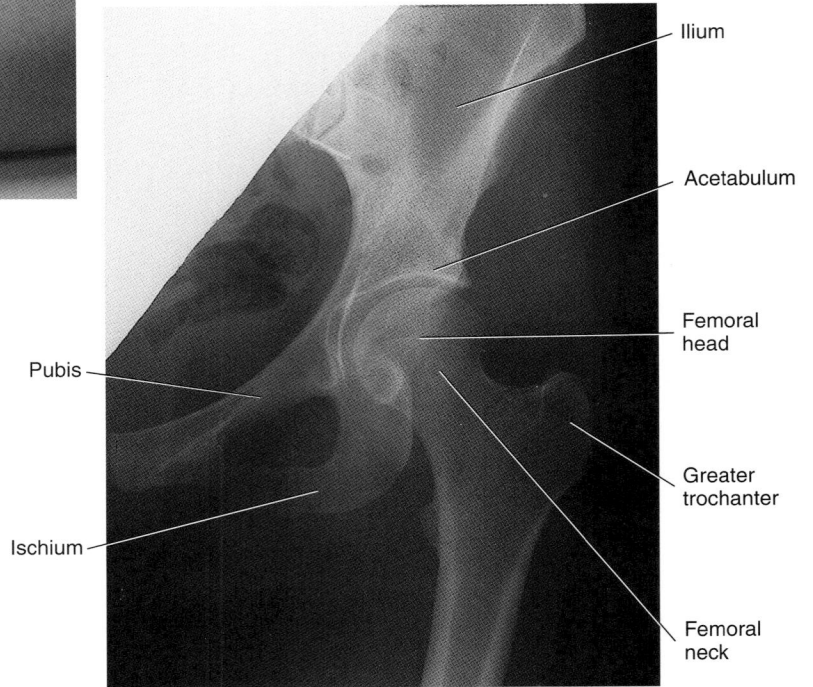

Ilium

Acetabulum

Femoral head

Pubis

Greater trochanter

Ischium

Femoral neck

AP hip radiograph.

Frog-leg lateral hip position.

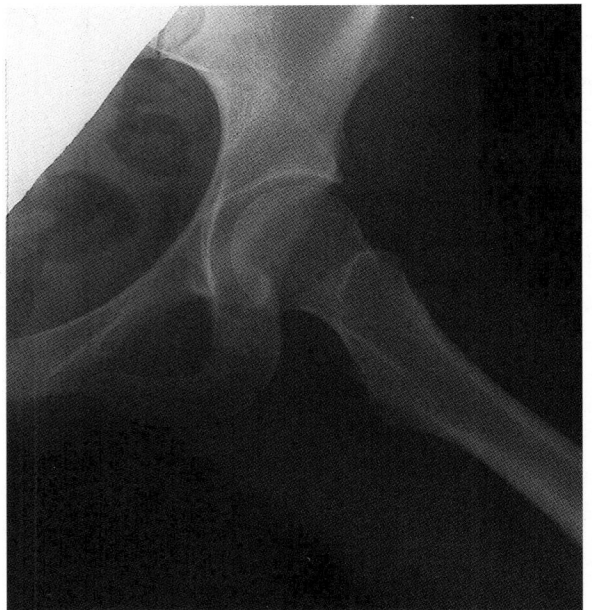

Frog-leg lateral hip radiograph.

Alternative Projection

When a lateral projection is needed in cases of known or suspected recent hip fracture, the true lateral hip view (inferosuperior projection) is substituted for the frog-leg lateral. This radiograph is taken without moving or rotating the affected leg.

The hip and knee of the unaffected extremity are flexed 90° and supported above the table. A 10 × 12-inch (25 × 30-cm) grid cassette is used or a stationary grid is attached to a regular cassette. The grid and cassette are supported on the tabletop on edge with their long dimension in contact with the table (crosswise). The cassette and grid are aligned to the angle of the femoral neck and the medial margin of the cassette is placed solidly into the soft tissues just proximal to the iliac crest. The x-ray tube is aligned horizontally, perpendicular to the center of the film through the patient's groin.

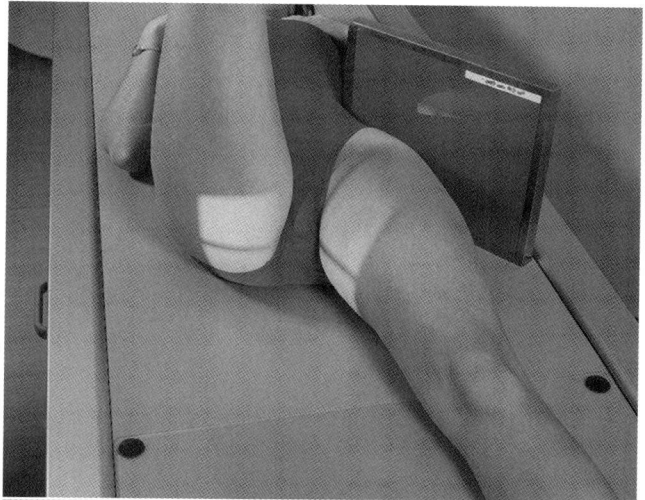

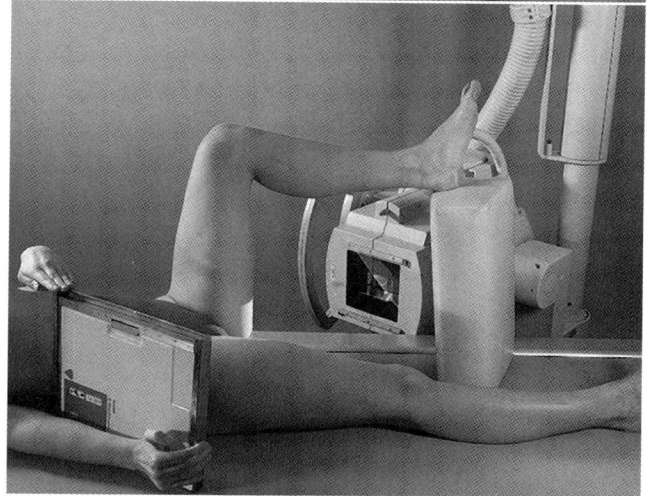

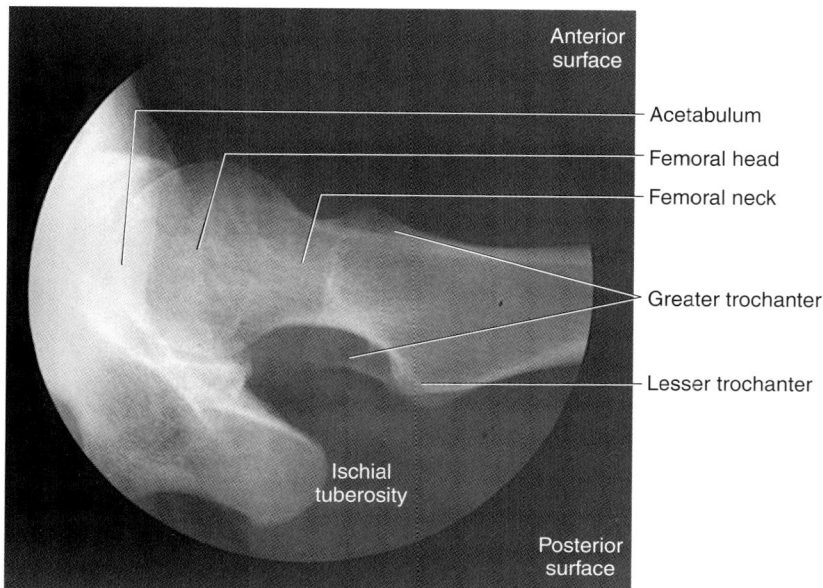

Inferosuperior (true lateral) hip position.

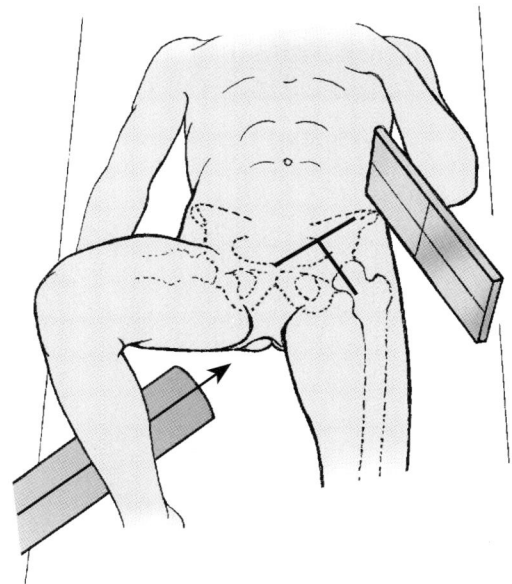

Inferosuperior (true lateral) hip radiograph.

Fig. 14-12 Stress fracture of third metatarsal *(arrows)*.

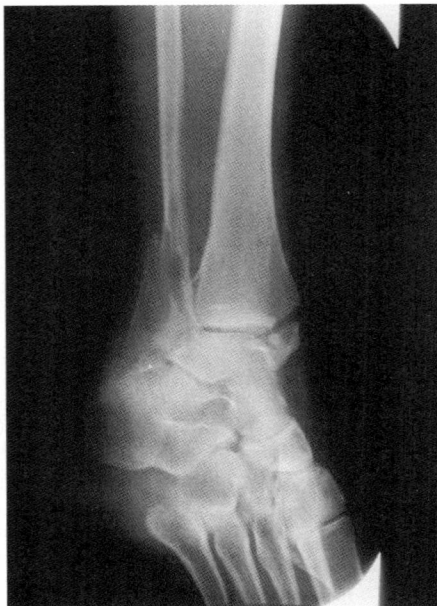

Fig. 14-13 Bimalleolar fracture.

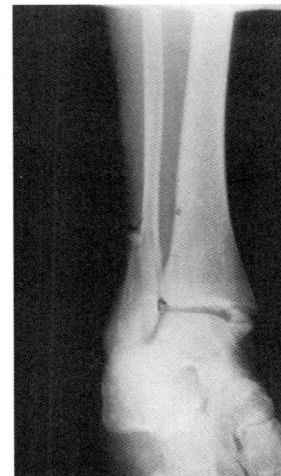

PATHOLOGY

Probably the most significant pathology affecting the lower extremity from the radiographer's viewpoint is trauma. Fractures and other conditions affecting this portion of the anatomy may vary greatly, and only those most commonly seen in radiography are discussed here.

Common Fractures

Stress fractures are often seen in the lower extremity, the result of stress to the bone from repeated injuries that would not cause fractures if they occurred only once. Stress fractures of long bones are usually simple undisplaced fractures (Fig. 14-12). They are common in the metatarsals and in the calcaneus as a result of running, jogging, or marching. They also occur in the tibia, fibula, femoral shaft, femoral neck, ischium, and pubis.

Sufficient force to cause a fracture of the tibia places a great strain on the fibula, often resulting in a fibular fracture as well. The associated fibular fracture may be in the same general region as the tibial fracture. One example is the bimalleolar fracture (Fig. 14-13). A similar injury in which the inferior posterior lip of the tibia is also fractured is called a trimalleolar fracture (Fig. 14-14). On the other hand, a distal tibia fracture may be associated with a fracture of the fibula at its weakest point, the proximal shaft, just distal to the head. When the shaft of the tibia is fractured with a twisting injury (common among skiers), the result is often a spiral fracture (Fig. 14-15).

Knee fractures in healthy individuals are relatively uncommon because the bones of the knee are very strong. When excessive force is applied to the knee joint, the result is more likely an injury to the meniscus cartilage and/or to one or more of the ligaments that connect the tibia to the femur. These soft tissue

Fig. 14-14 Trimalleolar fracture.

injuries are not visible on routine radiographs. Special imaging techniques such as arthrograms (joint studies involving injection of contrast media into the joint capsule) and magnetic resonance imaging (MRI) studies are used to evaluate soft tissue injuries to the knee.

The common fractures of the femur occur in the shaft, the neck, or the intertrochanteric region (between the trochanters). Fractures of the proximal femur (head, neck, and intertrochanteric region) are generally referred to as hip fractures (Fig. 14-16). Hip fractures associated with weakened bone from osteo-porosis are common among the elderly, particularly elderly women. Most fractures of the femur are treated by means of internal fixation, surgical application of hardware to hold the bones in place. Fig. 14-17 shows internal fixation of an intertrochanteric hip fracture. When there is severe degeneration of the hip joint, the treatment may be a total hip replacement (Fig. 14-18). In this case, the **prostheses** (anatomic replacements) for both the femoral head and the acetabulum are made of metal. It is not uncommon, however, for the acetabulum prosthesis to be made of a plastic material. Since the plastic is not visible on the radiograph, a radiopaque wire embedded in its rim permits it to be localized radiographically.

Because the pelvis as a whole is a rigid ring-like structure, fractures of the pelvis often occur in pairs. The reason for this is apparent if you consider how unlikely it would be to break a Life-Saver candy in only one place. The stress that causes one fracture creates an opposing stress, and two fractures result.

Fig. 14-15 Spiral fracture of tibia.

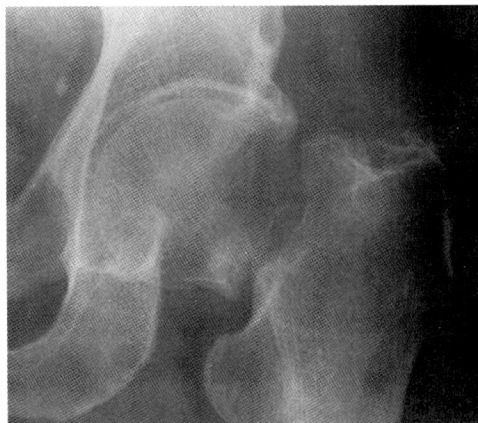

Fig. 14-16 Femoral neck fracture.

Fig. 14-17 A, Intertrochanteric hip fracture. **B,** Internal fixation of intertrochanteric fracture.

Fig. 14-18 Total hip replacement.

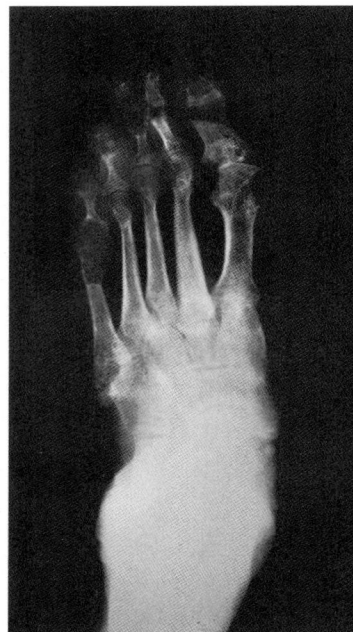

Fig. 14-19 Gout affecting the foot, particularly the great toe and first metatarsal.

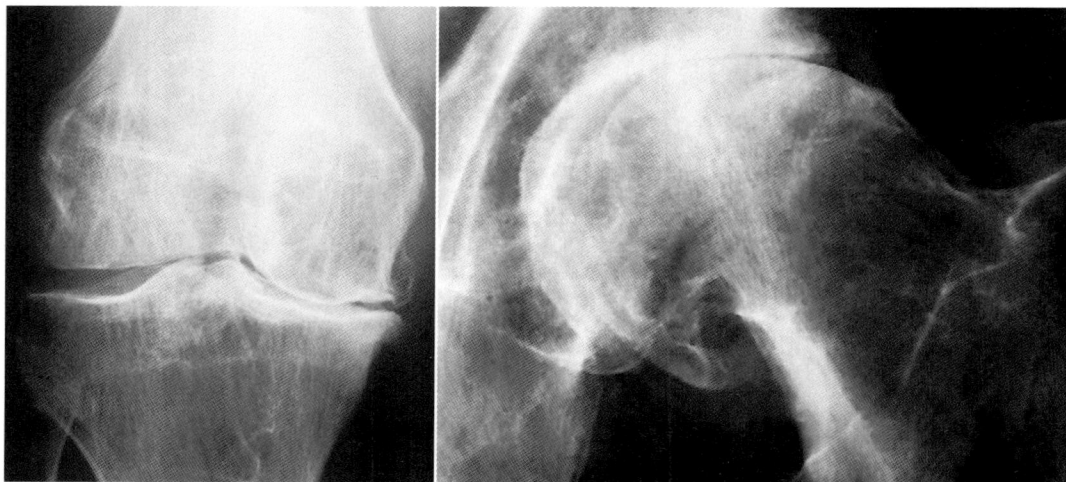

Fig. 14-20 Osteoarthritis. **A,** Knee; **B,** hip.

Nontraumatic Conditions

As previously mentioned, arthritis may affect any joints of the body, and there are a number of different types. Rheumatoid arthritis, discussed in Chapters 12 and 13, may also affect the joints of the feet. Gouty arthritis is a joint condition caused by gout, a systemic disorder that increases the uric acid content of the blood. Gouty arthritis commonly affects the feet, particularly the joints of the great toe (Fig. 14-19), although it may also involve the hands. Osteoarthritis

may cause degeneration of any of the joints of the lower extremity but is most common in the knee and the hip (Fig. 14-20). This condition is often associated with osteoporosis. Note the irregular contours of the articular surfaces and the bony hypertrophy at the margins of the joints.

Osteomyelitis, as introduced in Chapter 13, is an infection of the bone. In the acute phase of the disease, there is bony destruction. With healing, however, there is considerable new bone formation (Fig.

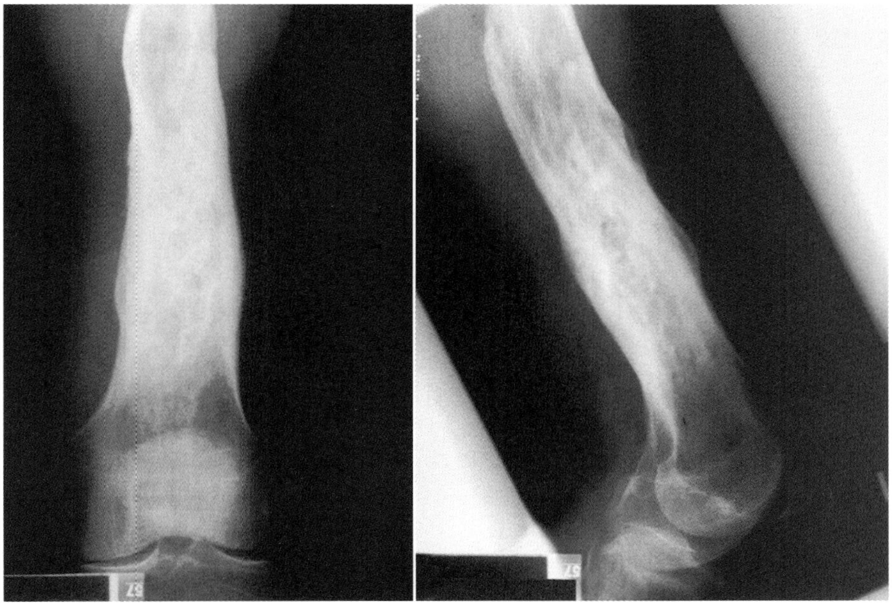

Fig. 14-21 Healed osteomyelitis of distal femur.

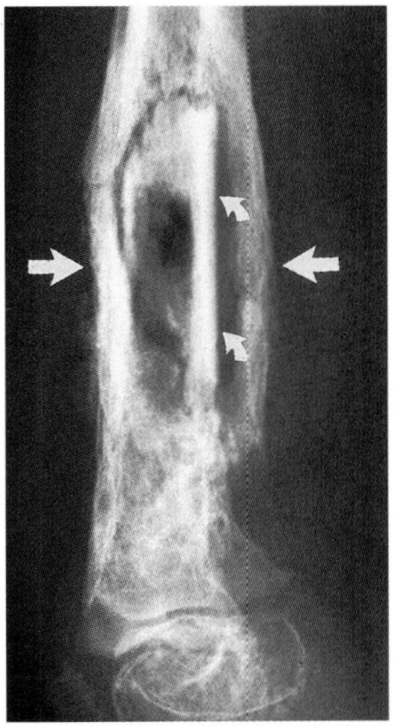

Fig. 14-22 Chronic osteomyelitis results in the formation of an involucrum *(straight arrows)*, a bony shell surrounding the sequestrum *(curved arrows)*, necrosed bone that is separated from the surrounding tissues.

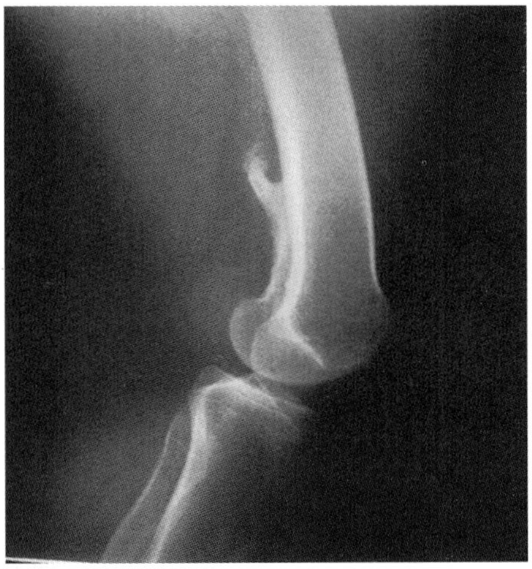

Fig. 14-23 Exostosis (osteochondroma) of distal femur.

14-21). When the condition is chronic, significant destruction and new bone formation are both present (Fig. 14-22).

Neoplastic and metastatic bone diseases may also affect the bones of the lower extremity. Fig. 14-23 illustrates exostosis of the tibia. Exostosis is also called osteochondroma, a benign tumor arising from cartilaginous tissue. Fig. 14-24 is an example of osteogenic sarcoma, one of several types of malignant bone tumors that occur in the lower extremity. The typical lesions of osteogenic sarcoma occur in the distal ends of long bones and are both destructive and sclerotic (thickened and hardened). They are associated with a tumor mass within the soft tissues. The bony spicules (needle-like formations) that extend into the soft tissue mass create the classic sunburst pattern of this disease. Fig. 14-25 shows osteoblastic metastases of the pelvis and proximal femurs, secondary to carcinoma of the urinary bladder.

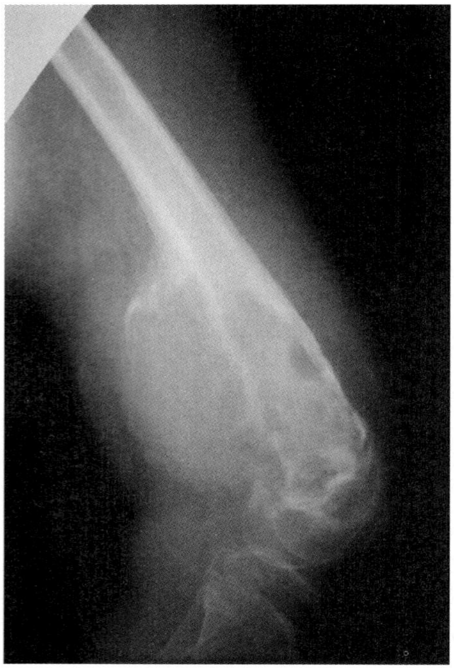

Fig. 14-24 Osteogenic sarcoma of the distal femur.

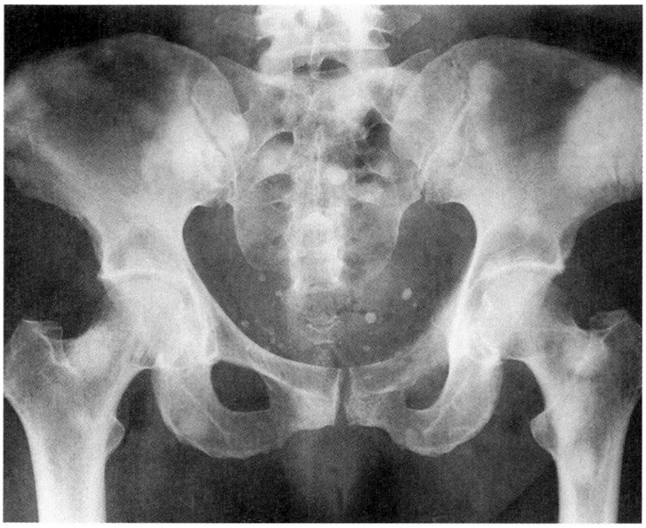

Fig. 14-25 Osteoblastic metastatic lesions of the pelvis and proximal femurs.

SUMMARY

The bones of the foot include the phalanges, metatarsals, and tarsals. The tibia and fibula form the lower leg, articulating at the ankle with the talus. The femur articulates with the tibia at the knee, and it forms the hip joint where it articulates with the innominate bone at the acetabulum. The two innominate bones—each consisting of ilium, ischium, and pubis—form the pelvis. Important palpable bony prominences include the medial and lateral malleoli of the ankle, the condyles and greater trochanter of the femur, the iliac crest, the anterior superior iliac spine, and the symphysis pubis.

Radiography of the foot, heel, and ankle is done tabletop using extremity cassettes. The patient is seated or recumbent on the radiographic table. Multiple views are often done on a single cassette. Examinations of the femur and pelvis, on the other hand, are done using rapid cassettes and grids or buckys with the patient recumbent. Radiography of the knee may be done either with or without a grid.

The most common radiographic pathology occurring in this portion of the anatomy involves trauma, particularly fractures. Nontraumatic conditions, such as arthritis, osteomyelitis, and neoplastic disease, are also seen.

▪ Review Questions ▪

1. How many phalanges are there in the great toe? The second toe?
2. Is the fibula medial or lateral to the tibia?
3. Name the bones that form the knee joint.
4. Name and point to three bony prominences on your own pelvis.
5. List two ways in which an examination of the foot differs from an examination of the ankle.
6. Describe the position of the leg for an oblique projection of the ankle.
7. Name two special projections of the knee.
8. If there is suspicion of fracture of the patella, what projections should be taken? What precautions should be taken?
9. How does a routine hip examination differ from an examination for possible hip fracture? Why?
10. List and describe four specific types of fractures of the lower extremity and hip.
11. List three general types of nontraumatic pathology that may affect the bones of the lower extremity or pelvis.

15

Spine

Learning Objectives

At the conclusion of this chapter, the student will be able to:

- Name the regions that make up the spine and identify each on an anatomic diagram and on a radiograph
- Draw or identify on a diagram the parts of a typical vertebra
- Identify significant positioning landmarks for the spine by palpation
- Demonstrate correct body and part positioning for routine projections and common special projections of the spine
- Correctly evaluate radiographs of the spine for positioning accuracy
- Describe and recognize on radiographs abnormalities and pathology common to the spine

Key Terms

atlas
axis
cervical spine
coccyx
dens
facet
intervertebral disk
kyphotic curve, kyphosis
lamina (pl. laminae)

lordotic curve, lordosis
lumbar spine
pedicle
sacrum
scoliosis
stenosis
thoracic spine
vertebra (pl. vertebrae)

Most radiography of the spine may be accomplished successfully in either the upright or the recumbent position. In medical practices and hospitals, radiography of the spine is usually done in the recumbent position. In chiropractic practices, spine radiography is almost always done in the upright position. This chapter provides instruction and illustration for both methods.

ANATOMY

The spine (Fig. 15-1) is the central portion of the skeletal system. It provides the supporting framework for the body. It also surrounds and protects the spinal cord. The spine is called the vertebral column because it is made up of many irregularly shaped bones called **vertebrae.** The spine is divided into five regions: cervical spine, thoracic spine, lumbar spine, sacrum, and coccyx. The vertebrae are named according to spinal region and are numbered from the top down. For example, the third vertebra from the top of the thoracic region is simply called the third thoracic vertebra, abbreviated T3.

When viewed from the front, the normal spinal column is relatively straight. When seen from the side, however, the spine has four curves (Fig. 15-2). It arches anteriorly and posteriorly to provide a spring-like flexibility that absorbs shock as we walk and run. A curvature that is convex (bowing outward) anteriorly is called a **lordotic curve,** or **lor-**dosis. One that is convex posteriorly is called a **kyphotic curve,** or **kyphosis.** An abnormal lateral curvature is called **scoliosis** (Fig. 15-3).

A typical vertebra is illustrated in Fig. 15-4. The block-like anterior portion is called the body. It consists of cancellous bone with a thin cortex. Posterior to the body is a ring of bone called the vertebral arch.

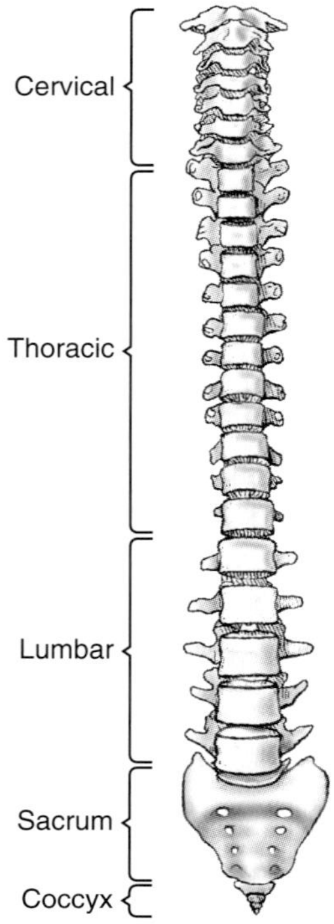

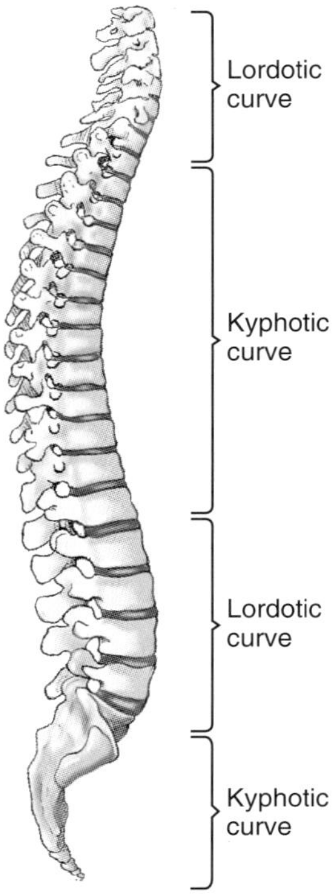

Fig. 15-2 Lateral aspect of spine.

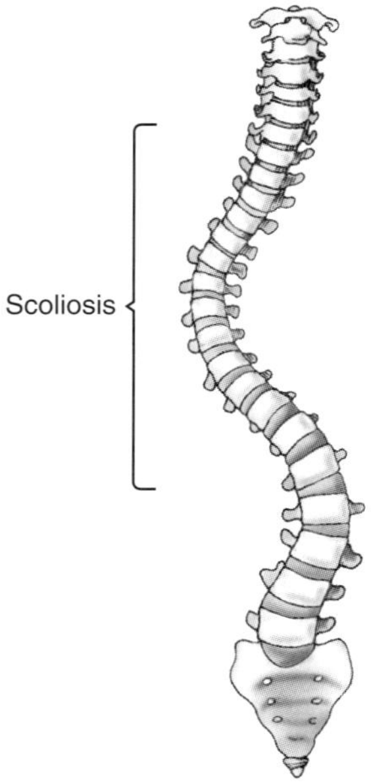

Fig. 15-1 Anterior aspect of spine.

Fig. 15-3 Scoliosis: abnormal lateral spine curvature.

It is formed by the pedicles, which attach to the body on either side, and by the **laminae** posteriorly. The hole in the ring is called the vertebral foramen. It is the passage for the spinal cord. Two lateral projections, extending from the sides of the vertebral arch, are called the transverse processes. The right and left laminae come together in the midline to form the spinous process, which projects posteriorly. Four articular processes extend superiorly and inferiorly from the vertebral arch. The articular surfaces of these processes are called **facets.** They articulate with facets on the articular processes of the vertebrae above and below, forming the zygapophyseal joints. The zygapophyseal joints are diarthrodial joints of the gliding type. Fig. 15-5 illustrates typical joints of the spine.

The vertebrae are connected by ligaments. They are cushioned anteriorly, between the bodies, by pads of fibrocartilage called **intervertebral disks** (Fig. 15-6). These disks have a tough outer covering, the annulus fibrosus, and a soft, pulpy center called the nucleus pulposus.

Cervical Spine

The **cervical spine** is the most superior region of the vertebral column. It supports the head and the structures of the neck. The cervical spine consists of seven vertebrae and has a lordotic curve.

The first two cervical vertebrae differ in form from the others to accommodate the support and rotation of the skull. C1 is called the **atlas** (Fig. 15-7). It is a ring-like structure with no vertebral body and a very short spinous process called the posterior tubercle. The atlas consists of two lateral masses connected by an anterior arch and a posterior arch. Each lateral mass has superior and inferior articular processes. The superior articular processes articulate with the base of the skull and the inferior ones form joints with similar processes on the superior aspect of C2. The transverse processes project laterally and slightly downward from the lateral masses. On the anterior surface of the anterior arch is a rounded process called the anterior tubercle.

C2 is called the **axis** (Fig. 15-8). It is the vertebra on which the atlas rotates, allowing the head to turn

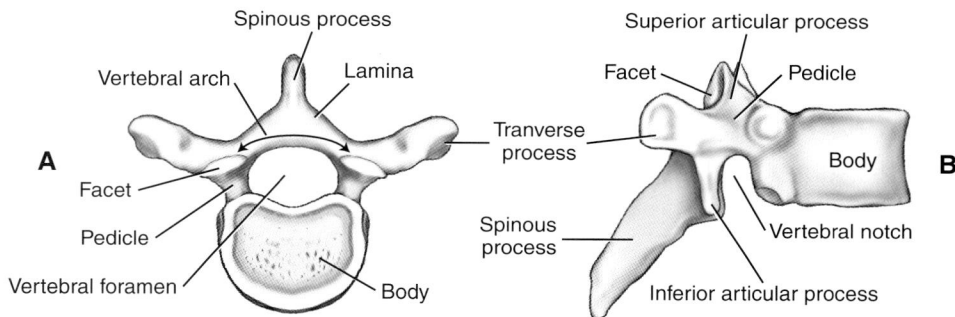

Fig. 15-4 Typical vertebra. **A,** Superior aspect. **B,** Lateral aspect.

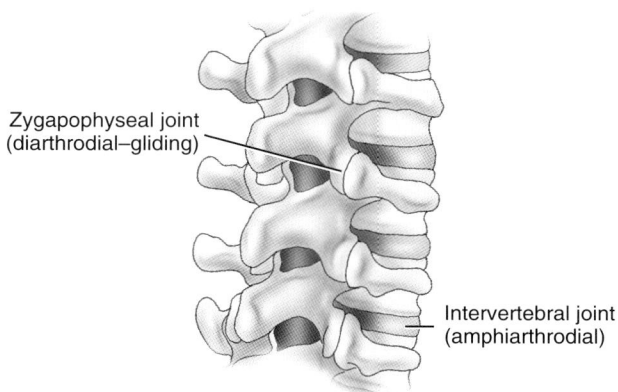

Fig. 15-5 Spinal joints. Intervertebral joint is between the bodies anteriorly, and the zygapophyseal joints are between the articular processes posteriorly.

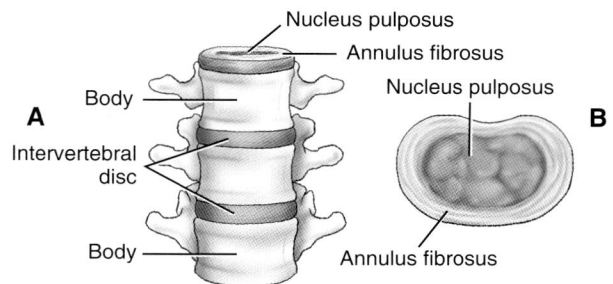

Fig. 15-6 Intervertebral disk. **A,** Anterior aspect. **B,** Superior aspect.

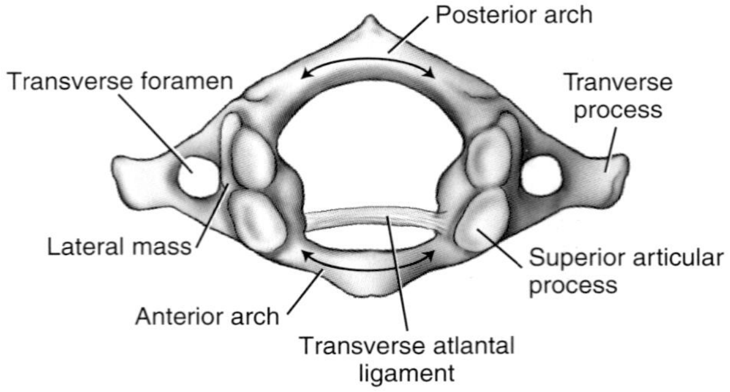

Fig. 15-7 Superior aspect of atlas (C1).

from side to side. Superior to the body of the axis is a tooth-like projection called the **dens,** or odontoid process. It projects into the anterior portion of the ring of the atlas and acts as a pivot between the two vertebrae.

C3 through C7 are typical cervical vertebrae (Fig. 15-9). The articular processes extend superiorly and inferiorly from a point posterior to the transverse process at the junction of the **pedicle** and the lamina. Together, the articular processes form a column of bone called the articular pillar. The cervical spinous processes are bifid—that is, they are split into two

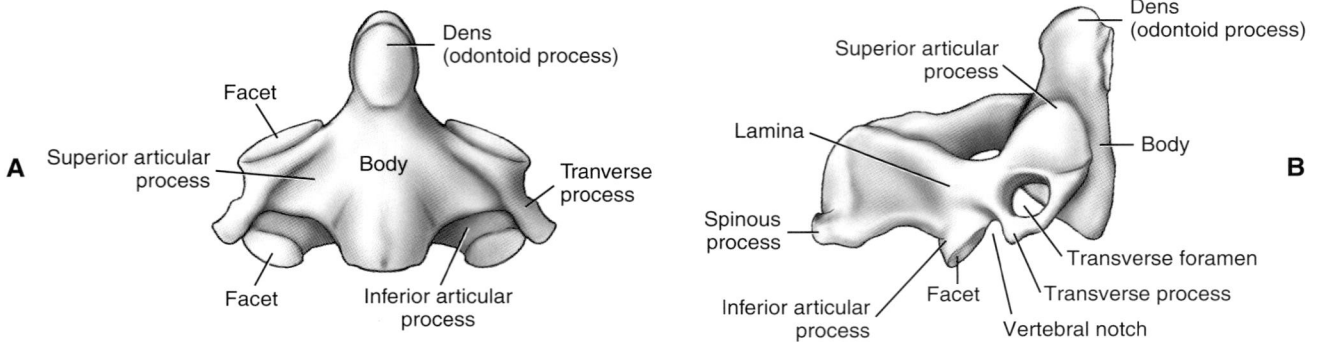

Fig. 15-8 Atlas (C2). **A,** Anterior aspect. **B,** Lateral aspect.

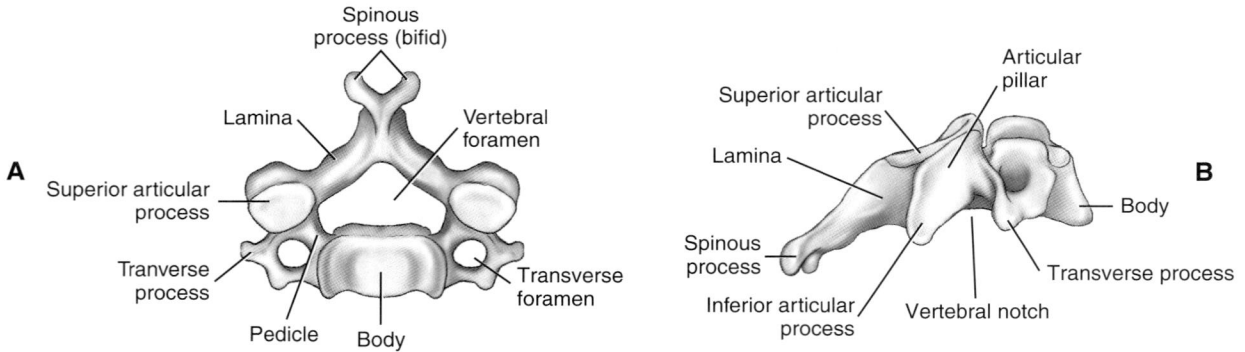

Fig. 15-9 Typical cervical vertebra. **A,** Superior aspect. **B,** Lateral aspect.

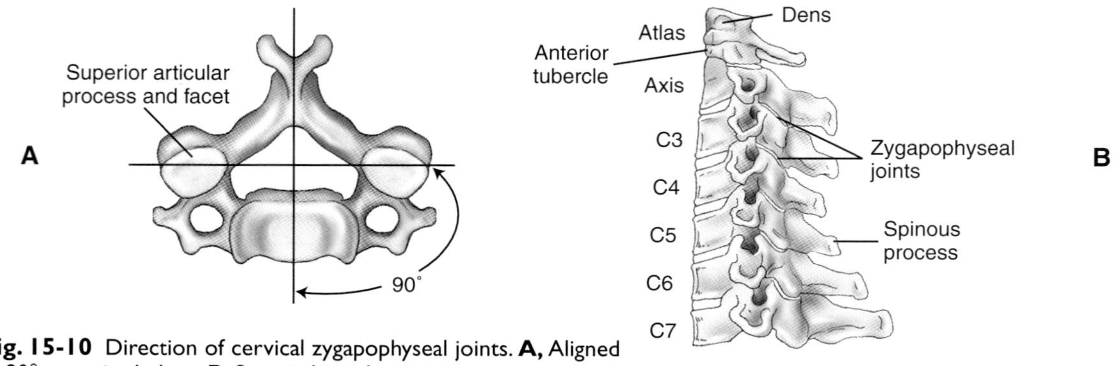

Fig. 15-10 Direction of cervical zygapophyseal joints. **A,** Aligned at 90° to sagittal plane. **B,** Seen in lateral projection.

posterior projections, forming a shape somewhat like a fish's tail. The spinous process of C7 is larger than the others and is easily palpable at the base of the neck. It is a convenient reference point for the location of other vertebrae.

With the exception of the C1-C2 articulation, the cervical zygapophyseal joints slope posteriorly and lie in the sagittal plane, so they are best seen from the lateral aspect (Fig. 15-10). The C1-C2 articulations differ in position and direction and are best seen in the AP projection.

Note the location of the vertebral notch, seen posterior to the body of the vertebra in Fig. 5-9, *B.* The space between each vertebral notch and the pedicle of the vertebra below it is called the intervertebral foramen (IVF). The intervertebral foramina are passages for nerves from the spinal cord to the upper torso and the upper extremity. They are set at an angle to the sagittal plane of 45° and 50° (Fig. 15-11).

Each cervical transverse process (including those of C1 and C2) features a hole called the transverse foramen. The transverse foramina form passages on each side for the vertebral artery and vein.

Thoracic Spine

The **thoracic spine** is sometimes also referred to as the dorsal spine. It consists of twelve vertebrae and has a kyphotic curve. The vertebrae vary somewhat in shape from one end of this spinal region to the other, but all have facets and/or demifacets that articulate with the ribs (Fig. 15-12). These joints are called costal or costovertebral joints and are diarthrodial joints of the gliding type.

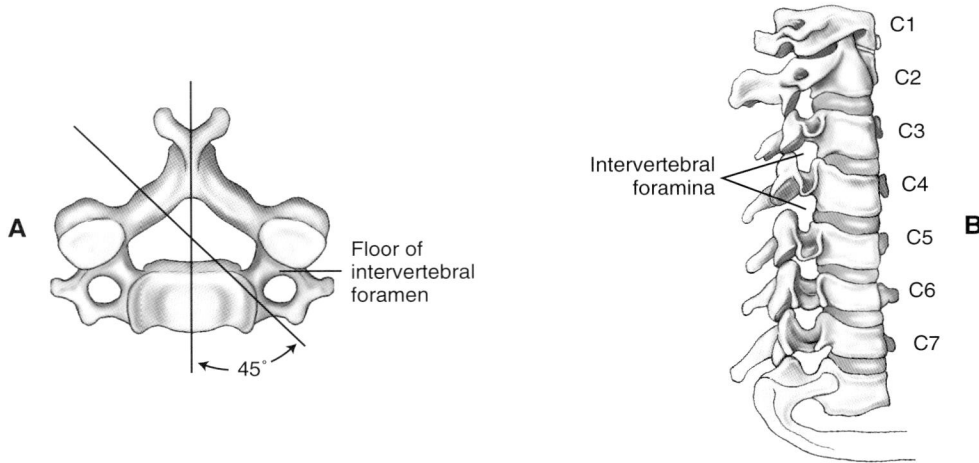

Fig. 15-11 Direction of cervical intervertebral foramina. **A,** Aligned at 45° to sagittal plane. **B,** Seen in oblique projections.

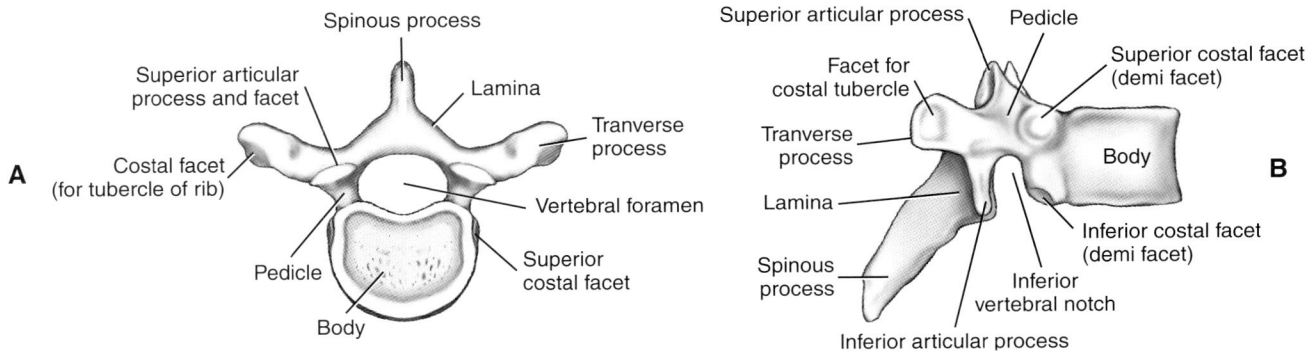

Fig. 15-12 Thoracic vertebra. **A,** Superior aspect. **B,** Lateral aspect.

The zygapophyseal joints of the thoracic spine are aligned at an angle of 20° posterior to the coronal plane (Fig. 15-13). The intervertebral foramina lie at an angle of 90° to the sagittal plane and so are seen from the lateral perspective (Fig. 15-14).

Lumbar Spine

The **lumbar spine** consists of five vertebrae and has a lordotic curve. The typical lumbar vertebra (Fig. 15-15) has a large, rounded body and a rather large, flat spinous process. The intervertebral foramina form an angle of 90° to the sagittal plane and are seen from the lateral perspective (Fig. 15-16). The zygapophyseal joints lie at an angle of 45° posterior to the coronal plane (Fig. 15-17). The narrow segment of bone between the superior and inferior articular processes is called the pars interarticularis.

When radiographed in the oblique projection, the lumbar vertebrae demonstrate a configuration that resembles a Scottie dog (Fig. 15-18). The superior articular processes form the ears of the dog, and the inferior articular processes form the front legs. The pars interarticularis represents the dog's neck.

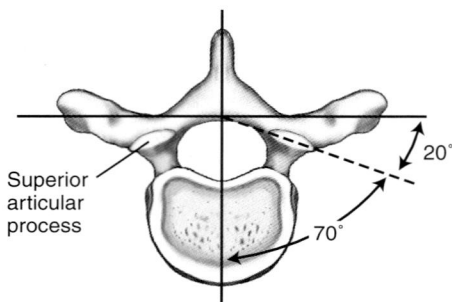

Sacrum and Coccyx

At birth, the **sacrum** consists of five sacral vertebrae. In the adult, they are fused into a solid bony structure (Fig. 15-19). The sacrum articulates with the ilia of the pelvis on either side, forming the sacroiliac joints. Its broad, flat superior surface is called the sacral base. The lateral portions of the first sacral segment are wing-like structures called the ala. The four pairs of sacral foramina are passages for nerves.

TABLE 15-1

Average Sacral Base Angulation

Body Position	Sacral Base Angle in Males	Sacral Base Angle in Females
Standing	35°	40°
Supine with legs extended	30°	35°
Supine with knees flexed	25°	30°

Fig. 15-13 Thoracic zygapophyseal joints are aligned at 70° to the sagittal plane and are seen in oblique projections.

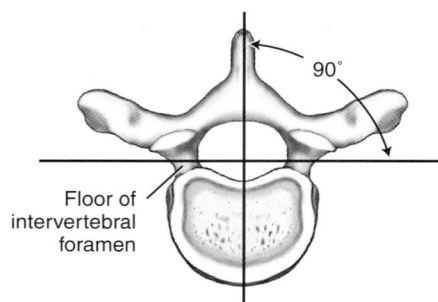

Fig. 15-14 Thoracic intervertebral foramina are aligned at 90° to the sagittal plane and are seen in lateral projection.

Fig. 15-15 Lumbar vertebra. **A,** Superior aspect. **B,** Lateral aspect.

The **coccyx,** the most inferior portion of the spine, is approximately the size of the fifth finger. In lay terms, it is called the "tailbone." The coccyx usually consists of four small vertebral segments, but it is not unusual for there to be three or five segments. The coccygeal segments tend to fuse in the adult. Two small bony projections extend superiorly from the posterior aspect on each side of the first coccygeal segment. These are called the coccygeal cornua (singular cornu, which means horn). They are joined to similar projections from the posterior inferior aspect of the sacrum, called the sacral cornua.

Together, the sacrum and coccyx form a kyphotic curve. This curvature is more pronounced in females than in males. The sacral base slopes downward anteriorly and the degree of slope is called the sacral base angle. This angle is greatest in females. It is greater when standing than when recumbent and is least when supine with the knees flexed. Average sacral base angles for males and females are listed in Table 15-1.

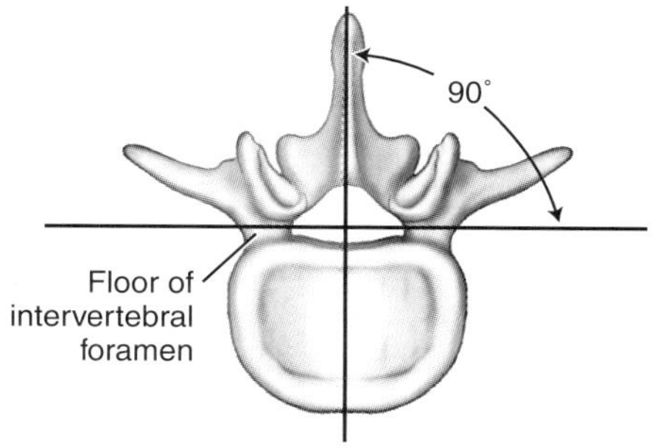

Fig. 15-16 Lumbar intervertebral foramina are aligned at 90° to the sagittal plane and are seen in lateral projection.

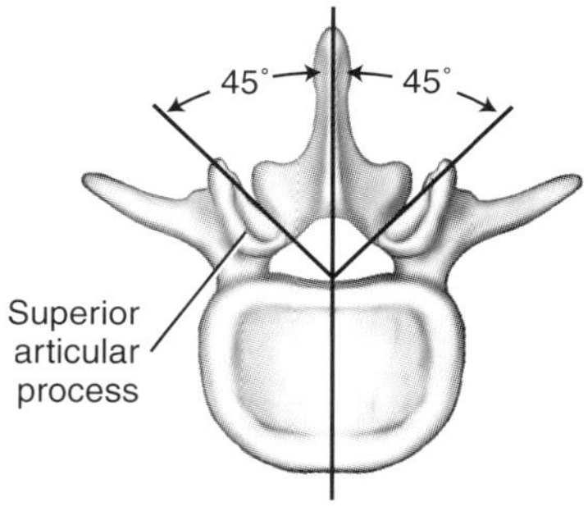

Fig. 15-17 Lumbar zygapophyseal joints are aligned at 45° to the sagittal plane and are seen in oblique projections.

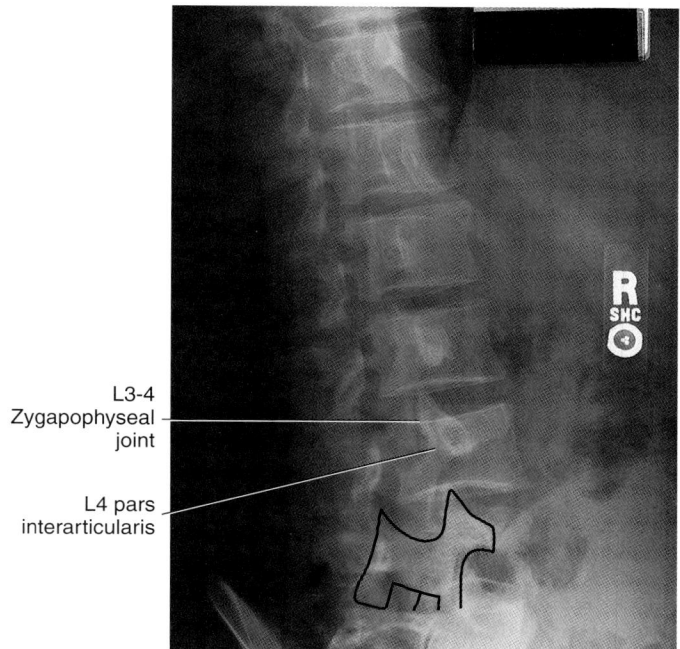

Fig. 15-18 Oblique lumbar spine radiograph showing Scottie dog configuration.

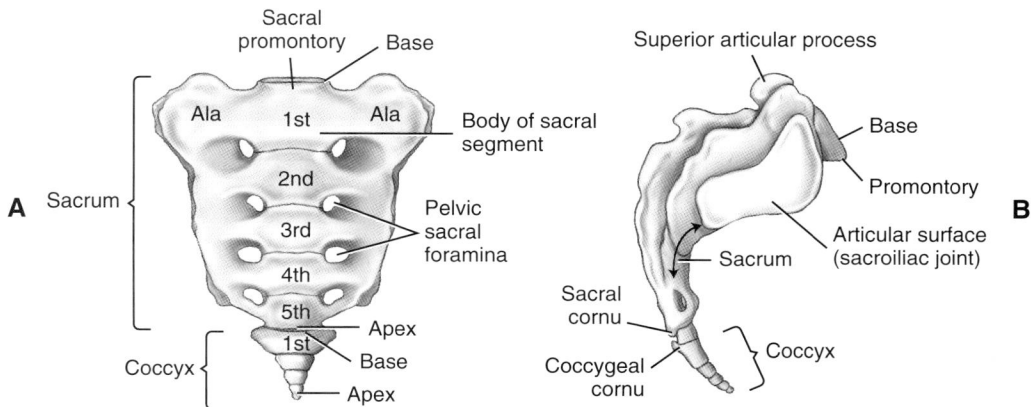

Fig. 15-19 Sacrum and coccyx. **A,** Anterior aspect. **B,** Lateral aspect.

POSITIONING AND RADIOGRAPHIC EXAMINATIONS

Many landmarks are used in positioning and alignment for various aspects of spine radiography. Fig. 15-20 illustrates the landmarks of the cranium and face that are helpful for radiography of the cervical spine. Fig. 15-21 shows the topographic anatomy that corresponds to specific vertebral levels of the spine. Memorizing these locations will enhance your ability to position accurately.

Spot films may be requested for better visualization of specific areas of the spine. As discussed in Chapter 9, the use of a small radiation field, centered on the area of clinical interest, improves contrast while minimizing the negative effects of distortion and parallax. The most common area for spot film radiography is the lumbosacral joint, but spot films may be helpful in any area of the spine. The radiographer must be able to correctly identify the location of any vertebra when a spot film is needed. When taking a spot film of a vertebra that does not have a precise palpable landmark, it is helpful to have reference to routine radiographs. The radiographer can measure the distance from a palpable landmark to the vertebra of clinical interest on the film and use this information to center the spot film.

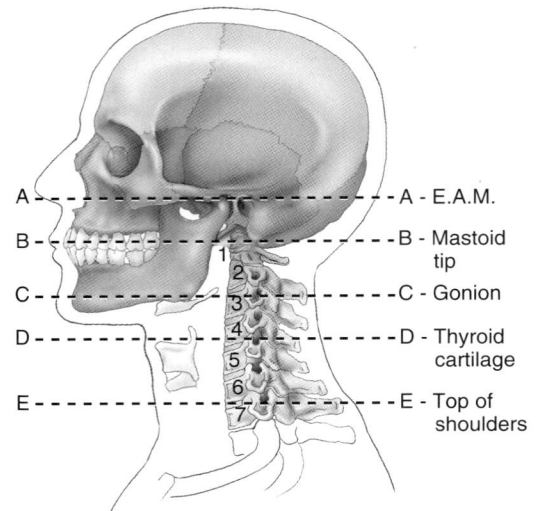

A - E.A.M.
B - Mastoid tip
C - Gonion
D - Thyroid cartilage
E - Top of shoulders
1
2
3
4
5
6
7

Fig. 15-20 Palpable landmarks for cervical spine positioning.

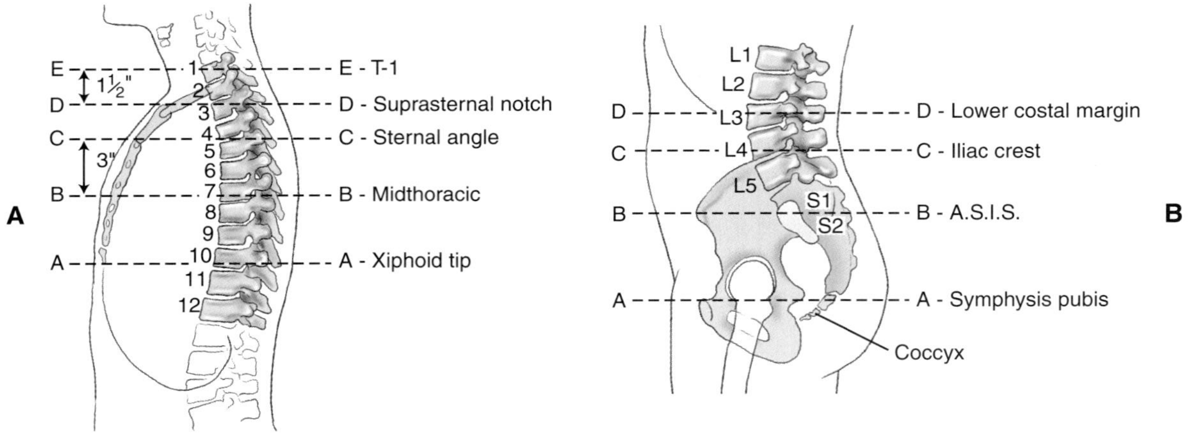

A
E - T-1
D - Suprasternal notch
C - Sternal angle
B - Midthoracic
A - Xiphoid tip
1½"
3"
1
2
3
4
5
6
7
8
9
10
11
12
B
D - Lower costal margin
C - Iliac crest
B - A.S.I.S.
A - Symphysis pubis
Coccyx
L1
L2
L3
L4
L5
S1
S2

Fig. 15-21 Palpable landmarks for spine positioning. **A,** Thoracic region. **B,** Lumbar region.

Two films are needed to demonstrate the entire cervical spine in the AP projection. The lower cervical view demonstrates C3 through C7, but the lower jaw and the teeth are superimposed over the atlas and axis. To demonstrate the upper cervical vertebrae, a second AP projection is taken through the open mouth. The AP upper cervical projection is sometimes referred to as the "AP open-mouth" or the "odontoid view."

Since it is not practical to measure the patient through the open mouth, the exposure factors for the upper cervical region are derived from those used for the lower cervical spine. More exposure is needed for the upper cervical area. The exposure factors from the lower cervical area are modified by adding 4 kVp or increasing the mAs by 30%.

For the lateral projection, the inferior margin of the film must be below the level of the upper surface of the shoulder in order to demonstrate all of C7. This results in a large OID between the neck and the film. To minimize magnification and improve definition on this projection, a 72-inch SID is used. Definition is also enhanced by use of the small focal spot. Although a grid is sometimes used for the lateral projection, the large OID constitutes an air gap (see Chapter 9) and the grid is not required.

Before cervical spine radiography, the patient must remove eyeglasses, earrings, hairpins, and necklaces as well as any clothing that has fasteners that might fall within the radiation field. Dentures and hearing aids should also be removed.

Basic Examination

The basic examination of the cervical spine includes the AP lower cervical, AP upper cervical, and lateral views.

AP LOWER CERVICAL

Film: 8 × 10 inch (20 × 25 cm), lengthwise

Screens: Rapid

Grid: Yes

SID: 40 inches

Body position:
Seated, standing, or supine

Part position:
Midsagittal plane of both body and head are aligned perpendicular to center of film, with patient facing tube. Head position is adjusted so that a line between mental point and base of skull makes an angle of 15° with horizontal plane. When patient is recumbent, patient's head rests on table. Head is placed firmly against cassette holder when upright. If desired position cannot be attained in this way, a radiolucent wedge sponge is placed under/behind head for stability.

> **TIP:** The upper margin of the collimator light field will fall across the patient's face at an angle of 15°, simplifying the adjustment of the head position.

Central ray:
Centered to film at angle of 15° cephalad through thyroid cartilage

Collimation:
Lengthwise, to film size. Crosswise, to soft tissue margins.

Patient instruction:
Stop breathing. Do not move.

Structures seen:
Vertebrae C3 through T2, including bodies, articular pillars, and intervertebral disk spaces

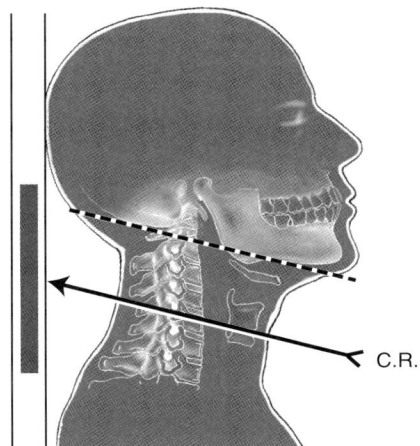

Position for AP lower cervical spine. Chin is projected over base of skull. Angled x-ray beam is parallel to cervical disk spaces.

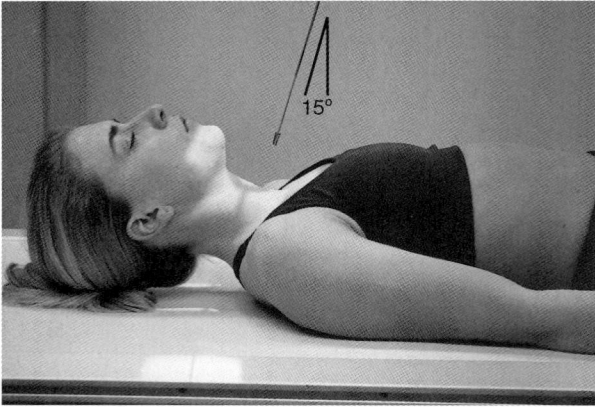

Recumbent position for AP lower cervical spine.

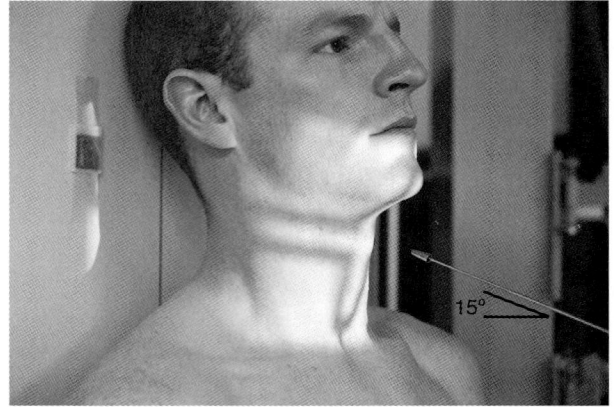

Upright position for AP lower cervical spine.

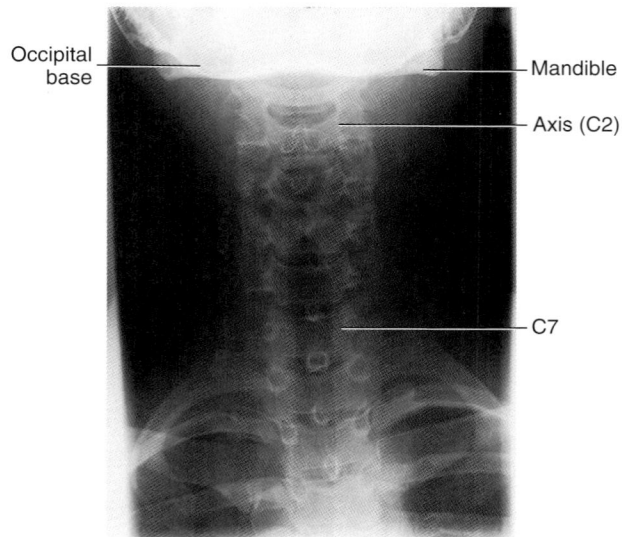

AP lower cervical spine radiograph.

AP UPPER CERVICAL

Film: 8 × 10 inch (20 × 25 cm), lengthwise

Screens: Rapid

Grid: Yes

SID: 40 inches

Body position:
Seated, standing, or supine

Part position:
Patient is facing tube with midsagittal plane of both body and head perpendicular to center of film. Position of head is adjusted so that a line between lower surface of upper teeth (occlusal plane) and base of skull is parallel to horizontal plane. When patient is upright, patient's head is placed firmly against cassette holder or a radiolucent wedge sponge for stability.

Collimation:
Lengthwise, to include lips when mouth is open. Crosswise, to include mastoid tips. Close collimation improves image quality and avoids unnecessary exposure to thyroid gland and eyes.

Patient instruction:
Open mouth as wide as possible. Stop breathing. Do not move.

> **TIP:** If the patient has closed the mouth following positioning and must reopen it prior to the exposure, it is wise to instruct the patient to "drop the lower jaw" as far as possible. When instructed to "open wide," patients may tend to extend the neck, causing incorrect position of the head.

Structures seen:
Lateral masses and transverse processes of atlas, dens, and upper half of body of axis (seen between upper and lower teeth)

> **TIP:** If the base of the skull is superimposed on the atlas and the dens, this indicates that the patient's neck was extended too far. If the upper teeth superimpose the atlas and the dens, the patient's neck was flexed too much. If the lower teeth superimpose the upper half of the axis and the base of the skull is in the proper position, the patient's mouth was not open far enough.

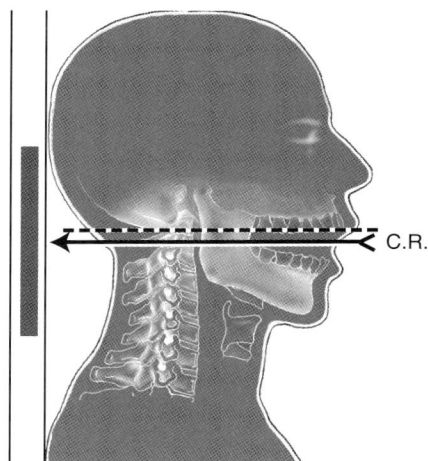

Head position for AP upper cervical spine. Upper teeth are projected over base of skull. With mouth wide open, atlas and axis are projected between upper and lower teeth.

AP open-mouth position for upper cervical spine, recumbent.

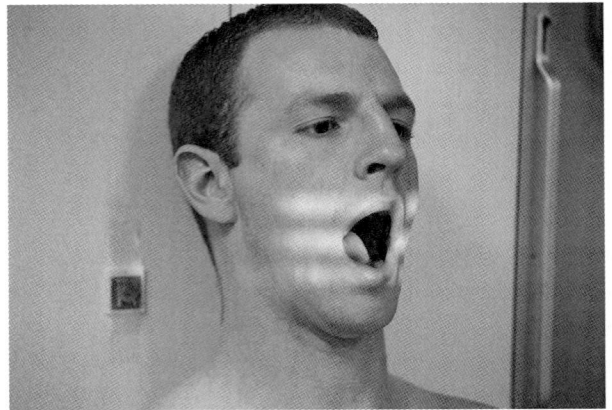

AP open-mouth position for upper cervical spine, upright.

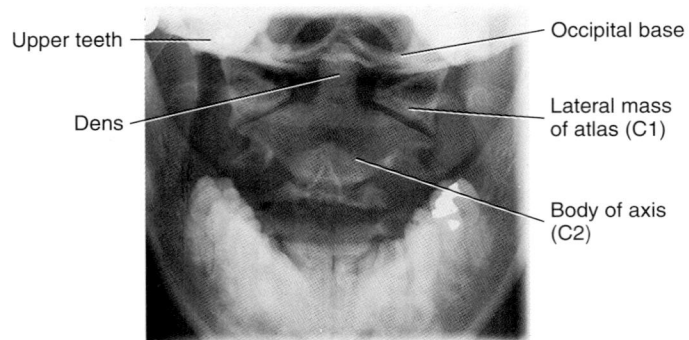

Upper teeth

Dens

Occipital base

Lateral mass
of atlas (C1)

Body of axis
(C2)

AP open-mouth projection of upper cervical spine.

LATERAL

Film: 8 × 10 inch (20 × 25 cm), lengthwise

Screens: Regular (rapid)

Grid: Optional

SID: 72 inches

Body position:
Seated or standing

Part position:
Midsagittal planes of body and head are parallel to film with infraorbitomeatal line parallel to floor. Shoulders must be relaxed and depressed. Film is positioned so that upper margin is level with top of ear.

> **TIP:** Patients with high, square shoulders may better achieve this position with sandbags of equal weight (10 pounds) suspended from the wrists. Alternatively, patient may stand on the center of a long strap, grasping the two ends to maintain downward tension on the shoulders.

Central ray:
Perpendicular to center of film through body of C4. Central ray enters neck at a point in line with mastoid process and 1 inch inferior to level of angle of mandible.

> **TIP:** Place your finger on the tip of the C7 spinous process and note the location of its shadow in the collimator light beam. It should be within the posterior margin of the film and at least 2 inches above its inferior margin.

Collimation:
Lengthwise, just inside film margins. Crosswise, to include spinous process of C7 and anterior soft tissues of neck. Eyes should be excluded from field.

Patient instruction:
Stop breathing. Do not move.

> **TIP:** Do *not* instruct patient to "take a deep breath," since this tends to elevate the shoulders.

Structures seen:
All seven cervical vertebrae and soft tissues of anterior neck, including spinal alignment, bodies, disk spaces, spinous processes, and zygapophyseal joints

> **NOTE:** When a good effort has been made to lower the shoulders for the lateral cervical spine projection but the radiograph fails to demonstrate C7, it is necessary to supplement the examination with the lateral projection of the cervicothoracic region, described later in this chapter.

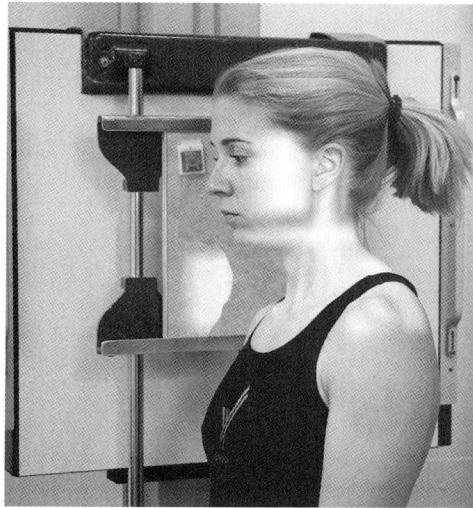

Lateral cervical spine position.

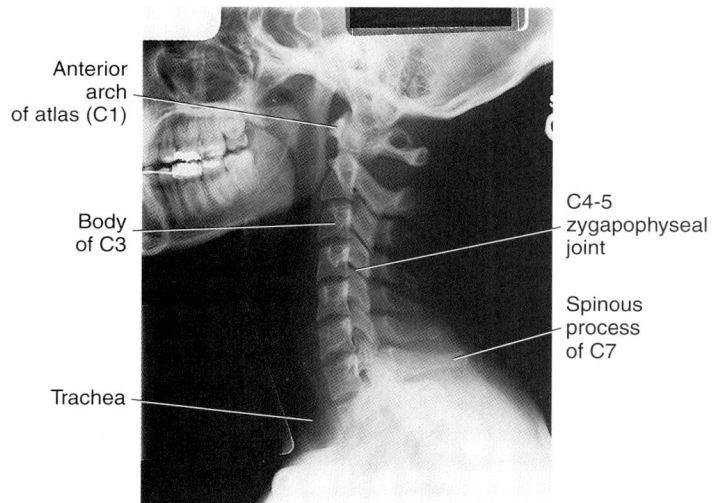

Lateral cervical spine radiograph.

Supplemental Views

FLEXION AND EXTENSION

Flexion and extension views of the cervical spine are taken in the lateral projection to evaluate intersegmental stability. Both views use the same general method as for the routine lateral projection.

NOTE: *When there has been recent trauma to the cervical spine, the routine lateral projection should be evaluated by the physician before proceeding with flexion and extension views.* If there is a fracture or significant vertebral subluxation (displacement), these positions may cause further damage to the spine or injury to the spinal cord.

Position for flexion:

Patient is positioned as for routine lateral projection. Patient is then instructed to first "tuck" chin close to neck and then to flex neck, attempting to look at a spot at midsternum.

Position for extension:

Patient is positioned as for routine lateral projection. Patient is then instructed to extend neck, looking at a spot on ceiling directly above head.

For both of these views, take care that there is no motion of the thoracic spine, which should remain straight. *The radiographer must not force these positions.* The desired degree of flexion or extension is the fullest extent that is tolerable for the patient. Double-check the head position to be certain that no rotation or lateral bending occurs with the flexion or extension.

Lateral cervical spine position with flexion.

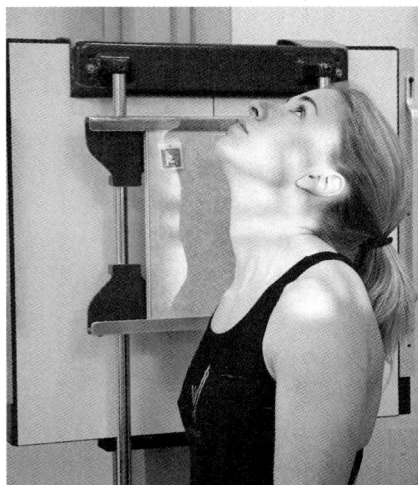

Lateral cervical spine position with extension.

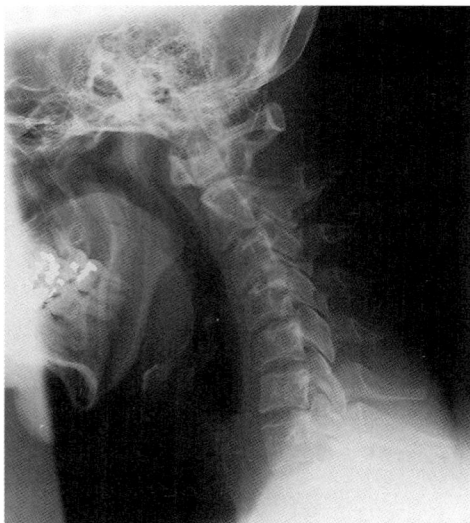

Lateral cervical spine radiograph with flexion.

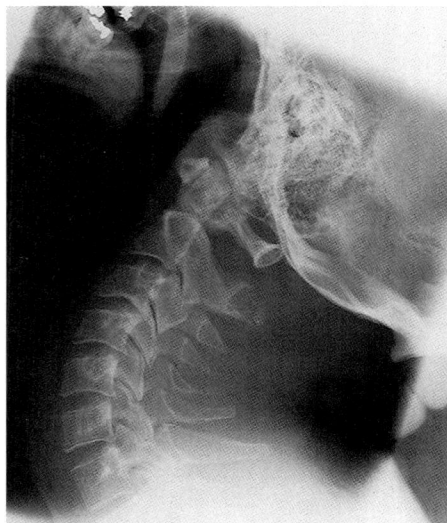

Lateral cervical spine radiograph with extension.

OBLIQUE PROJECTIONS

Oblique projections are taken in left/right pairs. The two basic approaches to oblique cervical projections are modifications of either the AP or the lateral projection. In the first case, the radiograph is taken either upright or recumbent at 40 inches SID, using the bucky. Alternatively, when done upright, a 72-inch SID may be used with or without a grid.

Oblique projections may be done PA (RAO and LAO) or AP (RPO and LPO). The available equipment and the preferences of the physician may dictate both the method and the positions used.

Film: 8 × 10 inch (20 × 25 cm), lengthwise

Screens: Regular (rapid)

Grid: Optional

SID: 40 inches or 72 inches

Body position:
Seated, standing, or recumbent

Part position:
Coronal plane of body forms angle of 45° to 50° with plane of film. Head is rotated so that sagittal plane of skull is parallel to film.

Central ray:

PA obliques: Angled 15° caudad to center of film through body of C4. Central ray enters neck at a point in line with mastoid process at level of angle of mandible.

AP obliques: Angled 15° cephalad to center of film through body of C4. Central ray enters neck at a point in line with mastoid process at level 2 inches inferior to angle of mandible.

Collimation:
Lengthwise, just inside film margins. Crosswise, to include soft tissues of neck. Eyes should be excluded from field.

Patient instruction:
Stop breathing. Do not move.

Structures seen:
PA obliques demonstrate intervertebral foramina on side nearest film. AP obliques demonstrate intervertebral foramina on side farthest from film.

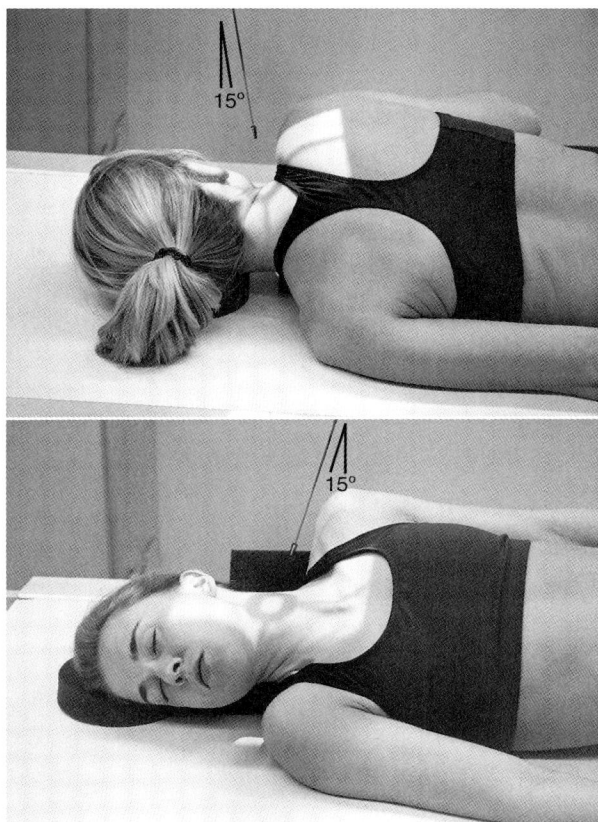

PA oblique cervical spine position, recumbent *(top)*. AP oblique cervical spine position, recumbent *(bottom)*.

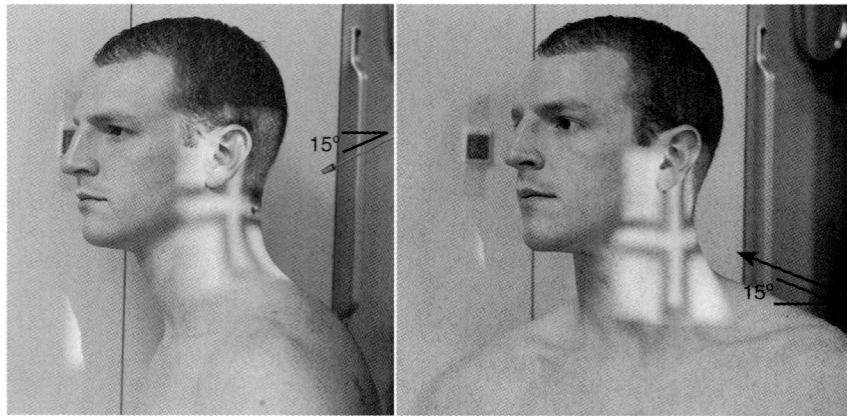

PA oblique cervical spine position, upright *(left)*. AP oblique cervical spine position, upright *(right)*.

Oblique cervical spine radiographs. **A,** AP oblique at 40 inches SID using the bucky. **B,** PA oblique at 72 inches SID without grid.

LATERAL CERVICOTHORACIC REGION

The lateral projection of the cervicothoracic area is commonly called the "swimmer's lateral projection." It is used when routine lateral projections of either the cervical or thoracic spine fail to demonstrate this area adequately. The shoulder positions create a small "window" between the shoulders, and the cervicothoracic spine is projected into this relatively open area.

Film: 8 × 10 inch (20 × 25 cm) or
              10 × 12 inch (24 × 30 cm), lengthwise

Screens: Regular (rapid)

Grid: Yes

SID: 40 inches or 72 inches

Body position:
Seated, standing, or recumbent

Part position:
Sagittal planes of body and head are parallel to film. Arm nearest film is raised above head and shoulder is rounded anteriorly. Opposite shoulder is depressed and slightly posterior.

Central ray:
Perpendicular to film at C7-T1 interspace. Central ray enters at base of neck in midcoronal plane at level of C7 spinous process.

Collimation:
Just inside film margins

Patient instruction:
Stop breathing. Do not move.

Structures seen:
Vertebrae C6 through T3 (C5 through T5 with larger film) in lateral projection without significant rotation. Bodies, disk spaces, spinous processes, and zygapophyseal joints are demonstrated between shoulders.

> **TIPS:**
> - The pressure of the body against the shoulder nearest the film helps to keep the shoulder anterior to the spine.
> - Take care when positioning the arms so that the sagittal plane of the body remains parallel to the film. The most common error associated with this position is rotation of the body so that the spine is oblique rather than lateral.
> - The clavicle nearest the film is projected across the spine on this projection, usually superimposing the body of T1. To avoid superimposition when T1 is of primary clinical interest, the central ray may be angled approximately 5°. A cephalad angle projects the clavicle over T2; a caudad angle projects the clavicle over C7.

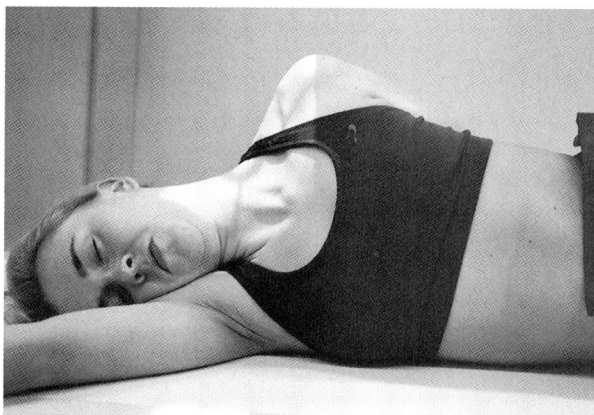

Recumbent position for swimmer's lateral projection of cervicothoracic region.

Upright position for swimmer's lateral projection of cervicothoracic region.

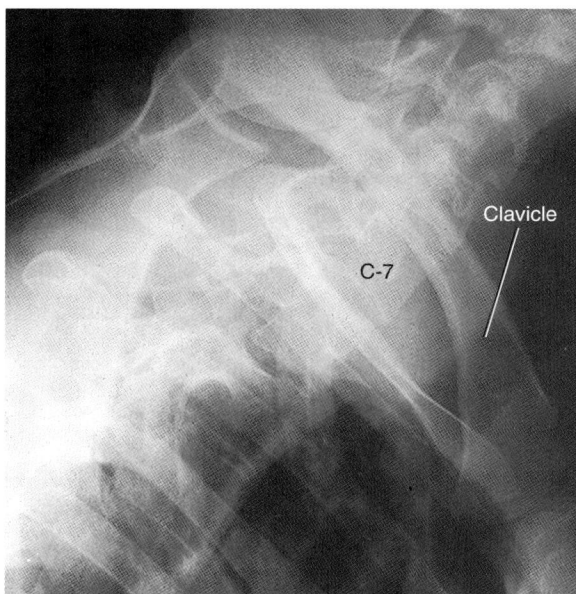

Swimmer's lateral projection of cervicothoracic region.

There is significant tissue density variation between the extreme ends of the thoracic spine. Near the neck there is much less tissue to penetrate than at the level of T12 in the upper abdominal region. For this reason, it is desirable to utilize the anode heel effect (see Chapter 4). For recumbent studies, the patient should be instructed to lie on the table with the head toward the anode end of the x-ray tube. If there is significant difference in thickness between the two ends of the thoracic spine, a #2 or #3 wedge compensating filter may be placed in the anode portion of the x-ray beam so that the shadow of its thin edge is approximately in the center of the radiation field.

In the lateral projection, the density variation is reversed. The proximal thoracic spine is more difficult to penetrate because of the bone and muscle mass of the shoulders, and there is little lung tissue in this area. The inferior portion is relatively easily penetrated because its mass is largely air-containing lung. For this reason, it may be desirable to reverse the position of the compensating filter for the lateral projection. In any case, the first three thoracic vertebrae are seldom visualized well on the lateral projection. When the area of clinical interest includes the upper thoracic vertebrae, it is usual for the examination to include a swimmer's lateral projection of the cervicothoracic region. This projection is explained and illustrated in the previous section. In some facilities, the swimmer's lateral projection is a routine part of the basic thoracic spine examination.

For this examination, the patient should undress down to the waist and don a gown that opens in the back. This will facilitate visualization and palpation of the spine. Any jewelry that would be in the radiation field should be removed. For standing examinations, the shoes should also be removed.

Basic Examination

The basic examination of the thoracic spine includes the AP and lateral views.

AP

Film: 14 × 17 inch (35 × 43 cm) or
             7 × 17 inch (18 × 43 cm), lengthwise

Screens: Regular (rapid)

Grid: Yes

SID: 40 inches

Body position:
Seated, standing, or recumbent

Part position: Midsagittal plane of body is perpendicular to film and centered to it, with patient facing tube. Superior border of film is aligned 1 inch above spinous process of C7. When patient is supine, it is helpful to place a bolster under knees. When patient is standing, feet should be shoulder-width apart with equal weight bearing, and patient's back should be firmly against cassette holder.

Central ray:
Perpendicular center of film at T7. This point is in midline at approximate midpoint of sternum.

Collimation:
Lengthwise, just inside film margins. Crosswise, 6 to 8 inches wide. Close collimation improves visualization and reduces patient dose. When there is significant scoliosis, a wider field may be needed.

Gonad shielding:
Lead half-apron

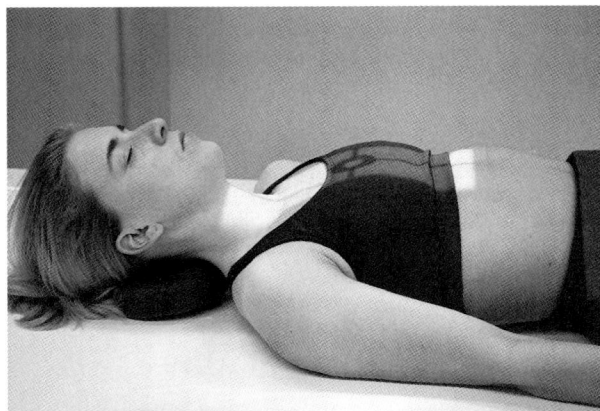

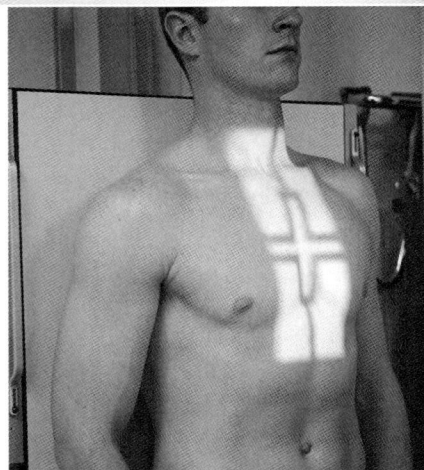

Recumbent AP thoracic spine position *(top)*. Upright AP thoracic spine position *(bottom)*.

Patient instruction:

Do not move. Suspend breathing on expiration.

Structures seen:

All twelve thoracic vertebrae, particularly the bodies, disk spaces, and transverse processes. C7 and at least a portion of L1 are usually also seen.

AP thoracic spine radiograph.

LATERAL

Film: 14 × 17 inch (35 × 43 cm), lengthwise

Screens: Regular (rapid)

Grid: Yes

SID: 40 inches

Body position:
Seated, standing, or recumbent

Part position:
Sagittal plane of body is parallel to film. Arms may be raised overhead or anterior to body with shoulders rounded anteriorly. Take care that entire length of thoracic spine is parallel to film. When patient is recumbent, this may require support of a radiolucent sponge under waist and/or hips.

Central ray:
Perpendicular to center of film at level of T7. Central ray enters at inferior angle of scapula in posterior axillary line.

Collimation:
Lengthwise, just inside film margins. Crosswise, 8 to 10 inches. A wider field may be needed if there is exaggerated thoracic kyphosis.

> **TIP:** Place a strip of lead or a lead rubber mask behind the patient so that its margin is aligned to the shadow of the patient's back in the collimator light. This absorbs backscatter and improves visualization of the spinous processes.

Patient instruction:
Do not move. Shallow breathing during exposure.

Structures seen:
T3 through T12 with blurring of ribs and lung markings when breathing technique is used

> **TIP:** A low mA setting that provides the desired mAs with an exposure time of 1 to 3 seconds is needed for best results with breathing technique.

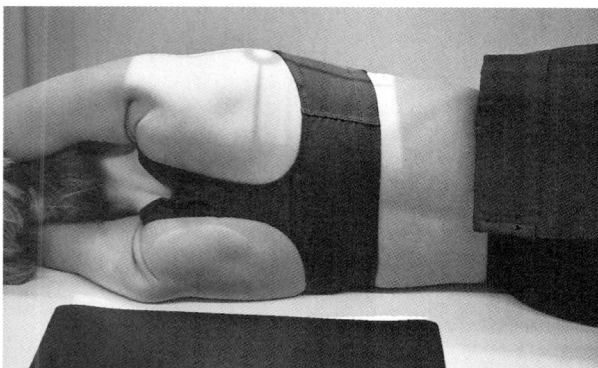

Upright lateral thoracic spine position.

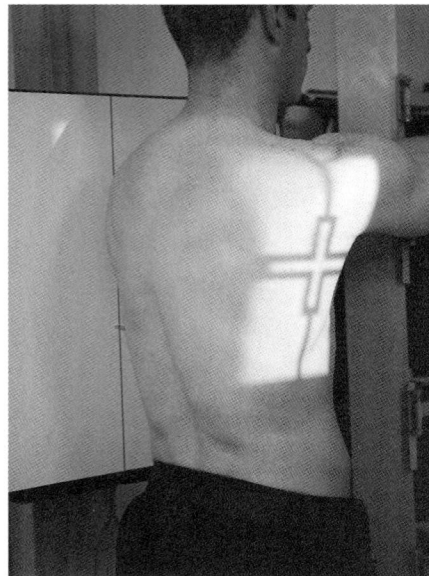

Lateral thoracic spine radiograph. Note blurred image of ribs and lung features caused by breathing technique.

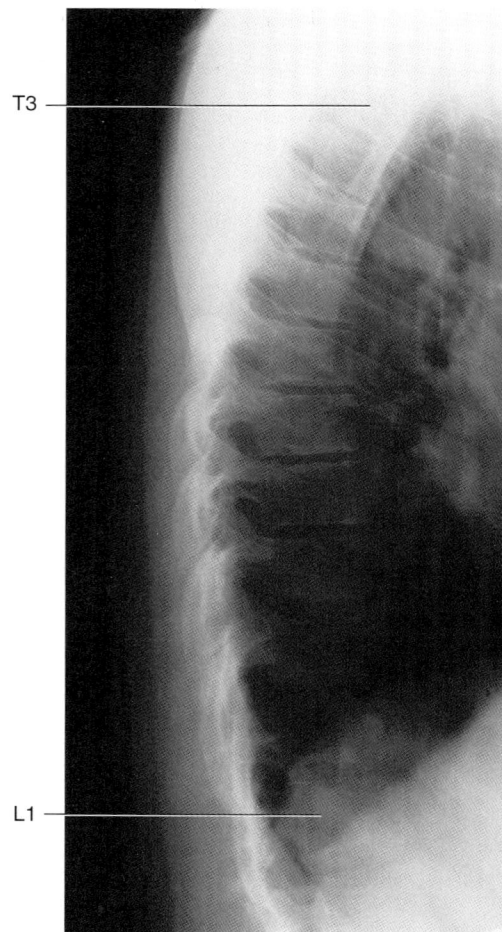

Recumbent lateral thoracic spine position.

Depending on the preferences of the physician, the routine frontal and lateral projections of the lumbar spine may be done using 11 × 14-inch (30 × 35-cm), 7 × 17-inch (18 × 43-cm), or 14 × 17-inch (35 × 43-cm) film. When the long dimension of the film is 17 inches, the center point is at the level of the iliac crest and the spine is visualized from T12 through the coccyx. When an 11 × 14-inch (28 × 35-cm) film is used, the center is 1½ inches superior to the level of the iliac crest to ensure that L1 is included. When the shorter film is used, the area of L1 through S2 is demonstrated.

For this series, patients should remove outer clothing from the torso and don a gown opening in the back. Female patients must remove bras. For upright studies, shoes are also removed.

Basic Examination

The basic examination of the lumbar spine includes the frontal projection (AP or PA) and lateral views.

FRONTAL PROJECTION (AP OR PA)

Film: 14 × 17 (35 × 43 cm), 7 × 17 inch (18 × 43 cm.), or 11 × 14 inch (28 × 35 cm), lengthwise. See introductory discussion.

Screens: Regular (rapid)

Grid: Yes

SID: 40 inches

Body position:
Standing or recumbent

Part position:

AP: Patient is facing tube with midsagittal plane perpendicular to film and centered to it. When patient is standing, feet are shoulder-width apart with equal weight bearing and torso is stabilized against upright cassette holder. When patient is supine, knees are flexed and supported with a bolster.

PA: Patient stands facing film or lies prone, with midsagittal plane perpendicular to film and centered to it.

Central ray:

14 × 17-inch (35 × 43-cm) or 7 × 17-inch (18 × 43-cm) film: Perpendicular to center of film through L4, in midline at level of iliac crest

11 × 14-inch (28 × 35-cm) film: Perpendicular to center of film through L3, in midline 1½ inches superior to level of iliac crest

Shielding:
Gonad shielding for males. Shield females only if shield will not interfere with purpose of examination. Consider PA projection for reduced ovarian dose.

Patient instruction:
Do not move. Suspend breathing on expiration.

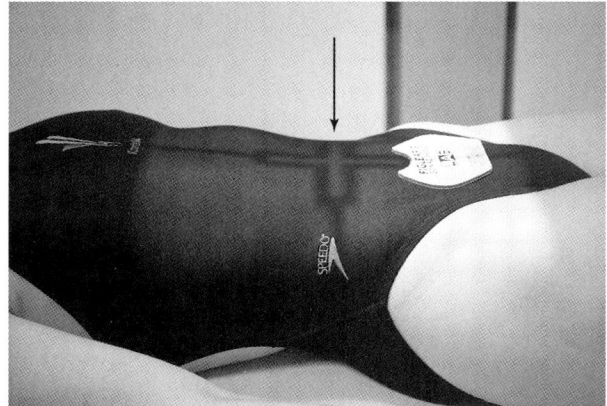

Recumbent AP lumbar spine position.

Upright AP lumbar spine position.

Structures seen:

All five lumbar vertebrae, intervertebral disk spaces, proximal portion of sacrum, and sacroiliac joints. This projection demonstrates the bodies, disk spaces, and transverse processes. The pedicles are seen "on end." When using a 14 × 17-inch (35 × 43-cm) film, central pelvis and hip joints maybe visualized if collimation is not too close. Visualization of hip joints is particularly important to demonstrate pelvic tilt when an upright AP projection is taken.

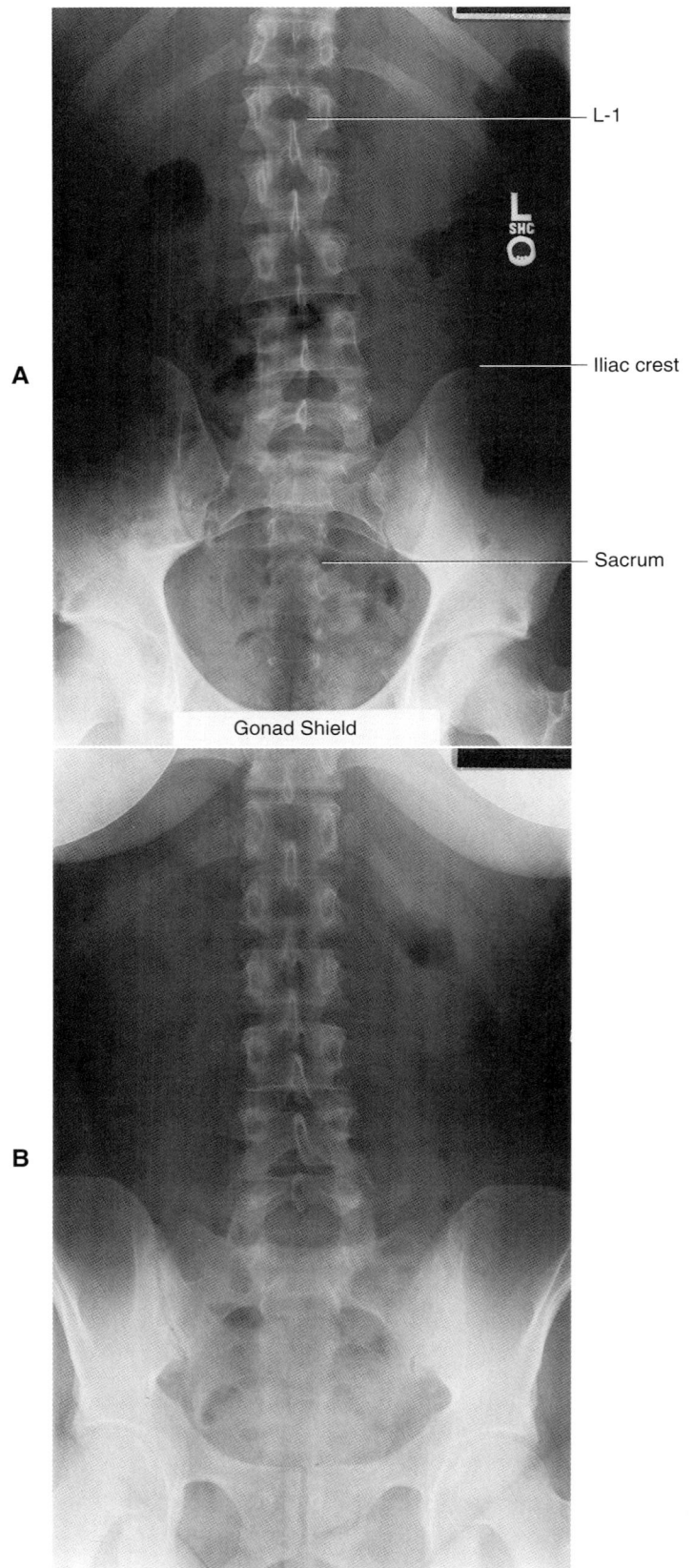

A

L-1

Iliac crest

Sacrum

Gonad Shield

B

A, AP lumbar spine radiograph, taken recumbent with knees flexed. **B,** PA lumbar spine radiograph, taken on the same patient. Note widening of disk spaces and magnification of sacrum.

LATERAL

Film: 14 × 17 inch (35 × 43 cm), 7 × 17 inch (18 × 43 cm), or 11 × 14 inch (28 × 35 cm), lengthwise. See introductory discussion.

Screens: Regular (rapid)

Grid: Yes

SID: 40 inches

Body position:
Standing or recumbent

Part position:
Sagittal plane is parallel to film.

Recumbent: In lateral recumbent position, spine is aligned parallel to center of bucky with arms anterior to body. Radiolucent sponges may be used to elevate waist and/or hip to keep spine level. Knees are flexed. A pad between knees helps keep pelvis lateral and maintain lateral position of spine.

Upright: Feet are shoulder-width apart with equal weight bearing and torso is stabilized against upright cassette holder. Arms are crossed over chest with hands supported on shoulders. Alternatively, arms may be supported out of radiation field by having patient grasp a pole.

Central ray:

14 × 17-inch (35 × 43-cm) or 7 × 17-inch (18 × 43-cm) film Perpendicular to center of film through L4, in midaxillary line at level of iliac crest

11 × 14-inch (28 × 35-cm) film Perpendicular to center of film through L3, in midaxillary line 1½ inches superior to level of iliac crest

> **TIP:** Place a strip of lead or a lead rubber mask behind the patient so that its margin is aligned to the shadow of the patient's back in the collimator light. This absorbs backscatter and improves visualization of the spinous processes.

Patient instruction:
Do not move. Suspend breathing on expiration. Respiratory phase is particularly important on the lateral projection. If exposed on inspiration, the posterior lung fields will superimpose the body of L1.

Structures seen:
All five lumbar vertebrae and superior half of sacrum, including intervertebral foramina, spinous processes and profile of the bodies, and intervertebral disk spaces

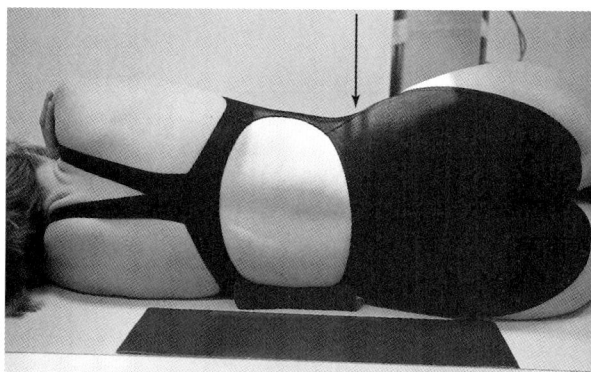

Recumbent lateral lumbar spine position.

Upright lateral lumbar spine positions.

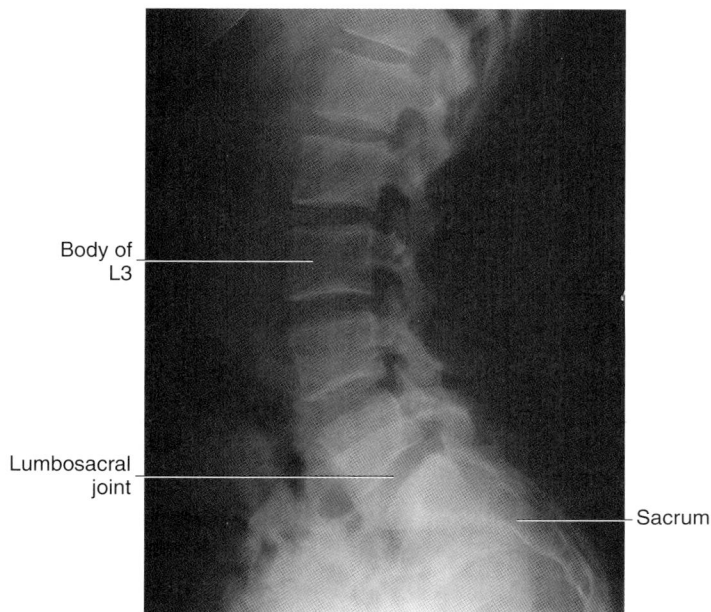

Lateral lumbar spine radiograph.

Supplemental Views

AP OBLIQUE PROJECTION

Bilateral oblique projections are taken for comparison. AP obliques (RPO and LPO) are most commonly done, since they demonstrate the zygapophyseal joints and pars interarticularis of the side nearest the film, providing better detail.

Film: 10 × 12 inch (24 × 30 cm) or
                 11 × 14 inch (28 × 35 cm), lengthwise

Screens: Regular (rapid)

Grid: Yes

SID: 40 inches

Body position:
Standing or recumbent

Part position:
Sagittal plane is aligned at angle of 45° to film.

Recumbent: From supine position, patient is rotated 45° toward side being radiographed. Position may be supported by a large 45°-angle radiolucent sponge. Take care that there is no torsion (twist) of spine.

Upright: From AP position, patient is rotated 45° toward side being radiographed. Feet are shoulder-width apart with equal weight bearing and torso is stabilized against upright cassette holder.

Central ray:
Perpendicular to center of cassette through L3. Central ray enters at point 2 inches medial to ASIS farthest from film and 1½ inches superior to iliac crest.

Shielding:
Use gonad shielding with precision so that shield does not cover areas of clinical interest.

Patient instruction:
Do not move. Suspend breathing on expiration.

Structures seen:
All five lumbar vertebrae and upper portion of sacrum, including zygapophyseal joints and pars interarticularis on the side nearest the film

Recumbent *(top)* and upright AP oblique lumbar spine position.

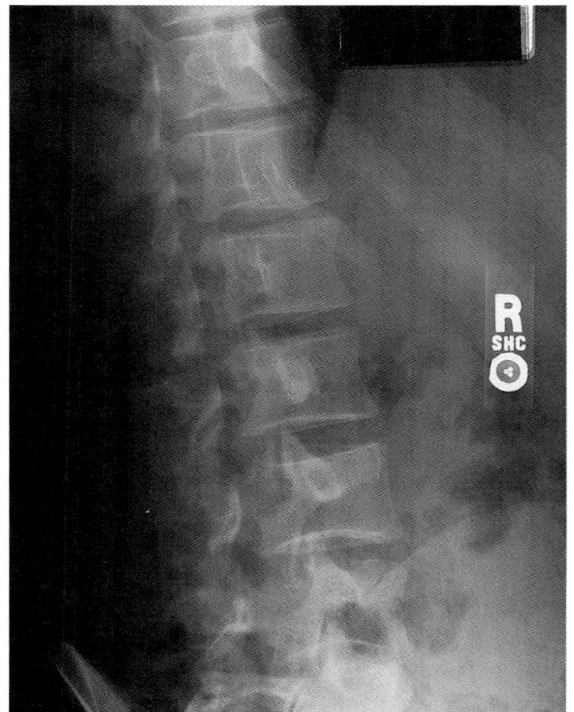

Oblique lumbar spine radiograph.

LATERAL LUMBOSACRAL JOINT

A "spot film" of the lumbosacral joint in the lateral projection is helpful when there is poor visualization of this area on the routine lateral projection. This may occur as a result of insufficient penetration of this dense area, or it may simply be desirable to enhance visualization, since this is the most common site of chronic problems in the lumbosacral spine. While this view may be taken with the patient upright, the result is usually superior when the patient is recumbent.

Body position:
As for routine lateral lumbar projection, taking care that spine is parallel to film. An 8 × 10-inch (20 × 25-cm) film is used with the field collimated to 5 × 5 inches.

Central ray:
Directed perpendicular to center of film through lumbosacral joint. This centering point is localized 1 inch inferior to iliac crest on coronal line midway between ASIS and posterior prominence of sacrum.

Position for lateral lumbosacral joint.

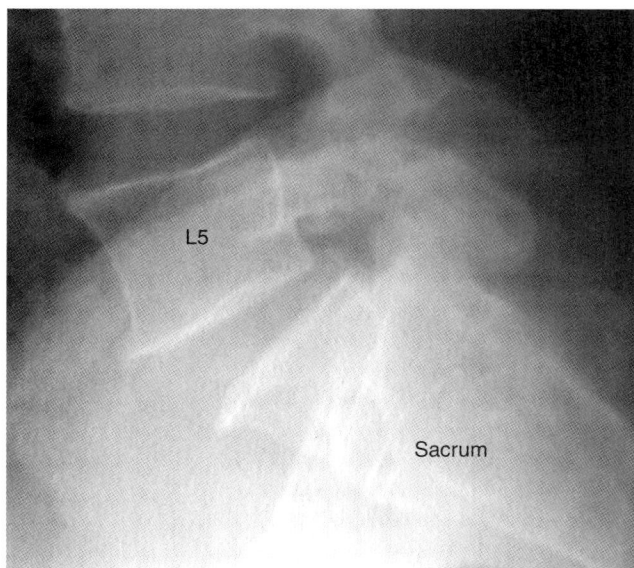

Lateral spot film of lumbosacral joint.

AP AXIAL PROJECTION OF LUMBOSACRAL AND SACROILIAC JOINTS

Because of the sacral base angle, the lumbosacral joint is not well visualized on the routine AP lumbar projection. The AP axial projection directs the central ray parallel to the sacral base. See Table 15-1 for variations in sacral base angle.

Film:　　8 × 10 inch (20 × 25 cm), crosswise

Body position:
Supine, as for AP recumbent lumbar spine, with knees flexed and supported

Central ray:
Angled 25° cephalad for males and 30° cephalad for females. It is directed to center of film through lumbosacral joint. Central ray enters in midline, midway between level of ASIS and symphysis pubis.

Shielding:
Gonad shielding for males. Ovarian shielding would interfere with purpose of examination.

Structures seen:
Lumbosacral joint in frontal projection, sacral ala, and sacroiliac joints

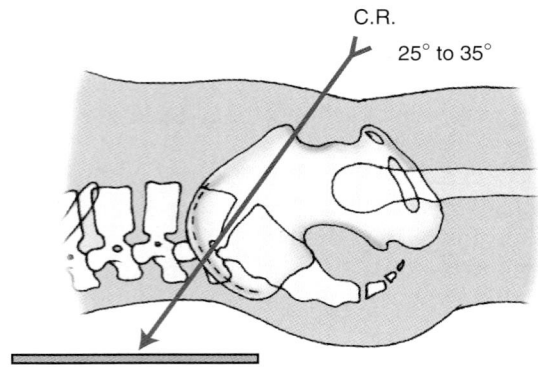

Alignment of central ray for AP axial projection of lumbosacral joint and sacroiliac joints.

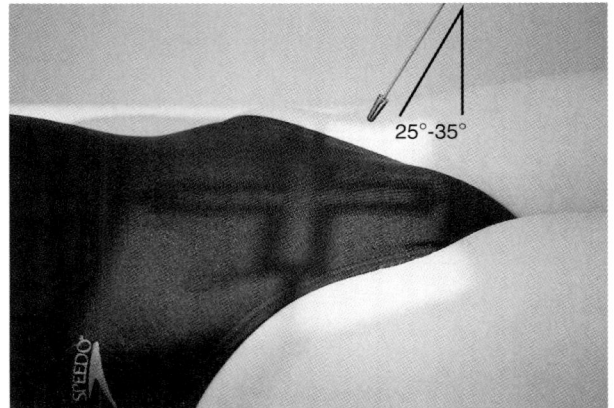

Position for AP axial projection of lumbosacral and sacroiliac joints.

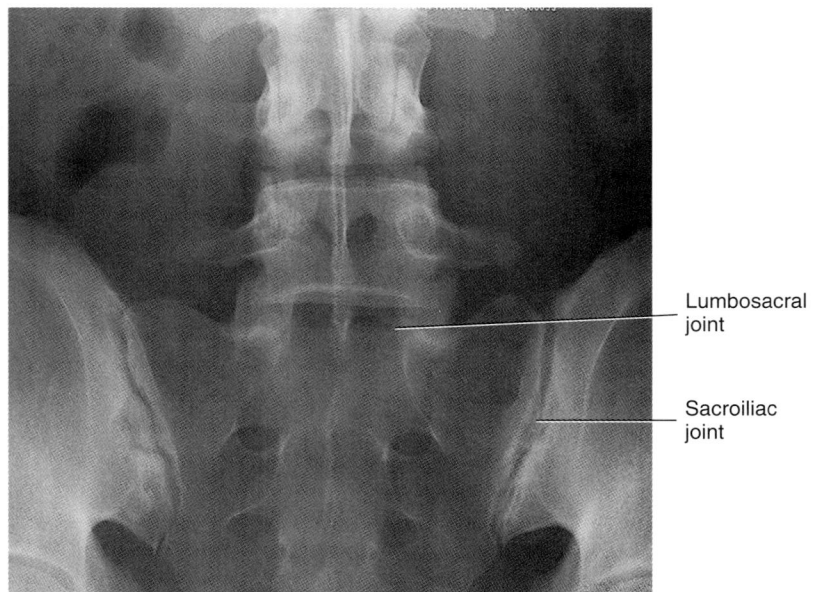

Lumbosacral joint

Sacroiliac joint

AP axial radiograph of lumbosacral and sacroiliac joints.

For this series, patients should remove outer clothing from the torso and don a gown opening in the back. For upright studies, shoes are also removed.

Basic Examination

The basic examination of the sacrum includes the AP axial and lateral views.

AP AXIAL

Film: 8 × 10 inch (20 × 25 cm), lengthwise

Screens: Regular (rapid)

Grid: Yes

SID: 40 inches

Body position:
Recumbent or supine

Part position:
Midsagittal plane is perpendicular to film and centered to it. Knees are flexed and supported with a bolster.

Central ray:
Angled 15° cephalad to center of film through midsacrum. Central ray enters body at midline, midway between symphysis pubis and level of ASIS.

Collimation:
Just inside film margins

Shielding:
Shield males with precision. Do not shield females.

Patient instruction:
Stop breathing. Do not move.

Structures seen:
Entire sacrum and sacroiliac articulations

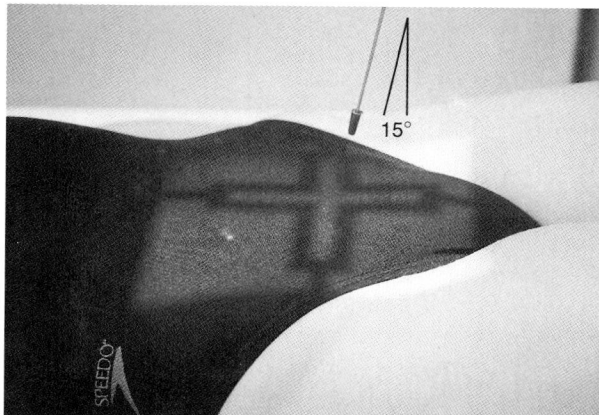

AP axial sacrum position.

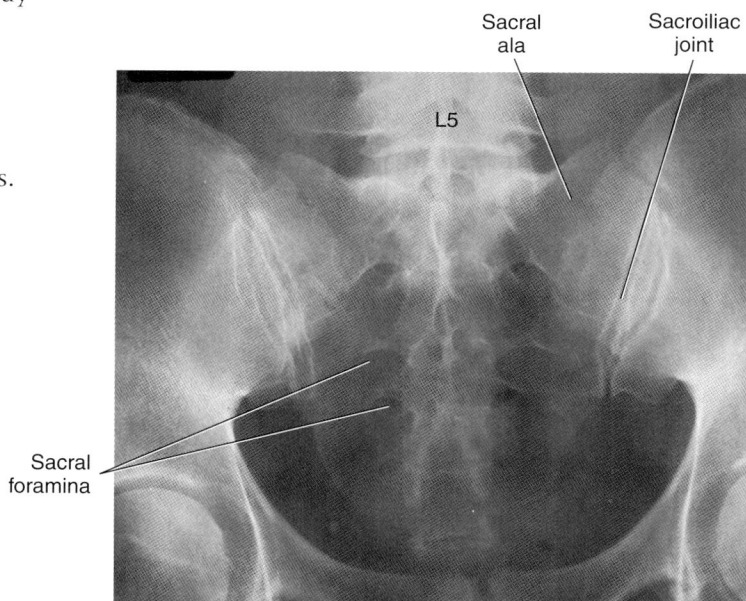

AP axial sacrum radiograph.

LATERAL

Film: 8 × 10 inch (20 × 25 cm), lengthwise

Screens: Regular (rapid)

Grid: Yes

SID: 40 inches

Body position:
Standing or recumbent

Part position:
Sagittal plane is parallel to film.

Recumbent: Spine is aligned parallel to center of bucky with arms anterior to body. Radiolucent sponges may be used to elevate waist and/or hips to keep spine level. Knees are flexed. A pad between knees helps keep pelvis lateral and maintain lateral position of spine.

Upright: Feet are shoulder-width apart with equal weight bearing and torso is stabilized against upright cassette holder. Arms are crossed over chest with hands resting on shoulders, or arms may be supported by having patient grasp a pole.

Central ray:
Perpendicular to center of film through center of sacrum. Central ray enters at point 2½ to 4 inches posterior to ASIS, depending on patient size; 3½ inches works well for average adult patient.

> **TIP:** Place a strip of lead or a lead rubber mask behind the patient so that its margin is aligned to the shadow of the patient's back in the collimator light. This absorbs backscatter and improves visualization of the sacrum.

Patient instruction:
Do not move. Suspend breathing on expiration.

Structures seen:
Entire sacrum and lumbosacral joint. Coccyx is usually also visualized.

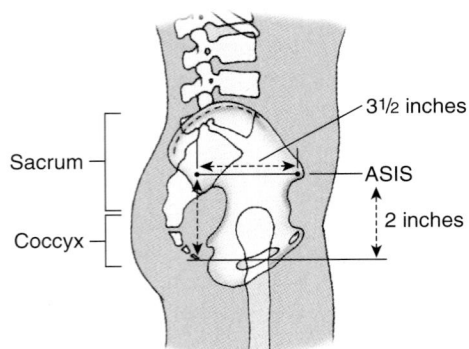

Localization of sacrum and coccyx in relation to palpable landmarks.

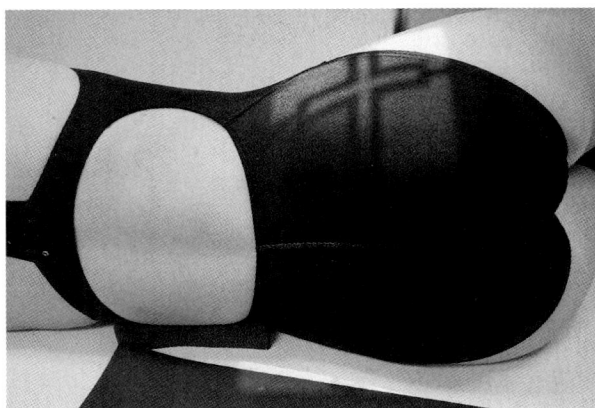

Lateral sacrum position.

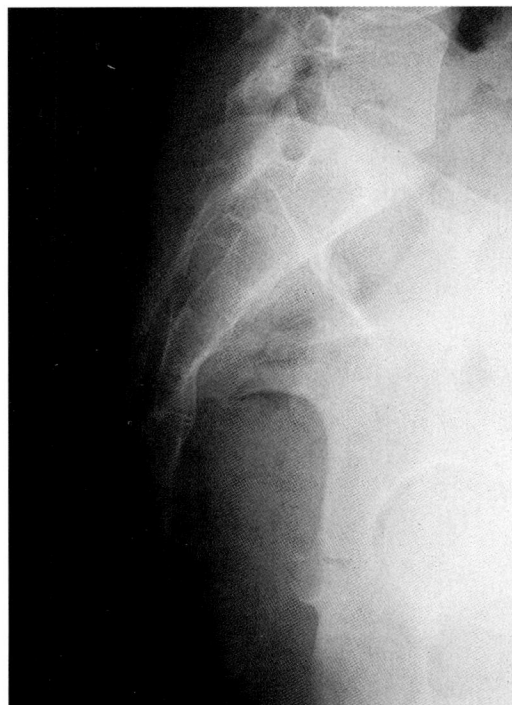

Lateral sacrum radiograph.

Supplemental View

OBLIQUE SACROILIAC JOINTS

Bilateral oblique views are usually taken for comparison.

Film:　　10 × 12 inch (24 × 30 cm), lengthwise

Screens: Regular (rapid)

Grid:　　Yes

SID:　　40 inches

Body position:
Recumbent

Part position:

AP obliques (RPO, LPO): From supine position, body is rotated so that coronal plane is aligned at angle of 25° to 30° to film. Side being radiographed is side that is elevated from film. Position may be supported by radiolucent sponge under hip and lumbar area of elevated side. Take care that there is no torsion of spine.

PA obliques (RAO, LAO): From prone position, body is rotated so that coronal plane is aligned at angle of 25° to 30° to film. Side being radiographed is side nearest film. This position may be supported by radiolucent sponge under hip and abdomen area on opposite side. Take care that there is no torsion of spine.

Central ray:

AP obliques (RPO, LPO): Perpendicular to center of film through point 1 inch medial to ASIS farthest from film

PA obliques (RAO, LAO): Perpendicular to center of film through point 1 inch medial to ASIS nearest film

Shielding:
Shield males with precision. Do not shield females.

Patient instruction:
Do not move. Suspend breathing on expiration.

Structures seen:

AP obliques: Sacroiliac joint farthest from film

PA obliques: Sacroiliac joint nearest film

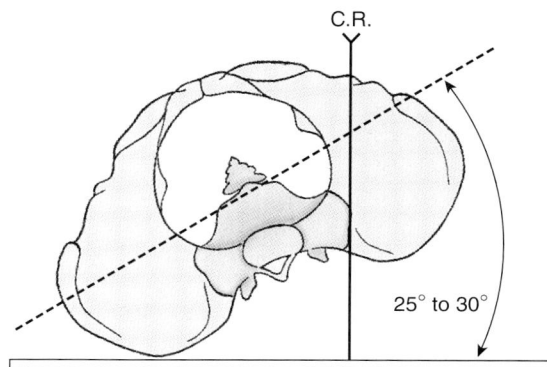

When coronal plane is aligned to film at 25° to 30° angle, perpendicular central ray passes through sacroiliac joint on elevated side.

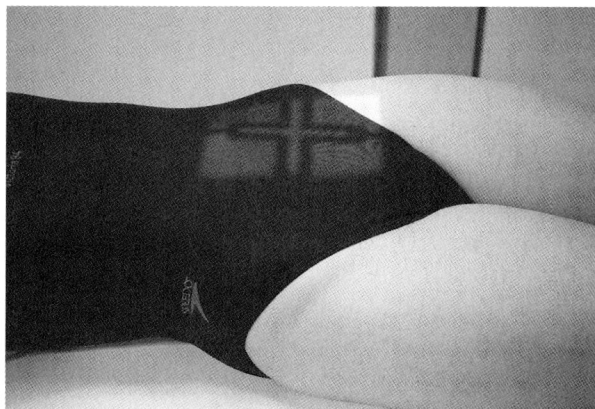

Oblique sacroiliac joint position.

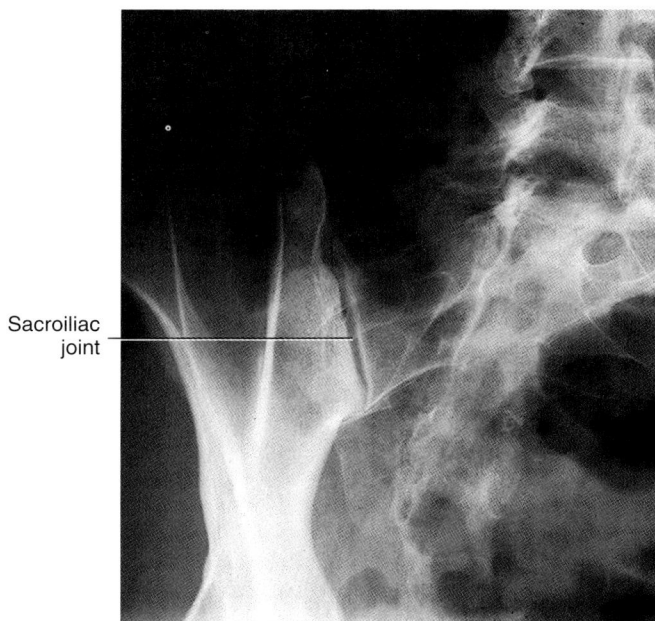

Oblique sacroiliac joint radiograph.

It is wise to use a short scale of contrast for coccyx radiography. The bones of the coccyx are tiny and are situated in a large mass of soft tissue, providing little subject contrast. A kVp setting of 65 to 75 kVp will improve visualization.

Preparation is the same as for sacrum examination.

Basic Examination

The basic examination of the coccyx includes the AP axial and lateral views.

AP AXIAL

Film:　8 × 10 inch (20 × 25 cm), lengthwise

Screens: Regular (rapid)

Grid:　Yes

SID:　40 inches

Body position:
Recumbent

Part position:
Coronal plane is parallel to film with patient facing tube. Midsagittal plane is centered to midline of bucky. When patient is supine, knees are flexed and supported with a bolster.

Central ray:
Angled 10° caudad and centered to film. Beam enters body in midline, midway between symphysis pubis and level of ASIS.

Collimation:
Lengthwise, 6 inches; crosswise, 4 inches. Close collimation is required for adequate visualization.

Shielding:
Precise gonad shielding for males. Do not shield females.

Patient instruction:
Do not move.

Structures seen:
Entire coccyx and distal portion of sacrum

AP axial coccyx position.

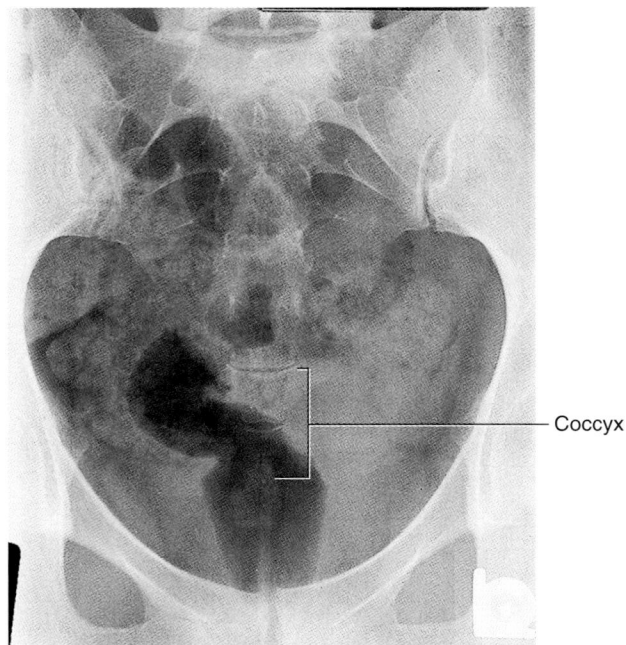

AP axial coccyx radiograph.

LATERAL

Film: 8 × 10 inch (20 × 25 cm), lengthwise

Screens: Regular (rapid)

Grid: Yes

SID: 40 inches

Body position:
Recumbent

Part position:
Sagittal plane is parallel to film. When patient is recumbent, knees are flexed and may be separated by a sponge or cushion.

Central ray:
Perpendicular to center of film through center of coccyx. Central ray enters at point 4 inches inferior to ASIS level and 2½ to 4 inches posterior to ASIS, depending on patient size; 3½ inches works well for average adult patient.

> **TIP:** Place a strip of lead or a lead rubber mask behind the patient so that its margin is aligned to the shadow of the patient's back in the collimator light. This absorbs backscatter and improves visualization of the coccyx.

Shielding:
Precise gonad shielding for males. Do not shield females.

Patient instruction:
Do not move.

Structures seen:
Entire coccyx and distal portion of sacrum

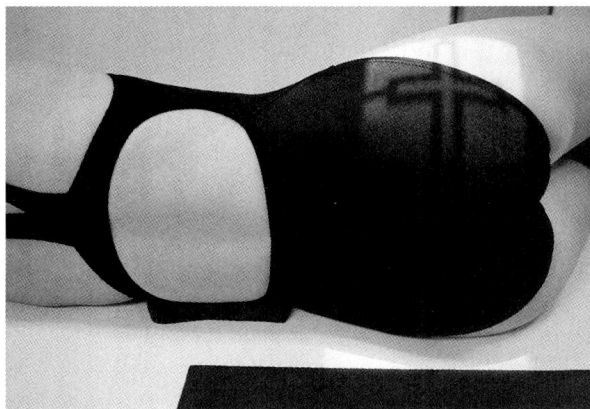

Lateral coccyx position.

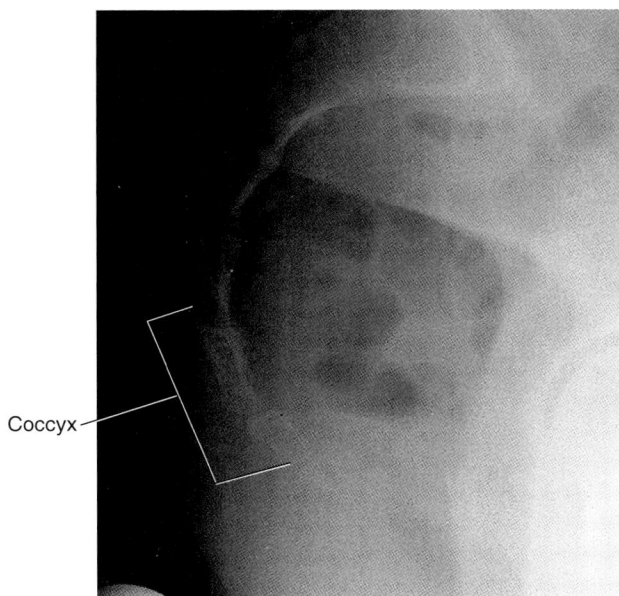

Lateral coccyx radiograph.

It is sometimes desirable to demonstrate the spinal column as a whole. This is useful for the evaluation of scoliosis. It is also used by some chiropractic physicians for evaluation of the entire spine with weight bearing. The frontal projection often constitutes a complete examination. The lateral projection is less frequently done but may be desirable for evaluation of abnormal spinal curvatures that involve both planes of the body. Lateral full-spine projections are also used in some chiropractic evaluations. For both frontal and lateral projections, exposure factors are adjusted for appropriate radiographic density in the lumbar region at the required distance.

Because of significant differences in thickness and tissue density between the cervical region and the lumbar region, some means is needed to provide even radiographic density throughout the spine with a single exposure. This is best accomplished by using compensating filters to reduce the exposure to the cervical and upper thoracic areas. Alternatively, cassettes are available that have gradient screens. These screens are slow at one end and fast at the other, providing appropriate exposure to the film at all levels. When using cassettes with gradient screens, the radiographer must ensure that the slow end of the cassette is uppermost. The use of compensating filters is preferred because this method provides significantly reduced exposure to the tissues in the region of the neck and upper chest as compared to the use of gradient screens.

Because patients who require this procedure must often be radiographed repeatedly, care is taken to shield sensitive tissues such as the gonads, the breasts, the thyroid gland, and the eyes. Idiopathic scoliosis (scoliosis of unknown cause) most often affects females, beginning with the onset of puberty. The prognosis for successful treatment is determined in large part by the degree to which the girl is still growing. This is evaluated by assessment of the growth plate along the rim of the iliac crest. For this reason, collimation in the pelvic region must be wide enough to include the iliac crests bilaterally.

FRONTAL PROJECTION

Film: 14 × 36 inch (35 × 90 cm)

Screens: See introductory discussion

Grid: Yes

SID: Usually 72 inches, but in some cases 84 inches

Position:
Upright as for routine standing AP or PA lumbar spine. Height of grid and cassette is adjusted to include area from top of patient's ears to level of greater trochanters. Head is positioned as for AP open-mouth projection of upper cervical spine, with neck position adjusted approximately 3° to allow for angulation of diverging x-ray beam at upper extremity of field. Neck is flexed slightly for AP projections and extended for PA projections.

Central ray:
Aligned perpendicular to center of film, entering at midline approximately at level of xiphoid process

Shielding:
Upright shield stands or shadow shield devices that attach to collimator

Collimation:
Wide enough in pelvic region to include iliac crests bilaterally

Patient instruction:
Hold mouth wide open. Stop breathing. Do not move.

PA full spine position.

AP full spine position showing use of upright shield stand.

LATERAL PROJECTION

Film: 14 × 36 inch (35 × 90 cm)

Screens: See introductory discussion

Grid: Yes

SID: Usually 72 inches, but in some cases 84 inches

Position:
Same as for upright lateral projection of thoracic spine. Shoulders are rounded anteriorly and arms are extended anterior to body and supported.

Shielding:
Upright shield stands or shadow shield devices that attach to collimator

Patient instruction:
Stop breathing. Do not move.

Lateral full spine position.

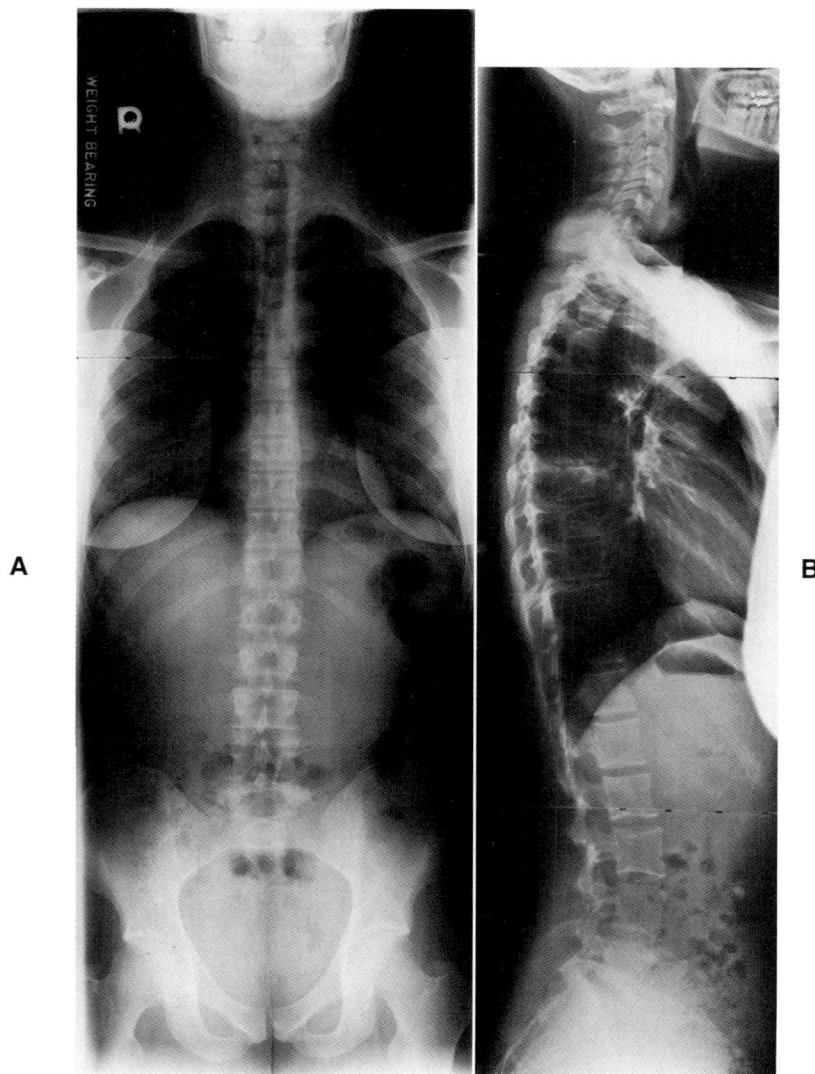

Full spine radiographs. **A,** AP; **B,** lateral.

PATHOLOGY

Congenital Anomalies

The spine is a common site for congenital anomalies and deformities. A transitional vertebra occurs when a vertebra of one spinal region takes on characteristics of the adjacent region. For example, anomalous ribs sometimes occur on C7 (Fig. 15-22) or L1. Sometimes one or both spinous processes of L5 become fused to the sacrum (Fig. 15-23). This is called sacralization of L5. Occasionally there will be an extra vertebra in one region. This may or may not be accompanied by the lack of a vertebral segment in the adjacent region. For example, some individuals have six lumbar vertebrae. Often, this is the result of "lumbarization" of the first sacral segment resulting from its failure to fuse with the remainder of the sacrum.

Another congenital deformity of the spine is spina bifida, which results when the posterior portions of the neural arches fail to close during development of the embryo. Usually spina bifida is relatively insignificant and produces no symptoms, in which case it is called spina bifida occulta. It is most commonly seen at L5 (Fig. 15-24). Less commonly, the defect may be quite large, leaving the spinal cord unprotected; the condition is then termed spina bifida vera (Fig. 15-25). This condition is sometimes accompanied by protrusion of the meninges or a portion of the spinal cord (Fig. 15-26).

Fig. 15-22 Bilateral cervical ribs *(arrows)*.

Fig. 15-24 15-3 Spina bifida occulta of L5 *(arrows)*.

Sacralized spinous process   Normal spinous process

Fig. 15-23 Unilateral sacralization of L5. There is a normal transverse process on the left, but on the right the transverse process is enlarged and fused with the sacrum.

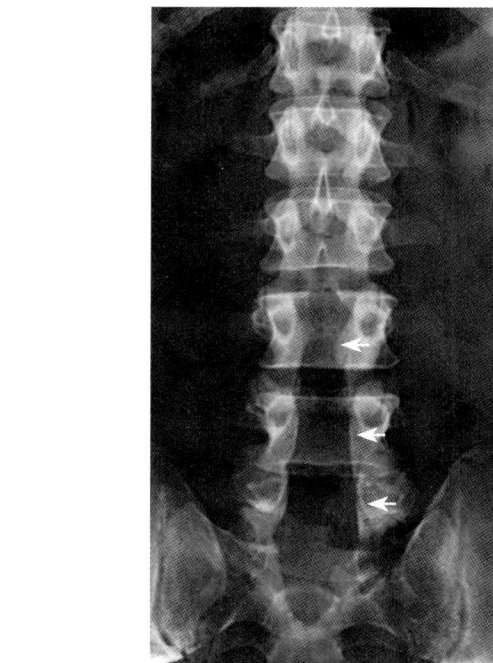

Fig. 15-25 15-1 Spina bifida vera *(arrows)*. The dark area involving of L3-L5 and the proximal sacrum is caused by congenital absence of portions of the posterior elements of the spine.

Spinal Fractures

Spinal fractures may result from trauma. Fractures of vertebral bodies are often compression fractures with anterior wedging. Fig. 15-27 shows a lumbar fracture that involves both the body and the posterior elements. The patient was a pedestrian who was struck by a car.

In bones weakened by disease processes, fractures may occur with little or no trauma. These are called pathologic fractures. Pathologic compression fractures of the thoracic spine due to osteoporosis are often seen in elderly women (Fig. 15-28).

Several fractures of the cervical spine are illustrated in Figs. 15-29 through 15-32. These radiographs also illustrate the importance of demonstrating all of C7 on lateral projections and of obtaining a high-quality radiograph of the atlas and axis in the AP projection. Unstable or displaced fractures of the cervical spine may cause pressure on the spinal cord. Spinal cord pressure, particularly in the cervical region, may cause paralysis and may also be life-threatening.

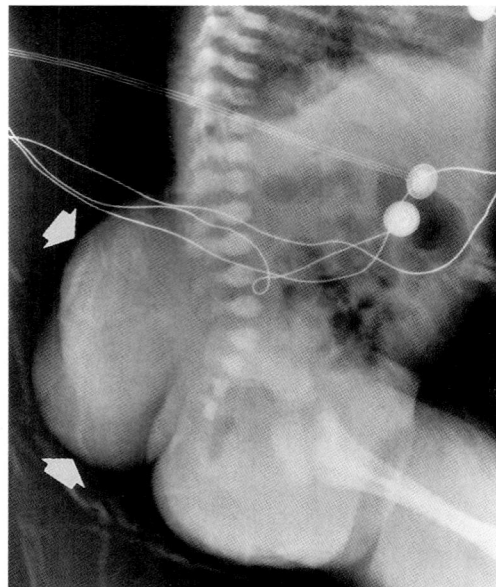

Fig. 15-26 15-2 Meningomyocele *(arrows)* is a herniated mass of neural tissue associated with spina bifida. Note absence of posterior elements in lower lumbar and sacral vertebrae.

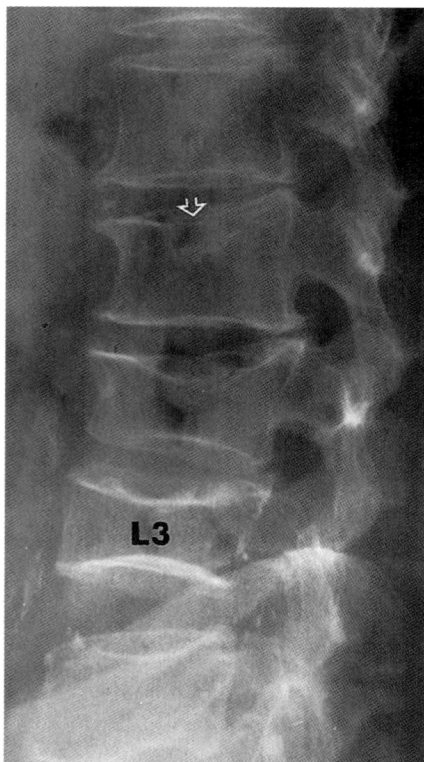

Fig. 15-27 Traumatic fracture of L1 with displacement

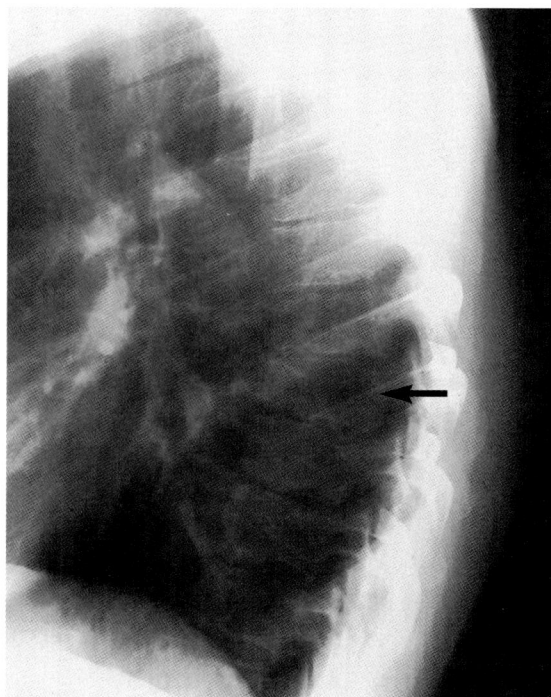

Fig. 15-28 Senile osteoporosis with partial collapse of T8 and T10.

Fig. 15-29 Clay shoveler's fracture, an avulsion of the spinous process of C7. **A,** AP projection shows the classic "double spinous process" sign *(arrows).* **B,** Avulsed fragment is clearly seen on the lateral projection *(arrow).*

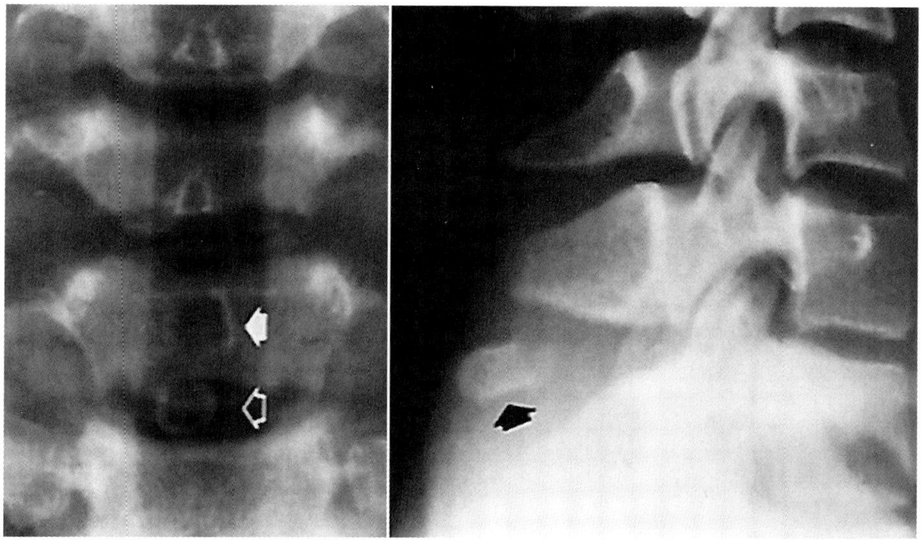

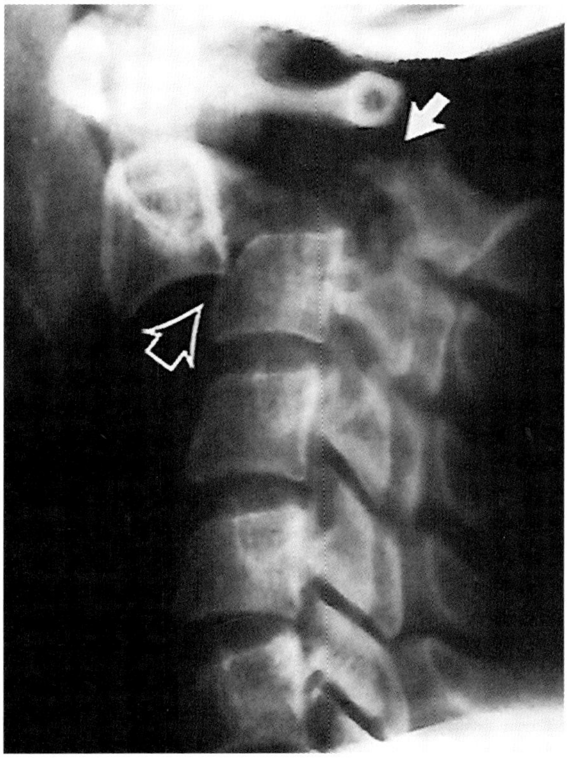

Fig. 15-30 Hangman's fracture is a fracture of the neural arch of C2 *(solid arrow)* with associated subluxation of C2-3 *(open arrow).*

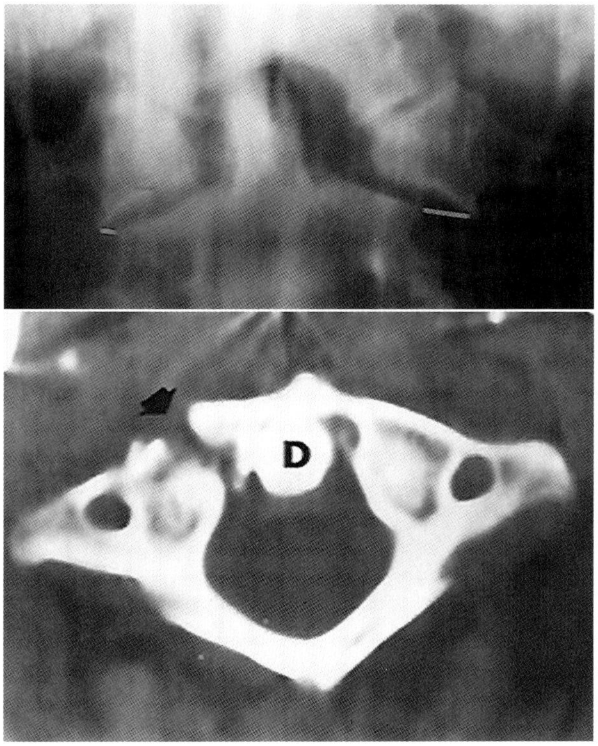

Fig. 15-31 Jefferson fracture is a fracture of the anterior arch of C1. **A,** Frontal projection tomogram shows lateral displacement of the lateral masses. **B,** CT scan clearly shows the break in the arch *(arrow). D,* Dens.

Fig. 15-32 Fractures demonstrated in AP upper cervical projection. **A,** Combined transverse and oblique fracture of the odontoid process *(arrrows)*. **B,** Fracture of C2 *(arrows)* at the base of the dens, separating it from the body.

Spondylosis, Spondylitis, Spondylolysis, Spondylolisthesis, and Spondyloschisis

Spondylosis, spondylitis, spondylolysis, spondylolisthesis, and spondyloschisis refer to very different conditions, but the words look and sound very much alike. They all have the same root, "spondylo-," which simply means vertebra.

Spondylosis refers to fixation or fusion of vertebrae.

Spondylitis has the suffix "-itis," which you may recall indicates inflammation. This term is often applied specifically to tuberculous disease of the vertebrae, which is also called Pott's disease. Spondylitis is also seen with rheumatoid arthritis.

Spondylolysis refers to the breakdown of the structure of the bone. This occurs with osteoporosis, with some metastatic lesions, and with other conditions that cause atrophy and bony destruction.

Spondylolisthesis refers to the anterior displacement of one vertebra on another (Fig. 15-33). It occurs most commonly at the lumbosacral joint and is usually caused by a defect or a fracture of the pars interarticularis or of the pedicle.

Spondyloschisis is the term for a congenital fissure (split or cleft) in the neural arch. Spina bifida occulta, discussed and illustrated earlier in this chapter, is an example of spondyloschisis.

Disk Pathology

The pulpy center of intervertebral disks is normally gel-like, semiliquid, and very flexible. Its mass shifts to change the shape of the disk with the pressure of various spinal movements. With advancing age and repeated minor traumas to the spine, the disks tend to degenerate. The nucleus may dry out and become

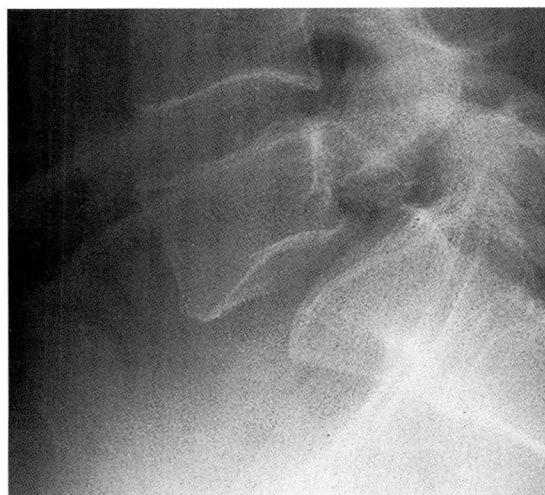

Fig. 15-33 Spondylolisthesis of L5-S1.

atrophied, causing narrowing of the disk space. Without adequate cushioning, the joint becomes inflamed and the surrounding bony structures show the characteristic signs of degeneration: sclerotic (hardened), irregular bone margins with hypertrophic lipping and spurring (Fig. 15-34). This condition is called degenerative disk disease (DDD) and is usually associated with osteoarthritis.

Disk herniation or herniated nucleus pulposus (HNP) is the condition often called, in lay terms, a "slipped disk." The annulus fibrosus ruptures and the nucleus is forced into the area posterior to the disk space (Fig. 15-35). The displaced nucleus causes pressure on the spinal cord and/or the nerve roots in this area.

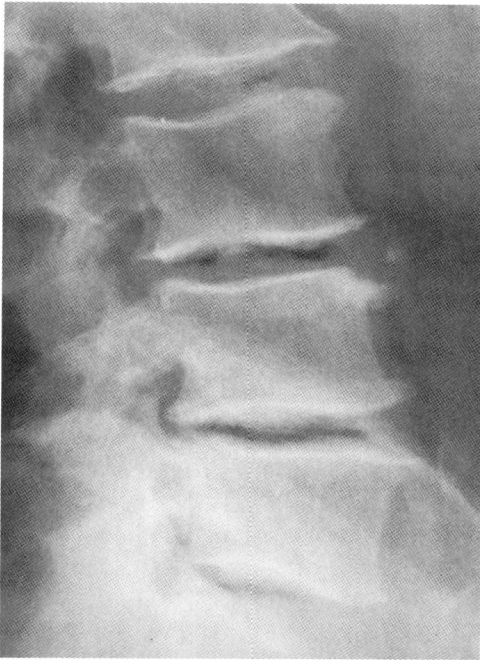

Fig. 15-34 Degenerative disk disease with associated arthritic changes. Note disk space narrowing and hypertrophic spurs on the anterior vertebral bodies. The dark linear shadows overlying two of the disks represent the "vacuum phenomenon" sometimes seen with severe disk degeneration.

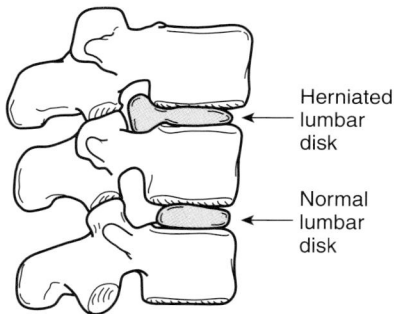

Fig. 15-36 Magnetic resonance image of the cervical spine in the sagittal plane shows herniation of the C4-5 disk. Note impression on the spinal cord.

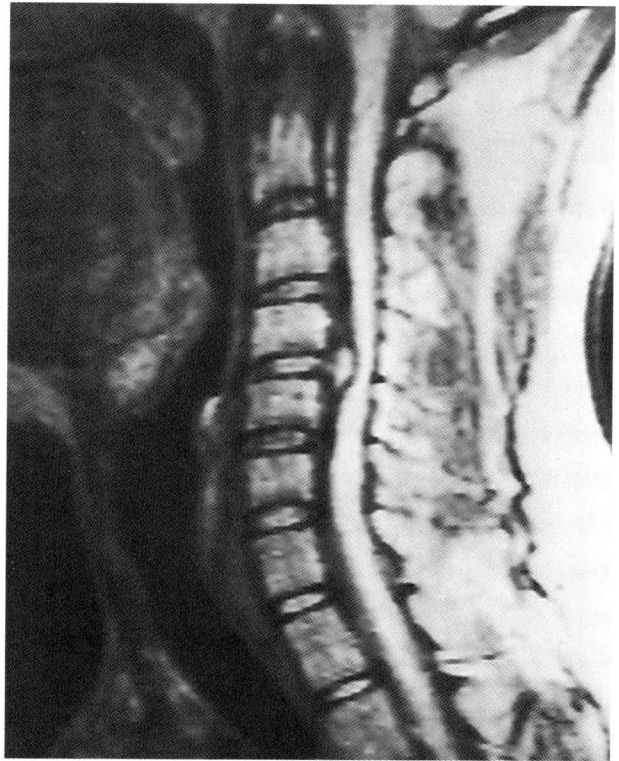

Herniated lumbar disk

Normal lumbar disk

Fig. 15-35 When an intervertebral disk herniates, the annulus ruptures and the nucleus pulposus is forced out posteriorly and/or laterally. The herniated nucleus then occupies space in the spinal canal or intervertebral foramen, causing nerve pressure.

Disk herniation is caused by trauma to the disk. In the lumbar area, it may be caused by pressure from lifting a heavy object. Herniated cervical disks are often the result of motor vehicle accidents. Disk herniation may cause acute pain, chronic discomfort, or recurrent painful episodes involving the site of the herniation. There may also be pain, numbness, or altered sensation in areas remote from the spine.

While the radiographic examination may show decrease in the height of the disk space, special imaging techniques are needed to demonstrate disk herniation definitively. Myelography, discography, CT,

and MRI studies (Fig. 15-36) may be used to identify disk pathology.

Remote Symptoms of Spine Pathology

The nerves communicate messages of motion and sensation between the brain and all parts of the body by way of the spinal cord. The spinal cord is surrounded by the vertebral foramina and nerves pass from the spinal cord to other parts of the body by way of the intervertebral foramina. This explains why changes in the vertebrae may cause symptoms in parts of the body remote from the spine. Such symptoms are indications of pressure or irritation to the nerve roots.

Nerve root insult may cause pain, numbness, or altered sensation. Sciatica, for example, is pain along the path of the sciatic nerve in the buttock, posterior thigh, and leg. It is caused by nerve irritation in the lumbar region. Damage to nerve roots may cause weakness or paralysis. The function or control of organs may be affected as well. For example, nerves in the upper cervical region control vital functions such as breathing, and nerves in the lumbar region control bowel and urinary bladder function.

There are many possible causes of nerve root compression. Hypertrophic arthritic changes, such as bony spurs on the vertebrae, may cause **stenosis** (narrowing) of the intervertebral foramina. Misalignment of vertebrae, subluxation, or spondylolisthesis may cause crowding of the nerve pathways. Disk herniation is also a common cause of remote nerve symptoms.

SUMMARY

The vertebral column surrounds the spinal cord and is the supporting structure for the body. It consists of 33 vertebrae or spinal segments: seven cervical, twelve thoracic, five lumbar, five sacral, and four coccygeal. Most vertebrae consist of an anterior body, a posterior vertebral arch, two lateral projections called transverse processes, and a posterior projection called the spinous process. The joints between the vertebral bodies are cushioned by intervertebral disks. Zygapophyseal joints between the posterior elements facilitate motion. Each region of the spine has either a kyphotic or lordotic curvature.

Radiographers must be familiar with the landmarks used to locate center points and individual vertebrae in each region of the spine. The zygapophyseal joints and the intervertebral foramina of each spine region vary in their relationships to body planes. The radiographer must be familiar with these relationships in order to demonstrate these structures accurately.

Most spine radiography may be done with the patient either upright or recumbent. A grid is required for all but the lateral and oblique cervical projections. All of the basic spine examinations consist of at least frontal and lateral projections. Oblique and axial projections and spot films are frequently taken to supplement routine examinations.

Congenital anomalies are common in the spine. Those most frequently seen include extra vertebrae, transitional vertebrae, and anomalous ribs. Many pathologic conditions of the spine are diagnosed radiographically. Trauma may require radiographs to identify fractures or significant displacements. Degenerative, inflammatory, and neoplastic diseases also affect the spine and are evaluated with radiography. Some spine conditions cause nerve symptoms in areas of the body remote from the spine.

▪ Review Questions ▪

1. List the regions of the spine and state number of vertebrae or vertebral segments in each.
2. Which spinal segments have a kyphotic curve? Which have a lordotic curve?
3. How do the atlas and axis differ from the other cervical vertebrae?
4. On your own body, indicate the location of the mental point, mastoid process, angle of mandible, laryngeal prominence, and jugular (sternal) notch.
5. An AP projection of the upper cervical spine is unsatisfactory because the patient's upper teeth are superimposed over the atlas and the dens. How should you adjust the position for a satisfactory radiograph?
6. Name and describe positions that will demonstrate each of the following structures: left cervical intervertebral foramina, cervical zygapophyseal joints, lumbar intervertebral foramina, left lumbar zygapophyseal joints, and sacroiliac joints.
7. An order for radiographic examination of the cervical spine includes a request for lateral flexion and extension views. The patient was in a car accident this morning. What precautions are needed? Why?
8. An AP projection of the thoracic spine appears to be quite dark in the region of T1-T4 and a bit too light in the region of T7-T12. List possible causes and suggest solutions.
9. How would you instruct a female patient to prepare for a lumbar spine examination?
10. List and describe three common congenital anomalies of the spine.
11. What type of spinal fracture is common among older women with osteoporosis?
12. List two possible causes of nerve root compression and four possible symptoms.

16

Bony Thorax, Chest, and Abdomen

Learning Objectives

At the conclusion of this chapter, the student will be able to:

- Name the bones that make up the bony thorax and identify each on an anatomic diagram and on a radiograph
- Name and identify on an anatomic diagram the principal organs located within the thoracic cavity
- Name and identify on an anatomic diagram the principal organs located within the abdominal cavity
- Identify significant positioning landmarks in the thoracic and abdominal areas by palpation
- Demonstrate correct body and part positioning for routine projections and common special projections of the bony thorax, chest, and abdomen
- Correctly evaluate radiographs of the bony thorax, chest, and abdomen for positioning accuracy
- Describe and recognize on radiographs pathology that is common to the bony thorax, chest, and abdomen

Key Terms

aorta	mediastinum
atelectasis	parietal membrane
bronchus (pl. bronchi)	peritoneum
cardiophrenic angle	pleura
carina	pleural effusion
colon	pneumoconiosis
costophrenic angle	pneumonia
diaphragm	pneumothorax
duodenum	sphincter
emphysema	sternum
esophagus	thorax
ileum	trachea
jejunum	vena cava
KUB	visceral membrane

Radiography of skeletal anatomy requires an approach that is quite different from that used for visualization of the soft tissues and organs. Examinations of the bony thorax and chest involve the same general part of the body, but they differ from each other considerably. This chapter details the positioning and techniques required for best demonstration of both types of thoracic tissue.

Although chest and abdomen radiography both involve soft tissues and organs, they are quite different. As explained in Chapter 6, subject contrast reflects tissue density differences within the body part. Differences in subject contrast require variations in technique to obtain optimum visualization on the radiograph. The subject contrast of the chest differs greatly from that of the abdomen. Subject contrast within the chest is provided by the radiolucency of the air in the lungs. The abdomen has tissues with very similar densities and therefore has relatively little subject contrast.

Chest radiographs provide diagnostic information about the heart, lungs, and other organs that lie in the area around the heart. While chest radiography may seem to be the simplest and most familiar of all radiographic examinations, it is also the most likely to provide life-saving information for the patient's care. For this reason, the ability to take high-quality chest radiographs is an essential skill for radiographers.

Radiography of the abdomen, on the other hand, is less likely to be required of the limited radiographer. In some states, abdomen radiography is beyond the scope of limited practice. Even where there is no restriction on abdomen radiography, these examinations are not often performed in an outpatient setting. If you are using this book as a text in a formal course in radiography, the abdomen may not be included in the curriculum. Basic information on the anatomy, positioning, and pathology of the abdomen is included here so that the content will be comprehensive for those who need this information.

Learning the meanings of the following word roots will aid in understanding and remembering some of the terminology in this chapter:

- *cardio-* heart
- *chole-* bile
- *chondro-* cartilage
- *costo-* rib
- *gastro-* stomach
- *hepato-* liver
- *nephro-* kidney
- *phreno-* diaphragm
- *pleuro-* lung membrane
- *pneumo-* air, lung
- *pulmo-* lung
- *thoraco-* chest

ANATOMY

Bony Thorax

The term **thorax** refers to the upper portion of the trunk, the chest. The thoracic cage (Fig. 16-1) is the bony structure that surrounds the organs of the chest and upper abdomen. It consists of the twelve thoracic vertebrae, twelve pairs of ribs, and the breastbone, which is called the **sternum.**

The sternum is a slender, flat bone located in the midline of the anterior thorax. It has three parts: the manubrium, the gladiolus or body, and the xiphoid process (Fig. 16-2). The manubrium is the superior portion. Its upper margin is indented to form the jugular notch, also called the manubrial notch or

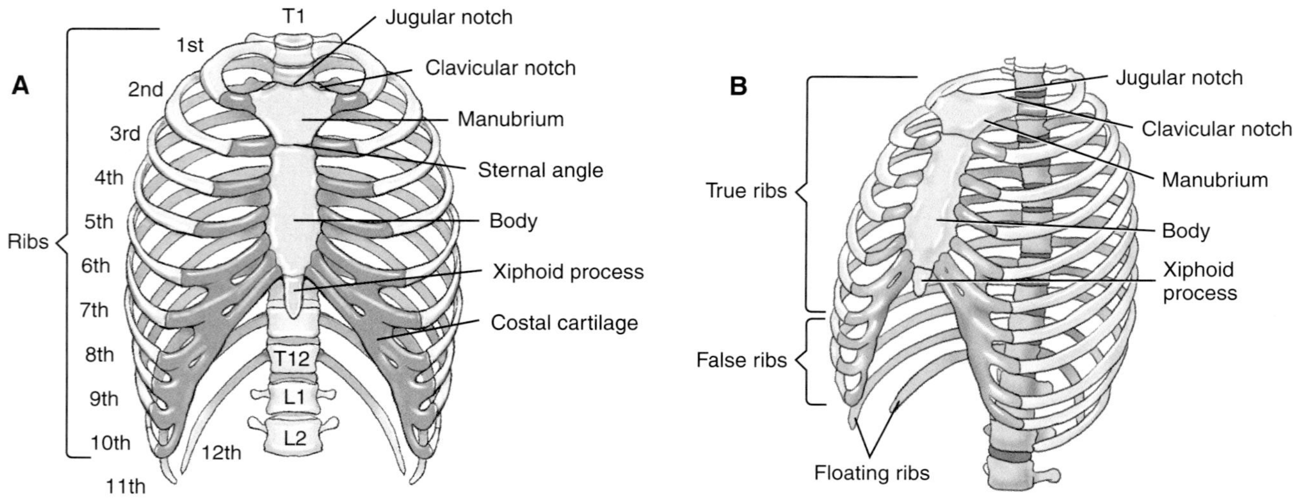

Fig. 16-1 Bony thorax. **A,** Anterior aspect. **B,** Anterolateral oblique aspect.

suprasternal notch, which is a useful positioning landmark. The gladiolus is the long, middle portion of the sternum. The junction of the manubrium and the gladiolus forms a palpable bony ridge called the sternal angle. The xiphoid process is the distal tip and is also a useful positioning landmark. In childhood the xiphoid process is formed only of cartilage; it ossifies in adulthood.

The ribs are numbered 1 through 12, from the top down. They are attached posteriorly to the thoracic vertebrae, forming the costovertebral articulations. Ribs are long, flat, slender, curved bones, and most have costal cartilage at their anterior ends. The connection between the rib and its associated cartilage is called the costochondral articulation. The anterior end of each rib is approximately 3 to 4 inches inferior to its

posterior end. When one is breathing, the rib articulations move gently, allowing expansion of the thorax with inspiration and contraction with expiration.

The first seven pairs of ribs are called "true ribs." They attach anteriorly to the sternum, forming cartilaginous amphiarthrodial joints. The lower five pairs of ribs are called "false ribs" because they do not completely surround the thorax. Ribs 8, 9, and 10 attach to bands of costal cartilage that attach to the sternum. Ribs 11 and 12 are also called "floating ribs." They have no cartilage and are not attached anteriorly.

Cavities of the Trunk

The interior of the trunk of the body is divided into two main cavities, the thoracic cavity and the abdominopelvic cavity (Fig.16-3). The two cavities are separated beneath the lungs by the **diaphragm.** The diaphragm is a large sheath of muscle that expands and contracts with breathing. It forms an arching curve from front to back.

The body cavities are lined with connective tissue called serous membrane. Serous membrane also provides an outer covering for organs. The cavity lining is called the **parietal** membrane and that which covers the organs is called the **visceral** membrane.

Chest

The thoracic cavity (Fig. 16-4) is divided into three parts: two pleural cavities that contain the lungs and the space between the lungs, which is called the **mediastinum.** The principal structures within the mediastinum are the heart with its associated great

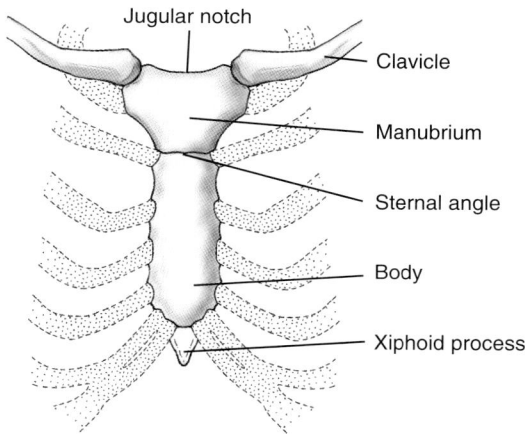

Fig. 16-2 Anterior aspect of sternum.

Labels: Jugular notch, Clavicle, Manubrium, Sternal angle, Body, Xiphoid process

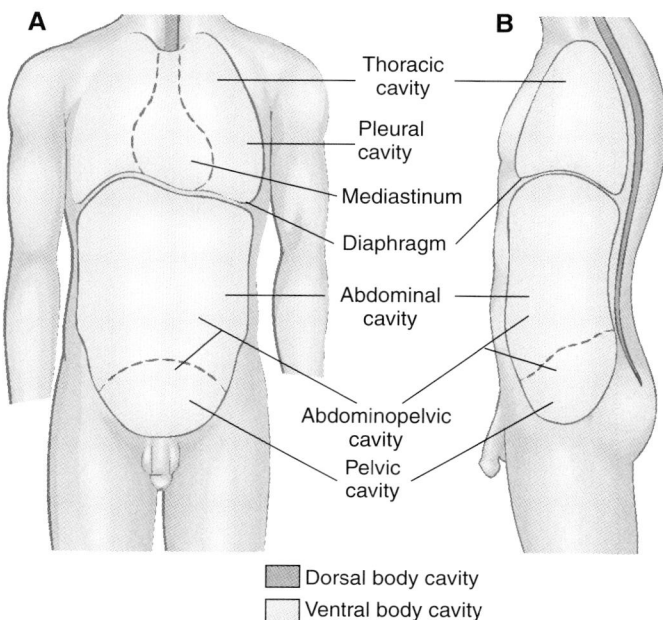

Fig. 16-3 Body cavities. **A,** Anterior aspect. **B,** Lateral aspect.

Labels: Thoracic cavity, Pleural cavity, Mediastinum, Diaphragm, Abdominal cavity, Abdominopelvic cavity, Pelvic cavity, Dorsal body cavity, Ventral body cavity

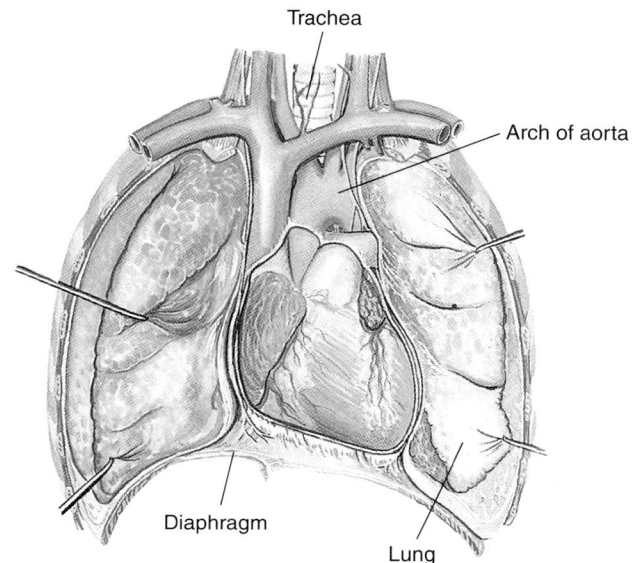

Fig. 16-4 Thoracic cavity is divided into three sections: right and left pleural cavities and mediastinum.

Labels: Trachea, Arch of aorta, Diaphragm, Lung

vessels, the **trachea** or "windpipe," and the **esophagus.** The trachea is a part of the respiratory system and connects the throat to the lungs. The esophagus is part of the digestive system and connects the throat to the stomach. Portions of the lymphatic system (the thymus gland and many lymph nodes) are also located within the mediastinum.

The heart (Fig. 16-5) is the principal organ of the circulatory system, introduced in Chapter 12. It occupies the inferior portion of the mediastinum in a sac of serous membrane called the pericardium. It is about the size of a fist and consists mainly of heart muscle tissue called myocardium. The heart is divided into four hollow chambers: the right and left atria, which are receiving chambers, and the right and left ventricles, which are discharging chambers. Blood enters the atria through veins and is pumped out into arteries via the ventricles. Valves between the chambers regulate blood flow.

The term "great vessels" refers to the veins and arteries that carry blood to and from the heart. The **vena cava** is the large vein that brings oxygen-depleted blood from the body to the right atrium. Venous blood contains carbon dioxide (CO_2), a gas that is a waste product of oxygen use. The blood then flows into the right ventricle and is pumped into the pulmonary arteries that carry it to the lungs. In the capillaries of the lungs, carbon dioxide is exchanged for oxygen and the carbon dioxide is exhaled. Oxygenated blood returns from the lungs to the left atrium via the pulmonary veins. From the left atrium it flows into the left ventricle, which pumps it back out to the body via the **aorta.** The aorta is the largest artery of the body. It leaves the heart in a superior direction and makes a "U-turn" called the aortic arch. Arteries branching from the aortic arch supply blood to the head and upper body. The descending aorta passes in an inferior direction posterior to the heart, through the diaphragm, and through the abdomen. Its many branches supply oxygenated blood to the remainder of the body (Fig. 16-6).

Respiratory System

The respiratory system was introduced in Chapter 12. The principal organs of the respiratory system are the lungs (Fig. 16-7). The lungs are divided into sections called lobes. The right lung has three lobes: superior, middle, and inferior. The left lung has only two lobes: superior and inferior. Each lung is shaped somewhat like a tall pyramid. The broad lower surface is called the base and the angle at the top is called the apex. The inferior lateral corners are called the **costophrenic angles.** The inferior medial corners are called the **cardiophrenic angles.** The left lung is slightly smaller and is narrower at the base than the right lung.

The other organs of the respiratory system include the mouth and nasal passages, the pharynx (throat), the trachea, and the **bronchi** (sing., bronchus) (Fig. 16-8). A small valve between the pharynx and the

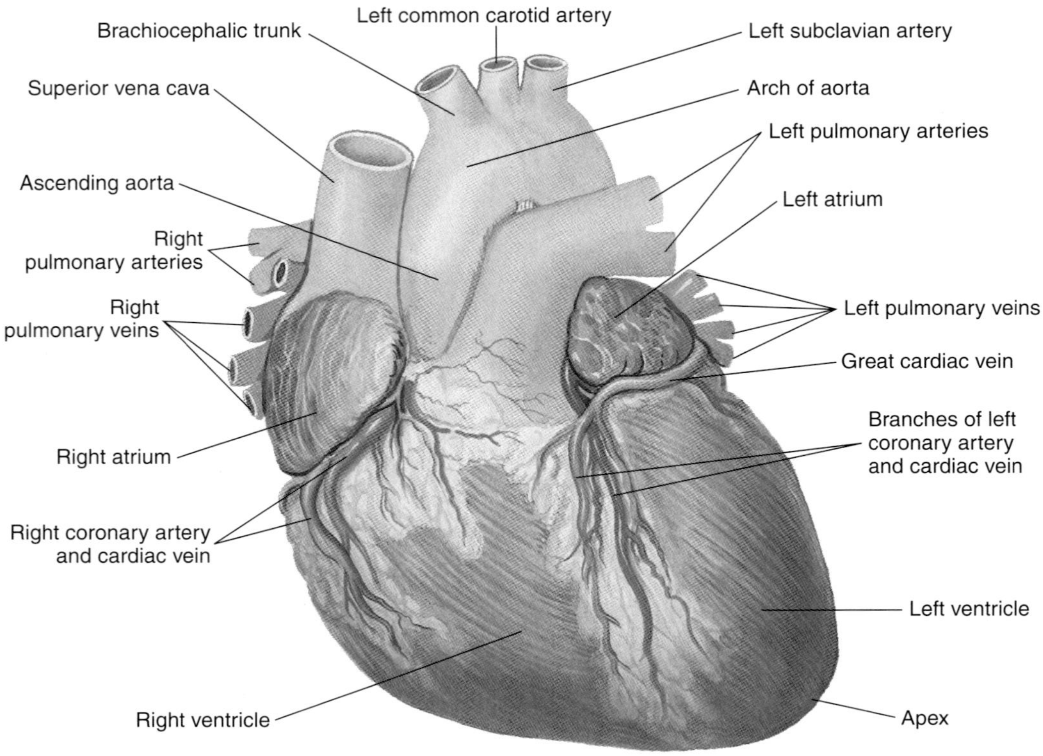

Fig. 16-5 Anterior aspect of heart.

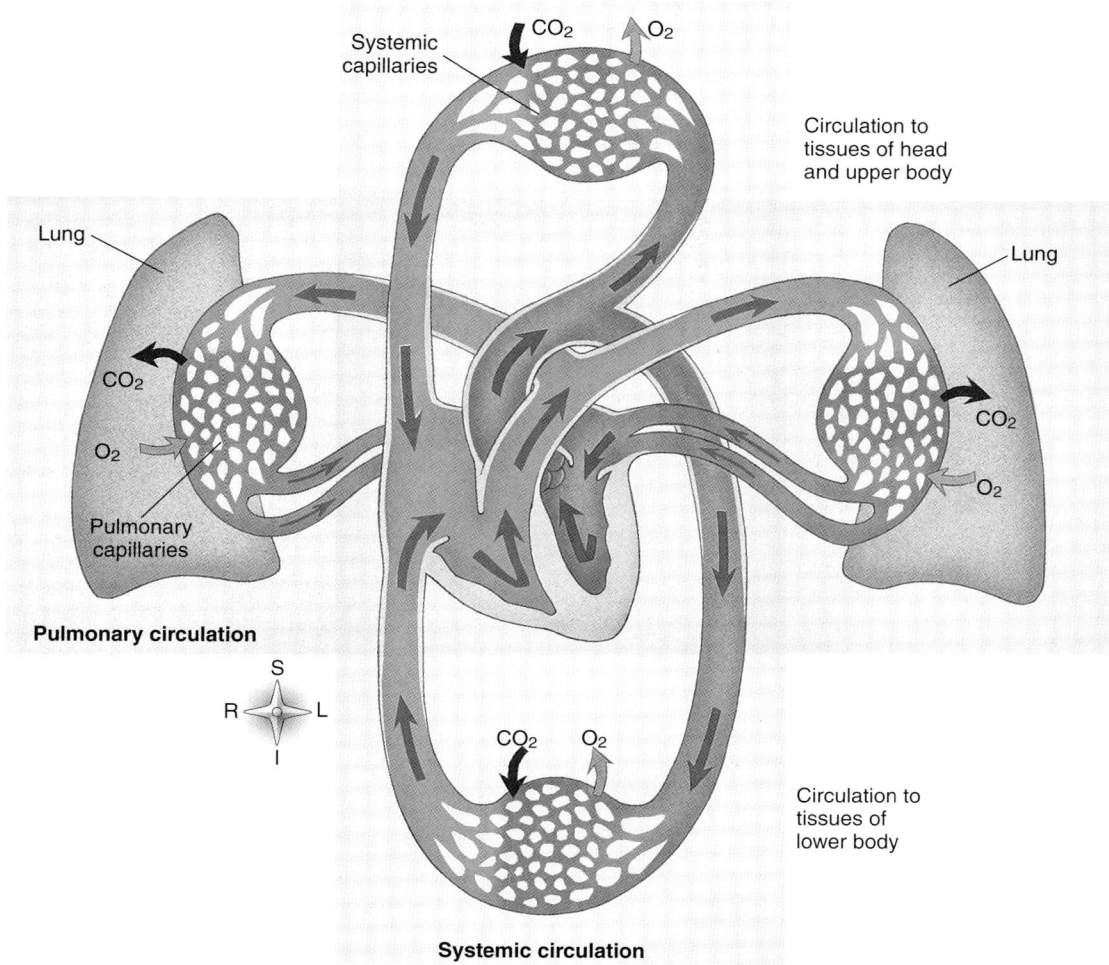

Fig. 16-6 Blood flow through circulatory system.

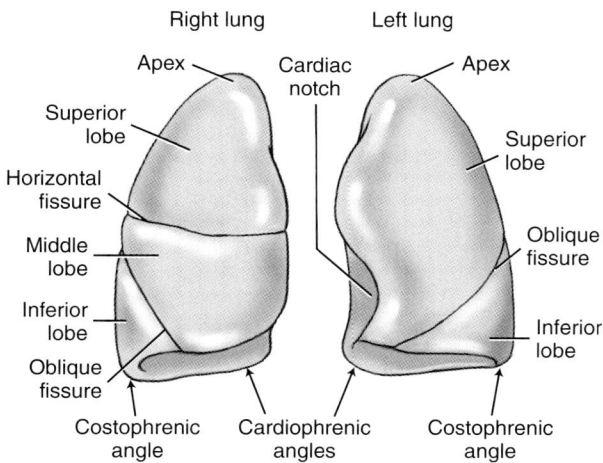

Fig. 16-7 Anterior aspect of lungs.

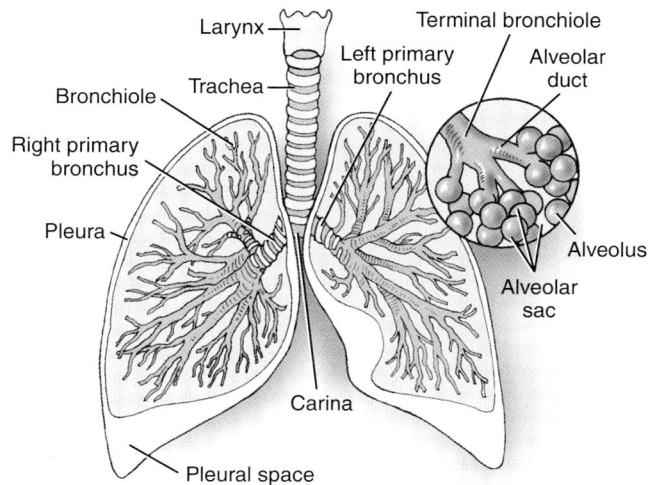

Fig. 16-8 Organs of respiratory system within thoracic cavity.

trachea, the epiglottis, closes off the trachea when we swallow so that food passes into the esophagus rather than the trachea. The trachea divides into right and left primary branches at the **carina,** about the level of the sternal angle and the T4-5 interspace. Each primary branch enters one lung and then divides into secondary branches, one for each lobe of the lung.

The membranes that line the pleural cavities and cover the lungs are called **pleura.** They secrete serous fluid that moistens and lubricates the surfaces so that the lungs can move smoothly against the walls of the pleural cavity with breathing motion. The space between the lungs and the cavity walls is called the pleural space.

Abdomen

The abdominopelvic cavity is divided into two sections: abdominal cavity and pelvic cavity. The abdominal cavity is the larger section, extending from the diaphragm into the upper portion of the bony pelvis. It contains the principal organs of the digestive tract: stomach, small and large intestine, liver, gallbladder, and pancreas. The spleen, which is part of the lymphatic system, is also located within the abdominal cavity. Its location is immediately beneath the diaphragm on the far left and behind the stomach.

The abdominal organs are contained in a double-walled serous membrane sac called the **peritoneum.** Parietal peritoneum lines the walls of the cavity and visceral peritoneum is positioned over and around the organs in folds. The folds between the organs, the mesentery, hold them in position. The anterior fold of the visceral peritoneum is called the omentum.

The abdominopelvic cavity also contains the urinary system. The kidneys and ureters are located in the retroperitoneal space, posterior to the visceral peritoneum against the posterior wall of the abdominal cavity.

The pelvic cavity is inferior to the abdominal cavity and situated within the bones of the pelvis. It contains the urinary bladder, the distal portion of the large intestine, and the internal parts of the reproductive system.

There are two systems for dividing the abdominopelvic cavity into parts so that locations within it can be readily identified. The simplest system is that which divides the area into four quadrants (Fig. 16-9). More specific localization is provided by a system that divides the abdomen into nine regions (Fig. 16-10). The radiographer must learn the names of the quadrants and regions and the principal structures that lie within each. This knowledge will help you to communicate clearly the site of a patient's pain or wound or to locate the area of clinical interest on a radiograph.

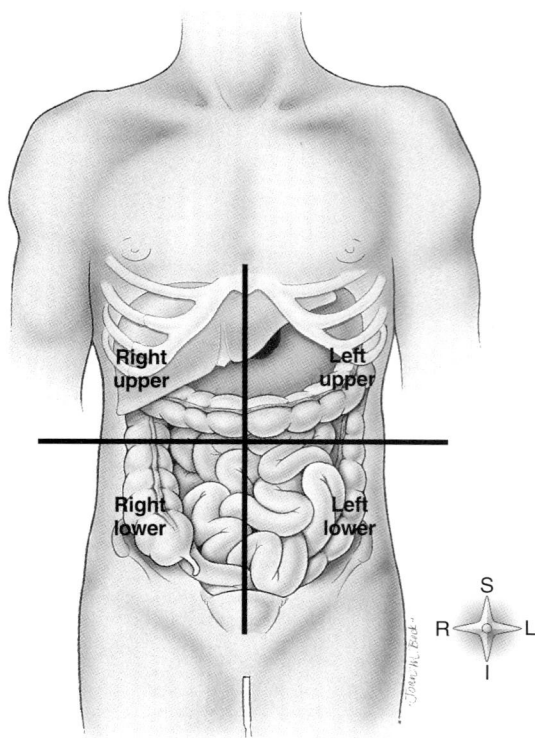

Fig. 16-9 Abdominopelvic cavity divided into quadrants.

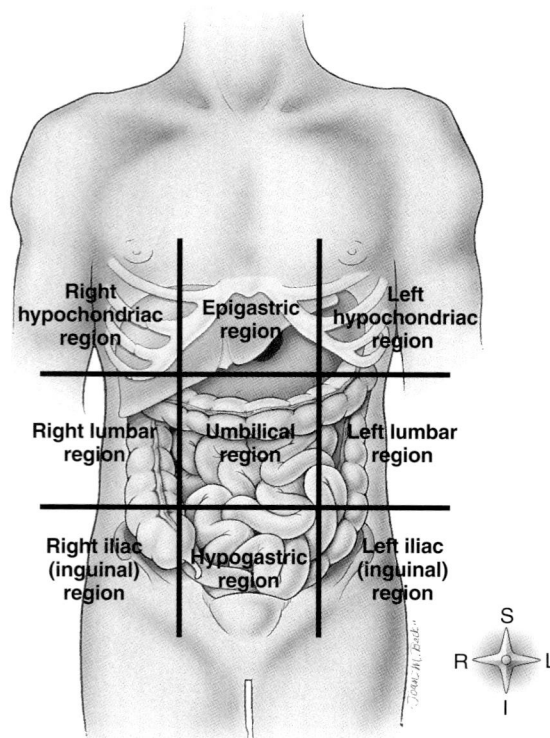

Fig. 16-10 Nine abdominal regions.

Alimentary Canal

The digestive system was introduced in Chapter 12. The portions of the alimentary canal that are within the abdominopelvic cavity are the distal end of the esophagus, the stomach, the small intestine, and the large intestine, also called the **colon** (Fig. 16-11).

From the mouth, food passes through the pharynx and into the esophagus at the epiglottis. The esophagus passes through the mediastinum and the diaphragm and into the abdominal cavity. Between the esophagus and the stomach is a round muscle called a **sphincter** that opens and closes the opening to the stomach.

The stomach (Fig. 16-12) is in the left upper quadrant of the abdomen. Its rounded upper portion is called the fundus. The large central curved portion is called the body. The lateral surface of the body is called the greater curvature and the medial one, the lesser curvature. The narrow distal portion of the stomach is called the pylorus. Another sphincter, the pyloric sphincter, joins the pylorus to the first portion of the small intestine.

Digestion and absorption of food occur in the small intestine. The adult small intestine is 20 to 22 feet in length. Its diameter ranges between 1½ inches at the proximal end to about 1 inch at the distal end.

The small intestine has three parts. The proximal portion is the **duodenum.** It is about 8 to 10 inches long and forms the shape of the letter C (Fig. 16-13). The rounded segment just distal to the pyloric sphincter is called the duodenal bulb. There is no obvious division between the second and third portions of the small intestine. The first 8 feet or so past the duodenum are called the **jejunum,** and the remainder is called the **ileum.** The distal portion of the ileum is called the terminal ileum. The ileum and jejunum loop back and forth across the central and lower part of the abdominal cavity and are framed by the colon (Fig. 16-14).

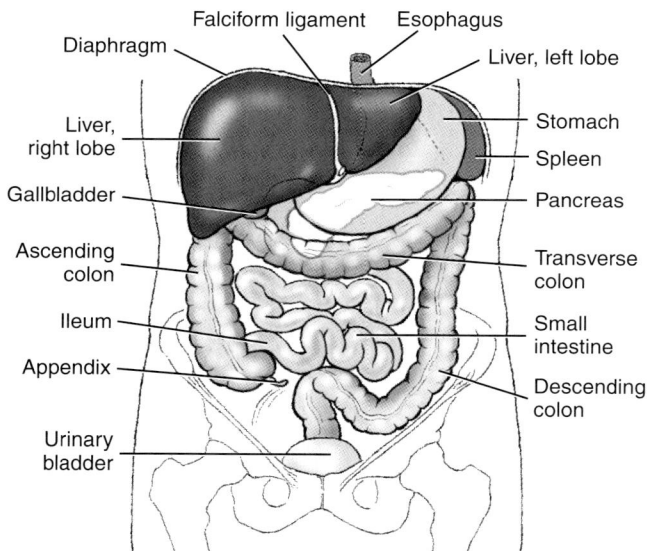

Fig. 16-11 Abdominal organs of digestive system.

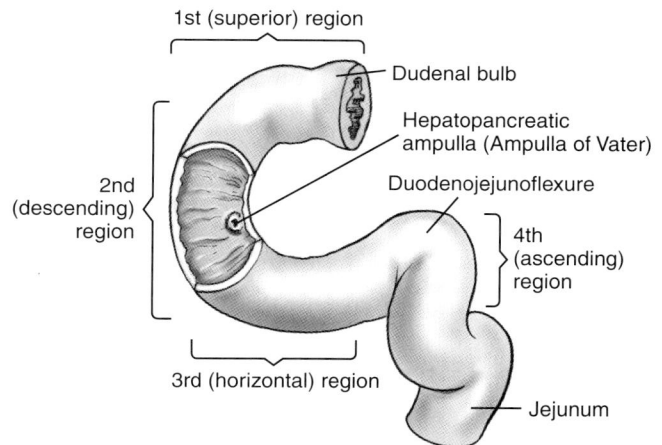

Fig. 16-13 Duodenum, first portion of small intestine.

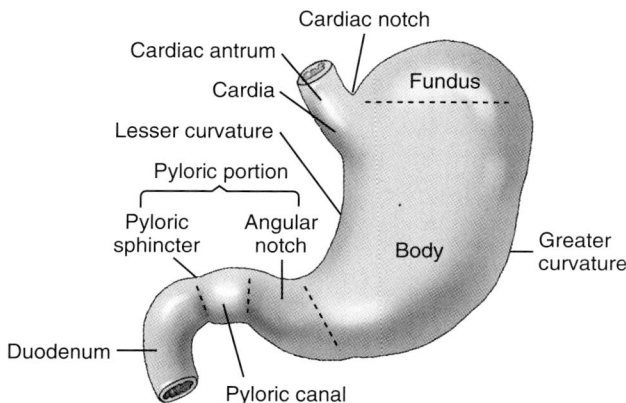

Fig. 16-12 Stomach.

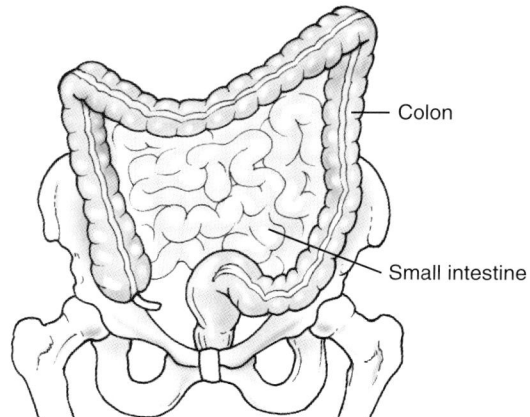

Fig. 16-14 Small intestine. Loops of ileum and jejunum are clustered in lower abdominal and upper pelvic cavity and are framed by colon.

The large intestine or colon is illustrated in Fig. 16-15. Its principal functions are to reclaim water from the intestinal contents and to eliminate solid food waste from the body. The first portion, the cecum, is located in the right lower quadrant. The terminal ileum attaches to the cecum at the ileocecal valve. The appendix is a small pouch attached to the cecum. From the cecum, the ascending colon extends superiorly to the right upper quadrant where it makes a right-angle turn in the medial direction. This turn is called the right colic flexure or the hepatic flexure, indicating that it is in the region of the liver. The transverse colon extends across the upper abdomen and turns inferiorly on the left side at the left colic flexure, also called the splenic flexure because it is located adjacent to the spleen. From this flexure it extends inferiorly and is called the descending colon. In the left lower quadrant, the colon forms an S-shaped curve. This portion is called the sigmoid colon for the Greek letter sigma (Σ). The sigmoid colon extends toward the midline where it turns inferiorly again and is called the rectum. The terminal portion of the rectum is called the anal canal. Food waste exits the body at the anal sphincter.

Other Digestive System Organs

The other organs of the digestive system that lie within the abdominal cavity are the liver, gallbladder, and pancreas (Fig. 16-16). The liver and gallbladder and their associated ducts are called the biliary system.

The liver is a large organ in the right upper quadrant. It fills the right hypochondriac region and part of the epigastric region. It is wedge-shaped, with its apex to the left of the midline. The liver has several important functions. It is a storehouse for energy and it removes toxins from the blood. Its most important function from a radiographic standpoint is the production of bile. The liver manufactures and secretes bile, which the body uses to digest fats. The hepatic ducts of the liver collect the bile and come together to form the common hepatic duct.

The gallbladder is a storage sac for bile and is located on the undersurface of the liver. The cystic duct is the passage for bile between the gallbladder and the common hepatic duct. The common hepatic and cystic ducts come together to form the common bile duct. Following a meal that contains fats, the gallbladder contracts, releasing bile. The bile flows through the common bile duct and into the duodenum at the hepatopancreatic ampulla, also called the ampulla of Vater.

The pancreas is an elongated gland that lies transversely in the upper abdomen. It is attached to the posterior abdominal wall. The wide end, which is called the head, lies in the curve of the duodenum. The tail lies adjacent to the spleen. The midportion is called the body. The pancreas manufactures insulin and glucagon, enzymes that are essential to sugar metabolism. The pancreas also secretes pancreatic juices that aid in digestion. The pancreatic juices flow through the pancreatic duct and empty into the duodenum at the hepatopancreatic ampulla.

BODY HABITUS

Accurate radiography of the chest and abdomen requires an awareness of body habitus, that is, the general shape of the patient's body. Organs vary greatly

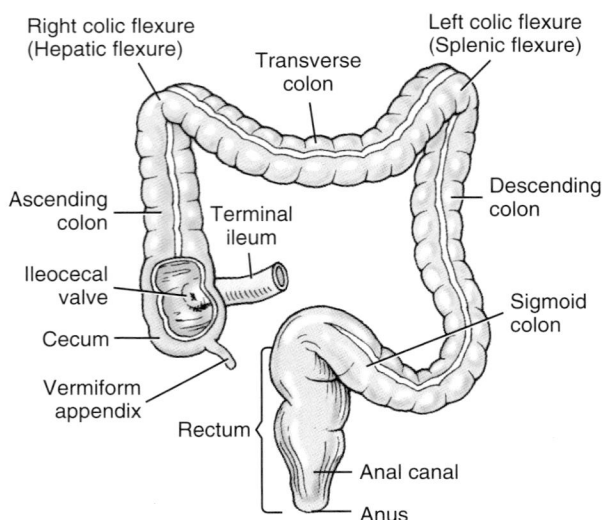

Fig. 16-15 Colon (large intestine).

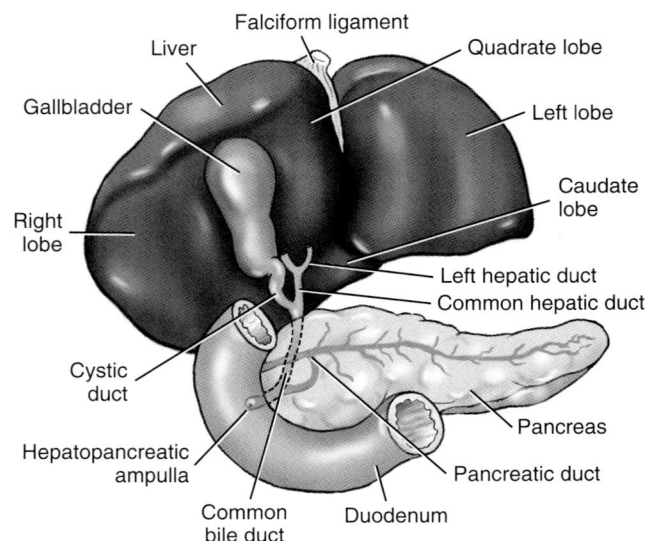

Fig. 16-16 Liver, gallbladder, and pancreas in relation to duodenum.

in size, shape, and location according to body types. The four basic types of body habitus are called hypersthenic, sthenic, hyposthenic, and asthenic and are illustrated in Fig. 16-17.

The sthenic body habitus is considered to be normal. About 50% of the population has this body type. The size, shape, and location of organs correspond to classic textbook descriptions and illustrations.

The hyposthenic body type might be thought of as "slender normal." About 35% of the population has a hyposthenic build. The organs tend to be longer, narrower, and more vertical in position.

The asthenic body type is tall and very slender. The organs are long and narrow and the abdominal organs are located much lower in the body. About 10% of the population has this body type.

The hypersthenic body type is a massive, stocky build. About 5% of the population has this body type. The thorax is short, broad, and deep. The organs tend to be high and more horizontal in position.

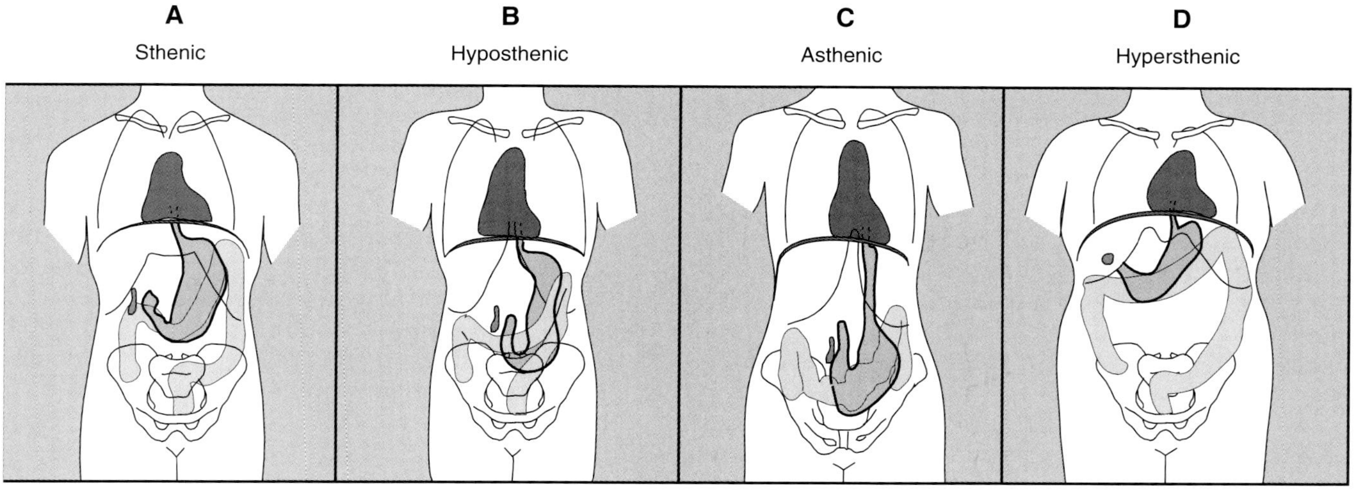

A	**B**	**C**	**D**
Sthenic	Hyposthenic	Asthenic	Hypersthenic

Fig. 16-17 Organ shape and location in relation to body habitus. **A,** Sthenic. **B,** Hyposthenic. **C,** Asthenic. **D,** Hypersthenic.

POSITIONING AND RADIOGRAPHIC EXAMINATIONS

Rib studies may be done with the patient recumbent or upright, and both methods are illustrated here. Upright positions are usually most comfortable for patients with painful rib injuries. Upright positions may also be safer, decreasing the possibility that pressure on rib fragments may puncture the lung during positioning.

Rib examinations vary considerably, depending on the specific area of clinical interest. Since the entire rib cage of most adults will not fit on a single 14 × 17-inch (35 × 43-cm) film, it is usual for a rib examination to be limited to the hemithorax, that is, the right or left half of the rib cage. When it is desirable to demonstrate the entire rib cage, both right and left studies are performed.

A rib examination usually consists of two projections, frontal and oblique. Some physicians prefer to include chest radiographs, because rib fractures may cause significant soft tissue injury that must be evaluated for proper patient care and treatment.

Both anterior and posterior ribs that are within the radiation field are seen on either AP or PA rib projections. The ribs nearest the film, however, are visualized with better definition. Thus, it is usual to take AP projections when the area of clinical interest is primarily posterior and to take PA projections for demonstration of anterior ribs. Variations are acceptable when the usual position would be too painful for the patient.

Frontal rib projections demonstrate the anterior and posterior portions of the ribs, which are in the coronal plane. The axillary (lateral) portions of the ribs are more or less in the sagittal plane and are seen "on end." Lateral projections are not useful because they result in the superimposition of right and left ribs. For these reasons, oblique projections are used to demonstrate the axillary portions of the ribs.

For posterior ribs, the AP oblique projection is taken in the "same side" oblique position, placing the side of clinical interest nearest the film. That is, the RPO position is used for right ribs and the LPO position for left ribs. For PA oblique projections of anterior ribs, however, the spine is projected over the ribs nearest the film. For anterior ribs, the correct oblique position is that in which the side of interest is farthest from the film. The RAO position demonstrates the left ribs and LAO position demonstrates the right ribs.

The ribs below the diaphragm require more exposure than the ribs above the diaphragm. This is because the air-containing lung tissues above the diaphragm are much more radiolucent than the dense abdominal tissues below the diaphragm. Separate exposure charts are used for the two areas. With deep inspiration, the lungs expand and the diaphragm is at its most inferior position. The posterior portions of ribs 1 through 9 and the anterior portions of ribs 1 through 7 are seen above the diaphragm on inspiration. On expiration, the lungs deflate and the diaphragm rises approximately two rib levels. Thus posterior ribs 8 to 12 and anterior ribs 6 to 12 are below the diaphragm on expiration. The position of the diaphragm in relation to the ribs and respiratory phase is illustrated in Fig. 16-18.

For all of these reasons, there are actually eight different possible rib examinations, depending on the location of the area of clinical interest. The radiographer must determine whether the site of interest is left or right, anterior or posterior, and above or below the diaphragm. The correct examinations for all possible combinations are summarized in Table 16-1.

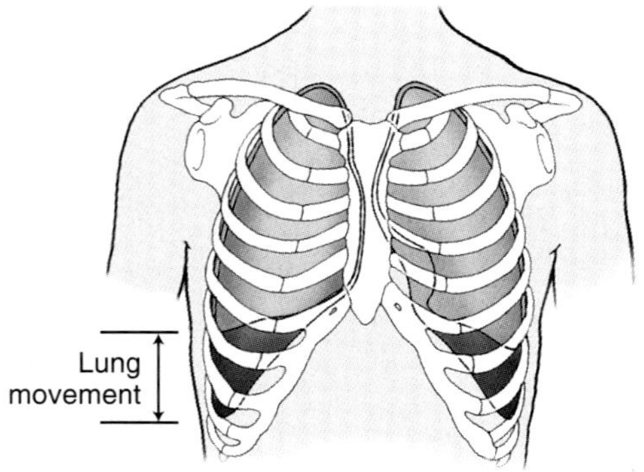

Fig. 16-18 Lung movement during respiration. Lungs expand during inspiration and contract on expiration.

TABLE 16-1

Rib Examinations According to Area of Clinical Interest

	Posterior Ribs 1-9	Posterior Ribs 8-12	Anterior Ribs 1-7	Anterior Ribs 6-12
Film Size	14 × 17 inch	10 × 12 inch	14 × 17 inch	10 × 12 inch
Projections for				
Right Ribs	AP, RPO	AP, RPO	PA, LAO	PA, LAO
Left Ribs	AP, LPO	AP, LPO	PA, RAO	PA, RAO
Respiratory Phase	Inspiration	Expiration	Inspiration	Expiration
Exposure	Above diaphragm	Below diaphragm	Above diaphragm	Below diaphragm

Basic Examination

UPPER POSTERIOR RIBS: AP AND AP OBLIQUE (AFFECTED SIDE NEAREST FILM)

Film: 14 × 17 inch (35 × 43 cm), lengthwise

Screens: Rapid

Grid: Yes

SID: 40 inches

Body position:
Standing or recumbent

Part position:

AP: Supine on table or upright with posterior surface of chest against upright grid cabinet. Coronal plane is parallel to film. Upper margin of film is 1½ to 2 inches above level of spinous process of C7.

AP oblique: RPO for right ribs or LPO for left ribs. Coronal plane forms an angle of 45° with film plane. Upper margin of film is 1½ to 2 inches above level of spinous process of C7.

Central ray:

AP: Perpendicular to center of film. Center point should be in midclavicular line at approximate level of axillary fold.

AP oblique: Perpendicular to center of film. Central ray enters at a point on midline of anterior surface at approximate level of axillary fold.

Patient instruction:
Do not move. Stop breathing on inspiration.

Structures seen:
Ribs 1 through 9. Posterior portions are best seen on AP projection; axillary portions are best seen on oblique projection.

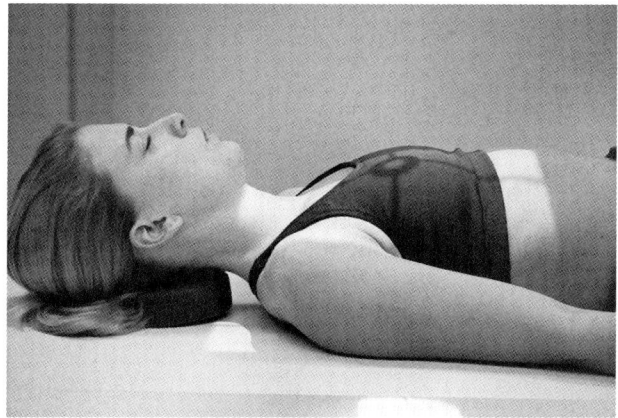

Recumbent AP position for right upper posterior ribs.

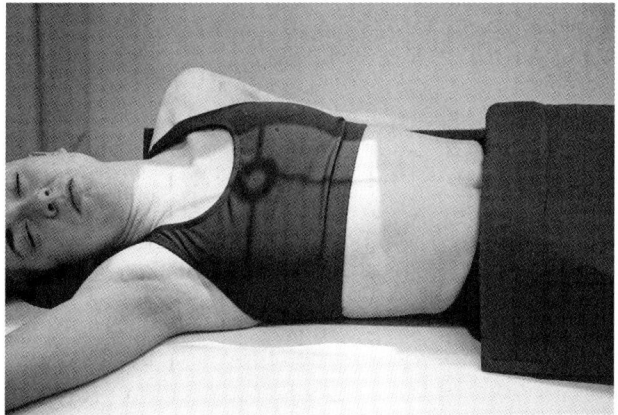

Recumbent AP oblique position for right upper posterior ribs (RPO).

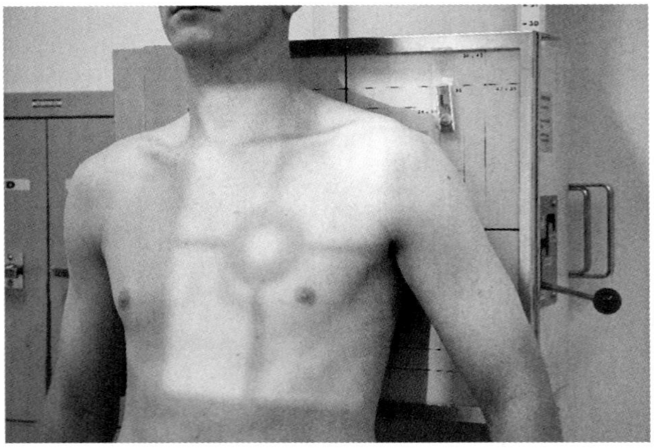

Upright AP position for left upper posterior ribs.

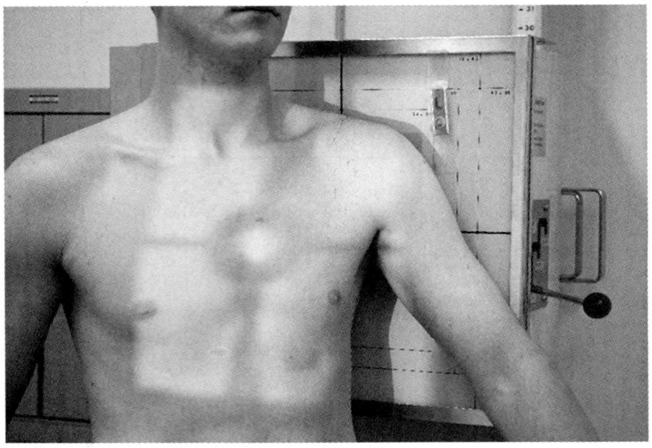

Upright AP oblique position for left upper posterior ribs (LPO).

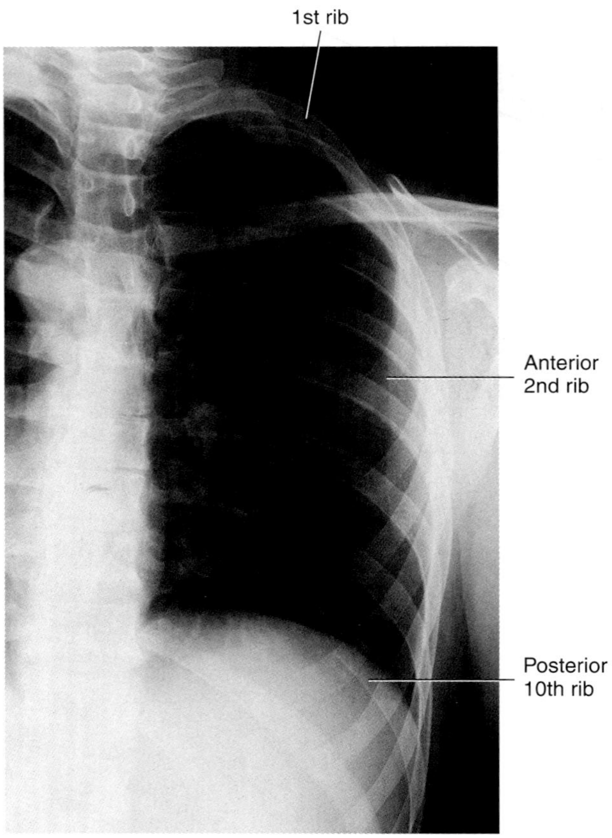

1st rib

Anterior 2nd rib

Posterior 10th rib

AP radiograph of right upper posterior ribs.

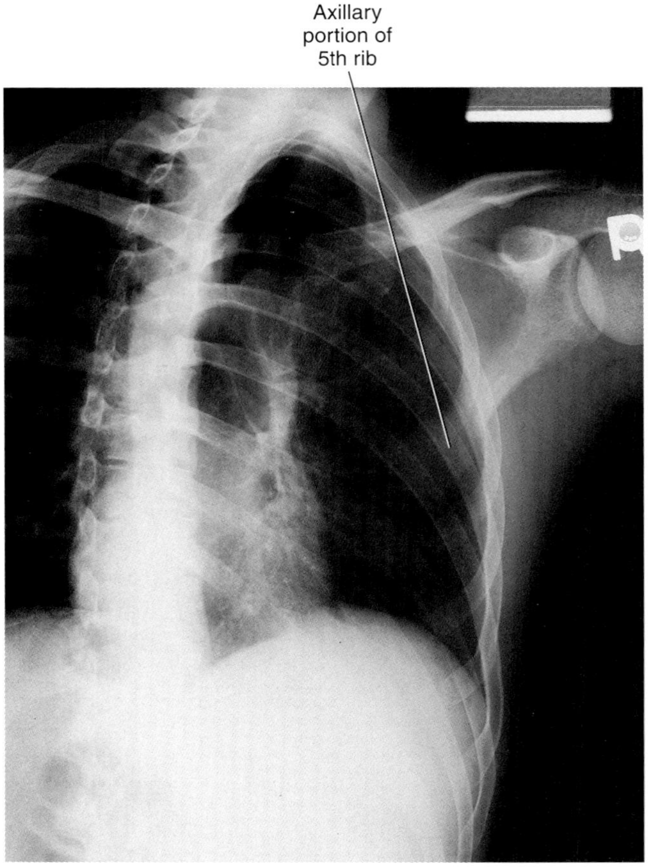

Axillary portion of 5th rib

AP oblique radiograph (LPO), left upper posterior ribs.

UPPER ANTERIOR RIBS: PA AND PA OBLIQUE (AFFECTED SIDE FARTHEST FROM FILM)

Film: 14 × 17 inch (35 × 43 cm), lengthwise

Screens: Rapid

Grid: Yes

SID: 40 inches

Body position:
Standing or recumbent

Part position:

PA: Prone on table or upright with anterior surface of chest against upright grid cabinet. Coronal plane is parallel to film. Upper margin of film is 1½ to 2 inches above level of spinous process of C7.

PA oblique: RAO for left ribs or LAO for right ribs. Coronal plane forms angle of 45° with film plane. Upper margin of film is 1½ to 2 inches above level of spinous process of C7.

Central ray:

PA: Perpendicular to center of film. Center point should be in midclavicular line at approximate level of axillary fold.

PA oblique: Perpendicular to center of film. Central ray enters at point on posterior surface midway between spine and midaxillary line of affected side at approximate level of axillary fold.

Patient instruction:
Do not move. Stop breathing on expiration.

Structures seen:
Ribs 1 through 7. Anterior portions are best seen on PA projection; axillary portions are best seen on oblique projection.

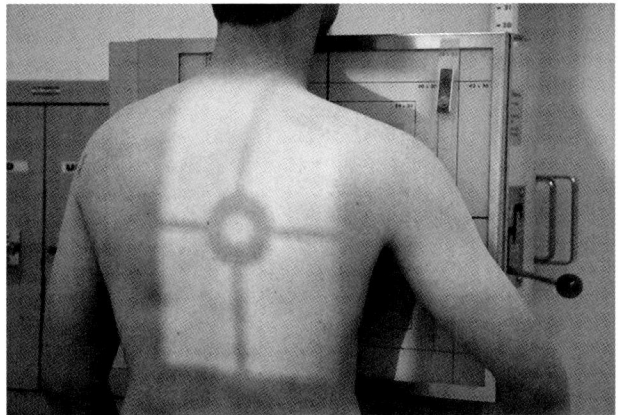

Upright PA position for right upper anterior ribs.

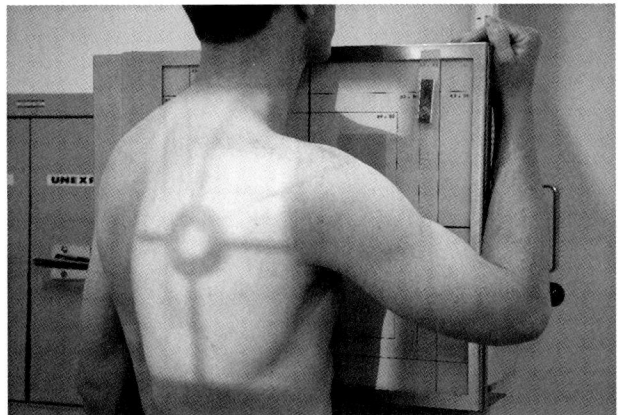

Upright PA oblique position (LAO) for right upper anterior ribs.

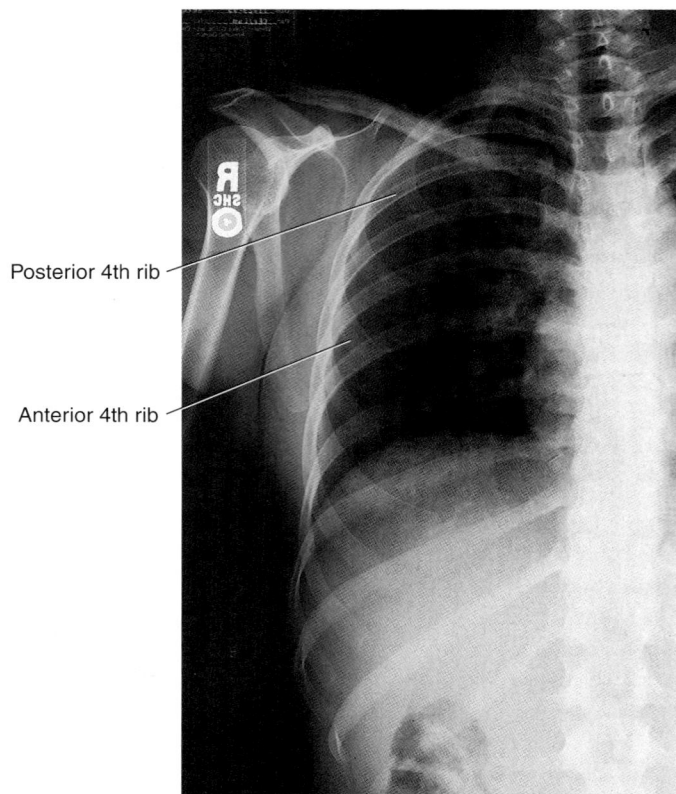

PA radiograph, right upper anterior ribs.

Posterior 4th rib

Anterior 4th rib

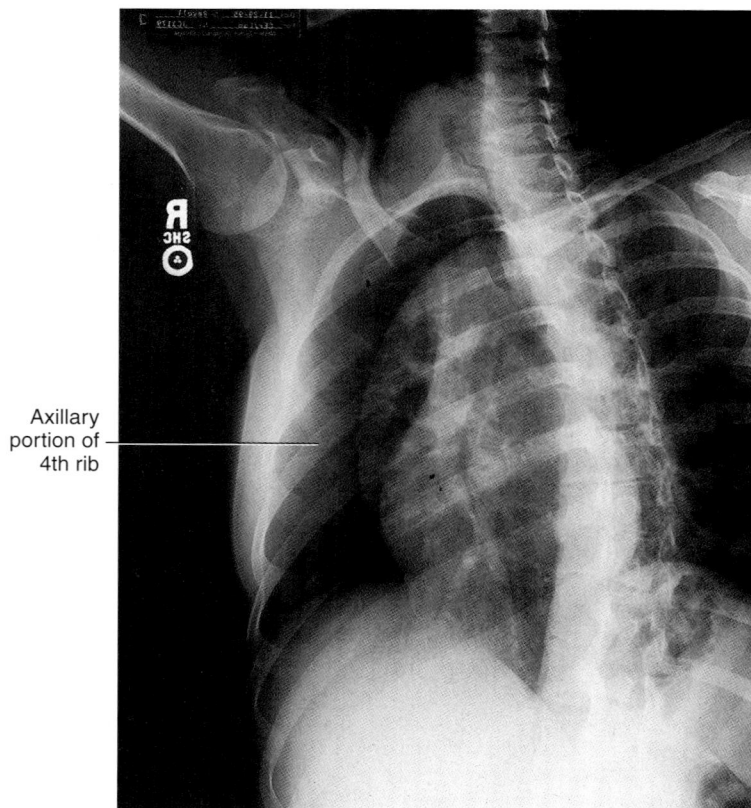

PA oblique radiograph (LAO), right upper anterior ribs.

Axillary portion of 4th rib

LOWER POSTERIOR RIBS: AP AND AP OBLIQUE (AFFECTED SIDE NEAREST FILM)

Film: 10 × 12 inch (24 × 30 cm), lengthwise or crosswise, depending on body habitus

Screens: Rapid

Grid: Yes

SID: 40 inches

Body position:
Standing or recumbent

Part position:

AP: Supine on table or upright with posterior surface of chest against upright grid cabinet. Coronal plane is parallel to film. Lower margin of film is at level of iliac crest.

AP oblique: RPO for right ribs or LPO for left ribs. Coronal plane forms angle of 45° with film plane. Lower margin of film is at level of iliac crest.

Central ray:

AP: Perpendicular to center of film. Center point should be in midclavicular line at approximate level of tip of xiphoid process.

AP oblique: Perpendicular to center of film. Central ray enters at point on midline of anterior surface at approximate level of tip of xiphoid process.

Patient instruction:
Do not move. Stop breathing on expiration.

Structures seen:
Ribs 8 through 12. Posterior portions of ribs are best seen on AP projection; axillary portions are best seen on oblique projection.

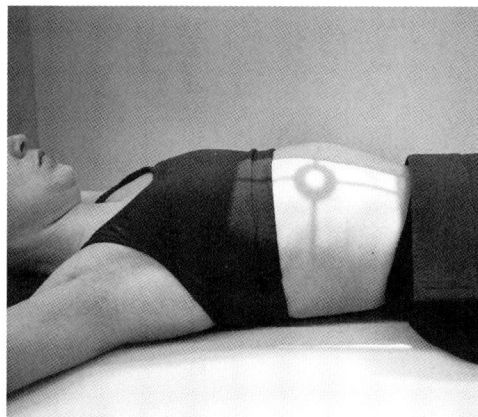

Recumbent AP position for right lower posterior ribs.

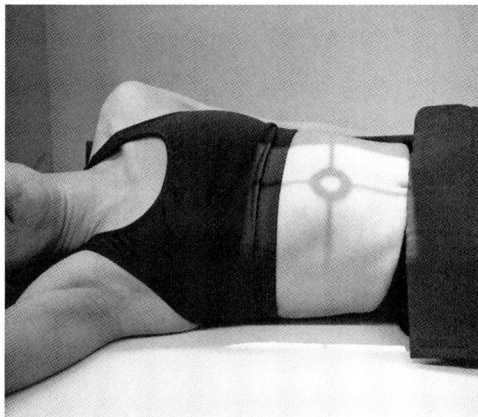

Recumbent AP oblique position (RPO) for right lower posterior ribs.

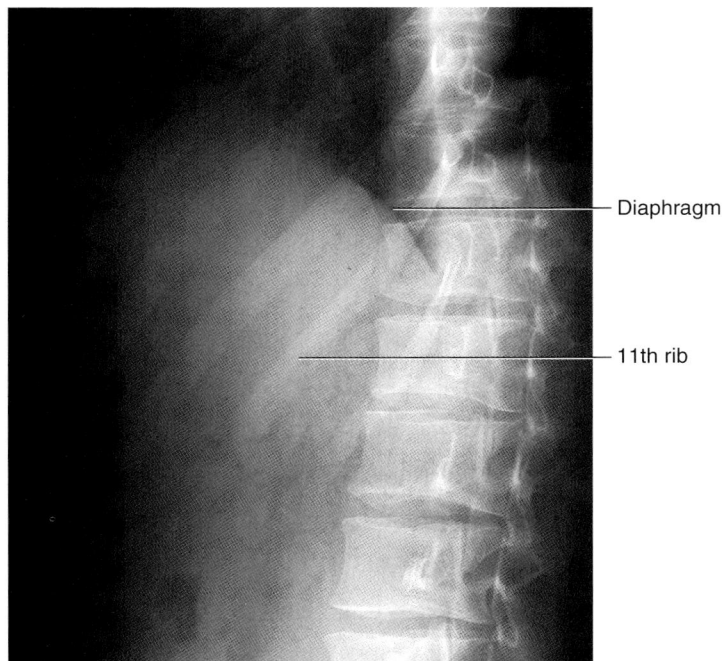

Oblique projection of ribs below diaphragm.

LOWER ANTERIOR RIBS: PA AND PA OBLIQUE (AFFECTED SIDE FARTHEST FROM FILM)

Film: 10 × 12 inch (24 × 30 cm), lengthwise or crosswise, depending on body habitus

Screens: Rapid

Grid: Yes

SID: 40 inches

Body position:
Standing or recumbent

Part position:

PA: Prone on table or upright with anterior surface of chest against upright grid cabinet. Coronal plane is parallel to film. Lower margin of film is at level of iliac crest.

PA oblique: LAO for right ribs or RAO for left ribs. Coronal plane forms angle of 45° with film plane. Lower margin of film is at level of iliac crest.

Central ray:

PA: Perpendicular to center of film. Center point should be in midclavicular line at approximate level of tip of xiphoid process.

PA oblique: Perpendicular to center of film. Central ray enters at point on posterior surface midway between spine and midaxillary line of affected side at approximate level of tip of xiphoid process.

Patient instruction:
Do not move. Stop breathing on expiration.

Structures seen:
Ribs 8 through 12. Anterior portions of ribs are best seen on PA projection; axillary portions are best seen on oblique projection.

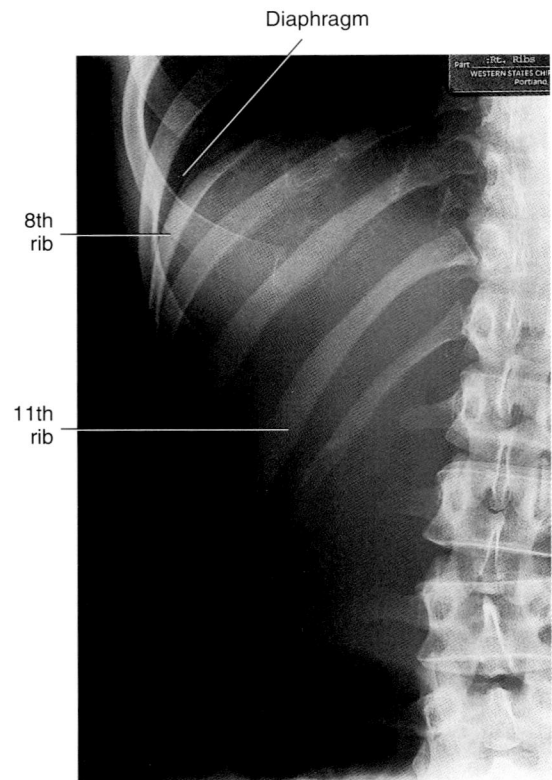

Frontal projection of ribs below diaphragm.

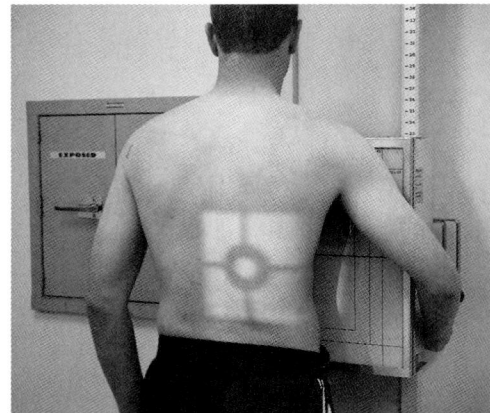

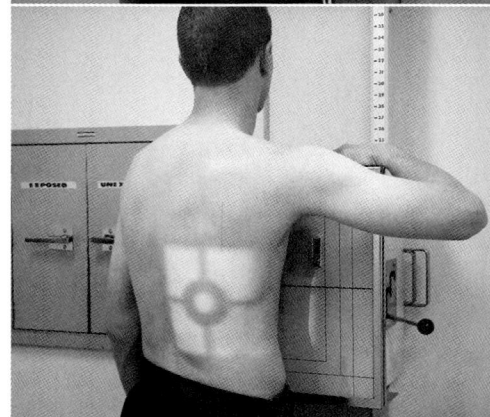

(A) Upright PA and **(B)** upright PA oblique (LAO) positions for right lower anterior ribs.

On first thought it might seem that a sternum examination should consist of PA and lateral projections. On second thought, however, it is apparent that a PA projection will cause superimposition of the thoracic spine over the sternum. The sternum is a thin, easily penetrated bone and cannot be seen on a PA projection. The solution is to rotate the body just enough to move the thoracic spine out of the way. A PA oblique projection with the body rotated 15° to 30° provides a nearly frontal view of the sternum. The amount of rotation needed depends on the AP diameter of the chest—the distance between the spine and the sternum. A deeper chest requires less rotation (Fig. 16-19).

The RAO position is used for the oblique sternum because it projects the sternum over the heart. The heart is a relatively homogeneous structure, making a relatively clear background for visualization of the sternum. When the LAO position is used, the sternum is projected over the right hilum, an area of contrasting and shadows that makes the sternum difficult to see. Breathing technique with an exposure time of 1 to 2 seconds improves visualization of the sternum by blurring the overlying ribs and lung markings to some degree.

Sternum radiography may be done with the patient either recumbent or upright, and both methods are illustrated here. The upright positions are usually most comfortable for patients with acute injury.

In preparation for sternum examination, the patient is instructed to remove all clothing and jewelry from the area between the neck and the waist and to don a gown that opens in the back.

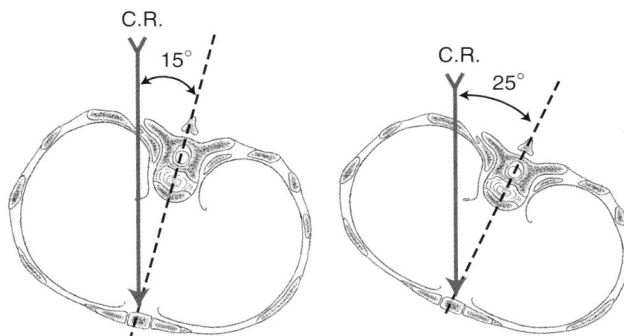

Fig. 16-19 Body rotation to project sternum free of spine depends on chest thickness. Greater rotation is required for thinner patients.

Basic Examination

The basic examination of the sternum includes the lateral and right anterior oblique views.

LATERAL

Film: 11 × 14 inch (28 × 35 cm), lengthwise

Screens: Rapid

Grid: Yes

SID: 40 inches (72 inches is sometimes used for upright lateral projections)

Body position:
Standing or recumbent

Part position:
Sagittal plane is parallel to film. For recumbent studies, patient's arms are extended over head. Upright, patient's arms are extended behind back and chest is thrust anteriorly.

Central ray:
Perpendicular to center of film through midpoint of sternum

Patient instruction:
Do not move. Stop breathing on inspiration.

Structures seen:
Entire sternum in profile

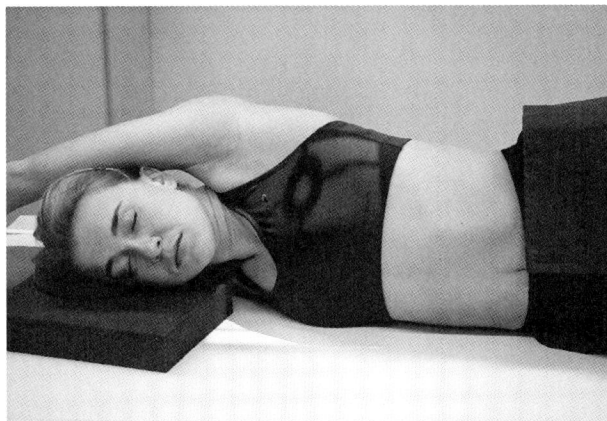

Recumbent lateral sternum position.

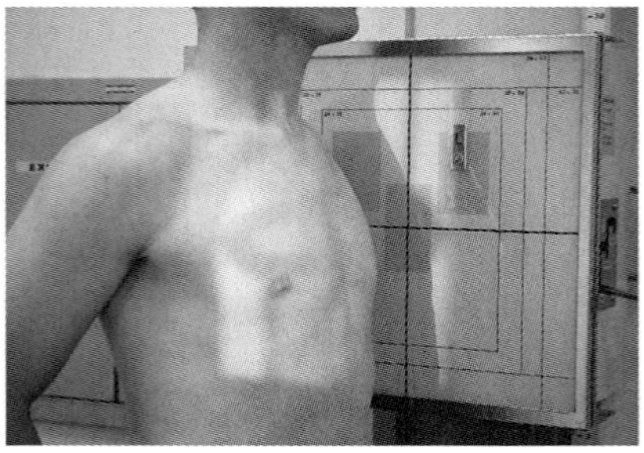

Upright lateral sternum position.

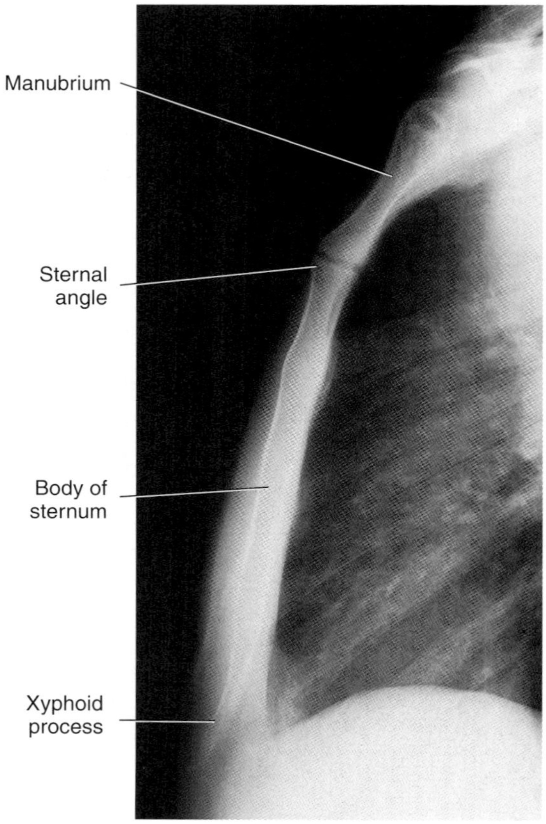

Manubrium

Sternal
angle

Body of
sternum

Xyphoid
process

Lateral sternum radiograph.

RIGHT ANTERIOR OBLIQUE

Film: 11 × 14 inch (28 × 35 cm), lengthwise

Screens: Rapid

Grid: Yes

SID: 40 inches

Body position:
Standing or recumbent

Part position:
Right anterior oblique position with coronal plane forming an angle of 15° to 30° with plane of film (see previous diagram and discussion). Sternum is aligned to center of film.

Central ray:
Perpendicular to center of film through midpoint of sternum. Seen from the back, sternum lies between spine and scapula when oblique position is correct.

TIP: The sternum is easily palpated in the lateral position, so it is easy to align the film and the tube to its center. When the lateral projection is done first, the film tray and tube are already correctly aligned for the oblique projection.

Patient instruction:
Do not move. Rapid, shallow breathing during exposure.

TIP: A low mA setting that provides the desired mAs with an exposure time of at least 1 second is needed to take full advantage of breathing technique.

Structures seen:
A nearly frontal projection of sternum, adjacent to spine

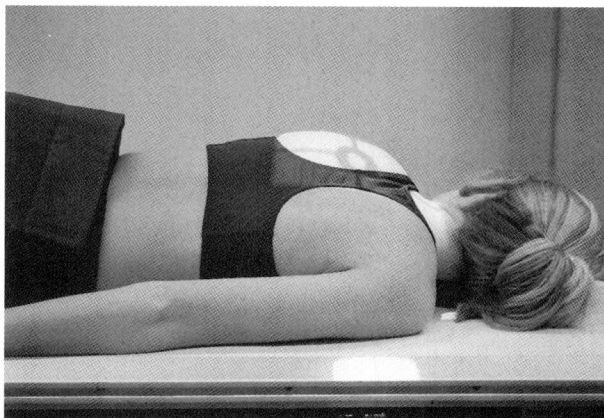

Recumbent oblique sternum position, RAO.

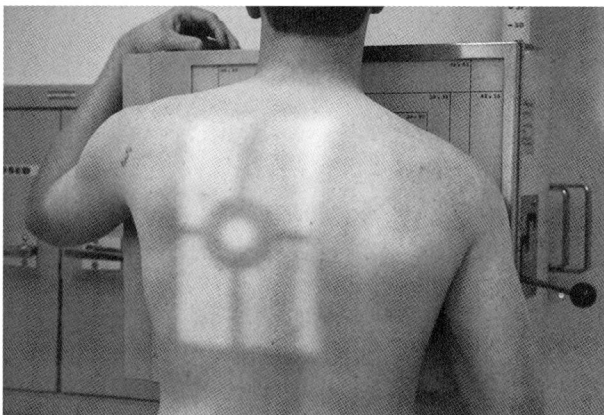

Upright oblique sternum position, RAO.

Oblique sternum radiograph, RAO position with breathing technique.

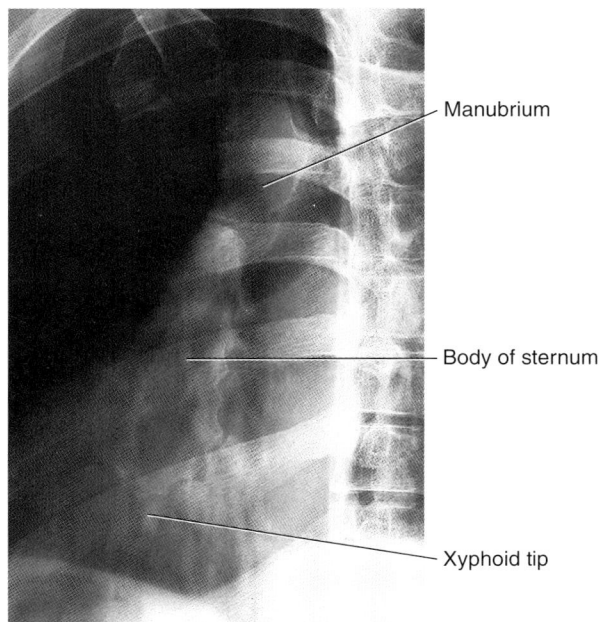

Manubrium

Body of sternum

Xyphoid tip

When physicians read chest radiographs, the diameter of the heart shadow is compared to the size of thoracic cage. This is an important diagnostic measurement. Chest radiography is done at 72 inches SID to minimize magnification of the image of the heart. For the same reason, PA and left lateral projections are preferred to AP and right lateral projections. Keeping the heart as close to the film as possible and using a 72-inch SID increases the accuracy of heart size measurements.

Exposures are made on deep inspiration to expand the lungs fields. It is a good practice to have the patient take two deep inspirations, holding the breath in on the second inspiration. This practice often produces a deeper inspiration for the exposure. In addition, if deep inspiration causes a cough reflex, the radiographer will be less likely to expose the film during a cough.

The upright position is important, both for maximum lung expansion and for the visualization of air-fluid levels. Air-fluid levels are discussed in the pathology section of this chapter. Fluid accumulations in the pleural space or within the lungs are important diagnostic signs.

The chest has a high degree of subject contrast because the lungs are air-filled and easily penetrated, while the mediastinum is a very dense structure. For this reason, chest radiography requires a high degree of technique latitude to avoid the appearance of black lungs and white mediastinum with little detail visible in either area. Latitude is obtained by using high kVp techniques (in the range of 110 to 130 kVp). A grid is used to reduce fog from the high-energy scatter radiation that is produced with high kVp. High mA settings are important to reduce exposure time because the heartbeat causes motion within the chest that blurs the image when the exposure is longer than $\frac{1}{20}$ second. Some facilities stock special film with wide latitude for chest radiography.

A lead apron that wraps around the pelvis, front and back, is the ideal gonad shield for chest radiography. When a half-apron is used, the question arises as to how it should be worn. Collimation and the posterior body structures provide gonad shielding from primary radiation, tube housing leakage, and collimator scatter. The quantity of collimator scatter and housing leakage at 72 inches SID is minimal. The principal source of potential gonad radiation in chest radiography is scatter from the grid cabinet and the wall. For this reason it is recommended that the half-apron be worn over the *front* of the patient.

In preparation for chest examinations, the patient is instructed to remove all clothing and jewelry from the area between the neck and the waist and to don a gown that opens in the back.

Basic Examination

The basic examination of the chest includes the PA and left lateral views.

Film: 14 × 17 (35 × 43 cm), lengthwise (crosswise placement for patients with hypersthenic body habitus)

Screens: Rapid

Grid: Yes

SID: 72 inches

Body position:
Standing, facing cassette holder

Part position:

PA: Anterior surface of chest is against upright grid cabinet with coronal plane parallel to film. Backs of hands are placed on hips and shoulders are rotated anteriorly. The purpose of arm position is to rotate scapulae out of the way so that they will not be superimposed on lungs. Film is aligned so that upper margin is 1½ to 2 inches above level of spinous process of C7.

Lateral: Both arms are raised overhead, with patient grasping opposite elbows. Left side of body is in contact with upright grid cabinet and midcoronal plane of thorax is perpendicular to center of film. Film and tube placement are unchanged from PA projection.

Central ray:

PA: Perpendicular to center of film. Center point should be in midline at approximate level of axillary fold.

Lateral: Perpendicular to center of film. Center point should be in midaxillary line, just inferior to axilla.

Patient instruction:

Do not move. Stop breathing on second deep inspiration.

Structures seen:

Heart, lungs, and mediastinum

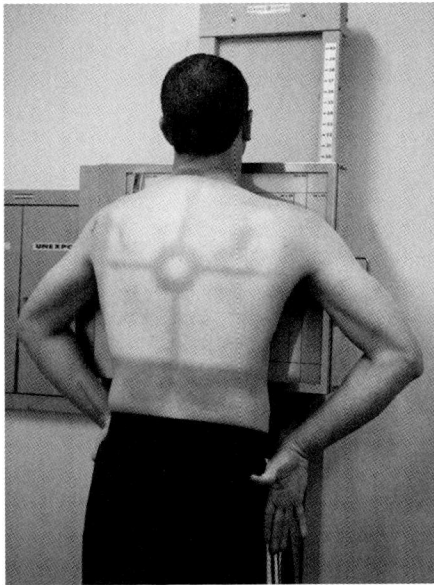

PA chest position.

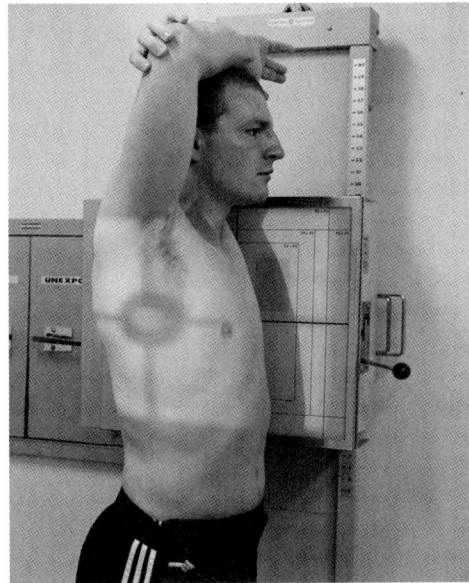

Left lateral chest position.

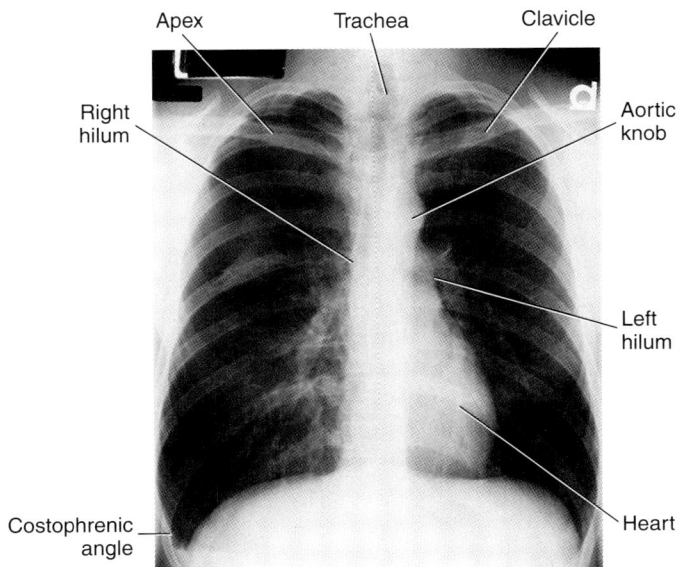

PA chest radiograph.

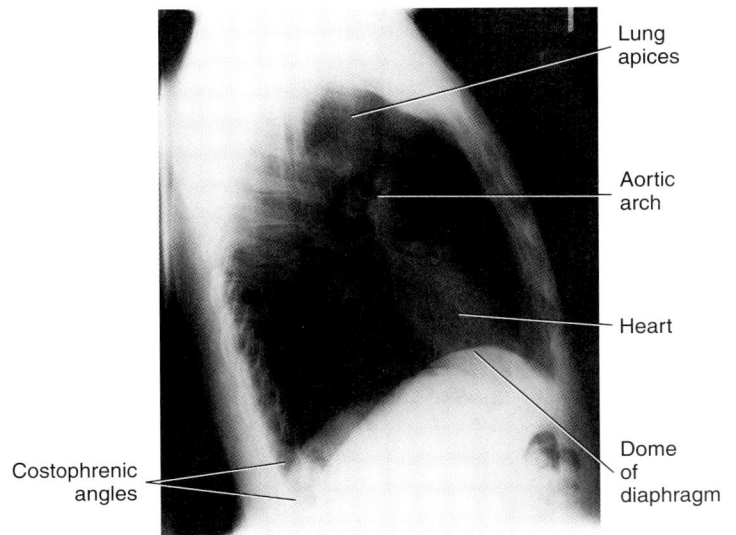

Left lateral chest radiograph.

Supplemental Projections

PA OBLIQUE

Film: 14 × 17 (35 × 43 cm), lengthwise

Screens: Rapid

Grid: Yes

SID: 72 inches

Body position:
Standing, facing cassette holder

Part position:
Left or right anterior surface of chest is against upright grid cabinet with body rotated so that coronal plane forms angle of 45° with film. Arm on side farthest from film is elevated and may be supported by grasping upper corner of grid cabinet. Arm nearest film is at patient's side but is abducted somewhat to avoid superimposition over soft tissues of chest. Upper margin of film is 1½ to 2 inches above spinous process of C7. Patient position is adjusted so that entire chest will be projected on film.

Central ray:
Perpendicular to center of film. Center point should be in midclavicular line of side farthest from film, approximately at level of axillary fold.

Patient instruction:
Do not move. Stop breathing on second deep inspiration.

Structures seen:
Heart, lungs, and mediastinum. LAO position demonstrates right chest primarily, and RAO, left chest. Spine is projected over lung on side nearest film.

Position variation for cardiac evaluation:
When cardiac evaluation is the primary objective of oblique chest radiography, the LAO projection is taken with the coronal plane of the body at an angle of 60° to the film. This position avoids superimposition of the spine over the cardiac silhouette.

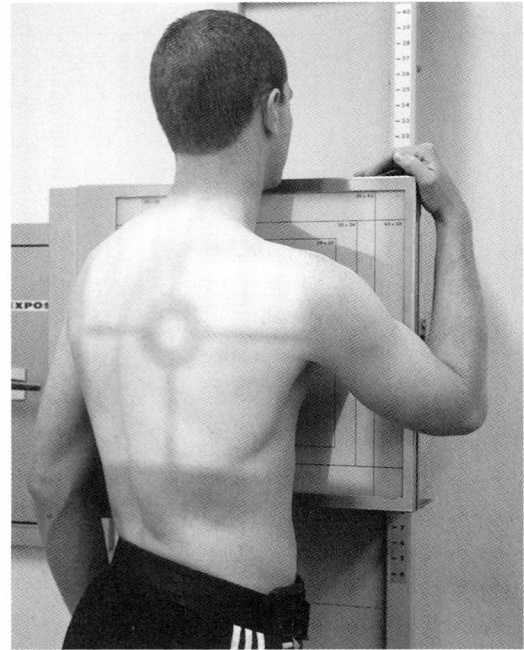

PA oblique chest position (LAO).

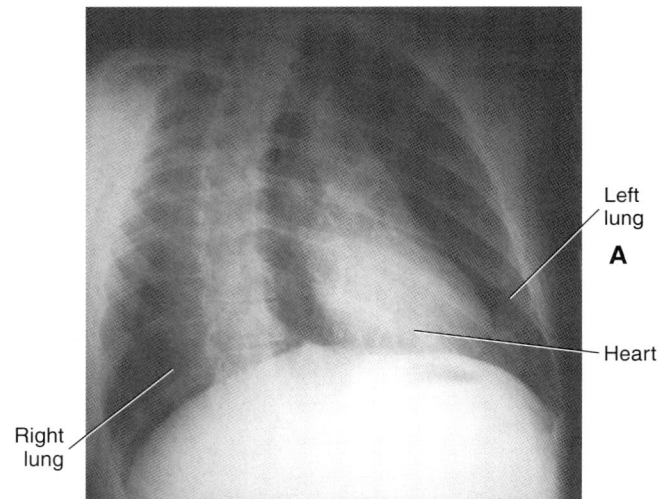

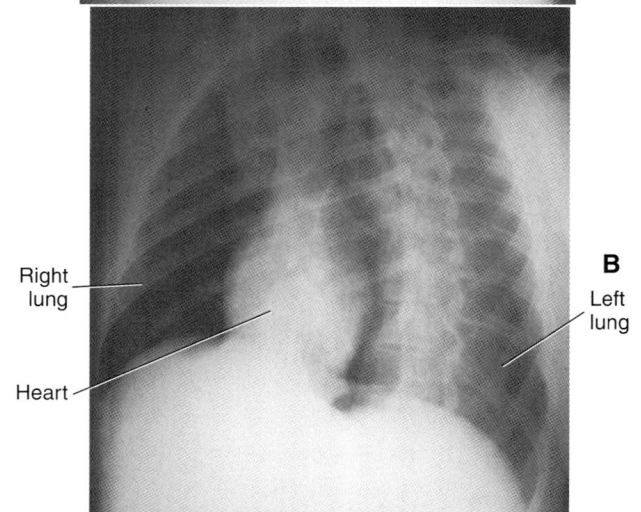

Oblique chest radiographs. **A,** RAO. **B,** LAO.

APICAL LORDOTIC PROJECTION

Film: 10 × 12 inch (24 × 30 cm), crosswise

Screens: Rapid

Grid: Yes

SID: 72 inches

Body position:
Standing

Part position:
Patient stands facing tube, 8 to 12 inches in front of upright grid cabinet. Patient then arches back and places shoulders against grid cabinet. Sagittal plane is perpendicular to film. Backs of hands are placed on hips and shoulders rotated anteriorly. The purpose of the arm position is to rotate scapulae out of the way anteriorly so that they will not be superimposed on lungs. Film is aligned so that upper margin is 2 inches above acromion process of scapula.

Central ray:
Perpendicular to center of film. Center point should be in midline at approximate level of suprasternal notch.

Patient instruction:
Do not move. Stop breathing onsecond deep inspiration.

Structures seen:
Apices of both lungs, free of superimposition by the clavicles

Position variation:
Patients who are weak or debilitated may not be able to assume and hold the lordotic position. In such cases, patient is positioned upright in AP position, standing or seated, and central ray is angled 30° cephalad through sternal angle. Film position is adjusted for alignment to central ray.

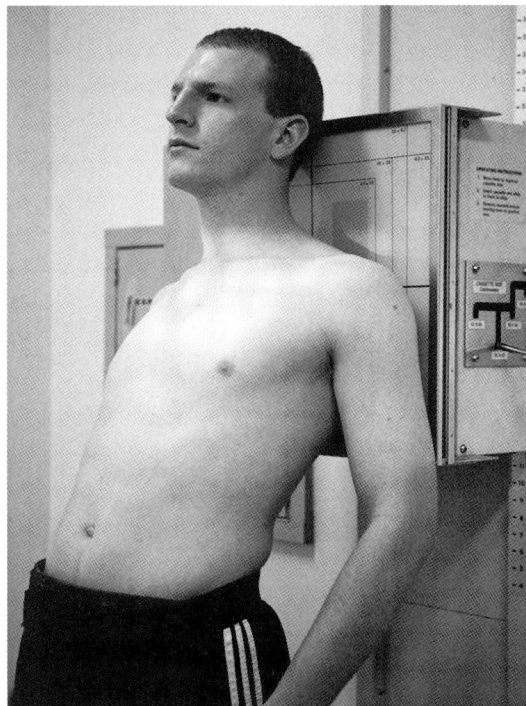

Apical lordotic chest position.

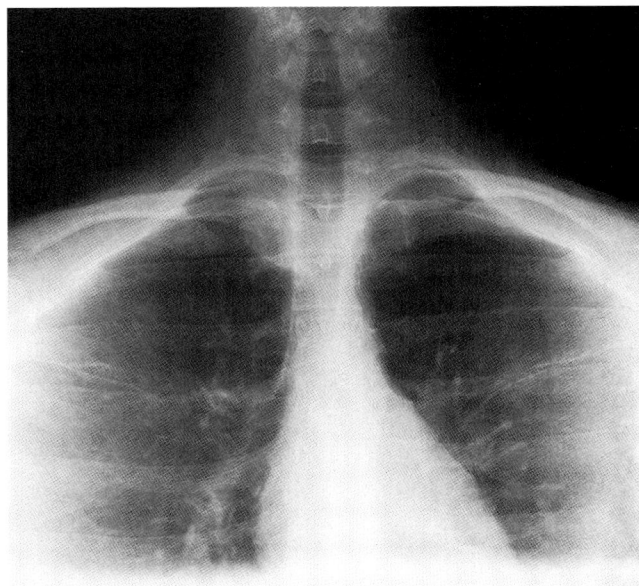

Apical lordotic chest radiograph.

Radiographic examinations of the abdomen may consist of one or more projections, depending on the purpose of the examination. The basic projection, which is a part of any abdomen study, is the AP supine projection, also called a **KUB.** Although KUB stands for "kidneys, ureters, and bladder," this projection is used in many cases where the urinary tract is not of primary clinical interest.

The basic examination for patients with acute abdominal pain should also include an AP upright projection. If the patient is unable to stand, the left lateral decubitus projection is substituted for the upright view. The purpose of including upright or decubitus projections is to demonstrate air-fluid levels in the intestines and to visualize free intraperitoneal air, if present. The significance of these findings is discussed in the pathology section of this chapter. To most effectively demonstrate these findings, the patient must maintain the upright or decubitus position for several minutes prior to the radiographer making the exposure. An upright PA projection of the chest is often included in examinations for acute abdominal pain.

In addition to the AP supine, upright, and decubitus projections described here, it may sometimes be desirable to take a lateral projection of the abdomen. The lateral projection is needed to evaluate the abdominal aorta and for the localization of abdominal foreign bodies that are not within the gastrointestinal tract. The lateral abdomen projection is taken like the lateral lumbar spine projection. The radiation field is expanded to cover the entire film and the mAs is somewhat reduced so as not to overexpose the abdominal area.

Gonad shielding is required for abdominal radiography of male patients. Properly placed, it does not affect visualization of the contents of the abdominopelvic cavity. Ovarian shields, on the other hand, obscure the pelvic cavity on females and should not be used for KUB studies. Shielding should be used for females for upright views, however, since the significant findings on upright views are in the upper and middle portions of the abdomen, and the pelvic portion is well demonstrated on the KUB.

Basic Examination

The basic examination of the abdomen includes the AP supine view (KUB) and the upright AP view.

AP SUPINE ABDOMEN (KUB)

Film: 14 × 17 inch (35 × 43 cm), lengthwise

Screens: Rapid

Grid: Yes

SID: 40 inches

Body position:
Recumbent or supine

Part position:
Sagittal plane is perpendicular to film and knees are flexed moderately and supported by a bolster. Check that inferior margin of film is at level of greater trochanter to ensure inclusion of floor of urinary bladder.

Central ray:
Perpendicular to center of film through a point in midline at level of iliac crest

Patient instruction:
Do not move. Stop breathing on expiration.

Structures seen:
Abdominal contents between diaphragm and pelvic floor. When exposure is correct, psoas muscles, liver margin, and kidney shadows should be visible.

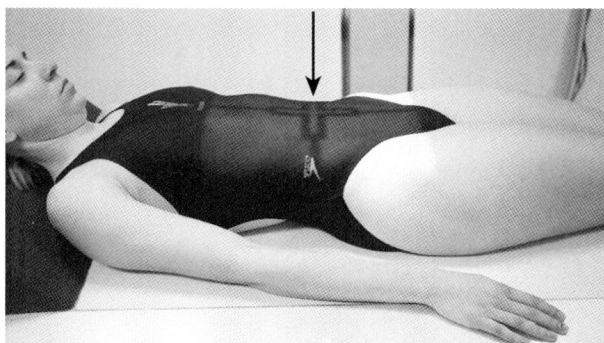

Position for AP supine abdomen (KUB).

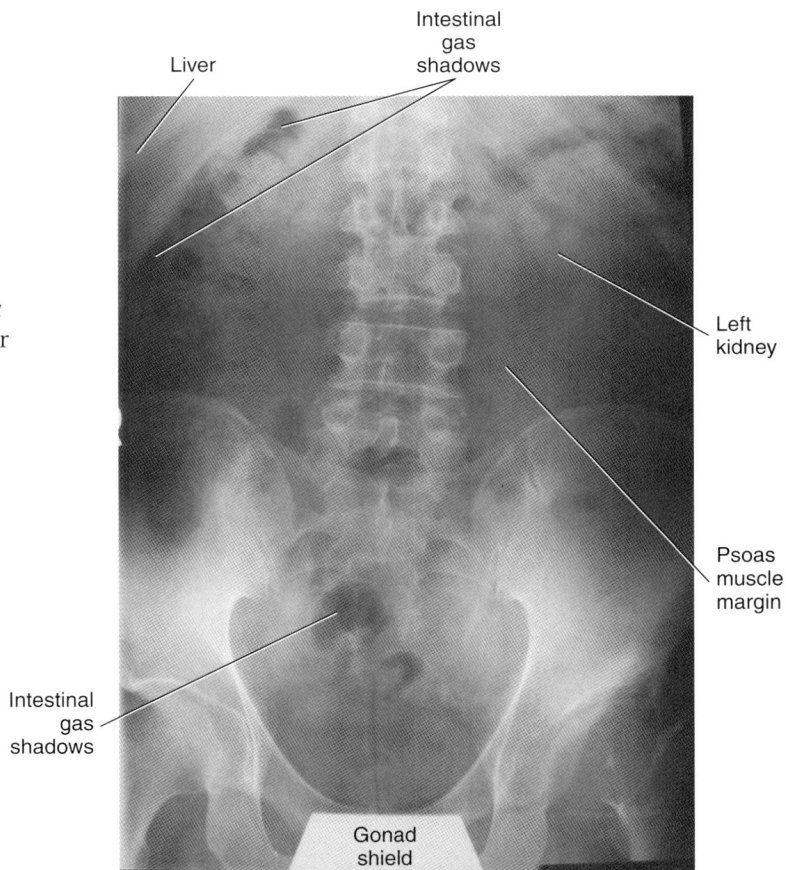

AP supine abdomen (KUB) radiograph.

UPRIGHT AP ABDOMEN

Film: 14 × 17 inch (35 × 43 cm), lengthwise

Screens: Rapid

Grid: Yes

SID: 40 inches

Body position:
Standing, facing tube

Part position:
Midsagittal plane is centered to film and perpendicular to it.

Central ray:
Perpendicular to center of film through a point in midline approximately 2 to 3 inches superior to level of iliac crest

Patient instruction:
Do not move. Suspend breathing on inspiration.

Structures seen:
Diaphragm must be seen at top of film. Air-fluid levels will be visible in intestines, if present. Free intraperitoneal air, if present, may be seen beneath diaphragm. See pathology section for discussion of air-fluid levels and free intraperitoneal air.

AP upright abdomen position.

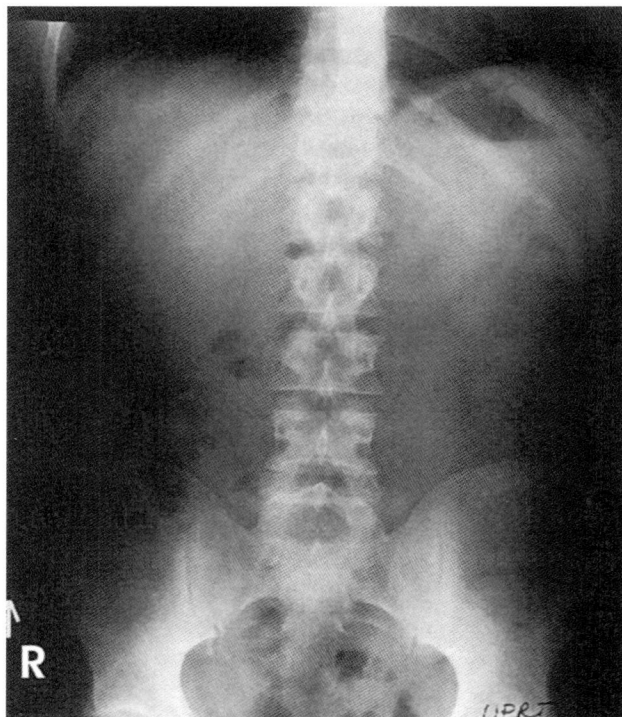

AP upright abdomen radiograph.

Alternative Examination

LEFT LATERAL DECUBITUS ABDOMEN

Film: 14 × 17 inch (35 × 43 cm), lengthwise

Screens: Rapid

Grid: Yes

SID: 40 inches

Body position:
Recumbent on left side on stretcher. Stretcher is placed in front of upright bucky or grid cabinet. Patient may be facing either tube or film.

Part position:
Midsagittal plane is centered to film and perpendicular to it.

Central ray:
Perpendicular to center of film through a point in midline approximately 2 inches superior to level of iliac crest

Patient instruction:
Do not move. Suspend breathing on inspiration.

Structures seen:
Diaphragm must be seen. Air-fluid levels will be visible in intestines, if present. Free intraperitoneal air, if present, will be seen near top of film, along right abdominal wall adjacent to liver. See pathology section for discussion of air-fluid levels and free intraperitoneal air.

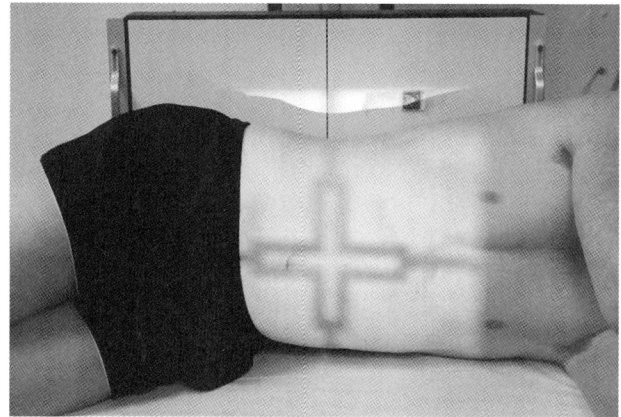

Left lateral decubitus abdomen position.

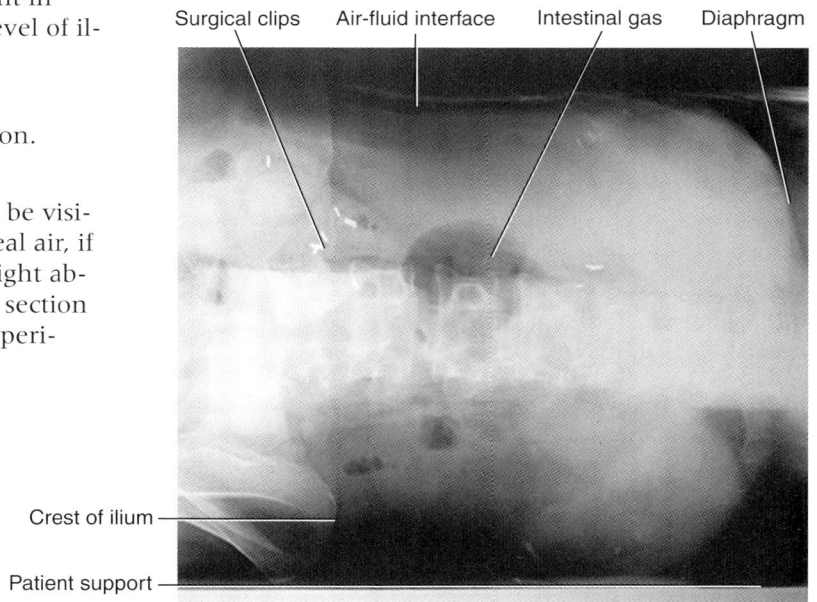

Left lateral decubitus abdomen radiograph.

PATHOLOGY

Bony Thorax

The most common reason for radiography of the bony thorax is trauma. The young boy in Fig. 16-20 was injured in a car accident. Trauma to the sternum is demonstrated in Fig. 16-21.

Nontraumatic pathology to the bony thorax includes malignant bone disease. Primary neoplastic lesions may occur in the bony thorax. Multiple myeloma is a fairly common example. Metastatic bone lesions also occur in the ribs, often secondary to lung tumors. Osteochondroma (exostosis), a benign bone tumor introduced in Chapter 14, is also seen in ribs.

Air-Fluid Levels

When fluid is present in a space normally occupied by air, it is an important diagnostic sign of pathology. The interface between air and fluid is usually clearly visible radiographically. The possibility of abnormal air-fluid levels exists in the chest, the abdomen, and the paranasal sinuses (Chapter 17).

Since fluid is heavier than air, the air will always be above the fluid, and the interface between the air and the fluid will be a horizontal line. *A horizontal x-ray beam is needed to demonstrate air-fluid levels.* This is easily understood when you look at a glass that contains water. Seen from the top, it is not possible to tell how much water the glass holds. Seen from the side, however, the horizontal line of sight makes the air-fluid level clearly visible (Fig. 16-22). For this reason,

upright and/or decubitus projections using a horizontal central ray are needed for demonstration of air-fluid levels.

Chest

Many acute and chronic conditions may affect the organs of the chest, and most can be evaluated radiographically. The scope of this text permits only a very limited introduction to this subject, offering examples of some of the common types.

Trauma to the chest may result in lung collapse, called **atelectasis.** Air escaping from the collapsing lung may occupy space in the pleural cavity formerly occupied by the lung. Air in the pleural cavity is called **pneumothorax.** Atelectasis is always present when there is pneumothorax. The lung markings (bronchioles and blood vessels that create the thread-like lines of decreased radiographic density in the lungs) are compressed with atelectasis. The radiographic appearance is such that the lung appears lighter in the collapsed area because the lung markings are very close together. Pneumothorax appears as an absence of lung markings in an area where lung is normally present. Pneumothorax is seen in Fig. 16-23.

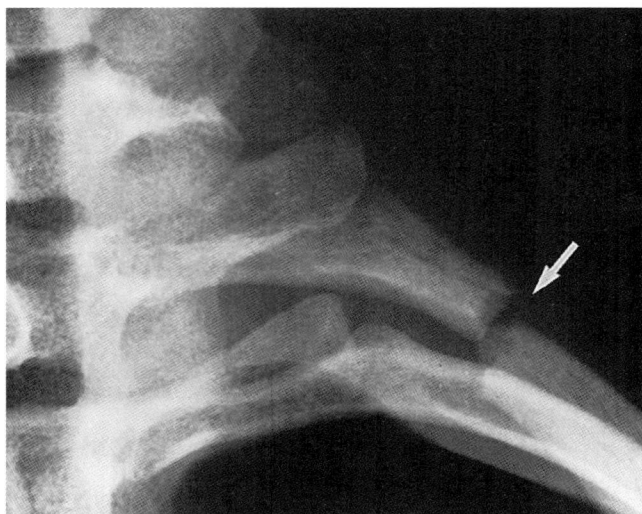

Fig. 16-20 Fracture of left first rib *(arrow)* posteriorly.

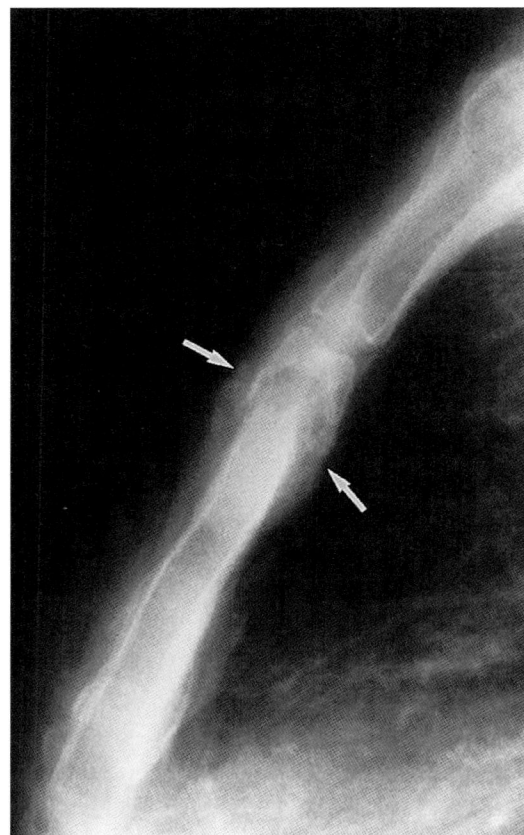

Fig. 16-21 Fracture of sternum *(arrows)* from striking steering wheel in car accident.

Atelectasis may affect a portion of a lung without causing pneumothorax (Fig. 16-24). The cause may be bronchial obstruction from foreign body or neoplasm. Abscess (localized infection), emphysema, and chronic bronchitis may also cause atelectasis.

Pneumonia is an inflammation of the lung that is usually caused by bacterial or viral infection. It may also be caused by the inhalation of chemical agents or the aspiration of vomitus. Acute pneumonia causes consolidation of the lung tissues, increased tissue density due to engorgement of the blood vessels, and fluid within spaces that are normally air-filled. Several types of pneumonia are illustrated in Fig. 16-25.

Fig. 16-22 Air-fluid levels are visualized with a horizontal line of view. **A,** Liquid in a glass. **B,** Radiograph of balloon containing air-fluid level.

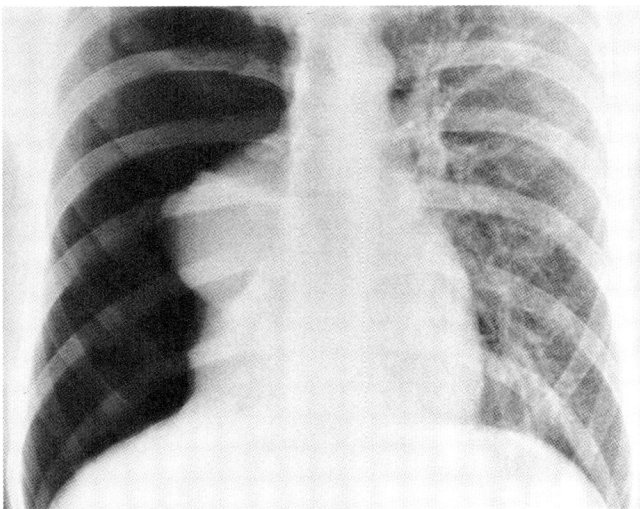

Fig. 16-23 Spontaneous pneumothorax. Complete collapse of right lung.

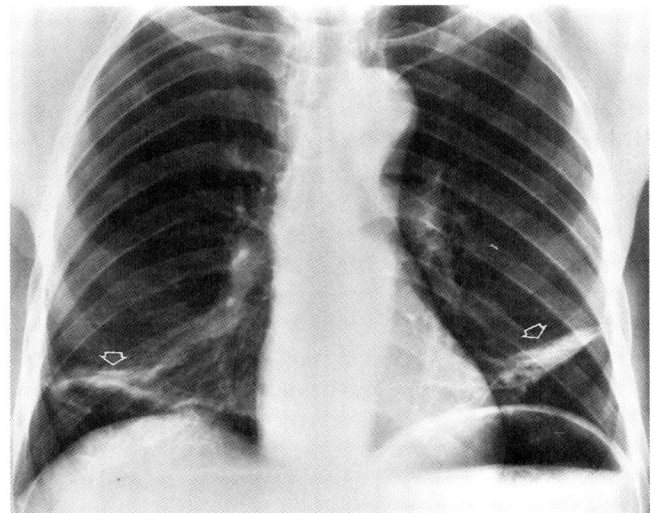

Fig. 16-24 Plate-like atelectasis *(arrows)* in both lung bases associated with chronic bronchitis.

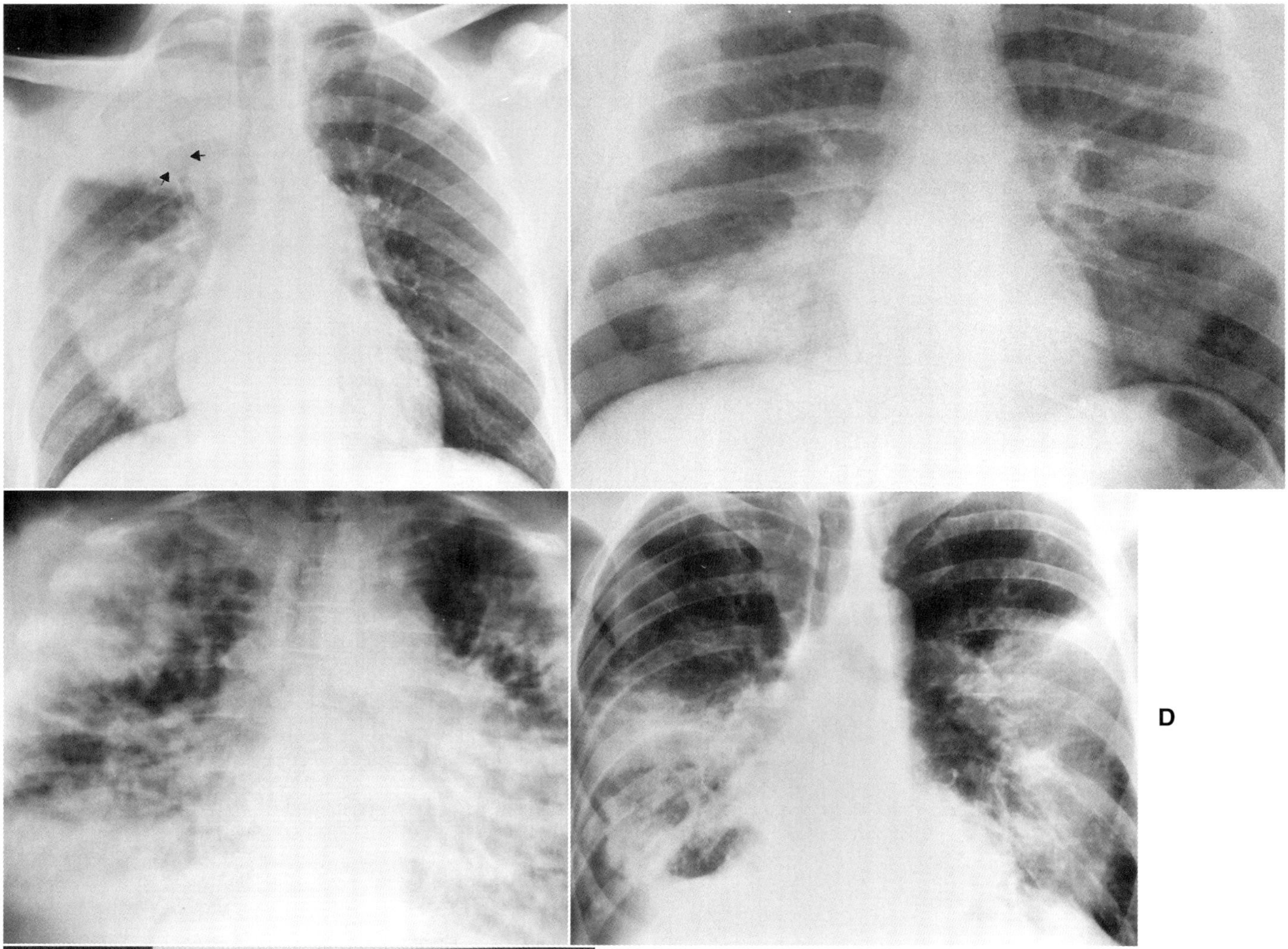

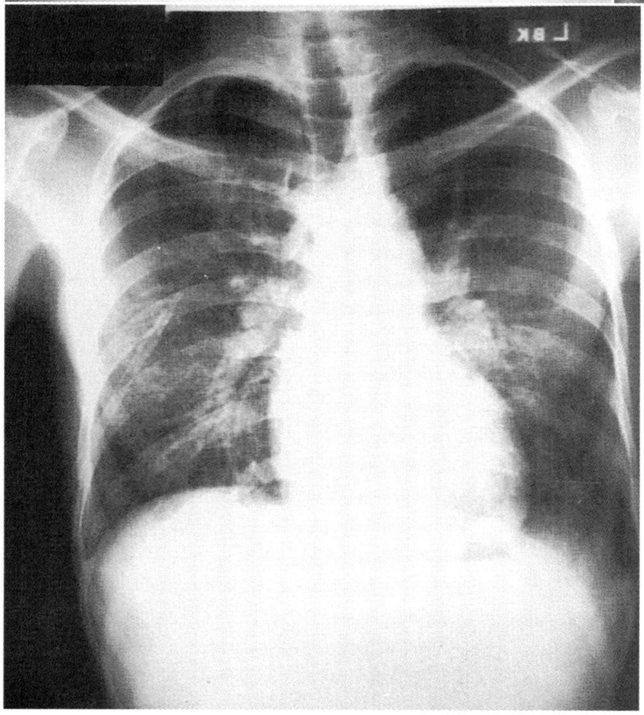

Fig. 16-25 There is great variation in the radiographic appearance of pneumonia. **A,** Bacterial pneumonia with consolidation of right upper lobe and medial and posterior segments of right lower lobe. The air bronchogram *(arrows)* is the air contrast of the upper lobe bronchus seen through the consolidated lung, a diagnostic sign. **B,** Bacterial bronchopneumonia producing ill-defined consolidation in right lung base. **C,** Example of viral pneumonia showing a diffuse infiltrate that obscures heart border ("shaggy heart" sign). There is also a patchy infiltrate in the right upper lobe. **D,** Aspiration pneumonia. Bilateral consolidation due to aspiration of vomitus. **E,** *Pneumocystis carinii* pneumonia, a common secondary infection associated with AIDS.

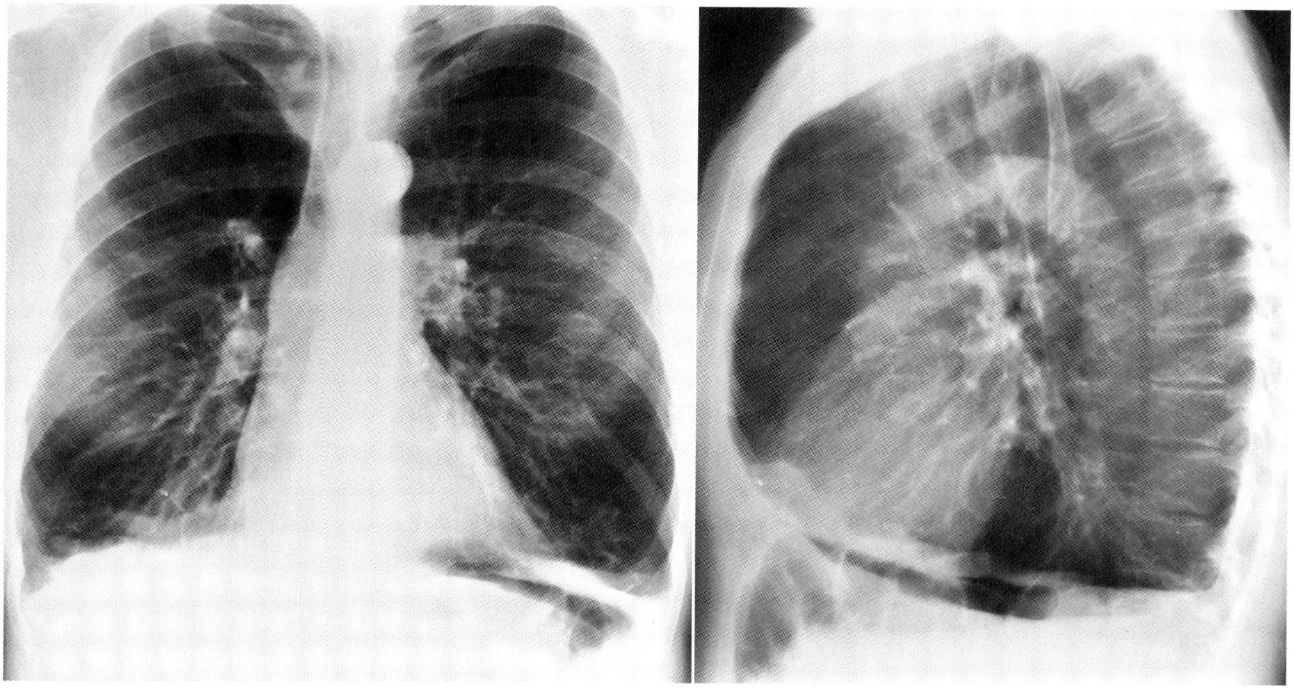

Fig. 16-26 Emphysema, a form of chronic obstructive pulmonary disease (COPD) characterized by over-inflation of the lungs.

Emphysema, a chronic lung condition, is a type of chronic obstructive pulmonary disease (COPD). It is characterized by obstruction and destruction of the small airways and alveoli of the lungs, resulting in overinflation of the lungs and the inability to effectively exhale stale air. Chronic overinflation of the lungs increases the AP diameter of the chest and causes depression and flattening of the diaphragm (Fig. 16-26). It is important for radiographers to recognize the outward physical signs of emphysema prior to radiography so that the exposures for chest films may be adjusted appropriately. Patients with emphysema are often described as "barrel-chested." The difference between the PA and lateral chest measurements will be only 8 cm or less. There may be retraction of the costal muscles near the neck with prominence of the clavicles. Patients may practice positive pressure breathing, pursing their lips as they exhale. The large chest measurement does not require a large exposure because the chest volume consists predominantly of air. Typically, these patients have a smaller-than-normal percentage of muscle and fat tissue and are very easily penetrated. The mAs for chest radiographs must be reduced by 30% to 60% from the usual mAs for patients of similar measurement. The actual percentage of mAs reduction needed depends on the severity of the disease.

Tuberculosis (TB) is an infectious lung disease (Fig. 16-27). Most persons exposed to tuberculosis do not develop active disease. The body's immune system walls off the infection so that it becomes dormant. In this inactive stage, the individual has no symptoms and cannot transmit the disease to others. Tuberculosis may become active, however, when the immune system is depressed. Malnutrition, HIV infection, other disease processes, or advanced age may allow tuberculosis to become active. When dormant tuberculosis becomes reactivated, it is called secondary tuberculosis. It often affects the apices of the lung. An apical lordotic projection of the chest may be ordered to better visualize this area of the lungs. The primary screening test for TB is the tuberculin skin test. Anyone who has ever contacted the disease will have developed antibodies to the TB organism and will test positive. When the skin test is positive, chest radiographs are taken to rule out active disease. Sputum cultures for tuberculosis bacteria may be needed to confirm the diagnosis of active disease. Precautions to prevent the spread of active TB are discussed in Chapter 21.

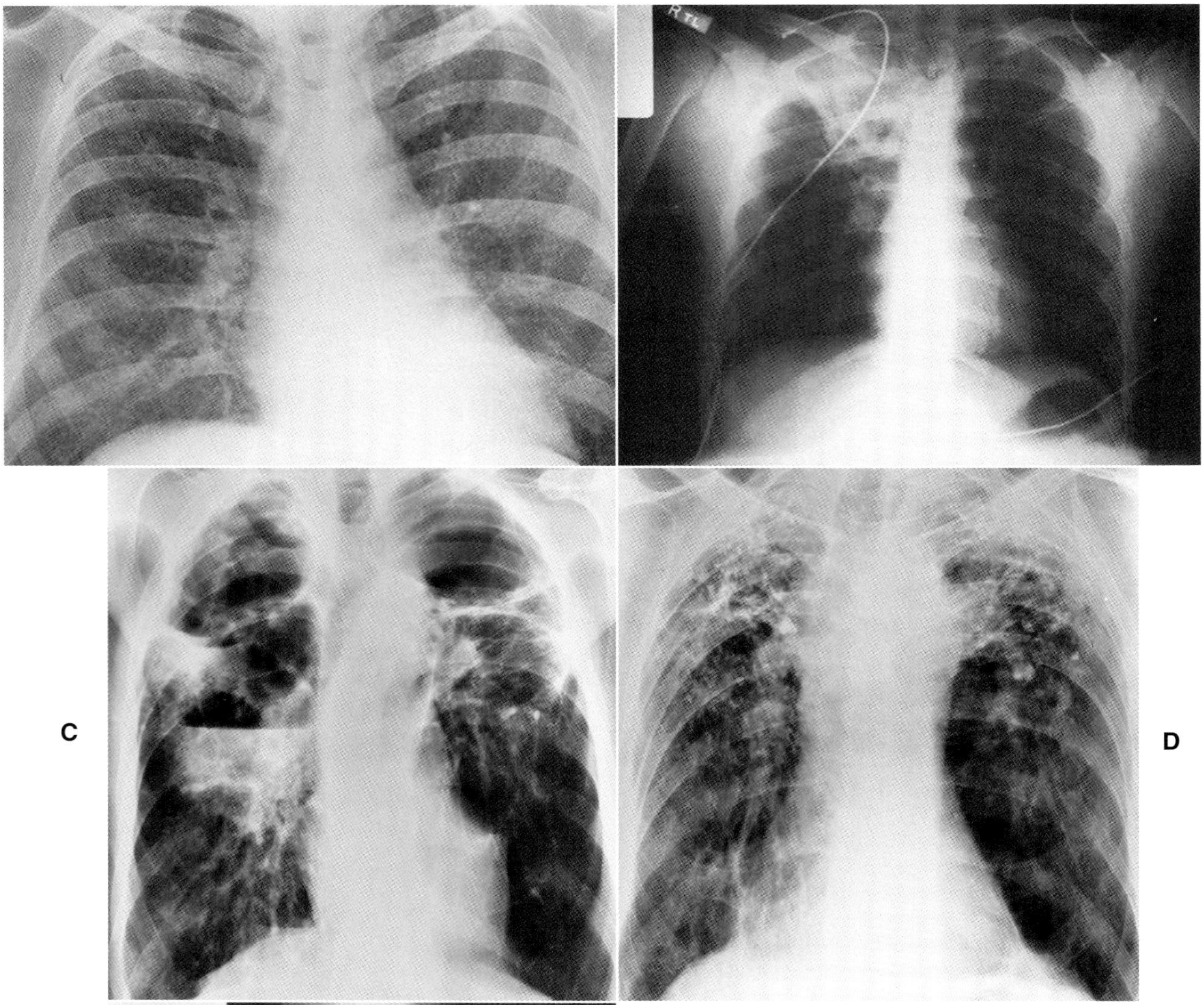

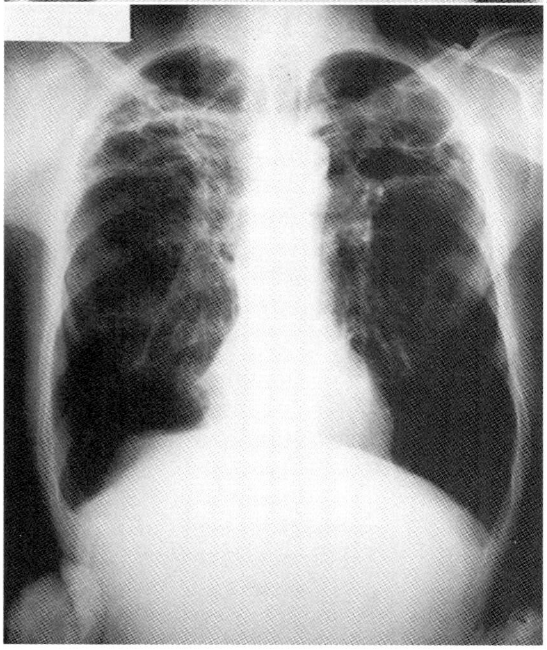

Fig. 16-27 Tuberculosis has many different radiographic appearances. **A,** Miliary tuberculosis, an acute form of the disease with many tiny foci of infection that look like millet seeds. **B,** Active, primary tuberculosis in right upper lobe. **C,** Secondary tuberculosis with fibrocalcific changes in both apices. **D,** Advanced tuberculosis with many large cavities exhibiting air-fluid levels. Note also chronic fibrous changes and upward retraction of the hila. **E,** Scarring and calcification from healed tuberculosis.

The term **pneumoconiosis** refers to a group of chronic occupational lung diseases caused by the inhalation of irritating dust. One type is silicosis (Fig. 16-28), which results from the inhalation of silicone dioxide. It affects workers in mines, foundries, and sandblasting operations. Other examples of pneumoconiosis include asbestosis and coal miner's disease (black lung).

Lung cancer may arise from the lung tissue, the pleura, or a bronchus. The radiographic appearance differs greatly, depending on the type of tissue involved and the stage of the disease. Fig. 16-29 provides examples of neoplastic lung disease.

Many different malignant diseases metastasize to the lungs. Metastatic lesions may be solitary or multiple and may display a wide variety of radiographic appearances. Fig. 16-30 provides examples of metastatic lung disease.

Heart disease may be caused by congenital malformations, hypertension (high blood pressure), stenosis of the coronary arteries that supply the heart muscle, or disease of the heart valves. There may also be calcification of the walls of the coronary vessels or the thoracic aorta.

When the walls of arteries become calcified, they also tend to become roughened and narrow. Fatty deposits on vessel walls, called plaque, also cause narrowing and roughening of vessels. A blood clot that forms on a rough area of a vessel wall is called a thrombus. The thrombus may cause stenosis of the vessel, impeding or occluding blood flow. A thrombus may also break free of the vessel wall and become a free-floating blood clot called an embolus. The embolus will flow through the bloodstream until it becomes

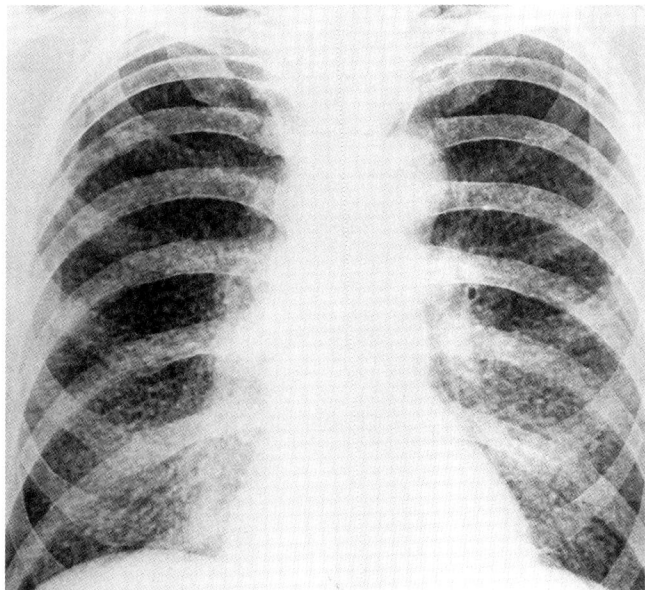

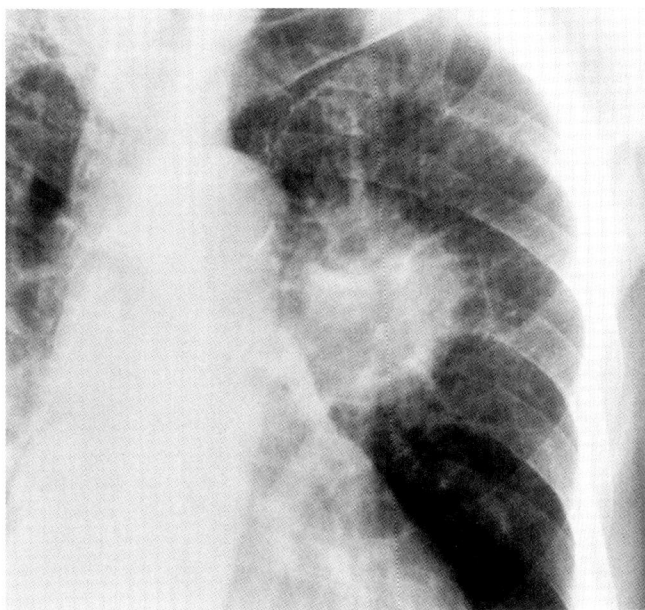

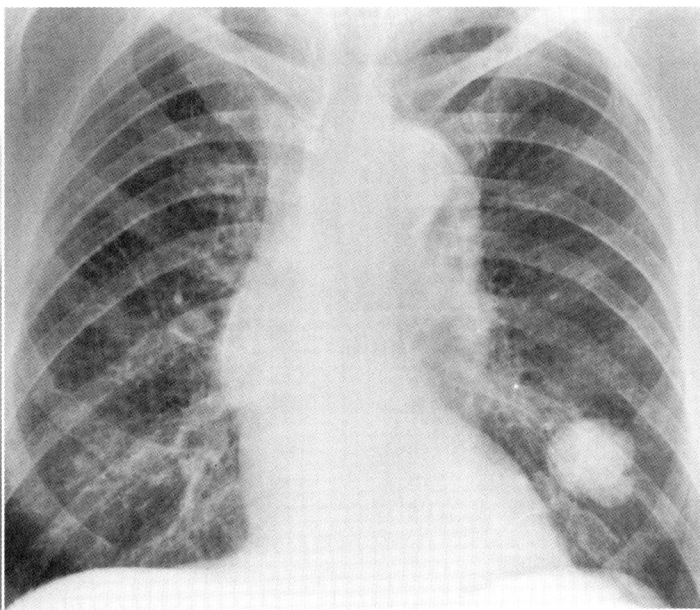

Fig. 16-28 Silicosis, characterized by miliary calcific nodules scattered throughout both lungs.

A

B

Fig. 16-29 Examples of neoplastic lung disease. **A,** Bronchogenic carcinoma, a highly malignant tumor. Its fuzzy, ill-defined margins are indicative of invasive growth into surrounding tissue. **B,** Peripheral bronchial adenoma in left lower lobe. This tumor has a low level of malignancy, which is typical of neoplasms with smooth, rounded margins.

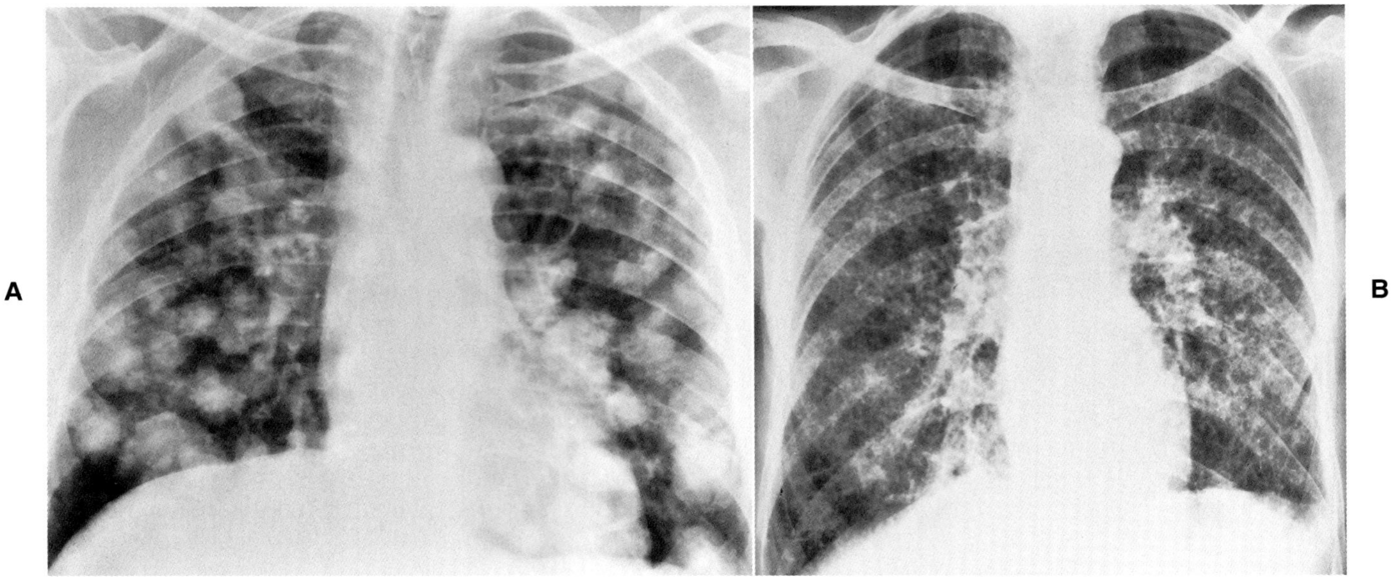

Fig. 16-30 Metastatic lung disease. **A,** Hematogenous metastasis (spread via blood) are well-defined lesions scattered throughout the lungs. The lesions may be any size from tiny to very large, depending on the site of the primary cancer. **B,** Lymphangitic metastases spread via the lymphatic system. In this case, the disease arises from primary cancer of the stomach.

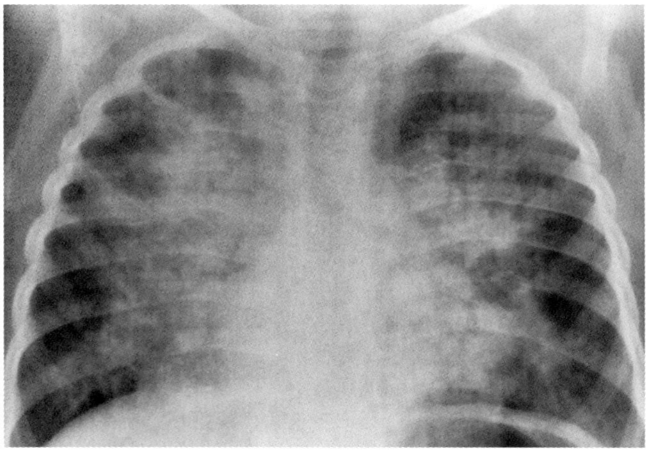

Fig. 16-31 Congestive heart failure. Decreased radiographic density in the lungs is due to pulmonary edema.

stuck in a vessel that is too narrow for it to pass. The embolus prevents blood flow to tissues, causing necrosis. A coronary occlusion (blockage of a coronary artery) causes myocardial infarction, a heart attack, which causes death of heart muscle tissue.

Congenital malformations, hypertension, and valvular disease may all cause inefficient heart function. The heart tends to become enlarged and the aorta may become elongated and tortuous. Cardiac insufficiency causes pulmonary edema, the collection of fluid in the tiny spaces within the lung. Advanced cardiac insufficiency with pulmonary edema is called congestive heart failure (CHF) (Fig. 16-31).

Pleural effusion is a collection of fluid in the pleural space (Fig. 16-32). This is a nonspecific radiographic finding with many possible causes. The most common causes are neoplastic disease, congestive heart failure, pulmonary embolism, infection (particularly tuberculosis), and pleurisy (inflammation of the pleura). When one is in the upright position, the fluid collects in the bottom of the pleural space and gives the appearance of blunting or rounding of the normally sharp costophrenic angles. This is an important reason why both costophrenic angles must always be included on PA chest radiographs.

Abdomen

Many abdominal structures, and particularly the internal contours of abdominal organs, cannot be seen radiographically without the use of contrast media. A contrast medium is a special compound that absorbs radiation to a greater extent than the tissues. A liquid suspension of barium sulfate is either swallowed or administered rectally to outline the internal structures of the gastrointestinal tract. Special contrast media tablets are available to provide contrast for examinations of the gallbladder. Intravenous in-

Fig. 16-32 Pleural effusion. **A,** Upright PA projection. **B,** Left lateral decubitus projection.

jections, or injections directly into the organs of interest, are used to provide contrast in the urinary tract and the biliary system. Procedures using contrast media are beyond the scope of this text.

The abdominal features that are visible on "plain films"(noncontrast media studies) are the outer contours of such abdominal structures as the liver and the kidneys. Since gas in the stomach and intestines provides radiographic contrast, portions of the gastrointestinal tract that contain gas are also visible on abdomen radiographs. The psoas muscles appear on abdomen radiographs as dense diagonal structures on either side of the lumbar spine. They are not abdominal organs, but their visibility is one indication that the exposure factors are correct (Fig. 16-33).

Conditions that may be diagnosed by means of abdominal plain films include enlargement of the liver, enlargement and/or displacement of the kidneys, abnormal gas patterns in the bowel, calcifications such as kidney stones and gallstones, and calcifications in arteries, particularly the abdominal aorta.

An acute abdomen series consists of AP supine and AP upright or decubitus abdomen films. An upright PA chest radiograph is also often included. The acute abdomen series is used in cases of suspected bowel obstruction or ruptured hollow viscus (organ). Bowel obstruction appears on the supine film as multiple

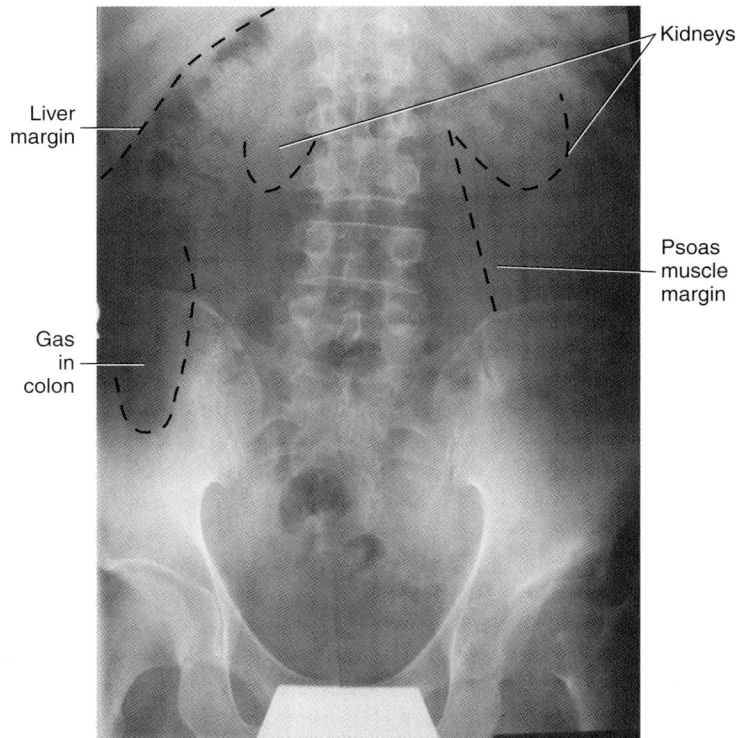

Fig. 16-33 AP abdomen radiograph showing kidney shadows, liver margin, and psoas muscles.

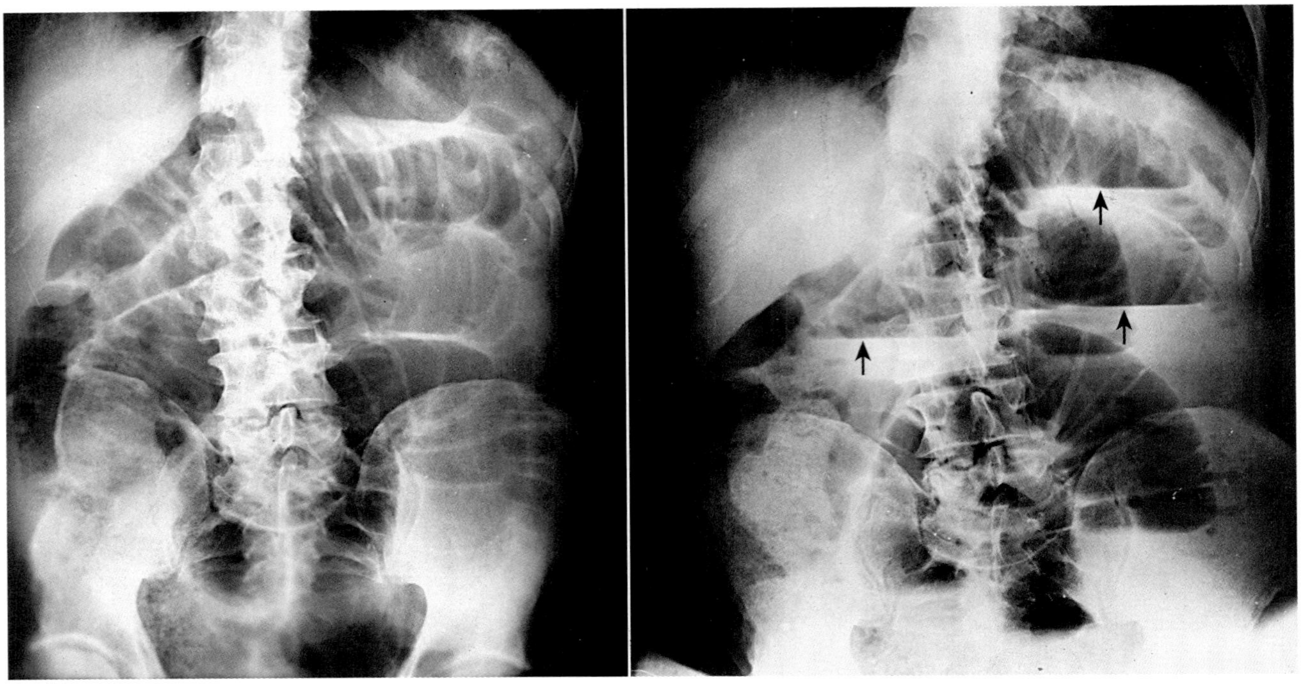

Fig. 16-34 Bowel obstruction. **A,** Supine view shows distended loops of small intestine. **B,** Upright view on the same patient shows air-fluid levels *(arrows)* within small intestine.

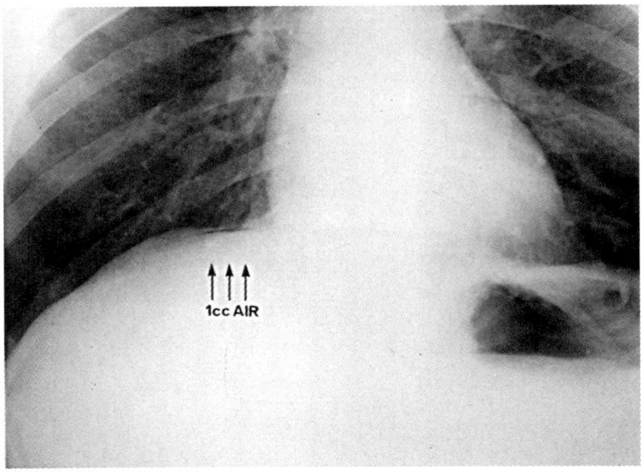

Fig. 16-35 Free air in the peritoneal cavity is seen as a thin, black horizontal line under the diaphragm on upright PA chest radiograph.

loops of distended, gas-filled small intestine. On the upright abdomen film, air-fluid levels are present (Fig. 16-34). The presence of large quantities of gas results in increased radiographic density. For this reason, it is desirable to decrease the mAs somewhat for abdomen radiography in cases of bowel obstruction.

The term "ruptured hollow viscus" refers an opening between the gastrointestinal tract and the peritoneal cavity. Ruptured appendix and perforated ulcer are examples. When this occurs, gas and in-

testinal contents leak into the peritoneal cavity, causing inflammation and a potentially life-threatening infection called peritonitis. Ruptured hollow viscus is diagnosed radiographically by the appearance of gas, or "free air," in the peritoneal cavity outside the gastrointestinal tract. Since gas rises, it is seen on upright projections just beneath the diaphragm (Fig. 16-35). On the left lateral decubitus projection, free air rises to the right abdominal wall and is projected over the liver.

The accumulation of fluid in the peritoneal cavity is called ascites. When ascites is present, the abdomen tends to be distended and rigid. Ascites is a nonspecific finding that is indicative of serious disease. Examples of conditions that may be associated with ascites include cirrhosis of the liver, neoplastic disease of the intestine, or pelvic inflammatory disease (PID), which is an infection involving the female reproductive organs. Ascites increases the tissue density of the abdomen, requiring an increase in mAs to obtain an adequate exposure.

An upright or decubitus PA chest film is often ordered for patients with acute abdominal pain. A chest film may be included routinely in an acute abdomen series. The chest radiograph is important when free intraperitoneal air is suspected because free air is often better seen on chest radiographs than on abdomen films. In addition, acute abdominal pain may result from problems related to the chest or diaphragm that are seen only on chest radiographs.

SUMMARY

The bony thorax consists of the ribs, the sternum, and the thoracic spine. Ribs and sternum are radiographed using relatively low kVp techniques to avoid overpenetrating these small bones. Rib examinations consist of both frontal and oblique projections, and procedures vary depending on the region of clinical interest. Different exposures are required, depending on whether the ribs of interest are above or below the diaphragm. Lateral and RAO projections are used to demonstrate the sternum. Both rib and sternum examinations are often done upright for patient comfort and safety. These examinations are usually performed to demonstrate fractures, but other abnormalities of these skeletal structures may be demonstrated radiographically as well.

Chest radiographs demonstrate the heart, lungs, and mediastinum and are commonly taken by limited radiographers. Chest films may provide life-saving information about the condition of vital organs. Routine chest radiography consists of PA and left lateral projections taken upright at 72 inches SID to minimize magnification of the heart. They are exposed on deep inspiration using a grid and high kVp exposure factors. Many inflammatory and neoplastic diseases are evaluated by means of chest radiography, as are conditions caused by trauma or heart problems.

Plain films of the abdomen are less common for limited radiographers. The AP supine (KUB) projection demonstrates the outlines of some abdominal organs and may also reveal calcifications in the urinary or biliary system or abnormal gas patterns in the intestines. Upright or decubitus projections are used to demonstrate air-fluid levels that indicate intestinal obstruction or free intraperitoneal air that is diagnostic of ruptured hollow viscus.

▪ *Review Questions* ▪

1. Name the parts of the sternum and point to each on your own body.
2. Make a simple drawing of a lung and indicate the apex, the hilum, the costophrenic angle, and the cardiophrenic angle.
3. Name four structures located within the mediastinum and state the body system to which each belongs.
4. Name two organs found in each quadrant of the abdomen.
5. Name the projections that constitute a routine examination of the left upper anterior ribs. The right lower posterior ribs.
6. Should ribs below the diaphragm be exposed on inspiration or expiration?
7. List as many differences as you can between rib radiography and chest radiography.
8. If a patient with acute abdominal pain cannot stand for an upright AP abdomen, what projection should be substituted? Why is this important?
9. List three conditions that involve inflammation of the lungs.
10. Name two common radiographic findings in cases of congestive heart failure.
11. What radiographic findings are typical of intestinal obstruction?
12. State two reasons why a chest radiograph might be important for a patient who has severe abdominal pain.

17

Skull, Facial Bones, and Paranasal Sinuses

Learning Objectives

At the conclusion of this chapter, the student will be able to:

- Name the principal bones that make up the cranium and the face and identify each on an anatomical diagram and on a radiograph
- Name and identify the four sets of paranasal sinuses on an anatomical diagram and on radiographs
- Identify significant positioning landmarks of the skull and face by palpation
- Demonstrate correct body and part positioning for routine projections and common special projections of the skull, facial bones, and paranasal sinuses
- Correctly evaluate radiographs of skull, facial bones, and paranasal sinuses for positioning accuracy
- Describe and recognize on radiographs pathology that is common to skull, facial bones, and paranasal sinuses

Key Terms

acanthion
blow-out fracture
cerebral concussion
contrecoup injury
cranium
external auditory (acoustic)
  meatus (EAM)
external occipital
  protuberance (EOP)
foramen magnum
glabella
gonion
lacrimal bone
mandible

maxilla
mental protuberance (point)
multiple myeloma
nasal concha
nasion
orbit
osteoma
Paget's disease
palatine bone
sella turcica
suture
vomer
zygoma

Skull radiography is much less common than it once was. Today, the first choice for diagnostic imaging in cases of head trauma is the computed tomography (CT) scan, which provides information about the condition of the brain that is not possible with routine radiography. CT is now also the principal imaging modality for the paranasal sinuses and some types of facial injuries.

On the other hand, radiography of the bones of the head is still a useful diagnostic tool. When a CT scanner is not immediately available, radiography can provide much valuable information. In addition, some types of skull pathology and facial injuries are still best evaluated radiographically.

The anatomy of the bones of the skull and face is complex. A comprehensive understanding of these structures requires diligent study. You will find it helpful to use a model of the skull that allows you to see both the internal and external structures in three-dimensional perspective.

ANATOMY

Cranium

The term **cranium** refers to the bones that surround the brain, the brain case (Fig. 17-1). It consists of eight bones: frontal, occipital, right and left parietal, right and left temporal, sphenoid, and ethmoid. Its outer shell is called the calvarium, and the portion within the skull is called the cranial base or floor.

The frontal bone is the bone of the forehead and also constitutes the anterior portion of the top of the skull. The bony prominence on the frontal bone between the eyebrows is a palpable positioning landmark called the **glabella.** On either side of the glabella, the frontal bone forms the superior portions of the **orbits** (eye sockets).

The occipital bone is at the lower part of the back of the skull and also forms the posterior portion of the cranial base. In the approximate center of the occipital bone is a palpable bony prominence called the **external occipital protuberance (EOP).** The large round hole in the anterior portion of the occipital bone is called the **foramen magnum.** It is the passage for the spinal cord between the skull and the spine. Between the frontal and the occipital bones are the two parietal bones.

The ethmoid bone forms the anterior floor of the cranium. It articulates with the underside of the frontal bone, extending inferiorly and posteriorly behind the nose. The ethmoid bone also forms a portion of each orbit.

The sphenoid bone is a complex bone that makes up part of the floor of the cranium and also portions of its lateral outer shell. The sphenoid bone is posterior to the ethmoid bone and is shaped somewhat like a bat (Fig. 17-2). The rounded fossa in the center of its anterior superior surface, called the **sella turcica,** is the location of the pituitary gland. The bilateral inferior projections are called the pterygoid processes.

The temporal bones (Fig. 17-3) also form portions of both the floor and the lateral outer surface of the cranium. They articulate with the sphenoid bone anteriorly and the occipital bone posteriorly. On the inferior lateral border is the **external auditory meatus,** also called the **external acoustic meatus.** Both terms are abbreviated **EAM.** The EAM is the opening to the ear canal and is an important positioning landmark. Extending medially from the EAM area is a dense pyramid of bone called the petrous portion, which contains the middle and inner ear structures. The mastoid process of the temporal bone contains many small air cells and is palpable just posterior to the earlobe. Just anterior to the EAM is the mandibular fossa, the socket that articulates with the lower jaw. Just superior to the EAM bilaterally are long slender horizontal bony projections that can be palpated on the sides of the face anterior to the ears. They are called the zygomatic arches and articulate anteriorly with the cheek bones.

The joints that connect the bones of the cranium are synarthrodial (immovable) joints called **sutures** that have individual names. The parietal bones are separated by the sagittal suture. Between the frontal bone and the parietal bones is the coronal suture. Between the parietal bones and the occipital bone is the lambdoidal suture. The joint between the parietal and temporal bones is the squamosal suture.

Facial Bones

The palpable bones of the face include the **maxilla** (upper jaw), **mandible** (lower jaw), **zygomas** (cheek bones), and nasal bones (Fig. 17-4).

The maxilla is actually two maxillary bones fused in the center beneath the nose. The maxilla is the largest immovable bone of the face and articulates with all of the other facial bones except the mandible. Its upper margins form the inferior medial orbital rims. Its inferior margin, called the alveolar process, contains the roots of the upper teeth. The junction of the two maxillary bones forms a superior prominence called the anterior nasal spine. This point, located at the junction of the nose and the upper lip, is a positioning landmark called the **acanthion.**

The nasal bones are situated between the orbits and articulate laterally with the maxillary bones and posteriorly with the ethmoid.

The zygomas are also called the malar bones or cheek bones. They form the superior lateral structure of the face. Each zygoma articulates superiorly with the frontal bone, inferiorly with the maxilla, and posteriorly with the zygomatic process of the temporal

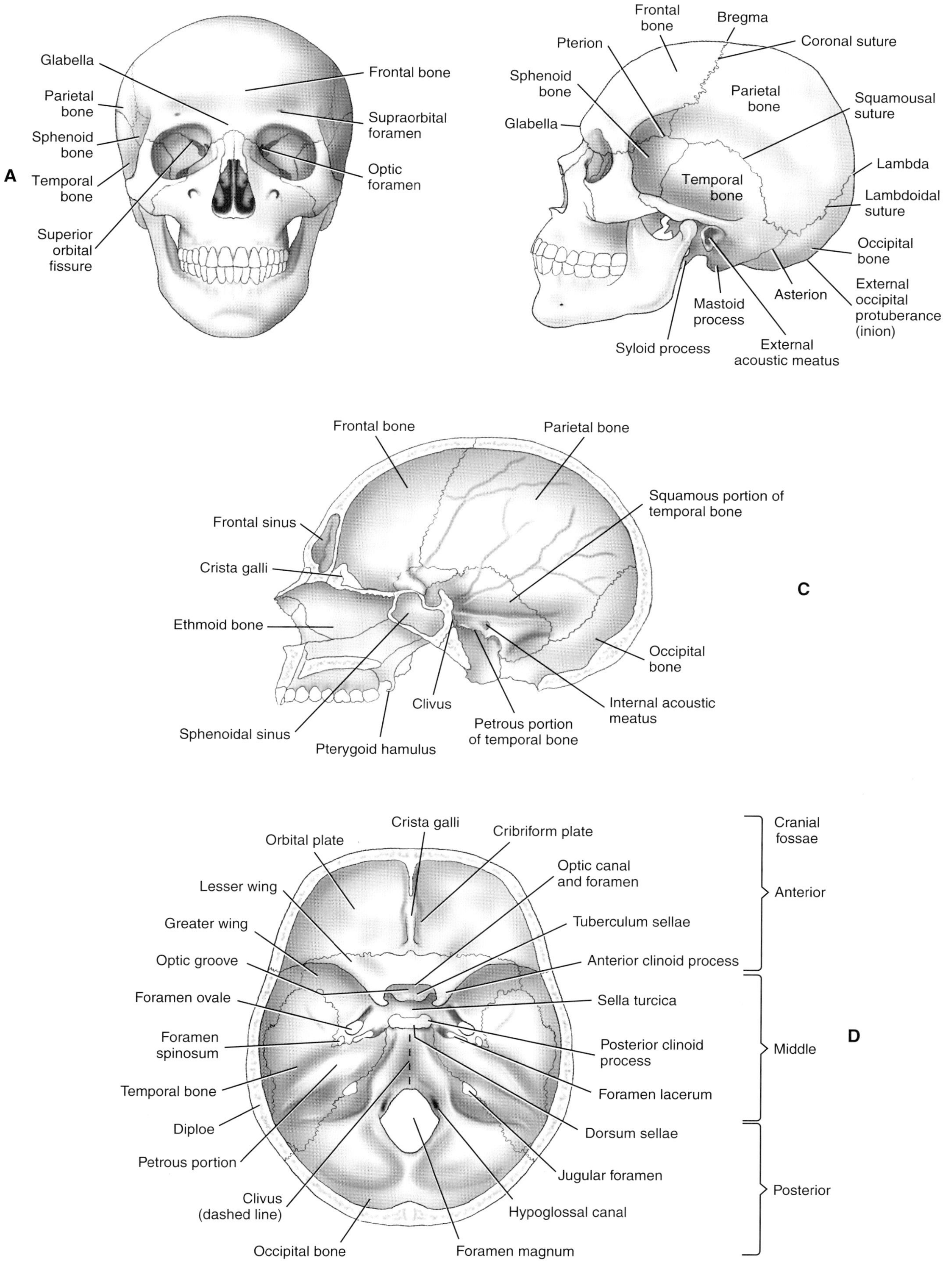

Fig. 17-1 Cranium. **A,** Anterior aspect; **B,** lateral aspect; **C,** lateral aspect of interior of cranium; **D,** superior aspect of cranial base.

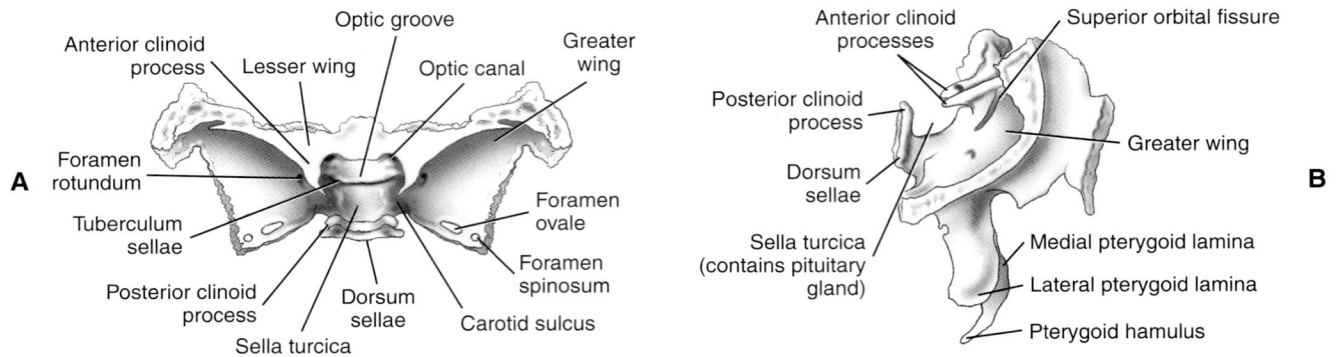

Fig. 17-2 Sphenoid bone. **A,** Superior aspect; **B,** lateral aspect.

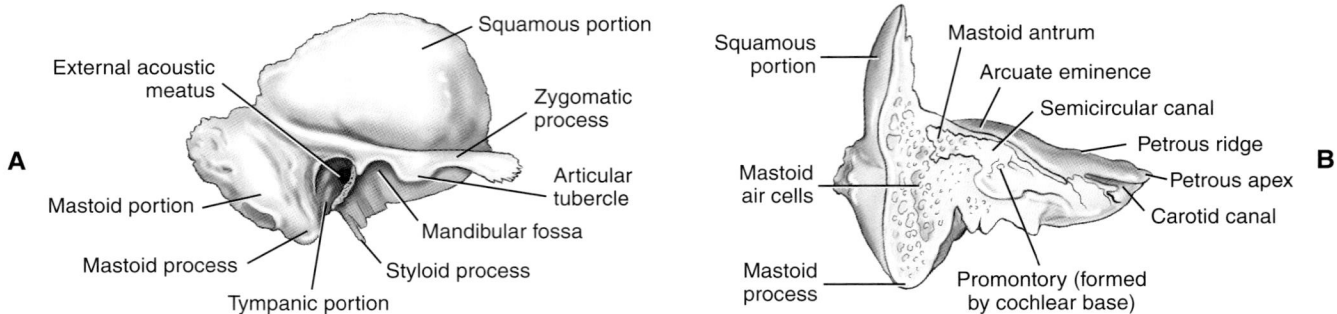

Fig. 17-3 Temporal bone. **A,** Lateral aspect; **B,** coronal section through mastoid and petrous portions.

bone. The anterosuperior border forms the lateral rim of the orbit.

The mandible, or lower jaw, is the only movable bone of the face. The horizontal medial portion is called the body. Its superior margin is the alveolar process that contains the roots of the lower teeth. The prominence in the center of its lower margin is called the **mental protuberance** or mental point and is a common positioning landmark. Extending superiorly from each end of the mandibular body is a large vertical projection called mandibular ramus (pl. rami). The right angle formed by the contour of the inferior posterior ramus is called the angle of the mandible, or **gonion,** and is a common positioning landmark. The superior margin of the ramus forms a concave curve called the mandibular notch. On the anterior end of the notch is a pointed projection, the coronoid process. The rounded projection on the superior posterior ramus, the mandibular condyle, articulates with the mandibular fossa of the temporal bone.

Seven additional small bones that are not externally palpable make up the remainder of the bony structure of the face: the **vomer,** two **palatine bones,** two inferior **nasal conchae,** and two **lacrimal bones.**

The vomer is posterior to the acanthion at the floor of the nasal cavity. It forms the inferior portion of the nasal septum, the wall that divides the nasal cavity. The two palatine bones, together with the maxilla, form the hard palate or roof of the mouth. Three pairs of nasal conchae—superior, middle, and inferior—are thin, curved bony projections that divide the nasal cavity, forming air passages lined with mucous membrane. The superior and middle conchae are projections of the ethmoid bone. The inferior conchae are separate bones that articulate with the maxilla on either side. The lacrimal bones are small thin bones that form a portion of the medial wall of each orbit.

Paranasal Sinuses

The paranasal sinuses are air-filled cavities within the ethmoid, frontal, and sphenoid bones and within the maxilla (Fig. 17-5). They serve as resonating chambers for the voice and help to warm and moisten inhaled air. The sinuses develop during childhood and are not fully formed until age 16 to 18. At maturity, passages connect the sinuses to each other and to the nasal cavity.

The maxillary sinuses are also called the maxillary antra or the antra of Highmore. They are the largest paranasal sinuses and are located within the body of the maxilla on either side of the nasal cavity.

The frontal sinuses are the second largest paranasal sinuses and are located in the anterior frontal bone, superior to the nasal cavity. They are divided by a central septum into right and left compartments and are

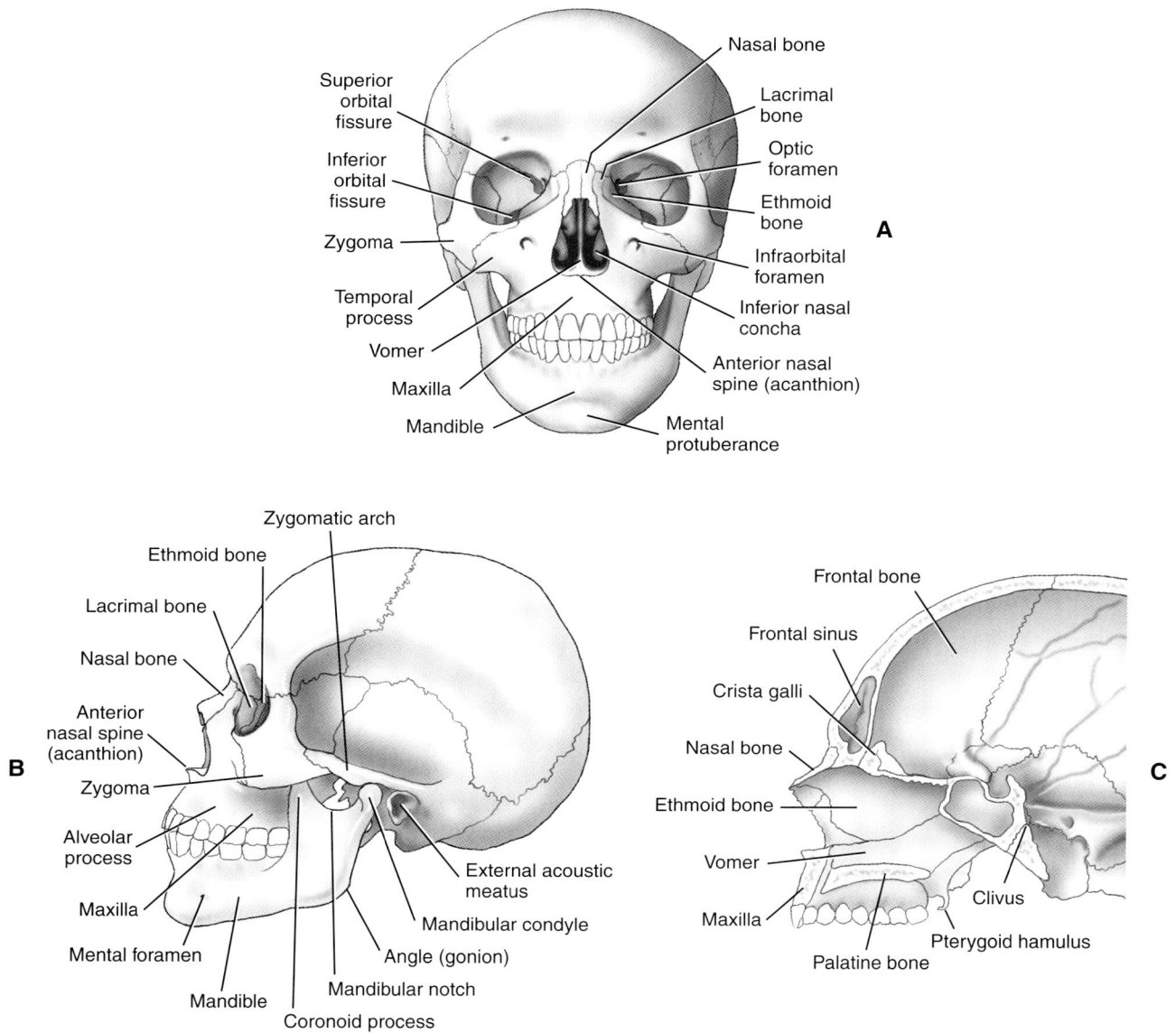

Fig. 17-4 Facial bones. **A,** Anterior aspect; **B,** lateral aspect; **C,** interior of facial bones, lateral aspect.

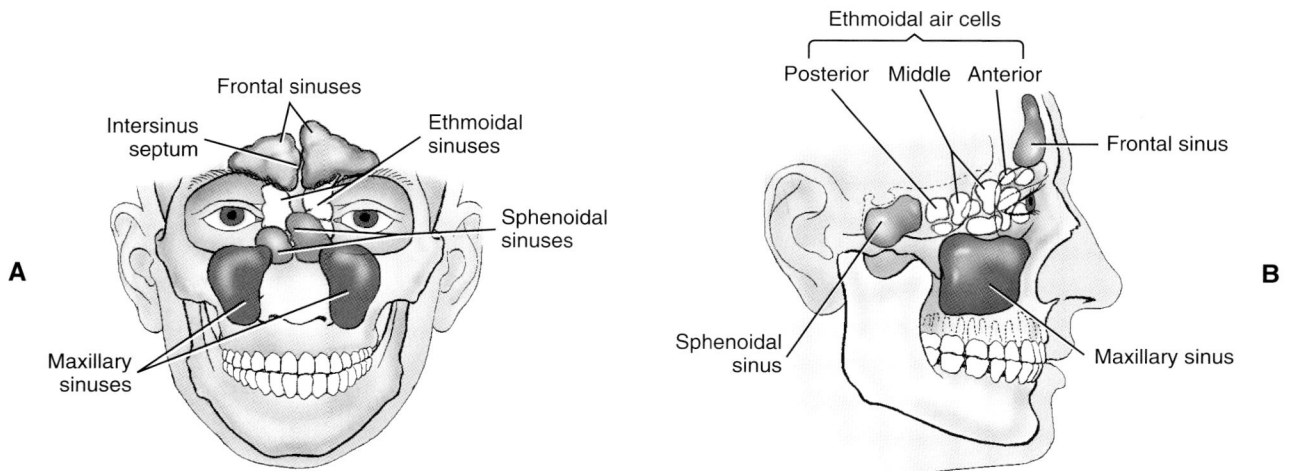

Fig. 17-5 Paranasal sinuses. **A,** Anterior aspect; **B,** lateral aspect.

usually further subdivided. They are not symmetrical and may vary greatly in size and shape. Absence of frontal sinuses is a normal variant.

The sphenoidal sinuses occupy most of the body of the sphenoid bone and are located immediately inferior to the sella turcica. Although they normally occur as a pair of chambers, it is not unusual for only a single chamber to be present.

The two ethmoidal sinuses are located within the lateral masses of the ethmoid bone and consist of a varying number of small air cells. They are situated between and behind the orbits and anterior to the sphenoid sinuses.

POSITIONING AND RADIOGRAPHIC EXAMINATIONS

Fig. 17-6 illustrates the landmarks used for radiographic positioning of the cranium, facial bones, and paranasal sinuses. Note the locations of land-

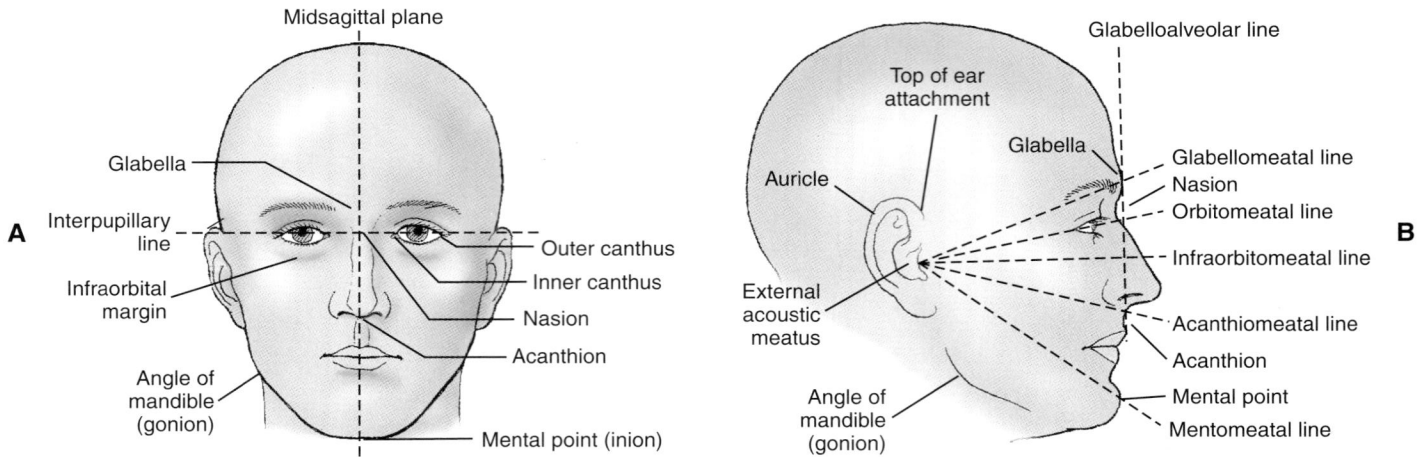

Fig. 17-6 Common landmarks for positioning of the cranium and facial bones. **A,** Anterior aspect; **B,** lateral aspect.

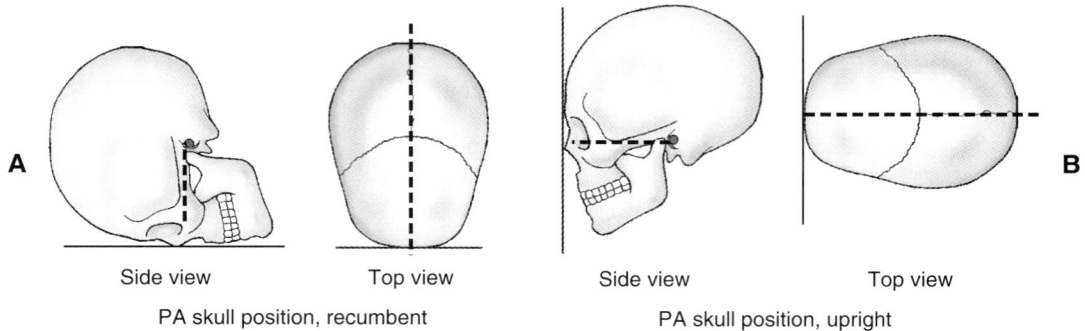

Fig. 17-7 PA skull position. **A,** Recumbent. *Left,* lateral aspect; *right,* superior aspect. **B,** Upright. *Left,* lateral aspect; *right,* superior aspect.

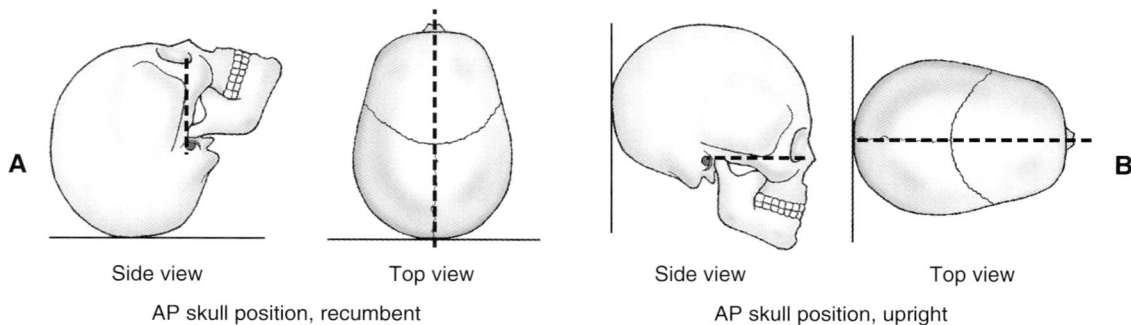

Fig. 17-8 AP skull position. **A,** Recumbent. *Left,* lateral aspect; *right,* superior aspect. Lateral aspect (left). Superior aspect (right). **B,** Upright. *Left,* lateral aspect; *right,* superior aspect.

marks mentioned previously in this chapter: glabella, acanthion, gonion, and mental point. Another significant landmark is the **nasion,** the anterior depression in the midline of the skull between the orbits. Note also the positioning lines used to judge the degree of flexion/extension of the neck that is appropriate for skull positions. Those lines that are used in this text are the orbitomeatal line (OML), the infraorbitomeatal line, and the mentomeatal line.

Dozens of positions have been described in the literature for demonstration of various aspects of the skull. It is possible, however, to demonstrate most of the structures of the cranium and face using only five basic positions: PA (Fig. 17-7), AP (Fig. 17-8), lateral (Fig. 17-9), submentovertical (Fig. 17-10), and parietoacanthial (Fig. 17-11). Variations in the direction and location of the central ray are used, together with adjustments in film size and exposure factors, to modify the images provided with these five positions.

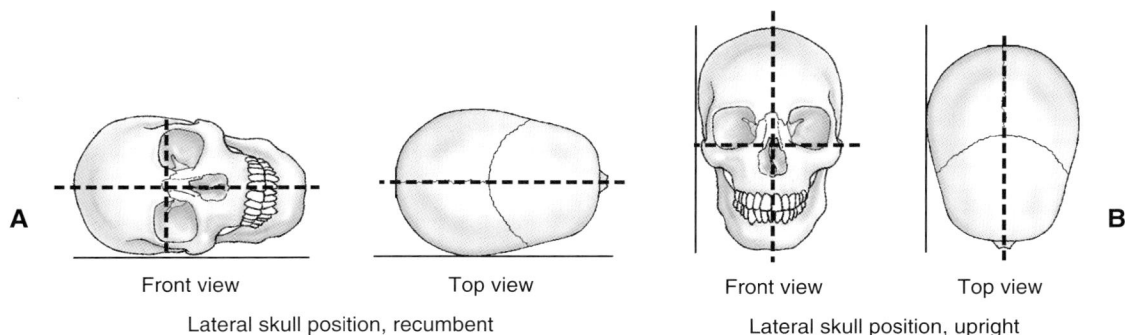

Front view   Top view

Lateral skull position, recumbent

Front view   Top view

Lateral skull position, upright

Fig. 17-9 Lateral skull position. **A,** Recumbent. *Left,* lateral aspect; *right,* superior aspect. **B,** Upright. *Left,* lateral aspect; *right,* superior aspect.

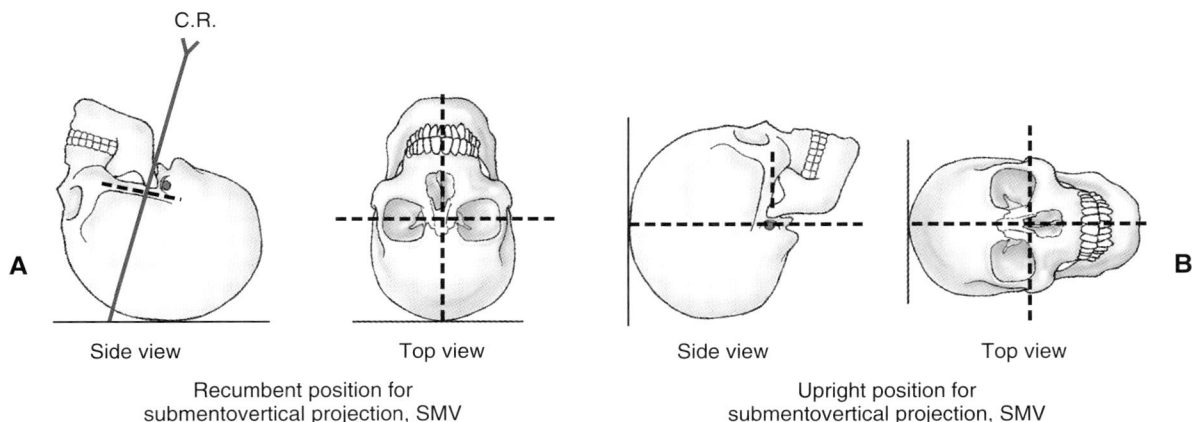

C.R.

Side view   Top view

Recumbent position for submentovertical projection, SMV

Side view   Top view

Upright position for submentovertical projection, SMV

Fig. 17-10 Submentovertical (SMV) skull position. **A,** Recumbent. *Left,* lateral aspect; *right,* superior aspect. **B,** Upright. *Left,* lateral aspect; *right,* superior aspect.

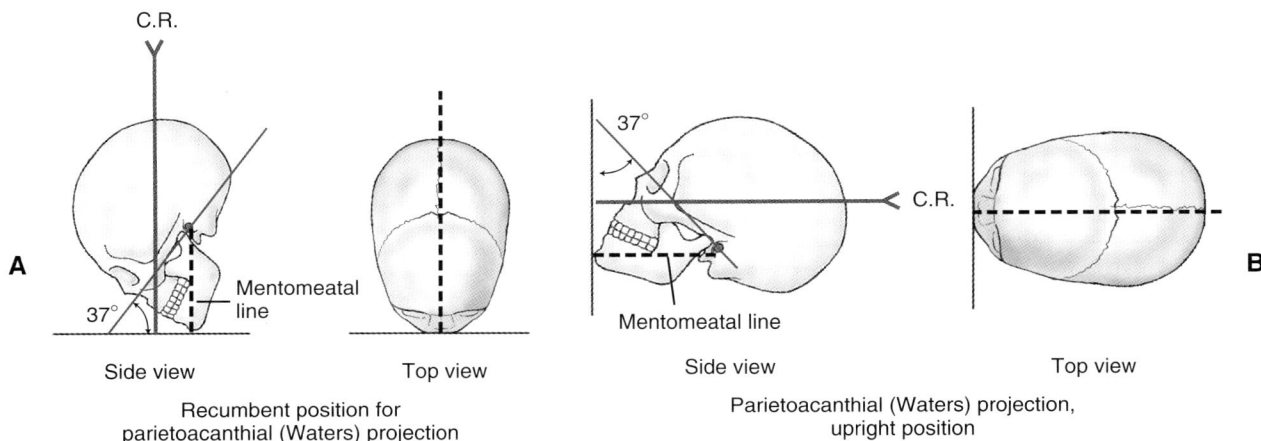

C.R.

37°   Mentomeatal line

Side view   Top view

Recumbent position for parietoacanthial (Waters) projection

37°   C.R.

Mentomeatal line

Side view   Top view

Parietoacanthial (Waters) projection, upright position

Fig. 17-11 Parietoacanthial (Waters) skull position. **A,** Recumbent. *Left,* lateral aspect; *right,* superior aspect. **B,** Upright. *Left,* lateral aspect; *right,* superior aspect.

Accurate positioning for examinations of the head requires precise attention to all three body planes: coronal, sagittal, and transverse. The coronal and sagittal planes are adjusted by body position and rotation of the head. The transverse plane alignment depends on the flexion or extension of the neck. To determine the alignment between body planes or positioning lines and the film plane, it is helpful to use a tool called an Angligner® (Fig. 17-12). In the absence of such a tool, a simple protractor or triangles cut from cardboard provide a useful guide. The cardboard liners from film boxes are suitable for this purpose.

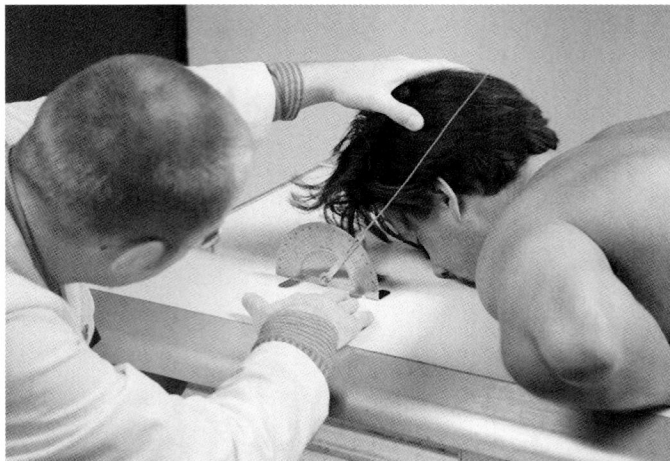

Fig. 17-12 Angligner® assists in positioning skull with baselines correctly aligned to film.

Depending on the preferences of the physician, the basic examination of the cranium may include either the PA projection or the PA axial (Caldwell) projection. Since the side nearest the film is best demon-strated on the lateral projection, some physicians prefer a basic series that includes both right and left lateral projections.

Basic Examination

The basic examination of the cranium includes the PA and PA axial (Caldwell), AP axial (Towne), and lateral views.

PA AND PA AXIAL (CALDWELL)

Film: 10 × 12 inch (24 × 30 cm), lengthwise

Screens: Rapid

Grid: Yes

SID: 40 inches

Body position:
Prone or seated facing upright bucky

> **TIP:** When the prone patient crosses the arms under the chest, this elevation of the thorax allows greater flexion of the neck. This facilitates accurate positioning and relieves pressure on the nose.

Part position:
Sagittal plane of skull is perpendicular to center of film with forehead and nose resting on table or against upright bucky. Orbitomeatal line (OML) is perpendicular to film.

Central ray:

PA: Perpendicular to center of film through nasion

PA axial (Caldwell): Angled 15° caudad to center of film through nasion

Patient instructions:
Do not move. Stop breathing.

Structures seen:
Frontal bone and outer contours of cranium from frontal perspective. When perpendicular central ray is used, petrous pyramids are projected within orbits. Using a 15°-caudad angle with Caldwell method, petrous pyramids are projected over inferior orbital rims and orbital margins are more clearly demonstrated.

Position variation:
A modification of the Caldwell method using a 30°-caudad angle projects petrous pyramids below orbital rims. This projection is sometimes used as part of the preliminary evaluation for patients who are to have MRI studies of the head.

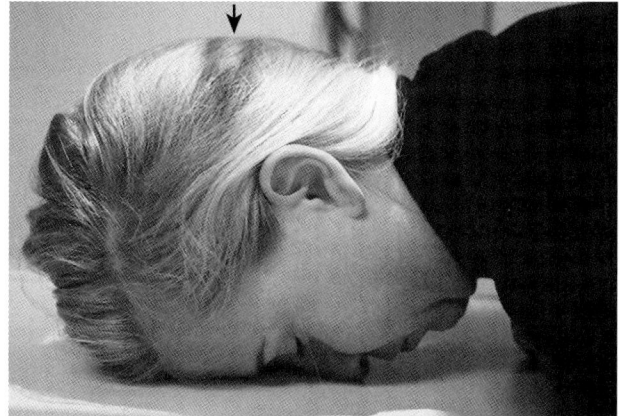

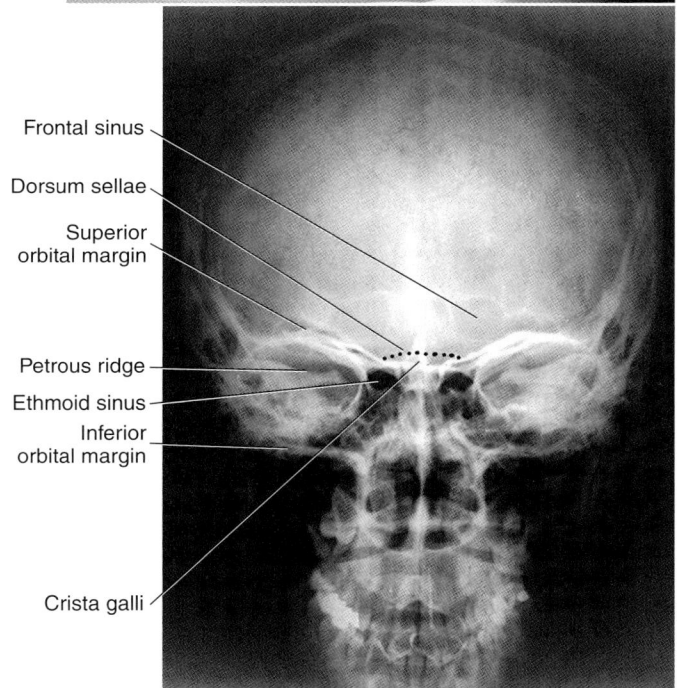

Frontal sinus
Dorsum sellae
Superior orbital margin
Petrous ridge
Ethmoid sinus
Inferior orbital margin
Crista galli

PA skull position and radiograph.

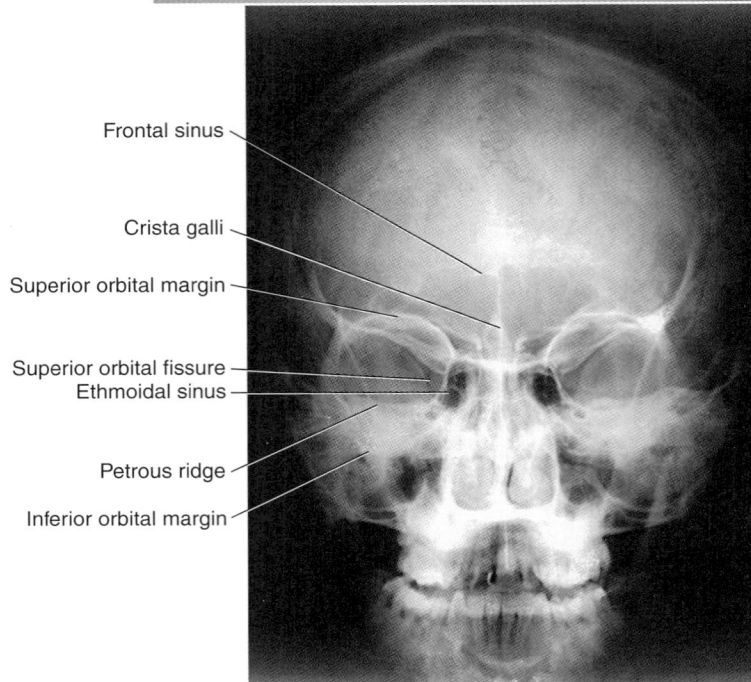

Frontal sinus

Crista galli

Superior orbital margin

Superior orbital fissure
Ethmoidal sinus

Petrous ridge

Inferior orbital margin

PA axial (Caldwell) skull position and radiograph.

AP AXIAL (TOWNE)

Film: 10 × 12 inch (24 × 30 cm), lengthwise

Screens: Rapid

Grid: Yes

SID: 40 inches

Body position:
Supine or seated

Part position:
Sagittal plane of skull is perpendicular to film with back of head resting on table or against upright bucky. OML is perpendicular to film.

> **TIP:** If patient is unable to flex neck sufficiently to get OML perpendicular to film, a wedge sponge under head may assist in attaining correct position.

Central ray:
Angled 30° caudad to center of film through external occipital protuberance. Central ray enters skull in midsagittal plane at approximate location of natural hairline, approximately 2½ inches superior to glabella.

> **TIP:** If patient is unable to flex the neck sufficiently to get the OML perpendicular to the film, the central ray may be angled 30° caudad to the plane of the OML. For example, if the OML is aligned 5° off-vertical, the central ray is angled 35° caudad.

Patient instructions:
Do not move. Stop breathing.

Structures seen:
Occipital bone, posterior parietal bones, foramen magnum, and petrous portions of temporal bones

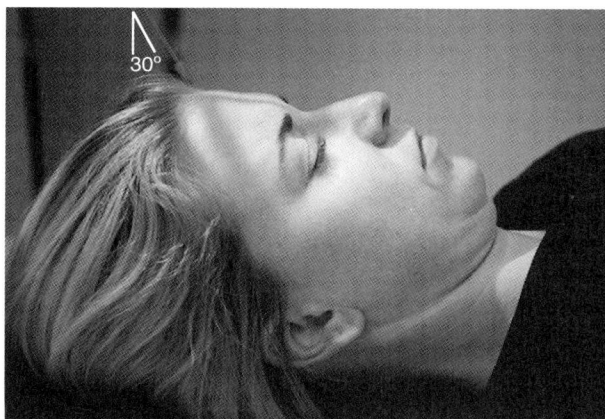

AP axial (Towne) skull position.

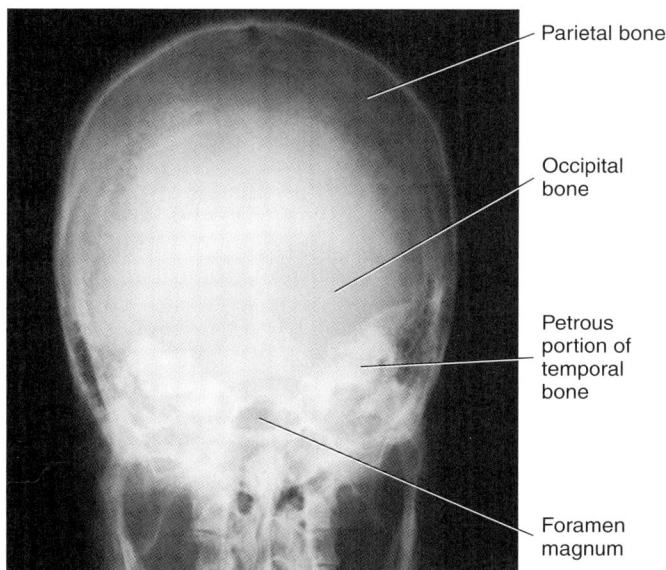

AP axial (Towne) skull radiograph.

LATERAL

Film: 10 × 12 inch (24 × 30 cm), crosswise

Screens: Rapid

Grid: Yes

SID: 40 inches

Body position:
Recumbent or seated in PA oblique body position with side of interest nearest film

Part position:
Sagittal plane of head is parallel to film and inter-pupillary line is perpendicular to it. Support under mandible may assist in maintaining this position.

Central ray:
Perpendicular to center of film through a point mid-way between external occipital protuberance and glabella. Central ray enters approximately 2 inches superior to EAM.

Patient instructions:
Do not move. Stop breathing.

Structures seen:
Lateral image of entire cranium. Sella turcica is seen in profile. There should be no rotation or tilt of cranium and paired structures should be superimposed. Side nearest film is most clearly seen.

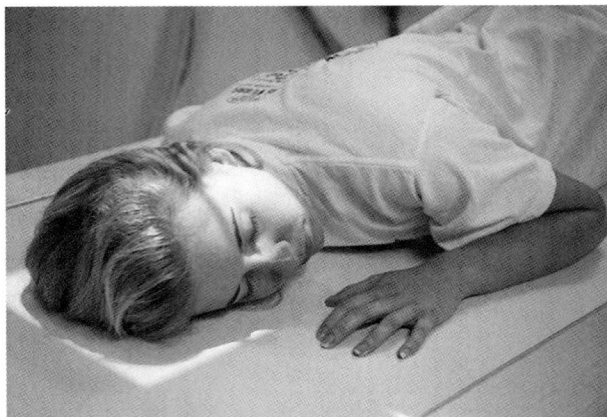

Lateral body and skull position.

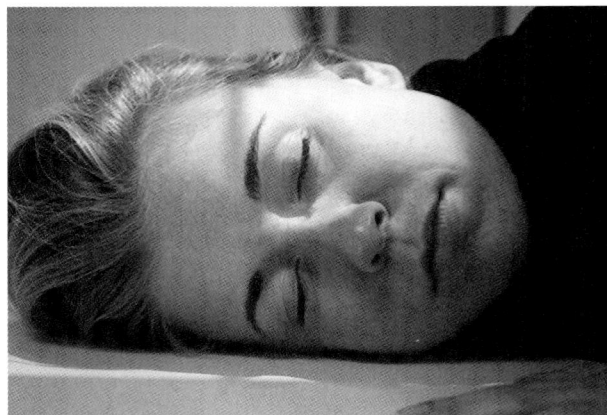

Lateral skull position.

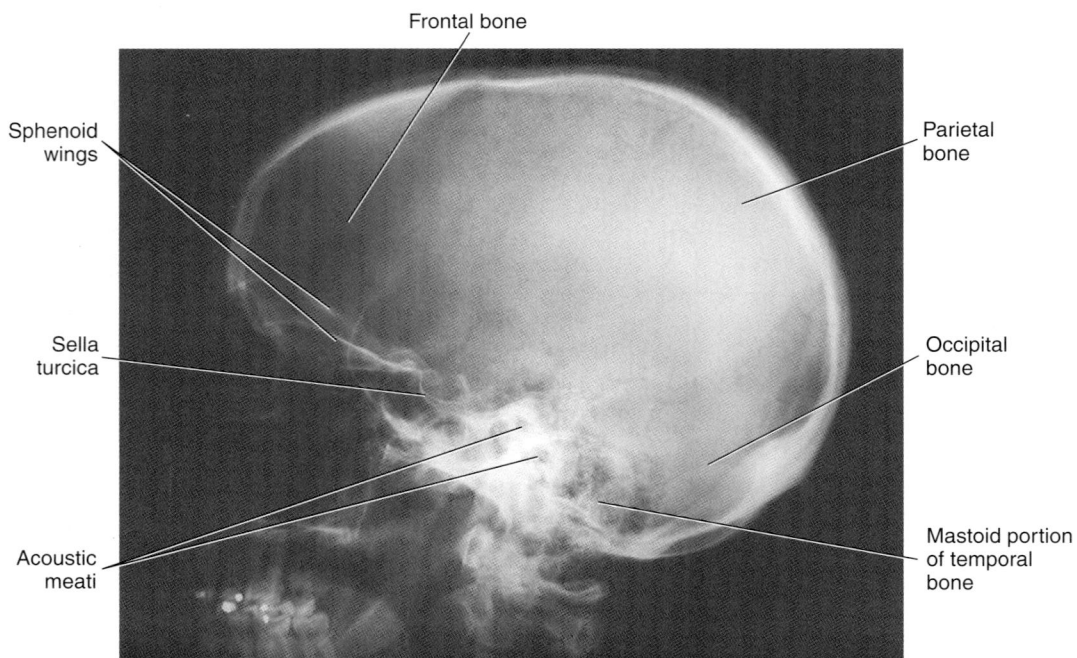

Lateral skull radiograph.

Alternative Projections

AP AND AP AXIAL (REVERSE CALDWELL)

When obesity or injury makes it difficult to position the patient prone, AP projections may be substituted for the PA and/or PA axial (Caldwell) projections. The structures demonstrated are the same, but the orbits and other anterior structures are magnified considerably when taken AP. Radiation dose to the eyes and thyroid gland is also increased as compared to the PA projections.

Film: 10 × 12 inch (24 × 30 cm), lengthwise

Screens: Rapid

Grid: Yes

SID: 40 inches

Body position:
Supine or seated

Part position:
Sagittal plane of skull is perpendicular to film with back of head resting on table or against upright bucky. OML is perpendicular to film.

> **TIP:** If the patient is unable to flex the neck sufficiently to get the OML perpendicular to the film, a wedge sponge under the head may assist in maintaining the correct position.

Central ray:
Perpendicular to center of film through nasion

> **TIP:** If the patient is unable to flex the neck sufficiently to get the OML perpendicular to the film, the central ray may be angled parallel to the OML.

Central ray alternative:
Angled 15° cephalad through nasion

> **TIP:** If patient is unable to flex the neck sufficiently to get the OML perpendicular to the film, a similar result is obtained by placing the *infraorbitomeatal line* perpendicular to the film and angling the central ray 7° cephalad.

Patient instructions:
Do not move. Stop breathing.

Structures seen:
Frontal bone and outer contours of cranium from frontal perspective. When a perpendicular central ray is used, petrous pyramids are projected within orbits. With reverse Caldwell method, using a 15°-cephalad angle, petrous pyramids are projected over inferior orbital rims and orbital margins are more clearly demonstrated. The orbits and other anterior structures are magnified in comparison to PA projections and the radiation to dose to eyes and thyroid gland is increased.

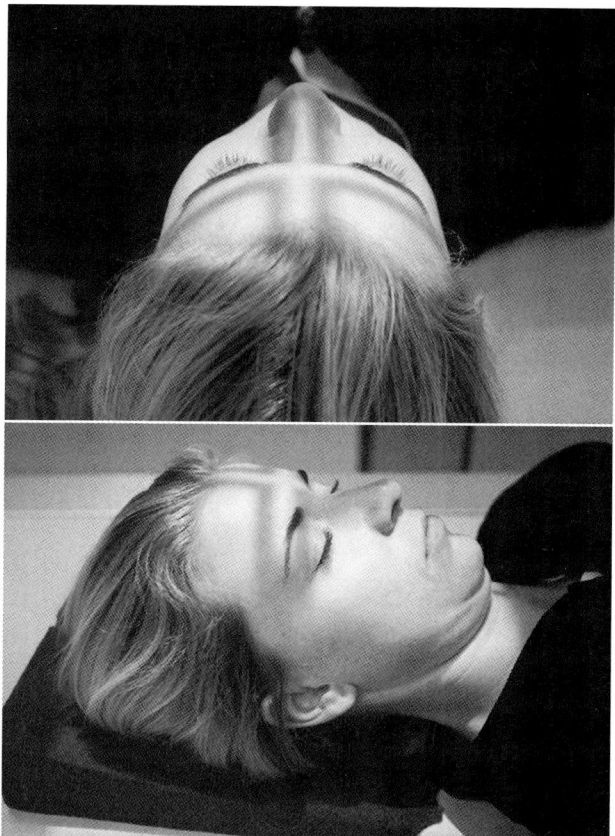

AP skull position.

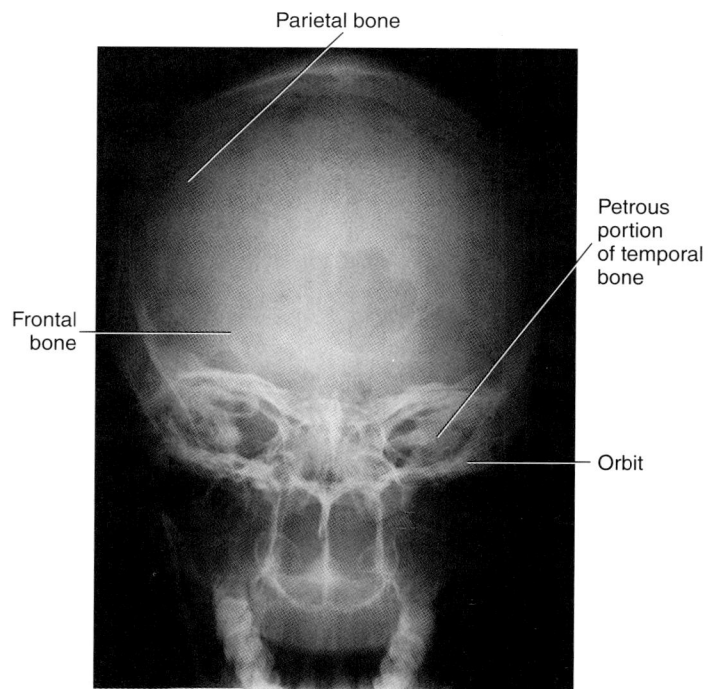

AP skull radiograph.

PA AXIAL, HAAS METHOD (REVERSE TOWNE)

This projection may be used instead of the AP axial (Towne) projection. It is useful when obesity or exaggerated kyphosis of the thoracic spine makes it difficult to assume a supine AP position with the OML perpendicular to the film or nearly so. Both the patient position and the tube angle are reversed. The resulting radiograph is similar to the AP axial (Towne) projection, but detail of the posterior structures is somewhat compromised due to the increased OID.

Film: 10 × 12 inch (24 × 30 cm), lengthwise

Screens: Rapid

Grid: Yes

SID: 40 inches

Body position:
Prone or seated facing film

Part position:
Sagittal plane of skull is perpendicular to film with forehead and nose resting on table or against upright bucky. OML is perpendicular to film.

Central ray:
Angled 30° cephalad to center of film through external occipital protuberance. Central ray exits skull in midsagittal plane at approximate location of natural hairline, 2½ inches superior to glabella.

Patient instructions:
Do not move. Stop breathing.

Structures seen:
Occipital bone, posterior parietal bones, foramen magnum, and petrous portions of temporal bones

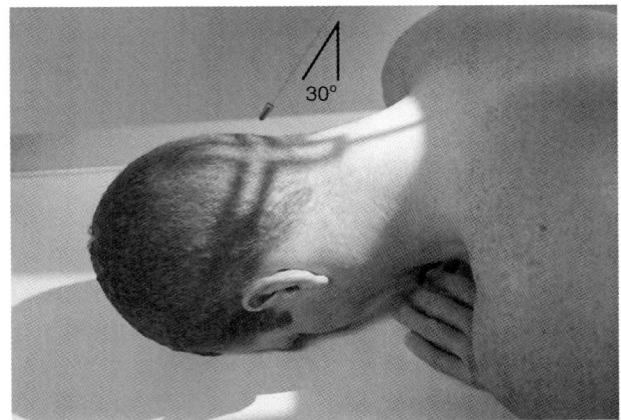

PA axial (reverse Towne) skull position.

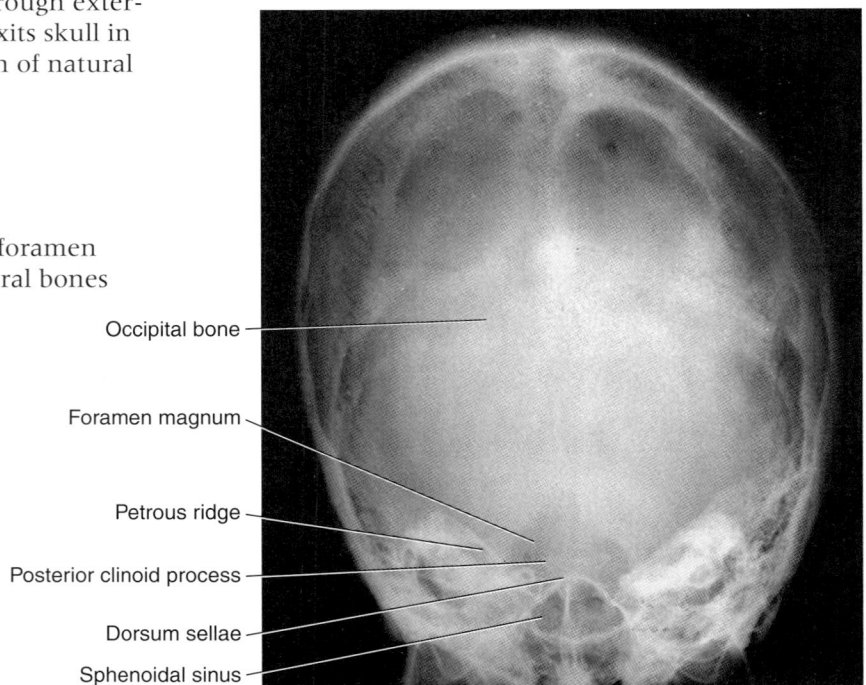

PA axial (reverse Towne) skull radiograph.

Supplemental Projection

CRANIAL BASE, SUBMENTOVERTEX (SMV)

This projection is added to the basic examination when it is desired to demonstrate the structures of the cranial base more completely than they are seen with the Towne method. It is especially helpful for demonstration of the sphenoid bone and the cranial foramina.

Film: 10 × 12 inch (24 × 30 cm), lengthwise

Screens: Rapid

Grid: Yes

SID: 40 inches

Body position:
Supine with shoulders and thorax elevated and supported, or seated with back to film and back arched to place top of head against upright bucky

Part position:
Sagittal plane of head is perpendicular to film. Neck is extended as much as possible with head resting on its vertex. OML is parallel to film or as nearly so as possible.

Central ray:
Perpendicular to OML through midline and passing through EAM

Patient instructions:
Do not move. Stop breathing.

Structures seen:
Cranial base, including occipital bone, foramen magnum, sphenoid and ethmoid bones, and petrous portions of temporal bone. Foramina ovale and spinosum are also demonstrated on this projection.

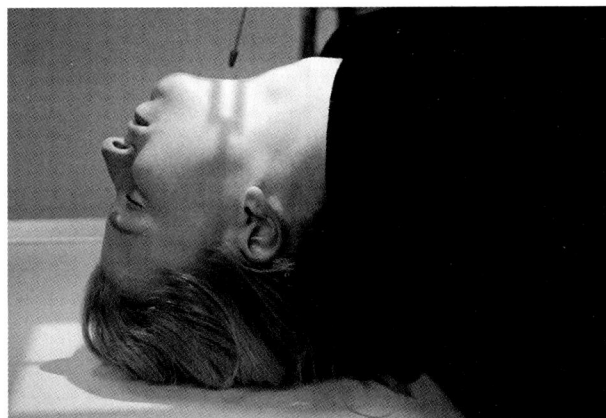

Axial (SMV) skull position.

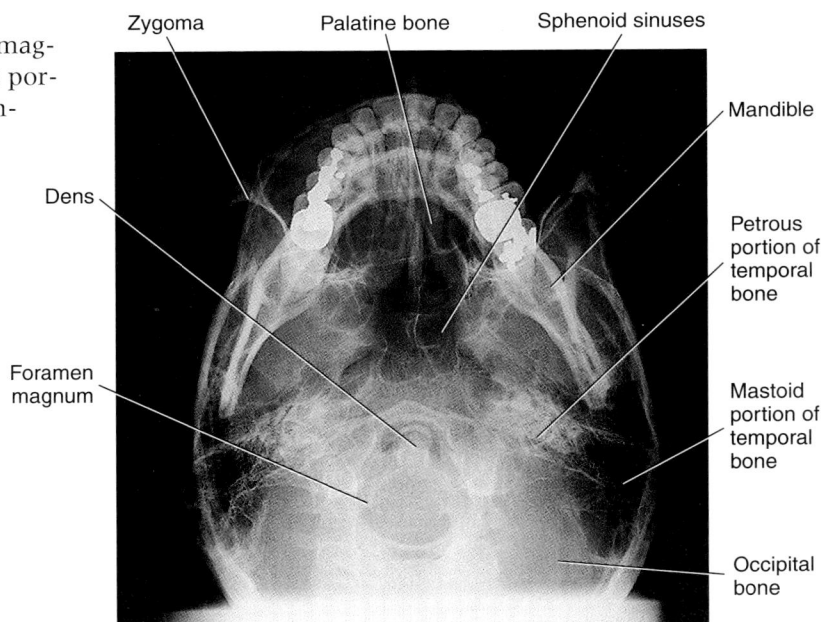

Radiograph of cranial base, submentovertical (SMV) projection.

Basic Examination

The basic examination of the facial bones includes the PA axial (Caldwell), parietoacanthial (Waters), and lateral views.

PA AXIAL (CALDWELL)

This projection is taken the same as for the Caldwell projection of the cranium described previously except that an 8 × 10-inch (18 × 24-cm) film is used with a smaller radiation field. It demonstrates the orbits and the zygomas.

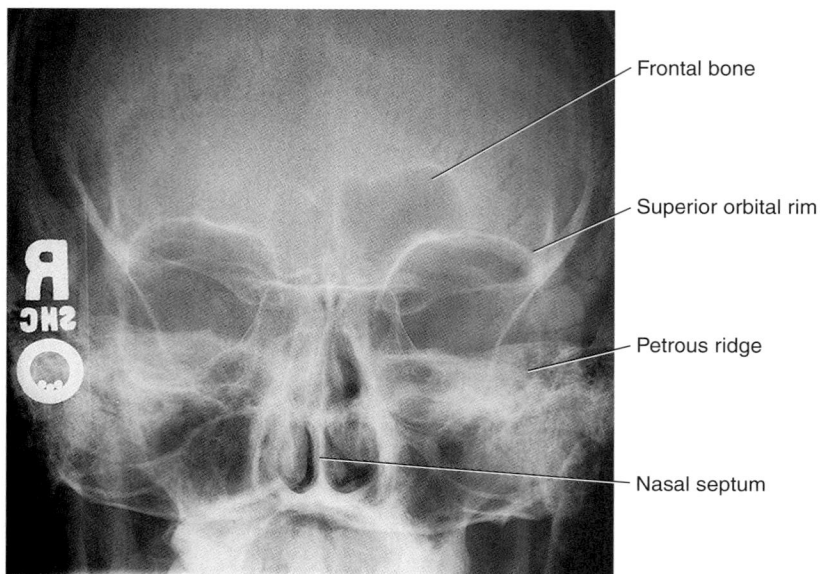

PA axial (Caldwell) radiograph of facial bones.

PARIETOACANTHIAL (WATERS)

Film: 8 × 10 inch (18 × 24 cm), lengthwise

Screens: Rapid

Grid: Yes

SID: 40 inches

Body position:
Standing, seated, or recumbent, facing film

Part position:
Neck is extended with chin resting on table or upright bucky. Mentomeatal line is perpendicular to film and infraorbitomeatal line forms 37° angle to film. Sagittal plane is perpendicular to film.

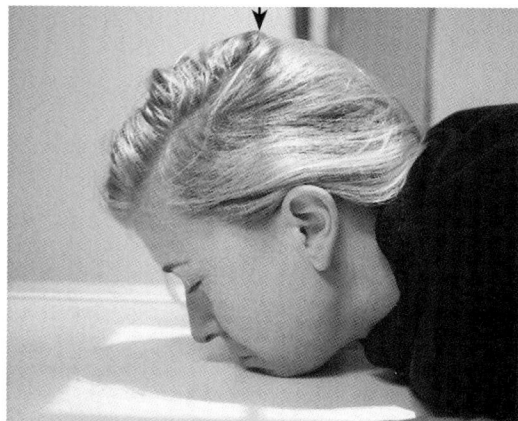

Position for parietoacanthial (Waters) projection of facial bones.

TIP: When patient is correctly positioned, the distance between table or upright bucky and the tip of the nose should be about the width of your thumb.

Central ray:

Perpendicular to center of film through acanthion

Patient instructions:

Do not move. Stop breathing.

Structures seen:

Maxilla, orbits, and nasal septum

TIP: The petrous portion of the temporal bone should be projected beneath the maxillary sinuses. If it is superimposed over the floor of the sinuses, the film should be repeated with a greater degree of neck extension.

Position variation:

A common modification of the parietoacanthial projection is taken with the neck flexed somewhat so that the OML is aligned at an angle of 55° to the film. This projection is sometimes used for demonstration of the orbits and zygomas, both of which are less distorted with this method than on the classic Waters view.

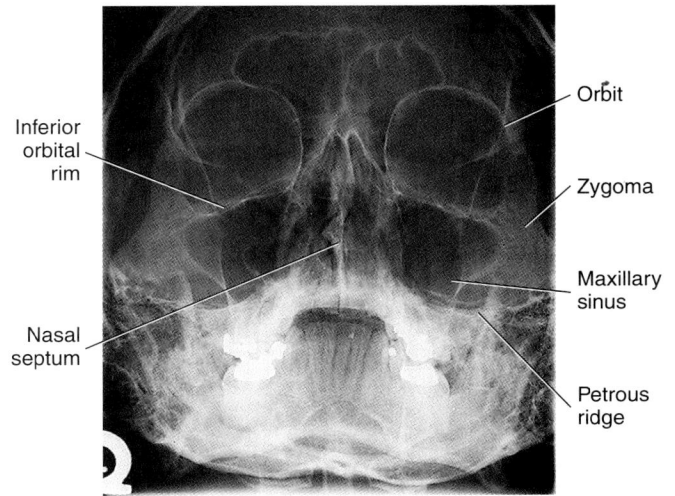

Parietoacanthial (Waters) projection of facial bones.

LATERAL

This projection is taken exactly the same as for the lateral projection of the cranium described previously except that an 8 × 10-inch (18 × 24-cm) film is used with a smaller radiation field and the center point is at the lateral orbital margin.

Lateral position for facial bones.

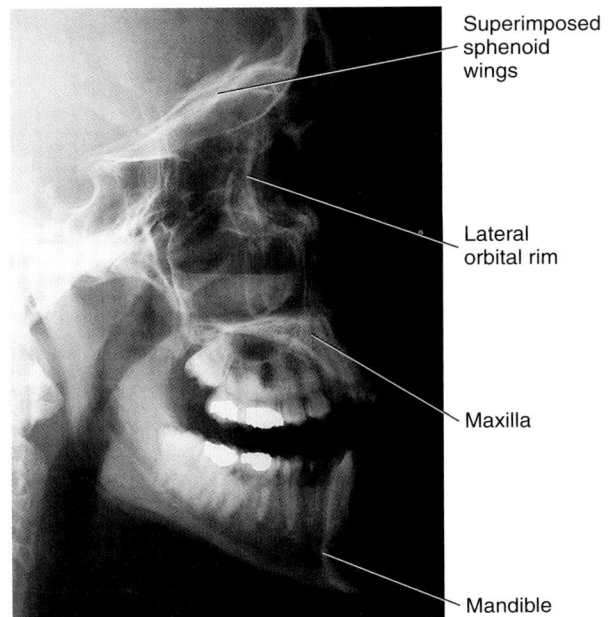

Lateral projection of facial bones.

Supplemental Projections

AXIAL PROJECTIONS OF ZYGOMATIC ARCHES

The zygomatic arches are seen on the submentovertical projection of the skull, positioned as for this projection of the cranial base. The exposure is reduced so as not to overpenetrate the zygomatic arches.

Alternatively, the position may be reversed to provide a verticosubmental (VSM) projection. With the patient prone or facing the upright bucky, the patient's neck is extended and the chin rests on the film, similar to the position for the parietoacanthial (Waters) projection. The central ray is angled caudad, perpendicular to the OML and passing through its center.

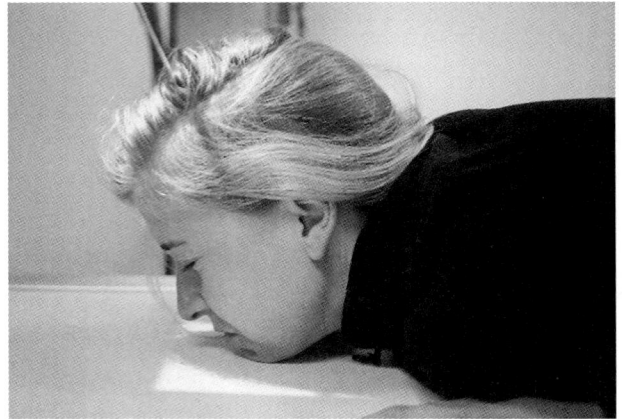

Position for axial verticosubmental (VSM) projection of zygomatic arches.

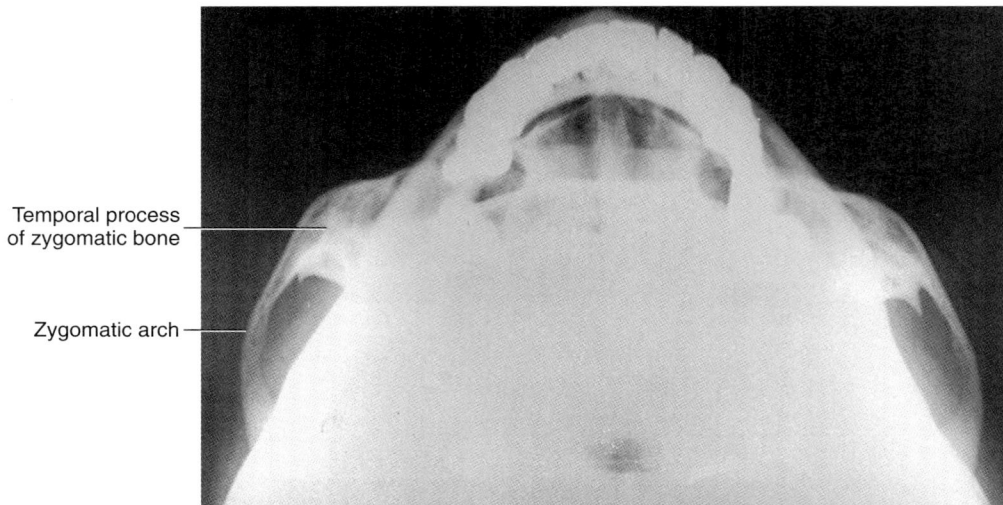

Temporal process of zygomatic bone

Zygomatic arch

Axial (VSM) projection of zygomatic arches.

BASIC EXAMINATION OF NASAL BONES

This examination is the same as the basic examination of the facial bones except that the lateral projection is specific to the nasal bones. The PA axial (Caldwell) projection demonstrates the orbital region surrounding the nasal bones and provides a frontal view of the nasal bones and nasal septum. The parietoacanthial (Waters) projection provides a semiaxial projection of the nasal cavity, nasal septum, and surrounding structures.

The lateral projection of the nasal bones is taken using an 8 × 10-inch (18 × 24-cm) detail (extremity) cassette on the tabletop (non-bucky). The body and part position are the same as for other lateral projections of the skull and face, and the central ray is aligned to the midpoint of the nasal bones. Close collimation is desirable, but the field should include the region of the acanthion to also demonstrate the anterior nasal spine of the maxilla.

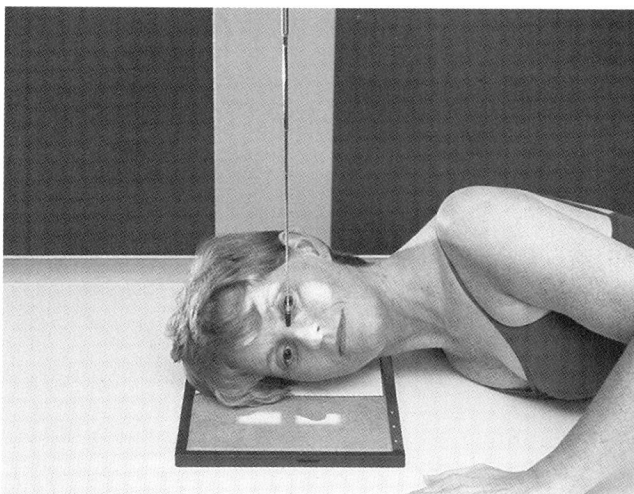

Lateral position for nasal bones.

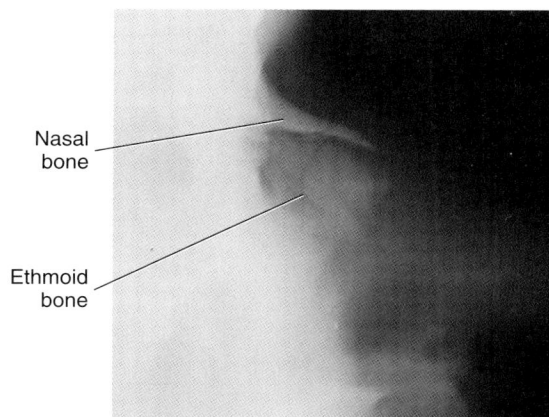

Lateral projection of nasal bones.

PA PROJECTION OF MANDIBLE

Film: 8 × 10 inch (18 × 24 cm), lengthwise

Screens: Rapid

Grid: Yes

SID: 40 inches

Body position:
Prone or seated facing film

Part position:
Sagittal plane of skull is perpendicular to center of film with forehead and nose resting on table or against upright bucky. OML is perpendicular to film.

Central ray:
Perpendicular to center of film through acanthion

Patient instructions:
Do not move. Stop breathing.

Structures seen:
Entire mandible. Rami are symmetrical.

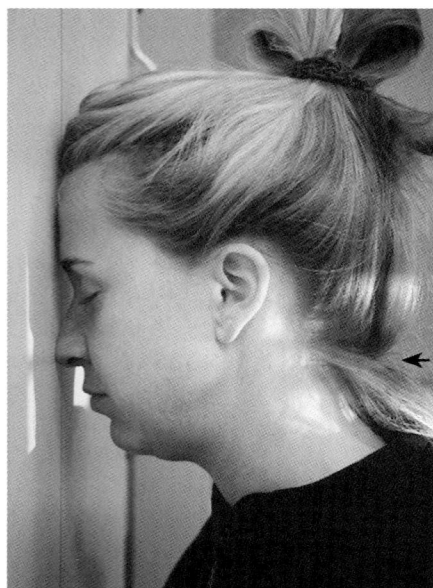

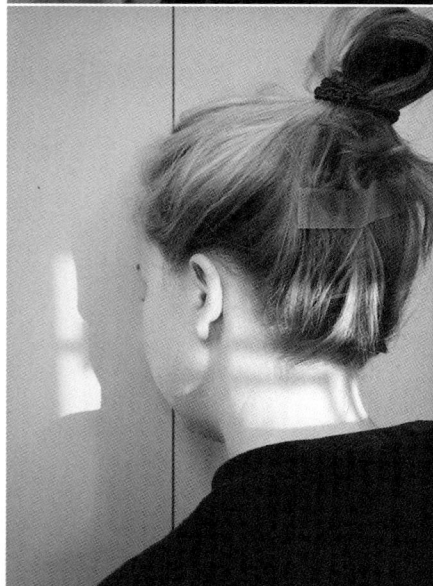

PA mandible position, upright.

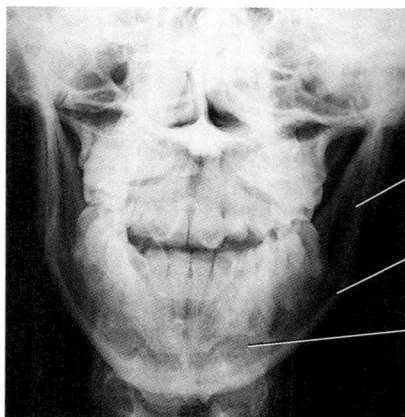

PA projection of mandible.

Mandibular ramus

Angle of mandible

Body of mandible

AXIOLATERAL PROJECTION OF MANDIBLE

Film: 8 × 10 inch (18 × 24 cm), lengthwise

Screens: Rapid

Grid: Yes

SID: 40 inches

Body position:
Seated, facing film with coronal plane of body somewhat oblique, as for lateral projection

Part position:
Neck is extended somewhat and flexed laterally so
that sagittal plane of skull forms 15° angle to film

Central ray:
Angled 15° cephalad through the mandible

Patient instructions:
Do not move. Stop breathing.

Structures seen:
Mandibular ramus and portion of body nearest film.
Mandibular condyle should be well demonstrated.

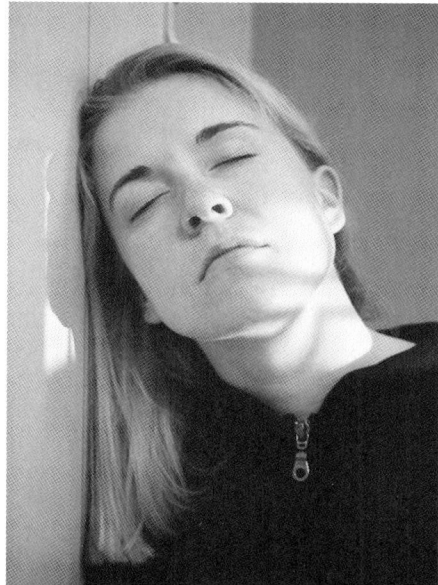

Axiolateral position for mandible.

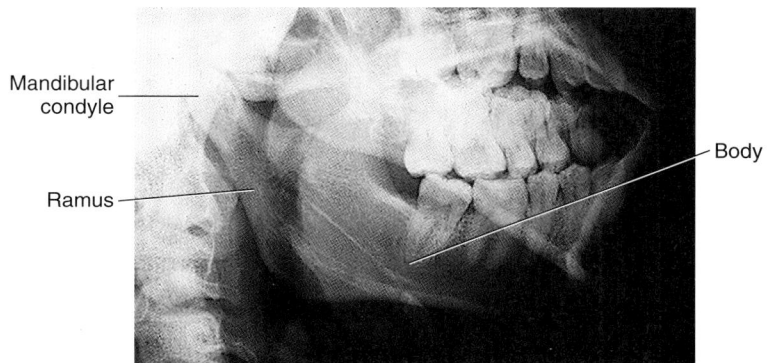

Axiolateral mandible radiograph.

AP AXIAL PROJECTION OF MANDIBLE

The AP axial (Towne) projection may be used to demonstrate the mandibular condyles and temporomandibular joints. The procedure is the same as for the AP axial (Towne) projection of the cranium except that an 8 × 10-inch (16 × 24-cm) film is placed crosswise and the central ray is aligned so as to pass through the mandibular rami.

To study the temporomandibular joints, this projection may be taken in both open-mouth and closed-mouth positions. For the closed-mouth position, the posterior teeth are in contact. For the open-mouth position, the jaw is opened wide, taking care that it stays in the same plane, that is, the mandible does not protrude anteriorly.

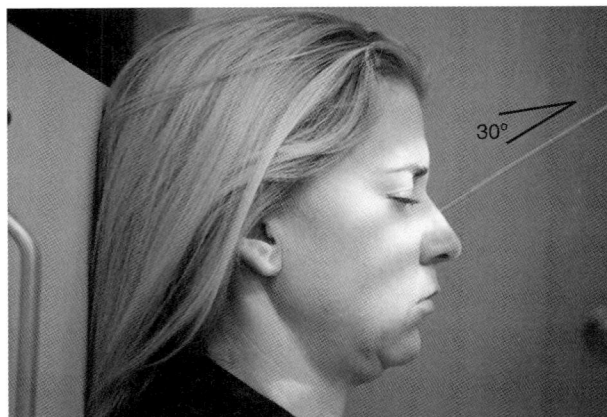

Position for AP axial projection of mandible.

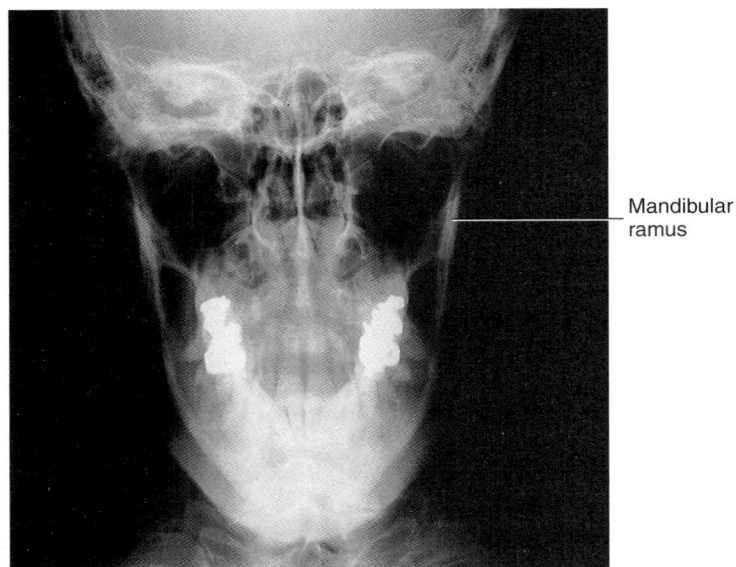

AP axial view of mandible.

Basic Examination

The basic examination of the paranasal sinuses includes the PA axial (Caldwell), parietoacanthial (Waters), axial (SMV or VSM), and lateral views.

A routine examination of the paranasal sinuses consists of four projections, all of which have been introduced in the positioning for the cranium and facial bones. *For sinuses, it is important that all views be done upright to demonstrate air-fluid levels if they are present.*

The PA axial (Caldwell) projection is the same as this projection of the facial bones in the upright position. It demonstrates the frontal and ethmoid sinuses.

The parietoacanthial (Waters) projection is the same as this projection of the facial bones in the upright position. It demonstrates the maxillary and ethmoid sinuses.

Upright position for PA axial (Caldwell) projection of paranasal sinuses.

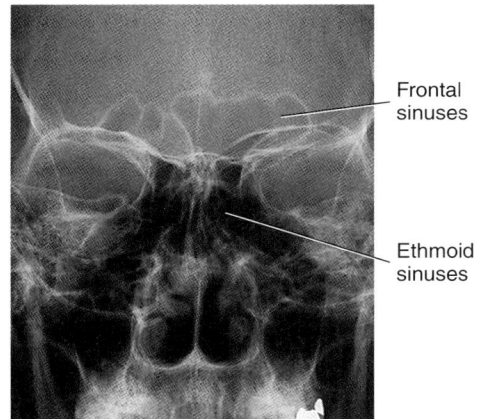

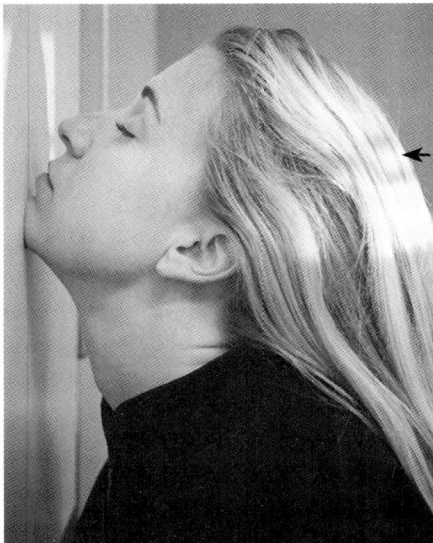

Frontal sinuses

Ethmoid sinuses

PA axial (Caldwell) projection of paranasal sinuses.

Upright position for parietoacanthial (Waters) projection of paranasal sinuses.

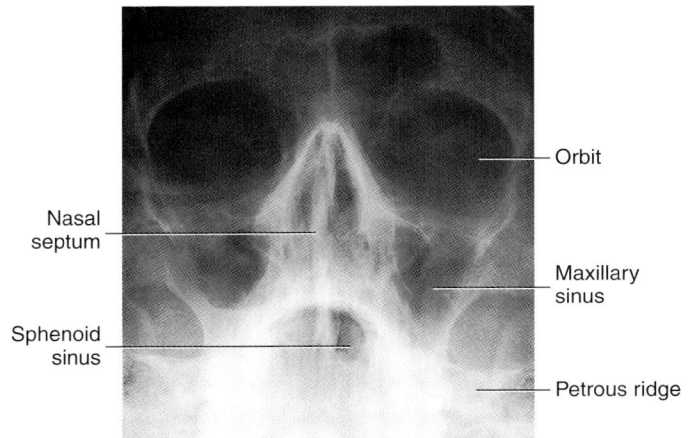

Orbit

Nasal septum

Maxillary sinus

Sphenoid sinus

Petrous ridge

Parietoacanthial (Waters) projection of paranasal sinuses.

The lateral projection is the same as the lateral projection of the facial bones in the upright position. Collimation is restricted anteriorly just beyond the glabella and posterior to the EAM. The lateral projection demonstrates all of the paranasal sinuses, with right and left chambers superimposed on each other.

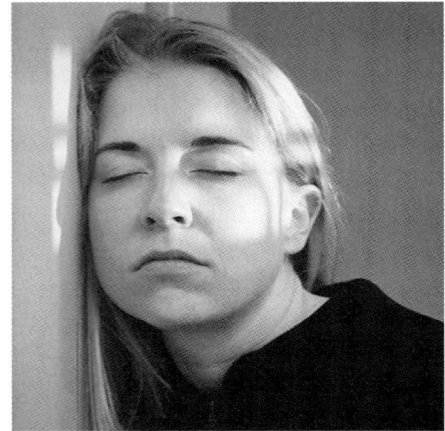

Upright lateral position for paranasal sinuses.

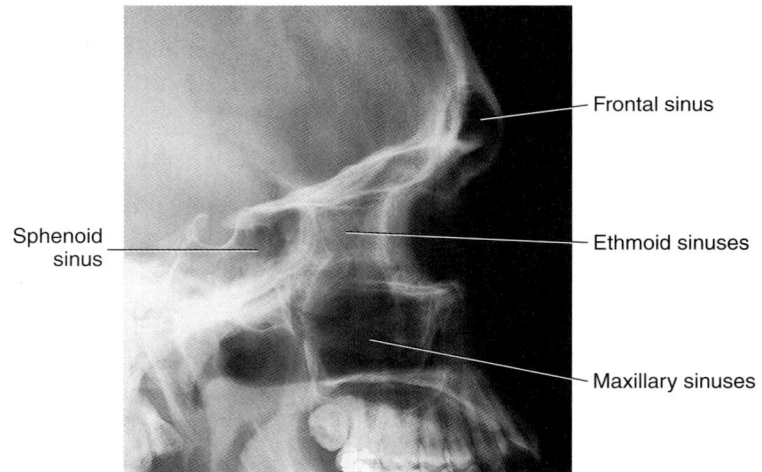

Lateral projection of paranasal sinuses.

The axial projection for the sinuses is taken upright in the same position as for either the submentovertical projection of the cranial base or the verticosubmental projection of the zygomatic arches. It demonstrates the sphenoid and ethmoid sinuses. The exposure is the same as that used for the cranial base.

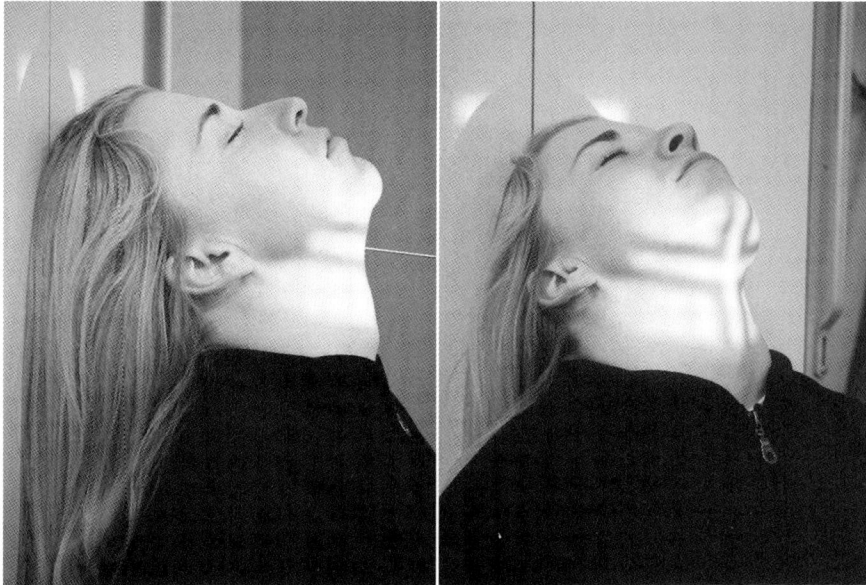

Upright axial (SMV) position for paranasal sinuses.

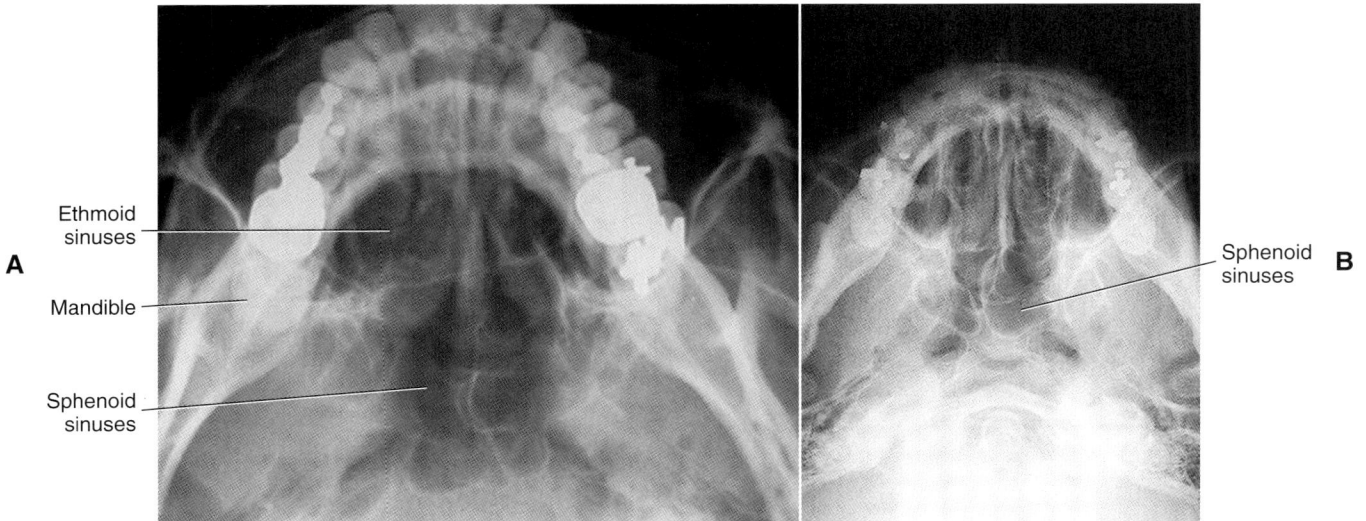

Ethmoid sinuses

A

Mandible

Sphenoid sinuses

Sphenoid sinuses

B

A, Axial projection (SMV) of paranasal sinuses. **B,** Axial projection (VSM) of paranasal sinuses.

PATHOLOGY

Trauma

A blow to the head that causes brief unconsciousness or disorientation, sometimes described as "seeing stars," is called a **cerebral concussion.** A severe blow to the head may cause the brain to move within the cranium, causing brain injury on the side opposite the location of the blow. This is called a **contre-coup injury.**

Cranial fractures may or may not represent serious injuries. Their principal significance is that they indicate sufficient trauma to cause potentially serious injury to the brain. Fig. 17-13 illustrates a linear fracture of the frontal bone caused by a head-on fall onto a hard surface. Fig. 17-14 is a depressed skull fracture, a type that occurs with a blow by an object, in this case, a baseball bat. Fractures of the cranial base are more difficult to see radiographically. They are sometimes diagnosed by the presence of air-fluid levels in the sphenoid sinuses, resulting from leakage of blood and/or spinal fluid into the sinus cavity. For this reason, when basal skull fractures are suspected, upright lateral or AP decubitus views of the skull may be requested.

The most common facial fracture is that of the nasal bones. The typical fracture caused by a blow to the nose is a transverse fracture of both nasal bones with depression of the fragments (Fig. 17-15).

The weakest areas of the orbit are the medial and inferior walls, posterior to the orbital rims. A blow to the eye may cause sufficient pressure to fracture the fragile bones of the orbit. Fracture of the orbital floor is called a **blow-out fracture** (Fig. 17-16). Bone fragments and soft tissues of the eye are forced into the maxillary sinus.

Blows to the side of the face may fracture the zygomatic arches (Fig. 17-17), and a direct blow near the center of the face sometimes causes multiple fractures (Fig. 17-18). Fractures of the mandible may

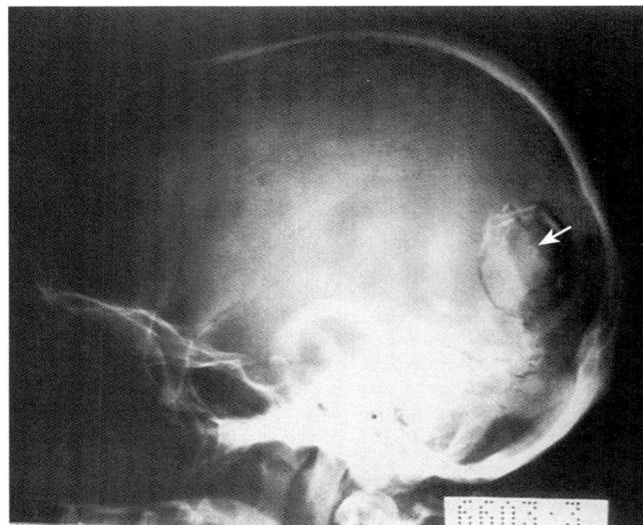

Fig. 17-14 Depressed skull fracture *(arrow).*

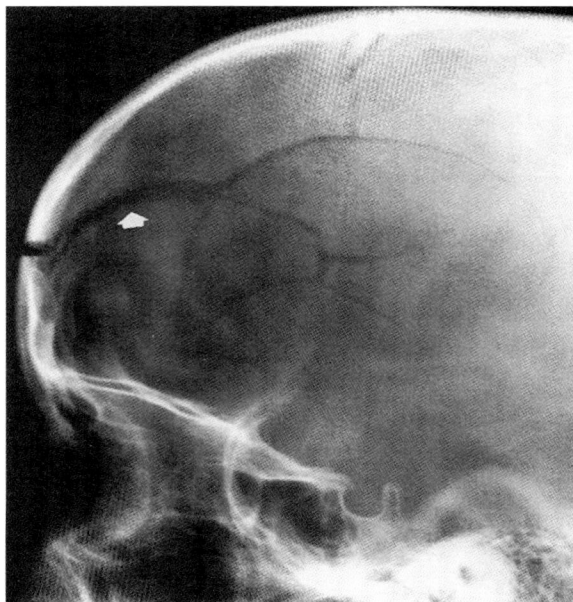

Fig. 17-13 Linear skull fracture *(arrow)* with separation, branching posteriorly.

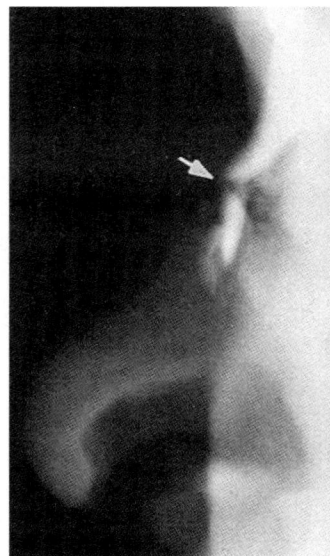

Fig. 17-15 Depressed fracture of nasal bones *(solid arrow).* Note demonstration of nasal spine.

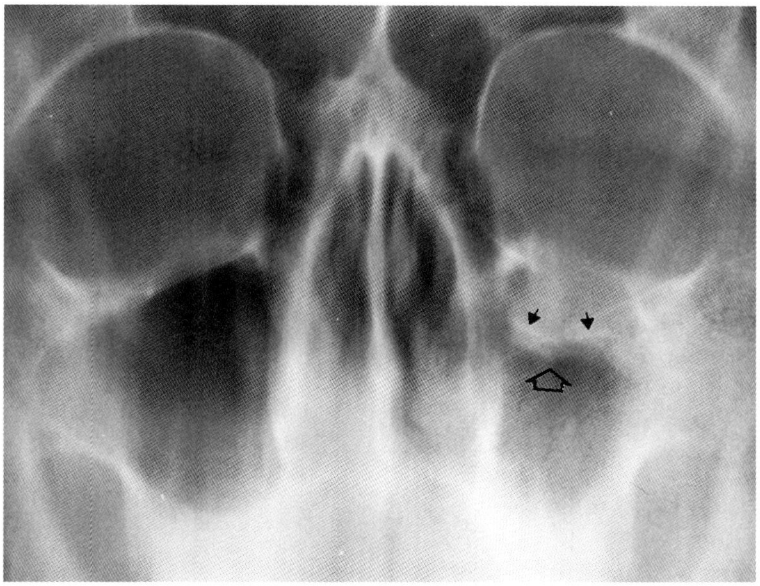

Fig. 17-16 Blow-out fracture of left orbit. Note soft tissue shadow in superior portion of left maxillary sinus *(solid arrow)*. The air in the orbit *(hollow arrow)* escaped from the sinus and is trapped in the orbit.

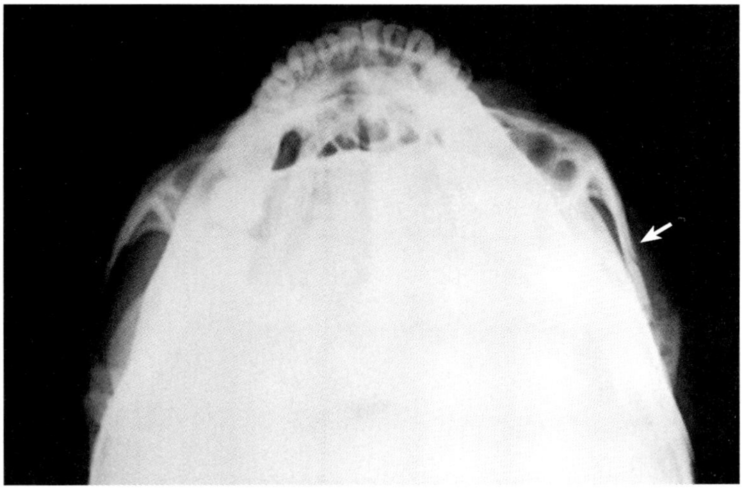

Fig. 17-17 Fracture with depression of left zygomatic arch.

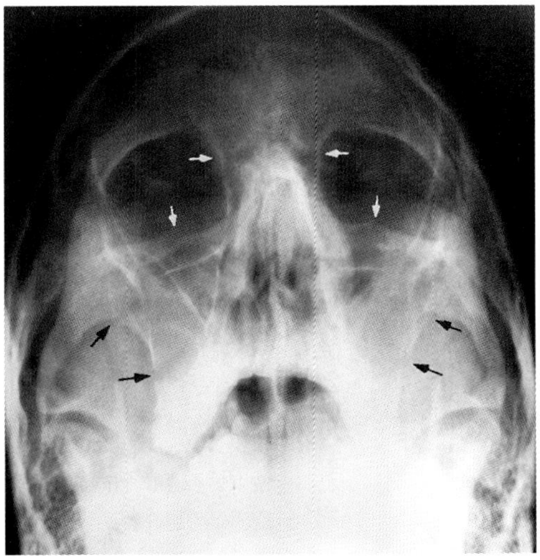

Fig. 17-18 Multiple facial fractures resulting in a large fragment of the face separated from the cranium. This combination of fractures, called LeFort II, results from a powerful frontal blow.

occur in the body, the ramus, or the narrow superior portion of the ramus just below the condyle. The structure of the mandible is such that sufficient force to cause a fracture in one area causes a wrenching stress in other portions of the bone. For this reason, mandible fractures often occur in pairs (Fig. 17-19).

Nontraumatic Conditions

Allergies or upper respiratory infections may cause inflammation of the paranasal sinuses. Since the sinuses communicate with the outside of the body via the nasal passages, they are common places for infection within the skull. Fig. 17-20 shows air-fluid levels in the maxillary sinuses and is an example of acute sinusitis. Fig. 17-21, on the other hand, shows the thickening of the walls of the maxillary antra that is typical of chronic sinusitis. Fig. 17-22 illustrates the destruction caused by osteomyelitis of the bones of the cranium. In this case, the infection spread from severe sinusitis.

Tumors within the brain are usually diagnosed by means of CT or MRI scans. Tumors that affect the bones of the cranium, however, may be evaluated by radiography. Fig. 17-23 illustrates multiple lytic lesions of **multiple myeloma,** a malignant bone disease that may involve many bones of the body. Fig. 17-24 shows destruction of the sella turcica caused by pituitary adenoma, a tumor of the pituitary gland. Fig. 17-25 is an example of an **osteoma,** a benign bone tumor.

Paget's disease affects the skeletons of older persons. It results in softening and destruction of the bone, followed by thickening and irregular calcification as the bone is repaired. While Paget's disease may affect many bones, it is most commonly seen in the bones of the cranium (Fig. 17-26) and of the pelvis and lower extremities.

SUMMARY

Radiography of the skull, facial bones, and paranasal sinuses is less common than it once was, having been replaced in many cases by CT studies and sometimes MRI. Radiography of the head is most common today for evaluation of bone disease in the skull, for screening for facial fractures, and for evaluation of the paranasal sinuses when CT is not readily available.

The bones that make up the cranium and facial structures are complex and require careful study. All are immovable, with the exception of the mandible, and are joined by synarthrodial joints called sutures. Proper positioning requires attention to all three body planes and the use of bony landmarks and positioning lines. The most common positioning line for alignment of the transverse plane is the orbitomeatal line (OML) between the midpoint of the lateral or-

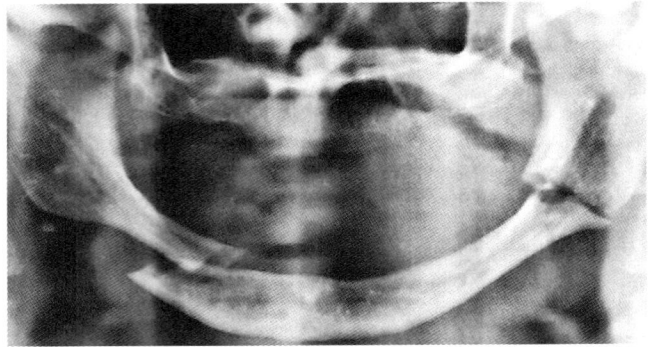

Fig. 17-19 Mandible fractures often occur in pairs.

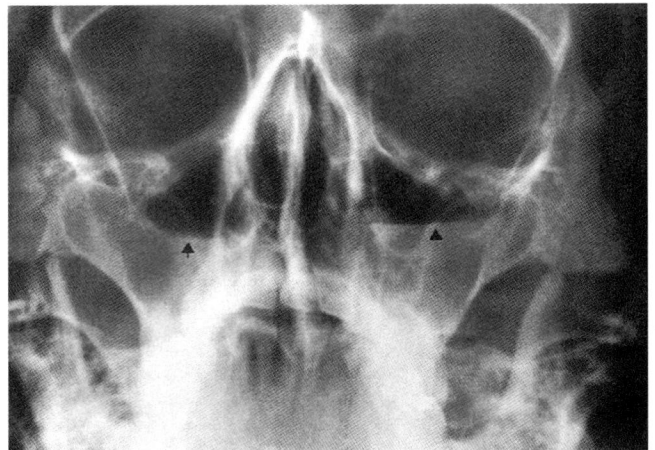

Fig. 17-20 Acute sinusitis. Air-fluid levels are demonstrated in both maxillary sinuses *(arrows)*.

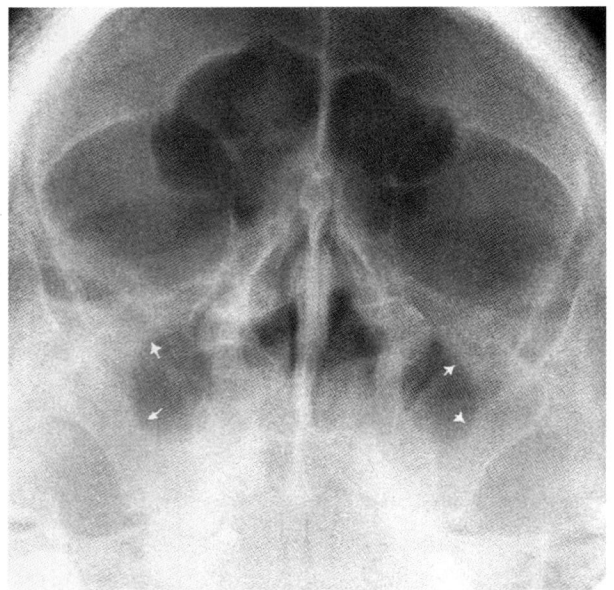

Fig. 17-21 Chronic sinusitis shows characteristic thickening of the walls of the maxillary sinuses *(arrows)*.

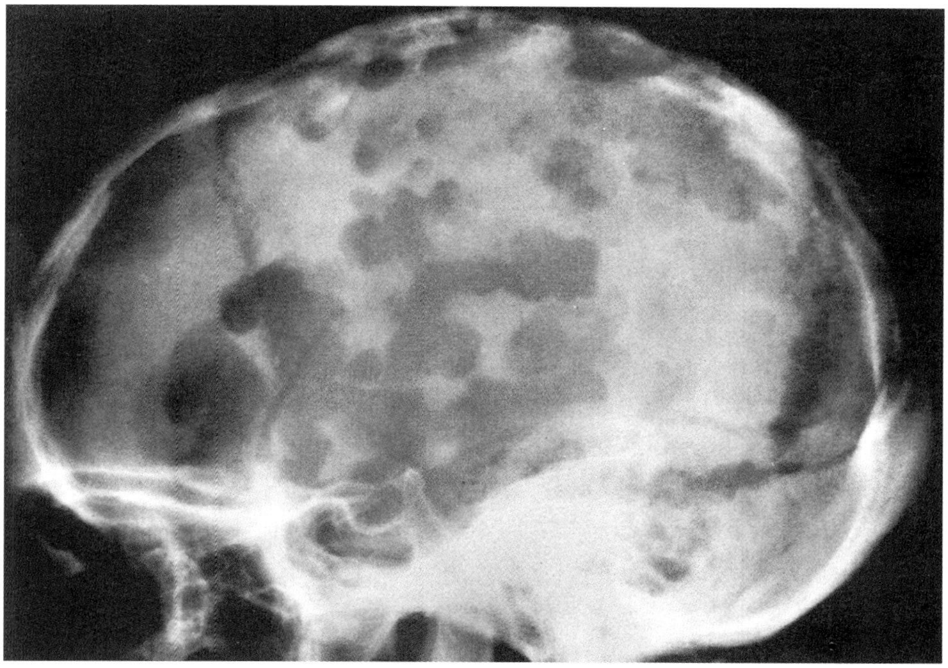

Fig. 17-22 Osteomyelitis of the skull.

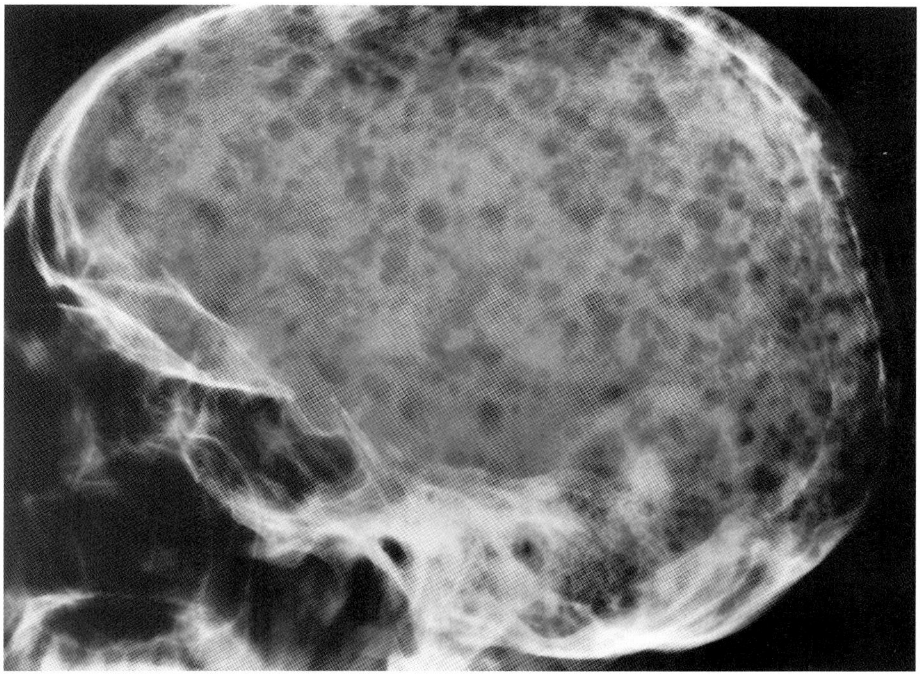

Fig. 17-23 Multiple myeloma as manifested in the skull.

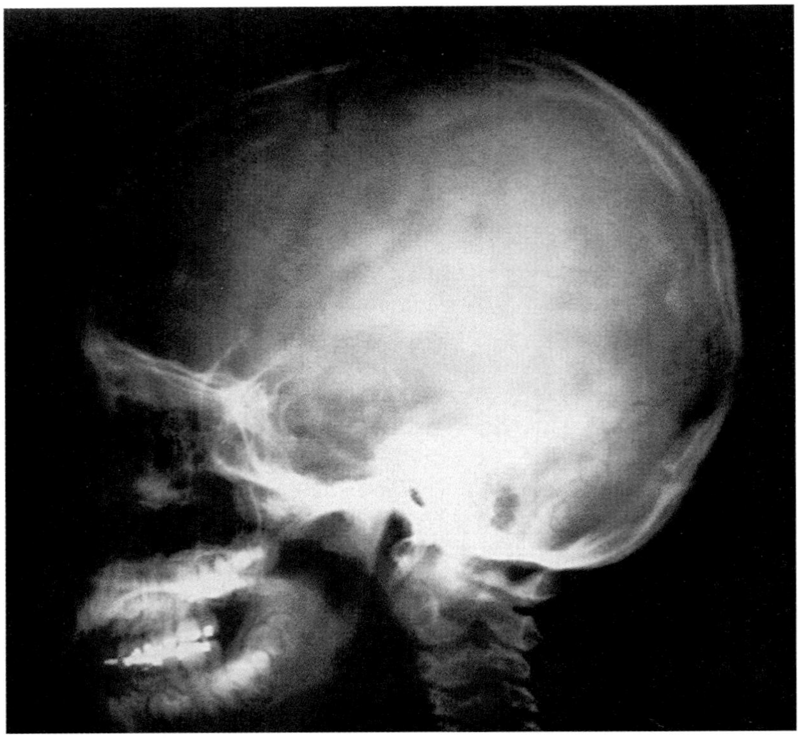

Fig. 17-24 Destruction of the sella turcica caused by pituitary adenoma.

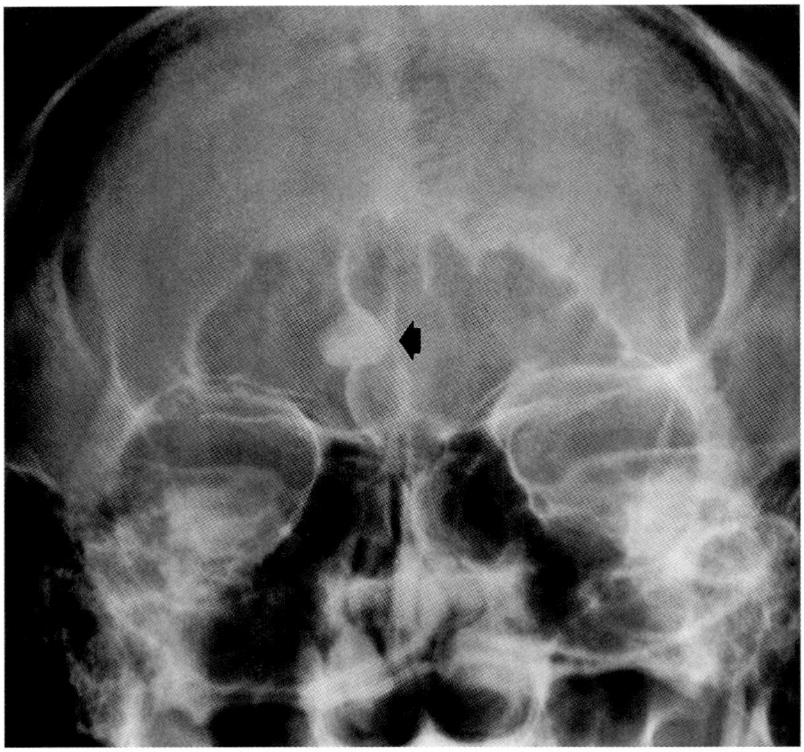

Fig. 17-25 Osteoma of the frontal sinus *(arrow).*

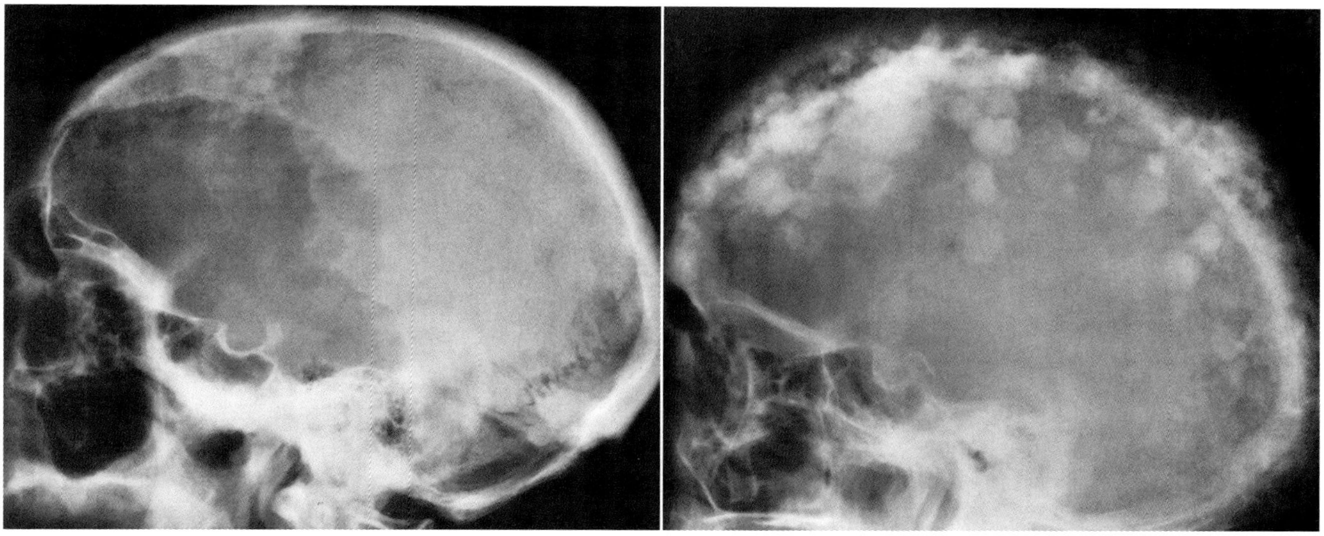

A   B

Fig. 17-26 Examples of Paget's disease in two patients. **A,** Bone destruction during early phase. **B,** "Cotton-wool" appearance produced by sclerosis in reparative phase.

bital margin and the external auditory meatus (EAM).

The basic positions for radiography of the skull, face, and sinuses are AP, PA, lateral, axial, and parietoacanthial (Waters). Variations in tube angle are used to project structures of interest with least distortion and free of superimposition by structures that could compromise visualization. The axial projec-

tions (SMV and VSM) are used to demonstrate the cranial base, the zygomatic arches, and the sphenoid sinuses.

There are four sets of paranasal sinuses, named for the bones in which they are located: maxillary, frontal, ethmoid, and sphenoid. The sinuses are radiographed with the patient in the upright position to visualize air-fluid levels.

▪ Review Questions ▪

1. Name the bones that make up the cranium.
2. Which cranial bones contain the auditory canals?
3. List the bones that make up the orbit.
4. List the bones that contain paranasal sinuses.
5. Name a projection that demonstrates the cranial base.
6. Compare the procedure for an AP axial (Towne) projection with the procedure for demonstrating the same structures with the patient prone.
7. Name two projections that demonstrate the maxillary sinuses.
8. How does the procedure for a lateral projection of the nasal bones differ from that for a lateral projection of the facial bones?
9. Describe the patient/part position for a parietoacanthial (Waters) projection of the facial bones and sinuses.
10. If the petrous ridge is projected over the floor of the maxillary sinuses on the parietoacanthial (Waters) projection, how should the position be modified to clearly demonstrate this area?
11. List three types of facial fractures and state which projection is most likely to provide a clear demonstration of each.
12. Name three types of pathology that may be diagnosed by radiography of the cranium.

18

Radiography of Pediatric and Geriatric Patients

Learning Objectives

At the conclusion of this chapter the student will be able to:

- Demonstrate appropriate levels of communication with children of any age
- Immobilize an infant or toddler for a radiographic examination
- Compare the characteristics of the developing skeleton with that of the mature skeleton
- Formulate exposures for pediatric radiographic techniques
- Identify pediatric radiographic examinations that vary in method from adult examinations
- List considerations that improve communication and compliance when dealing with older patients
- Describe changes that occur to the skeleton and the soft tissues as a result of aging
- Adjust radiographic exposures appropriately for patients with osteoporosis and/or advanced age

Key Terms

Alzheimer's disease
aspiration
battered child syndrome
decubitus ulcer
demineralization
diverticulitis
geriatrics

nonaccidental trauma (NAT)
organic brain syndrome
osteopenia
osteoporosis
Parkinson's disease
pediatrics
valid choice

The term **pediatrics** refers to the care of children. **Geriatrics** refers to the care of elder adults. This chapter is divided into two parts, one devoted to the special needs of infants and children and the other to the requirements of older adults. Pediatric and geriatric patients have the same needs as other patients: confidence and reassurance, safety and security, comfort and competent care. It is in the way these needs are met that their requirements differ.

This chapter offers strategies for effective communications with both pediatric and geriatric patients in the clinical setting. It also covers variations in the skeletons and soft tissues of these patients and the exposure adjustments required for successful radiography. Instructions are provided for immobilizing infants and small children. Pediatric procedures that vary from the methods used for adults are addressed as well. Pathology unique or especially common to the very young and the very old is also presented.

PEDIATRICS

Communication

Relating to a child can be difficult for those who have little experience with children. Even successful parents sometimes have problems relating to other people's children. This is a skill that improves with practice. Experience with children without the stress of the workplace may help. Consider spending an afternoon at the park with children of a relative or friend. Make it a point to get acquainted with a neighbor child. When introduced to a child, make a special effort to relate in a comfortable way.

Children tend to be more intuitive than adults. They can usually sense when you really care about them. If your real concern is whether you will get off duty on time or whether your employer will be satisfied with the films, the child will not be fooled by your pretended interest. To gain a child's trust, your concern must be genuine.

Effective communication is age-appropriate and shows both respect and concern. It is both verbal and nonverbal. Research indicates that about 70% of communication received by adults is nonverbal. That is, we learn far more from posture, body language, facial expression, and tone of voice than from what is actually said. For example, the radiographer in Fig. 18-1 might be saying, "I'm so glad to see you," but that is certainly not what the listener is likely to hear. In Fig. 18-2, the same radiographer radiates warmth with her smile, her body language, and her touch. Positive touch is both firm and gentle. It is an important aspect of nonverbal communication. There are no available figures for the percentage of communication received by children that is nonverbal, but it is presumed to be much higher than the percentage for adults.

Children are more likely to have a positive attitude about radiography if they perceive your facility to be a child-friendly place. Some small furniture and an assortment of books and games (Fig. 18-3) will help them feel welcome and will keep them occupied if they must wait. At the very least, it is helpful to have a few toys or interesting objects that can capture the attention of children of different ages (Fig. 18-4). One inexpensive item that appeals to a wide age range is stickers. They may be used as a get-acquainted gift or as a reward for good behavior or both.

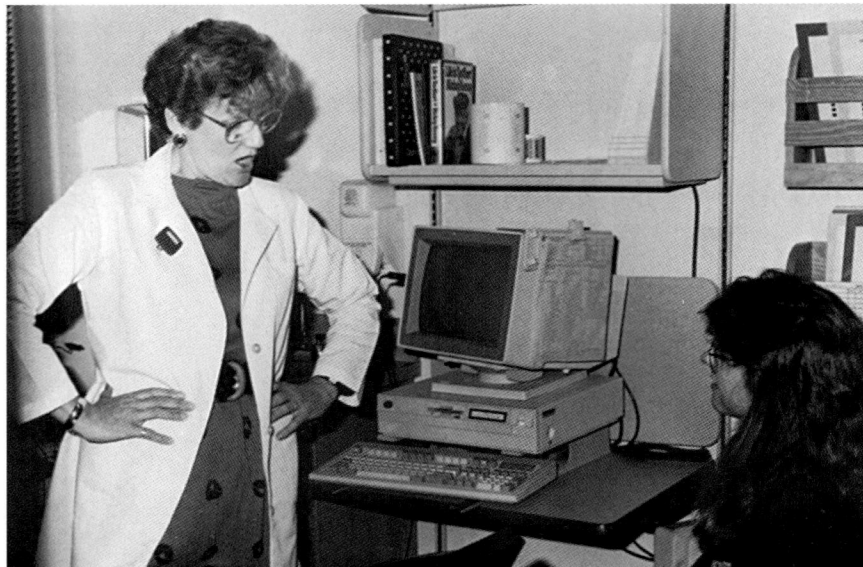

Fig. 18-1 More than 70% of communication received by adults is nonverbal.

Infants There are four basic things to remember if you wish to relate successfully to infants:

1. They like to be bundled up. They feel more secure when they are warm and snugly wrapped.
2. They like to be held firmly and gently. It may help to settle them if you rock them or walk around with them.
3. They relate to faces. They like eye contact, but they cannot focus very far away, so closeness is good.

4. They respond to the sound of voices long before they can understand words. Talk to them softly while you work.

Toddlers Toddlers may present your greatest challenge in dealing with children. They perceive themselves to be the center of the universe, and they have learned to be assertive. They may say, "Me want . . ." or "Me do it" or "NO!" and their social skills do not yet include cooperation. While they may understand what you say, their attention span is short. They need lots of reminders and lots of patience.

Fig. 18-2 Warmth is not conveyed with words.

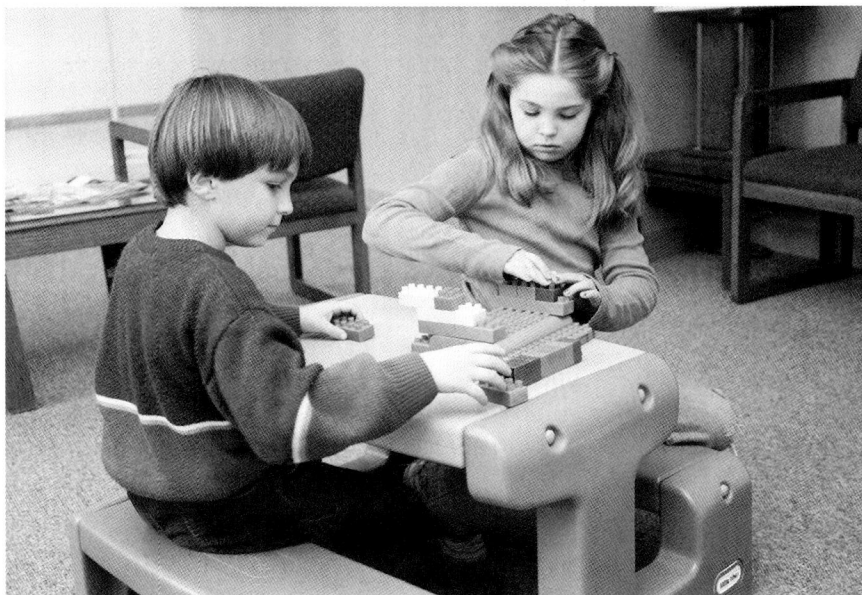

Fig. 18-3 Children feel welcome when the clinic waiting room has an area for them to play.

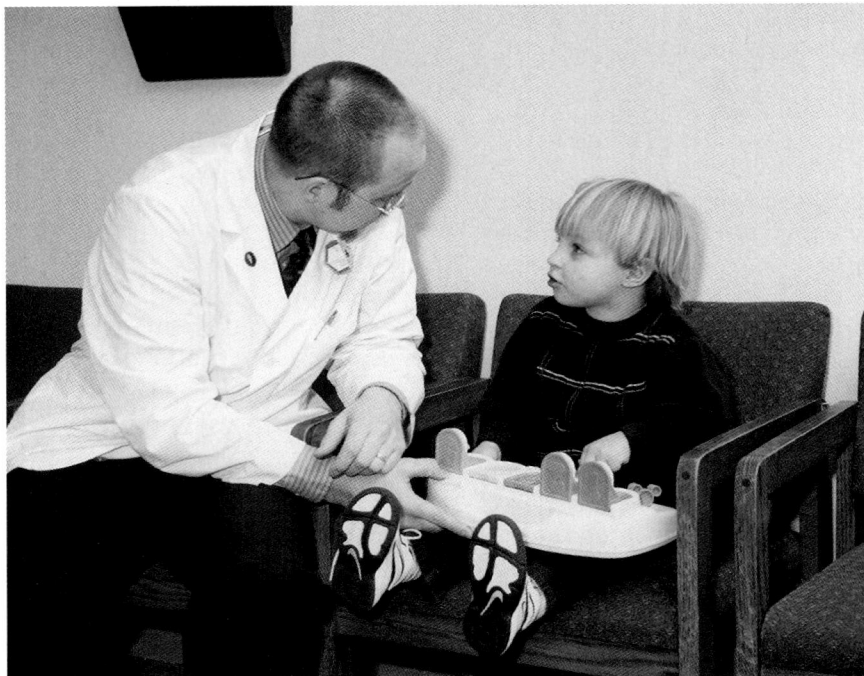

Fig. 18-4 Toys or other interesting objects may serve to distract a frightened child.

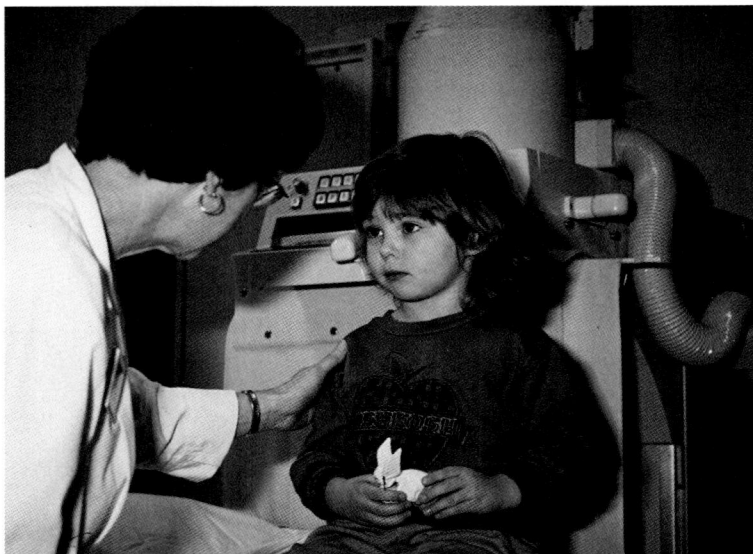

Fig. 18-5 Use positive touch and eye contact at the child's eye level. A familiar object to hold is reassuring.

There are some helpful clues for dealing with toddlers in Fig. 18-5. First, note that the child is holding a toy. Toddlers feel more secure when they can hold onto something familiar. It may be a blanket or a teddy bear or mother's car keys. Second, the radiographer is at the child's eye level. Raise the child up to your level or stoop down to the child's level. If both of you are standing, you will loom over the child like a giant. Unfamiliar adults are often intimidating to chil-

dren because of their stature. And finally, remember to make eye contact and use positive touch.

Time spent getting acquainted (Fig. 18-6) can pay big dividends in cooperation later. Toddlers and small children may be frightened by strange surroundings and equipment, but they are probably familiar with lights and cameras. Explain that you have a very special camera to take their picture. Let them practice saying "Cheese!" They may be allowed to turn on the

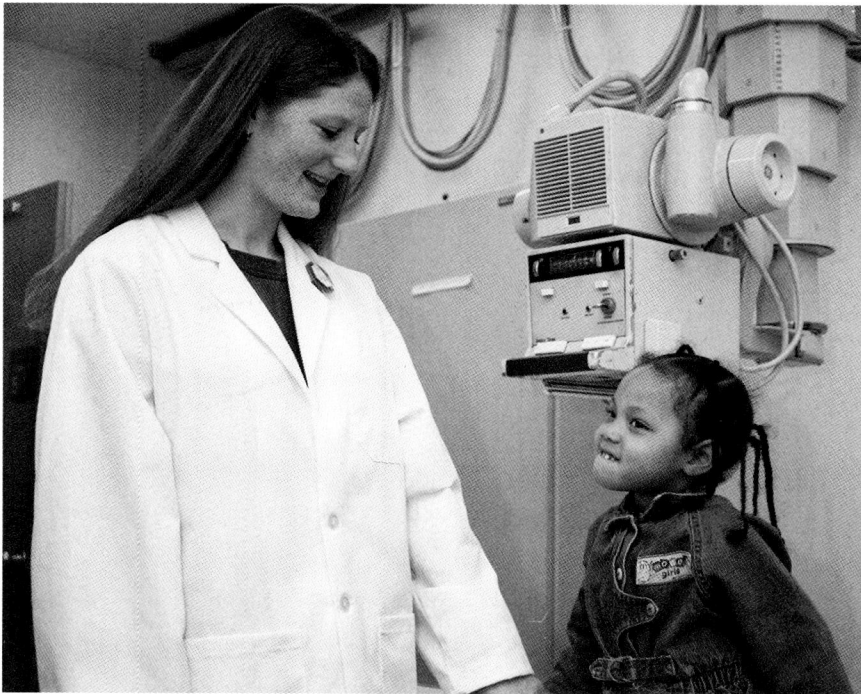

Fig. 18-6 Take time to get acquainted.

collimator light, which will make the equipment seem less mysterious and help them feel in control (Fig. 18-7).

An important way that toddlers learn is by imitating adults. Show them what you want them to do, and encourage them to copy your actions. Toddlers respond to both praise and disapproval *when used appropriately.* If you give undeserved praise, they will quickly identify you as a "soft touch" and assume that anything they do is all right with you. If your disapproval takes the form of anger or impatience, they may become fearful or decide that you are impossible to please and quit trying. If the child is not cooperating, don't smile. Look him in the eye and calmly say, "No. Hold still." Look for something that is really positive to praise, no matter how small. Positive expectations and appropriate praise are powerful motivators for good behavior (Fig. 18-8).

Remember that crying and muscular activity are natural responses to anxiety in a small child. When the situation seems unbearable, kicking and squirming are quite normal. If the child is crying, the two best strategies are patience and distraction. You might show interest in a toy or turn on the collimator light. A funny face or a funny noise may provide a distraction. Raising your voice over a child's screams is almost never effective, but sometimes a whisper will capture a distressed child's curiosity and attention. Any effort you make to gain a child's cooperation is worthwhile as long as you are making progress. If, after a patient and reasonable attempt,

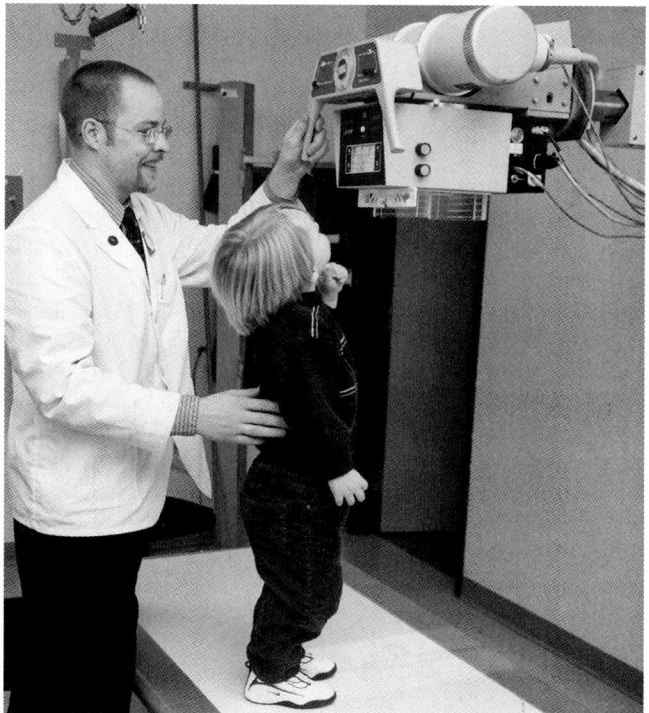

Fig. 18-7 A child may be allowed to turn on the collimator light.

you are not succeeding, conclude that cooperation is not possible. Immobilize the child gently but firmly and complete the procedure as quickly as you can. Immobilization techniques are explained later in this chapter.

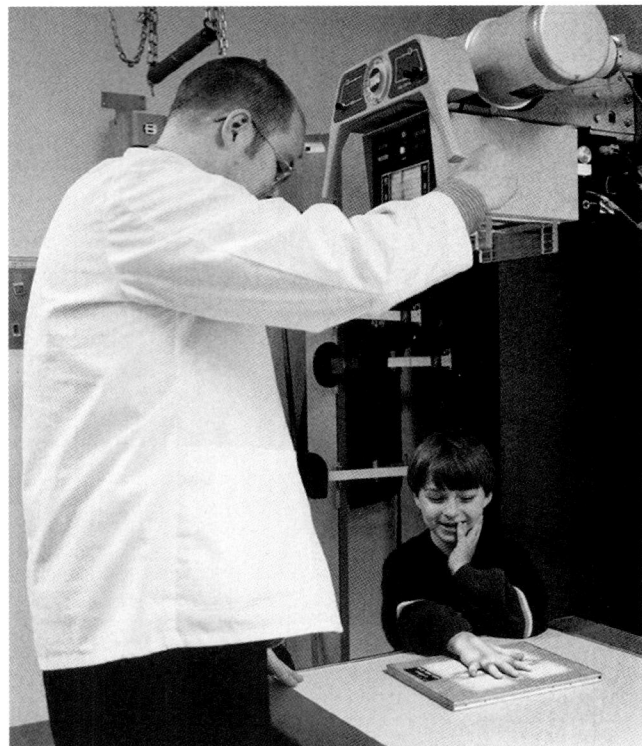

Fig. 18-8 When a positive relationship is established, children are more likely to cooperate.

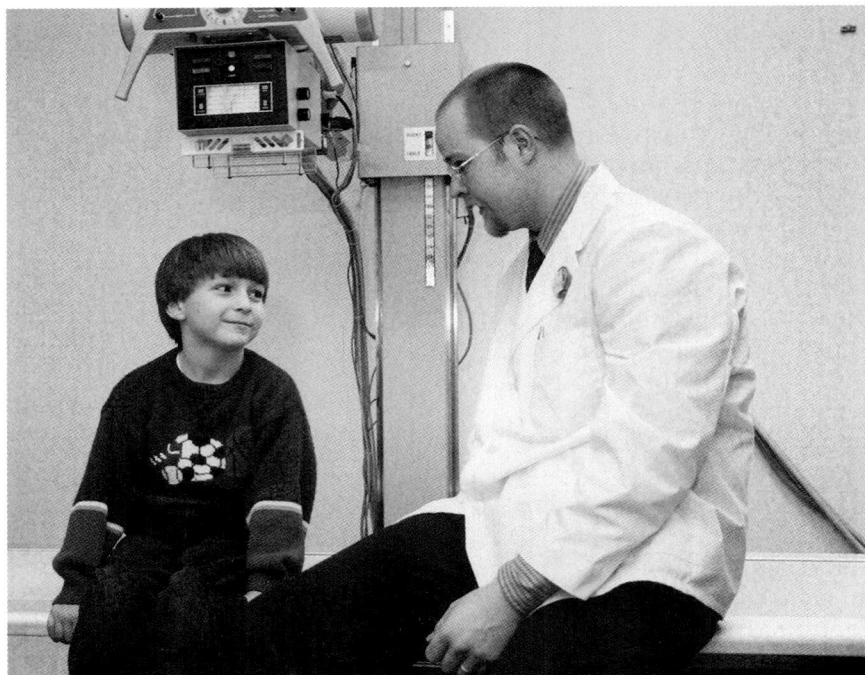

Fig. 18-9 Keep explanations simple, direct, and honest.

Young Children Children aged 4 to 7 (Fig. 18-9) are usually less of a challenge than toddlers. They are less likely to be intimidated by strange surroundings and are better able to understand what you tell them. They may behave like small adults or toddlers, depending on their level of confidence in themselves and their trust in you. You can encourage both by offering **valid choices.** A valid choice is one where either alternative is acceptable to you. For example you might ask, "Would you like to wear a blue gown or a red one?" or "Would you like to get up on the table by yourself, or would you like me to help you?" Asking, "Would you like to lie down here?" is *not* a valid choice. If the child must lie down for the procedure, there is no choice involved, and if the child answers "No," you have placed yourself in an awkward position.

Children of this age want to understand what is happening. They are curious and will probably ask questions. Keep your explanations simple, direct, and honest. Too much detail may be frightening or boring, especially if they do not understand all you say. Avoid stating how many pictures you plan to take. These children will count, and if additional films are needed, you may lose credibility. Honest praise is a good motivator with this group also.

Older Children Children aged 8 to 12 are usually easy to deal with (Fig. 18-10). They relate quickly to their circumstances and understand what is going on. They want to be brave and are usually willing to help. Most respond readily to humor. If pain or fear causes them to revert to the behavior of a younger child, the techniques for dealing with younger children may be applied. Valid choices, positive expectations, and honest praise will usually ensure success.

Early Adolescence Young teens usually behave much like adults. On the other hand, the hormonal changes of puberty make them subject to mood swings. They may behave somewhat like toddlers when hurt or frightened. In this case, they may act self-centered and their attention span may be short. When this happens, appropriate praise and/or disapproval are effective strategies.

One important characteristic of most young teens is an exaggerated sense of modesty. They are in the process of coming to terms with the physiologic changes of puberty and can be easily embarrassed by any attention to their bodies. Girls may feel "naked" if asked to remove their bras. The x-ray may be perceived as an "all-seeing eye," ready to reveal their

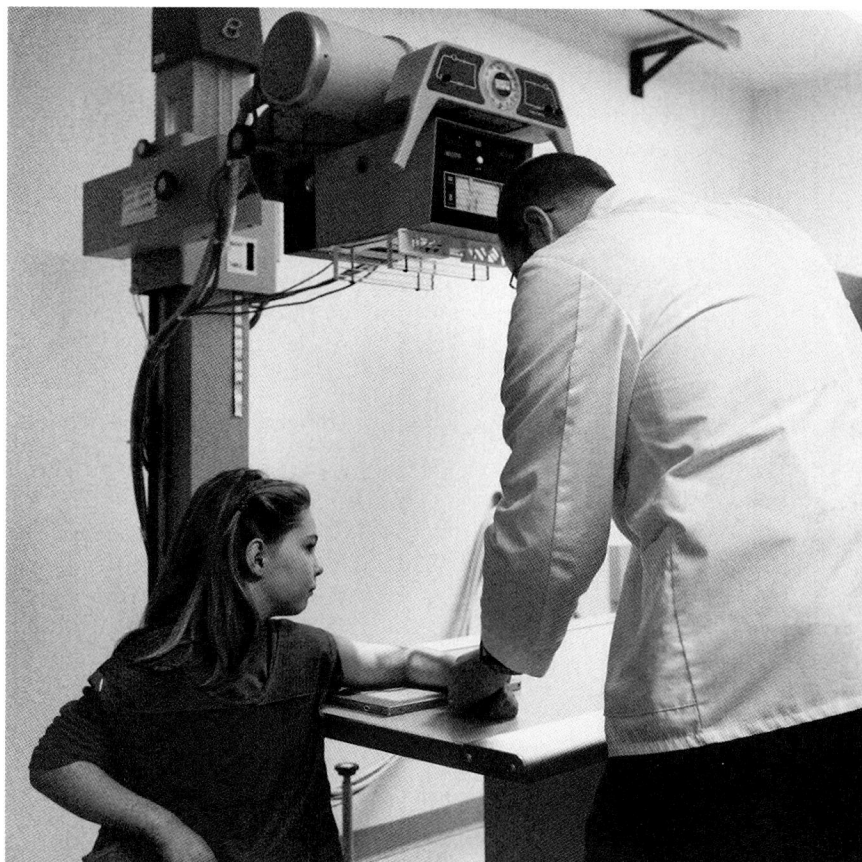

Fig. 18-10 Older children are usually quite cooperative.

innermost secrets. Special sensitivity is needed. If undressing is required, provide one or more gowns (Fig. 18-11) so that the patient is modestly covered during the examination. If you must inquire about sensitive subjects such as bowel habits or menstrual periods, treat these topics in a matter-of-fact way, and do it privately.

Parents in the X-ray Room

One question that radiographers must answer when small children are x-rayed is whether or not to allow parents in the x-ray room. Experts disagree on this subject, and there are good arguments on both sides of the issue.

The principal argument in favor of a parent's presence is that separating parent from child creates anxiety in both the child and the parent. Having the parent close may be reassuring to the child. In addition, the parent may have skills for calming this particular child and gaining cooperation with the procedure. If help is needed to hold a child in position during an exposure, a parent is the logical choice.

The opposing argument holds that a parent's presence may create anxiety for both parent and child. The child may appeal to the parent for "rescue" from the procedure, creating a dilemma for the parent: rescue and comfort the child, or support the continuation of the procedure? This argument tends to support the idea that if the parent is not visible, the child is more likely to accept the procedure. If the child is distressed and must be immobilized, the parent will be less upset if waiting out of sight and out of earshot. The parent can comfort the child as soon as the procedure is completed.

Each viewpoint has merit under certain circumstances. The duty of the radiographer is to evaluate the circumstances and make a judgment call. Consider the age of the child. Babies may be calmed by hearing their mother's voice nearby. Toddlers that usually get their way with their parents may behave better in the absence of their parents. Consider the state of mind of both the parent and the child. Even the best parents sometimes behave irrationally when their children are sick or injured. The parent may feel helpless and have a strong need to control everything. The presence of a parent who is not calm is likely to be upsetting to the child. When both parents are present, it is usually best for only one to accompany the child into the x-ray room. If you choose to have a parent present, try to select the one with the most matter-of-fact attitude.

At this point, there are two important things to remember, and they are both good news: You can change your mind, and you can ask for help. Whatever you decide, if you make the wrong decision, it is reversible. If a parent is a problem in the room, you can say, "I think little Sara and I can work this out by ourselves. Please wait in the waiting room and I'll bring her out just as soon as we're finished." And if you begin without the parent and change your mind, you can say, "I think Jason needs his Mommy. Would you mind giving me a hand?" Finally, if the situation seems unmanageable, don't hesitate to ask for help. A more experienced staff member can sometimes save the day. As a last resort, consult the physician, who may decide to postpone the procedure or to sedate the patient.

Immobilization

Infants and small children can be immobilized for most examinations without the need for someone to hold them. Whenever possible, mechanical immobilization of some type is the best answer. This section introduces some of the commercial devices available for this purpose. If pediatric patients are frequently seen in your facility, it is wise to invest in commercial immobilization devices to meet your needs. Noncommercial devices and the use of items commonly found in the x-ray room or the clinic are also illustrated in this section.

When circumstances require that someone hold a child during an exposure, remember that this is *not* to be done by a radiographer. Occupationally ex-

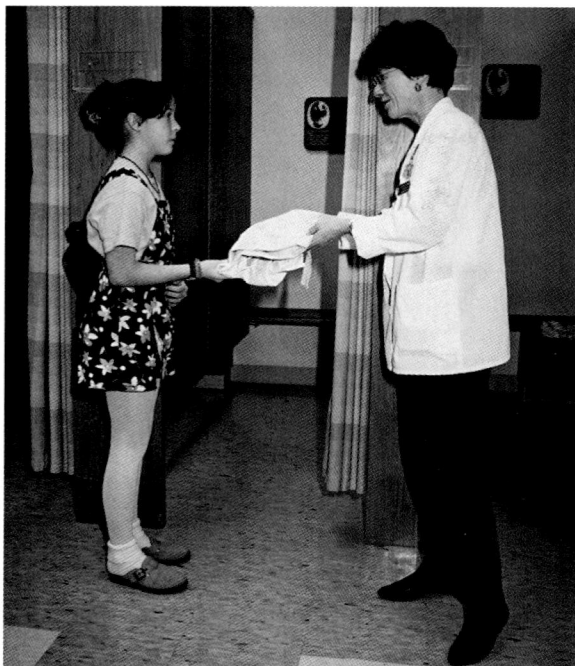

Fig. 18-11 Sensitivity to privacy improves relationships with young teens.

posed persons such as radiographers are prohibited from holding either patients or cassettes during exposures. A nonoccupationally exposed person must be recruited for this duty, and the best candidate is usually the child's parent, provided that person is not pregnant. Provide a lead apron. Lead gloves should be worn if the hands will be in the radiation field. Demonstrate precisely how the child should be held and how to sit or stand to minimize exposure to primary radiation. Using extended arms so that the holder's body is at arm's length from the child will reduce exposure from scatter radiation (Fig. 18-12). To avoid the need for repeat exposures, take care to ensure that the holder has a comfortable and *firm* grip on the child in correct position.

Commercial Immobilization Devices Table restraints are common accessories for radiographic tables. These wide bands attach to the sides of the table and may be adjusted for placement along its length. Wide strips of Velcro secure the bands around the patient. Similar devices called compression bands consist of a single band of cloth that is secured to rails on both sides of the table. It is tightened using a ratchet roller. While originally designed to provide abdominal compression for specific procedures, compression bands are also useful for immobilization and to provide security from falls.

Many radiographic devices designed for pediatric immobilization are modifications of the original design for circumcision boards. They are frames to which the child is attached by Velcro straps at strategic locations. Figs. 18-13 through 18-15 illustrate a variety of these products. Some come in several sizes; others are adjustable in size. Each has some advantages and disadvantages. For example, the Octastop® restrainer has octagonal end pieces that allow the child to be placed in eight different positions. Once the child is secured in the device, AP, PA, right and left lateral, and four oblique positions of the torso are obtainable without readjustment of the attachments. The disadvantage of this device is the OID, a distance between the patient and the film that is greater than usual.

Radiography of the skull, face, and neck requires precise immobilization of the head. The devices described above have head straps that aid in this process, but specific devices for head immobilization can be very helpful if they are available. Fig. 18-16 illustrates head clamps for this purpose.

The Pigg-O-Stat® (Fig. 18-17) is a unique device for upright chest radiography of infants and small children. Its saddle-like seat is mounted in a disk that rotates for various projections. An adjustable clear plastic sheath surrounds the child's upper body and is fastened securely with leather straps behind the waist and the head. This plastic portion supports the body upright and holds the arms overhead. The unit incorporates a film holder and gonad shielding. The Pigg-O-Stat is not inexpensive, but it is a worthwhile investment where there is a high volume of pediatric chest radiography.

Noncommercial Devices and Methods for Immobilization The principal objective of most immobilization is to prevent motion from flailing arms and kicking legs. When the extremities are under control, it is nearly impossible for a child to turn over or move about. In the absence of a commercial immobilization device, the extremities can be controlled by using a "mummy wrap." A sheet is used to secure the arms at the sides of the body and to hold the legs together. This technique is illustrated in Fig. 18-18. Infants wrapped in this way, with a lead apron placed over the pelvis, are both immobilized and shielded. Older children may still be able to "buck," flexing their knees and necks to bounce the torso up and down. Fig. 18-19 shows the use of the mummy wrap in conjunction with the table restraint strap across the knees, which is very effective.

Fig. 18-12 Take appropriate precautions for radiation safety when holding is required.

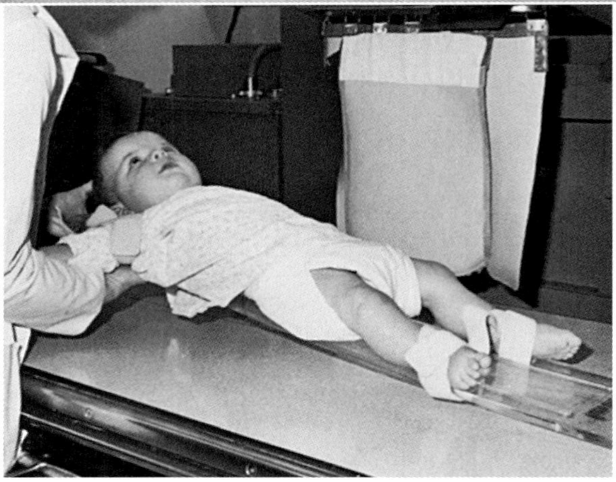

Fig. 18-13 Tame-Em® adjustable infant restraining device. **A,** Device is made of Lucite with Velcro straps. **B,** Device in use.

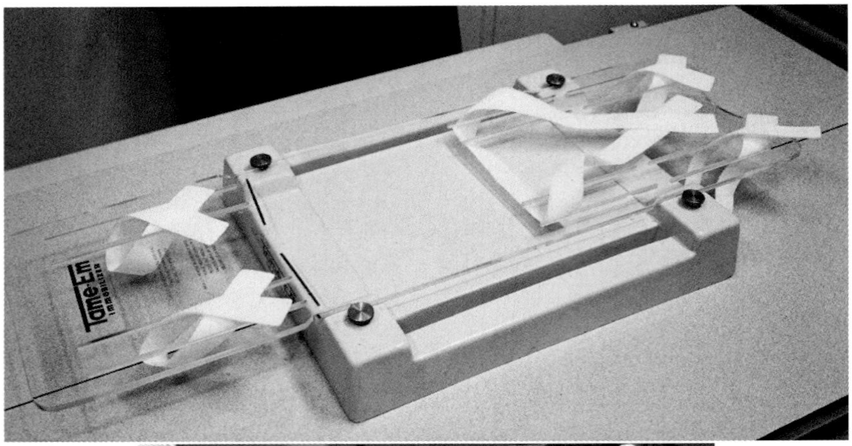

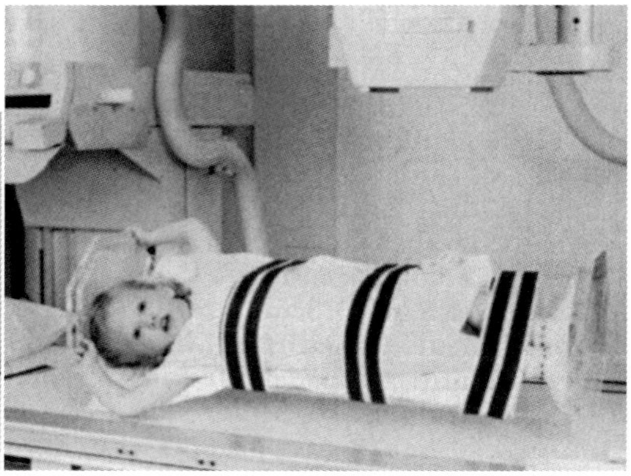

Fig. 18-14 Octastop restraint board. **A,** Device features octagonal end plates and Velcro restraining straps. **B,** Device holds child securely in oblique position shown here plus seven other positions.

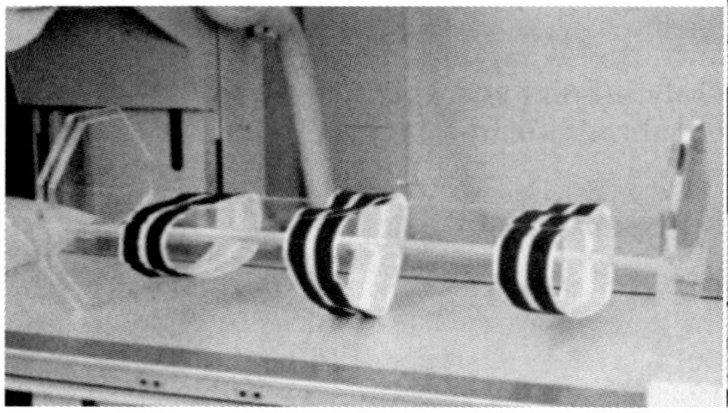

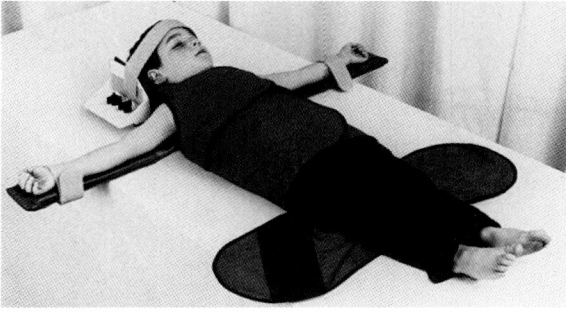

Fig. 18-15 Papoose Board provides selective restraints for children 2 to 6 years of age. Wide fabric straps with Velcro secure the child's torso. (Courtesy Olympic Medical, Seattle.)

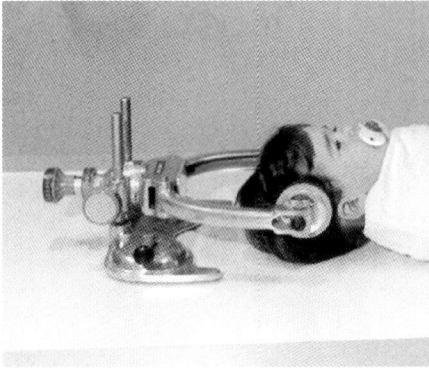

Fig. 18-16 Adjustable head clamp secures skull positions with cushioned contacts.

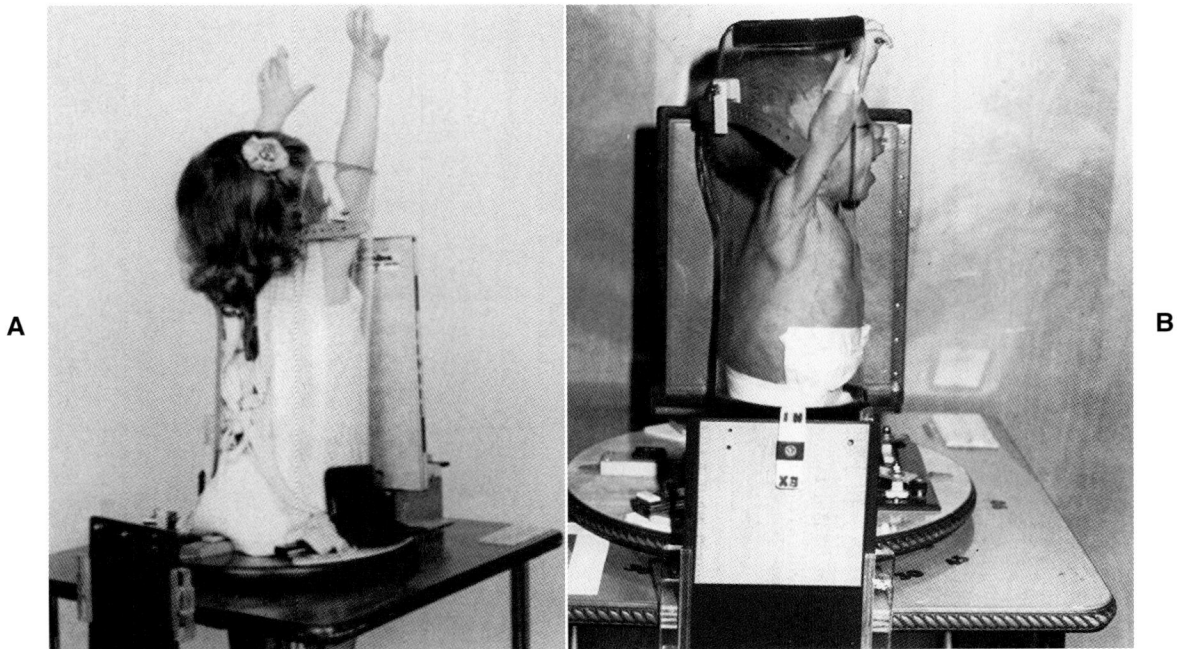

A

B

Fig. 18-17 Pigg-O-Stat positioning chair for upright chest radiography. **A,** Toddler positioned for PA chest radiograph. **B,** Infant positioned for lateral chest radiograph.

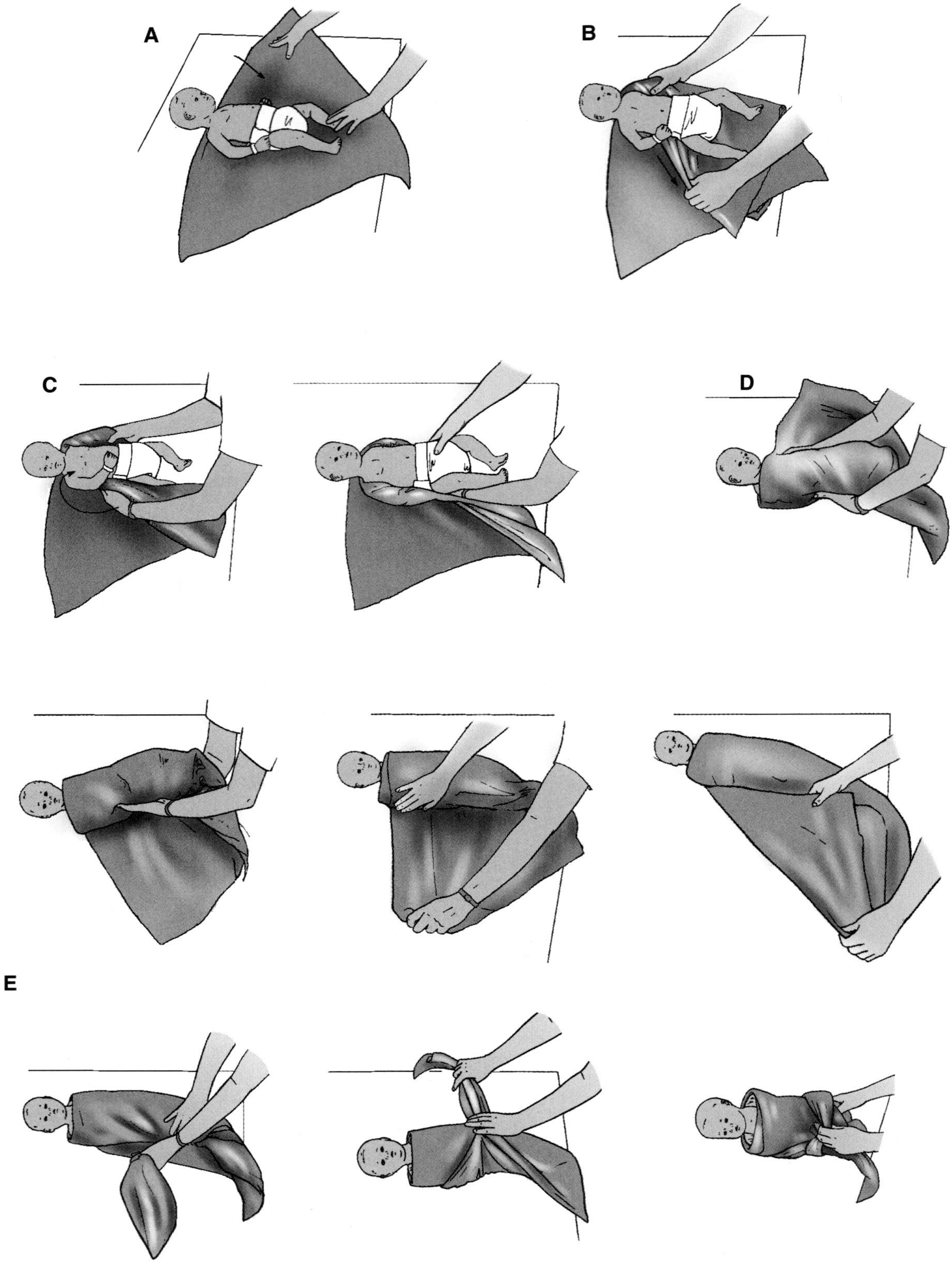

Fig. 18-18 Mummy wrap technique. **A,** Fold the sheet on the diagonal to make a triangle. Place the child on the sheet with wide edge under neck. **B,** Wrap one corner up over an arm, tuck it under the body and pull it through. **C,** Wrap the first corner over the second arm and tuck it under the body. **D,** Wrap the second corner over the chest and secure it under the body. **E,** Complete mummy wrap by securing the second corner around the child.

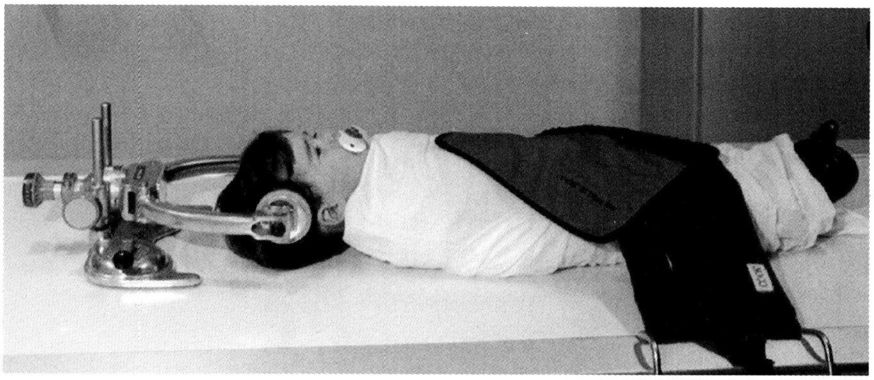

Fig. 18-19 Mummy wrap used in combination with other immobilization methods.

The use of tape for maintaining position is illustrated in Fig. 18-20. Note that the adhesive surface of the tape is not in contract with the patient's skin. It may be twisted so that the nonadhesive side is against the skin, or a gauze pad may be placed between the tape and the skin. Tape is not effective if applied to the tabletop or the cassette surface. It must be wrapped around the edge of the cassette or table so that lifting pressure does not loosen it.

Stockinet is a tubular knitted fabric placed on extremities before the application of a cast. It is also useful for securing arms or legs together (Fig. 18-21). Velcro straps may also be used to hold the legs together and in position. Sandbags, too, can be used to hold extremities in place. Another convenient way to hold extremities in place is to use a flexible sheet of Plexiglas (Fig. 18-22). This plastic material is available at reasonable cost from plastics dealers.

When the head must be precisely positioned and head clamps are not available, tape is not the only answer. Two or three large books may be placed on each side of the head. Radiolucent sponges are placed between the books and the head, and the books are moved close enough to hold the head firmly (Fig. 18-23). Another option is to use large, heavy "bookends" made from angle iron in place of the books. These can be custom made for relatively small cost at metal shops that do iron work. Similar devices are available commercially.

Pediatric Radiographic Procedures

Radiographic studies involving infants or children may differ in approach from procedures used for adults. Pediatric variations may be due to the inability of the patient to stand, the requirements of immobilization, or the technical modifications desirable for body parts that are small.

Gonad shielding is especially important in pediatric radiography. Immature reproductive cells are more vulnerable to mutation than those of adults. While studies of the head or foot may not require gonad shielding on adults, all body parts on pediatric patients are closer to the gonads and shielding should be used for all examinations. Children with chronic problems may need many examinations, and since gonad dose is cumulative, shielding is imperative. Small shields for infants and children may be cut from lead rubber or lead vinyl stock material. They may also be purchased from your x-ray supplier.

Film and field sizes for pediatric studies are smaller than for adults because of the smaller size of the body parts. Select a film that is slightly larger than the area of clinical interest. Studies that would involve several different examinations on an adult are sometimes combined for children. For example, the entire lower extremity might be included in a single exposure on an infant, whereas a study of the lower extremity would involve separate exposures of the femur and the lower leg on an adult.

Bilateral studies of extremities are often done on children. Comparison of the two sides is frequently helpful in determining whether variations in the growth centers of young bones are the result of normal development or injury. Years ago it was routine to take bilateral comparison views for all extremity examinations on children under age 12. With the exception of hip and clavicle examinations, comparison views of the unaffected extremity are now taken only when they are needed, and only on the specific order of the physician. This practice avoids unnecessary radiation exposure to the patient.

Many examinations that are done with a grid or bucky on adults can be done better without the grid on small children. The deciding factor with respect to grids is the thickness of the body part. When the part thickness is less than 12 cm., a grid is not required. It is advantageous to avoid using a grid whenever possible for several reasons. First, grid use requires more exposure. Nongrid exposures lower patient dose and

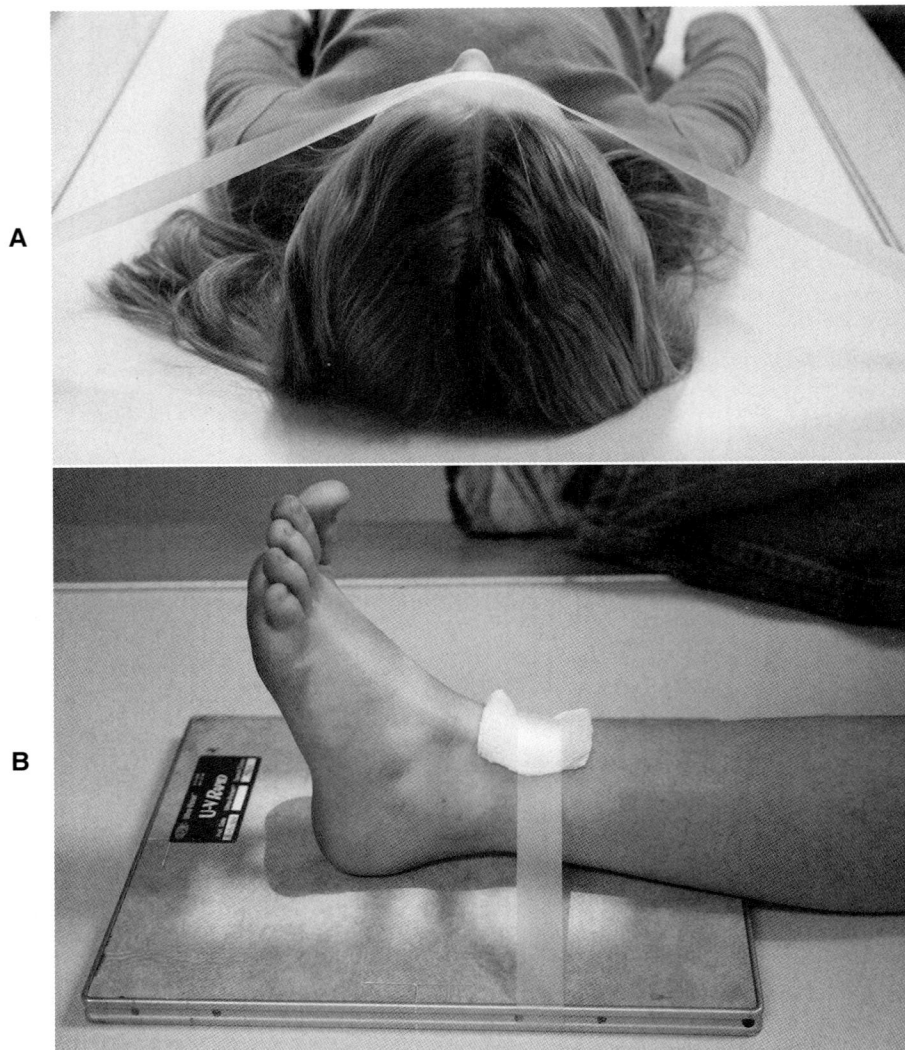

Fig. 18-20 Tape is often used to maintain pediatric positions. **A,** Tape is twisted to avoid adhesive contact with skin. **B,** Gauze is used to prevent adhesive contact with skin.

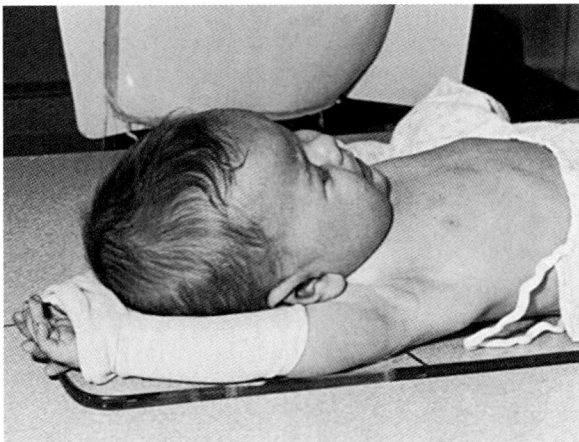

Fig. 18-21 Stockinet may be used to secure position of arms or legs.

Fig. 18-22 A simple sheet of clear Plexiglas aids in immobilizing a child's hand.

Fig. 18-23 Books and sponges may be used to position the child's head.

Fig. 18-24 Chest and abdomen studies on infants may be done tabletop at 40-inch SID. (In practice, cassette should be covered with a soft, warm blanket.)

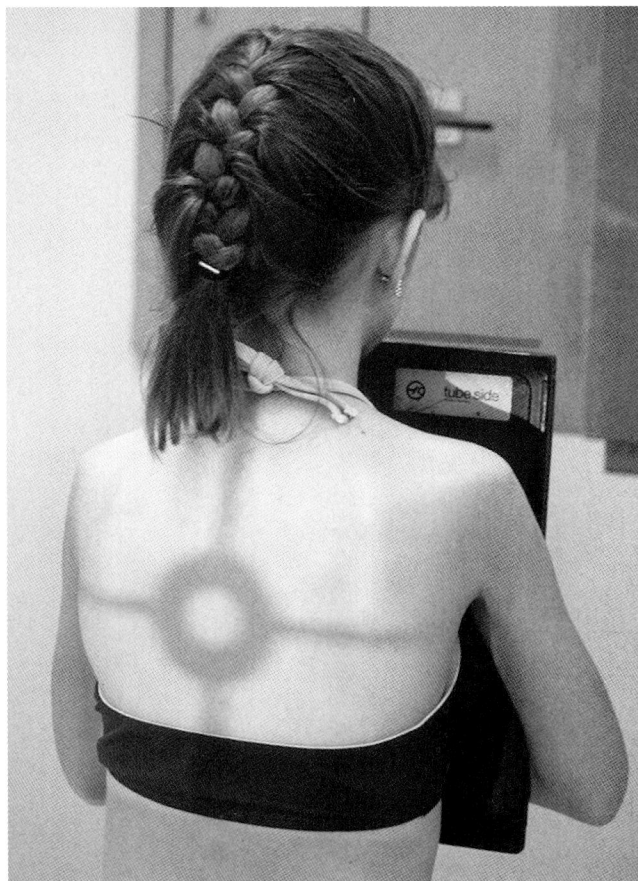

Fig. 18-25 When a cassette holder is not available, a cooperative child may hold the cassette.

permit shorter exposure times. When dealing with an active child it is helpful to have the cassette on the tabletop so that you can clearly see that the body part is correctly centered.

Chest radiographs on children do not require a grid. Unless you have a Pigg-O-Stat, infant chest radiographs will probably be done recumbent at 40 inches SID (Fig. 18-24). Older children who are cooperative may sit or stand at an upright nongrid cassette holder (see Fig. 18-12) or may sit at the end of the radiographic table and hold the film (Fig. 18-25). These methods allow for upright projections at 72 inches SID.

Clavicle radiography differs significantly between adults and children. Adult studies involve only one clavicle and are usually done in the PA projection using the bucky, often upright. Pediatric clavicle studies are usually bilateral and are done recumbent in the AP projection with the film on the tabletop (Fig. 18-26). An axial projection may be taken in the same position with the central ray angled 30 degrees *cephalad*. This is a frequent pediatric examination, since clavicle fractures are quite common in children.

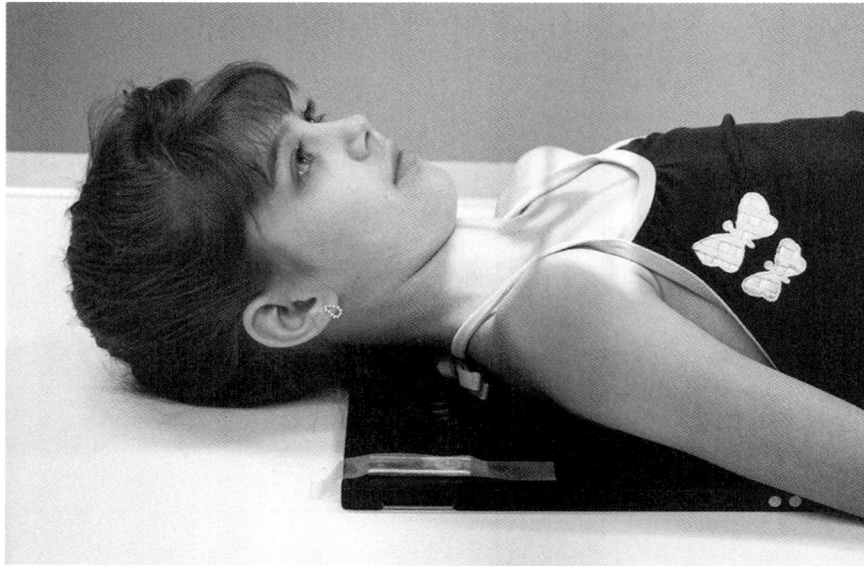

Fig. 18-26 Position for bilateral clavicle examination on a child.

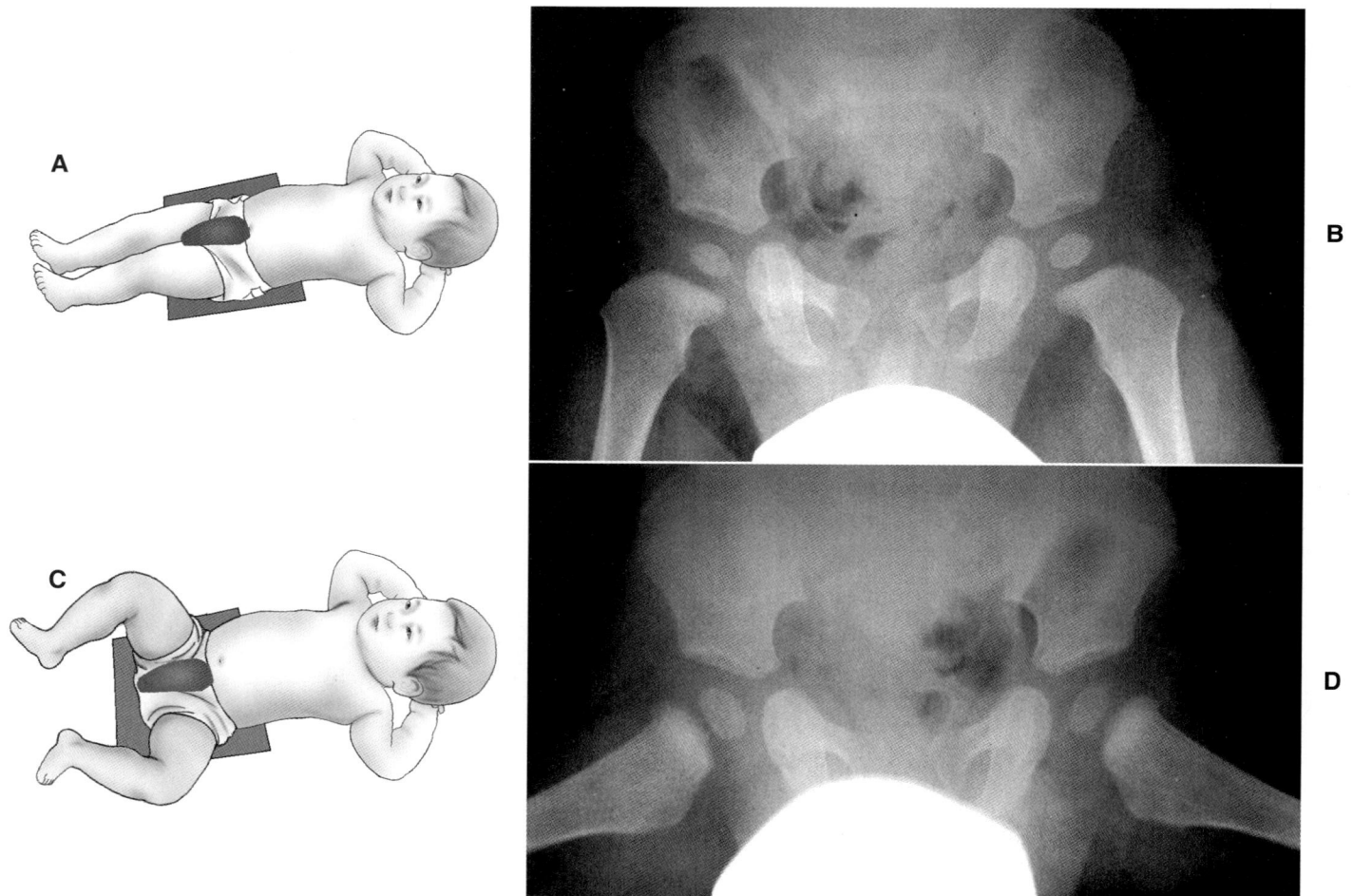

Fig. 18-27 Positions for bilateral hip examination on a small child. **A,** AP position; **B,** AP radiograph. **C,** Frog-leg position; **D,** frog-leg radiograph.

Hip studies also are done bilaterally. The routine examination consists of AP and frog-leg views of the pelvis and proximal femurs (Fig. 18-27). Congenital hip dislocation is a fairly common condition that requires radiography for diagnosis and continued evaluation until treatment is complete.

Pediatric Anatomy

Children are not just small versions of adults. Their anatomy is similar, but there are some definite differences. An infant's head is much larger in relation to body size than that of an adult. Adult spines have multiple curvatures, as discussed and illustrated in Chapter 15. An infant spine forms a C-shaped curve. The curve straightens and then re-curves as the child grows.

Fig. 18-28 illustrates an infant skeleton. Note that the skull and the joints are not completely ossified. With the exception of the anterior rib cage, the dark areas in this illustration that represent cartilage will become bone as the child develops. For this reason, the radiographic appearance of children's bones and joints is much different from that of adults. Fig. 18-29 is a typical example. It provides a radiographic comparison of the ankle and heel of a small child with that of an adult.

The structure of children's bones is less solid than that of adults and is more easily penetrated by the x-ray beam. Note the difference in bone density in the two radiographs in Fig. 18-29. Children's soft tissues also differ from those of adults. "Baby fat" is a thick layer of subcutaneous (under the skin) fat that develops in the first 4 months of life and begins to disappear between the ages of 3 and 4. Children with baby fat are more easily penetrated than older children with the same measurements because more of their soft tissue is fat and less is muscle.

There is an exception to this rule, however, with respect to chest radiography. Babies and toddlers have smaller lungs and more fat in the chest. When the chest measurements are similar, a 5-year-old child will generally require less exposure for a chest radiograph than a 3-year-old. This is because the older child's chest has little fat and is almost entirely made up of air-containing lung.

Even after the baby fat is gone, children's muscles tend to be small and underdeveloped. Their soft tissues are softer and more easily penetrated than those of an adult. This is true even when the child appears to be very strong.

Formulating Techniques for Pediatric Exposures

Unless you work in a children's hospital or a pediatrician's office, you are unlikely to have a comprehensive pediatric x-ray technique chart. Pediatric ex-

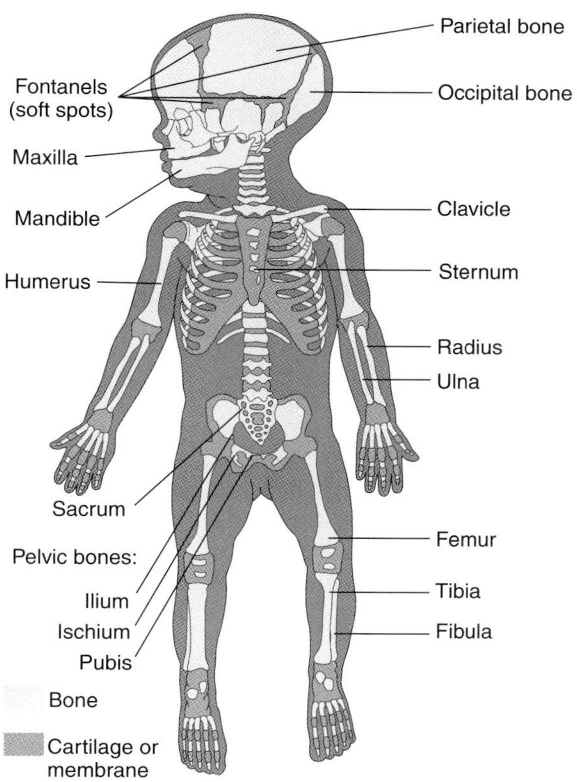

Fig. 18-28 An infant's skeleton has many bones that are not completely ossified.

posure references often consist of a few penciled notes in the margins of an adult chart.

One method of arriving quickly at a pediatric technique is to compare the body part size to parts of similar size on an adult. For example, a child's wrist might be about the same thickness as an adult finger, or a baby's pelvis might be similar in thickness to an adult elbow. Using an adult technique for a body part of similar size is usually successful provided that the same cassettes, SID, and grid or nongrid techniques are used. This method provides a starting point for deriving pediatric exposures, especially for examinations of the extremities.

Another method of calculating pediatric exposures is also based on an adult technique chart. This formula uses the exposure for the smallest adult listing on the chart and modifies it as follows:

- Reduce kVp by 2 kVp for each centimeter of difference between the chart measurement and the patient measurement
- Use 80% of the mAs suggested on the adult chart

Example: Suppose you wish to take an AP lumbar spine on a 9-year-old patient who measures 13 cm. Your technique chart states that an adult measuring 18 cm would require 20 mAs and 80 kVp. The

Fig. 18-29 Ankle radiographs show anatomic changes that occur with skeletal maturation. **A,** Age 3; **B,** adult.

difference in measurement is 18 cm – 13 cm, or 5 cm. Multiply this number by 2 to obtain the kVp change:

$$5 \text{ cm} \times 2 \text{ kVp/cm} = 10 \text{ kVp}$$
$$80 \text{ kVp} - 10 \text{ kVp} = 70 \text{ kVp}$$

Now multiply the mAs by 80%:

$$20 \text{ mAs} \times 0.8 \ (80\%) = 16 \text{ mAs}$$

The new technique is 16 mAs at 70 kVp.

Another method of formulating pediatric techniques is to use the Supertech® Calculator introduced in Chapter 10. This slide rule has separate windows for both adult and pediatric measurements. It produces appropriate exposures for all body parts for children under age 12, as well as for adults. It also provides multiple options for pediatric chest radiography: grid and nongrid, 40 inches and 72 inches SID. The Supertech® Calculator may be used to create an entire technique chart or may be kept in the x-ray room to derive individual exposures as needed.

When formulating pediatric techniques, a short exposure time is always an advantage. A high mA setting that permits the least possible exposure time will help to avoid repeat exposures caused by patient motion. If there is any likelihood that the patient will move during the exposure, do not lengthen the exposure time by using the small focal spot or the slow-speed detail cassettes for pediatric studies. Even slight motion defeats these efforts to provide fine detail.

Pediatric Pathology

Children tend to put things in their mouths, so it is not unusual for them to swallow foreign bodies.

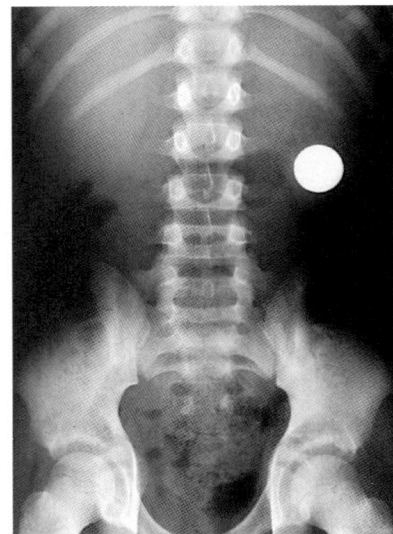

Fig. 18-30 AP abdomen radiograph shows foreign body (coin) in child's stomach.

While swallowed objects may be alarming to a parent, smooth objects, such as coins, do not often cause a serious problem. They are likely to be located radiographically in the stomach (Fig. 18-30) or the intestines and will pass without incident in a few days. When the child is in distress and has difficulty swallowing, the object may be lodged in the upper esophagus (Fig. 18-31). AP and lateral projections to include the neck and chest area are indicated in these cases.

Foreign bodies in the mouth or nose may also be **aspirated,** that is, inhaled into the trachea or a bronchus.

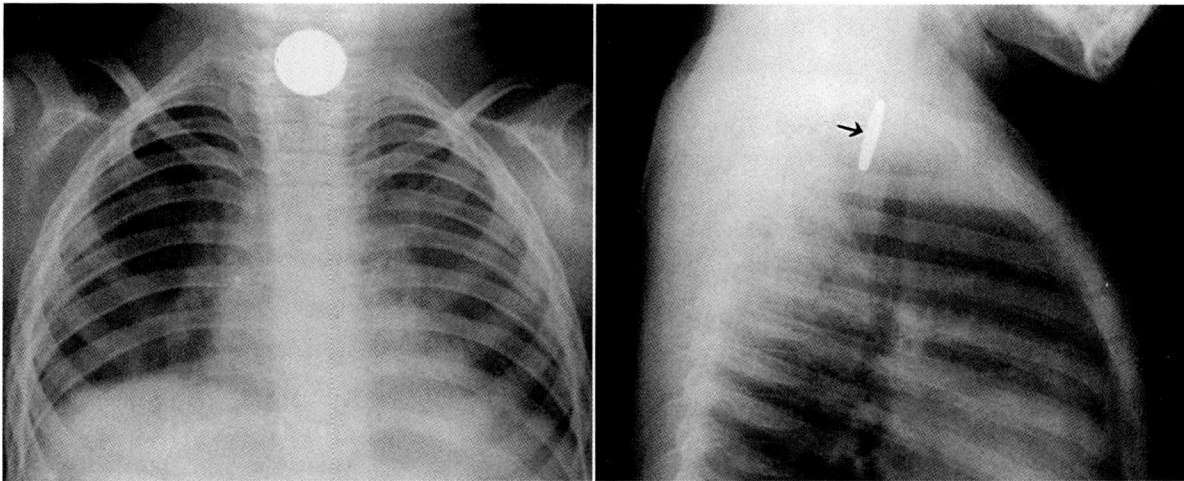

Fig. 18-31 Frontal and lateral chest radiographs demonstrate foreign body (coin) in child's lower cervical esophagus.

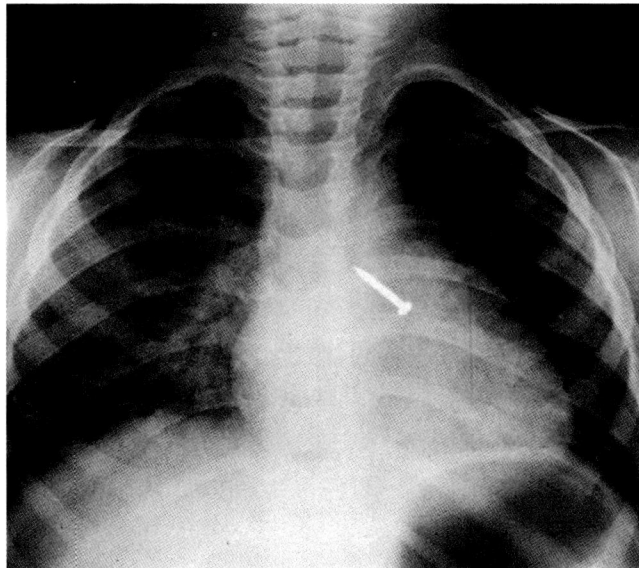

Fig. 18-32 PA chest radiograph shows foreign body (nail) aspirated into left main bronchus.

Fig. 18-32 illustrates an opaque foreign body in the left main bronchus. Nonopaque items such as plastic beads or peanuts may also be aspirated and are more difficult to locate. Frontal chest radiographs in both inspiration and expiration are sometimes used to demonstrate failure of a lung segment to expand. This helps to identify the location of a bronchial blockage, even when the aspirated object cannot be seen on the radiograph.

Children's long bones are more flexible than those of adults, so they are far more likely to have incomplete fractures. The periosteum ruptures and the cortex separates on one side of the bone, but the other side remains intact. This is called a greenstick fracture. Fig. 18-33 illustrates a greenstick fracture of the humerus in a 3-year-old girl.

Problems with the endocrine glands may affect the growth of a child. When endocrine pathology is suspected, x-rays may be used to evaluate the degree of maturity of the skeleton. Different sites may be radiographed, depending on the actual age of the child, but the wrist is common because it has many different growth centers that mature at different ages. Fig. 18-34 illustrates the wrist of a child with severe hypothyroidism. The skeletal maturity is typical of a child 2 to 3 years old, but this child's chronologic age was 11.

Fig. 18-35 is an example of a condition known as slipped capital femoral epiphysis. The growth plate of the femoral heads has separated and the epiphysis is displaced medially. This condition usually

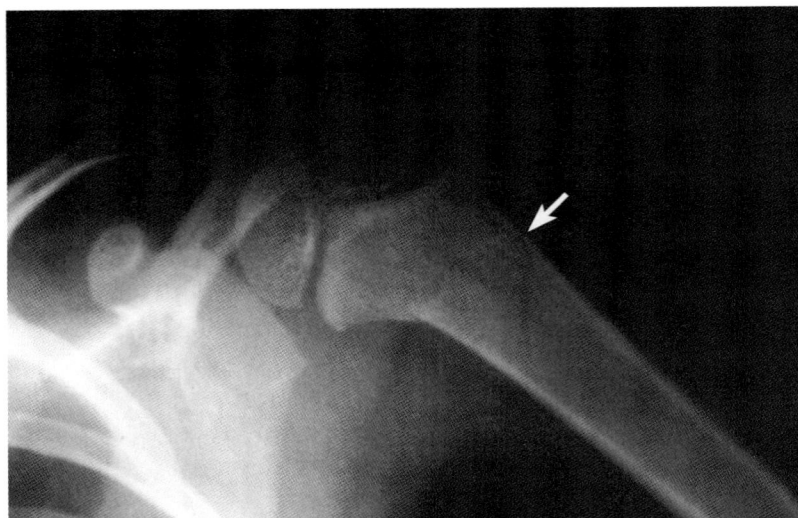

Fig. 18-33 Greenstick fracture of humerus in a 3-year-old girl.

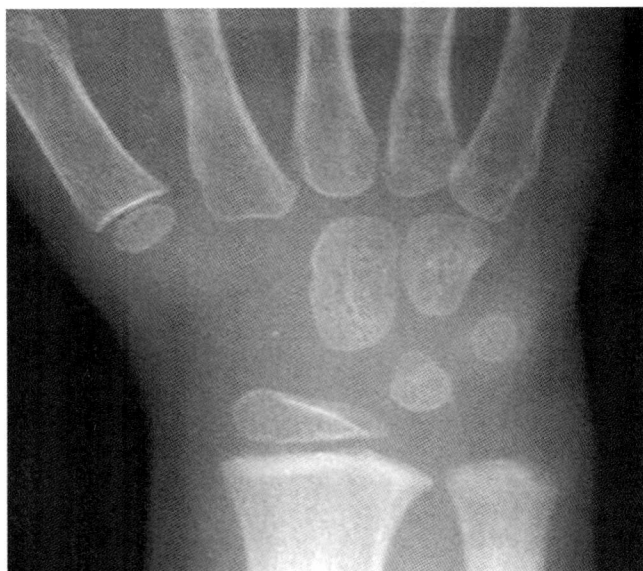

Fig. 18-34 Bone age study of wrist on 11-year-old child reveals severely retarded skeletal maturation due to hypothyroidism. This wrist radiograph resembles that of a 2-year-old child.

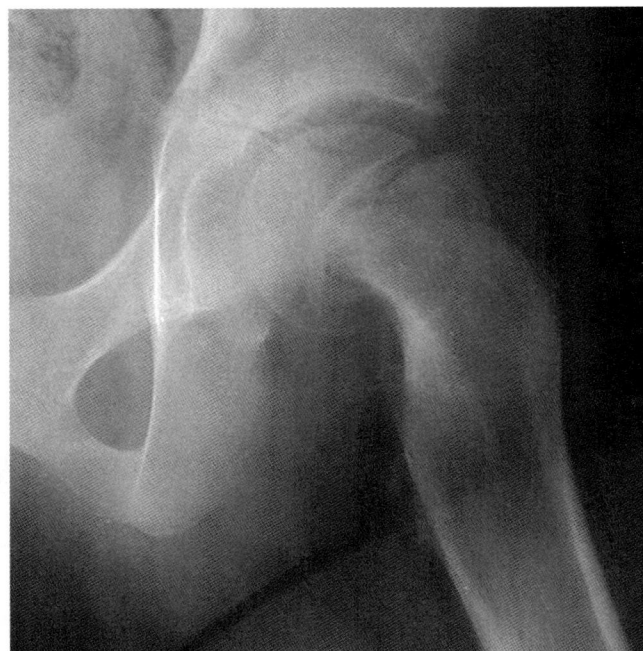

Fig. 18-35 Bilateral slipped capital femoral epiphysis in overweight boy, age 11.

occurs in overweight children, who place too much stress on the growth plate before it is strong enough to support the excess weight.

Whenever an injured child is brought for treatment, physicians and other health care personnel should be alert for indications of **battered child syndrome,** the characteristics of child abuse. Battered child syndrome is sometimes also referred to as **nonaccidental trauma,** abbreviated **NAT.** There are four signs that should raise concern about this possibility:

1. Multiple injuries
2. Evidence of chronic or repeated injury with no other explanation

3. Injuries that are not consistent with the parents' report of the trauma
4. Failure to seek prompt treatment for serious injury

Fig. 18-36 illustrates a classic case. The radiographs of the lower legs show periosteal reaction from repeated bruising of the bone. This type of injury is sometimes referred to as CML, classic metaphyseal lesion. Because abuse was suspected, a skeletal survey was done. That is, radiographs were taken of the skull, extremities, and thorax. The chest and rib study showed highly suspicious fractures of five ribs, which had partially healed. Four months after the skeletal survey the child was again treated, this time for a

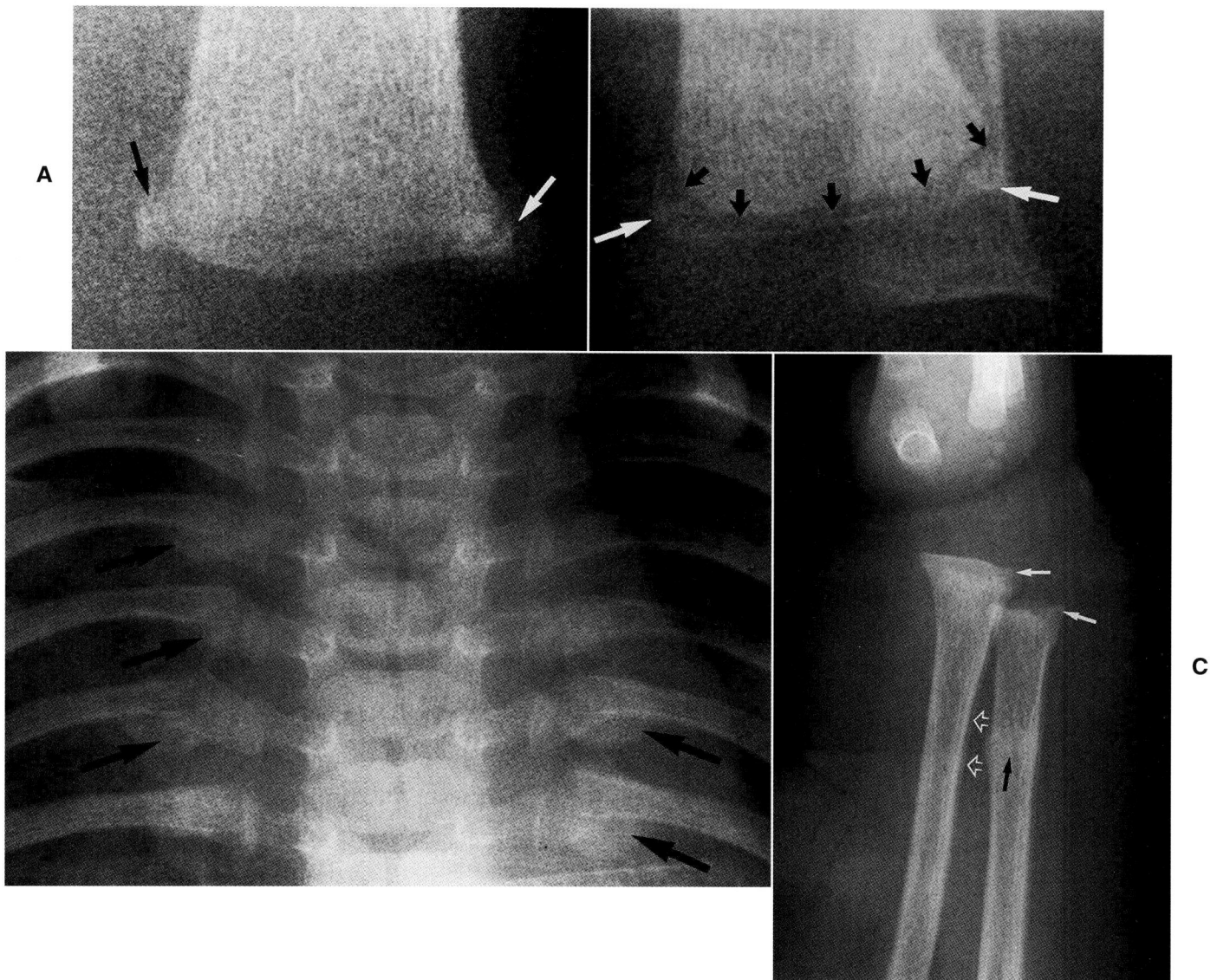

Fig. 18-36 Battered child syndrome. **A,** Distal tibias reveal periosteal reaction from repeated bruising of the bone. **B,** Skeletal survey revealed five healing rib fractures *(arrows)*. **C,** Four months later, patient was treated for fracture of the forearm. Radiographs revealed evidence conflicting with parent's report of trauma.

fracture of the forearm. The parent stated that this injury had occurred on the previous day, but the radiographs showed evidence of healing at the fracture site, indicating that this injury had been present for a longer time. CML was also noted at the wrist. The diagnosis of battered child syndrome was undeniable.

GERIATRICS

Over the next two decades the size of the geriatric population will increase dramatically in the United States. The youngest members of the "baby boom" generation reached the age of 55 in the year 2000 and will soon reach retirement age. While age 55 is no longer considered to be "old," it is common for persons in this age group to begin experiencing health problems associated with aging and for the level and frequency of health care to increase. This generation is better educated and more self-directed than the generation it follows and will place ever-increasing demands on the health care system.

Communication

Most older adults are mentally alert and have little sensory impairment. Treated with respect, they present few problems for radiographers. On the other hand, aging is sometimes accompanied by the gradual loss of hearing, vision, mobility, or mental acuity, and radiographers must be alert to the possibility of these problems. It is important to evaluate the needs of older patients on an individual basis (Fig. 18-37).

A typical attribute of aging is a tendency to proceed at one's own pace, which is often slower than that of younger adults. Most older patients do not respond well to being pushed or hurried.

Hearing Loss   Patients with hearing loss may display levels of impairment that vary from the need to use a high-intensity hearing aid to a mild difficulty hearing voices in a high or low register. Do not assume that patients who are hard of hearing cannot communicate. Expecting others to listen and speak for them may be offensive. Some older adults deny that their hearing is failing or are embarrassed to admit that they cannot hear clearly. They may guess at what is being said or pretend to understand, so it is important for you to validate that you have been understood correctly. The following list of suggestions will help to improve communication with patients whose hearing is impaired:

- Have their attention before you begin to speak.
- Face the person, preferably with light on your face. Lip reading may be an important supplement to their hearing.
- Hearing loss is frequently in the upper register, so speak lower as well as louder.
- Speak clearly at a moderate pace. Do not shout.
- Avoid noisy background situations.
- Rephrase when you are not understood.
- Avoid potential misunderstandings by asking open-ended questions. Validate understanding by asking patients to repeat instructions.
- Be patient.

When in doubt, ask the patient for suggestions to improve communication. Allow the patient who wears a hearing aid to retain it as long as possible. When a hearing aid must be removed for an examination, give all instructions before placing the aid in a safe location. Since visual clues are more important when hearing is impaired, do not remove the patient's glasses until necessary.

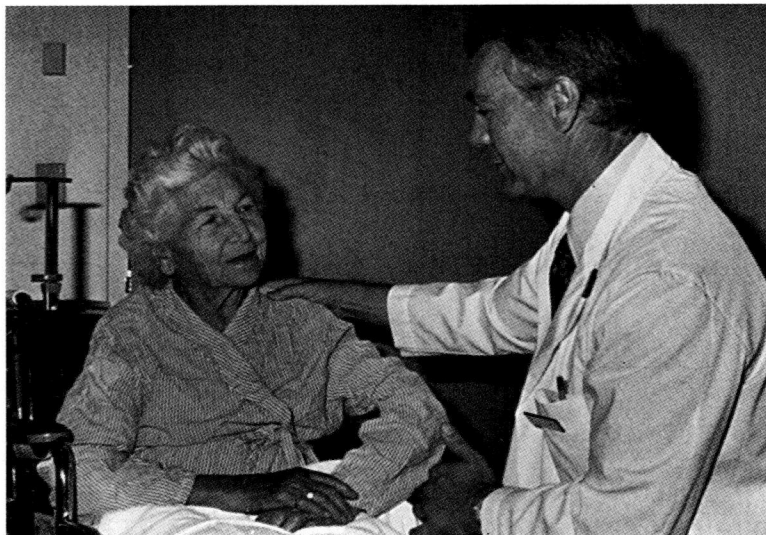

Fig. 18-37 Evaluate the needs of older patients on an individual basis.

Failing Vision   Persons with failing vision usually see better in bright light. They may be able to walk about and to recognize faces but be unable to read fine print. Offer to read written material without waiting for the patient to request this assistance. Patients who manage quite well with glasses may be disoriented and unable to walk around safely when their glasses are removed. Allow these patients to retain their glasses as long as possible, and stand by to assist when they must move about without them.

Impaired Mental Function   Loss of mental function is a part of the aging process for some individuals. **Organic brain syndrome** is a term that refers to a large group of disorders associated with brain damage or impaired cerebral function. In older persons the cause is often either **Alzheimer's disease** or circulatory impairment. Alzheimer's disease is a specific type of brain tissue deterioration that causes memory loss and gradual deterioration of mental function. It is often indistinguishable from other forms of organic brain syndrome. Circulatory impairment may be caused by a major stroke or a series of small strokes. Mental impairment may involve orientation, memory, intellect, judgment, and/or insight. Mood and personality may also be affected. Medication, illness, or injury can cause similar symptoms.

While these patients may be confused about where they are or why they have come, their memory of the past may be quite clear. It sometimes helps them to focus if you converse with them about their early life. Short-term memory is often diminished, so you may need to repeat instructions. Keep them simple and give them one at a time. Using valid choices and treating aged patients with the respect due any adult will help them maintain their sense of identity.

Physical Changes That Accompany Aging

As the skeleton ages, the bones tend to lose their calcium content, becoming porous and more radiolucent. This process is variously referred to as **demineralization, osteopenia,** or **osteoporosis.**

The soft tissues also undergo change with age. The muscle tissues tend to atrophy and become fatty, causing them to be more radiolucent than the muscles of younger adults. The subcutaneous fat layer that cushions the skin of younger persons is gradually lost with age and may be nonexistent in the elderly. The skin also loses its elasticity. For these reasons the skin of older persons may be very fragile. Any shear pressure may cause it to tear and bleed. The skin of the feet and legs is especially delicate on patients whose circulation is compromised. Veins are also fragile, causing older patients to bruise easily. Pay special attention to avoid bumping the extremities as you position the patient, and avoid wearing jewelry on your hands or wrists that could harm a patient during the process of moving or positioning. Sliding across the table surface may injure the skin, so it is best to place a sheet on the table and slide the sheet to move the patient.

Older or debilitated patients may develop ulcerated areas over bony prominences when pressure is exerted for even a short period of time, especially on the hard surface of a radiographic table. These lesions are called **decubitus ulcers** or bedsores. They are caused when pressure on a limited area inhibits circulation, depriving the cells of oxygen and nutrition. When pressure is not relieved and circulation restored promptly, the cells in the central portion may die, causing the beginning of ulcer formation. Pressure for as little as 10 minutes may be sufficient to initiate this process in a high-risk patient. Over the course of a few days, the tissue breaks down and the ulcer becomes apparent on the skin surface. Since the damage is not visible immediately, the true origin of these lesions may not be recognized. High-risk patients are weak or debilitated and may be in a poor nutritional state with impaired circulation. When this is the case, the ulcers may not heal well and may even require skin grafting.

Decubitus ulcers can be prevented by padding bony prominences with radiolucent sponges (Fig. 18-38). Pay particular attention to the midthoracic area, the sacrum, and the heels when the patient is supine. In the lateral position, provide cushioning under the trochanters, knees, and ankles. Alternatively, a full-size table pad may be used. Radiolucent pads with plastic covers are available to fit radiographic tables and should always be used when the procedure is likely to be prolonged. Failure to protect the skin also increases the likelihood of decubitus ulcer formation.

Formulating Techniques for Geriatric Exposures

Because the bones and soft tissues of elderly persons are more easily penetrated than those of younger adults, exposures must be adjusted to allow for these differences in tissue density. Demineralization of the bone produces less subject contrast on radiographs, so the best result is obtained by reducing the kVp. A reduction of 4 to 6 kVp is usually appropriate for patients over 70 years of age. Thin patients over age 80 may require a reduction of up to 10 kVp.

Geriatric Pathology

The aging process is generally accompanied by aches and pains. Osteoarthritis was discussed in Chapter 15. It is very common in the elderly, causing joints to be stiff and painful. There is no cure for it, and the usual treatment is a nonprescription pain medication such as Tylenol or Advil. But pain may also have other causes that require a different treatment, so

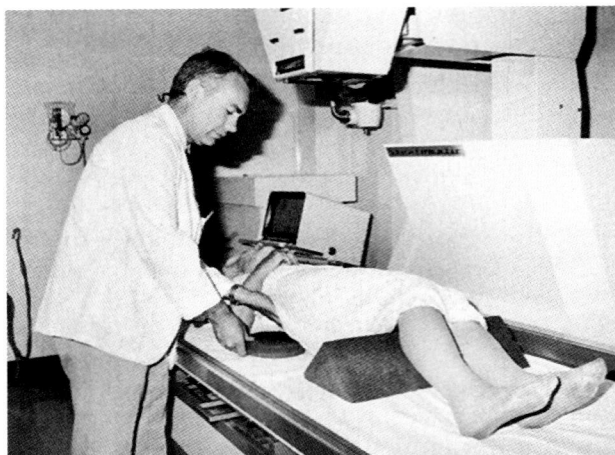

Fig. 18-38 Positioning with radiolucent sponges under bony prominences provides both comfort and safety.

complaints of pain must be investigated to rule out other possibilities. Radiography is often a significant part of this evaluation process.

Heart disease is seen frequently in older patients. Chest radiography is an important aspect of the diagnosis and continued evaluation of patients with heart conditions.

Many older patients have gastrointestinal complaints. A common cause is **diverticulitis,** a degenerative inflammatory disease of the colon. Small sacs or pouches called diverticuli form in the walls of the colon and feces tends to stagnate within them, causing inflammation. The symptoms of this condition are constipation and/or diarrhea with abdominal pain or cramping. The radiographic diagnosis of diverticulitis requires fluoroscopy with a contrast medium instilled rectally. This examination is called a barium enema or lower gastrointestinal series. It is beyond the scope of this text, since it is not usually performed by limited radiographers. Severe, chronic diverticulitis may lead to sufficient thickening of the colon walls to cause bowel obstruction. Also, diverticuli may rupture, allowing gas and intestinal content to leak into the peritoneal cavity and causing peritonitis. These conditions cause acute abdominal pain and may be diagnosed by plain films of the abdomen. The procedure for radiography of acute abdominal conditions is discussed in Chapter 16.

Parkinson's disease is a degenerative condition of the nervous system that sometimes attacks in middle age but is far more common in the elderly. The characteristic symptom is a fine tremor that may begin in a hand or foot and gradually spreads to involve all parts of the body. As the disease advances, the body becomes weak and rigid; the patient has a peculiar gait and a lack of facial expression. Parkinson's disease is not evaluated with radiography, but patients with this condition may require radiographs for other reasons. The tremors complicate radiography because these patients may be unable to hold still during exposures. These patients will be better able to relax and minimize motion if they are resting comfortably in a recumbent position. A solid object for them to grip may help to reduce hand tremors. A high mA setting with a short exposure time will help to avoid motion blur on the films.

SUMMARY

Compared to young adults, both pediatric and geriatric patients may exhibit limitations in their ability to cooperate with radiography. Respectful and age-appropriate communication will help to smooth the way, especially when accompanied by a generous amount of patience. It is important not to make assumptions based on age stereotypes, young or old. Each patient has a unique personality. Patients will respond best when approached with a genuine interest and concern for them as individuals.

Immobilization of infants and small children without the aid of human hands is highly desirable and usually possible. A variety of excellent commercial devices is available for this purpose, but creative use of available equipment and supplies will allow you to improvise when such devices are not available. Radiographers must never hold patients during exposures. If a child must be held in position, this duty should be delegated to a parent or other person who is not occupationally exposed. These persons should be provided with lead apparel and instructed to position themselves to minimize exposure.

Because of differences in bone and soft tissue density, both pediatric and geriatric patients require modification of exposures found on adult technique charts. Guidelines based on the adult chart are provided in this chapter, and the Supertech Calculator provides suitable exposures as well.

An understanding of the special requirements of pediatric and geriatric patients allows the radiographer to function confidently and competently when radiography of these patients is needed. Working with the very young and the very old will undoubtedly provide some of your most delightful and satisfying experiences as a radiographer.

Review Questions

1. List three things you might do to calm an infant.
2. Demonstrate examples of appropriate praise and disapproval when dealing with a toddler.
3. Explain what is meant by "valid choice" and give two examples.
4. Describe the concerns that might face a young teen who is anticipating radiography.
5. Using a doll and a sheet, demonstrate the "mummy wrap" technique.
6. List four ways in which the anatomy of young children differs from that of adults.
7. If the technique for a PA skull radiograph on an adult patient measuring 20 cm is 12 mAs and 82 kVp at a 40-inch SID, calculate exposure factors for an 8-year-old child measuring 15 cm.
8. List five considerations that might improve communication with patients who have some degree of hearing loss and two considerations for patients with failing eyesight.
9. List three possible causes of confusion or memory loss in older patients.
10. List three physical changes that usually accompany advanced age.
11. Describe what is meant by the term "decubitus ulcer." State the cause of this condition and two precautions for its prevention.
12. If the routine technique for an AP hip radiograph on an adult patient measuring 18 cm is 8 mAs and 80 kVp, suggest exposure factors for a thin 84-year-old patient of the same size.
13. State the most common cause of the aches and pains that accompany aging.
14. List three symptoms of Parkinson's disease and two considerations for successful radiography of patients with this condition.

Film Critique

Learning Objectives

At the conclusion of this chapter the student will be able to:

- Describe optimum conditions for viewing and evaluating radiographs
- Place radiographs on viewboxes with the correct orientation
- Demonstrate a systematic review of a radiograph for diagnostic, technical, and esthetic quality
- Recognize artifacts and technical errors on radiographs and state their causes
- Suggest appropriate changes in technique or procedure when film quality is less than optimum
- List appropriate criteria for determining whether a radiograph should be repeated

Much of the information in this chapter is not new. Facts from many chapters are brought together here and considered with respect to the evaluation of film quality.

Critical film review is significant to your patients and to the physician who will read the films. It is also essential to your continuing education as a radiographer. Film critique is the process by which you determine whether each film is correctly identified and marked and whether it has sufficient diagnostic quality, that is, whether it meets the minimum requirements of the order or must be repeated. When a film must be repeated, your review provides information to ensure that the repeat is satisfactory.

Up to this point, the text has dealt primarily with the science of radiography and the basic procedures involved in the application of that science. Radiography is also an art, and your ability to take consistently high-quality radiographs will only develop with time and practice. Every radiograph you take is an opportunity to learn; each film you review can teach you something that will improve your ability. Your skills will develop most rapidly if you pay close attention to the results of your work, always striving for excellence.

VIEWING RADIOGRAPHS

Viewing Conditions

The first consideration in viewing radiographs is the x-ray illuminator, or "viewbox." Radiographs cannot be accurately evaluated by holding them up to a ceiling light or window. A suitable viewbox must be used. It is highly desirable that all viewboxes in your facility have the same type and number of fluorescent tubes. This ensures that the brightness and color of the viewing lights are consistent. Films viewed in the physician's office will have the same appearance as when viewed by the radiographer in the x-ray department. Viewbox surfaces must be kept clean and free of artifacts. Window cleaner and a soft cloth should be available for this purpose.

It is not necessary to view films in a dark room, but a low light level in the viewing area is needed. Too much light causes the pupils of the eyes to contract, admitting less light to the eye when viewing, and causing radiographs to appear dark. A dark mask made of black film can be used to surround a small film on the viewbox, enhancing your ability to perceive its detail accurately.

Most x-ray departments have a high-intensity light, sometimes called a "hot light," that is used to view details in portions of a film that are too dark to be seen clearly on the viewbox. These lights are literally hot and will damage the film when held too close for too long. While such lights are convenient in certain circumstances, the radiographer must not rely on

them for routine viewing. Radiographs should be repeated when they are so dark that the principal details can only be appreciated with a hot light.

Film Orientation

Radiographs are generally viewed "right side up," with the most superior aspect of the anatomy at the top. Regardless of the patient's position when the film is exposed, it is customary to view frontal radiographs, both AP and PA, as if the patient were facing the viewer in anatomic position. The patient's right side is toward the viewer's left. Oblique radiographs are viewed using the same general rules. Lateral radiographs are usually viewed in the same position as they are taken; for example, a left lateral projection is hung on the viewbox with the patient facing toward the viewer's right.

Exceptions to these basic rules are made for examinations of the distal upper extremity and the foot and for decubitus projections. Radiographs of the fingers, hand, wrist, forearm, and foot are usually viewed with the distal aspect at the top. Decubitus projections are often viewed horizontally, in the same position as they are taken.

The radiographs in this text are presented using these conventional rules for viewing.

Some chiropractic physicians prefer to view frontal and oblique spine radiographs from the same perspective used to examine the patient, as if they were facing the patient's back. In this case, the film is placed on the viewbox so that the patient's left side is on the viewer's left.

SYSTEMATIC FILM REVIEW

With practice, you will learn to critique a film quite rapidly. It is important, however, not to rush this process or significant details may be overlooked. An accurate assessment of film quality requires a systematic approach. You may find it helpful to use the acronym, **I AM ExPERT.** These letters stand for identification, anatomy, marking, exposure, processing, esthetic considerations, radiation safety, and troubleshooting.

- **Identification.** First, check the film identification. Is it clear and complete? Does it match the identification on the requisition?
- **Anatomy.** Is the pertinent anatomy included on the radiograph and clearly visible? Was the patient properly positioned? Look for evidence of rotation or superimposed structures that indicate improper position.
- **Marking.** Is the correct side marker clearly visible and in the proper location? If your facility's protocol requires additional markers for this film, such as flexion, weight bearing, or

upright, check for placement and clear visibility of additional markers as well.

- **Exposure.** Were the exposure factors appropriate? Is the film too dark? Too light? Does it have sufficient contrast? Too much contrast? Are the essential features of the radiograph sharp and distinct?
- **Processing.** Does the radiograph show evidence of darkroom fog or handling artifacts? If the film is too dark, too light, or lacking in contrast, is it possible that these problems are attributable to processing?
- **Esthetic considerations.** Does this film have artistic merit? Do artifacts, misalignments, or other features of the image detract from its general appearance?
- **Radiation safety.** Is there evidence of collimation on at least three margins of the film? Where appropriate, is shielding apparent and correctly placed?
- **Troubleshooting.** Identify the cause of any problems noted on the film. Are they so significant as to require a repeat? If so, what changes are needed for the repeat to be successful? Regardless of whether a repeat is required, what steps are needed to improve film quality in the future?

Each of these aspects of film critique is discussed in greater detail later in this chapter.

FILM IDENTIFICATION AND MARKERS

The first thing to check when reviewing a radiograph is the patient identification. This information must be complete and legible. Certainly, it is a good practice to be aware of the need for accurate film identification and to cultivate a habit of identifying films properly. Errors sometimes occur, however. When films are not properly labeled, your facility should have some method of rectifying the mistake. It is not necessary to repeat a film that has been mislabeled. The usual practice is to apply a gummed label with the correct information over the incorrect label. This method is definitely preferable to marking out the original label and writing directly on the film.

Check to be certain that the correct side marker (and any other required marker) is clearly visible and properly located. Correct marker placement is discussed in Chapter 12. Many facilities stock gummed labels with the words "right" and "left." These labels can be applied in place of a missing or incomplete marker or to cover an inaccurate marker. Of course, it is best to mark films properly when they are exposed, but using a gummed label is preferable to writing on the film.

It is a common misconception that identification and markers added after the film has been processed are "not legal." You may hear that radiographs with added markings are not admissible as evidence in court. In fact, any radiograph is admissible in court, but radiographs where the identification and markers are applied using standard procedures may have more credibility as evidence. Some facilities have policies that require radiographers to initial and date any changes made to radiographs after they have been processed.

ANATOMY AND POSITIONING ERRORS

Exclusion of Significant Anatomy

Standard film sizes and center points are used in radiography to ensure that the required anatomic structures will be included on the film. Even when these standards are followed, some significant anatomic details may be excluded due to inaccurate collimation or variations in patient size. When positioning, check to be certain the outer margins of the structures to be included are within the film margins and within the radiation field. The image should include the surrounding soft tissues as well as the bony structures. It is helpful to remember that all radiographic images are magnified to some degree. Double check that the shadow of the outer margin of the part is seen within the collimator light field, since this shadow is magnified to approximately the same extent as the radiographic image.

Fig. 19-1 shows incomplete anatomic demonstration of the scapula caused by excessive collimation. The hand image in Fig. 19-2 is incomplete because the fingertips are not included. In both cases the film should be repeated unless the physician is fully satisfied that the area of clinical interest is adequately shown. The lower leg radiograph in Fig. 19-3 is incomplete because it does not demonstrate the proximal tibia and the knee joint. Depending on the specific area of clinical interest, this examination may be sufficient. If not, the best approach would be to take a separate film of the knee area to include the proximal tibia.

Fig. 19-4 shows incomplete anatomic demonstration of the sacrum caused by poor alignment of the x-ray beam to the film. There are three possible causes for this error: failure to secure the cassette in the film tray, failure to push the film tray all the way into the table or upright bucky, and/or failure to align the x-ray tube to the center of the grid. In this case it is apparent that grid alignment is not a problem because there is no evidence of grid cut-off (see Chapter 9).

Incorrect Positioning

Routine positions are established to ensure a comprehensive evaluation of the anatomy. Errors in diagnosis can occur when structures are not well visu-

Fig. 19-1 Lateral scapula with incomplete anatomy demonstration resulting from improper collimation.

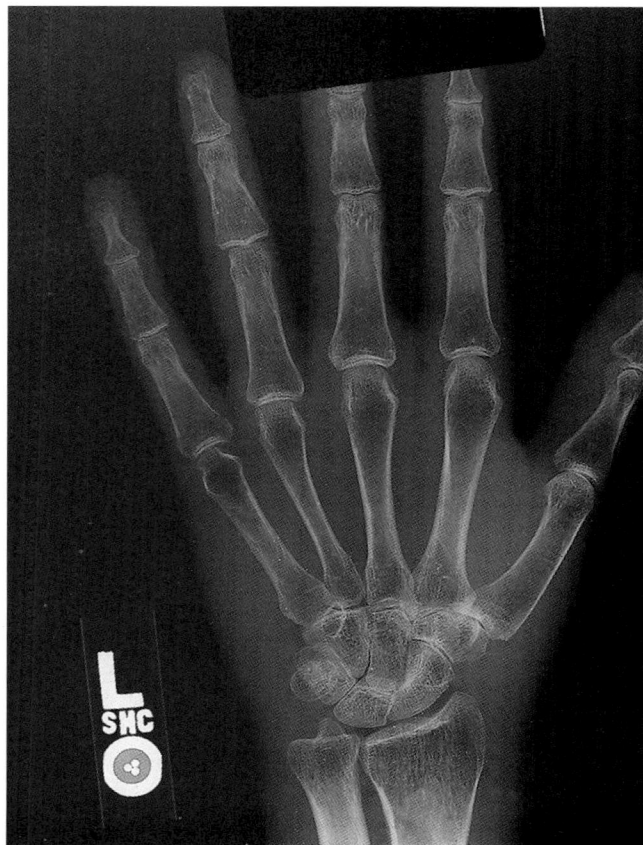

Fig. 19-2 Hand examination is incomplete without inclusion of fingertips.

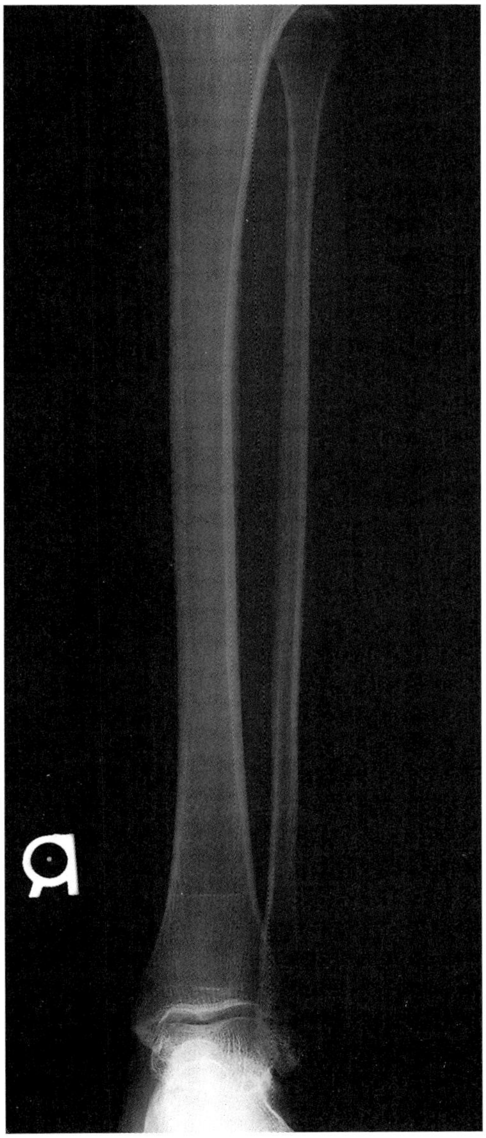

Fig. 19-3 Lower leg examination fails to demonstrate proximal tibia and knee joint.

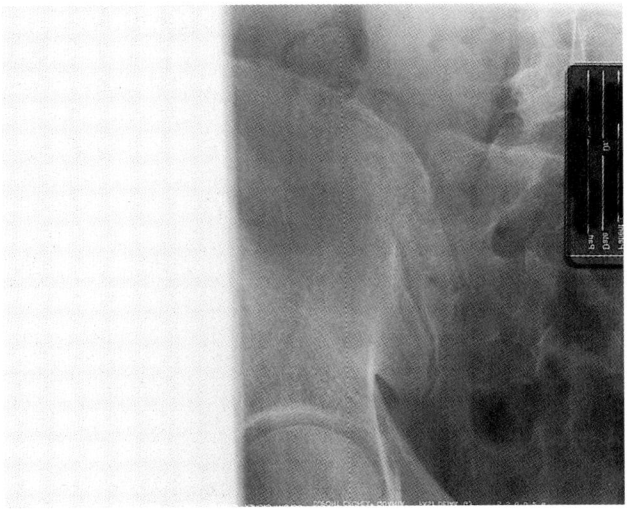

Fig. 19-4 Radiograph is incomplete because film was not centered to the x-ray beam. This may occur when the film is loose in the tray or the tray is not fully inserted in the table.

Fig. 19-5 Visualization of elbow joint is compromised by improper position. Humerus was not parallel to the film.

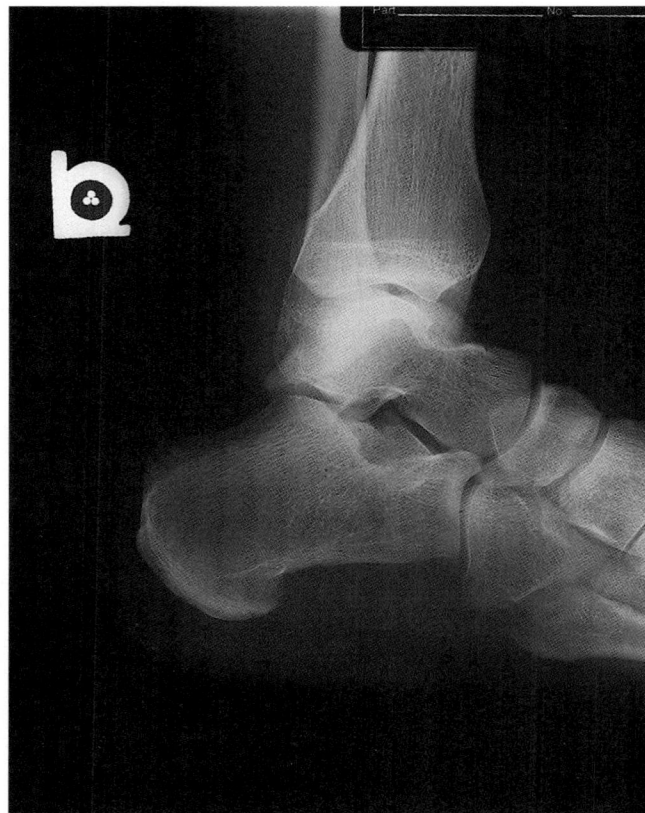

Fig. 19-6 Visualization of ankle joint is compromised by rotation of leg. Sagittal plane was not parallel to the film.

alized or when incorrect positioning prevents comparison with the usual normal radiographic appearance. If you cannot readily identify the errors on the radiographs in this section or on radiographs that you have taken, compare them to the normal radiographs illustrated in Chapters 13 through 17.

Figs. 19-5 and 19-6 illustrate common examples of incorrect extremity positioning. In Fig. 19-5, the elbow joint is not well visualized because the radiographer failed to position the patient with the axilla at table level. Distortion of the joint in this case is due to the fact that the humerus is not parallel to the film. In Fig. 19-6, visualization of the ankle joint is compromised because of part rotation. The sagittal plane of the ankle and lower leg was not parallel to the film.

Figs. 19-7 and 19-8 are examples of the errors that occur in the AP projection of the upper cervical spine when the skull is positioned with incorrect alignment of the transverse plane. Fig. 19-7 was taken with too much *flexion* of the neck, projecting the teeth over C1 and the dens. Fig. 19-8, on the other hand, is an ex-

ample of too much neck *extension,* projecting the base of the occipital bone over C1 and the lower teeth over the upper portion of C2. Both of these radiographs are unsatisfactory and must be repeated.

Fig. 19-9 is a lateral lumbar spine radiograph that is obviously unsatisfactory because the patient's arm is projected over the spine. In this case it is clear that the radiographer failed to consider the location of the patient's arms when positioning. The arms should be positioned across the chest with the hands resting on opposite shoulders or supported anterior to the body and out of the radiation field.

Fig. 19-10 illustrates two common errors in positioning for the PA projection of the chest. First, the patient's torso is somewhat rotated; that is, the coronal plane of the body is not parallel to the film. This is apparent when checking the location of the proximal ends of the clavicles in relation to the spine. When the position is correct, the clavicles will appear symmetrical with the spine centered between them. In addition, the scapulae are superimposed on the lat-

Fig. 19-7 Teeth obscure the atlas and the dens due to neck flexion.

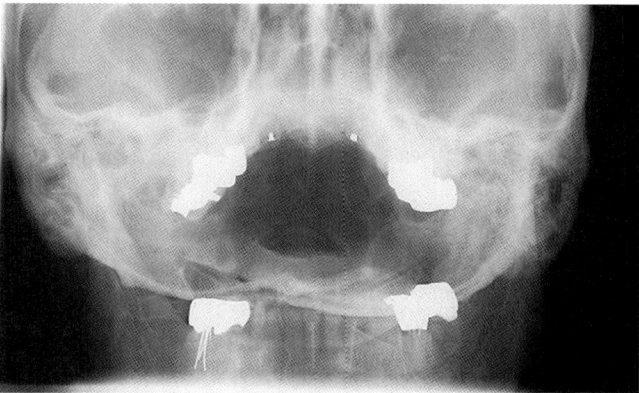

Fig. 19-8 Occipital bone obscures the atlas and the dens, and lower teeth obscure the upper body of C2 due to neck extension.

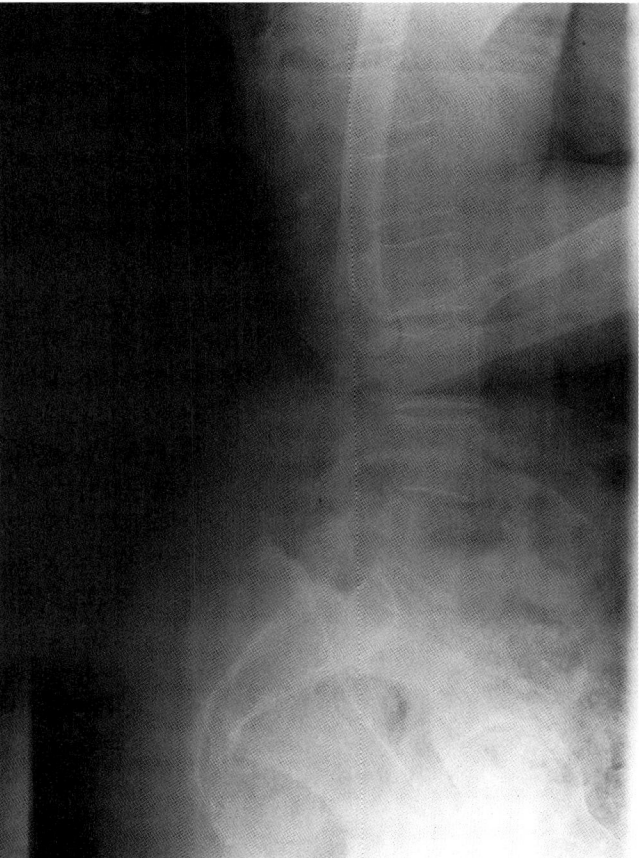

Fig. 19-9 Poor attention to positioning details resulted in arm projected over lumbar vertebrae.

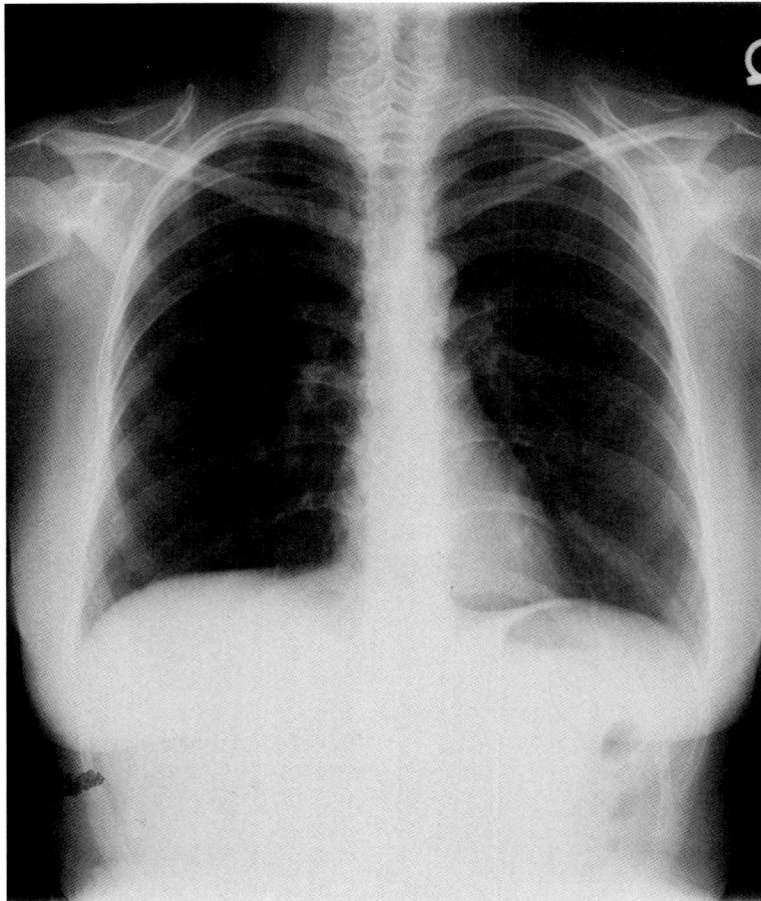

Fig. 19-10 PA chest shows rotation and projection of scapulae over lateral lung fields.

Fig. 19-11 Asymmetry of PA skull due to rotation.

eral lung fields because of inadequate anterior rotation of the patient's shoulders.

Fig. 19-11 illustrates poor position due to rotation of the skull for the PA projection. Note that the right and left halves of the image are not symmetrical. When the skull is properly positioned, any lack of symmetry of cranial structures is an important diagnostic sign. When the position is rotated, cranial symmetry cannot be adequately evaluated and a significant aspect of diagnostic quality is lost. For PA projections, there are few visual clues to assist in determining whether the skull is rotated, so the ability to *feel* whether the alignment is correct is an essential positioning skill.

Some patients cannot assume standard, routine positions for radiography. Compromise is required when deformity, injury, or discomfort prevents patients from maintaining the usual position. The positioning sections of Chapters 13 to 17 provide some alternative positions for common situations where patients may be unable to comply with routine positioning. For example, the PA thumb projection may be substituted for the AP projection when the patient is unable to assume a satisfactory AP position. Trauma views of the shoulder and upper humerus may be used when the patient is unable to rotate the arm for a routine shoulder or humerus examination. Where no standard position can be used, the radiographer should strive to obtain at least two different projections of the area of interest that are as close to standard as possible. All extremity examinations must include at least one joint, and the same joint should appear on all views. Radiographs taken in nonstandard positions should be shown to the physician before concluding that the examination is complete.

EXPOSURE FACTORS

This term in the I AM ExPERT acronym is a reminder to consider the image quality factors: radiographic density, contrast, definition, and distortion.

Radiographic Density Problems

First, consider the overall density of the radiograph. Is it too dark or too light? If so, to what degree? When using a reliable technique chart to provide exposures for patients within the usual size range, it should not be a common occurrence that films are significantly over- or underexposed. When films are grossly too light or too dark without other explanation, double check to be certain that the technique chart has been followed with respect to the source-image distance, cassette type, and use of a grid or bucky. Using a regular (rapid) cassette with an exposure calculated for detail screens will produce an image that is much too dark as seen in the phantom ankle radiograph (Fig.

19-12, *A*). The opposite is also true; using rapid screen exposure factors with detail cassettes will cause underexposure (Fig. 19-12, *B*). Using bucky techniques for tabletop exposures produces films that are too dark, while tabletop techniques used with a grid will cause underexposure. Improper SID also affects radiographic density; when the SID is too great, the film will be too light and vice versa.

Another reason why a film may be too light is that the radiographer released the exposure switch before the exposure was complete. When this occurs, the exposure is terminated before the preset exposure time has elapsed, resulting in underexposure. To avoid this possibility, always watch the exposure indicator and hold both the rotor and the exposure switches firmly until you are certain that the exposure is complete (see Chapter 5). If your finger "stutters" on the exposure switch, causing it to be momentarily released and then depressed again, the result will be a double exposure. In this case, the film will be too dark and may also exhibit image blur due to patient motion.

When a film is repeated because of incorrect exposure factors, mAs is the factor that is altered to adjust radiographic density. On control panels where mA and time are set separately, the mAs is usually adjusted by changing the exposure time. Fig. 19-13 shows a radiograph that needs to be repeated because it is too dark. In this case the PA projection of the chest was accidentally taken using the exposure factors for the lateral projection. The mAs must be reduced by approximately 65% to provide an acceptable image. The radiograph in Fig. 19-14 is too light and must be repeated with an mAs increase of approximately 100%, twice the original amount. In this case, the patient was a decathlon athlete with extremely dense muscle tissue, which was not taken into account by the radiographer. Chapter 10 explains changes in radiographic density. The density change produced by various percentages of adjustment in mAs is illustrated in Fig. 10-6.

Contrast Problems

Next, evaluate the radiographic contrast. If some portions of the image are properly exposed while other portions are too dark or too light, the film exhibits too much contrast. Fig. 19-15 shows a radiograph of a phantom ankle with too much contrast and one with kVp adjustment according to the 15% rule (see Chapter 10) to provide a longer, more acceptable, scale of contrast.

Fig. 19-16 is another example of a contrast problem. Note that while the upper lumbar region is appropriately exposed, the area of the lumbosacral joint is too light. In this case, the problem results from extreme subject contrast. The patient was slender in the mid-

A

B

Fig. 19-12 A, Overexposure occurs when detail screen technique is used with rapid screens. **B,** Underexposure occurs when rapid screen technique is used with detail screens.

Fig. 19-13 Overexposure. Correction requires mAs reduction of approximately 65%.

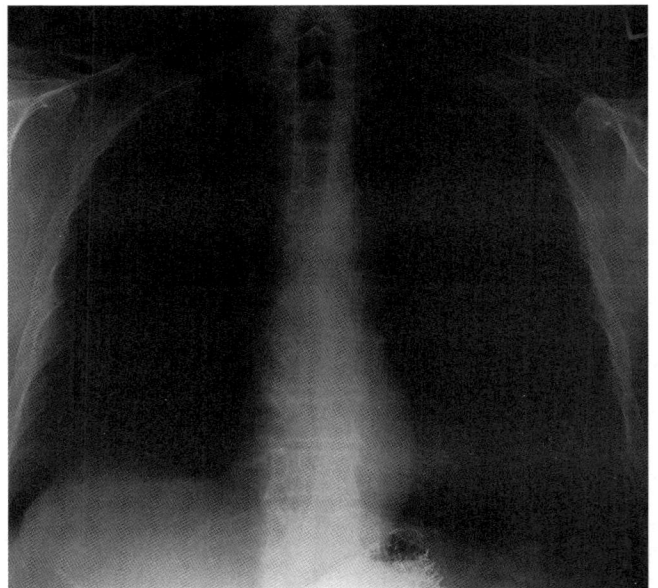

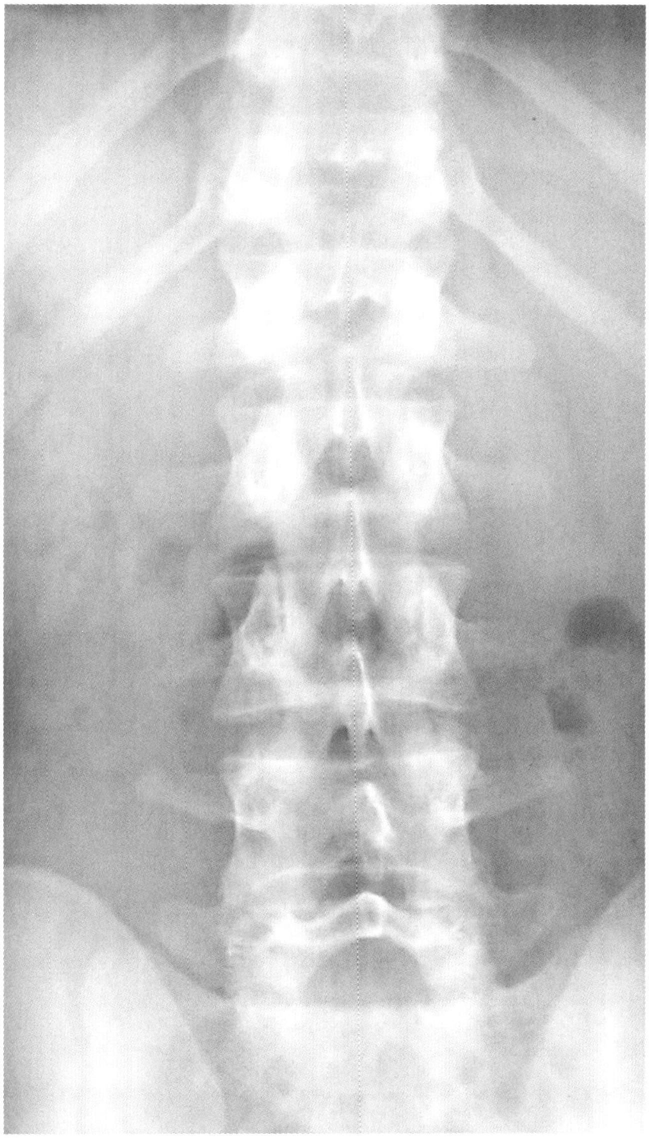

Fig. 19-14 Underexposure due to dense muscle tissue. Correction requires mAs increase of 100%.

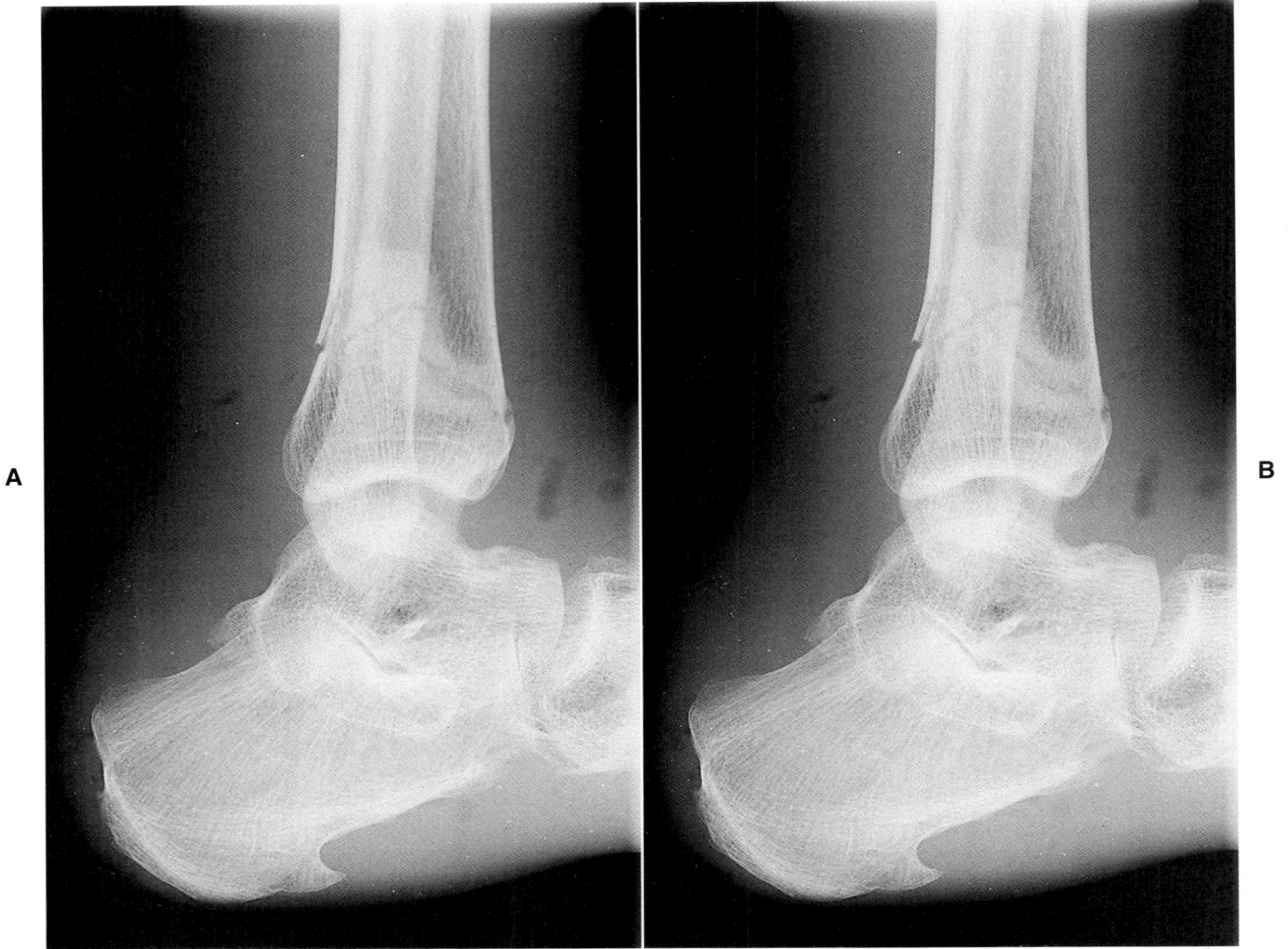

Fig. 19-15 A, Excessive contrast results in poor visualization of detail in portions of the image. **B,** Scale of contrast is improved by increasing kVp and decreasing mAs according to the 15% rule.

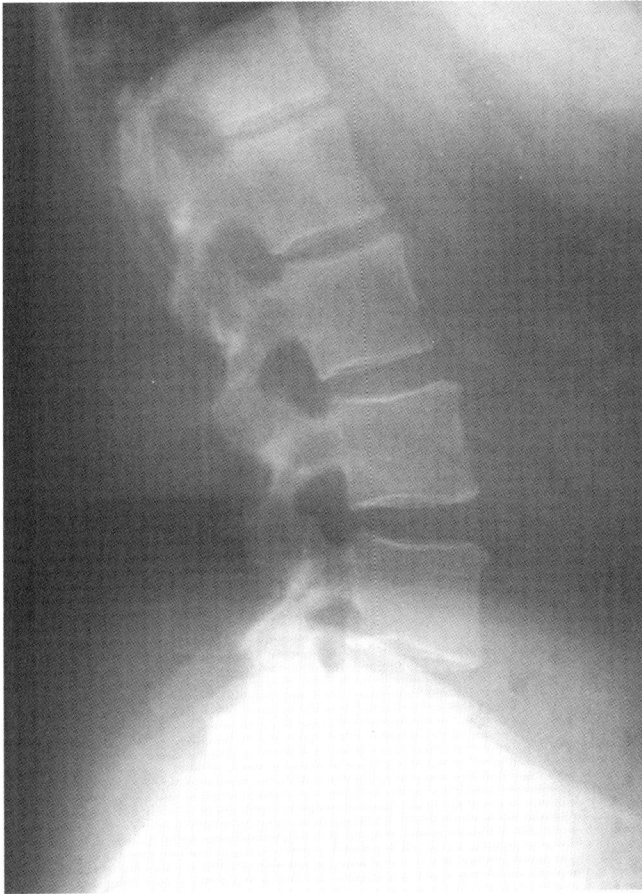

Fig. 19-16 Lumbosacral joint is underexposed due to extremes in tissue density. The problem is best corrected by taking a spot film of the lumbosacral joint.

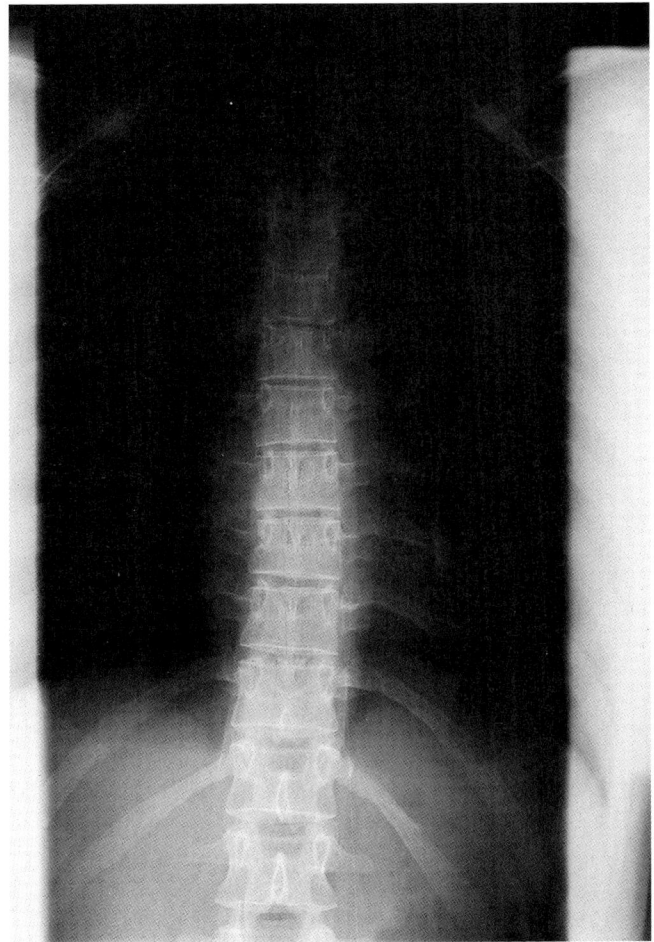

Fig. 19-17 This AP thoracic spine was exposed with the patient's head toward the cathode end of the tube. Reversing this position would utilize the anode heel effect properly, providing more uniform density.

portion of the torso but broad in the hips. There are several possible approaches to this problem. As in the previous example, the kVp could be increased using the 15% rule to provide less contrast overall and greater penetration of the lumbosacral region. Another solution would be to increase the mAs for sufficient radiographic density in the lumbosacral area while using a compensating filter attached to the collimator to prevent overexposure of the other lumbar vertebrae. Perhaps the best solution would be to take a spot film of the lumbosacral region with an increase in mAs of approximately 150%, 2½ times the original exposure.

Fig. 19-17 also represents an example of uneven radiographic density. The upper portion of the thoracic spine is too dark, while the remainder is appropriately exposed. While this case could be approached as a contrast problem as in the previous examples, the real problem here was failure to take the anode heel effect into account (see Chapter 4). If the patient were reversed on the table, placing the thinner upper por-

tion toward the anode end of the x-ray tube, the radiographic density would be more uniform.

A different contrast problem exists where contrast is insufficient to adequately demonstrate anatomic details. Fig. 19-18, *A*, illustrates such a radiograph of the lumbar spine on a large patient. Note that the details of the vertebrae are difficult to see, especially the transverse processes. In this case, the exposure was made using 100 kVp. Fig. 19-18, *B*, is a radiograph of the same patient using 80 kVp with an appropriate adjustment in mAs (see Chapter 10). It shows considerable enhancement of detail visibility attributable to increased contrast. In such cases it is also important to collimate as much as possible.

A low level of contrast is typically seen on radiographs of large body parts, such as the lumbar spine, on very large patients. This is especially true when a low grid ratio permits excessive scatter radiation fog, causing further degradation of contrast. If the use of an 8:1 or 10:1 ratio grid causes frequent contrast

Fig. 19-18 A, AP lumbar spine on a large patient was exposed at 100 kVp and lacks sufficient contrast. **B,** This radiograph on the same patient was exposed at 80 kVp. Contrast is improved with decreased kVp and increased mAs.

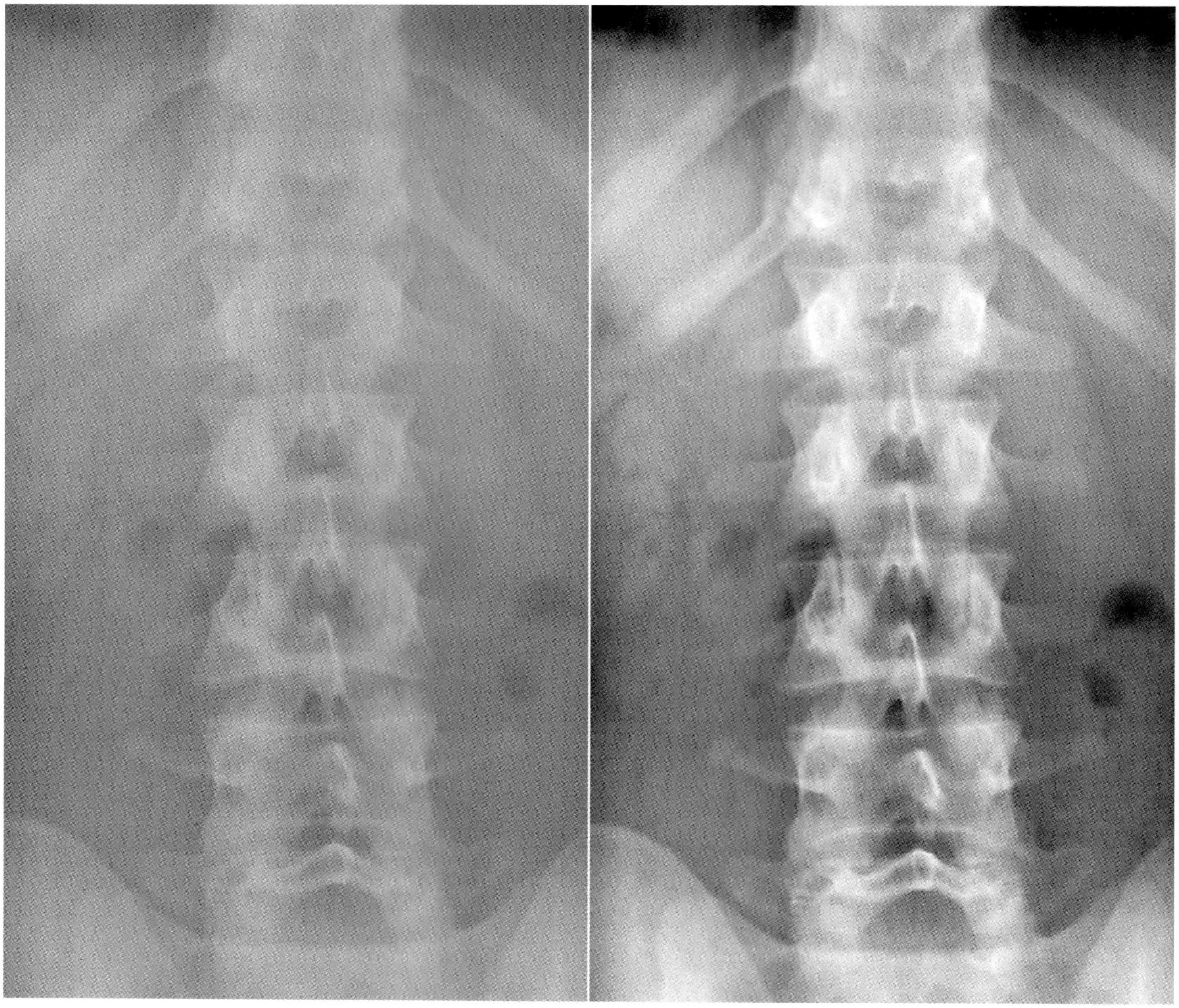

Fig. 19-19 Lateral cervical spine taken at 40-inch SID using the large focal spot has poor definition. Poor positioning and artifacts also need to be corrected when this radiograph is repeated at 72-inch SID with the small focal spot.

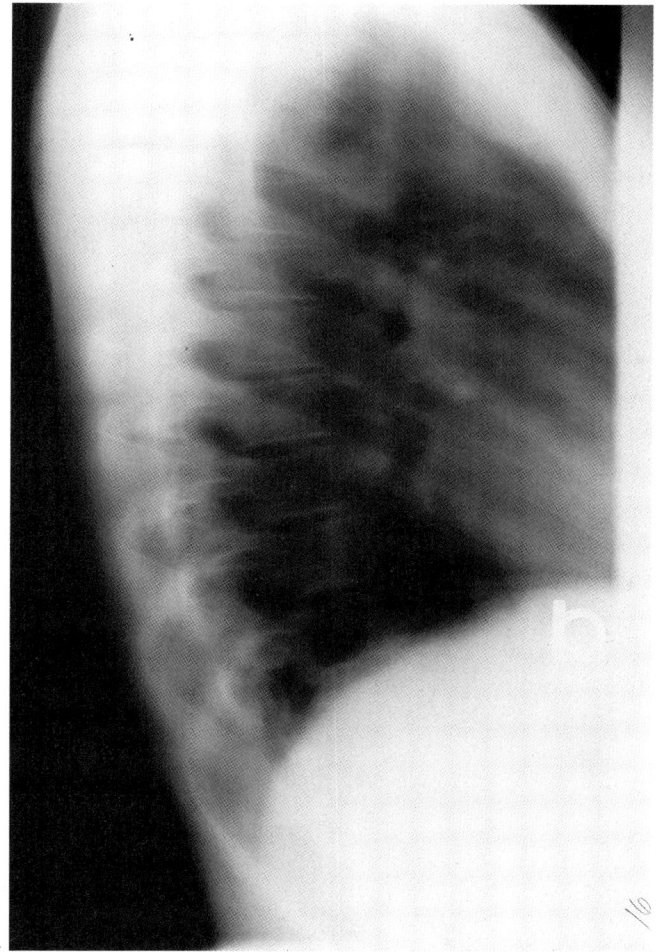

Fig. 19-20 Lateral thoracic spine radiograph exhibits motion blur.

problems with large patients, consider purchase of a 12:1 grid (see Chapter 9).

Lack of Definition

When evaluating definition on the radiograph, note whether the margins and fine lines of the image are sharp and clear. Chapter 6 explains the factors that affect sharpness of detail, which include screen speed, object-image distance (OID), focal spot size, parallax, and motion. Fig. 19-19 illustrates a lateral cervical spine radiograph that was taken at a 40-inch SID using the large focal spot. For these reasons, the film exhibits magnification distortion and poor definition. Because of the large OID involved in lateral cervical spine radiography, a 72-inch SID and the small focal spot should be used. This radiograph also shows that the patient's shoulders were not depressed, resulting in failure to

properly demonstrate the C6 and C7 vertebrae. Failure to remove the patient's glasses degrades image quality with distracting artifacts. This radiograph must be repeated to correct all of these errors.

Since screen speed, SID, and focal spot size are usually established in advance to provide the required degree of definition, blurring or unsharpness is most likely to be caused by patient motion during the exposure. Fig. 19-20 is an example of a blurred image resulting from patient motion. In this case, breathing technique (see Chapter 12) was used in the hope of blurring only the ribs and lung markings, but the patient was unable to avoid movement of the entire torso. If a radiograph must be repeated because of motion, take care that the patient's position is stable and secure. Provide clear instructions and adequate time for the patient to comply. Use sandbags or a compression band to aid in immobilization, if necessary. Use the highest possible mA setting and the shortest exposure time that will provide the desired mAs (see Chapter 10).

PROCESSING PROBLEMS

Film critique involves review for possible processing problems and film-handling errors. Information on these topics is included in Chapter 8. This portion of the critique should also include a check for evidence of screen dirt or other screen artifacts. These are discussed in Chapters 7 and 8.

Whenever a film is too light, processing should be considered as a possible cause. Fig. 19-21 illustrates two radiographs of a hand phantom taken in the same way with the same exposure factors. The first was processed just a few minutes after the processor was turned on, before it had reached operating temperature. The second was processed under standard conditions with the developer temperature at 94° F. It is apparent that processor operation can have a significant impact on film quality. To differentiate between film density problems caused by exposure and those

caused by processing, it is helpful to note aspects of the film that are not a part of the radiographic image. In this case, the film density of the background area, outside the soft tissue margins, is much lighter than usual. In addition, there is a lack of radiographic density in the identification portion of the film.

Film fog caused by age or improper storage affects all of the films in a box. These problems and those involving safelights are usually detected when performing routine sensitometric quality control evaluations (see Chapter 8). Darkroom fog diminishes radiographic contrast and, as with processor problems affecting density, can be seen in areas outside the image.

Some fog is mysterious in origin, and the source must be identified to avoid such problems in the future. Usually the film itself provides clues. Fig. 19-22, for example, shows fog over the upper third and a dark, dotted-line artifact. This problem was traced to daylight exposure of the unopened inner envelope of the film package. The radiographer had removed the envelope from the box, unaware that the envelope was not light-tight. The fog and the dark dotted line

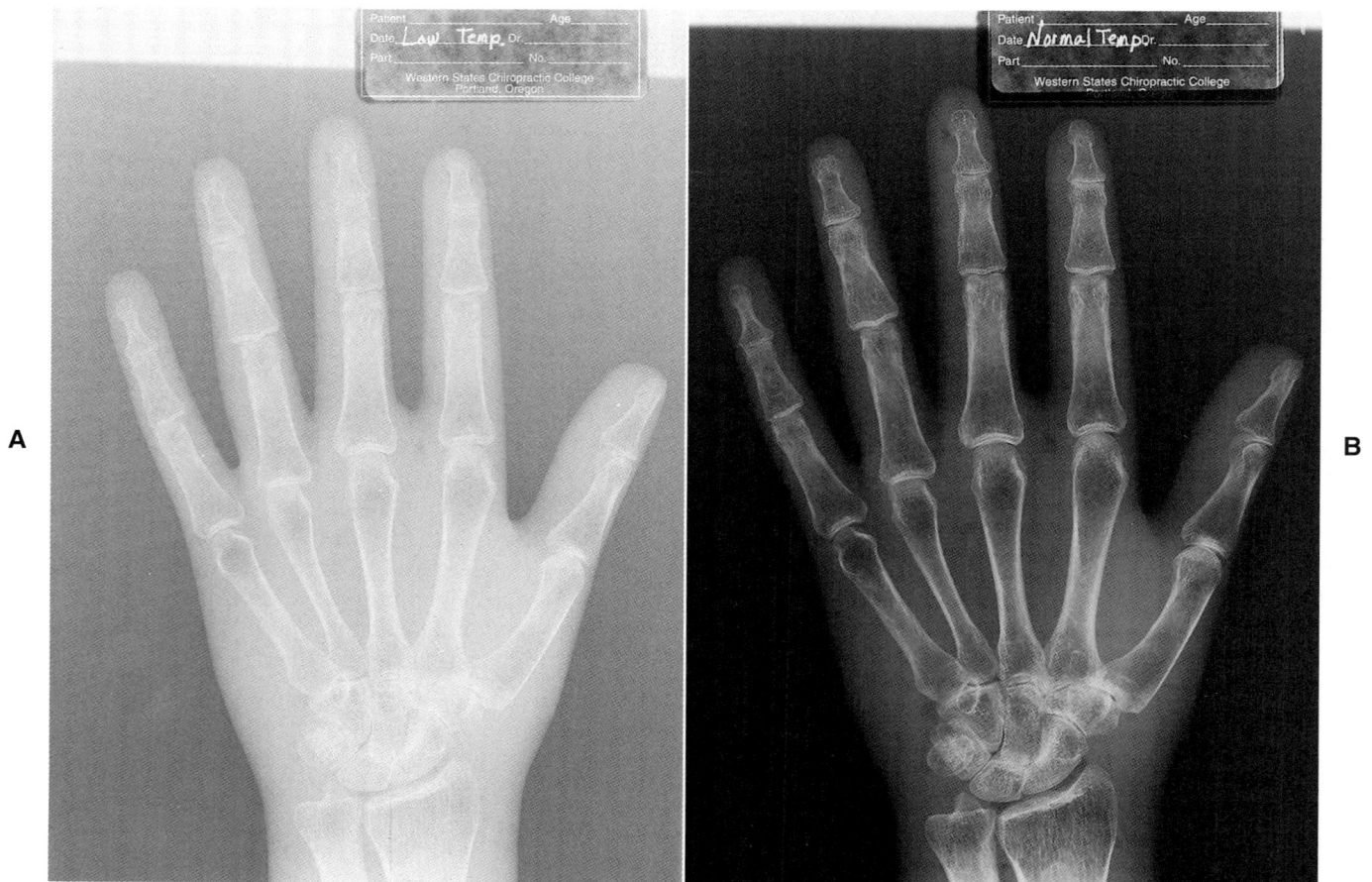

Fig. 19-21 A, Phantom hand radiograph lacks density and contrast because it was processed before the processor reached operating temperature. **B,** The same subject exposed with the same factors and processed at standard temperature.

were caused by light leaking through the envelope's perforations. In this case, only the top film was affected, but it was necessary to process additional films from the front and the back of the envelope to ensure that cassettes were not loaded with fogged film.

Fog artifacts caused by leaky cassettes and failure to properly load cassettes are illustrated in Chapter 8.

Fig. 19-23 illustrates a film that was processed in contact with another. The irregular artifacts indicate areas where processing chemicals seeped between

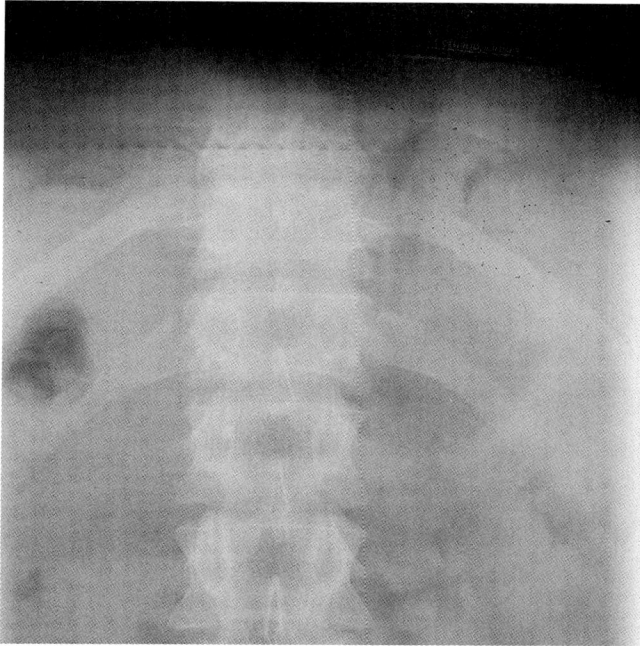

Fig. 19-22 Thoracolumbar spine radiograph exhibits fog artifact and dotted-line artifact due to daylight exposure through inner envelope of film package.

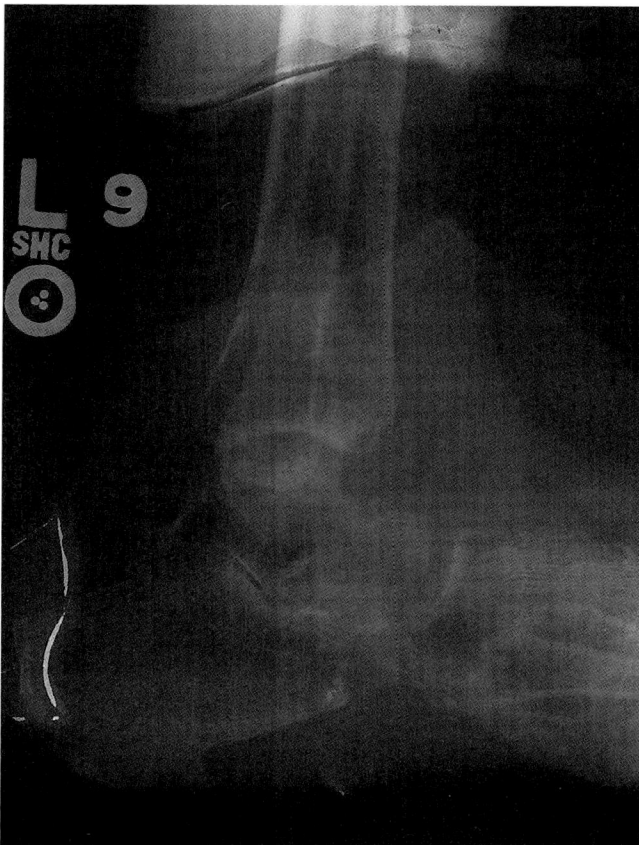

Fig. 19-23 Phantom ankle radiographs that were processed in contact with each other.

the films. This error obviously requires that the films be repeated. It may also cause the processor to jam, causing potential damage to other films that are in the processor at the time. Overlapping of films in the processor is avoided by being certain that the film is moving into the processor when it is fed. Otherwise, it may still be lying on the feed tray when the next film is fed. Be sure to wait for the signal before feeding the next film.

ESTHETIC QUALITY

The term "esthetic quality" refers to eye appeal of the radiograph. Esthetic quality is sometimes called "artistic merit," a value that is placed on the excellence of the film's general appearance.

Radiographs that lack esthetic quality have a careless or sloppy appearance, even though they may meet the standards required for diagnosis. *Esthetic quality may also have diagnostic significance.* Informal research studies have shown that radiographs with a high de-

gree of artistic merit are more likely to be interpreted accurately than diagnostic films with poor esthetic quality. Lack of attention to detail in the production of a radiograph may cause the physician to lack confidence in the information it provides. Artifacts and irregularities in the image may distract the viewer, affecting ability to focus on the significant details.

While esthetic considerations alone seldom require repeat exposures, it is important for the radiographer to assess the esthetic quality of radiographs and, when artistic merit is absent, to strive for improvement at the next opportunity.

What is the first thing you notice when you look at the radiograph in Fig. 19-24? Patient preparation for radiography involves removing outer clothing, jewelry, and other artifacts from the area that will be included on the film. The radiographic images of eyeglasses, hearing aids, snaps, buttons, and pocket change detract from the image and may obscure important anatomic details. Artifacts such as safety pins or coins projected over the abdomen may raise a question as to whether the item has been swallowed or is actually outside the body. Note also the patient's hair. While hair is not normally seen on radiographs, heavy locks of wet hair, braids, or dense masses of hair may interfere with the image. Fig. 19-25 shows a hair artifact that was originally suspected to be a soft tissue mass in the patient's neck.

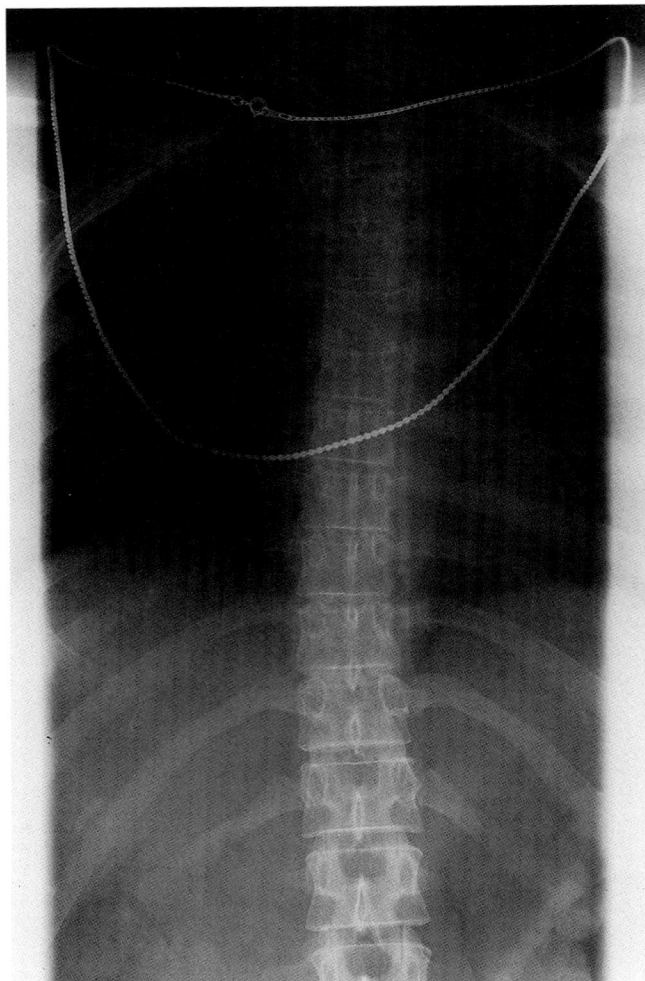

Fig. 19-24 Artifacts are distracting and may obscure significant anatomic detail.

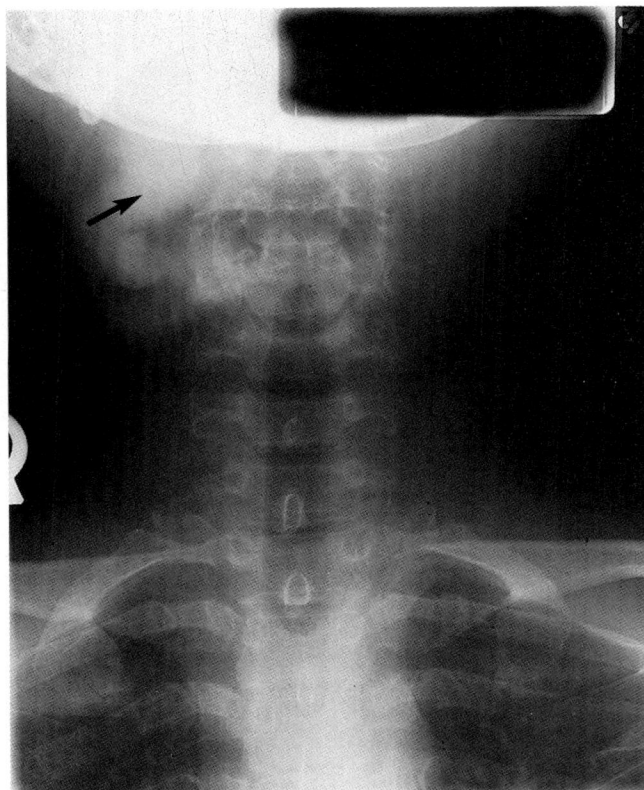

Fig. 19-25 AP cervical spine with hair artifact (*arrow*).

Another aspect of esthetic quality is the alignment and centering of the image. Long bone studies are sometimes done with the anatomy placed diagonally on the film. Usually, however, the long axis of the anatomy should be aligned to the long axis of the film. Inattention to alignment and centering produces images with a careless, sloppy appearance. Careful centering ensures that the collimation will show equal margins on each side and on the top and bottom of the film, giving it an orderly appearance. This is especially important when multiple exposures are made on a single film. The spaces between the exposures should be uniform and the anatomy should be similarly aligned on all views. Fig. 19-26 shows two radiographs with multiple exposures that illustrate a significant difference in esthetic quality.

Images that are too dark, too light, or lacking in contrast also lack esthetic quality, even when essential structures can be clearly seen. Careful attention

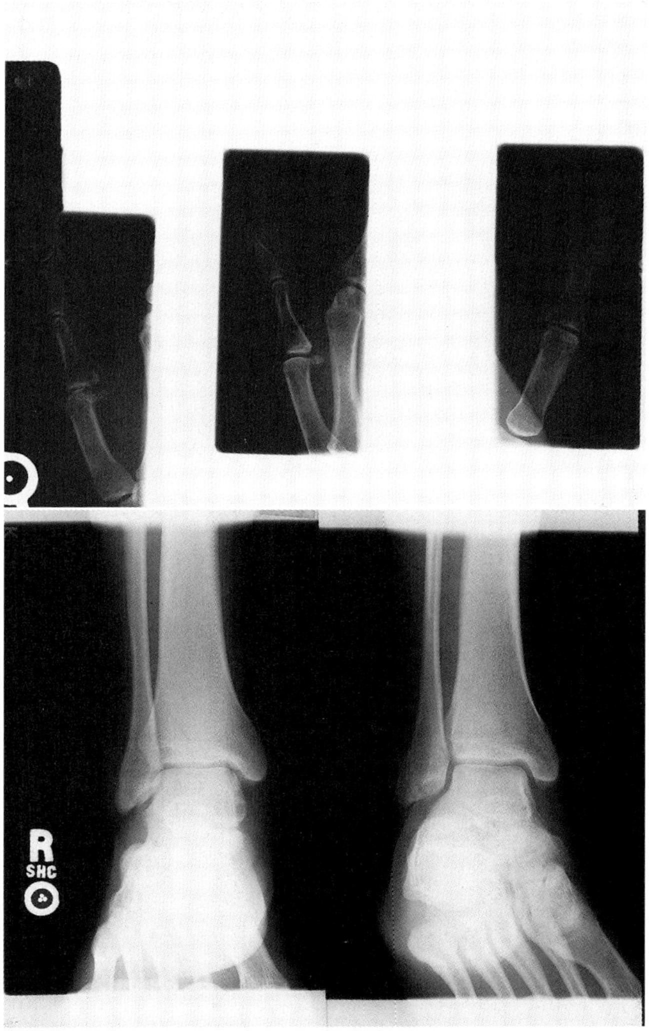

Fig. 19-26 Compare the esthetic quality of these multiple-exposure radiographs.

to the exposure factors and processing problems discussed here will aid in assuring that your radiographs have artistic merit.

RADIATION SAFETY FACTORS

While radiation safety involves many considerations, only two can be assessed by evaluating radiographs. These are collimation and shielding. Both are typically reviewed by radiation control officers when inspecting x-ray facilities for safety. This awareness should motivate radiographers to ensure that films demonstrate respect for radiation safety regulations.

Collimation

The need for proper collimation applies to all radiographs. Appropriate field limitation reduces patient dose and also improves radiographic quality. Regulations usually require that the field size be limited, not only to the film size but to the area of clinical interest. Radiation control officers usually judge that these requirements have been met when collimation is evident on three sides of a radiograph. The radiographer's goal, however, is to show evidence of collimation on all four sides of the film and to narrow the field further, when possible, so that nonessential anatomy is excluded from the field.

The radiation field light indicator does not always indicate the radiation field precisely. Legally, there may be as much as ¾-inch difference between the light field margin and the radiation field margin. For this reason, it is wise not to collimate *too* closely, raising the possibility of excluding an essential aspect of the anatomy and requiring a repeat examination.

Shielding

As stated in Chapter 11, gonad shielding is required on patients under age 55 when the gonads are within the radiation field and a shield will not interfere with the purpose of the examination. Some states require special labeling of radiographs where shielding is omitted because it would interfere with the examination's purpose. When reviewing radiographs for quality, check to be certain that shielding is apparent where required and that the shield is properly positioned. When shields are omitted and labeling is required, check to be sure that the proper label is in place.

TROUBLESHOOTING

The final step in film critique is to determine the causes of any problems identified, to decide whether the film should be repeated, and if so, what steps should be taken to ensure the success of the repeat.

This process has been addressed specifically in the preceding sections. At this point it is important to emphasize that causes of film quality problems should always be accurately identified, whether a repeat is required or not. Understanding the reason for a problem is essential to preventing its recurrence.

When the cause of a problem is not immediately apparent, continue to investigate. Review the pertinent sections of this chapter and the chapters they refer to. When the opportunity arises, show puzzling films to more experienced radiographers and seek their advice. The local technical representative of the company that manufactures your film may be willing to assist with problems related to film and processing and can often help solve other technical problems as well. If you are not already acquainted with your film company's technical representative, contact can be made through your film dealer.

REPEATING RADIOGRAPHS

Few radiographs can accurately be described as perfect. While some may be excellent, some are so poor that they must be repeated without question. Others may be marginal, acceptable to some physicians and unacceptable to others. Deciding whether or not to repeat a film is a subjective judgment. Some radiographers would prefer to take films over and over, striving for a perfect radiograph every time. Unnecessary repeats waste time and resources and cause unnecessary radiation exposure to patients. On the other hand, some radiographers are very reluctant to repeat films, preferring to submit sloppy or substandard work than to tell the patient that a repeat is required. *If the film is worth taking in the first place, it is worth taking well, even if success requires one or more repeat exposures.* Radiation safety should not be used as an excuse to avoid repeating unsatisfactory films. A reasonable standard is established by asking the question: Does this film have sufficient quality to reveal or rule out any pathology that may be present?

The positioning sections of Chapters 13 through 17 provide general guidelines to assist you in determining whether the pertinent anatomy has been included on the film. The normal radiographs in the positioning sections generally represent an acceptable standard of positioning and technique. The pathology sections reveal some of the radiographic findings that should be visualized, if present. A thorough study of these chapters will aid in determining whether your radiographs are acceptable. General guidelines for acceptable standards should be established in consultation with the physician who reads the films. In the beginning, you may need to consult the physician or your supervisor for help in deciding whether a repeat is required. With experience, you will learn to anticipate their opinions and to trust your own judgment.

It is useful to keep a record or log of all repeat exposures. A simple notebook with columns can be used to list the type of examination, the view or projection that was repeated, the problem that required a repeat, and comments about the cause of the problem and any steps taken to prevent the problem in the future. Review the record periodically to see if the same mistakes are recurring. For example, if repeats are frequently required because the original films are too light, the technique chart may need attention. If repeats are often needed on cervical spines because of positioning errors, you should review your procedure for these positions.

It is also useful to note the percentage of the radiographs you take that must be repeated. If your repeat rate is less than 1%, you could be an exceptionally outstanding radiographer. More than likely, you are failing to repeat films that should be redone. If your repeat rate is greater than 10%, consider whether you are repeating films unnecessarily or are making the same mistakes over and over. Most experienced radiographers have a repeat rate between 3% and 8%. The number will vary with the types of patients seen and the types of examinations performed. A reduction in your repeat rate, with no reduction in quality standards, indicates that your skills are improving.

SUMMARY

Film critique is a critical review of all aspects of a radiograph and is an essential skill for all radiographers. This process is required to determine when repeat exposures are needed and how they should be done. It also aids the radiographer in developing skills and improving performance.

Radiographs should always be viewed on viewboxes that are in good condition. Standardization of viewboxes throughout the facility is desirable. Radiographs should be placed on the viewbox using conventional viewing standards, usually with the superior anatomy at the top and the patient's right side on the viewer's left.

A systematic review process ensures that important details are not missed. It should include attention to each of the following factors: film identification and labeling; anatomy and positioning; radiographic density, contrast, and definition; film handling artifacts, screen artifacts, and processing; artistic quality; collimation; and shielding. The causes of any problems should be determined and steps taken to correct them.

The decision to repeat a film should be based on reasonable diagnostic standards developed in consultation with the physician who reads the films. A log of repeated films aids the radiographer in solving problems and assessing progress toward excellence.

REVIEW

Using the guidelines provided in this chapter, critique the radiographs in Figs. 19-27 through 19-31. List the errors you identify and your suggestions for correcting them. Compare your list to the critiques of these radiographs in Appendix G.

Note that film quality and detail are always lost when radiographs are reproduced in books, so it is also important for you to practice your film critique skills on actual radiographs.

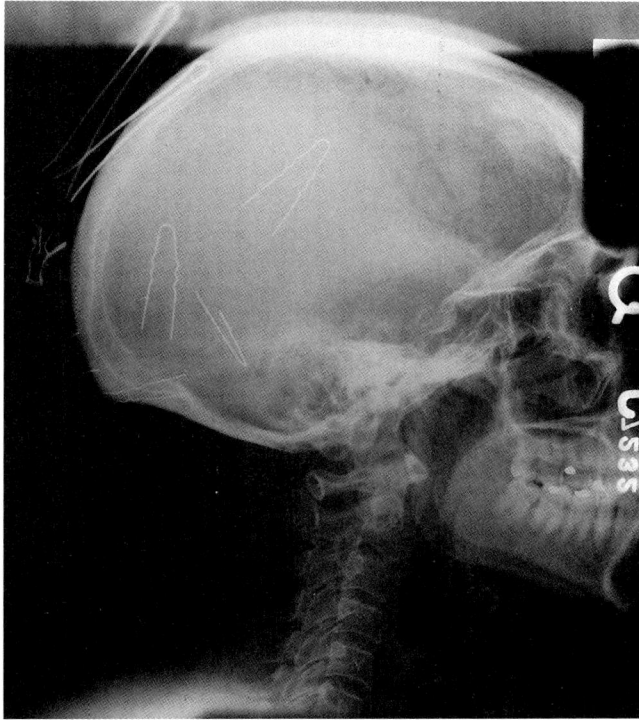

Fig. 19-27

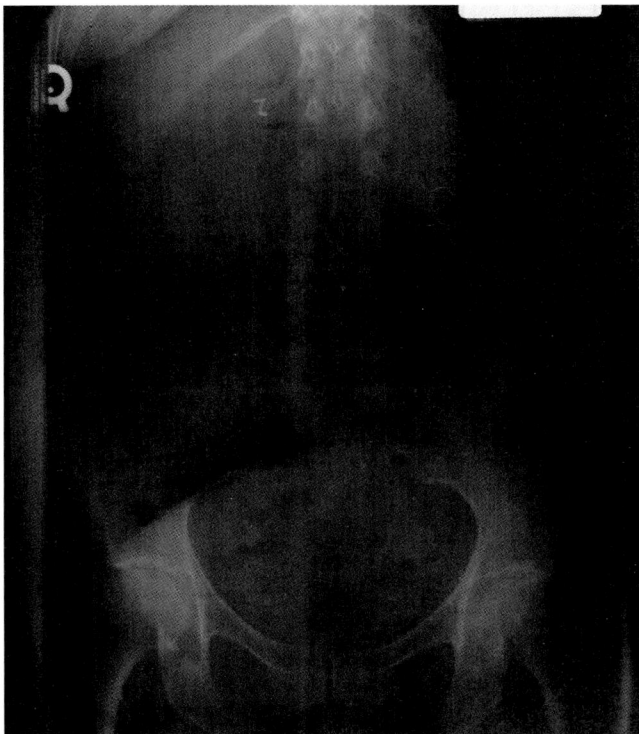

Fig. 19-28

Fig. 19-29

Fig. 19-30

Fig. 19-31

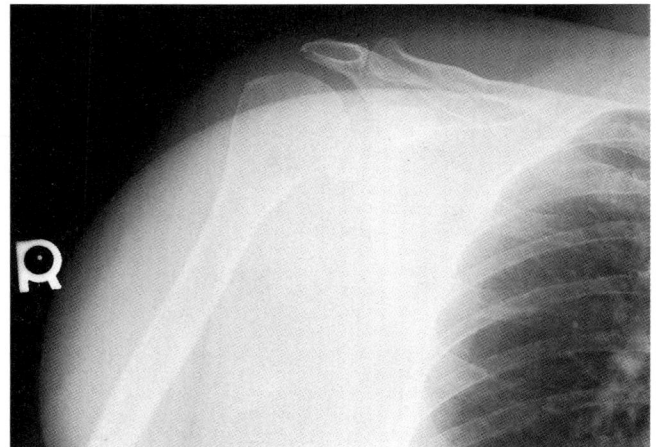

PART IV

Professionalism and Patient Care

20 Ethics, Legal Considerations, and Professionalism

Learning Objectives

At the conclusion of this chapter, the student will be able to:

- Discuss reasons why a study of professional behavior is important to the limited radiographer
- Apply ethical concepts to typical situations that arise in the health care setting
- Explain the rationale for confidentiality of professional communications and precautions for maintaining confidentiality
- Demonstrate respect for patient rights that the radiographer is responsible for protecting
- List specific acts of misconduct and malpractice that could occur in the practice of radiography and describe the most frequent circumstances causing patients to initiate litigation
- List aspects of self-care that demonstrate responsible behavior by the radiographer
- Demonstrate effective communications skills, including listening skills, nonverbal skills, and validation of communication; discriminate between assumed and validated statements
- Suggest positive strategies for both verbal and nonverbal communication with patients with hearing and/or visual impairments and patients from other cultures
- Demonstrate communication strategies that promote teamwork in the workplace
- Demonstrate professional skills in handling messages sent and received on paper and by telephone, voice mail, and fax
- Demonstrate the use of patient charts for both obtaining and recording information; state the essential characteristics of good medical records
- Explain requirements for maintaining radiographs and procedures for lending them

Key Terms

aggressive	intentional misconduct
assault	invasion of privacy
assertive	libel
battery	malpractice
chart	morals
charting	negligence
defamation of character	reasonably prudent person
empathy	*respondeat superior*
ethics	rule of personal responsibility
false imprisonment	slander
informed consent	values

ow does a profession differ from a job? Many different definitions have been advanced for a profession. Generally speaking, a profession is more than a field of study; it is the application of specialized knowledge in a way that benefits others and carries a high degree of responsibility to the community it serves. A profession is organized to govern itself; to effectively set standards of professional behavior, education, and qualification to practice and to enforce those standards within its ranks. Having a peer review journal or publication is also expected of a profession. This allows the profession to advance and to continually review and challenge the basis of knowledge on which it functions.

As stated in Chapter 1, limited radiographers are taking the first steps to becoming a profession and have not yet attained true professional status. On the other hand, their work is closely associated with that of physicians, nurses, and other health care professionals. The public does not usually distinguish between limited radiographers and professional radiologic technologists. For these reasons, a high level of professionalism in both attitude and behavior is expected of the limited radiographer. Strict adherence to professional standards by limited radiographers will hasten the day when professional status is achieved. As a radiographer, your work must be focused on the patients in your care and your efforts must be devoted to providing quality of service. It is a primary goal of this text to assist you in this effort.

ETHICS

Correct behavior or "right action" may be dictated by moral, legal, or ethical considerations. **Morals** are right actions based on religious teachings. Most religions have similar guidelines for the proper conduct of life and relationships. Such concepts as honesty, fairness, and compassion are cultural standards based on moral principles. When principles are in conflict— for instance, in a situation where it may not seem compassionate to be honest—**values** will determine which concept prevails. Values refer to the priority that is placed on the significance of moral concepts. Values vary with individuals, and morals are matters of individual choice as dictated by conscience.

Laws, on the other hand, are legal requirements for behavior. In this way, the government can control the behavior of groups and individuals. Laws govern health care delivery, the practice of radiography, and certain interpersonal interactions, as discussed in Chapter 1 and in the section on legal considerations that follows.

Professional **ethics** are rules that apply values and moral standards to activities within a profession. They define what is meant by professional behavior. A code of ethics is a hallmark of a profession because it signifies high principles of professional behavior and willingness by the profession to control its own conduct. Limited radiographers have yet to develop a code of ethics that applies specifically to this group. This does not mean, however, that there are no applicable standards. State laws and licensing boards may limit the scope of practice and prohibit "unprofessional conduct." As the employee of a physician or health care organization, the radiographer must strive to support and uphold the ethics that apply to all health care personnel.

In general, the ethics of health care require that *all* patients receive respectful, competent, and compassionate care. The dignity and confidentiality of each patient must be respected by those who provide health care.

Standard of Ethics for Radiographers

The Code of Ethics for the profession of radiologic technology (Box 20-1) is developed and jointly adopted by the American Society of Radiologic Technologists and the American Registry of Radiologic Technologists (see Chapter 1). The purpose of this code is to establish a high standard of professional conduct and to assist the members of the profession to practice ethical principles. While the code and the rules that follow apply directly to those who are certified by ARRT, they are very important to limited radiographers as well. Until such time as limited radiographers have established themselves professionally and have their own code of ethics, *this is the standard by which all who practice radiography will be judged.*

The ten principles of the ARRT/ASRT Code of Ethics are self-explanatory, and some of these concepts are expanded upon in other sections of this chapter. Principles 3, 5, 7, and 9, however, deserve additional attention.

Principle 3 requires radiographers to put aside all personal prejudice and emotional bias when rendering professional services. This is more difficult than it may at first appear. Most of us can easily identify prejudice in others, but our own biases or judgments are beyond our awareness or are seen by us as being fully justified or "only common sense." All of us have some natural preferences that may result in discriminatory treatment if we are not fully aware of them. With what patients do you feel most comfortable? Men? Women? Those of your own race? Those over 16? Under 65? Middle class? Do you feel greater compassion for a patient with a heart problem than for one with a sexually transmitted disease? Once we identify those areas in human relationships where we are most at ease, it becomes apparent that we are less comfortable in some situations or would prefer to avoid them altogether. It is instructive to pay attention to how we deal with patients who are outside

BOX 20-1

American Society of Radiologic Technologists Code of Ethics

1. The radiologic technologist conducts himself or herself in a professional manner, responds to patient needs, and supports colleagues and associates in providing quality patient care.
2. The radiologic technologist acts to advance the principal objective of the profession to provide services to humanity with full respect for the dignity of mankind.
3. The radiologic technologist delivers patient care and service unrestricted by concerns of personal attributes or the nature of the disease or illness, and without discrimination, regardless of sex, race, creed, religion, or socioeconomic status.
4. The radiologic technologist practices technology founded upon theoretical knowledge and concepts, utilizes equipment and accessories consistent with the purpose for which they have been designed, and employs procedures and techniques appropriately.
5. The radiologic technologist assesses situations; exercises care, discretion, and judgment; assumes responsibility for professional decisions; and acts in the best interest of the patient.
6. The radiologic technologist acts as an agent through observation and communication to obtain pertinent information for the physician to aid in the diagnosis and treatment management of the patient, and recognizes that interpretation and diagnosis are outside the scope of practice for the profession.
7. The radiologic technologist utilizes equipment and accessories, employs techniques and procedures, performs services in accordance with an accepted standard of practice, and demonstrates expertise in minimizing the radiation exposure to the patient, self, and other members of the health care team.
8. The radiologic technologist practices ethical conduct appropriate to the profession and protects the patient's right to quality radiologic technology care.
9. The radiologic technologist respects confidences entrusted in the course of professional practice, respects the patient's right to privacy, and reveals confidential information only as required by law or to protect the welfare of the individual or the community.
10. The radiologic technologist continually strives to improve knowledge and skills by participating in educational and professional activities, sharing knowledge with colleagues, and investigating new and innovative aspects of professional practice. One means available to improve knowledge and skills is through professional continuing education.

Courtesy of The American Society of Radiologic Technologists and The American Registry of Radiologic Technologists, Copyright 1994.

our "comfort zone." Sometimes we tend to act more friendly or solicitous to cover up feelings. At other times we may remain aloof, appearing to be preoccupied because we are in a hurry. Lack of interest and concern is unacceptable; feigned concern or pretended interest is never the same as the real thing. Faithfulness to the spirit of Principle 3 requires a high degree of self-awareness and presents a serious professional challenge to the radiographer.

Principle 5 deals with the question of professional responsibility. It implies that radiographers are sufficiently educated and experienced to be capable of independent discretion and judgment. Within the scope of their professional activity, they are expected to be both capable of making decisions and accountable for the decisions they make. A very important aspect of assuming this responsibility is awareness and acceptance of one's limitations. Although responsibilities may vary with the working environment, regular duties should be specified in job descriptions and must be consistent with the permitted scope of practice. It is in no one's best interest to perform tasks without adequate knowledge or to undertake a responsibility without adequate qualification. This principle also holds the radiographer accountable for errors committed under the orders of another

person, if the responsible radiographer knew, or *should have known*, that the order was in error.

Principle 7 requires that radiographers adhere to accepted practices and make every effort to protect themselves and all patients and staff from exposure to unnecessary radiation. The ethical implications of this issue are very important. When a radiographer violates this principle, there is no telltale evidence. The negative consequences of other breaches of ethics might be immediate, but the latent and genetic effects of unnecessary radiation to patients may not be apparent for 10 to 20 years, even for several generations. Making every effort to minimize radiation exposure—even when the patient is difficult to handle, even when you are really in a hurry, and even when no one is watching—requires both good habits and a strong ethical commitment to radiation protection.

Principle 9 relates to the confidentiality of information in a health care setting, which is one of the cardinal concepts in all codes of ethics relating to health care. The confidentiality of conversations between patients and their physicians is considered so important that, along with communications to lawyers and the clergy, it is protected by "legal privilege." This means that the professional cannot be required to divulge such information, even when to do so might be of material value in a court of law.

Radiographers are often privy to conversations between patients and their physicians as well as to confidential information contained in patient records. They may witness circumstances where patients are unable to preserve their dignity and may behave in ways that would cause them shame or embarrassment if known to friends or family. Many patients do not want it known that they are ill. Some may wish to keep their diagnosis confidential. Information that may seem of no consequence to you may constitute a very sensitive issue for the patient. Any breach of confidence, even if no names are mentioned, may rightly be interpreted by others as an indication that the radiographer does not respect professional confidence. Betrayals of confidence cause individuals to lose faith in health care providers and may prevent them from revealing facts essential to their care.

The patient's right to confidentiality is not violated by appropriate communications among health care workers when the information is pertinent to the patient's care. It is justifiably assumed in such a case that the transfer of information is for the patient's benefit and that all personnel involved are bound by the ethics regarding confidentiality. Appropriate communications are those directed privately to those who have need of the information. Conversations about patients must never be held in public areas such as waiting rooms, elevators, or cafeterias.

The ethics of patient-staff communication also require the exercise of sound judgment and restraint to avoid exposing to patients the radiographer's personal concerns or the problems of the staff. Using the patient as a sounding board for complaints or gossip is inexcusable.

Rules of Ethics

In addition to the Code of Ethics, which lists general ethical principles, the standards of ethics for professional radiologic technologist as published by the ARRT also include Rules of Ethics. These are mandatory specific standards of minimally acceptable professional conduct for all registered radiologic technologists and applicants for certification by ARRT. Since ARRT does not issue credentials for limited radiographers, it has no means of enforcing most of these rules where limited radiographers are concerned. They do represent, however, established guidelines for acceptable behavior by radiographers. These rules are published in their entirety by the ARRT. They prohibit the practices summarized below:

- Using fraud or deceit to obtain employment or credentials
- Dishonest conduct with regard to the ARRT examination

- Conviction or no-contest plea with respect to a felony, gross misdemeanor, or misdemeanor except speeding or parking infractions
- Failure to report to ARRT that legal or ethical charges are pending against the person in any jurisdiction
- Failure or inability to practice the profession with reasonable skill and safety
- Engaging in any professional practice that is illegal or contrary to prevailing standards or that creates an unnecessary danger to the patient
- Delegation, or acceptance of delegation, of professional functions that might create an unnecessary danger to a patient
- Actual or potential inability to practice radiologic technology safely by reason of illness, use of alcohol or drugs, or any physical or mental condition; adjudication of mental incompetence, mental illness, chemical dependency, or of being a person dangerous to the public
- Revealing a privileged communication except as permitted by law
- Knowingly engaging in or participating in abusive or fraudulent billing practices
- Improper management of patient records, such as failure to maintain records, or actions that may result in a false or misleading record
- Assisting a person to engage in the practice of radiologic technology without current and appropriate credentials
- Violating an administrative rule of a state board or violating any state or federal law governing the practice of radiologic technology or the use of controlled substances
- Providing false or misleading information directly related to the care of a patient
- Practicing outside the scope of practice authorized by one's credentials
- Making a false statement to ARRT or failing to cooperate with an investigation conducted by ARRT or the Ethics Committee
- Engaging in dishonest or misleading communication regarding one's education, experience, or credentials
- Failing to report to ARRT any violation or probable violation of any Rule of Ethics by any Registered Technologist or applicant for certification by ARRT

Patient Rights

A major aspect of ethical concern for radiographers is to protect patient rights at all times. Considerable emphasis is placed on consumer advocacy in our society, and this value is especially significant in the field of health care. An example of a patients' rights statement is in Box 20-2. Many of these concepts apply specifically to physicians or hospitals, but sev-

BOX 20-2

A Patient's Bill of Rights

*Bill of Rights**

1. The patient has the right to considerate and respectful care.

2. The patient has the right to and is encouraged to obtain from physicians and other direct caregivers relevant, current, and understandable information concerning diagnosis, treatment, and prognosis.

   Except in emergencies when the patient lacks decision-making capacity and the need for treatment is urgent, the patient is entitled to the opportunity to discuss and request information related to the specific procedures and/or treatments, the risks involved, the possible length of recuperation, and the medically reasonable alternatives and their accompanying risks and benefits.

   Patients have the right to know the identity of physicians, nurses, and others involved in their care, as well as when those involved are students, residents, or other trainees. The patient also has the right to know the immediate and long-term financial implications of treatment choices, insofar as they are known.

3. The patient has the right to make decisions about the plan of care prior to and during the course of treatment and to refuse a recommended treatment or plan of care to the extent permitted by law and hospital policy and to be informed to the medical consequences of this action. In the case of refusal, the patient is entitled to other appropriate care and services that the hospital provides or transfer to another hospital. The hospital should notify patients of any policy that might affect patient choice within the institution.

4. The patient has the right to have an advance directive (such as a living will, health care proxy, or durable power of attorney for health care) concerning treatment or designating a surrogate decision maker with the expectation that the hospital will honor the intent of that directive to the extent permitted by law and hospital policy.

   Health care institutions must advise patients of their rights under state law and hospital policy to make informed medical choices, ask if the patient has an advance directive, and include that information in the patient records. The patient has the right to timely information about hospital policy that may limit its ability to implement fully a legally valid advance directive.

5. The patient has the right to every consideration of privacy. Case discussion, consultation, examination, and treatment should be conducted so as to protect each patient's privacy.

6. The patient has the right to expect that all communications and records pertaining to his/her care will be treated as confidential by the hospital, except in cases such as suspected abuse and public health hazards when reporting is permitted or required by law. The patient has the right to expect that the hospital will emphasize the confidentiality of this information when it releases it to any other parties entitled to review information in these records.

7. The patient has the right to review the records pertaining to his/her medical care and to have the information explained or interpreted as necessary, except when restricted by law.

8. The patient has the right to expect that, within its capacity and policies, a hospital will make reasonable response to the request of a patient for appropriate and medically indicated care and services. The hospital must provide evaluation, service, and/or referral as indicated by the urgency of the case. When medically appropriate and legally permissible, or when a patient has so requested, a patient may be transferred to another facility. The institution to which the patient is to be transferred must first have accepted the patient for transfer. The patient must also have the benefit of complete information and explanation concerning the need for, risks, benefits, and alternatives to such a transfer.

9. The patient has the right to ask and be informed of the existence of business relationships among the hospital, educational institutions, other health care providers, or payers that may influence the patient's treatment and care.

10. The patient has the right to consent to or decline to participate in proposed research studies or human experimentation affecting care and treatment or requiring direct patient involvement, and to have those studies fully explained prior to consent. A patient who declines to participate in research or experimentation is entitled to the most effective care that the hospital can otherwise provide.

11. The patient has the right to reasonable continuity of care when appropriate and to be informed by physicians and other caregivers of available and realistic patient care options when hospital care is no longer appropriate.

12. The patient has the right to be informed of hospital policies and practices that relate to patient care, treatment, and responsibilities. The patient has the right to be informed of available resources for resolving disputes, grievances, and conflicts, such as ethics committees, patient representatives, or other mechanisms available in the institution. The patient has the right to be informed of the hospital's charges for services and available payment methods.

A Patient's Bill of Rights was first adopted by the American Hospital Association in 1973. This revision was approved by the AHA Board of Trustees on October 21, 1992.

*These rights can be exercised on the patient's behalf by a designated surrogate or proxy decision maker if the patient lacks decision-making capacity, is legally incompetent, or is a minor.

eral of them are especially pertinent to the work of radiographers.

Foremost is the right to considerate and respectful care. This statement is self-explanatory and applies to every patient, regardless of current status. This is essentially the same professional behavior prescribed by Principles 2 and 3 of the ARRT/ASRT Code of Ethics.

The patient also has a right to information, but this does *not* place an obligation on the radiographer to provide any and all information that may be requested. Radiographers must be prepared to offer explanations of radiographic procedures and to identify themselves and the physicians with whom they work. *Questions regarding diagnosis, treatment, and other aspects of care must be referred to the physician.*

The right of privacy implies that the patient's modesty will be respected and that every effort will be made to assist the patient in maintaining a sense of personal dignity. The radiographer must remember that many health care procedures may threaten the patient's modesty and dignity. Patients are likely to be much more sensitive in these situations than the health care workers, for whom the procedures are an everyday occurrence. Somewhat related to this right is the generally accepted practice to ensure that a patient and a physician or other health care worker of opposite sexes are not left alone together in a setting that requires undraping of the patient or examination of the genitals or female breasts. A "chaperone," preferably of the same sex as the patient, should be present if at all possible. The objective of this practice is not to prevent the health professional from violating ethical principles, although this may be a consideration. The main purpose is to ease the patient's mind if he or she fears such an encounter and to provide a witness in case the patient later claims to have been assaulted or touched in an unprofessional manner. Many health care facilities have policies that apply in these situations, and many physicians prefer to be chaperoned, even when no such policy exists. The radiographer should be aware of any such policies and sensitive to others' needs in this regard. Similar considerations may affect decisions on whether to allow parents to observe the care provided to their minor children.

Note that students and others not required for a procedure must have the patient's permission to be present. The taking of photographs, other than for the sole purpose of the patient's care, also requires consent.

The right of privacy also includes the expectation of confidentiality, discussed earlier in this chapter.

The right to refuse treatment also implies the right to refuse examination. If a patient chooses to exercise this right, the radiographer must not proceed with the study. Signing an informed consent does not invalidate the patient's right to refuse treatment once

the procedure has begun. Consent may be revoked at any time during the procedure. If this occurs, take time to explore the reason why the patient is unwilling to continue. This may be a response to a temporary discomfort and not an objection to the procedure itself. Experience with these situations will allow you to respond with tact and concern, calming the patient so that the examination can be resumed. If the patient still refuses to complete the procedure, comply gracefully and notify the physician. When patients wish to leave a clinic or outpatient facility, they must not be prevented from doing so.

Informed Consent

Although patient consent to routine procedures is implied by the continued acceptance of care, **informed consent** is necessary for any procedure that is considered experimental or that involves substantial risk. Certain imaging procedures require that the patient receive an explanation of both the procedure and the potential risk and sign a consent form. This is particularly true for procedures involving the use of contrast media. When informed consent is required for complex procedures or surgery, the physician will provide the information and obtain the consent. For patients undergoing more routine procedures, a staff member may provide the necessary form and explanation (Fig. 20-1).

Most procedures commonly performed by limited radiographers do not require informed consent. When it is your duty to obtain informed consent, be sure that you are prepared with a full understanding of the procedure and its risks so that you can give an adequate explanation to the patient and answer any questions. If the patient asks a question that you are not prepared for, seek the correct answer before continuing. An improper response can invalidate the

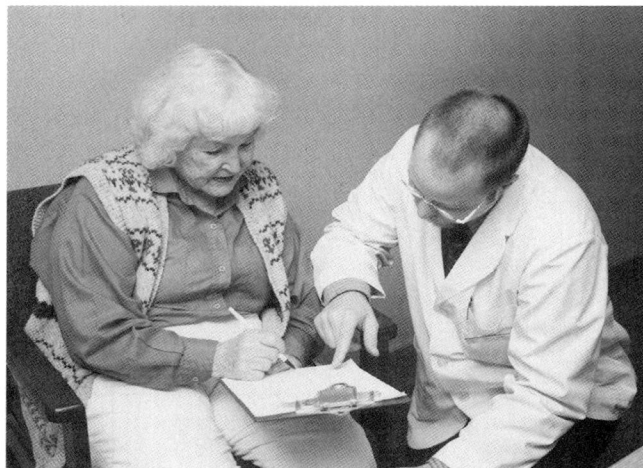

Fig. 20-1 Radiographer obtaining informed consent.

consent. The legal implications of informed consent cannot be overemphasized. Successful lawsuits against health care providers have been based on lack of compliance with the following guidelines:

- Patients must receive a full explanation of the procedure and its risks and sign the consent form before being sedated or anesthetized.
- A patient must be legally competent in order to sign an informed consent.
- Only parents or legal guardians may sign for a minor.
- Only a legal guardian may sign for a mentally incompetent patient.
- Consent forms must be completed before being signed. Patients should never be asked to sign a blank form or a form with blank spaces "to be filled in later."
- Only the physician named on the consent form may perform the procedure. Consent is not transferable from one physician to another, even an associate.
- Any condition stated on the form must be met. For example, if the form states that a family member will be present during the procedure, the consent is not valid if the family member is not in attendance.
- Informed consent may be revoked by the patient at any time after signing.

It is the radiographer's responsibility to be aware of which procedures require written consent and to be certain that these forms are in order before proceeding with the examination.

LEGAL CONSIDERATIONS

Violation of Local and Institutional Standards

As discussed in Chapter 1 and mentioned earlier in this chapter, it is essential to maintain all required credentials. In states that require a current license or permit to practice radiography, practicing outside the legal requirements may result in fines, loss of credentials, or even imprisonment. Failure to maintain the qualifications required by your employer may result in termination of your employment. Infractions of laws or professional rules may make it impossible for you to obtain professional standing and/or employment as a radiographer in the future.

Intentional Misconduct

Personal injury lawsuits are becoming more and more common in the field of health care. They usually fall into one of two categories: **intentional misconduct** or **negligence.** The types of intentional misconduct that may occur in a health care setting include assault, battery, false imprisonment, invasion of privacy, libel, and slander (defamation of character).

Assault may be defined as the *threat* of touching in an injurious way. The person need not be touched in any way for assault to occur. If the patient feels threatened and is caused to believe that he or she will be touched in a harmful manner, justification may exist for a charge of assault. To avoid this, the radiographer must explain what is to occur and reassure the patient in any situation where the threat of harm may be an issue. *Never use threats to gain a patient's cooperation.* This statement applies to pediatric patients as well as to adults.

Battery consists of an unlawful touching of a person without consent. If the patient refuses to be touched, that wish must be respected. Even the most well-intentioned touch may fall into this category if the patient has expressly forbidden it. This should not prevent the radiographer from placing a reassuring hand on the patient's shoulder, as long as the patient has not forbidden it, when there is no intent to harm or to invade the patient's privacy. On the other hand, a radiograph taken against the patient's will or on the wrong patient could be construed as battery. This emphasizes the need for consistently double checking patient identification and being certain that proper informed consent has been obtained for procedures that require it.

False imprisonment is the unjustifiable detention of a person against his or her will. This becomes an issue when the patient wishes to leave and is not allowed to do so. Inappropriate use of physical restraints may also constitute false imprisonment. Reasonable judgment must be used to decide whether restraints are necessary. Physical restraints that tie down an adult patient's hands or legs are applied only on the order of a physician.

Invasion of privacy charges may result when confidentiality of information has not been maintained or when the patient's body has been improperly and unnecessarily exposed or touched. The significance of confidentiality is re-emphasized here. Health care facilities, physicians, and their employees may be liable if they disclose confidential information obtained from a patient or contained in the medical record. If the information disclosed reflects negatively on the patient's reputation, one may also be liable for **defamation of character.** Protection of the patient's modesty is vitally important and is noted throughout this text as it pertains to specific procedures. Liability can result if photographs are published without a patient's permission.

Libel and **slander** refer to the malicious spreading of information that results in defamation of character or loss of reputation. Libel usually refers to written information, and slander is more often applied to information spread verbally. It should be clear that any breach of confidentiality rules is not only unethical but could also cause the radiographer to be sued for slander.

Negligence and Malpractice

Negligence refers to the neglect or omission of reasonable care or caution. The standard of reasonable care is based on the doctrine of the **reasonably prudent person.** This standard requires that a person perform as any reasonable person would perform under similar circumstances. In the relationship between a professional person and a patient or client, the professional has a duty to provide reasonable care. An act of negligence in the context of such a relationship is defined as professional negligence or **malpractice.** The radiographer is held to the standard of care and skill of the "reasonable radiographer" in similar circumstances.

Malpractice lawsuits against physicians, hospitals, and health care workers are becoming increasingly common. As a result, rates for malpractice insurance coverage have soared, and this topic is a serious concern to all health professionals.

There has long been a tendency to place legal responsibility on the highest authority possible. For instance, according to the legal doctrine of *respondeat superior* ("let the master respond"), the employer is liable for employees' negligent acts that occur in the course of their work. In recent years, however, the **rule of personal responsibility** has been increasingly applied. This means that each person is liable for his or her own negligent conduct. Under this rule, the law does not allow the wrongdoer to escape responsibility, even though someone else may be legally liable as well.

There has been much discussion concerning whether radiographers should carry malpractice insurance. Health care facilities carry liability insurance, which covers them in the case of negligence by employees acting in the course of their employment. This coverage may also apply to employees individually. Radiographers should learn the extent of provisions for malpractice coverage provided by their employers.

Malpractice lawsuits have resulted in unfavorable judgments against radiographers as individuals. In rare cases, insurers who have paid malpractice claims have successfully recovered damages from negligent employees by filing separate suits against them. Some believe these are sufficient reasons for radiographers to be protected by their own liability insurance policies. Others argue that the potential for a large insurance settlement is an incentive to sue, and that if the radiographer has no means of paying a large claim, no suit will be filed. The possibility of losing personal assets, such as one's home,

---

> ### BOX 20-3
>
> ### *The Seven C's of Malpractice Prevention**
>
> - **Competence.** Knowing and adhering to professional standards and maintaining professional competence reduce liability exposure.
> - **Compliance.** The compliance by health professionals with policies and procedures in the medical office and hospital avoids patient injuries and litigation.
> - **Charting.** Charting completely, consistently, and objectively can be the best defense against a malpractice claim.
> - **Communication.** Patient injuries and resulting malpractice cases can be avoided by improving communications with patients and among health care professionals.
> - **Confidentiality.** Protecting the confidentiality of medical information is a legal and ethical responsibility of health professionals.
> - **Courtesy.** A courteous attitude and demeanor can improve patient rapport and lessen the likelihood of lawsuits.
> - **Carefulness.** Personal injuries can occur unexpectedly on the premises and may lead to lawsuits.

*The Seven C's are reprinted here with the permission of David Karp, loss prevention manager for the Medical Insurance Exchange of California.

may provide motivation for considering the purchase of malpractice insurance.

Lawsuits can result in conflict, expense, professional embarrassment, and loss of public confidence, even when the patient is denied any award. Thus a great need for caution exists, both in the interest of quality patient care and in the avoidance of possible malpractice claims. Research indicates that lawsuits are most likely when patients feel alienated from the people providing their care. When a trusting professional relationship is established, suits are less likely to occur. With this in mind, David Karp has developed The Seven C's of Malpractice Prevention (Box 20-3), a list of considerations to reduce the likelihood of lawsuits and legal liability for negligence.

Proper patient identification, accuracy in medication administration (Chapter 23), and compliance with patient safety requirements (Chapter 21) are positive steps the radiographer can take to avoid malpractice suits. Harm may result when medications or contrast media are administered without proper precautions or when drug reactions are not immediately identified and appropriately treated. Poor radiographs pose a potential for misdiagnosis that may have serious consequences for both patient and radiographer. The potential for harmful error is often greatest in stressful situations. Appropriate responses in an emergency ensure the least possible risk. You must understand and accept that an appropriate response depends on your level of experience and education. Do not hesitate to ask questions and receive help when needed.

The radiographer can also protect patients and the employer by reporting illegal or unethical professional activities to the proper authority. In such a situation the radiographer must take care to be neither too zealous nor too hesitant. A simple, written statement that includes the facts (dates, times, names, places) but avoids judgments or conclusions should be prepared as soon as possible after the occurrence. This statement should be submitted to the appropriate person, probably your immediate supervisor unless he or she was involved in the incident. The supervisor receiving such a report is responsible for seeing that it is given to the proper authority, who must then follow up by investigating. A single report may not produce change, but it may add strength to other reports or lead to increased supervision where needed.

PROFESSIONAL ATTITUDE

Think for a moment of a physician, nurse, or other health care worker that you have encountered as a patient. If this person had a neat, appropriate appearance, spoke in a way that was friendly and concerned but not too familiar, and projected an air of confidence and competence, you probably perceived that this person projected a professional attitude. Patients expect professionalism in health care workers and will be most likely to respond with cooperation and confidence when these expectations are met. A professional attitude also promotes positive relationships with co-workers and employers.

Self-Care

Health is a state of physical, mental, and social well-being, and to be healthy implies that you are a person capable of promoting health. Health professionals are responsible for their own well-being and are also expected to serve as health role models for their patients and members of the community. Take care not to project an attitude that announces, "Do as I say, not as I do." This attitude undermines credibility in the eyes of the very people we most want to help.

A radiographer who is not healthy is not a good health role model and cannot function effectively for both physical and psychologic reasons. To help others, we must first meet our own physical and mental needs. Certain needs are common to everyone and can be listed and ranked in importance (Fig. 20-2). Any unmet need imposes stress and prevents us from achieving our optimum state of well-being.

Needs on the most elemental level are foremost until they are adequately satisfied. As satisfaction occurs on one level, the needs of the next higher level

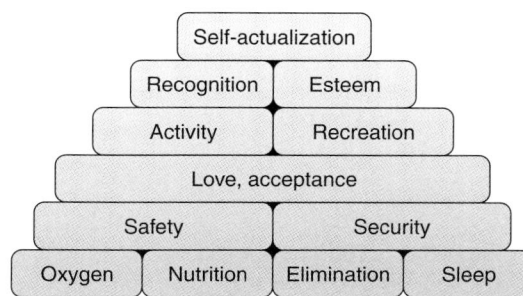

Fig. 20-2 Hierarchy of needs.

occupy the individual's attention. Self-actualization is the state in which a person welcomes tension and effort as a stimulus to creativity and self-expression. Constructively meeting your needs for well-being, knowledge, and self-esteem enables you to function more fully, free from self-preoccupation at those times when the patient requires full attention.

Since many patients have a lowered resistance that makes them especially vulnerable to infection, the radiographer who is ill should stay at home. However, since you are counted on to be present when scheduled, it is even better to prevent the onset of illness. Everyone experiences grief and acute anxiety occasionally, and such stresses can make you susceptible to illness. Whenever possible, you should stay at home to deal with such problems until some resolution is reached. Anxiety and stress can prevent you from properly fulfilling your responsibilities.

Don't overlook the importance of good nutrition and exercise habits. These practices pay off in less time lost from work and an increased sense of well-being. Knowledge and practice of the principles of body mechanics (Chapter 21) will help avoid the types of injuries that can result from lifting and moving patients or equipment.

Preventive health measures are equally important. For example, health care facilities are required to offer hepatitis B vaccine to employees who are at risk of contracting this disease in the course of their work. If your work places you at risk of infection from needle sticks or other exposure to potentially contaminated blood, you are responsible for taking advantage of this important protection. The infection control precautions discussed in Chapter 21 have been developed to help prevent transmission of diseases from patient to patient and from patient to you. Your health and that of your patients depend on your commitment to understand and use these principles. Minimizing the spread of disease in a health care setting is a primary responsibility of all health care personnel.

Radiation exposure over the course of a career can have serious health consequences if proper precautions are not observed. Chapter 11 provides instruction in radiation protection. Adhering to radiation safety practices is another important aspect of self-care for radiographers.

Job Satisfaction

Most health care workers enter the field with a desire to help and to provide excellent patient care. Sometimes, however, the demands of clinical practice may tend to overshadow your best intentions. Patient needs may be overlooked in the stress of coping with highly technical material unless you make a conscious effort to learn, from the beginning, to handle both at once.

Your work will be the most satisfying when your contributions and the personal contacts they involve are genuine and sincere. When you do things because you *want* to do them and you enjoy doing them, you will do them well and your work will be much less stressful (Fig. 20-3).

Appearance

Everyone we meet forms an instant impression of who we are. What does your appearance tell patients about you? Appearance can communicate how we feel about our work. Uniforms are worn by radiographers to present a simple, neat appearance. They are washable and plain to make them easy to keep clean. They should fit comfortably and be worn with simple, appropriate accessories. Jewelry on hands or arms that could injure patients should be avoided. Although fads and fashions change over time, a professional image will continue to be conservative.

Personal cleanliness and grooming are essential. Fingernails should be kept reasonably short and smooth. Shoes should be comfortable and quiet to

Fig. 20-3 With a high level of job satisfaction, a radiographer begins the day with a positive outlook.

walk in and must be kept clean. Fragrances should be light and used sparingly. Some patients have an allergic response to perfumes; others find heavy fragrances offensive. Strong fragrances can cause nausea in patients who are ill.

The appearance of the x-ray room and the public areas of your work environment is equally important. An untidy, cluttered room does not show respect for patients. It suggests that personnel may be too pressured, too disorganized, or too uncaring to perform competently.

Teamwork

Teamwork is defined as the cooperative effort by the members of a group to achieve a common goal. The pressures of work may sometimes make this concept hard to apply. Our goal is the best patient care we are able to provide, and it becomes easier when we remember that by working cooperatively with the others we *can* accomplish that goal.

Teamwork is a two-way street. If you appreciate help when schedules are tight, you need to be sensitive to the needs of your co-workers as well. The essence of teamwork is found in good communication among co-workers. Communication that promotes teamwork is discussed later in this chapter.

Empathetic Care

As stated earlier in this chapter, patients are entitled to considerate and respectful care. This statement applies to every patient, regardless of current status. Dealing effectively with clinical situations involves several abilities. One is the ability to show **empathy,** a sensitivity to the needs of others that allows you to meet those needs constructively, rather than merely sympathizing or reacting to their distress. Understanding and compassion are accompanied by an objective detachment that enables you to provide an appropriate response. For example, you could express sympathy for the victim of a tragic accident by crying or by smothering him with expressions of pity. On the other hand, a more productive expression of empathy would show concern and care while quickly and accurately providing the films that could aid in rapid diagnosis and treatment.

Beginners in the care of patients often express concerns such as "What shall I do if the patient vomits? I just know I'll get sick too!" or "I faint at the sight of blood." As you gain experience and confidence, you will learn to cope by focusing on the patient rather than on yourself. Thoughts of how you can meet the patient's needs will enable you to project a calm, reassuring attitude.

A focus on patient needs will also help you to respond calmly and assertively when the actions of a patient are inappropriate. It may seem strange or frightening that some individuals respond to stress and anxiety by becoming hostile or even threatening. These are often people whose coping mechanism depends on being in control of the situation. Preserving a calm, objective attitude is most effective in dealing with such patients.

Overt expressions of sexuality by patients are encountered very infrequently. Such events are usually a reflection of anxiety by patients who no longer feel functional as sexual human beings because of their current physical state. With this in mind, you can be less judgmental while setting limits on patient behaviors. In other words, you can refuse to accept the behavior while continuing to reassure and care for the patient.

Care of Supplies and Equipment

Health care facilities must stock large quantities of supplies to function effectively. In such an environment, it is easy to assume that free access implies free use. In truth, however, someone must pay the bill. To the purchase price of each inventory item one must add on an overhead factor that may be two or three times the item's basic value. For example, to the cost of a film that is accidentally exposed one must add the proportional costs involved in ordering, accounting, shipping, and storage. Medical equipment is expensive, and proper care is required to ensure that its value is preserved and that it is available for use when needed. The misuse of equipment or supplies, or their diversion for personal use, wastes funds and increases health care costs. The radiographer who avoids such waste is demonstrating a high standard of professional behavior.

Continuing Education

In radiography, as in any rapidly changing technologic field, continuing education is essential to be aware of current trends and maintain competencies. This also is an important professional responsibility.

Radiographers must place a priority on the acquisition of additional skills and regularly expand their knowledge. Those who are content with the status quo quickly find themselves behind the times. Standard practice changes rapidly, and today's knowledge will soon be out of date. Textbooks often contain information that is valid when the manuscript is completed but outdated by the time the book is published. Formal and informal continuing education helps us to maintain interest in our work and to avoid the boredom and routine that are detrimental to emotional health.

Many opportunities for continuing education are available to the radiographer. Hospitals, colleges, and professional organizations all provide educational opportunities that meet the needs of radiographers to stay abreast of current developments and expand their skills. Education may take the form of courses, classes, workshops, seminars, and other group experiences, but there is also a variety of materials available for individual learning and self-study.

Some states require continuing education as a condition of license or permit renewal. When you are required to provide evidence of continuing education, be sure to determine in advance whether the education you plan to receive is approved and accredited for this purpose. Also, be sure to keep an accurate record of your continuing education activities and any documentation of participation that you receive. Documentation is valuable even if it is not immediately required. It may help in qualifying for a promotion or a new position or assist your professional advancement in unforeseeable ways.

Failure to maintain competency and required certifications places the employer and the employee at risk and may result in loss of employment and professional reputation. Knowing the credentials required in a given situation and maintaining current credentials are important professional responsibilities.

PROFESSIONAL COMMUNICATION

To communicate means to convey information, to express oneself clearly, and to have an interchange of ideas and information with others. How we communicate also involves our attitude and our manners. Attitude is a state of mind or an opinion that is sometimes revealed by body position, tone of voice, or other nonverbal signals. Manners are customs that express respect, and they are sometimes referred to as the oil that makes daily contacts run smoothly.

Accurate communication is essential to quality patient care. The ability to give instructions depends on the speaker being clear and precise. The listener, on the other hand, is equally responsible for attentive and receptive behavior. The rapport we establish with both patients and coworkers by listening attentively and responding in a meaningful way can easily be overlooked under the pressures of a busy schedule. Stress in the workplace can easily accelerate when interpersonal communication breaks down and good manners are neglected.

While language is the primary means of communication, nonverbal behaviors also reveal to others how we feel about them. Literature provides us with interpretation of many nonverbal behaviors. Everyone can recognize some of the more common ones. We usually perceive frowns or pursed lips as disapproval. Refusal to look directly into an individual's face while speaking conveys avoidance, submission, or rejection, while clenched teeth or fists suggest angry feelings under rigid control. Patients in pain may present a tight, protective posture. Leaning forward while listening shows interest in the subject being discussed. Touch is another mode of communication. An abrupt or tentative touch may be perceived as distaste or as reluctance to care for the individual. Positive touch is both firm and gentle, assuring the patient that you are both capable and caring. As a rule, a positive and caring attitude naturally results in nonverbal behaviors that are also positive. The reverse is also true. A negative attitude will be unconsciously revealed, no matter what is said.

Clear, distinct speech habits help to ensure accurate communication. Try to tailor the content of your speech to the comprehension level of the listener. Chapter 18 provides specific suggestions for communication with children and older adults. Use good eye contact and speak face to face. This approach assures others that they have your full attention and concern.

It is an asset to cultivate an ability to be **assertive.** This does not imply that you need to be **aggressive.** Expressions of aggression involve anger or hostility, whereas assertion is the calm, firm expression of feelings or opinions. You have the right to be assertive when you require assistance in a patient care situation that is beyond your ability. Employers may be assertive in requiring employees to maintain the level of competence required by their job descriptions. In dealing with patients who are reluctant to cooperate, pleasant assertiveness is the attitude that is most productive in obtaining compliance.

Listening Skills

How do you feel when you are interrupted or when the listener looks out the window while you attempt to make your point? Are you irritated when others "put words in your mouth" or change the subject without responding to what you have just said? Good communication is a two-way street. A good listener does more than await his or her turn to speak. Listening skill involves the ability to focus on the speaker and to project an attentive attitude. If you focus on the speaker, you can respond to what has been said rather than making a quick switch to the next item on your own agenda. In discussion, patients often give us clues about a physical problem that can be easily overlooked if we rush to get on to the next question.

Validation of Communication

Good communication requires validation of understanding. An informal response such as a smile, nod, or brief "okay" may be perfectly satisfactory in a social situation, but when essential information is being presented, the response must reflect clear understanding. This is particularly true with all instructions that involve your professional activities. As a listener, you can be sure that you have understood the message by reflecting or repeating back the essential elements of the speaker's statement in your response. When you are imparting information and do not receive a validating response, continue the conversation by asking your listener to restate the information. The conversation that follows is an example of a valid communication between a physician and a radiographer:

Physician: I think Mrs. Kirkland may have a right scaphoid fracture. Please take the routine exam and a Stecher view and let me see them right away.

Radiographer: That's a routine right wrist plus a Stecher view on Mrs. Kirkland and I'll bring them to you as soon as they're ready.

Physician: Right. Thank you.

The lesson for both speakers and listeners is that messages must be both clear and complete and that comprehension of the message must be confirmed. Without such validation, neither party can be certain that all elements of the message have been understood.

Communication under Stress

Any situation that disturbs our everyday activities imposes stress. Most health care involves some anxiety and often proves stressful to patients, families, and health care workers alike. This is especially true in a crisis when speed is a factor or when a complex situation causes disagreement about priorities.

Stress interferes with our ability to process information accurately or appropriately. A classic example is the victim of a house fire who flees with the object closest to hand, such as a rubber plant, rather than essential papers or treasured family possessions. In a stressful situation accurate communication can be difficult. The principles of communication already discussed are always important, but these additional suggestions can improve your effectiveness under stress:

- Lower your voice and speak slowly and clearly.
- Be nonjudgmental in both verbal and nonverbal communication.
- Do not allow the inappropriate actions or speech of an upset individual to goad you into a similar response.
- When you are uncertain whether the listener has understood you, request an answer. For example, "Did you read the consent form? What did it ask you about allergies?"

Communication with Patients

The first contact with a patient is usually an introduction. In many social situations today, given names are used as soon as introductions are made. Although this may seem to project an air of friendliness and informality, it also poses certain problems. "Good morning, Mr. Robles. I'm Lisa McCall, the radiographer," is more than an example of good manners. It shows respect and concern and allows the patient to choose how he wishes to be addressed. In an effort to show friendliness, some staff may address adults as "honey" or "sweetie" instead of calling them by name. Others, who are focused on the work routine, may refer to "the diabetic in room 2" or "that sprained ankle in the hall." Talking down to adults or treating them impersonally diminishes their self-esteem and raises feelings of resentment. Such feelings can diminish the ability of the patient to understand and follow directions. They may also prevent retention of information and could actually hinder recovery. Resentment is destructive of the trusting relationship that is essential to quality care.

A helpful way to show respect and elevate patients' self-esteem is to involve them in their own care by giving them opportunities to make choices. Offering a valid choice, as discussed in Chapter 18, requires some thought, but the rewards in terms of patient satisfaction are well worth the effort. The choice does not need to be an earth shaking decision. Questions such as, "Would you like a blanket over your knees?" or "Would you like to stop in the restroom before we begin?" can reassure patients who would like to feel capable of making decisions and who need to have a share in their own care. Treating patients as individuals, allowing them to make valid choices, and using good nonverbal skills are tools that alleviate fear and promote cooperation.

In determining why patients fail to follow instructions, one factor frequently encountered is the assumption that the patient understands the procedure. Such an assumption is really just a guess. For example, we could assume that since you are reading this text, you are a student in a radiography or medical assisting program. That could be true, or perhaps you are a nurse who is expanding professional skills. You might also be an instructor or the proofreader for this book. Making assumptions about patients implies that you might also guess about physical status or mental ability, or even about willingness to cooperate. Can you assume that Mr. White, who may have broken his ankle, can be positioned flat on the radiographic table? No. While his ankle injury might not prevent such positioning, he may have emphysema, which would interfere with his ability to breathe while lying supine.

Conversing with patients allows you to use your powers of observation. How alert or confused is the patient? How well does the patient hear? Is English comprehension a problem? From observation, you can often make a tentative assessment of the patient's ability to get on and off the examination table, walk

unassisted to the bathroom, and so forth. Patient assessment is discussed in depth in Chapter 22, but assessment begins with communication. You should learn from this chapter that good communication with patients can help you establish a spirit of trust and cooperation that will assist in both patient assessment and patient care.

Special Circumstances in Communications

Deafness The deaf patient presents a set of problems unlike those of patients with a hearing loss, as discussed in Chapter 18. Many totally deaf individuals live in a cultural setting that has its own social structure, language, and even "inside" jokes. Certain cues help in differentiating between the patient with a hearing loss and the deaf patient, especially in an emergency. You may become aware that a seemingly alert patient is totally deaf when he or she:

- Does not respond to noises or words spoken out of the range of vision
- Uses lip movements without making a sound or speaks in a flat monotone
- Points to the ears and mouth while shaking the head in a negative motion
- Uses gestures or writing motions to express the need for paper and pencil

Some deaf people are adept at lip reading and are able to speak, at least to a limited degree. More often the deaf are educated in American Sign Language (ASL), which is the most common sign language and is distinctly different from English. It has a unique grammar, syntax, and rules. Learning a few basic signs may aid in establishing rapport with deaf patients. A card showing the alphabet and some common useful signs in ASL should be available through your local hospital's nursing service department or from community service agencies that assist the deaf. An interpreter is essential in any situation that requires complex instruction or an exchange of important information. Deaf patients have the right to choose the most preferred method of communication, which might be pencil and paper. Be sure that writing materials are available and that the patient's writing arm is free.

The health care setting can seem overwhelming, especially when the patient is a deaf child. If possible, allow the child and parents to tour the area before the examination begins. Take time to explain fully the procedures and activities so the parents can help the child understand what to expect. If the child is distressed, you might consider allowing a parent to stay in sight or near the child while following appropriate radiation safety precautions.

Blindness Most of us depend greatly on our eyes to become familiar with our surroundings and ensure our safety as we move about. Vision enables us to recognize individuals and locate items of daily living. The ability of a blind person to accomplish these same tasks without vision can seem astounding. They rely on hearing and touch to a much greater extent than sighted persons. With the aid of a cane or guide dog, many blind persons lead very independent lives. Having learned to work outside the home, use public transportation, and maintain their own households, these patients may be insulted by attitudes that are too solicitous. They may be quite capable of proceeding confidently after a quick description of a room and the obstacles in it. You might say, "This is a square room, Mrs. Lord. The x-ray table is about 5 feet in front of you, and a chair is at 7 o'clock. After you're on the table, I'll be in a booth to your left." On the other hand, they may welcome some special help in a strange environment. Some patients prefer to follow you by listening to your footsteps and using a cane, whereas others may wish to place a hand on your shoulder or elbow. Those who are infirm may prefer your arm around their waist while you reassure and direct them verbally. Take care that obstacles such a step-stools do not present a safety hazard as they move about. None of these approaches applies to all blind persons. Good communication is the key to determining which form of help is acceptable and appropriate.

Remember that loss of the ability to see, hear, or speak is a communication impairment and not a reflection of the individual's intelligence or ability to think. Patients with sensory deprivation challenge us to be more flexible and innovative in the way we offer explanations and reassurance.

Impaired Mental Function   Special sensitivity is needed when dealing with adult patients who are mentally or emotionally handicapped. Such patients may include those with congenital defects such as trisomy 21 (Down syndrome), accident victims, those with illnesses affecting the brain, and those with severe emotional disorders that affect comprehension. As with children, you must assess the patient's ability to understand and follow instructions, since this ability may vary from a near infantile response to a functional capability close to normal. In general the same clear, simple, and direct instructions offered to children are appropriate. You may need to repeat instructions if the attention span is short. Avoid the sort of approach you might use when talking to toddlers. Use the adult form of address, and treat these patients with the respect and dignity due anyone their age.

Varied Languages and Cultures

Non–English-Speaking Patients   Language barriers are not handled very effectively in the United States.

Recent federal legislation has addressed the patient's right to understand and communicate effectively in health care situations regardless of language barriers. Most large hospitals now have a service that will arrange for an interpreter when necessary. In a clinic setting, translation is most often handled by family members. If patients who do not speak English are commonly seen in your facility, you should become familiar with the policies in place for ensuring that communication meets the needs of these patients.

The difference between a certified interpreter and a friend or family member who assumes this role may be significant. The interpreter is trained to translate only what has been said, both by the patient and to the patient, and not to explain what is implied. Family and friends may tend to add extraneous information or to edit the conversation in an effort to be cooperative or to save time. For example, a complete explanation of positioning and when to stop and start breathing may be abbreviated in translation to, "It's okay, Mama. Just hold still." Family members may hesitate to reveal information about the patient that they believe is private or embarrassing. The patient may hesitate to reveal personal information through family or friends. Family members whose command of English is limited may have good intentions but be unable to provide adequate translation of complex information. The services of a trained interpreter provide a professional bridge in difficult communication situations. As with deaf clients, official interpreters must be used when complex and important information is being exchanged so that interpersonal relationships will not interfere and the parties to the conversation can be certain that the translations are accurate.

When using an interpreter, look directly at the patient and speak as though the patient were able to understand you. The interpreter will translate as you speak or as soon as you have finished a sentence. Speaking to the interpreter directly tends to make the patient feel left out or talked about rather than involved in the process.

If a translator is unavailable, use demonstrations or pencil sketches to validate whether the individual understands and make extensive use of nonverbal encouragement. A friendly smile and a warm touch may be worth many words.

Culture and Communication   The relationship between culture and communication is an integral part of our everyday life. Our reactions and habits are learned from our parents, are passed down to our children, and largely govern the way we conduct our activities of daily living. Each society develops unwritten rules regarding such ordinary things as how close we stand when talking to another person, where we touch another person in public, and even how frequently we touch. The differences in such nonverbal behaviors are highly significant. For example, a Vietnamese may smile to cover up disturbed feelings. Repeated head nods may indicate respect for the person who is speaking rather than agreement with the subject under discussion.

It is difficult to generalize about the cultural practices of any individual, since variations within the group depend on so many factors. The appearance of individuals often has no relationship to how extensively they have achieved cultural integration into the mainstream of American life. The recent arrival from Eastern Europe who comes in wearing high-top shoes, a baseball cap, and blue jeans may speak little or no English, while the patient wearing a turban or a dashiki may have been born in Chicago of ancestors who have been here for generations.

Even when we speak the same language, we don't always mean the same thing. In England when a young lady is shopping for new "pants," she will be in the underwear section. If you ask, "Where can I find a lift?" while in London, you will be directed to the elevator, not to the cab stand. If we don't always understand when we speak the same language, it is clearly important to clarify and validate instructions when our patients speak limited English.

The way we welcome, avoid, or perceive touch has powerful cultural roots. Touching while providing health care is perceived more positively by women than by men. Women find it reassuring and comforting, while men find personal touch less positive. Touch should never be forced on a patient. The brief hug around the shoulder, so reassuring to some Americans, may be an upsetting invasion of personal space and privacy to those whose culture does not include a casual embrace. When touch is required to assist or position the patient, tell the patient in advance what you are going to do and then use a firm, appropriate touch. This approach is least likely to give unintentional offense.

Eye contact in the United States is considered a positive behavior. To look an individual in the eye while speaking is usually perceived as an expression of interest, concern, or honesty. Many Asian and Native American societies, however, are non–eye-contact groups and may resent direct eye contact as being impolite and an invasion of personal space. In countries with a high population density, touch is less common among adults than in the United States. It is important in many Asian societies to avoid conflict or placing another person in an embarrassing position. Harmony is to be promoted, and loud or aggressive behavior is considered a sign of poor manners. Such patients may respond more positively to a soft, quiet tone of voice than to the brisk, assertive commands so easy for many Americans to adopt when in a hurry. When apprehensive or nervous, many Asian patients become reticent and unsociable, which can hinder effective communication.

Individuals of Hispanic descent have played a large part in the exploration and settlement of the United States. Spanish is the second most common language spoken in the United States. In Hispanic culture embracing, touching, and close proximity are easily accepted from familiar people. This may seem to contrast with a strong sense of modesty that can be demonstrated during physical examinations. It is important to provide both men and women with ample gowns and covering during radiographic studies.

An old superstition of Mediterranean origin is occasionally seen among Hispanic clients. The "evil eye," or *mal ojo,* is thought to bring bad luck or illness if children are praised or admired without also being touched. Eye contact with adults is perfectly acceptable, but when praising a child it is wise to give a touch or pat while expressing admiration. While the parents may no longer *express* belief in *mal ojo,* the ability to cause illness in a child by looking admiringly without touching is a very strong old superstition.

Many Native Americans avoid direct eye contact, considering it a mark of disrespect. Pointing directly at an individual can also be considered insulting. Personal space is important, and while patients may embrace or touch others with whom they feel close, touching should be confined to that needed to provide health care.

The best way to understand people of another culture is to learn to know them personally. While this may not always be practical, books are available on transcultural health care. If your geographic area has a significant number of individuals from another culture, you can enrich your life and provide better care by learning as much as possible about ethnic groups with which you come in frequent contact.

Nonverbal behaviors such as eye contact and touching are not interpreted the same by all those within our own society, let alone those of a different culture. This text can touch only briefly on the subject of culture, but perhaps this short discussion has heightened your awareness, not only to those of different ethnic backgrounds but to differences within American culture as well. The more sensitive you are to the reactions of all your patients, the more comfortable your interpersonal contacts will become.

Communication with Patients' Families

When we are sick or injured, the presence of those who care about us is very reassuring and may be essential to our ability to cope. It is natural that family members rush to the emergency room after an accident, visit patients during hospital admissions, and accompany patients to their appointments. You may have to deal with family members who want to hold the patient's hand during a radiographic examination or who eagerly await the results of a diagnostic procedure. When you are busy and the patient is your primary concern, family members may appear as obstacles to your work. Dealing sensitively with families is often necessary and helps your patient in ways that may not be apparent.

Your communication with families often involves the transfer of practical information. Those waiting for a patient want to know how long the procedure will take, and they appreciate an update from you when a delay occurs. When the wait is prolonged, your attention to the waiting family's comfort might include directions to needed services such as the restrooms, cafeteria, or telephone.

If the patient is a minor, is incompetent, or is sedated, you may need to provide instructions to a family member regarding preparations or follow-up care. Be sure you are speaking to the person who will actually assist the patient, since information can be lost when it is passed from person to person.

Questions often arise regarding the immediate presence of family members during a procedure. The family must usually stay outside the room, preferably in a waiting area or lobby that is out of hearing range. This is done not only because of radiation safety but also because it allows the staff to proceed without interruptions from concerned family members who may not understand what is happening and may require explanations and reassurance. Procedures that involve patient discomfort or some blood loss may be very unsettling to loved ones. If families are waiting nearby, you should be aware of this and avoid making statements within hearing range that might alarm them or betray a professional confidence.

Occasionally a family member may need to stay with the patient in the procedure room, as with the deaf child mentioned earlier. In these situations, only one family member should be selected, and this person should receive a clear explanation before the procedure. You should answer questions at this point and clarify the role of family members. Provide radiation protection as necessary.

Sometimes dealing with families can be especially difficult. In an emotionally charged situation, we all use different means to cope with our anxiety. Some of us become dependent and wait for others to make decisions and give us instructions. Others maintain self-control by withdrawing or denying the importance of the situation. Anxiety causes some individuals to be quite aggressive or controlling when communicating about patients who are dear to them. Fear frequently engenders anger. If you can understand aggressive demands for service and attention as being an expression of fear, you can concentrate on reassuring rather than responding with anger yourself. Although you should refer inquiries about diagnosis or prognosis directly to the physician, an expression

of concern can demonstrate empathy. "I know how worried you must be about Cynthia, Mr. Roth. I've let the doctor know you're waiting for the results."

Communication with Co-Workers

The ability to relay information to other health professionals is essential. Many problems we encounter when dealing with patients are also met when communicating with professional workers. The pressures of time and work load may compound the personality conflicts encountered in any group. Good interpersonal relationships are built on the ability to make others feel good about themselves (Fig. 20-4). The nonverbal behaviors that we use with patients, such as touch and appearance, are equally effective with co-workers. Be a good listener. Use praise and appreciation as positive reinforcements when work is well done or when others go out of their way to offer assistance. Demonstrate respect for your co-workers as individuals by avoiding cliques and gossip. Especially avoid revealing personal information about your employer and your co-workers to patients or others.

One other concern regarding your interpersonal relationships with co-workers has ethical and legal implications. The pressure of work may make it difficult to find time to exchange general information. For this reason, break or lunch time is often used to catch up on recent developments and share information on difficult situations. *Never* discuss patients in a public setting. It is acceptable to talk about changes in schedules, the holiday party, or the new computer system, but discussions of interesting cases, celebrity patients, or possible treatment errors can be overheard and used in damaging litigation. Such conversations are an invasion of the personal rights and privacy of the patient.

In a modern health care facility a great amount of information is exchanged. Although much of the

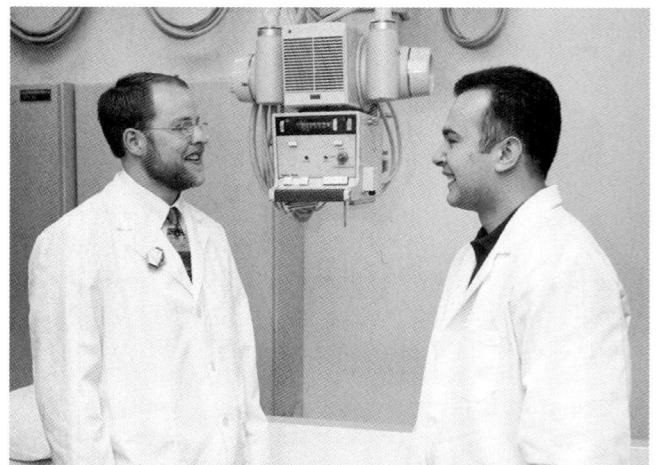

Fig. 20-4 Good interpersonal relationships are built on the ability to make others feel good about themselves.

technical communication is conveyed using charts and forms, it can be equally important to relay informal messages accurately. Attention to details, such as adding your name and the date to telephone message forms and notes for the bulletin board, can help keep information retrieval pertinent.

Many businesses and health care facilities have voice mail systems to facilitate messaging when personnel are away from their telephones. When you receive a voice mail message that is important, it is courteous to call back and confirm that you have received the message. Playing "telephone tag" can be very frustrating to the caller, and the problem is worsened when there is no way to know if the message has been delivered. When leaving a voice mail message (Fig. 20-5), be sure to identify yourself and your position or facility. State the date and time and your telephone number clearly. If you want your call returned so that you can talk to the person, suggest a convenient time to return the call. Here are some additional guidelines for avoiding problems with telephone communications:

- Be familiar with your telephone system, including forwarding and "hold" functions.
- Identify yourself and your facility or department when calling or answering a call.
- Keep paper handy and make notes during the call to avoid losing details.
- Use a pleasant, receptive tone of voice.
- Validate the message before concluding the call.
- When receiving a call for someone else, avoid disclosing personal information. Simply state that the person is unavailable or out of the office and offer to take a message. Be sure the message is relayed to the proper person or department promptly.

Any written communication is valuable only if the recipient receives it in good time. Whether this is a telephone message, a personal note, or a change in schedule, try to see that such messages are given directly to the individual or posted in plain sight in a predetermined spot. Be sure to identify yourself, your department (if appropriate), and the time sent or received. If a response is needed, remember to include the return address and/or telephone number.

Fax transmissions are often used to aid in the rapid transmission of information between health care facilities. Physicians may use the convenience of fax transmission to avoid the problems that can occur with verbal telephone orders. If you are responsible for faxing information to another institution, remember to fill out the cover sheet first. Confusion about where the information is needed or who is to receive it can cause needless delays in patient care. When sensitive, confidential information regarding a patient is to be sent by fax, it should be preceded by a phone call to alert the recipient. Confidentiality is difficult to preserve at best, and such information should be treated in a responsible manner. Most fax machines print out a record of the fax transmission. If it is your duty to transmit information by fax, know the procedure for maintaining this record. It may be attached to the original copy of the document, or you may need to note the document's content on the fax record and file it separately.

MEDICAL INFORMATION AND RECORDS

Effective documentation of information about patients and their care marks the radiographer who recognizes efficient record keeping as a way to meet ongoing patient needs (Fig. 20-6). Attention to clerical details such as dates, account numbers, chart numbers, Social Security numbers, and similar data is necessary to your facility and the patients it serves. While many dif-

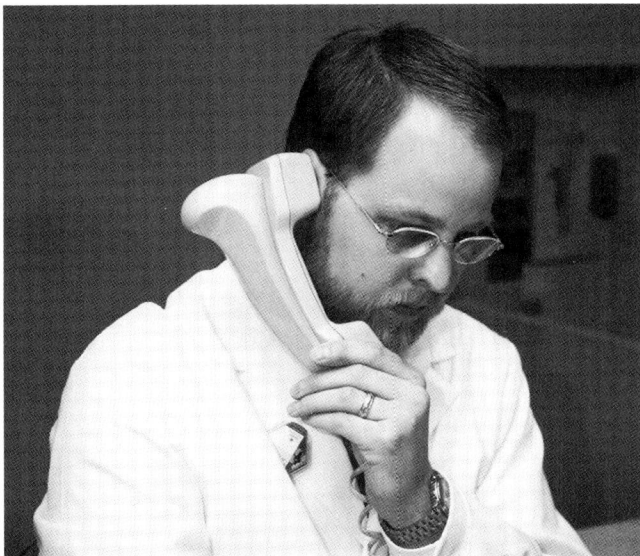

Fig. 20-5 A professional approach to telephone communication improves patient care.

Fig. 20-6 All medical records you initiate must be pertinent, accurate, and legible.

ferent forms and types of data may be used, certain terms are common. **Charting** refers to any records you are expected to add to a document. In most radiology departments the majority of the record keeping is done on requisition forms that usually have a limited area for charting. A **chart** refers to a more extensive compilation of information. Clinics keep a chart of pertinent information on each patient. These medical records include not only written information on the condition of the patient, medications, and treatments but also laboratory results, radiology reports, and any other information pertaining to the health and welfare of the patient.

Health care providers today use computers extensively for clerical functions. The storage of computer data on patients creates an individual file for each patient. Access to these files supplies basic information about the patient, such as full name, address, telephone number, and birth date. Computer files can be used for scheduling, generating requisitions, or entering charges when procedures have been completed. However, computers are only capable of using the information provided them. If a date is incomplete or a name misspelled, the computer is unable to right the error and may reject the entire entry. If your work requires access to the computer system, be sure that you receive an adequate orientation to the system.

Unlike the business world where each individual in an office has a personal computer, it is more common in the clinical setting for a few computers to be used by the entire staff. It is inappropriate to use the business computer for entertainment. It should be used for personal communications only in an emergency.

Health care agencies are also businesses. Proper record keeping is required to ensure that patients are billed accurately, that supplies are ordered to replace those used, and that insurance companies receive verification of the care given to their clients. Failure to process information accurately and promptly may inconvenience your co-workers and pose serious problems for patients.

The chief reason for keeping accurate, pertinent medical records is to provide data about the patient's progress and current status. Good records promote a systematic approach to therapeutic care, allowing for longitudinal comparisons that aid in a more comprehensive approach to extended health care.

The chart is a legal document that can substantiate or refute charges of negligence or malpractice and can also serve as a record of behavior. The course of treatment and quality of care are reflected in the chart.

Chart as a Resource

The need to validate impressions in assessing the status of patients is discussed in Chapter 22. The chart is frequently your most accessible resource. Although the organization of charts may vary somewhat, cer-

tain elements are consistent. The diagnosis or "impression" is found at the conclusion of the history sheet. The patient's current status is found in the physician's progress notes. Allergic sensitivities are stated in the history and may also be listed in a special place or on the cover. Laboratory reports, radiology reports, and results of other studies are found in a separate section. Medication records and other standard forms will be included.

To reduce the time involved in the charting process, as well as the volume of medical records, charting is a somewhat streamlined form of written communication. Many frequently used words are abbreviated, and comments are made in the form of brief phrases rather than complete sentences. Some practice is needed for beginners to translate this jargon accurately. The lists of abbreviations and terms in Appendix H are helpful to students learning to use medical charts.

Medical Recording by Radiographers

Requisitions and reports are forms of particular importance to radiographers. An x-ray requisition serves as the formal order for a diagnostic procedure. It includes patient data, a brief medical history, and specific instructions. The requisition may be part of a multicopy form that eventually includes the x-ray report. Both the requisition and the report are medicolegal records and may be filed with the films or separately. The original copies of reports become part of the patient's chart. Although radiographers are not responsible for initiating these records, they rely on requisitions for information about each examination they perform. They may also refer to previous reports for information about the patient's problem or recommendations for further studies.

Medical recording by radiographers varies greatly depending on the job description and place of employment. You must become familiar with your facility's requirements. Documenting certain information about patients is an essential part of the medicolegal record. This includes administration of contrast media or medications, changes in patient status, and reactions to contrast or medications, as well as any treatment received in the radiology department. An example of such treatment might be oxygen given to a patient who becomes short of breath.

In some facilities x-ray procedures are routinely charted by the radiographer. The information you chart should include the date, the time (using the 24-hour clock; e.g., 2:15 PM is charted as 1415), a specific statement of what occurred, and your signature. When charting observations or treatments on behalf of the physician, include the physician's name followed by a slash mark and your signature. When radiographers chart, they should use a full signature and a designation of their department or position unless their initials with their full name are recorded elsewhere in the chart for legal verification. Complete signatures are always needed for witnessing documents such as informed consents, consents to treatment, and incident reports.

Accountability is essential to the medicolegal aspects of patient care. Whether or not your duty involves making entries in patient charts, remember that all written records you initiate must be *accurate, pertinent,* and *legible.* Medical records should also be objective. For example, do not chart, "patient is confused" because this does not demonstrate how you came to this conclusion. This is an unvalidated clinical judgment. "Patient cannot relate why he is at the clinic or how he got here" is a clearer, more objective statement. Objectivity is particularly important when dealing with situations that have a strong potential for legal action.

Poor medical records are often a major contributing factor when a defensible court case is lost. Charting should be complete, objective, consistent, legible, and accurate. The list in Box 20-4 alerts you to some rules for avoiding mistakes frequently found when charts are audited.

Radiographs as Records

Radiographs are a part of the medicolegal record and are considered the property of the facility in which they are made. Patients often assume that because they have paid for the examination, the films belong to them. Tact is required when explaining that the charges cover the expense of the procedure and that every effort will be made to ensure the films are available when needed to assist in the patient's care.

State laws vary with respect to the length of time radiographs must legally be kept on file. Usually the retention period is 5 to 7 years, with the additional requirement that films on minors be kept 5 to 7 years after the patient reaches majority, or legal age (18 to 21 years, depending on the state).

Since charts can easily be photocopied, the original is never allowed to leave the health facility. Radiographs are more difficult and expensive to duplicate, however, and the quality of copies is never equal to that of the original. For this reason it is often helpful for original films to be made available for comparison or consultation outside the facility of origin. This is not necessarily true with digital imaging systems or computer-assisted modalities that permit more than one original copy of each film to be made.

Since rules governing confidentiality also apply to radiographs, the patient must sign a release form when films are needed by another provider. A written record of the date and the borrower's name and address meets the legal obligation to keep the films on file. Usually follow-up procedures exist to ensure return of films that have been loaned, but radiographs are sometimes checked out for indefinite periods, as when a patient with a chronic condition moves to another state.

It is usually recommended that films be sent directly to a consulting physician rather than allowing the patient to transport them. Sometimes, however, it is more convenient for the patient to carry the films. In this case a physician should view the films with the patient and answer any questions in advance, since the patient's curiosity may result in an attempt to interpret the films. This may lead to unnecessary confusion and anxiety. For example, patients have been known to assume that a heart shadow is a lung tumor or that a gas bubble in the stomach indicates a serious disease.

SUMMARY

Ethics refers to the application of moral standards to professional behavior. Ethical conduct for professional radiologic technologists is prescribed by the ARRT/ASRT Code of Ethics and the Rules of Ethics of the ARRT. Many state laws and regulations define the scope of practice and guidelines for professional conduct of limited radiographers. Ethical conduct is essential to good patient care. Such conduct safeguards patient rights and reduces the likelihood of medicolegal difficulties.

Our success in any endeavor depends largely on our ability to communicate with one another. In a health care setting, many factors cause anxiety in both patients and staff. When we take time to empathize, we find greater personal satisfaction in our work and improve the quality of care. The same principles of communication that increase the effectiveness of our relationships with patients and their families will aid in our relationships with co-workers as well.

Radiographers must respect the importance of medical records and strive to keep records that are pertinent, accurate, and legible. An understanding of patient charts provides the radiographer with a valuable resource to validate information. The ability to chart competently is essential if the job description requires making entries in these records. Radiographs belong to the facility in which they are taken and are a part of the legal medical record.

◫ BOX 20-4

Rules for Charting

- To delete an entry, simply draw a line through it; do not erase or use correction fluid.
- Always initial and date corrections.
- Never leave blanks on forms. Insert "NA" (not applicable) or "0."
- Never insert loose or gummed slips of paper.
- Always include the year when dating written materials.

▪ Review Questions ▪

1. Discuss ways in which adherence to ethical principles might help a radiographer to avoid an accusation of malpractice.
2. Using "A Patient's Bill of Rights" from this chapter, identify those patient rights for which a radiographer may have direct responsibility.
3. What is the standard of care that is used to legally define negligence?
4. List four aspects of self-care that demonstrate responsible behavior by radiographers.
5. List three reasons why limited radiographers should pursue continuing education, even if it is not required for the renewal of credentials.
6. Demonstrate three nonverbal behaviors that enhance communication.
7. Make two requests for a glass of water, demonstrating both aggressive and assertive attitudes.
8. Describe ways in which Asian culture, Hispanic culture, and Native American culture differ from the prevailing culture in the United States with respect to interactions that occur during health care.
9. List three things a radiographer can do to support the anxious relatives of an injured patient.
10. List three positive actions for promoting teamwork and cooperation in the workplace.
11. Who owns radiographs? What should you do if a physician calls from across town and requests films that are in your file?

21 Safety and Infection Control

Learning Objectives

At the conclusion of this chapter, the student will be able to:

- Recognize potential hazards in health care settings, such as those caused by fire, obstructions, and spills, and suggest appropriate responses
- List four complications that may arise from improper patient positioning and state the correct position in each case
- Demonstrate safe techniques for assisting patients to stand, sit, lie down, and walk using the principles of good body mechanics
- Demonstrate methods for immobilizing and restraining adult patients, and list precautions to be taken when these methods are used
- List and explain the four factors involved in the cycle of infection and state the most direct way to intervene in this cycle
- Describe the disease processes involved with infection by HIV, hepatitis, and tuberculosis and explain how to limit the transmission of these diseases
- Define *medical asepsis, disinfection,* and *sterilization* and give examples of the correct application of each
- List examples of personal hygiene that help prevent the spread of infection
- Demonstrate the technique for effective hand washing
- Demonstrate correct principles of medical asepsis in linen handling, disposal of contaminated items, and disinfection of radiographic tables and equipment
- Demonstrate correct techniques for establishing a sterile field, donning sterile gloves, removing contaminated gloves, and changing dressings

Key Terms

airborne contamination	microorganism
asepsis	normal flora
autoclave	opportunistic infection
biohazard symbol	orthopnea
carrier	orthostatic hypotension
disinfection	PASS
droplet contamination	pathogen
dyspnea	sharps container
emesis	spontaneous combustion
endospore	Standard Precautions
fomite	sterilization
microbe	vector
microbial dilution	vehicle

Whether your job description involves only radiography or includes other back office clinic functions, patient care skills will be essential to your work. In this chapter you will learn some basic principles of patient care and how to ensure the safety of your patients while also preventing injury to yourself or others.

Gathering places for the sick are often focal points for the transmission of disease. Anyone with a health problem is more susceptible to infection, and therefore infection control is of critical importance in patient care. It is your professional duty to follow established infection control policies. This will promote the safety of patients, yourself, and your co-workers.

Health care facilities that are affiliated with hospitals or government agencies are required to have policy and procedure manuals that provide protocols for many procedures discussed in this chapter. Small, private facilities may have less formal policies but should also have written protocols. It is your duty to be familiar with these protocols and to know where to find answers to questions about procedures.

As stated in Chapter 1, this text assumes that most limited radiographers will not be employed in hospitals and will not perform procedures involving contrast media. If your work involves hospital care or contrast procedures, a comprehensive text on patient care for radiographers is recommended.

HAZARD CONTROL

Fire Prevention

The first consideration in fire safety is fire prevention, since it is obviously preferable to practice prevention than to cope with a fire. An awareness of potential hazards is the first step toward prevention.

Three components must be present for a fire to burn: a flammable substance (fuel), oxygen, and heat (Fig. 21-1). Fire can be avoided by ensuring that these three elements never occur in the same place at the same time. Conversely, a fire can be stopped if one of the elements can be removed from the situation. We use this principle when we fight a fire by adding water (lowering the temperature) or by smothering (removing oxygen, as when we wrap a blanket around a person whose clothing has ignited). Most accidental fires are traceable to one of four causes: (1) spontaneous combustion, (2) open flames, (3) smokers, and (4) electricity.

Spontaneous combustion occurs when a chemical reaction in or near a flammable material causes sufficient heat to generate a fire. This is a relatively infrequent cause of fire in health care facilities, since safety standards and local safety regulations control the types of chemicals and cleaning products in general use. Paint products, solvents, and cleaning rags

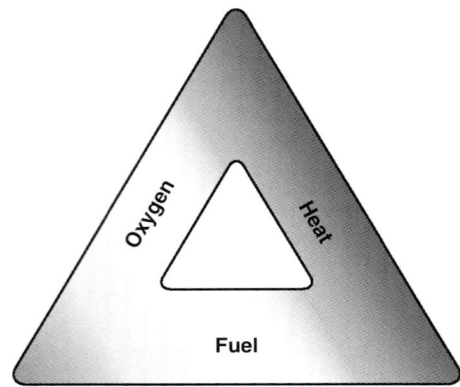

Fig. 21-1 Chemistry of fire.

capable of spontaneous combustion may be used during periods of renovation, however. Regulations require recycling or special disposal of hazardous chemicals, prohibiting them from being included with ordinary trash. For this reason, dangerous chemicals may be accumulated and stored with other recyclable items until it is convenient to dispose of them. Spontaneous combustion can occur when they are stored temporarily in a closed environment or too near a heat source. Oily or paint-soaked waste should be placed in tightly covered containers outside the building. Storage areas for dangerous products must meet the safety standards of the local health department.

Open flames that burn out of control are a common source of fires in homes but are unlikely to cause problems in health care facilities. Take appropriate precautions in kitchens or laboratories where open burners are used. Precautions include keeping flammable substances a safe distance from the burner, using strict standards of cleanliness in kitchen areas, and never leaving open burners untended when they are in use.

Smoking has essentially been eliminated from health care facilities. Health care providers promote positive health habits by prohibiting smoking, so most hospitals, clinics, and doctors' offices are designated as nonsmoking facilities. Prohibiting smoking reduces the incidence of this type of fire, but smokers may be tempted to smoke covertly. Smoking is especially dangerous when facilities are not equipped to accommodate smoking. Be alert for the smell of smoke. Direct smokers to the designated smoking area, which may be outside the building.

Radiographers use a wide variety of complex electrical equipment. Do not let familiarity with electrical items lull you into a false sense of security. Electrical fires are potential sources of fire hazard and are of special concern in radiology departments where there is much electrical equipment. For this reason, it is especially important for radiographers to be alert to electrical hazards. Box 21-2 later in this chapter lists precautions for avoiding electric hazards. Adherence to these rules greatly reduces the risk of electrical fire. These principles apply in any area where electrical equipment is used. When the building is not occupied, x-ray machines and all equipment that does not require constant power should be turned off or unplugged.

Short circuits in older x-ray control panels can generate enough heat to cause a fire. This is usually preceded by smoldering wire insulation, which causes smoke and an unpleasant odor and is usually readily detectable before an actual fire. If a short circuit occurs, turn off the electricity at the main power source, call for qualified assistance, and stand by with the proper fire extinguisher.

Oxygen by itself does not burn, but it does support combustion. Since the presence of oxygen greatly increases the fire hazard, it is important to exercise extreme care when oxygen is in use. There should be no smoking, no open flames, and no ungrounded appliances near areas where oxygen is in use.

Preparedness

Radiographers must be familiar with a fire plan for the facility. Be sure you know the evacuation route from your area and at least one alternate route. In addition, have a general knowledge of your facility's floor plan. Take special note of the location of fire alarms and fire extinguishers.

In the event of a fire, large facilities use a coded communication to notify the staff without alarming the patients. This is usually a code number or code name announced over the paging system: "Attention all staff, there is a code 100 in the west wing." The same code is commonly used for fire drills. Fire drills must be taken seriously. Take full advantage of fire drills and any in-service classes to gain confidence in evacuation procedures and the use of fire extinguishers. If you are ever involved in an actual fire, your preparation and self-confidence will allow you to function effectively and will reassure those around you.

In small offices there may not be a formal fire plan or fire drills. While the potential for fire within the facility may be small, there can be risk of fire from adjacent offices or buildings. Some sort of plan is essential. Local fire departments provide safety inspections and instruction in fire safety.

According to professional fire marshals, the most frequent infractions of fire safety rules include the following:

- Blocked fire exits
- Doors blocked open
- Equipment stored in corridors
- Improper storage of flammable items
- Improper use of extension cords

BOX 21-1

In Case of Fire

Remain Calm
1. Evacuate everyone from the immediate area.
2. Report the fire and precise location.
3. Close all doors.
4. Shut off oxygen valves.
5. Shut off all electrical equipment.
6. Go to a safe location and await further instruction.

Doors should never be blocked open. Closed doors help prevent the rapid spread of fire. Wheelchairs, carts, and other equipment must be placed to avoid obstructing passages and doorways. Pay particular attention to passages and doors that are not often used. They may be the only safe exit in case of fire. Corridors should not be used to store equipment. If necessity demands that some items be placed there temporarily, keep them all on the same side with room to pass easily. Ask yourself the question, "If we had to evacuate this area, would this equipment be a problem in this location?"

In Case of Fire

If you discover a fire, your primary responsibility is to evacuate everyone from the immediate area to a safe location. Second, report the fire and location, using the prescribed procedure. A small wastebasket blaze may be extinguished with a nearby pitcher of water or smothered with a pillow, but do not waste precious minutes in futile attempts. Your responsibility is the safety of patients and yourself. Box 21-1 lists the steps to follow in case of fire. These steps apply whether the fire is in your area or in another part of the building.

During *any* emergency it is important to remain calm and use a low voice. During a fire evacuation, try to avoid using the word "fire." Instead, you might say, "Mrs. Jensen, there is a little smoke in one of the rooms, and we are going to move you outside until we can see how serious it is."

No substitute exists for knowing the location of fire extinguishers and fire alarms. Familiarity with this information is your personal responsibility. Thorough knowledge of the procedure for reporting a fire is also essential.

Fire Extinguishers

Fire extinguishers are marked to indicate the class or classes of fire for which they are appropriately used. Class A fires involve combustibles such as paper or wood, Class B fires involve flammable liquids or gases, and Class C fires involve electrical equipment or wiring. A multipurpose, dry chemical extinguisher is suitable for all three classes of fires and is the type most often found in public buildings.

Fig. 21-2 shows a close-up view of a typical fire extinguisher mechanism. To use the fire extinguisher correctly, remember the acronym **PASS:**

Pull the pin.
Aim the nozzle.
Squeeze the handle.
Sweep. Use a sweeping motion from side to side.

While the temptation may be to aim the fire extinguisher steadily at the flame, a sweeping motion is more effective and covers a wider area. This decreases the likelihood that the fire will spread. Fire extinguishers have considerable force and are effective at a safe distance from the fire. Stand back so as not to endanger yourself (Fig. 21-3).

Fire extinguishers must be inspected regularly and recharged periodically. A tag attached to the unit should indicate the dates of the last inspection and the last recharge. The last inspection should be no longer than 1 year ago. When an extinguisher has been used, it must be recharged. It should be replaced immediately with a fresh unit.

Electric Shock

Electric shock may pose a serious hazard to both patients and personnel if safety precautions are not observed. This is especially true with x-ray equipment, which carries an electrical potential in excess of 100,000 volts. The hazard of lesser circuits should not be underestimated, however, since shocks from standard 110-volt outlets can prove fatal under certain circumstances. The rules for reducing the likelihood of electric shock are listed in Box 21-2. In addition, use extreme caution when using electricity around water. *Never stand on a wet floor or use wet hands to perform tasks involving the use of electricity.*

Falls and Collision Accidents

Reducing the risk of falls and collisions is a major safety concern. Caution is needed for the safety of both patients and personnel. Be especially conscious of hazards when moving wheelchairs and other mobile equipment, and do not store or "park" equipment

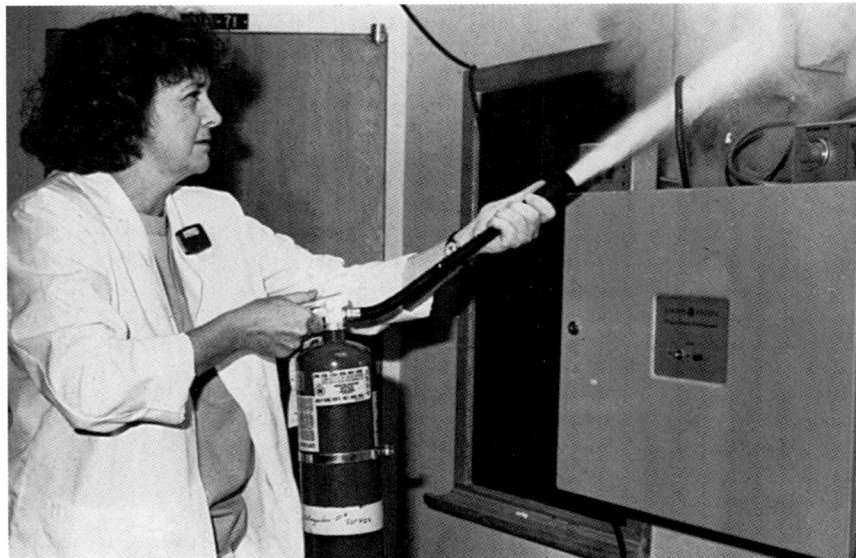

Fig. 21-2 Fire extinguisher mechanism.

1. Pull
2. Aim
3. Squeeze
4. Sweep (side-to-side)

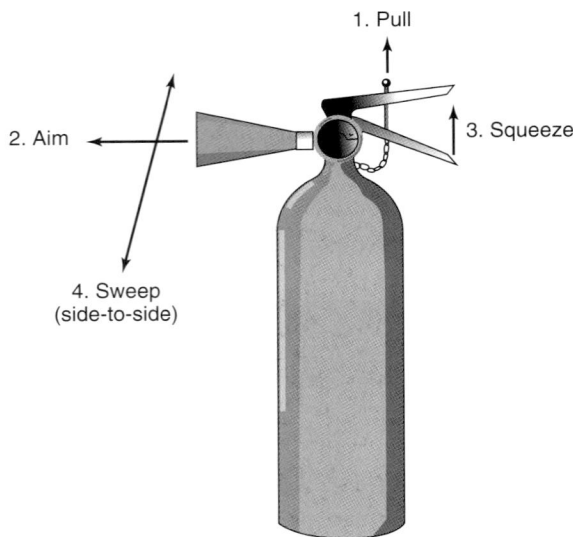

Fig. 21-3 Use a fire extinguisher from a safe distance with a sweeping motion.

where it might cause a problem. Too narrow a passageway is unsafe for patients who must use walkers or crutches. Equipment too close to a corner may be an unseen obstruction to someone hurrying from an intersecting hallway or carrying a bulky object.

Storage areas are common sources of accidents when items are not placed properly and secured when necessary. Heavy items should be placed near the floor. Do not be tempted to stack items precariously. Any item that could cause harm should be situated so that its position will not shift unexpectedly.

To access items that are above your reach, use a secure stool or stepladder. *Never* stand on tables or chairs with wheels.

Electrical cords should not be strung across doorways or other traffic patterns. Try to position equipment as close as possible to a suitable outlet. If a cord must cross a traffic path temporarily, secure it to the floor with tape to minimize the possibility of someone tripping over it. If hazardous, makeshift electrical connections are a common problem, discuss this with your supervisor or employer and suggest a safe, permanent remedy.

Spills

Spills deserve special attention. Depending on the nature of the substance, spills may pose a chemical hazard in addition to the risk of injury from falls. Household bleach or concentrated darkroom chemicals can cause eye damage or skin injury. Appropriate cleaning measures are needed to avoid potentially serious problems.

The Occupational Health and Safety Administration (OSHA), a federal agency governing safety in the workplace, requires that all chemicals be properly labeled and that Material Safety Data Sheets (MSDS) for all hazardous materials be on file and easily accessible to personnel. The MSDS for any chemical will spell out the required equipment and procedure for safe handling in the event of a spill. The following steps help to ensure safety when a spill occurs:

- Limit access to the area.
- Evaluate the risks involved.
- Obtain both the information and the equipment to clean up the spill safely.
- Clean up the spill immediately.
- If you lack the necessary skill or equipment, call your supervisor.

BODY MECHANICS

The principles of proper body alignment, movement, and balance are referred to as *body mechanics*. The application of these principles minimizes the energy required to sit, stand, and walk. Your effective strength is increased when you use these principles to perform tasks that require stooping, lifting, pushing, pulling, and carrying.

BOX 21-2

Electrical Safety

- All electrical equipment and appliances should be inspected and approved prior to purchase by a qualified testing agency such as the Underwriters Laboratories (UL).
- Follow manufacturer's instructions in using and caring for electrical equipment. Pay particular attention to whether a given item is washable or immersible. Improper cleansing may create a hazard.
- All electrical equipment used in contact with patients and in areas where there is water, such as the darkroom, should be equipped with grounded (three-pronged) plugs.
- Equipment with three-pronged plugs must always be used with properly grounded outlets. Never use plug adapters to avoid grounding an appliance.
- Electrical equipment must be inspected regularly. Arrange for prompt repair of any equipment that shows evidence of damage, frayed insulation, or loose plugs.
- Immediately discontinue use of any equipment that causes sparks or an unusual smell.
- Electrical circuits are protected from accidental overload by circuit breakers. If circuit breakers trip repeatedly, have an electrician check all equipment on the circuit to determine and remedy the cause of the overload. This may be a sign of faulty wiring or inadequate electrical supply.
- Always disconnect the power supply before exposing the electrical circuitry in any piece of equipment. Unplug portable units or shut off the main power switch on permanent installations.
- Never attempt to repair electrical equipment unless you have been specially trained to do so. Simply changing batteries or fuses may create a hazard if done incorrectly.
- If you must use an extension cord, use the shortest cord that will do the job. Be certain that all extension cords are labeled to show their amperage capacity and are approved for the intended use. Specifications must match or exceed the capacity of the original cord attached to the equipment. Do not bring cords from home for any purpose.
- In case of an electrical fire, use a fire extinguisher approved for Class C fires. Water and some chemical extinguishers increase the hazard of electrical shock.

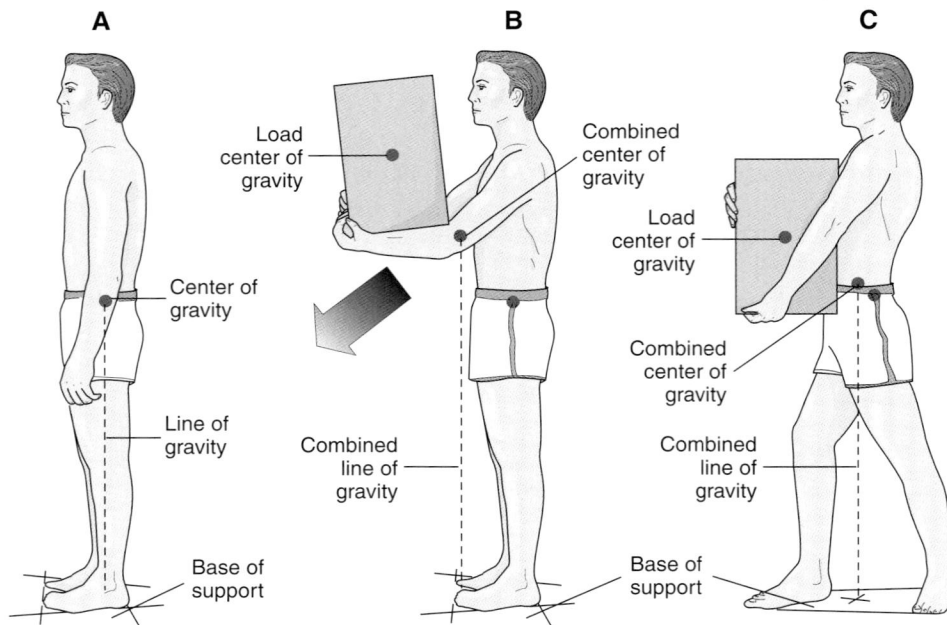

Fig. 21-4 Body mechanics. **A,** With good posture, the line of gravity bisects the base of support. **B,** When the load is held away from the body, the line of gravity does not bisect the base of support. **C,** A wide stance with the load held close to the body allows the combined line of gravity to bisect the base of support.

Applied body mechanics also prevents muscle and back strain. Such strains are a common problem among health care workers, causing much discomfort and reduced efficiency. When you injure yourself on the job, you place a greater burden on your co-workers. If you injure yourself while lifting a patient, you may injure the patient as well.

Three concepts are essential to understanding the principles of body mechanics (Fig. 21-4):

1. **Base of support**—This is the portion of the body in contact with the floor or other horizontal surface. A broad base of support provides stability for body position and movement.
2. **Center of gravity,** or center of body weight—This is the point around which body weight is balanced. It is usually located in the midportion of the pelvis or lower abdomen, but the location may vary somewhat depending on body build. The body is most stable when the center of gravity is nearest the center of the base of support.
3. **Line of gravity**—This is an imaginary vertical line passing through the center of gravity. The body is most stable when the line of gravity bisects the base of support.

Using these concepts, the principles of body mechanics can be stated in the five simple rules printed

> ## BOX 21-3
> ### *Rules of Body Mechanics*
>
> 1. Provide a broad base of support.
> 2. Work at a comfortable height.
> 3. When lifting, bend your knees and keep your back straight.
> 4. Keep your load well balanced and close to your body.
> 5. Roll or push a heavy object. Avoid pulling or lifting.

in Box 21-3. Memorize them and practice them, both at work and at home.

Bending and twisting the back while lifting is the most common cause of back strain (Fig. 21-5). A broad and stable base of support can be easily provided by standing with feet apart and one foot slightly advanced. Remember that your thigh muscles are among the strongest in your body. Good body mechanics use the combined strength of your legs, arms, and abdomen to protect the shorter, more vulnerable back muscles. Think ahead and use the tools available to you when anticipating a task that may cause muscle strain. Adjust the height of your work surface, use a cart to move a heavy load, and obtain help to lift heavy objects.

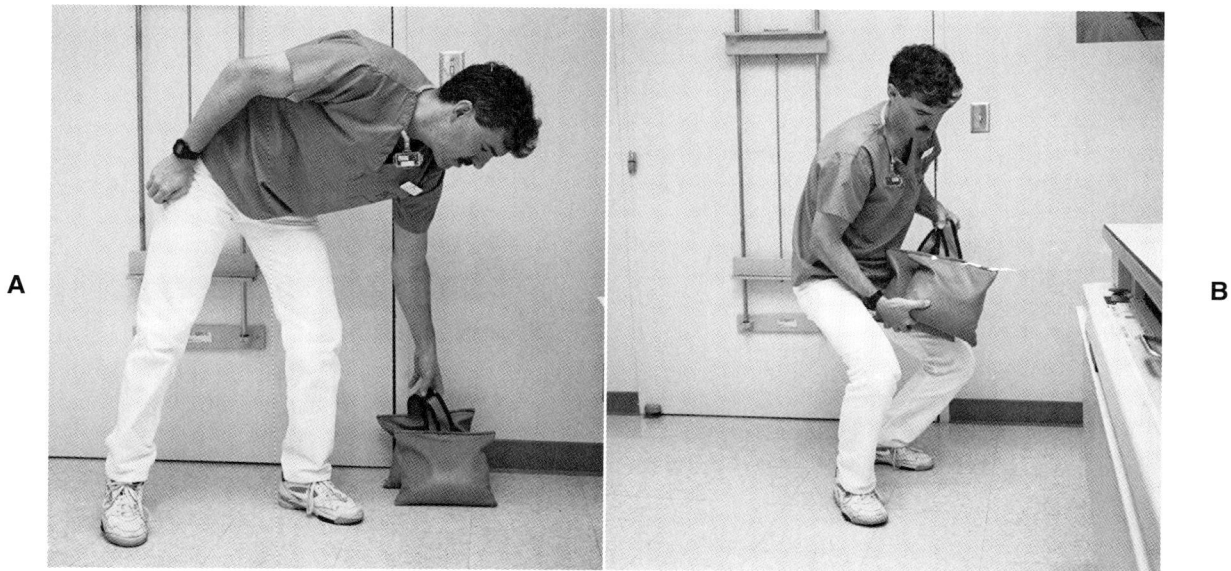

Fig. 21-5 Good body mechanics help avoid fatigue and prevent back strain. **A,** Wrong. Back is bent and twisted. **B,** Right. Knees are flexed and back is straight.

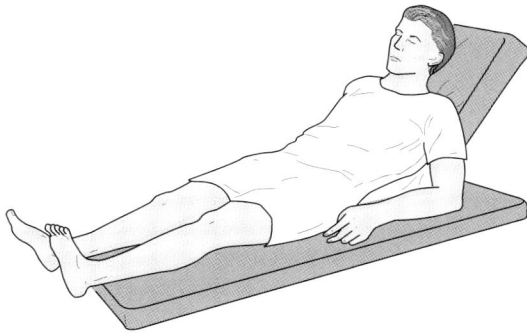

Fig. 21-6 Fowler's position.

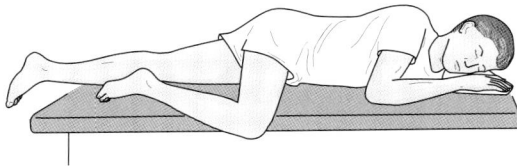

Fig. 21-7 Sims' position.

ASSISTING PATIENTS WITH POSITIONS AND MOVEMENTS

Body Positions

Common body positions have names. It is easier to communicate and to follow physicians' orders if you understand these terms.

You are already familiar with the terms for body positions used in radiographic positioning such as *recumbent, supine,* and *prone.* In addition, there are specific positions most commonly associated with pa-

tient care situations. Fowler's position (Fig. 21-6) is a modification of the supine position in which the patient's upper body is elevated. This position provides more comfort and safety than the supine position for patients who are short of breath or are nauseated. Sims' position is a recumbent oblique position (Fig. 21-7). In the Trendelenburg position (Fig. 21-8), the patient's head is lower than the feet. Usually the table is tilted approximately 15°. This position is used during some fluoroscopic procedures and is helpful in the treatment of patients suffering from shock.

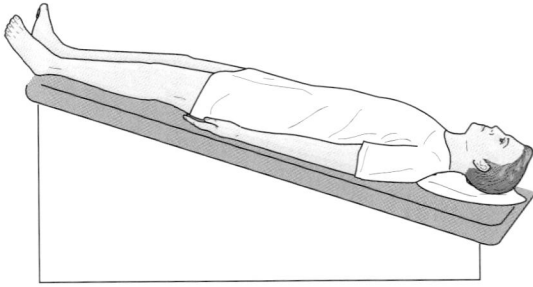

Fig. 21-8 Trendelenburg position.

Fig. 21-9 Assisting patient to stand.

Support and Padding

When lying on a hard surface such as a radiographic table, patients are more comfortable with radiolucent sponges or cushions strategically placed for support. If a cushion or pillow under the head will not be a hindrance to the examination, it can enable the patient to see what is occurring and thus help relieve apprehension. Elevation of the patient's head also relieves neck strain, allows easier breathing, and helps avoid the uncomfortable sensation that the head is lower than the feet.

A bolster under the knees of a supine patient relieves lumbosacral stress by straightening the lordotic lumbar curve. This is especially comforting to arthritic and kyphotic patients and to most elderly persons. It is essential for patients with spine injuries and those who have recently undergone spinal or abdominal surgery. Patients with abdominal pain must have the head elevated and a bolster under the knees to relieve strain on the abdomen.

The measures used to promote comfort are frequently the same interventions designed to prevent complications. When the body is supine, the weight of the abdominal contents pushes the diaphragm up into the thoracic cavity, making it more difficult to take a deep breath. This is no problem for most of us, but patients with **dyspnea** (difficulty breathing) or **orthopnea** (inability to breathe lying down) are unable to lie supine. A patient who becomes short of breath when supine must be assisted to sit up immediately.

Patients who are nauseated also need to have their heads elevated. This position helps control nausea and prevents aspiration of **emesis** (vomit) if the patient vomits. Patients who become nauseated and cannot be assisted to the Fowler's position should be rolled into a lateral recumbent position.

Padding placed under body prominences such as the sacrum, heels, or midthoracic curvature is important for several reasons. One reason is that if patients are reasonably comfortable, they are better able to maintain the positions needed for an effective examination, even on a hard surface. Another reason, explained in Chapter 18, is that older or debilitated patients may develop decubitus ulcers over prominent bony structures when pressure is exerted for even a short period of time.

Positioning is one area in which your learning will be enhanced by acting out the patient's role. Practice positioning with your classmates until comfort and positioning are part of the same action.

Assisting Patients to Change Position

Radiographers assist patients to sit, stand, lie down, and move about many times each day. The correct technique is least likely to cause discomfort or injury to either the patient or the radiographer.

When a patient who is seated needs help to stand up, stand facing the patient. Reach around the patient's upper body and place your hands firmly over the scapulae. The patient's hands may rest on your shoulders. If the patient has weakness on one side of the body, position yourself to brace the patient's weak leg with your knee as the patient stands. This will help to keep the patient's leg from bending and giving way with weight bearing. On your signal, lift upward while the patient rises to a standing position (Fig. 21-9). Remember to use a broad base of support and keep your back straight. This method may be re-

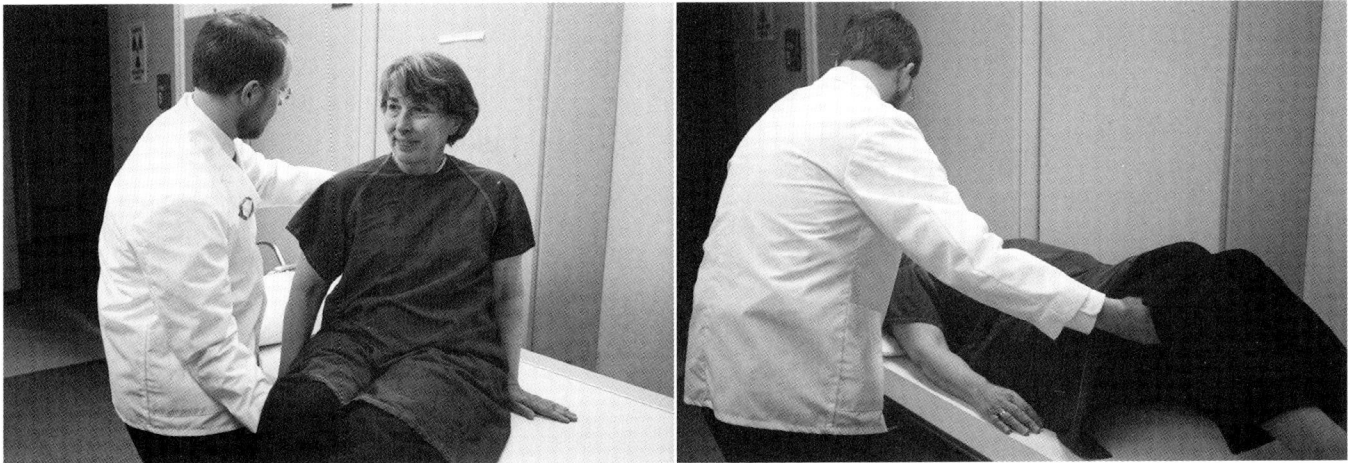

Fig. 21-10 Assisting patient to lie down. **A,** Place one arm around the shoulders and the other under the knees. **B,** Pivot the patient and lift the legs as you ease the upper body to a supine position.

versed when assisting a standing patient to sit down. Be certain that the seat is secure and will not move as the patient sits.

Patients seated on the side of the radiographic table often find it easier to lie down with some help. Place one arm under the knees and the other around the patient's shoulders. Lift the legs as you pivot the patient and rest the legs on the table (Fig. 21-10). At the same time, ease the shoulders down so that the patient is supine. This method is reversed when assisting a supine patient to sit up. It is much easier to sit up when the legs have been lowered somewhat than when they are extended on the table.

Patients suffering from recent spinal injury and those recovering from spinal surgery will find it difficult to lie down and sit up. Moving from a supine position to a sitting position or from sitting to supine places considerable stress on the spine. It is preferable for these patients to sit from the lateral recumbent position. When lying down, the patient should lie first on one side and then turn to the supine position with the knees flexed. Provide support and assistance to the patient while extending the legs, and place a bolster or pillow under the knees for support when supine.

Some patients have **orthostatic hypotension** when rising from a recumbent position. This condition is a temporary state of low blood pressure that causes patients to feel lightheaded or faint when they first sit up. A pause before assisting them to stand will give them an opportunity to regain their sense of balance.

Assisting Patients to Move About

Some patients who are weak or cannot bear their full weight easily on both legs may use a cane or a walker for support to move about. A walker is a lightweight metal frame with four legs and bars at the front and

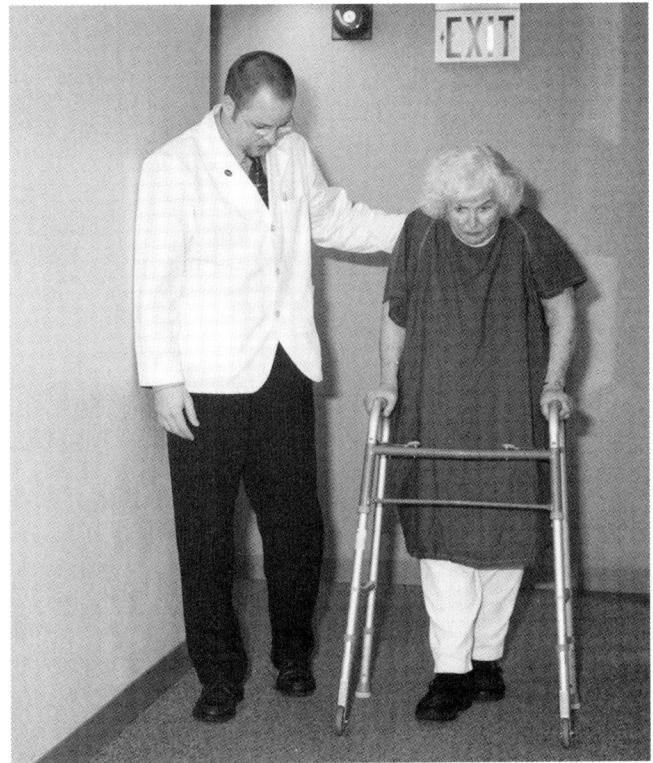

Fig. 21-11 Patient using walker.

sides on which the patient may lean for support (Fig. 21-11). The patient moves the walker ahead before taking each step. Patients with ample strength who cannot bear weight on one leg will usually walk with crutches (Fig. 21-12). Patients who are accustomed to using crutches or a walker do not need other assistance to walk. Your responsibility is to show them the way and to make sure there are no obstacles in the path.

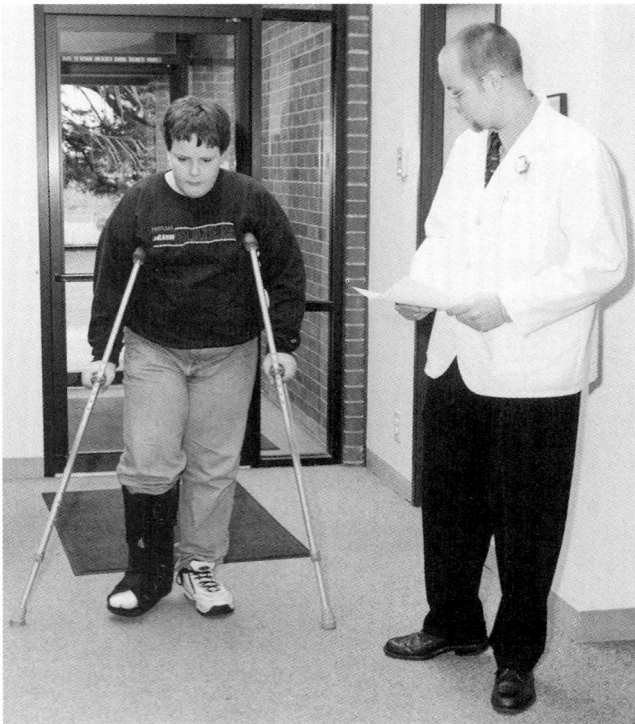

Fig. 21-12 Patients competent with crutches do not require assistance to walk.

Patients who cannot walk alone and who do not use a walker or crutches will need physical support to walk. Some patients have weakness of one side of the body. This is typical of stroke victims and those who have had injury or surgery to a lower extremity. Determine which is the patient's weak side and position yourself next to the weak side. Grasp the patient around the waist while the patient leans on your shoulder for support. Instruct the patient to lead with the strong leg. Supporting such patients for more than a few steps is a slow process and will be uncomfortable for both the patient and the radiographer. It may also be hazardous to the patient. When any distance is involved, these patients should be moved by wheelchair.

You may use the wheelchair to transport patients from cars, from seats in the waiting area, or from the radiographic table. The process of assisting a patient to or from a wheelchair may seem elementary, but it is a common cause of falls and accidents. The correct technique makes this procedure safer and easier. Start with the patient seated. Position the wheelchair beside the patient with wheels locked and footrests out of the way. Assist the patient to stand and pivot so that the front edge of the wheelchair seat is touching the back of the patient's knees. Ease the patient into a sitting position in the chair. Position the footrests and leg rests. If the patient is in an examination gown, cover the patient's lap and legs with a sheet or

light blanket. This provides warmth and comfort while protecting the patient's modesty.

If the patient's foot or leg is injured, elevate the leg rest of the wheelchair to support the injured leg during transport. Help the patient lift the leg when changing position. Take care that the injured limb is not twisted or bumped during transport.

The most common type of fall associated with wheelchair transfer occurs when the patient backs into the wheelchair to sit down. The patient may miss the edge of the seat, or tip the chair by sitting too near the edge. To avoid such an accident, be sure to lock the wheels of the chair and assist the patient until seated securely.

To move the patient from the wheelchair to the x-ray table, place the wheelchair near the table, lock the brakes, and move the footrests out of the way. At this point the procedure will vary depending on whether you are fortunate enough to have an x-ray table that is adjustable in height.

If the height of the x-ray table is adjustable, lower the table to chair height. In this instance the transfer to the x-ray table is the reverse of the transfer from any seat. Using the face-to-face assist explained previously, help the patient to stand and pivot with the back to the table. Then ease the patient into a sitting position on the edge of the table.

If table height is stationary, position a step stool with a tall handle nearby. The patient now places one hand on the stool handle while you provide support from the opposite side. Assist the patient to step up onto the stool and pivot with the back to the table. If one side is weak, instruct the patient to step up with the strong leg. Then ease the patient to a sitting position. This procedure is illustrated in Fig. 21-13. Once the patient is seated on the table, assist the patient into a recumbent position as described previously. If the patient is paralyzed or unable to stand for any reason, a two- or three-person lift from the wheelchair is required. Never hesitate to request assistance when there is any question of whether you can handle the patient safely.

Lifting Patients from a Wheelchair

Patients who are paralyzed or unable to stand will arrive in wheelchairs. Depending on the design of the chair and the requirements of the procedure, extremity examinations and chest radiography may be performed with the patient seated in the wheelchair. Most procedures, however, will require that the patient be placed on the radiographic table. This requires a two- or three-person lift.

If the patient is not too heavy, two people can lift the patient from the wheelchair to the table (Fig. 21-14). The stronger of the two is the primary lifter, and the second person assists. First, place the wheelchair parallel to the table and lock the wheels. Re-

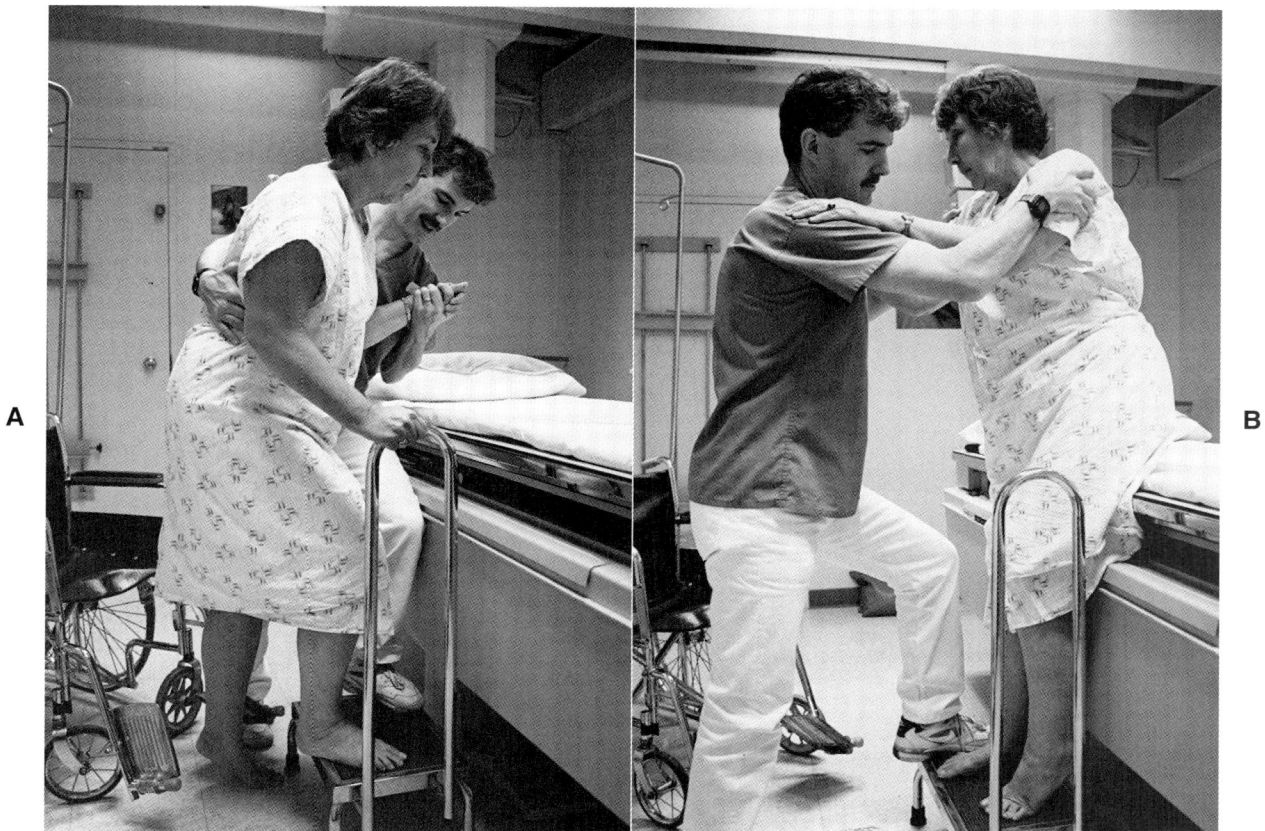

Fig. 21-13 Assisting patient onto x-ray table. **A,** Assist patient with step stool, if necessary. **B,** Assist patient to pivot and sit on the table.

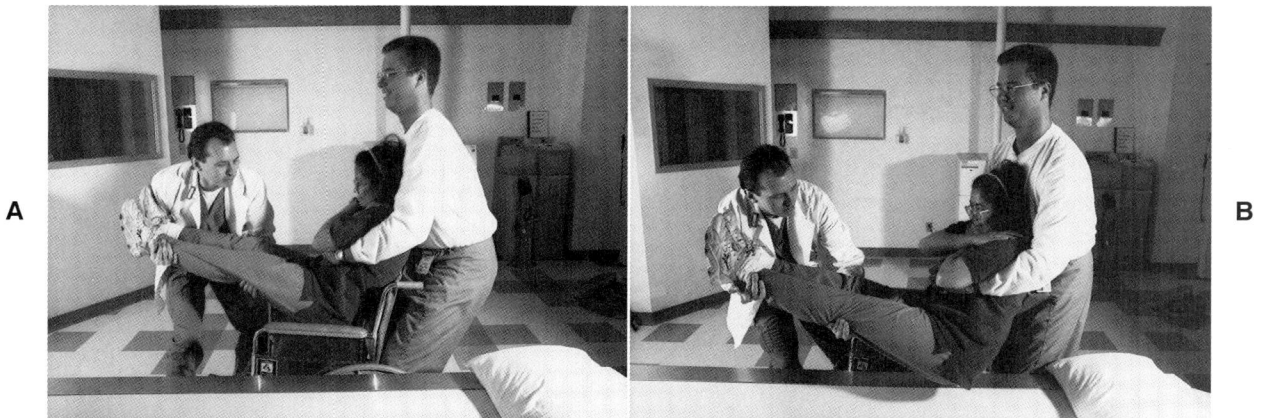

Fig. 21-14 Two-person lift. **A,** Primary lifter stands behind the chair and reaches around the patient, extending his arms through the patient's axillae and grasping her arms from the top. The assistant kneels on one knee, cradling the patient's thighs and legs. **B,** On signal, both lift together and place the patient gently on the table.

move the chair arm that is nearest the table, if possible. Instruct the patient to cross both arms across the chest. The primary lifter then stands behind the chair and reaches around the patient, extending his or her arms through the patient's axillae and grasping the patient's forearms from the top. The assistant kneels on one knee near the patient's feet and cradles the patient's thighs in one arm and lower legs in the other. On signal, both lift together and place the patient gently on the table.

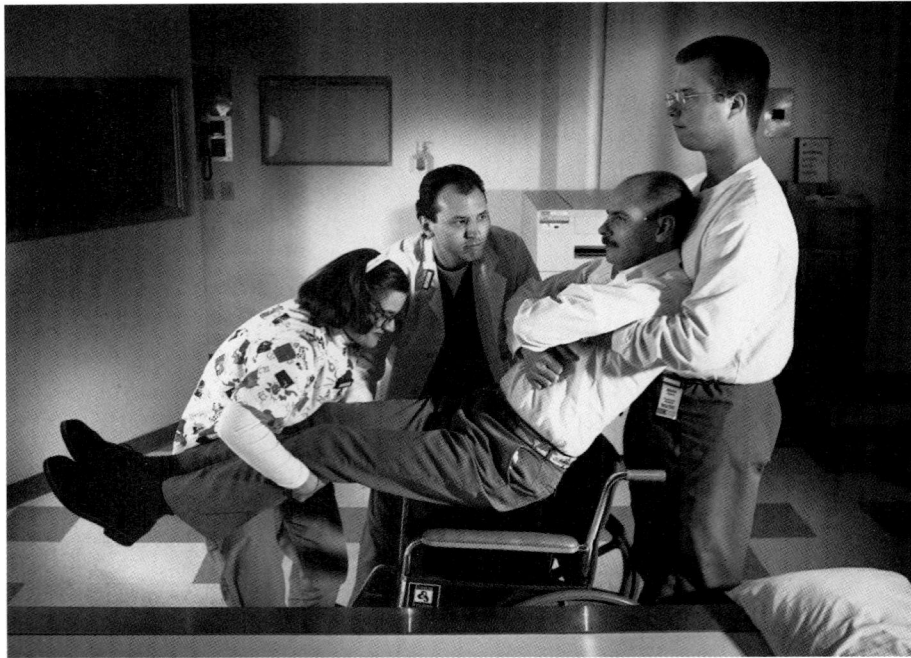

Fig. 21-15 The three-person lift is much like the two-person lift, with the third person helping to raise the patient's hips until they are clear of the chair.

The three-person lift is similar to the two-person lift and is safer if the patient's weight is beyond the ability of two people to lift easily (Fig. 21-15). In this case, remove both arms of the wheelchair and position the first two lifters as for the two-person lift. The third person kneels on one knee at the side of the chair that is farthest from the table. The third lifter places one arm around the patient's waist and the other under the buttocks. All lift together on signal. The role of the third lifter is primarily to assist in raising the patient from the chair. The wheelchair will block any forward motion of the third lifter, so the first two lifters will complete the transfer.

IMMOBILIZATION

Several methods can be utilized to aid in the immobilization of adults who have difficulty holding still in the desired position. When tremors such as those associated with Parkinson's disease complicate the procedure, it is useful to support the patient as comfortably as possible. A sandbag placed across an extremity proximal to the area of interest will stabilize the part during radiography.

Safety straps or compression bands may be used on the x-ray table for stabilization during radiography or during waiting periods as a precaution against falling (Fig. 21-16). Patients whose motion is restricted by safety devices must be monitored carefully. *Never leave a patient unattended who is unable to change position.* Difficulty breathing or the need to

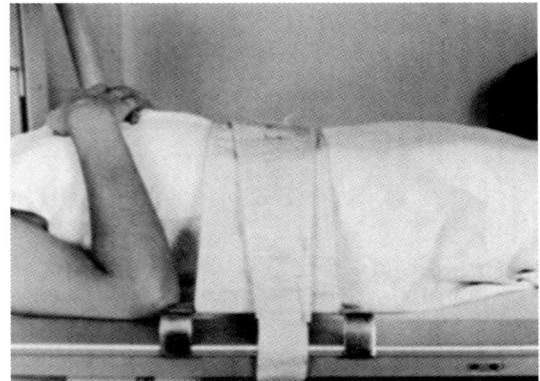

Fig. 21-16 Table restraint provides stabilization and security during radiography.

cough or vomit may require an immediate position change. The patient's inability to respond to this need may pose a serious hazard. If you must leave the room, another qualified person must be assigned to attend the patient.

The application of physical restraints to an adult patient without the patient's consent requires a physician's order. Liability for a charge of false imprisonment may result from the unauthorized use of restraints.

Immobilization of infants and children is discussed extensively in Chapter 18. Some of these methods may be applied to adults who are unable to cooperate. For example, the weight of a sandbag or of lead protective devices can be used to aid in maintaining position. Tape can also be used to maintain a position if it is not in direct contact with the patient's skin. Be certain that cloth or a tissue protects the skin from the adhesive surface of the tape, or twist the tape so that the nonadhesive side is against the skin.

When patients are coherent and cooperative and are neither sedated nor in distress, they may be left alone for brief periods. If delays occur, check with the patient frequently.

ACCIDENTS AND INCIDENT REPORTS

Any fall, accident, or occurrence that results in injury or potential harm must be immediately reported to your supervisor and/or your employer. As soon as the victim has been properly attended, an incident report, sometimes called an *unusual occurrence form,* must be completed. The reporting of incidents is essential whether the victim is a patient, a visitor, or a member of the staff. Do not hesitate to report incidents in which you are injured, even though the injury may seem minor at the time.

Incident reports are crucial to risk management. They aid in establishing or limiting liability for any injury, as well as documenting the need for changes that may improve safety practices in the future.

Occasionally a very minor incident may result in no harm and may not seem to require the formal procedure of an incident report. It is always a good idea to keep a record of *any* unusual occurrence in case it should later turn out to be of greater consequence than was originally apparent. Making a note of such events in the patient's chart or the requisition form provides a record that can be important if questions or consequences develop regarding the event.

The decision as to whether a particular occurrence merits an incident report is a judgment call. For example, a simple sneeze should not prompt an incident report, but a severe asthmatic attack is a reportable occurrence. A very mild asthmatic episode that is successfully self-treated by an outpatient with the patient's own medication is an example of a situation where judgment will vary with individuals. The ability to make these kinds of judgments develops with experience. Seek the counsel of a supervisor when such questions arise, and when in doubt, err on the side of caution by filing a report.

INFECTION CONTROL

Cycle of Infection

The four factors involved in the spread of disease are sometimes called the *cycle of infection* (Fig. 21-17). For infections to be transmitted, there must be an infectious organism, a reservoir of infection, a susceptible

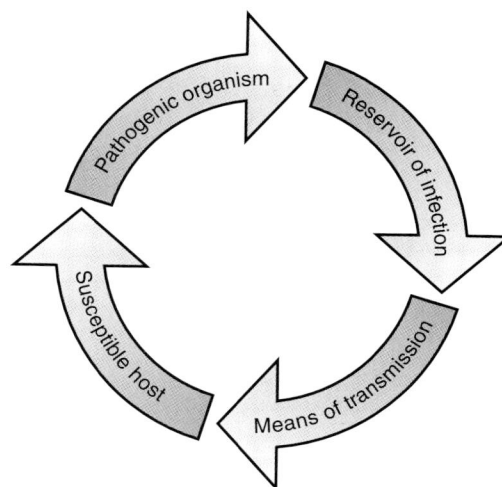

Fig. 21-17 Cycle of infection.

host, and a means of transporting the organism from the reservoir to the susceptible individual.

Microorganisms

Microorganisms or **microbes** are referred to in lay terms as "germs." They are living organisms too small to be seen with the naked eye. They include bacteria, viruses, protozoa, prions, and fungi. Most microorganisms do not cause disease and many are essential to our continued well-being. Some microorganisms live on or within the body without causing disease and are referred to as **normal flora.** They aid in skin preservation and digestion and protect us from infection. Microorganisms that cause disease are called **pathogens.** Normal flora may also be pathogenic when they are not confined to their usual environment. For example, *Candida albicans* may be found in the throats or gastrointestinal tracts of most healthy individuals. This organism can cause urinary or vaginal infections in females and sometimes causes respiratory disease in infants or in adults with compromised immune systems.

Bacteria

Bacteria are very small single-cell organisms with a cell wall and an atypical nucleus that lacks a membrane (Fig. 21-18). Bacteria grow independently and do not need a host cell to reproduce. They are classified according to shape. Most bacteria have one of three distinct shapes: spherical, called *cocci;* rod-shaped, called *bacilli;* and spiral, called either *spirilla* or *spirochetes* (Fig. 21-19).

Bacteria are able to adapt to new conditions and are also able to mutate, allowing them to resist and survive in the presence of antimicrobial drugs. Some

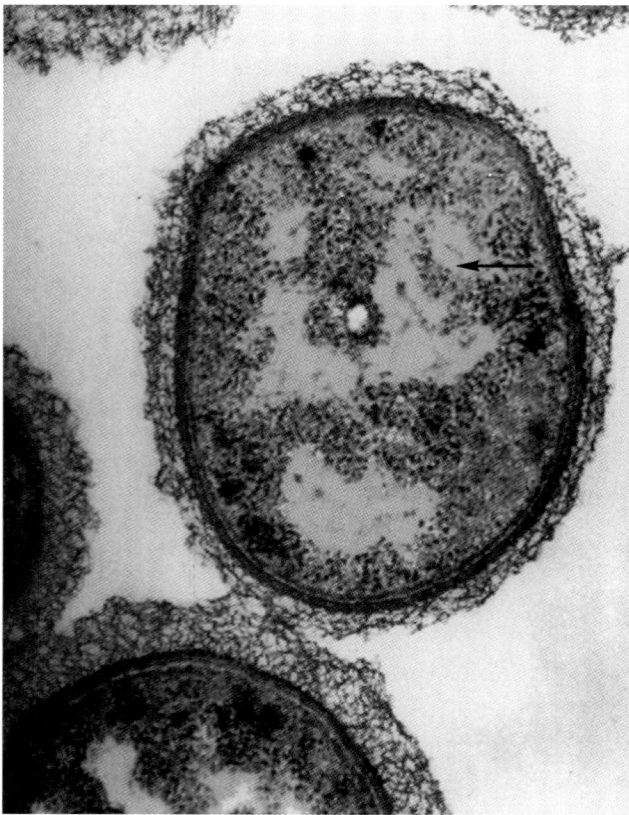

Fig. 21-18 Photomicrograph of a bacterium. Note the absence of a nuclear membrane.

types of bacteria have the ability to generate **endospores,** which are formed within the cell when environmental conditions are unfavorable. This bacterial form is resistant to heat, cold, and drying and can live without nourishment. Most endospore-forming bacteria live in the soil, but they can reside almost anywhere. When conditions improve, endospores germinate, revitalizing the bacteria. Endospore-forming organisms are responsible for tetanus, anthrax, and gas gangrene. These diseases are serious but are relatively uncommon.

Significant diseases caused by bacteria include tuberculosis and streptococcal pharyngitis (strep throat), as well as infectious diarrhea and kidney disease caused by a particular strain of *Escherichia coli* (*E. coli* 0157:H7).

Viruses

Viruses are subcellular organisms. They are among the smallest known disease-causing organisms. In order to be studied, they must be viewed with an electron microscope. Examples of viruses include influenza, human immunodeficiency virus (HIV), herpes, hepatitis, and rhinovirus. Rhinovirus causes the common cold. Other common viruses are Epstein-Barr virus, which causes infectious mononucleosis, and varicella, which causes chickenpox.

Viruses cannot multiply on their own. A virus invades a host cell, stimulating it to participate in the

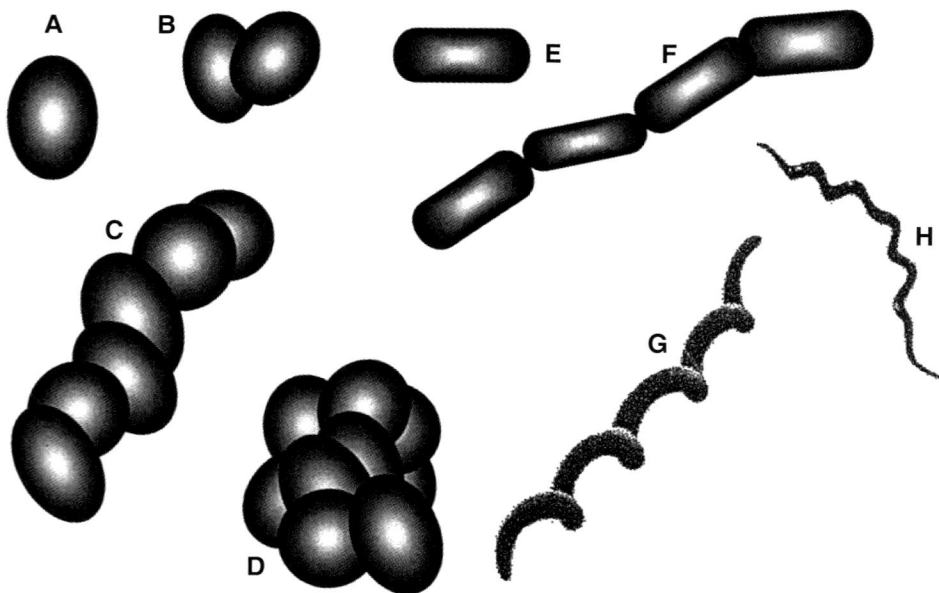

Fig. 21-19 *Bacterial forms: cocci.* **A,** Single coccus. **B,** Diplococcus. **C,** Chain formation *(Streptococcus).* **D,** Cluster *(Staphylococcus). Bacterial forms: bacilli.* **E,** Single bacillus. **F,** Chain. *Spiral bacteria.* **G,** Spirillum. **H,** Spirochete.

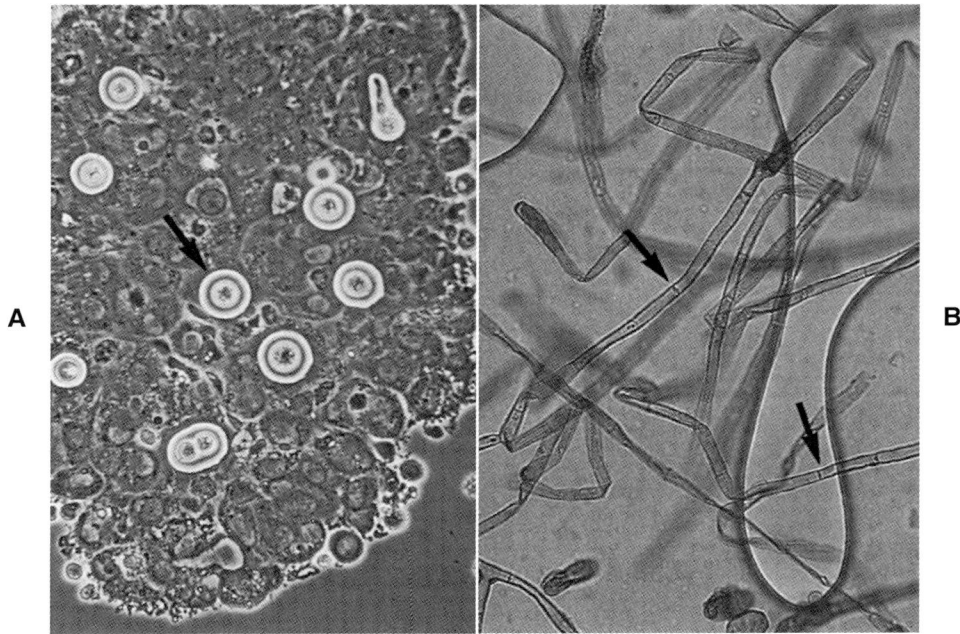

Fig. 21-20 Fungi. **A,** Single-cell yeasts. **B,** Molds are multicellular fungi.

formation of additional virus particles. Each type of virus is specific to a particular type of cell. For example, the hepatitis virus attaches to liver cells. Because viruses reside in the host cell and use it to replicate, it has been difficult to develop antiviral drugs that are not also harmful to the host cell. Only a few antiviral agents exist, and these are useful against only a limited number of viruses.

Fungi

Fungi (singular, fungus) occur as single-celled yeasts or as filament-like structures called *molds* that are composed of many cells (Fig. 21-20). There are more than 100,000 diverse species of fungi, and many have useful purposes. They assist in the production of alcoholic beverages, are responsible for the flavor of cheese, give bread its lightness, and produce the antibiotic penicillin. In humans, fungi cause skin infections such as athlete's foot and ringworm, respiratory infections such as histoplasmosis, and infections in individuals with compromised immune systems.

Prions

The smallest and least understood of all microbes is the prion, which was discovered in 1983. Scientists believe that prions may be infectious proteins. Their method of replication is not understood. They were first identified as the cause of scrapie, a degenerative disease affecting the nervous systems of sheep. It is thought that

prions are the cause of Creutzfeldt-Jakob (mad cow) disease in humans, and perhaps other conditions that are characterized by slow deterioration of the nervous system. There is early speculation that further research into the nature of prions may help us to better understand the cause of Alzheimer's disease.

Protozoa

Protozoa are complex, single-cell animals that generally exist as free-living organisms (Fig. 21-21). A few, however, are parasitic and live within the human body. Most parasitic protozoa produce some type of resistant form such as a cyst to survive in the environment outside the host. Other protozoa have complicated life cycles involving alternate existence in the human body and in insects. This is true of the protozoan that causes malaria. Protozoa can infect the gastrointestinal, genitourinary, respiratory, and circulatory systems.

Reservoir of Infection

The reservoir, or source, of infection may be any place where pathogens can thrive in sufficient numbers to pose a threat. Such an environment must provide moisture, nutrients, and a suitable temperature, all of which are found in the human body. A source of infection might be the patient with hepatitis, a radiographer with an upper respiratory infection (URI), or a visitor with staphylococcal boils.

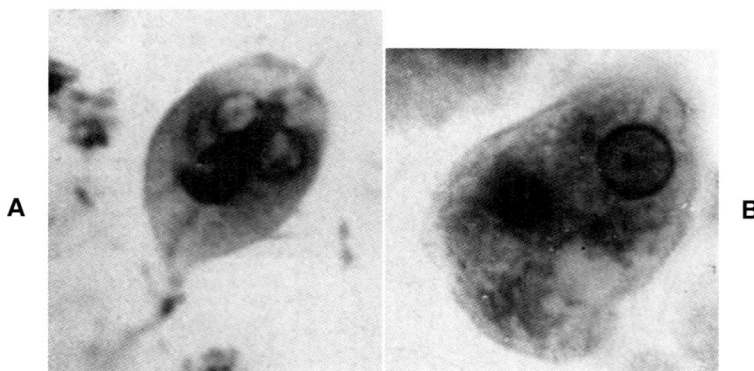

Fig. 21-21 Photomicrographs of protozoa. **A,** *Giardia lamblia,* a flagellate. **B,** *Entamoeba histolytica,* an ameba.

Since some pathogens live in the bodies of healthy individuals without causing apparent disease, a person may be the reservoir for an infectious organism without realizing it. These persons are called **carriers.** Many of us have throat cultures that are positive for *Staphylococcus aureus* ("staph"), but we do not have a sore throat. A susceptible patient with an open wound could contract a life-threatening infection if sufficiently contaminated with this organism. The classic example of a carrier of infection is Typhoid Mary, a "healthy" food handler. Hundreds of cases of typhoid fever were attributed to contamination of the meals she helped to prepare. Better sanitation and food-handling education have reduced the incidence of food-borne diseases. Today, a more common example of a carrier of infection is the asymptomatic individual infected with HIV who spreads the disease through sexual intercourse or by sharing contaminated needles with intravenous drug users.

Although the human body is the most common reservoir of infection, any environment that will support the growth of microorganisms has the potential to be a secondary source. Such sources include contaminated food or water or any damp, warm place that is not cleaned regularly.

Susceptible Host

Healthy individuals have a high level of natural resistance to infection. Fatigue, stress, malnutrition, illness, and injury tend to tax and weaken the immune response, reducing natural resistance. In a health care setting, susceptible hosts are frequently patients whose natural resistance to infection is diminished. In addition to the primary problems that caused them to seek care, they may develop iatrogenic (health care–related) infections.

Infections also pose a threat to health care workers because their work results in exposure to many pathogens. In a single day a radiographer may care for patients with pneumonia, hepatitis, and wound infections. Hepatitis B and C are the biggest concerns. Both are spread by blood and blood products and are most often transmitted to health care workers by accidental needle sticks. The radiographer who must work when resistance is low because of fatigue, stress, or a low-grade infection has increased susceptibility to infection. Maintain your resistance to infectious illness by keeping your body healthy and well rested.

Disease Transmission

The most direct way to intervene in the cycle of infection is to prevent transmission of the infectious organism from the reservoir to the susceptible host. To accomplish this successfully, you must understand the six main routes of transmission.

The first route is by means of direct contact. In this transmission mode the host is touched by an infected person in such a manner that the organisms are placed in direct contact with susceptible tissue. For example, syphilis and HIV infections may be contracted when infectious organisms from the mucous membranes of one individual are placed in direct contact with the mucous membranes of a susceptible host. Also, skin infections often occur among health care workers because of frequent contact with patients who have bacterial diseases. The five other principal routes of transmission are indirect and involve transport of organisms by means of fomites, vectors, vehicles, airborne particles, and droplet contamination.

An object that has been in contact with pathogenic organisms is called a **fomite.** Examples of fomites in the radiology department might include the x-ray table, upright Bucky, cassettes, calipers, and positioning sponges that are contaminated with infectious body fluids.

A **vector** is an arthropod (insect, spider, or similar form) in whose body an infectious organism develops or multiplies before becoming infective to a new host. Some examples of vectors are mosquitoes that transmit malaria or dengue fever, fleas that carry bubonic plague, and ticks that spread Lyme disease or Rocky Mountain spotted fever. In these examples, the bites of infected insects transmit disease to humans.

A **vehicle** is any medium that transports microorganisms. Examples include contaminated food, water, drugs, or blood.

Airborne contamination is spread by dust containing either endospores or droplet nuclei (small infectious particles from evaporated droplets that contain microorganisms). Contaminated dust may remain suspended in the air for long periods. These particles may be dispersed by air currents and may be inhaled by a susceptible host. Special air handling and ventilation are required to prevent airborne transmission of these infective particles. *Mycobacterium tuberculosis,* rubeola (measles), and the varicella (chickenpox) viruses are examples of airborne infections.

Droplet contamination often occurs when an infectious individual coughs, sneezes, speaks, or sings in the vicinity of a susceptible host. Droplet transmission involves contact of the mucous membranes of the eyes, nose, or mouth of a susceptible person with droplets containing microorganisms. These particles are relatively large and do not remain suspended in the air. They travel only short distances, usually 3 feet or less. Influenza, meningitis, diphtheria, pertussis (whooping cough), and streptococcal pneumonia are examples of illnesses spread by droplet contamination.

Although most microorganisms are fragile, requiring continuous warmth, moisture, and nutriment to exist, some are resistant to destruction and can remain viable for long periods of time. Bacteria that are capable of forming endospores may live in this form for many years. They are often carried in invisible dust particles in the atmosphere. Research has shown that some viruses can resist drying, remaining infectious for weeks. This is true of the viruses that cause herpes, both oral and genital. These examples indicate that some microorganisms can float through the air and lurk in dusty corners waiting for the opportunity to invade a susceptible host. This should emphasize to you the need for cleanliness as a defense against infection.

INFECTIOUS DISEASES

Disease Information

There are many new diseases in the world. Some old diseases are returning in epidemic proportions after years of low-level incidence. The wide and inappropriate use of broad-spectrum antibiotics has led to the development of drug-resistant infections in hospitals and in the community. Some of these infections are untreatable because they are resistant to the available drugs. As a worker in the health care field, you may be on the front line of exposure to infectious diseases. The Centers for Disease Control and Prevention (CDC) monitor and study the types of infections occurring in the nation, compile statistical data about these infections, and publish this information in both a weekly report and an annual surveillance summary report. The CDC also provides information about prevention and treatment of specific infections. When questions arise regarding current information on disease prevention and infection control, the best source for answers is the CDC.*

HIV and AIDS

The rapid spread of HIV and acquired immunodeficiency syndrome (AIDS) is a source of concern to everyone. As of 1998 the CDC estimated that 800,000 to 900,000 persons were living with HIV or AIDS in the United States. The cumulative number of AIDS-related deaths reported by June 2000 was greater than 400,000. Estimates as of December 2000 indicate that worldwide there are 36 million people living with this disease and more than 21 million have died from it. The number of infections increases by 16,000 every day! Some scientists predict that 75% to 90% of those infected with HIV will develop AIDS.

In the early 1980s, HIV was identified as the cause of AIDS. Two major types of HIV are known to infect humans. HIV-1 is the predominant type throughout the world, and HIV-2 is found primarily in heterosexual populations in West Africa.

HIV can infect many different cells in the body, including the cells of the central nervous system. It is the adverse affect of the virus on the cells of the immune system that suppresses the normal immune response and causes the manifestations of AIDS.

An HIV-infected individual can transmit the virus to others starting a few days after infection, even though evidence of the virus may not be detected in the blood for 3 to 6 months. Infected individuals pass through several phases of infection over a span of months or years before immunosuppression causes the symptoms of AIDS. Soon after infection there is often a brief period of flu-like symptoms, usually followed by years without symptoms. During this phase the virus is silently replicating in the body. At the end of the asymptomatic period, before the full development of AIDS, the individual will experience

---

*Centers for Disease Control and Prevention, 1600 Clifton Road, Atlanta, GA 30333. Telephone: (404) 639-3311. Toll-free hotlines are available on specific subjects. On the Internet, contact http://www.cdc.gov.

night sweats, oral infections, weight loss, enlarged lymph nodes, and a low fever. The appearance of AIDS is characterized by a significant decrease in the number of certain white blood cells (CD4 lymphocytes) and the occurrence of multiple infections and malignant diseases. The infections that occur with AIDS are called **opportunistic infections** because they occur when certain microorganisms take advantage of the opportunity to invade in the absence of an immune response. Some of the opportunistic infections observed are *Pneumocystis carinii* pneumonia, *Candida albicans* infection of mucous membranes, and widely scattered lesions from herpes virus and cytomegalovirus. There is also an increased risk of developing active tuberculosis. Kaposi's sarcoma, a malignancy of pigmented cells of the skin, is the most common form of cancer affecting AIDS patients.

Drugs have been developed that prolong the time required for HIV infection to progress to AIDS, but at this time there is no known cure. Research on vaccines is continuing but is hampered by the rapid mutation of the virus.

Fortunately, the AIDS virus is *not* acquired by casual contact. Touching or shaking hands, eating food prepared by an infected person, and contact with drinking fountains, telephones, toilets, or other surfaces do not result in transmission of HIV. The principal routes of transmission are through sexual contact and through contaminated blood or needles. HIV is also transmitted from mother to fetus via the placenta and from mother to child through breast milk.

As a health care worker in today's world, you must expect to encounter unidentified or undiagnosed cases of HIV and other blood-borne diseases. Controversy surrounds the patient's right to confidentiality regarding AIDS diagnosis or HIV status, even within the health care setting. This may prevent you from being informed about diagnosed cases. The CDC estimates that at least two undiagnosed cases exist for every case that is known.

Anxiety about HIV infection is typical and understandable among health care workers. However, the occupational risk is not great. The vast majority of health care workers infected with HIV were exposed as a result of activities unrelated to their work. The most common occupational exposure, by far, is the needle stick. According to the CDC the probability of infection following a needle stick injury with blood containing HIV is 3 out of 1000 exposures. Many thousands of needle sticks have been reported over the years, but as of 1999 only 192 health care workers with no other identified risk factors have been diagnosed as HIV-positive. Of these cases, 56 were sufficiently documented by testing to confirm that the infection was contracted as a result of work-related

exposure. The implications here are obvious. Although prevention at work is essential, self-care in terms of safe sexual practice is equally crucial.

Hepatitis

Hepatitis is an infectious disease of the liver. There are seven types of hepatitis, classified A through G. Hepatitis A and E are transmitted through food and water that is contaminated with feces. Hepatitis B, C, D, and G are blood-borne. Hepatitis E is uncommon in the United States, and little is known about hepatitis F and G. Hepatitis D appears only as a co-infection with hepatitis B.

Hepatitis A, B, and C are the types of primary concern with respect to transmission in the health care setting. Hepatitis B can be spread through contact with blood or blood products; with body fluids such as saliva, semen, and vaginal secretions; and through maternal-fetal contact. Hepatitis C is primarily spread by blood or blood products; the risk for sexual or maternal-fetal transmission is unknown.

The clinical features of all forms of hepatitis are similar: jaundice (bile pigment in the blood causing the skin and whites of the eyes to appear yellow), fatigue, abdominal pain, loss of appetite, nausea, vomiting, and diarrhea. Both hepatitis B and hepatitis C have the potential to develop into chronic infections, although the risk is greater with hepatitis C.

Health care workers can protect themselves against hepatitis B by taking a vaccine, which usually provides immunity for 7 to 10 years. Protection from hepatitis A and C can be achieved by following established infection control practices. Hepatitis A remains the most common form of the disease and is best controlled by practicing good personal hygiene, especially hand washing.

Tuberculosis

Tuberculosis (TB) is a disease of the lungs caused by the acid-fast bacillus *Mycobacterium tuberculosis,* also referred to as *tubercle bacillus.* Historically, this disease was called "consumption" because of the victim's tendency to "waste away." In the past the incidence of TB in the United States was spread across all economic levels of society. Today, however, the highest rate of active cases is seen among the homeless, recent immigrants, and immunosuppressed individuals. While the incidence of cases in this country is much lower now than it was in the years before 1950, recent outbreaks of TB have raised grave concern due to the appearance of drug-resistant strains of the bacteria.

Pulmonary tuberculosis is spread through airborne droplet nuclei that are generated when an infected person coughs or speaks. These particles are easily transmitted because they are extremely tiny (1 to 5 microns in size), and air currents keep them airborne. The probability that a susceptible person will be infected depends upon the concentration of the infectious droplet nuclei in the air.

The vast majority of those infected with tubercle bacilli do not develop a clinical disease and become infectious. Generally, within 2 to 10 weeks following infection, the body's immune system begins walling off the infection, preventing its multiplication and spread. The walled-off disease is inactive or dormant, but the infection can be reactivated at any time. Reactivation may occur with lowered resistance due to immune deficiency, malnutrition, other illness, or old age.

In a weakened or immunosuppressed state, such as with HIV, patients progress rapidly to active disease. Symptoms of active disease include productive or prolonged cough, fever, chills, loss of appetite, weight loss, fatigue, and night sweats. As the bacilli multiply, they cause tissue necrosis that results in cavities in the lung. These spaces are major reservoirs of the infection that can then be spread by coughing. Severe cases can be fatal. Examples of radiographs showing various stages of tuberculosis are shown in Chapter 16, Fig. 16-26.

Extrapulmonary TB, infecting bone or organs other than the lung, accounts for a small percentage of TB infections. Patients with extrapulmonary infection and no active pulmonary disease do not transmit the disease through airborne contamination.

The simplest and most common method of testing for TB infection is the tuberculin skin test. This test involves an intradermal injection on the anterior forearm. The injection produces a raised area on the skin, similar to an insect bite, that is called a *weal*. The weal is inspected 48 to 72 hours later to determine whether the individual has been infected with TB. A negative test indicates that the person has never been infected. A positive result indicates that a person has at one time been infected and has developed antibodies (resisting proteins) to the organism. Since few people develop clinical symptoms or become infectious, many people have a positive skin test without having active disease. This test is administered when a person is known to have been exposed to TB and has not already tested positive.

If the skin test is positive, or if symptoms are present, a chest radiograph is ordered to rule out active disease. When there are symptoms and/or positive radiographic findings, sputum smears and cultures are tested for acid-fast bacilli (AFB). Positive results are definitive proof of active disease and are an indication to begin treatment.

TB screening is often required for those who work in contact with vulnerable or high-risk populations.

For example, schoolteachers, corrections officers, and health care workers are often required to have pre-employment tuberculin skin tests.

According to the CDC, there were over 17,000 new cases of tuberculosis reported in the United States in 1999. This represents a decrease of nearly 5000 from 1995 and is the lowest number of reported cases since national reporting began in 1953. This decline is similar for both US-born and immigrant populations. Less than 3% of cases reported represent infected health care workers.

Early identification, isolation, and treatment are required to minimize transmission of TB. Health care workers are at risk of contracting this disease only when a patient is exhibiting symptoms of active TB.

In 1993 OSHA issued an enforcement policy on TB, based on CDC guidelines, to protect exposed workers against TB. According to this policy, infection control experts within the health care facility are to conduct an assessment to determine the actual risk for transmission of TB in inpatient and outpatient settings. If the findings reveal risk, they are to develop TB infection-control interventions. These include free TB skin tests.

PREVENTING DISEASE TRANSMISSION

Historical Perspective

When infectious disease was rampant, infected persons were often "quarantined," which meant that members of a household were prevented from leaving their home and others were excluded. This practice helped to confine the infection to one family rather than spreading it through an entire community. Later, hospitals developed policies that involved separating patients admitted with infectious diseases from other patients. Contact with other persons was rigidly controlled. This "isolation" was a logical outgrowth of the former practice of quarantine. These techniques provided for specialized methods of infection control when the danger of disease transmission was exceptionally great. Although isolation techniques were effective when used correctly, no mechanism existed for the prevention of serious diseases carried by asymptomatic individuals. This was certainly true of patients who were carriers of hepatitis B virus and HIV.

In acknowledgment that many patients with blood-borne infections are not recognized, the CDC introduced a system in 1985 known as Universal Precautions (UP). With this system, all patients are treated as potential reservoirs of infection. It is based on the use of barriers for all contacts with blood and certain body fluids known to carry blood-borne

BOX 21-4
Standard Precautions for All Patient Care

Wash hands often and well.	Wear protective gloves when likely to touch body substances, mucous membranes, or non-intact skin.	Wear plastic apron or a gown when clothing is likely to be soiled.	Wear mask and eye protection when likely to be splashed.	DO NOT RECAP — Place intact needle/syringe units and "sharps" in designated disposal container. Do not break, bend, or recap needles.

pathogens. The need to use barriers such as gloves and masks depends on the nature of the interaction with the patient rather than on the specific diagnosis.

Because UP placed emphasis on blood-borne infections and did not include precautions for contamination by feces, nasal secretions, sputum, urine, and vomitus (unless contaminated with visible blood), a new system was introduced in 1987 called Body Substance Precautions (BSP), also called Body Substance Isolation (BSI). This system focused on the use of barriers for all moist and potentially infectious body substances from all patients. The system was developed to protect health care workers from acquiring and transmitting infections from all pathogens. As of 1996, however, the CDC has recommended a new system that synthesizes the features of UP and BSP. This most recent system is called **Standard Precautions** and incorporates guidelines for isolation in hospitals. As applied in an outpatient setting, Standard Precautions are essentially the same as BSP.

Standard Precautions involve the use of barriers whenever contact is anticipated with any of the following:

- Blood
- All body fluids and wound drainage
- Secretions and excretions (except sweat), regardless of whether they contain visible blood
- Non-intact skin
- Mucous membranes

Standard Precautions, as applied in an outpatient setting, are summarized in Box 21-4. They are designed to reduce the risk of transmission of unrecog-

nized sources of pathogens in health care facilities, whether blood-borne or not. If unanticipated contact with any body substance occurs, thoroughly wash the contact area as soon as possible. Use gloves to wipe up after all blood spills and disinfect using 1 part bleach to 10 parts water.

Standard Precautions require that each individual use judgment in determining when barriers are needed. Each individual must establish his or her own standards for consistent use of barriers. These personal standards should be based on the individual's skills and the anticipated interactions involving the patient's body substances, non-intact skin, and mucous membranes. You will be making frequent decisions about when to take the extra time to protect both yourself and your patients. In the beginning, your level of precautions should be very high. Although you may observe more experienced workers taking fewer precautions, do not think you must follow their example. At this stage it is far better to take too many precautions than to take too few.

Be aware of the specific infection control policies in place at your facility. The key to effective protection is a consistent approach to *all contact* with *all body substances* of *all patients* at *all times.*

MEDICAL ASEPSIS

Medical **asepsis** is the process of reducing the *probability* of infectious organisms being transmitted to a susceptible individual. The healthy human body has the ability to overcome a limited number of infectious organisms. This resistance can be overwhelmed by a massive exposure. On the other hand, a reduced resistance caused by disease, cancer chemotherapy, immunosuppressants, or extremes in age may result in infection after only minimal exposure. The fewer organisms to which an individual is exposed, the more likely it is that he or she will resist infection. The process of reducing the total number of organisms is called **microbial dilution** and can be accomplished at several levels.

Simple cleanliness measures avoid transmitting organisms when proper cleaning, linen handling, and hand washing techniques are used. The second level is **disinfection** and involves the destruction of pathogens by chemical agents. The third stage is surgical asepsis, or **sterilization.** This involves treating items with heat, gas, or chemicals to make them germ-free. They are then stored in a manner that prevents contamination.

You can easily find examples of poor aseptic technique in most clinical settings. Unfortunately, the results of carelessness are seldom traced to the culprit. It is the patient acquiring an infection who suffers. Armed with the knowledge of disease transmission, how can you fight the spread of infection?

- Stay home when you are ill, if possible. Definitely avoid contact with immunocompromised patients.
- Cover your mouth when you sneeze or cough.
- Wear a clean uniform daily, and remove it immediately when you go home.
- Wash your hands frequently.
- Use established precautions when handling patients, linens, or items contaminated with body substances.
- Practice good housekeeping techniques in your work area.
- When in doubt about the cleanliness of any object, do not use it.
- Dispose immediately of linens, instruments, or other items that touch the floor because the floor is always considered contaminated.
- When patients are coughing or sneezing, provide tissues and ask them to cover their mouth and nose.

Hand Washing

The first three principles listed above are simple and self-explanatory. Hand washing also may seem obvious, but this is the rule most consistently ignored in many health care settings. *Frequent hand washing is the single best protection against disease.* Aseptic hand washing technique is both simple and effective (Fig. 21-22). It should be followed explicitly before and after work, before meals, and often during the day. Since a radiographer's duties frequently demand brief contacts with a series of patients, it is especially important that you always wash your hands between patients and after removing gloves.

Housekeeping

Good housekeeping in the workplace reduces the incidence of airborne infections and the transfer of pathogens by fomites. A clean, dry environment discourages the growth of all microorganisms. A custodian or cleaning service may do much of the cleaning in the office or clinic at night, but the radiographer is responsible for inspecting the work area regularly and maintaining high standards of medical asepsis.

Several general principles apply whenever cleaning is required:

- Always clean from the least contaminated area toward the more contaminated area and from the top down.
- Avoid raising dust.
- Do not contaminate yourself or clean areas.
- Clean all equipment that comes in contact with patients after each use using a cloth moistened with disinfectant.

The CDC recommends a diluted solution of sodium hypochlorite bleach (Clorox) as the preferred disinfectant for preventing the spread of HIV. Dilute bleach by mixing 1 part bleach with 10 parts water. Mix fresh bleach daily because its effectiveness declines rapidly when diluted.

Your facility may have written procedures with detailed instructions concerning preferred cleansing agents and the extent of responsibility for disinfecting rooms. Consult the policy and procedure manual.

Handling Linens

Objects or linens soiled with body secretions or excretions are considered contaminated and may serve as fomites even though stains may not be apparent. For this reason many clinics use disposable gowns and linens. Any linens used by patients should be handled as little as possible. To prevent airborne contamination, fold the edges to the middle without shaking or flapping and immediately place loosely balled linens in a hamper or a lined trash container. *Never use any linen for more than one patient.*

Handling and Disposal of Contaminated Items and Waste

A modern health care facility uses many disposable items, from simple objects such as paper cups and tissues to more complex items such as trays for minor

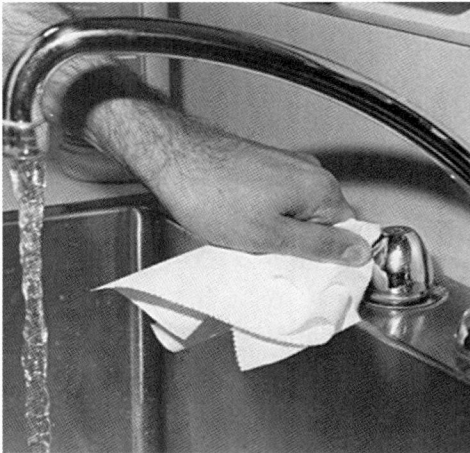

A, Sink is contaminated. Use paper towel to handle controls unless there are foot or knee levers.

B, Wet hands thoroughly. Keep hands lower than elbows so water will drain from clean area (forearms) to most contaminated area (fingers).

C, Apply antimicrobial soap.

D, Lather well. Rub hands and fingers together with firm rotary motion for 20 seconds. *Friction is more effective than soap in removing microorganisms from skin.* Rub palms, backs of hands, and areas between fingers.

E, Rinse, allowing water to run down over hands. Repeat steps to cleanse wrists and forearms.

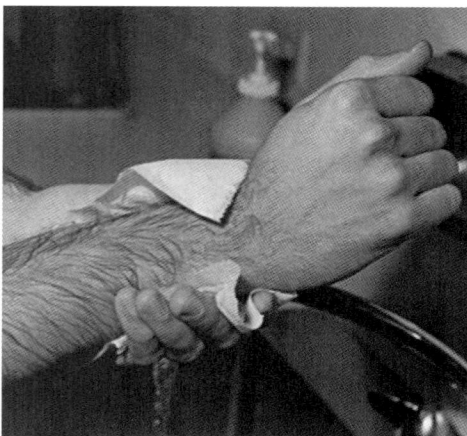

F, Use paper towel to dry thoroughly from fingertips to wrist.

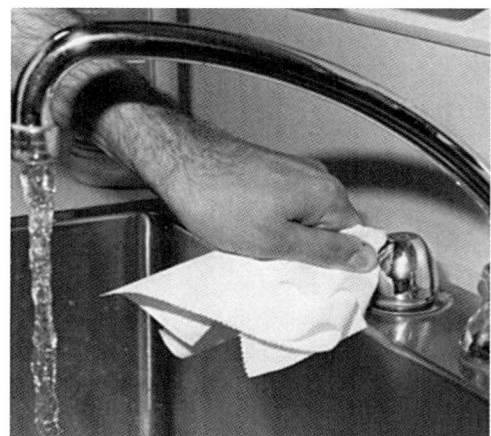

G, Turn off water with paper towel to avoid contaminating hands.

Fig. 21-22 Hand washing.

surgical procedures. Disposable items are designed to be used only once and then discarded. The only exception to this rule involves the immediate reuse of an unsterile item such as an emesis basin by the same patient.

Your facility will have a protocol for discarding disposable items. Some place glass, plastic, and paper into separate covered containers. Others place everything together. Regulations demand that objects contaminated with blood or body fluids be discarded in a suitable container and marked with the **biohazard symbol** (Fig. 21-23). Used bandages and dressings are assumed to be contaminated. They are handled with gloves and placed directly into waterproof bags, which are then sealed and discarded.

Needles, syringes, and contaminated items capable of puncturing the skin are disposed of in **sharps containers.** Sharps containers are designed to receive the syringe without recapping it. They are discussed further and illustrated in Chapter 24. *Never recap a needle.* This is how most finger punctures occur.

An accidental needle stick or skin broken by a contaminated object is cause for concern. If this occurs, allow the wound to bleed under cold water and wash with soap. An incident report must be filed, even though the injury seems insignificant. In addition to the incident report, a baseline blood sample should be drawn. Since HIV infection will not be apparent in the blood for approximately 3 months, this helps rule out infection acquired before the occupational exposure. At 6 months, another sample is tested for HIV.

Before specimens are sent to the laboratory, they should be placed in clean containers with secure caps and slipped inside a plastic bag labeled with a biohazard symbol.

Fig. 21-23 Biohazard symbol.

Always wear gloves when assisting patients with bedpans or urinals. Collect any specimen needed and empty the bedpan or urinal immediately. Rinse it well over the toilet, and discard it or put it in the proper place to be sterilized.

SURGICAL ASEPSIS

Earlier in this chapter, we defined *medical asepsis* as a method of reducing the number of pathogenic microorganisms in the environment and intervening in the process by which they are spread. *Surgical asepsis,* on the other hand, is the complete removal of all organisms and their spores from equipment used to perform patient care or procedures. The linens, gloves, and instruments used in surgery may be the first examples brought to mind, but many other procedures such as injections and the drawing of blood samples also require sterile equipment. In addition, some procedures require special skin preparation to prevent pathogens from entering the body.

Sterile items used in clinics and doctors' offices are usually disposable items such as syringes and needles. They are sterile when purchased and are protected by a paper or plastic wrap. Reusable items such as instruments and glass syringes are wrapped, sterilized, and stored in a clean, dry location.

Sterilization

Although the radiographer is seldom directly involved in the process of sterilization, it is helpful to understand the methods that may be used. Chemical, gas, and steam sterilization are most common.

Chemical sterilization involves the immersion and soaking of clean objects in a bath of germicidal solution. Sterilization depends on solution strength and temperature and the immersion time, all of which are difficult to control accurately. Contamination of the solution or the object being sterilized may occur and is not easily detectable. For this reason, chemical sterilization is one of the less satisfactory methods for providing surgical asepsis and is not recommended. If chemical sterilization must be used, be certain to follow the chemical manufacturer's instructions completely.

Items that would be damaged by moist heat are usually sterilized with a mixture of gases (Freon and ethylene oxide) heated to 135° F (57° C). Gas sterilization is used primarily for electrical, plastic, and rubber items and for optical ware. Telephones, stethoscopes, blood pressure cuffs, and other equipment may be sterilized in this manner. Hospitals have equipment for gas sterilization, but this method is not usually available in outpatient facilities. Items of value that have become contaminated may have to be sent out for gas sterilization. This treatment sterilizes very effectively but has one drawback: since the gases used are poisonous, they must be dissipated by means of aeration in a controlled environment. Aeration is a slow process, so it is important to send items for gas sterilization well in advance of the time they will be needed.

An **autoclave** is an electric steam chamber that seals tightly to achieve high temperatures under pressure. Autoclaving, or steam sterilization, is the quickest and most convenient means of sterilization for items that can withstand heat. Higher temperatures can be achieved under pressure, making this an extremely effective method. Clinics that need to sterilize reusable equipment will have a small autoclave for this purpose.

An advantage of both steam and gas sterilization is that indicators can be used to identify that a pack has been sterilized. Indicators are placed inside the pack, as well as outside, to show that the gas or steam has penetrated to all surfaces. Indicators change color when the required conditions have been met. Radiographers are responsible for correctly recognizing the sterilization indicators used in their clinical facilities.

Sterile Fields

If your job description involves assisting the physician with sterile procedures, you will need to know how to prepare a sterile field. A sterile field is a germ-free area prepared for the use of sterile supplies and equipment. The principles of surgical asepsis used in establishing and working with a sterile field are stated in Box 21-5. The first step in preparing a sterile field is to confirm the sterility of packaged supplies and equipment. Packages are considered sterile if they meet the following criteria:

- They are clean, dry, and unopened.
- Their expiration date has not been exceeded.
- Their sterility indicators have changed to a predetermined color, confirming sterilization.

Preparations must be made before starting a procedure that requires sterile technique. The radiographer may be responsible for assembling the needed equipment. Most procedures today use disposable equipment that is wrapped in paper or plastic. Directions on the packages are usually clear and precise. The time taken to read them well in advance increases self-confidence when assisting the physician. Nondisposable equipment that has been sterilized is double-wrapped in cloth or heavy paper and sealed with indicator tape. All packs are wrapped in a standardized manner and are always opened using the following method (Fig. 21-24):

- Place the pack on a clean surface within reach of the physician.
- Just before the procedure begins, break the seal and open the pack.

BOX 21-5

Standard Principles of Surgical Asepsis

- Any sterile object or field touched by an unsterile object or person becomes contaminated.
- Never reach across a sterile field. Organisms may fall from your arm into the field. Also, reaching increases the chance of brushing the area with your uniform.
- If you suspect an item is contaminated, discard it. This includes items that are damp (moisture permits the transfer of bacteria from the outside to the inside of a wrapped set) and items that have had the seal broken or on which the indicator tape has not assumed the correct color.
- Do not pass between the physician and the sterile field.
- Never leave a sterile area unattended. If the field is accidentally contaminated, for example, by a fly or a patient reaching for her glasses, no one would know.
- A 1-inch border at the perimeter of the sterile field is considered to be a "buffer zone," and is treated as if it were contaminated.

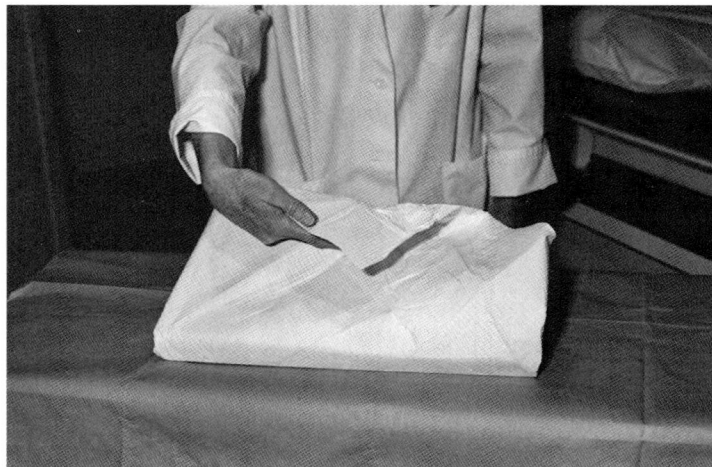

Fig. 21-24 Sterile field. **A,** After checking the sterilization indicator and expiration date, open the first corner away from you.

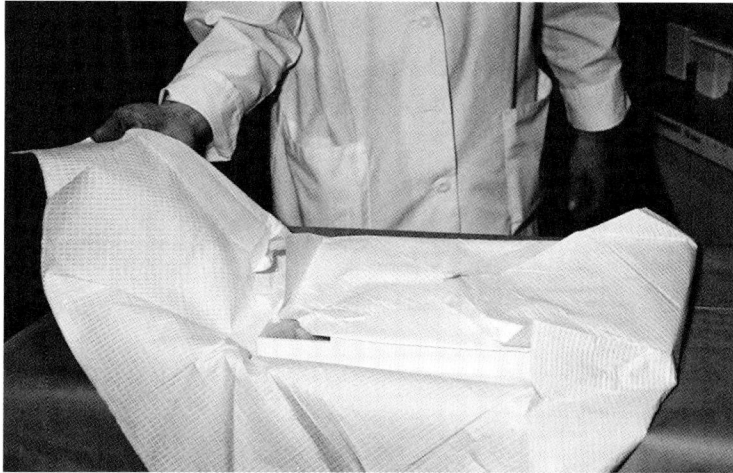

B, Open one side by grasping its corner tip.

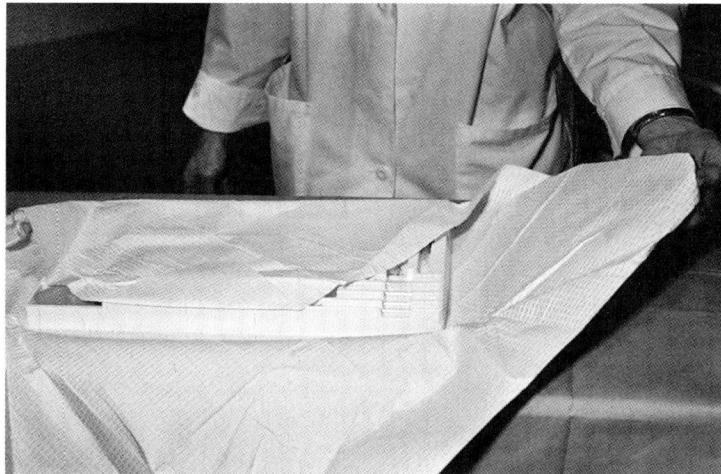

C, Open second side in same manner.

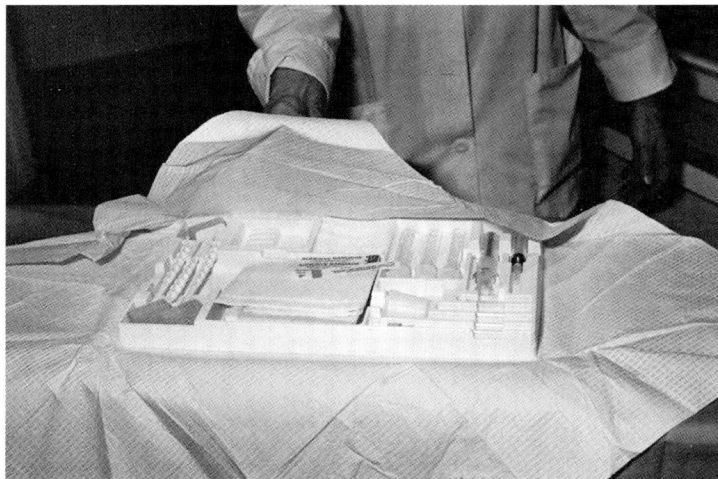

Fig. 21-24, cont'd D, Pull remaining corner toward you. If there is an inner wrap, open it in the same manner. A sterile field is now established.

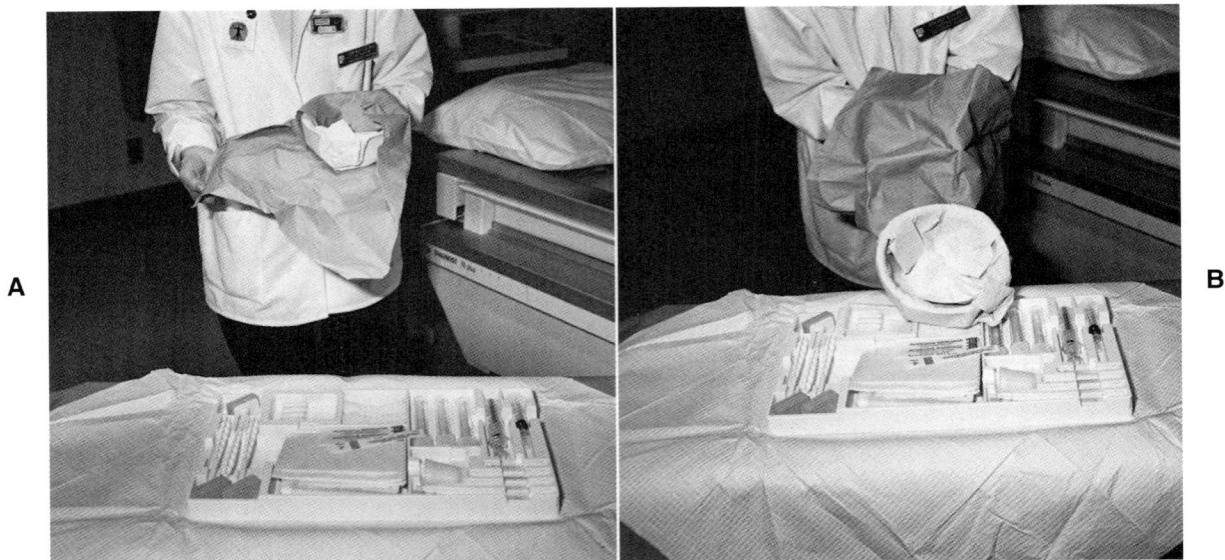

Fig. 21-25 Adding a double-wrapped item to the sterile field. **A,** Holding the item in the nondominant hand, open the outer wrap, opening the first fold away from your body. **B,** Avoid contamination of the field by holding the corners of the outer wrap while dropping the item onto the tray.

- Unfold the first corner away from you, and then unfold the two sides.
- Pull the front fold down toward you and drop it. Do not touch the inner surface.
- The inner wrap, if there is one, is opened in the same manner.
- You have now established a sterile field.

Non-disposable sterile items wrapped separately may now be added to the sterile field. Standing back from the table, grasp the object through the wrapping with one hand. With the other hand, unseal the wrappings, allowing them to fall down over your wrist. Hold the edges of the wrapper with your free hand, and drop the object onto the sterile field without releasing the wrapper (Fig. 21-25).

Disposable sponges, gloves, and other small items are supplied in "peel-down" paper wraps and may be added to the sterile field. Following the instructions, separate the paper layers, invert the package, and allow the object to fall onto the sterile field without contaminating the object or the sterile field (Fig. 21-26).

It may be necessary to add a liquid medium such as Betadine (a skin disinfectant) to a sterile tray. After double-checking the label, position the label toward your hand, open the spout and squirt the first few drops into the wastebasket or sink. Then pour the required amount into the sterile receptacle on the tray, show the physician the label, and close the spout. By discarding the first small amount poured, you "wash" the container's lip and avoid the possibility of contaminating the tray (Fig. 21-27).

When a radiographer must manipulate items in a sterile field without wearing sterile gloves, a sterile transfer forceps is used. Unwrap the forceps, grasping the handles firmly without touching the remainder of the instrument. Keep the forceps above your waist and in your sight at all times. After use, place the tips in a sterile field with the handles protruding so you can use them again. Do not reach across the sterile field.

If a procedure must be postponed, do not open the tray. If it is already open, cover it immediately with a sterile drape or discard it, since airborne contamination is just as serious as a break in sterile technique.

When the sterile procedure is completed, don protective gloves and thoroughly clean all reusable items to be sterilized. Items must be free of all residue so that the sterilizing agent can penetrate to all surfaces. Thorough cleaning is very important and is most easily accomplished when done promptly. Discard disposable items; place needles in the sharps container and the remainder in a biohazard bag. Remove your gloves and wash your hands.

Anyone whose work involves sterile fields must have a "sterile conscience." This refers to an awareness of sterile technique and the responsibility for telling the person in charge whenever you contaminate a field or observe its contamination by someone else. You may feel reluctant to speak out about apparent breaks in technique because of the inconvenience of reestablishing a sterile field. Physicians and co-workers may not seem to appreciate your challenge at the moment, but your professionalism and

Fig. 21-26 Adding a disposable item to a sterile field from a "peel-down" wrap. **A,** Separate the wrap according to package instructions. **B,** Invert the package, allowing the item to drop onto the field.

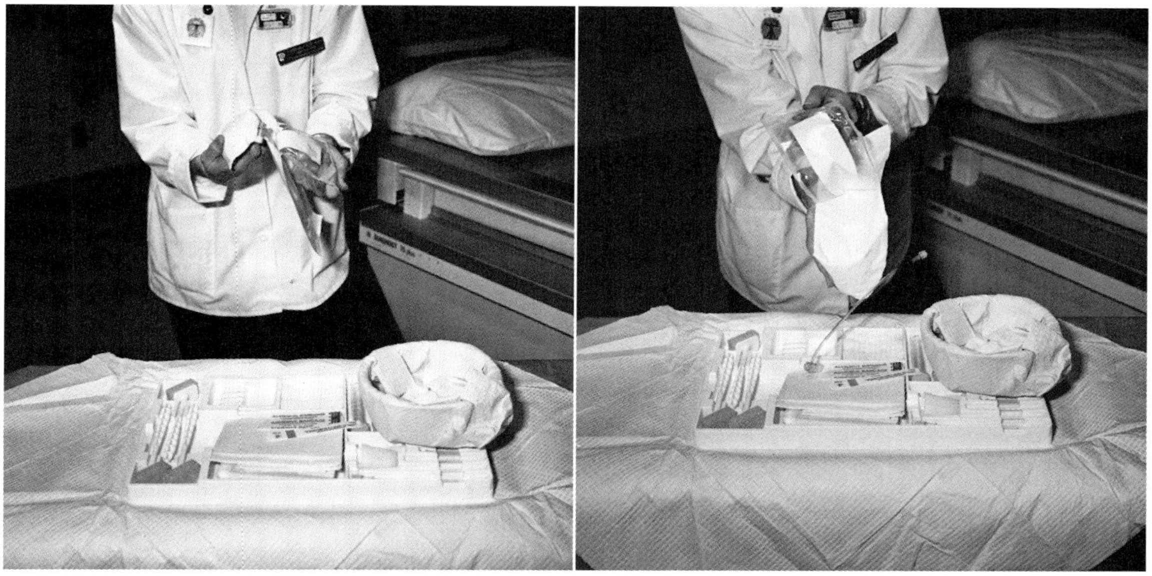

Fig. 21-27 Adding liquid to a sterile field. **A,** After checking the label, flush the tip of the container by squirting a small amount into a waste container. **B,** Pour the required amount into a receptacle on the tray, taking care not to contaminate the field.

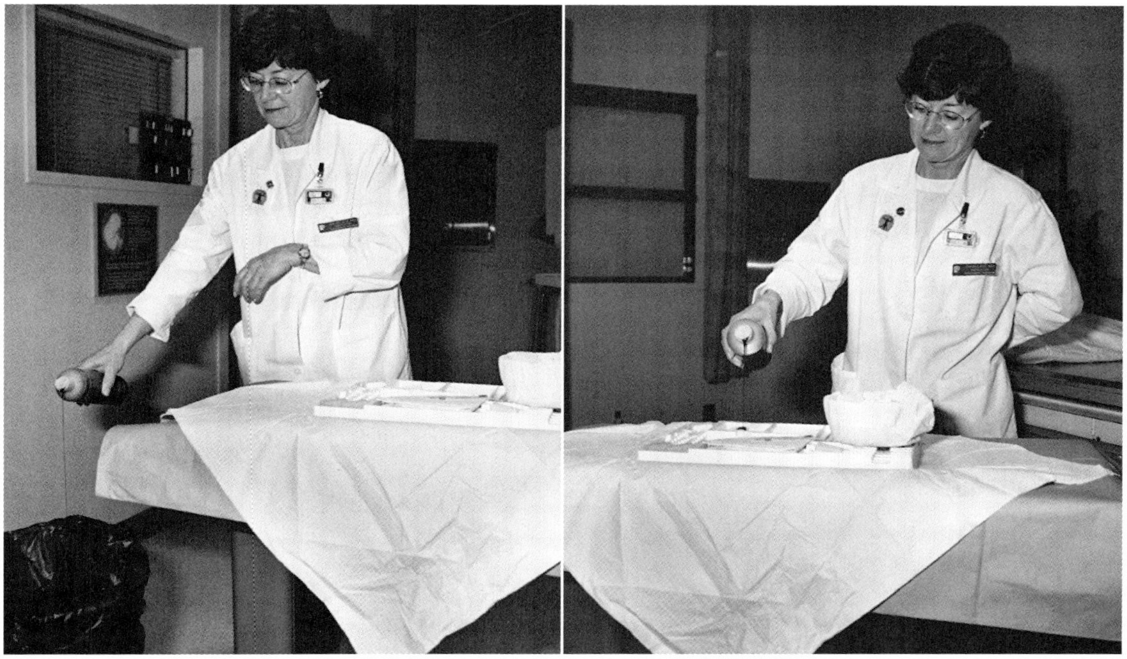

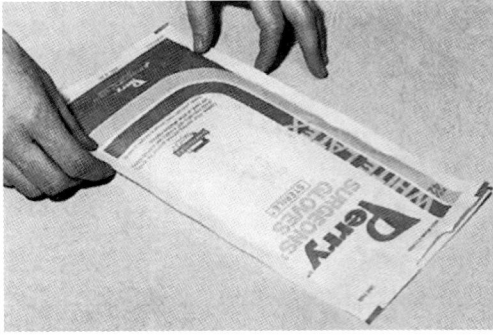

A, *Wash your hands.* Open the gloves and check for correct sizes.

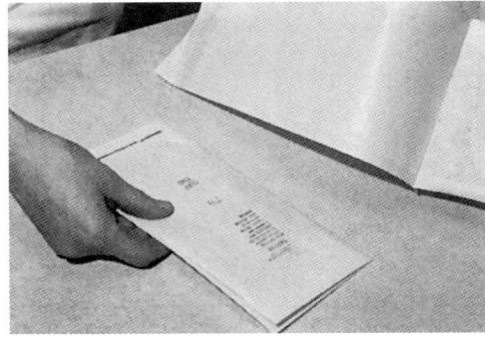

B, Open the outer wrap to expose the folded inner wrap.

C, Open the inner wrap, touching only the outer surface. Expose the gloves with the open ends facing you.

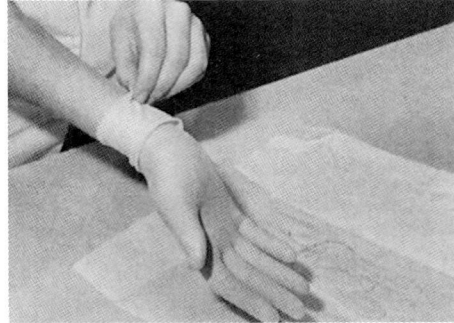

D, Put on the first glove, touching only the inner surface of the folded cuff.

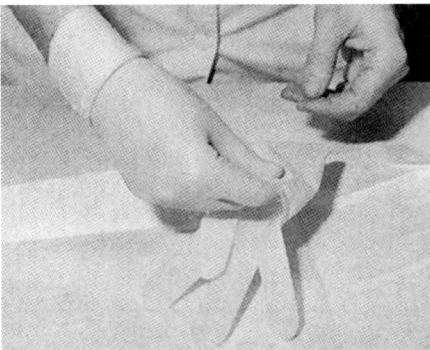

E, Using the gloved hand, grasp the second glove *under* the cuff.

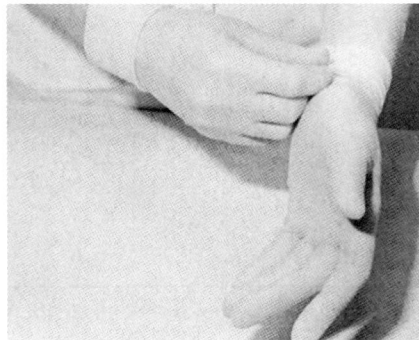

F, Put on the second glove and unfold the cuff.

G, Insert the fingers under the cuff on the first glove and unfold the cuff.

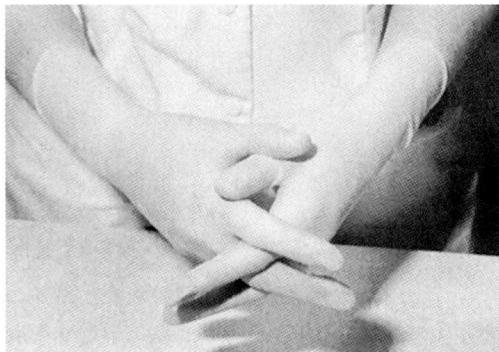

H, Gloving is complete. Keep your hands in front of your body at a safe distance from your uniform to avoid contamination.

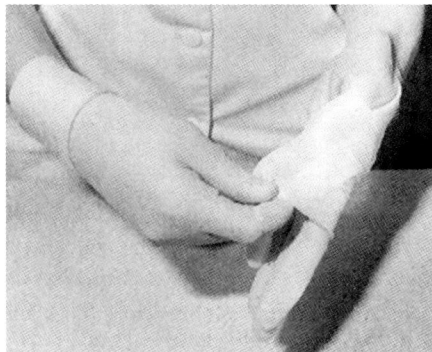

I, Remove the gloves by inverting them as you pull them off.

Fig. 21-28 Donning sterile gloves.

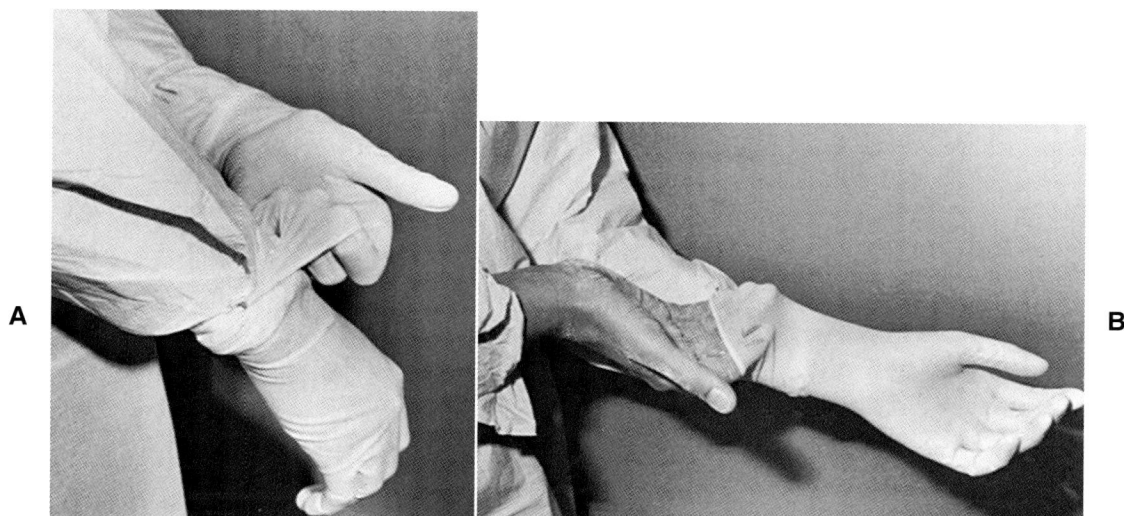

Fig. 21-29 Removing contaminated gloves. **A,** Grasp the first glove from the outside and pull it off. **B,** Insert your clean fingers inside the cuff of the second glove and remove it. Wash your hands.

concern for the patient's welfare will be reflected in the confidence they place in your aseptic technique.

GLOVING

Radiographers are seldom required to don sterile gloves. This skill may be important, however, if you are needed to assist with certain sterile procedures or to apply a sterile dressing. The technique is illustrated in Fig. 21-28.

The technique for removing contaminated gloves is shown in Fig. 21-29. This method avoids contamination of your arms or sleeves and also of your ungloved hand.

REMOVING AND APPLYING DRESSINGS

In many health care facilities today radiographers and medical assistants are called upon to perform skills that were once performed solely by nurses. For example, you may be directed by a physician to remove a patient's dressing. It may also be your duty to apply a fresh dressing when the examination has been completed.

When a dressing is to be removed, wash your hands, don protective gloves, and inform the patient of what you are about to do. Use care in removing the dressing to prevent cross-contaminating the wound and yourself. Remove the dressing gently to avoid hurting the patient. Place the soiled dressing in a plastic bag and seal it before adding it to the biohazard container. Remove your gloves and wash your hands.

The application of a new dressing requires sterile technique. Begin by preparing your supplies: sterile gloves, sterile drape, sterile gauze, and tape. You may

also need some normal saline solution to clean the area around the wound. When everything you will need has been assembled, proceed as follows:

- Tell the patient what you plan to do.
- Wash your hands.
- Tear several strips of tape to a convenient length.
- Open the sterile drape pack, placing the drape near the patient.
- Partially open the drape by pulling from the corners. This creates a small sterile field for your other sterile items.
- Open the dressing package and add the sterile dressing to the sterile field.
- If you will need to cleanse around the wound, drop sterile gauze sponges into your field for this purpose.
- To moisten the gauze sponges, open a small vial of sterile normal saline solution. Recheck the label and pour a small amount of the saline over the sponges. Do not allow liquid to soak through to the sterile towel. Check the label for the third time before discarding the vial.
- Don sterile gloves using the method described for sterile gloving.
- Use the moist sponges to clean gently around the wound.
- Allow the area to dry completely.
- Apply the dressing over the wound and secure it in place with tape.
- Cover the patient.
- Dispose of any waste.
- Remove your gloves and wash your hands.

SUMMARY

The principle underlying everything discussed in this chapter is safety. Radiographers must be alert to potential hazards from fire, falls, spills, and electric shock and prepared to respond appropriately when any of these hazards pose a threat. The safety of both the patient and the radiographer is assured when radiographers use correct techniques for positioning patients and for assisting them to move. The objective is to protect patients when they are unable to protect themselves, and to do so without personal injury.

Despite the "miracle drugs" developed over the past 60 years, infectious diseases are still a significant public health problem and some are growing alarm-ingly worse. There are still no medications to treat most viral infections. Other organisms are mutating rapidly, acquiring immunity against medications that were once highly effective. Asymptomatic carriers of HIV and hepatitis B virus pose a significant threat to members of the community, including health care workers, who may be exposed to infectious blood or body fluids without being aware that the infection is present. For all of these reasons, the safety of patients and health care workers requires conscientious infection control practices. Aseptic techniques begin with a commitment to proper practices and are implemented through the conscientious application of knowledge and skill.

▪ Review Questions ▪

1. Imagine you have just started a new job at a small facility that offers no formal orientation program. What information must you obtain immediately in order to respond appropriately in case of fire?
2. When investigating an unpleasant smell, you realize that it is coming from the x-ray control panel and there is also some smoke in the area. What should you do?
3. While preparing a dilute bleach solution to disinfect the radiographic table, you accidentally knock the open bleach bottle off the counter creating a large spill. What should you do?
4. List three situations when the patient's head should be elevated.
5. How should the technique for assisting a patient to lie down or sit up be modified for patients with back pain?
6. List three special considerations that should be applied when assisting a patient who has weakness on one side of the body.
7. List five categories of pathogenic microorganisms and explain how they differ from each other.
8. List the six routes of disease transmission and suggest methods for intervening in each.
9. How would you apply Standard Precautions when removing a dressing?

22

Assessing Patients and Managing Acute Situations

Learning Objectives

At the conclusion of this chapter, the student will be able to:

- List four personal comfort needs common to most patients and describe appropriate responses to meet these needs
- Assist patients who need to use bedpans or urinals
- Obtain and record a patient history
- Accurately observe patients' physical status and report status using appropriate terminology
- Take and record temperature (oral, rectal, and axillary), pulse rate (four common sites), respiration rate, and blood pressure; state normal values for each
- Administer oxygen or suction appropriately in an emergency
- Recognize acute life-threatening conditions such as shock, heart attack, respiratory arrest, and cardiac arrest and respond appropriately
- List the four levels of consciousness and discuss possible causes for changes in LOC
- Demonstrate correct handling of patients with extremity fractures and recently applied casts
- Assist patients who experience asthma attack, hyperventilation, nausea and vomiting, epistaxis, hypoglycemia, vertigo, seizures, and syncope

Key Terms

anaphylaxis
angina
anoxia
asthma
cerebrovascular
  accident (CVA)
cyanotic
diabetic coma
diaphoretic
diastolic
epistaxis
fibrillation
hemorrhage
hypertension

hyperventilation
hypoglycemia
hypotension
incontinence
shock
stridor
syncope
systolic
tachycardia
thready
transient ischemic
  attack (TIA)
urticaria
vertigo

469

This chapter addresses basic principles involved in meeting patient needs. Before these needs can be met, however, they must be clearly identified. Observation, evaluation, and assessment are the skills needed to determine patient needs. When these skills are consciously practiced in the clinical area, they increase your value as a radiographer. They help you become more sensitive to the safety of the environment and the condition of your patients.

The dictionary defines an *emergency* as a serious event that happens unexpectedly and demands immediate attention. Sudden deterioration in the status of any patient under your care is an acute situation requiring an appropriate response. Whether such a situation leads to a more serious problem may depend on whether you are prepared to act quickly and efficiently. Seen from this perspective, no patient problem can be considered trivial. Acute situations are bound to occur when you are dealing with patients who are ill or injured, and you must be prepared to cope in a way that will minimize the possibility of further injury or complication.

ASSESSING THE PERSONAL CONCERNS OF PATIENTS

Uncertainty about the coming procedure, fear of a possible diagnosis, or concern about the effect of illness on family members can cause varying reactions in patients. Sometimes these concerns are expressed as anger and demonstrated by inappropriate speech or rude behavior toward personnel. Other expressions of anxiety may be a need to talk constantly or, conversely, becoming quiet and withdrawn. You may observe fidgeting or other nervous mannerisms.

Anxiety can also be caused by a concern over modesty, especially when patients must undress for examinations or treatments. Reassure patients by providing ample cover, an explanation of the procedure, and a matter-of-fact attitude.

Your presence is comforting to the anxious patient. Touch patients reassuringly and tell them what to expect. Let them know when you leave the area and when you expect to return. Escort ambulatory patients to the bathroom or back to the waiting area. It can be very distressing to patients if they must wander about in an examination gown wondering where to go. Once you start a procedure, try to remain near the patient. If patients must wait in an x-ray room or dressing room, let them know that you are within hearing distance and that they may call on you for help. If a call button is available, show patients how to use it and assure them that someone will assist them promptly if they call. If no call signal is available, check with patients frequently while they wait.

Physical discomfort adds to tension as well. Remember that most patients will find it hard to remain

Fig. 22-1 Valuables are placed in a suitable container within sight of the patient.

still during a long procedure on a hard surface. This is especially difficult for a thin patient or an elderly person with kyphosis. Note whether an obese patient has difficulty breathing when lying flat on the table. Note skin temperature when you touch the patient, and inquire whether the patient is warm enough. If the patient feels chilled, provide a blanket and tuck it around the patient to provide both warmth and a sense of security. As you move briskly around the room the temperature may seem warm enough to you, but elderly or frail patients may not be active enough to keep warm. If the patient is coughing or sniffling, offer paper handkerchiefs and position a waste container within reach for the soiled tissues.

If dentures must be removed, provide a suitable disposable container and place it in a safe and visible location. Dentures slide in much more easily when wet, so add water to the container or direct the patient to a sink when dentures are replaced. Eyeglasses and hearing aids are also items essential to activities of daily living and are difficult and expensive to replace or repair. A bright-colored plastic box or basket is a useful container for these items and other small valuables (Fig. 22-1). Choose a safe location in view of the patient. Use the same place consistently and point out the location to the patient.

PHYSIOLOGIC NEEDS

Water

A dry mouth can be caused by medication but can also result from anxiety or simply from thirst. A drink of water, offered with a straw if the patient is lying down, may be very comforting. Since some tests may require that the patient have nothing by mouth, even water, it is wise to check that water is permitted.

Elimination

One concern that can greatly bother a patient is an urgent need to void. A full bladder may cause discomfort, irritability, and difficulty remaining still during the procedure. If this need is ignored in an older or debilitated patient, **incontinence** (loss of bladder control) may result, causing embarrassment for the patient and cleanup problems for the radiographer. Be especially sensitive to the need for bathroom facilities when procedures are prolonged.

Before a patient uses the bathroom, check to see if a specimen of urine or feces should be collected. If so, provide instructions and the correct container. Urine collection procedure is explained in Chapter 24.

When a patient needs to defecate or urinate and is unable to walk or be taken to the bathroom in a wheelchair, a bedpan or urinal is used. This is not a common requirement in most outpatient facilities, but assisting patients with bedpans or urinals is a basic clinical skill for all radiographers.

When a bedpan is needed, follow the procedure outlined in Box 22-1. Be sure that the patient is adequately covered for privacy. When a female patient is placed on the bedpan, the upper torso needs to be slightly elevated to prevent urine from running up her back. When a patient is restricted in mobility, two people may be needed to assist the patient onto the bedpan. If the patient is on the x-ray table, one person should stand on each side of the table to prevent the patient from falling.

When the patient is finished, you may have to assist with wiping. Wear gloves and have toilet tissue, a wet washcloth, and a dry towel conveniently placed. Assist the patient to lift the hips or roll away from you onto one side while you steady and remove the pan. Place it safely aside and, if necessary, help the patient by wiping from front to back with paper first, and then with a wet cloth, before drying. Offer the patient a disposable moist towelette or a clean wet cloth and towel to cleanse the hands.

Male patients may need to use a urinal. Usually this is simply a matter of providing the urinal and removing it again when the patient is finished. If he is unable to use it himself, don protective gloves and spread the patient's legs; lift the sheet with one hand and slide the penis into the urinal with the other. It may be advisable to hold the urinal in position until the patient is finished.

Patients may find it difficult to use a bedpan or urinal if they feel that they are under observation. If possible, you should remain out of their line of sight while staying close enough to meet safety requirements.

Empty the bedpan or urinal carefully into the toilet to avoid splashing. Remember to wash your hands after removing your gloves.

Placing a patient on the bedpan or offering the urinal is not a complex task. Once you are familiar with this procedure, the chief obstacle to overcome is embarrassment. A cheerful, matter-of-fact attitude will help you and the patient.

Sanitary Supplies

Occasionally a patient requires a sanitary napkin. Know where these are kept. If a soiled napkin is to be removed, direct the patient to a bathroom or place a paper bag within reach. Dispose of the bag with the soiled napkin in the appropriate container.

TAKING A HISTORY

It is important for both the radiographer and the physician who reads the films to know why an examination is being done. If the films are sent out to a radiologist, this information must accompany the films. If the requisition does not provide complete and accurate information about the patient's history and condition, you will need to obtain this information from the patient. The answers you receive may influence how the examination is conducted. The history also aids the radiologist in focusing the interpretation to meet the referring physician's needs. This does not need to be a detailed medical history, but rather a thoughtful consideration of the patient's

TABLE 22-1			
Guidelines for Taking a History			

Examination	Questions	Observations	Example of History*
Orthopedic, acute injury	How did the injury occur? When? Can you show me exactly where it hurts?	Swelling, deformity, discoloration, laceration, abrasion	Twisting injury, l. ankle, while skiing today; swelling & pain over lateral malleolus.
Orthopedic, not involving acute injury	Where does it hurt? How long has it been bothering you? Were you ever injured there? How was the injury treated (cast, surgery)? Has there been any recent change?	Deformity, scars, range of motion, weight bearing	Chronic pain, r. knee 2 yrs, worse since building fence Sat. Prev Rx c̄ cortisone inj. No known injury.
Neck	Did you injure your neck? How? When? Where does it hurt? Do you have any pain, numbness, or tingling of the shoulder or arm? Which side?	Range of motion	MVA 10/12/97; lower neck pain & l. shldr pain c̄ numbness & tingling, l. hand.
Spine	Did you injure your back? How? When? Do you have pain, numbness, tingling, or weakness of the hip or leg? Which side? Any bowel or bladder problems?	Gait, range of motion	Lifting injury 2 wks ago. LBP radiating to r. hip.
Head	Were you injured? When? How? Do you have pain? Where? Did you lose consciousness? For how long?	Speech: clarity, confusion; gait	Severe HA, blurred vision, dizziness, & gen'l weakness, 24 hrs. No known injury. Speech slurred.
Chest	Do you know why your doctor ordered this examination? Are you short of breath? Do you have a cough? Do you cough anything? Do you cough up blood? Have you had a fever? Do you have any heart problems?	Respirations, cough	SOB, wheezing, & r. chest pain since resp flu 4 wks ago. Moderate, non-productive cough.
Abdomen, gastrointestinal examinations	Do you know why your doctor ordered this examination? Do you have pain? Where? Do you have nausea? Diarrhea? Have you had any other tests for this problem (lab tests, ultrasound)? Do you know the results? Have you ever had abdominal surgery? When? Why?		LLQ pain, incr over past mo. ? of mass seen on US done here 10/21/97.
Urology	Do you know why your doctor ordered this examination? Do you have any pain? Where? For how long? Do you have trouble passing urine? Pain? Urgency? Frequency? Have you ever had this problem before? Do you have high blood pressure?		2 prior episodes of UTI; current malaise, fever, & mid back pain.

*To identify the common abbreviations used by radiographers (e.g., MVA, motor vehicle accident), see Appendix H.

current status and why this particular radiographic study is being done.

Patients sometimes complain that they feel like products on an assembly line. The process of taking a history presents an opportunity for you to give the patient individual attention and build rapport. In addition, your ability to gain the patient's confidence will influence the amount of relevant information you obtain. Remember to introduce yourself, call the patient by name, and deal with immediate patient concerns as soon as possible.

Begin the history by asking a general question about the nature of the problem such as, "Do you know why Dr. Chen wants you to have an x-ray of your chest?" Be realistic in the scope of your questions. Focus on expanding the information provided on the x-ray requisition. This is especially important when the request or order does not indicate the rationale for ordering the procedure. Most requisition forms have a place for this information to be provided, but in practice the information received is often absent or so limited that it may seem irrelevant without further explanation. The information you obtain is most useful when recorded on the requisition form. You may also need to enter it into the computer record.

History requirements vary with the nature of the examination. Table 22-1 provides history questions and observations pertinent to many patient complaints. You can use it to become familiar with the type of information that will be most useful in specific situations. The history examples in this table use common abbreviations that are also used in charting and other medical recording.

Patients may have been asked to complete a history questionnaire on their initial visit, but this information may be out of date. In facilities that perform procedures using contrast media, a special history questionnaire that includes allergy information may need to be filled out and signed prior to each procedure that uses a contrast medium.

Examinations for patients with chronic conditions or those receiving post-treatment follow-up may require a comparison to prior imaging studies. If these are not part of the current file, your history should contain information on previous relevant examinations, including when and where they were done.

Some medical assistants take preliminary histories before the physician sees the patient. Although the physician is responsible for taking the official medical history, a preliminary history can save time, allowing the physician to focus quickly on details of the patient's problem. If taking patients' preliminary histories is one of your accepted responsibilities, the standard format that follows can serve as a guide. Using this outline will allow you to elicit the greatest amount of information in the least amount of time and will help you avoid missing relevant facts.

Onset: How did it start? What happened? When did it first trouble you? Was it sudden or a complaint that gradually got worse?

Duration: Have you ever had it before? Has it been continuous? Does it bother you all the time? How long has this attack been bothering you?

Specific location: Where does it hurt (or where is the problem)? Can you put your finger on where it hurts the most? Does it hurt anywhere else?

Quality of pain: What does it feel like? Sharp, stabbing pain? Dull ache? Throbbing pain? How severe is it? Mild, moderate, severe? (Some like to use a pain scale of 1 to 5 or 1 to 10, with 1 being no pain at all and the highest number representing the worst pain the patient can imagine.) Does it wake you up at night?

What aggravates: When is it worse? What seems to aggravate it? Is it worse after meals (at night, when you walk)?

What alleviates: What has helped in the past? Does that still help? What seems to help now? Does the time of day (amount of rest, change in position) make a difference?

Tact and caution are required when obtaining a history. Anxious patients may read too much into your questions. Information regarding such serious matters as cancer, surgery, or heart attacks is best elicited in a general way rather than through blunt questions. "Do you know why your doctor ordered this examination?" is less threatening than "Is your doctor checking for cancer?" Victims of accidents where legal liability is in question may be reluctant to provide information that could increase their liability or jeopardize a legal settlement. Minors may be hesitant to reveal personal information in the presence of their parents. Where information is difficult to obtain, it is usually wise for the physician to take the complete history.

At this point the process of taking a history may seem complex and confusing. This is a skill that improves with practice. Role-playing with other students, including a critical observer, will improve your ability to take a history with sensitivity and confidence. As clinical practice provides additional knowledge and experience, you will find that your observation and history-taking skills become increasingly accurate and pertinent.

ASSESSING CURRENT PHYSICAL STATUS

Establishing a Baseline

The radiographer may be the first and primary observer of a significant change in the patient's current condition. In order to accurately assess change, you must first establish a baseline for your observations.

Before you start the procedure it is important to review the requisition. Unfortunately, the requisition may not have enough specific information, and this is a place where your skill in history taking will prove valuable. If you have access to a chart, read the diagnosis and the most recent progress notes. An order for the radiographic procedure should be there. If a recent notation reads, "Unable to stand to void," you can anticipate a need for help when transferring this patient to the x-ray table.

Some notations have special significance to the radiographer. Allergies are usually noted in red on the outside of the chart as well as in the history. A patient with a previous history of allergies, sometimes referred to as an "allergenic individual," is more likely to have an adverse reaction to medications, especially when administered by injection.

Physical Evaluation

In the context of this chapter, evaluation is an ongoing process of observation, assessment, and measurement to note and evaluate changes in patient condition. How do you know when the condition of a patient is changing for the worse? What do you look for?

The most important process is sometimes called "eyeballing the patient." This skill of acute observation compares the actions and appearance of *this* patient with those of similar patients you have seen. You also use this skill to compare the appearance of this patient *now* with the way he or she appeared earlier. Although this may seem intuitive, you are actually responding to subtle changes in the overall appearance of the patient.

One of the easiest signs to recognize is a change in skin color. Individual complexions vary, but when pale skin becomes **cyanotic** or olive skin becomes pale and waxen, the change is usually quite apparent. The term *cyanotic* denotes a bluish coloration in the skin and indicates a lack of sufficient oxygen in the tissues. This is most easily seen on the mucous membranes such as the lips or the lining of the mouth. Nail beds may also show a bluish tinge. For some patients with heart or lung conditions, this may be a chronic or usual state, but the patient who *becomes* cyanotic needs oxygen and immediate medical attention. Any patient who looks pale and anxious and does not feel well is subject to fainting and needs to sit or lie down immediately. Do not leave the patient! A patient who loses consciousness and falls to the floor may suffer injuries far more serious than the cause of the fainting.

We have discussed the importance of touch as a form of communication and reassurance, but contact with your hands also allows you to make physical observations. The acutely ill patient in pain may be pale, cool, and **diaphoretic** (perspiring), in what is frequently called a "cold sweat." Hot, dry skin may indicate a fever, whereas warm, moist skin may only be a response to the weather or the room temperature. Cool, moist skin may indicate acute anxiety. Wet palms and shaking hands are typical of the apprehensive individual who will need an unusual amount of reassurance. These patients may find it difficult to concentrate. They often need more frequent instructions during the procedure and should receive written directions for any required follow-up care.

If you note any of these signs, it is important to determine whether this is a new symptom. Has the patient just received any new medication? If so, notify your supervisor or the physician immediately. You may be observing the first signs of an impending allergic reaction.

Vital Signs

The next four procedures used for assessment are usually referred to as *vital signs.* They involve the measurement of temperature, pulse rate, respiratory rate, and blood pressure. The ability to take vital signs is a valuable clinical skill. Even if vital signs are not a part of your usual job description, you may need to assess them in an emergency. If you do not take vital signs routinely, keep your skills sharp by reviewing your technique frequently. When time allows, check your co-workers. We should all be aware of our own baseline vital signs, so your practice will benefit you, the person on whom you practice, and the patient who may need your skill in an emergency. Know the location of a blood pressure cuff and gauge (sphygmomanometer), a stethoscope, and other equipment that might be needed in an emergency. Even before you are proficient, you may be asked to obtain it for a nurse or physician.

Temperature   The first of the vital signs is temperature. Taking a temperature is a basic clinical skill that can also be used in your own home. An accurate temperature reading measures the body's basic metabolic state, the rate at which it uses energy.

Body temperatures vary during the day, being lowest in the morning and highest in the evening. Normal oral temperatures vary from 96.8° to 99.6° F (36.6° to 37.6° C). Rectal temperatures range from 0.5° to 1.0° higher than oral temperatures; axillary temperatures range from 0.5° to 1.0° lower. In addition, normal temperatures vary slightly from person to person. A tense, "high-strung," quick-moving individual is likely to have a higher basal temperature than the placid, slow moving person, all else being equal. What is your average temperature range?

Fever, also termed *pyrexia* or *hyperthermia,* is a sign of an increase in body metabolism, usually in response to an infectious process. For adults, fever commonly refers to any temperature of 100.4° F or higher when taken orally, or 101.4° F taken rectally.

Temperatures may be obtained by the oral, rectal, axillary, or tympanic routes. Alert, cooperative patients usually prefer the familiar oral route. It has long been held that the oral method is less accurate than the rectal method, but research does not confirm this belief. The oral route provides an accurate measure of changes in body core temperature when taken correctly with the bulb or probe of the thermometer well under the base of the tongue.

The oral method is *not* appropriate when the patient has recently had a hot or cold beverage, is receiving oxygen, or breathes through the mouth. In these situations the rectal or axillary method may be used. The rectal temperature is accurate and faster; axillary temperature is slower and somewhat less accurate. The axillary method is sometimes preferred, however, because it is less invasive. Rectal temperatures may be contraindicated in certain patients with cardiac conditions to avoid stimulating the vagus nerve.

Fig. 22-2 Digital thermometer used for oral, rectal, and axillary measurements.

Fig. 22-4 Tympanic thermometer is inserted into the auditory canal.

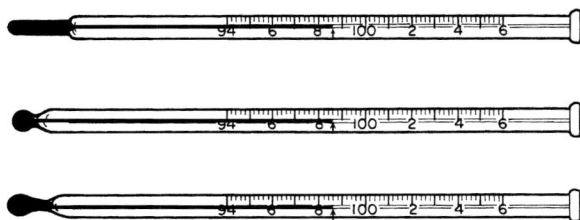

Fig. 22-3 Glass thermometers. Comparison of oral, stubby, and rectal thermometers (top to bottom).

Many health care facilities now use digital electronic thermometers with interchangeable oral and rectal probes (Fig. 22-2). These instruments can be read in 1 minute or less. They emit a short beep when the highest temperature is recorded. They are read from a digital display on the thermometer itself or on a handheld display unit. Disposable sleeves are used to cover a probe, which avoids the need to disinfect the thermometer after each use.

You are probably familiar with the use of glass thermometers. There are three common configurations: the oral type has a long, slender bulb; the rectal type has a short, round one; and the "stubby" type is similar to the rectal type, but with an even smaller bulb (Fig. 22-3). The rectal and stubby thermometer types are suitable for oral, rectal, and axillary use. The oral type may be preferred for both oral and axillary use because the longer bulb exposes more mercury to body heat, shortening the time necessary to obtain

a reading. To read, hold the glass thermometer horizontally and slowly rotate it until the mercury indicator is clearly visible. Glass thermometers may be stored dry or in antiseptic solution in containers labeled "oral" and "rectal." If your facility uses glass thermometers there will be a specific protocol for disinfecting them after use. Be familiar with the required procedure.

Tympanic thermometers (Fig. 22-4) are also used, primarily with pediatric patients. They measure temperature at the tympanic membrane (eardrum) in the ear. Disposable thermometers consist of a strip of temperature-sensitive paper with adhesive backing that may be attached to the forehead. These also are primarily used for children.

Rectal, axillary, or tympanic routes are best for children under the age of 6 years and for anyone who is confused or unable to follow directions. A general procedure to obtain accurate temperature readings for each common route is found in Box 22-2. Remember to tell your patient what you are about to do. When taking an oral temperature, remind the patient not to bite down and to keep the lips closed. *Never leave a patient alone with a rectal or axillary thermometer in place.*

Pulse   A pulse is the advancing pressure wave in an artery caused by the expulsion of blood when the left ventricle of the heart contracts. Since this wave occurs with each contraction, it is an easy and effective way to measure the rate at which the heart is beating.

A rapid pulse, called **tachycardia,** occurs when the heart beats more than 100 times per minute. This

BOX 22-2

Taking a Patient's Temperature

Digital Thermometer

Oral route
- Wash your hands and don disposable gloves.
- Cover the probe with a clean plastic sleeve.
- Actuate the thermometer and insert the probe under patient's tongue.
- Remove probe when the audible tone or flashing number indicates that the maximum temperature has been reached (about 1 minute).
- Note the temperature reading.
- Remove and discard plastic sleeve.
- Remove and discard gloves.
- Wash your hands.
- Make sure thermometer is off and return it to storage.
- Record the temperature.

Glass Thermometer

Oral route
- Wash your hands.
- Rinse the thermometer if it is stored in antiseptic liquid.
- Hold the thermometer firmly by the top and briskly shake the mercury down to 96° F.
- Place the thermometer bulb under the patient's tongue and leave in place at least 3 minutes.
- Instruct the patient to keep lips closed.
- Remove and read thermometer.

- Cleanse the thermometer and return it to be disinfected.
- Wash your hands and record results.

Axillary route
- Wash your hands.
- Obtain an oral thermometer.
- Shake the mercury down to 96° F.
- Place the thermometer bulb into the axilla so that skin folds are in direct contact with thermometer bulb.
- Instruct the patient to hold upper arm firmly to chest wall for 5 to 7 minutes.
- Remove and read thermometer.
- Return the thermometer to be disinfected.
- Wash your hands and record results.

Rectal route
- Wash your hands and don gloves.
- Obtain a rectal thermometer.
- Instruct the patient to lie in a lateral recumbent position. Cover the patient and expose the anus by raising the top fold of the buttock.
- Slowly insert the bulb until the anal sphincter is passed.
- Hold thermometer in place for 3 minutes. *Do not leave the patient.*
- Remove the thermometer slowly. Wipe it with a tissue and read.
- Cleanse the thermometer and return it to be disinfected.
- Remove gloves, wash your hands, and record results.

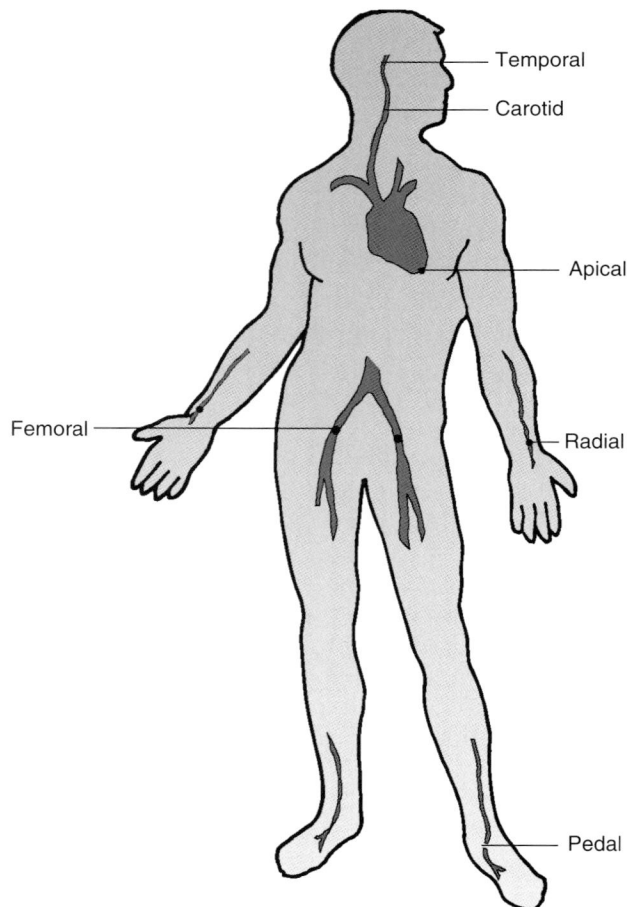

Fig. 22-5 Common pulse points.

may be temporary, as after exertion or when nervous or excited, or may be constant, caused by a damaged heart or an endocrine disorder. A rapid pulse rate may also result from interference with oxygen supply or from significant blood loss. In these cases the heart must beat faster to circulate the remaining blood and/or supply oxygen to the cells of the body.

Average normal pulse rates in adults vary between 60 and 100 beats per minute. The tense, nervous individual is more likely to be in the upper range, whereas athletes tend to have a slower rate.

In addition to rate, the pulse volume or quality may vary. A weak, rapid pulse is described as **thready.** A thready pulse may indicate that the heart is not pumping enough blood.

Common pulse points are shown in Fig. 22-5. The most common site for palpation of the pulse is the radial artery on the lateral aspect of the wrist at the base of the first metacarpal. Since your own thumb has a pulse, you cannot take an accurate pulse using your thumb. Place your fingers over the artery with your thumb on the back of the wrist. Compress gently but firmly. By compressing the artery against the radius, the pulse is easy to feel, especially if the patient's wrist is held palm down (Fig. 22-6).

If the pulse is weak or difficult to count, you can use the carotid artery. Place your fingers just below the angle of the mandible (Fig. 22-7). This site is easily accessible and is particularly important if a patient loses consciousness. If the pulse is not palpable at this site, the heart is not beating effectively and emergency measures are necessary. These measures are discussed later in this chapter.

The *dorsalis pedis* or pedal pulse is taken over the instep of the foot (Fig. 22-8). This pulse may be significant when there is a question of compromise in the peripheral circulation. For example, radiographers

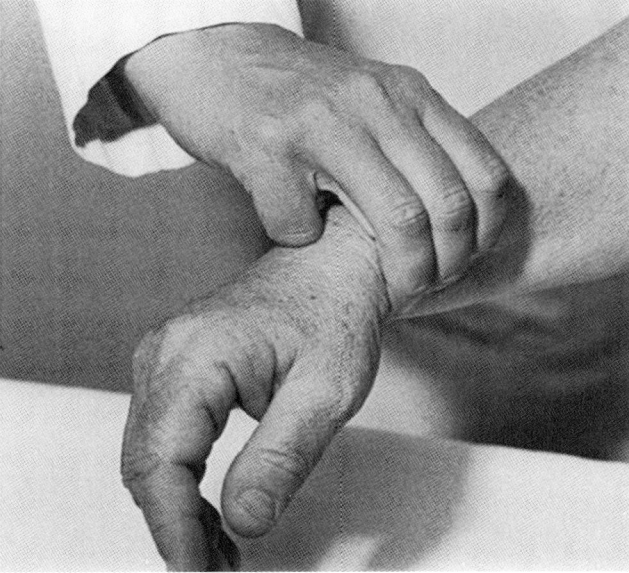

Fig. 22-6 Taking a radial pulse with the palm down.

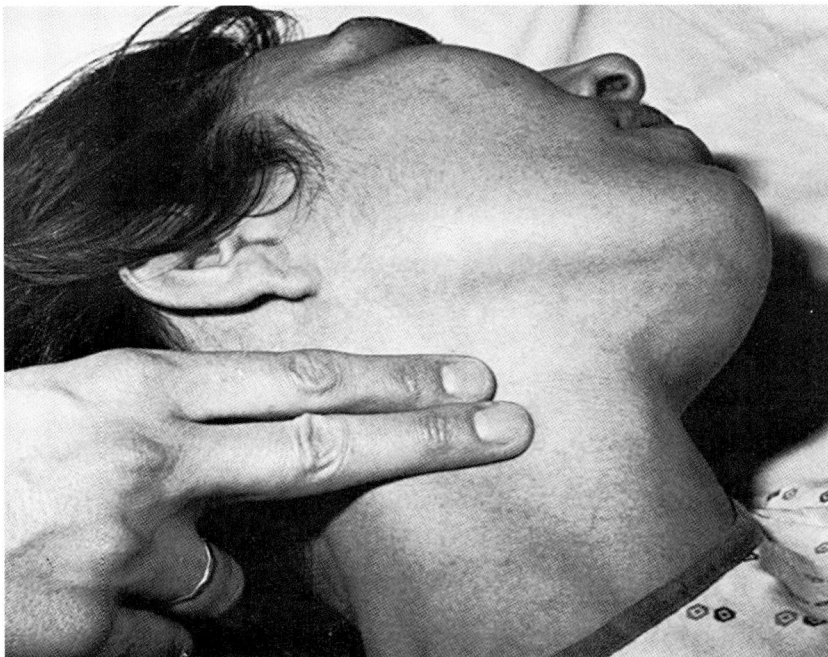

Fig. 22-7 Palpate for carotid pulse just below the angle of the mandible.

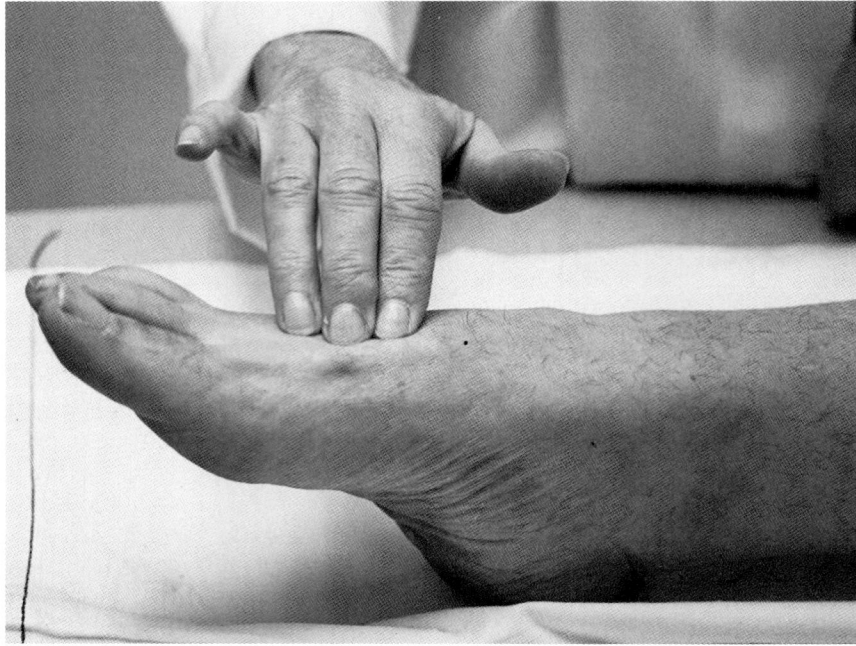

Fig. 22-8 Taking a pedal pulse.

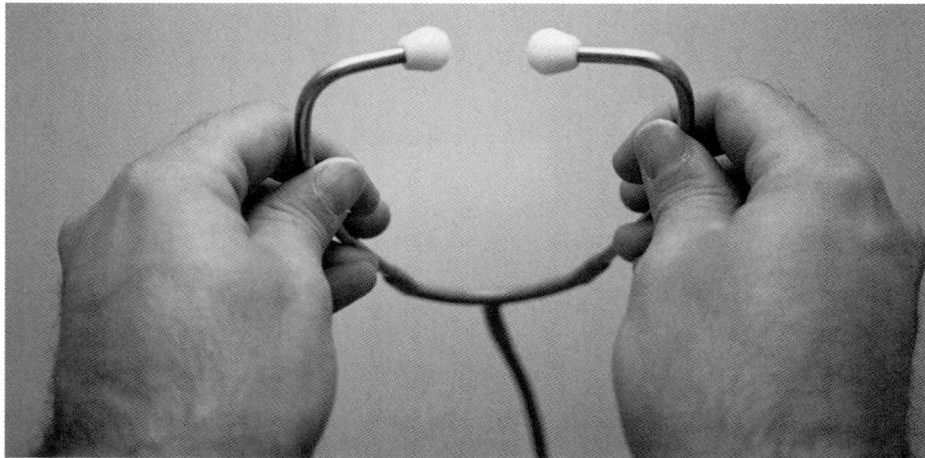

Fig. 22-9 Stethoscope ear tips should point up slightly. Correctly placed ear tips improve accuracy.

may be requested to check the pedal pulse following the application of a cast to the lower extremity. Since inability to feel this pulse may be an important diagnostic sign, you must practice until you are certain that you can detect the pulse when it is present.

If the pulse is slow or irregular, you may want to use a stethoscope to take an apical pulse. Look at the stethoscope carefully and become familiar with its use. It may have a bell as well as a diaphragm; less expen-

sive models may have a diaphragm only. On the bi-modal stethoscope, you can switch from bell to diaphragm. For most purposes, however, the diaphragm is preferred.

Hold the earpieces of the stethoscope horizontally in front of you. The ear tips should point up slightly (Fig. 22-9). Insert the tips in your ears and then tap the diaphragm gently with your finger to be sure you can hear. Now press the diaphragm firmly over

BOX 22-3

Measuring Blood Pressure

- Wash your hands and explain the procedure to the patient.
- The patient may be sitting or lying down, but the cuff should be at the level of the heart. Either arm may be used.
- Wrap the cuff snugly with the bottom edge above the antecubital space. Most cuffs are self-securing.
- Place the gauge where you can easily read the dial.
- Palpate the brachial artery pulse in the antecubital space.
- Place the stethoscope's ear tips in your ears and press its diaphragm over the brachial artery.
- Close the valve on the bulb pump and inflate the cuff rapidly to approximately 180 mm Hg.
- Open the valve on the pump and *slowly* release the pressure.

- Listen for the beat of the pulse while watching the gauge. Note the figure at which the pulse is first heard. Note this as the systolic reading.
- As the pressure is released, the sound increases in intensity and then, suddenly, becomes much softer. Note this point as the diastolic reading.
- Release the remaining pressure.
- If the situation permits, ask the patient to raise the arm and clench and release the fist. Lower the arm and repeat the procedure to check the results.
- Remove the cuff and record the results as systolic over diastolic (e.g., 140/86).
- Clean the ear tips and diaphragm of the stethoscope with alcohol and return the equipment to its storage place.

---

the apex, or tip, of the heart. This is normally found in the fifth anterior intercostal space at the left mid-clavicular line. Count the pulse for a full minute and record the rate and any irregularities.

When the radial pulse rate is taken routinely, it is common to count for 15 seconds and then multiply by 4. Whenever there is an irregular rate or rhythm, count for a full 60 seconds. If the apical rate is faster than the radial rate, the heart is not beating efficiently. This should be recorded on the chart or requisition. If the patient also shows signs of distress, report to the physician immediately.

Respirations When a patient shows evidence of respiratory distress, a measurement of respiratory rate will help in making an assessment. To count respirations, simply note the number of inhalations per minute. This is done while continuing to hold the wrist after the pulse has been counted, since some patients may force a change in the respiratory rate if aware that a count is being made. If you are having difficulty counting breaths, place one hand lightly over the patient's diaphragm. Compare your findings with the normal adult range of 12 to 20 breaths per minute. If a patient complains of difficulty in breathing (dyspnea) or has extremely rapid breathing (tachypnea), you should inform the physician and prepare oxygen equipment for immediate use if ordered. Oxygen administration is discussed later in this chapter.

Blood Pressure A blood pressure (BP) reading is usually expressed in two figures, such as 120/78. The top figure is the **systolic** pressure and is a measure of the pumping action of the heart muscle itself. The

bottom figure is the **diastolic*** pressure and indicates the ability of the arterial system to accept the pulse of blood forced into the system when the left ventricle contracts. If you are angry, afraid, or exercising, the top figure greatly increases. The diastolic figure may also rise, but to a lesser degree. What is a normal BP? As a rule of thumb, an average diastolic pressure greater than 90 mm Hg indicates some degree of **hypertension** (abnormally high BP). Less than 50 mm Hg indicates **hypotension** (abnormally low BP) and may indicate shock. A systolic pressure of 95 to 140 mm Hg is within normal limits, as is a diastolic pressure of 60 to 90 mm Hg. The acceptable range of BP varies depending on age, weight, and physical status.

In order to establish a baseline or average BP, at least two readings are customary. These may be taken over a period of time, and on some occasions are taken sitting, standing, and lying down. In an emergency, however, a single reading can be reported to the physician and should be compared with prior recorded results, if available. Take an additional reading as soon as possible.

Wall-mounted or rolling floor model sphygmomanometers that measure BP in millimeters of mercury (mm Hg) are the most accurate. Small, portable, aneroid manometers using air pressure are more often found in the radiology department and are sufficient for most purposes. While not measuring pressure in mm Hg directly, these manometers provide readings in the same units. The procedure is detailed in Box 22-3 and is illustrated in Fig. 22-10.

---

*An easy way to remember is to think, "D is for down."

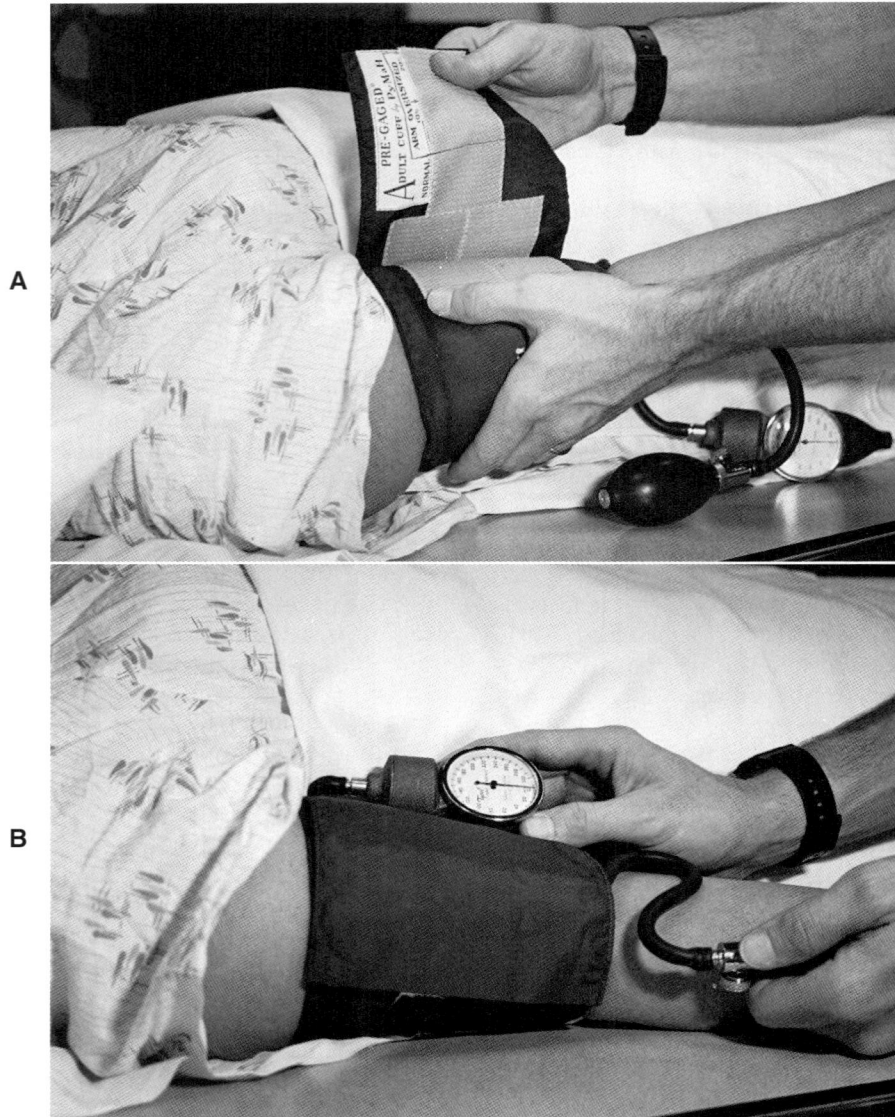

Fig. 22-10 Blood pressure procedure. **A,** Wrap the cuff snugly above the antecubital space. **B,** Place the gauge for easy visibility.

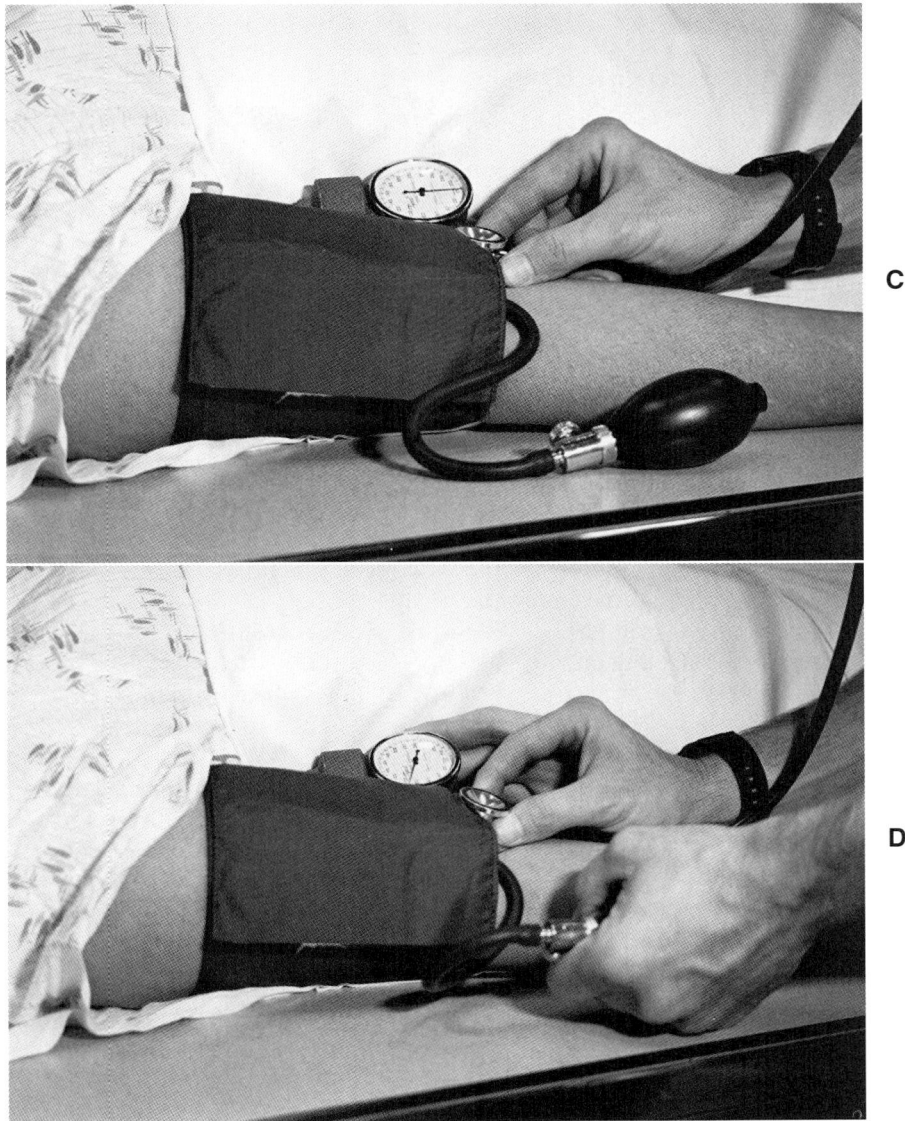

Fig. 22-10, cont'd C, Place the stethoscope over the brachial artery and inflate the cuff. **D,** Release pressure while noting pressures at start and end of audible pulse.

ACUTE SITUATIONS

You will see patients in widely varying states of health. Using assessment skills learned earlier in this chapter, you will be able to evaluate patients and note changes in their symptoms and conditions. Individuals suffering from prolonged illness, the onset of acute illness, or recent trauma may suffer a sudden change in status that could be life threatening. In obvious emergency situations, patients are usually taken directly to a hospital. Sometimes, however, the seriousness of a patient's condition may not be immediately recognized or the patient may be reluctant to go to the emergency room. These patients may seek care in a physician's office or an urgent care center.

Patients who have been involved in accidents are subject to sudden changes in condition and may go into physical or psychologic shock. Once the acute phase of an accident is over, many patients who were full of fortitude experience a delayed emotional reaction. This may take the form of uncontrollable crying or a compulsive urge to tell everyone about the accident. They may also have a physical reaction, such as fainting, trembling, or violent nausea. Your most positive action is to be available, offer nonverbal support, and watch carefully for any signs of deteriorating physical condition. Your ability to speak calmly and work competently under pressure is reassuring.

Patients experiencing the early onset of a heart attack sometimes seek immediate care in physicians' offices or clinics. If at any time a patient suddenly develops an irregular pulse; complains of feeling faint, weak or nauseated; or has a sudden onset of pain in the chest, shoulder or jaw, *a physician must be notified immediately* regarding the possible onset of a heart attack.

EMERGENCY SUPPLIES AND EQUIPMENT

All health care facilities are prepared for emergencies to some degree. The extent of preparation will depend on the facility and the likelihood of dealing with a life-threatening condition. For example, a cardiologist's office should be equipped to handle a heart attack or a cardiac arrest, but a chiropractic office might be equipped with only a first aid kit. In the rare event of a life-threatening emergency in a chiropractic office, the standard procedure would be to call 911.

Preparedness for an emergency usually includes certain essential items such as airways, emergency medications and the equipment for administering them, a blood pressure cuff, and a stethoscope. There may also be a board to slip under the patient when giving external cardiac massage, artificial ventilation equipment, and a defibrillator. Emergency items may be kept in a box or drawer or on an emergency cart. It should be possible to move the entire set quickly to any location within the facility where it might be needed. Emergency equipment and supplies should be inspected regularly to ensure that they are available for instant use. It is helpful to have an up-to-date list of contents with the kit so that precious time is not wasted searching for items that are not present.

It is your duty to know the extent of emergency equipment and supplies available and to know where they are kept. *Never borrow equipment or supplies from the emergency set for routine use.* This practice results in the absence of lifesaving items when they are most needed. When you use these items in an emergency, be sure that supplies are replenished and the kit is ready for use before returning it to storage.

Oxygen and Suction

As a radiographer you may encounter patients who need supplemental oxygen. The administration of oxygen is noninvasive. If a physician has not yet evaluated the patient, it is appropriate to provide a low flow rate of oxygen to any patient who experiences acute anxiety accompanied by a rapid heart rate and shortness of breath. Oxygen is prescribed for patients with a wide range of illnesses.

Since the need for oxygen and/or suction can be sudden and dramatic, you should be familiar with the mechanics of oxygen and suction systems available in your facility.

Oxygen Administration   Oxygen (O_2) can be administered by various means. A simple face mask (Fig. 22-11) is often used to provide oxygen temporarily. It is shaped to conform to the face and held in place by an elastic strap. The mask is used primarily for short-term therapy. It fits loosely and delivers O_2 concentrations that can vary from 30% to 50%, depending on the fit and the O_2 flow rate. Attach the mask to the O_2 supply and adjust the flow meter to deliver 3 to 5 liters per minute (L/min.). Place the mask over the nose and mouth and slip the elastic band over the patient's head. Most patients on long-term O_2 therapy use a portable tank with a nasal cannula (Fig. 22-12). When applying a nasal cannula, make certain that the O_2 is flowing through it before placing it on the patient. The O_2 should be delivered at a rate of 4 L/min. or less, since higher rates are drying to the nasal mucosa.

While some outpatient facilities have O_2 available from wall outlets, portable O_2 units are more commonly used. You must be familiar with the operation of these units and with the procedure for checking to ensure that they will be immediately available when needed. The O_2 tank has an on-off valve with a dial indicating how much gas remains (Fig. 22-13). A separate valve adjusts the rate of O_2 flow, and an associated flow meter shows the delivery rate in units of L/min. *Both valves must be turned on to provide oxygen to the patient.*

Fig. 22-11 Oxygen mask.

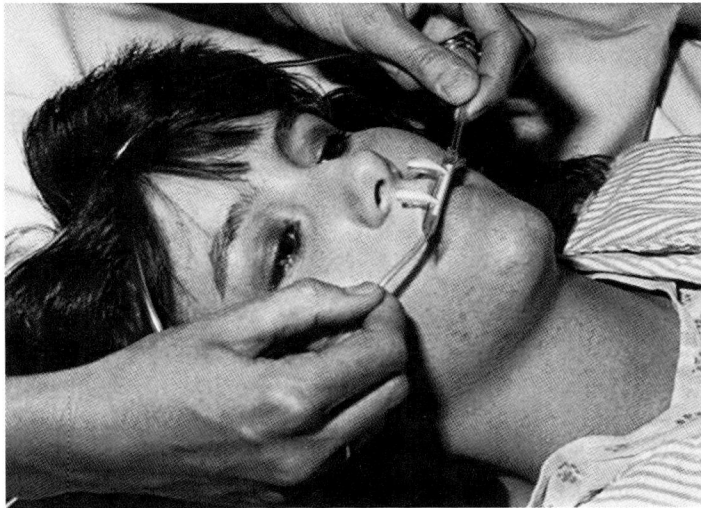

Fig. 22-12 Nasal cannula for oxygen administration.

Fig. 22-13 Portable oxygen tank controls.

The O_2 flow rate for many patients is 3 to 5 L/min. Severely compromised patients such as trauma victims in shock may receive O_2 at a much higher rate. Patients with emphysema or chronic obstructive pulmonary disease (COPD) (see Chapter 16) should receive O_2 at a slower rate, less than 3 L/min. These patients must not receive a higher rate of flow since the level of carbon dioxide in the blood controls their rate of respiration. If too much O_2 is administered, their respiratory rate may become too slow for adequate ventilation.

Suction Mechanical suction is used when a patient is unable to clear the mouth and throat of secretions, blood, or vomitus. If suction equipment is available for emergency use in your facility, you must assume responsibility for understanding and operating this equipment. If you are responsible to ensure that the suction system is operational, check to see that:

- The pump is working.
- The receptacle is connected to the pump.
- An adequate length of tubing connects the suction catheter to the receptacle.
- An assortment of disposable suction catheters is on hand.

Be alert to the need for suction whenever a patient becomes nauseated, is bleeding from the mouth or nose, or is unable to swallow and cope with secretions because of a low level of consciousness. If a patient does begin to aspirate mucus or vomitus, turn the patient immediately to the lateral recumbent position, don gloves and a face mask or goggles, and attempt to clear the airway manually. Remember to stand to one side when clearing an airway, since the sudden violent expulsion of the obstructing material may spray your face. If a reflex cough does not clear the airway at this point, suction is needed and your role may be to assist the nurse or physician with the procedure. Unwrap the suction tip attached to the suction apparatus and turn on the suction. At this point the nurse or physician proceeds with suctioning while you assist by holding the patient in position. When the emergency is over, check to be sure that you have cleaned or replaced the receptacle and replaced the disposable tip and tubing so that the suction unit will be ready for use when needed.

In the event that you must suction the patient yourself, call for help while you unwrap the catheter and turn on the suction. After you have cleared the mouth, pull the chin down and forward while inserting the suction catheter tip over the tongue in the midline. Do not insert the catheter forcibly, since you may injure the larynx (voice box). Any suctioning beyond the nasopharynx (back of the throat) should be done by a physician or someone trained in this procedure.

RESPIRATORY EMERGENCIES

Asthma

Asthma is difficulty in breathing caused by bronchospasm or constriction of the bronchi. Asthmatic attacks are frequently precipitated by stress, and if the radiologic procedure is new or frightening, the patient who is subject to asthma may experience dyspnea. The usual response to respiratory distress is to administer oxygen, but this is not the best treatment for patients with asthma. The objective is to relieve the bronchospasm. Most chronic asthmatic persons carry a nebulizer (mist applicator) with a bronchodilating medication. If so, they should bring this medication with them into the examining room. For an acute episode, the treatment is an injection of epinephrine, which is ordered by the physician. Although anxiety-producing to both patient and radiographer, a single asthmatic attack is seldom fatal.

Bronchial Obstruction

A common example of respiratory arrest caused by choking occurs when an older person enjoying a festive meal has difficulty chewing the meat. To avoid embarrassment, the meat may be swallowed whole and lodge at the larynx. The combination of alcohol, talking while eating, and poorly fitting dentures predisposes to such a "cafe coronary," which is not a heart attack at all. While this type of emergency is unlikely in a clinical setting, the ability to recognize this condition and respond appropriately may enable you to save a life.

When any foreign body lodges in the opening of the trachea, victims usually become quite agitated, their faces become congested, and they may tear at their collars or clutch their throats. Since the lungs hold more air than is normally used during respiration, the reserve supply can be used to help dislodge the foreign body by using a technique called the *abdominal thrust* or Heimlich maneuver. Ask, "Can you speak?" If the person does not answer, tell the person what you are about to do. Place both arms around the waist from behind. With your thumb on the outside of your fist, place the fist on the abdomen just below the sternum, and with the other hand grasping the opposite wrist, quickly and forcefully apply pressure upward against the diaphragm just below the ribs (Fig. 22-14). This will compress the lungs and will frequently expel the aspirated object. Never insert fingers into the mouth of a conscious patient in an effort to retrieve an obstructing object. Severe bite injury can occur, and the obstructing material can be forced farther into the throat during the struggle. If foreign material is visible in the mouth of an unconscious patient, use gloved fingers to grasp the tongue and lower jaw. Insert the index finger of the other

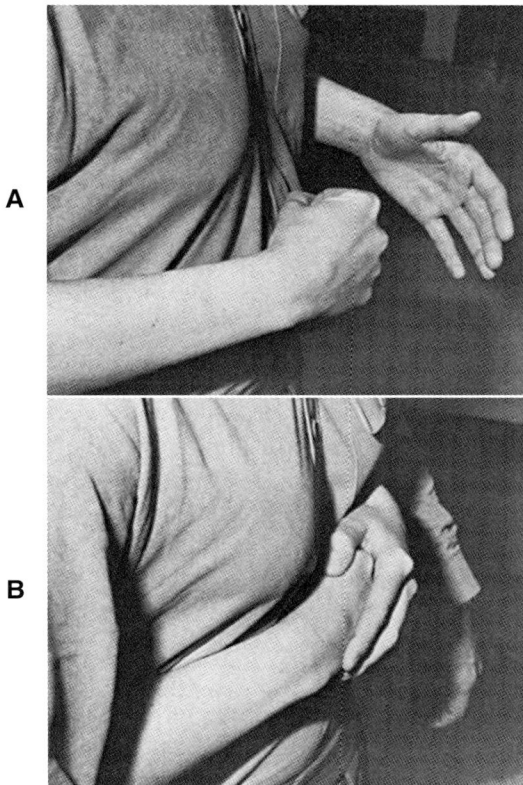

Fig. 22-14 Heimlich maneuver. **A,** Encircling victim from behind, place the knuckles of your right fist over the solar plexus. **B,** Grasp your right wrist with your left hand. Squeeze forcefully and quickly against diaphragm.

hand at the base of the tongue, and sweep it forward to clear the obstruction. This should never be done for children if the obstruction is not clearly visible in the mouth and easily accessible.

Children more frequently aspirate foreign objects such as large wads of chewing gum, buttons, small toys, or coins that have been held in the mouth. More than 65% of the deaths from foreign body aspiration occur in children under 1 year. Consumer product safety standards regulating minimum sizes of toys and toy parts for young children have markedly decreased the incidence of fatal aspiration incidents.

Foreign body obstruction should always be suspected in children when there is a sudden onset of coughing or **stridor** (a harsh, high-pitched sound). In infants, if respirations and coughing stop:

- Turn the infant prone with the head lower than the trunk.
- Support the head by firmly holding the jaw.
- Deliver up to five forceful back blows between the shoulder blades using the heel of the hand.
- While supporting the head and neck, turn the infant over, head lower than torso.

- Place the three middle fingers on the sternum just below the nipple line.
- Administer five thrusts over the sternum, taking care not to press over the liver.

In children over 1 year and under 8 years of age you may use the same Heimlich maneuver as in adults, with a series of quick thrusts beneath the diaphragm. Use good judgment in determining the amount of force, however, since young children have a relatively large and unprotected liver. The smaller the child, the greater the chance that damage to the liver could occur.

CARDIAC EMERGENCIES

Angina Pectoris

Angina is the term for chest pain that occurs when the coronary arteries are unable to supply the heart with sufficient oxygen to meet current needs. Episodes of chest pain are precipitated by exertion or stress and are usually relieved by rest or nitroglycerin tablets administered under the tongue. The discomfort caused by angina varies from a vague ache to an intense crushing sensation. Sometimes it is reported as severe indigestion, since it often presents as pain under the sternum. Patients who are known to suffer from angina usually carry their medication with them at all times. An emergency supply of nitroglycerin is usually stocked in medical health care facilities.

Heart Attack

If a patient complains of sudden, intense chest pain, often described as a crushing pain, you should assume until proven otherwise that the patient might be having a heart attack. Since the pain is caused when a portion of the heart wall becomes ischemic, you must prevent further damage by minimizing patient exertion. Patients may underestimate the importance of this type of pain and assume instead that the sudden onset is "terrible heartburn" or "indigestion from that peanut butter sandwich." Pain may be referred to the left arm, jaw, or neck. These patients often appear diaphoretic and pale. They may feel nauseated and short of breath and may have an irregular heartbeat. Stay with the patient, obtain help, and assist the patient to a comfortable position. If the patient has shortness of breath, raise the head and administer oxygen at 3 to 5 L/min.

Cardiac Arrest

One of the most anxiety-producing situations encountered by health care workers is to discover an unconscious patient or to observe a patient suddenly lose consciousness. When this occurs, it is important to initiate the "shake and shout" maneuver. Most patients who have simply fainted will respond if you call out their name and give them a gentle shake. If there is no response, feel for the carotid pulse and observe for respiration. If the patient has stopped breathing or no pulse is detected, the attention of a specially trained medical team is required. Summon help using the procedure prescribed by your facility for this type of emergency. You must be familiar with this protocol. If no protocol exists, call 911 and then summon the physician, telling him or her that emergency medical services are on the way.

Time is vital, since lack of effective circulation to the central nervous system can cause irreparable brain damage in 3 to 5 minutes. While awaiting the emergency medical team, you should proceed with cardiopulmonary resuscitation (CPR) only if you are trained and certified in this procedure.

CPR is the basic life support system that is used to ventilate the lungs and circulate the blood in the event of respiratory or cardiac arrest. Correct instruction and certification in CPR is important for all health care personnel. Your class should be given by a certified instructor, and you should take responsibility for updating your CPR card every 2 years. CPR instruction is not provided here, since recommendations are updated frequently and printed materials are easily obtained from the American Heart Association or the American Red Cross. Practice on a resuscitation dummy is essential. Your local college, fire department, and hospital in-service department may also offer courses taught by certified instructors that include lecture, film, practice time, and review and testing sessions.

Even if you are certified in adult CPR, do not attempt CPR on infants or children unless you are currently certified specifically in pediatrics as well. Courses for health care personnel usually include the techniques for infants and children.

Once the emergency medical team has arrived, you may no longer feel needed, but a radiographer can perform several important tasks. Record keeping is essential. Write down the time the attack started and when the emergency team responded. You may be asked to record times and amounts of medications. It may be necessary to obtain equipment, call for other personnel, or monitor a telephone.

During cardiac arrest and/or resuscitation, a rapid, weak and ineffective heartbeat may be present. This state is called **fibrillation** and is caused by interruption of the electrical signals that control the heart muscle. An electric shock to the chest can stimulate the heart to return to a normal rhythm. One recent development in treating cardiac arrest is the increasing use of automatic external defibrillators (AEDs). These devices are proven to be very effective when used by personnel with only a limited amount of training. They are often part of a clinic's emergency equipment. Learning to use an AED is much easier than learning to use a conventional defibrillator, or even learning CPR. Some of these defibrillators are completely automatic and have the ability to analyze cardiac rhythm, identify the need for defibrillation, and automatically deliver a shock. A semiautomatic version also exists in which the operator presses a button to start rhythm analysis. If the AED identifies the need for defibrillation, an indicator signal is given, instructing the operator to press the shock control.

The use of AEDs has been reduced to four simple steps:

1. Turn on the power.
2. Attach the adhesive pads to the victim's chest.
3. Turn on rhythm analysis.
4. Press the shock control to deliver the shock, if indicated.

While different brands and models have a variety of features, including different monitors, paper strip readouts, and instructions to the operator, application of the device is simple. If your facility has an AED, periodic instruction should be provided. Your participation in both CPR and AED classes can help you to be both prepared and effective in cardiac emergencies.

TRAUMA

Head Injuries

Patients who have received a blow to the head may have sustained serious injury, even when no external signs of trauma are present. Damage may occur with or without a skull fracture. The brain is soft, has a rich blood supply, and is suspended in cerebrospinal fluid within the skull. A blow to the head may cause a concussion or contrecoup damage as discussed in Chapter 17. If bleeding or swelling occurs inside the skull, a rise in intracranial pressure (pressure within the skull) may cause seizures, loss of consciousness, or respiratory arrest. Brain tumors can also result in increased intracranial pressure, causing patients to exhibit similar symptoms.

Four levels of consciousness (LOC) are generally recognized and may be described as follows:

1. Alert and conscious
2. Drowsy but responsive
3. Unconscious but reactive to painful stimuli
4. Comatose

The patient who is alert and oriented on arrival and then becomes increasingly incoherent and drowsy may be showing signs of increased intracranial pressure. The earliest signs of increasing pressure may be irritability and lethargy, often associated with a slowing pulse and slow respirations. Notify the physician immediately if you suspect a change in LOC.

Extremity Fractures

Trauma involving the long bones of the body may be classified in two categories: (1) compound fractures, in which the splintered ends of bone are forced through the skin; and (2) closed fractures. Compound fractures are usually partially reduced and a dressing applied before radiographic examination.

There are many ways of temporarily immobilizing extremity fractures. The two legs may be fastened together for stability during transportation (self-splinting), or a stiff object such as a board or rolled-up magazine may serve as a splint. Splinting devices should not be removed except under the physician's direct supervision.

When you must position a fractured extremity that is not supported by a splint, maintain a *gentle* traction (pull) while supporting and moving the arm or leg. Two people may be required to support and position patients with a potential long bone fracture, since the extremity must be supported at sites both proximal and distal to the injury. It is important to minimize motion of the fracture fragments. This helps avoid unnecessary pain and, more importantly, the initiation of a muscle spasm that could interfere with the physician's attempt to reduce and immobilize the fracture more permanently. Movement of fracture fragments may tear surrounding soft tissues, nerves, and blood vessels, seriously complicating the patient's condition.

Special care is required when positioning extremities following application of a plaster cast. Undue pressure against a fresh (wet) cast may cause it to change shape. Lift the casted extremity by placing your open hands underneath it; never grasp it from above. Observe the patient's fingers or toes for evidence of impaired circulation. They should be warm, pink, and sensitive to touch and pressure. Coldness, numbness, or lack of normal coloration should be reported to the physician at once. Swelling within the cast and subsequent pressure can cause permanent nerve and tissue damage if not relieved promptly.

Wounds

Patients with open wounds have usually been treated before you see them in the radiology suite. Bleeding has been controlled and dressings applied. The radiographer's primary responsibility regarding open wounds is to maintain the dressings and to report promptly any significant amount of fresh bleeding. This is usually considered to be the amount of bright-red blood sufficient to soak through a fresh dressing. If a laceration or incision opens, causing severe **hemorrhage** (continuous abnormal blood flow), apply direct pressure to the site of bleeding while summoning immediate assistance.

MEDICAL EMERGENCIES

Drug Reactions

The administration of any medication has the potential to cause a medical emergency. The reaction can range in severity from a sudden bout of nausea and vomiting to a cardiac arrest. The nature of the symptoms will determine the appropriate treatment. Treatment for the various conditions seen in response to drug administration is found throughout this chapter. Individuals who have been sensitized to an over-the-counter medication may be just as prone to an allergic reaction as those who are receiving prescription medications. Reactions may occur to medications administered orally or by injection. When drugs are administered intravenously (directly into a vein) the effects will appear more rapidly.

A moderate allergic reaction is characterized by erythema, **urticaria** (hives), and/or dyspnea. The physician should be notified while you remain with the patient. The treatment for these patients is administration of an antihistamine medication such as diphenhydramine intravenously or intramuscularly, depending on the reaction's severity. Some physicians prefer to give epinephrine at this point. A severe allergic reaction is called **anaphylaxis** or anaphylactic shock. This life-threatening condition may result in respiratory or cardiac arrest and, less often, in seizures. The early symptoms of anaphylaxis include a sense of warmth, tingling, itching of palms and soles, difficulty swallowing, constriction in the throat, a feeling of doom, an expiratory wheeze, and then progression into laryngeal and bronchial edema. Treatments for shock, respiratory arrest, and cardiac arrest are appropriate and are outlined elsewhere in this chapter. At the onset of anaphylaxis the radiographer should maintain the patient's airway, alert the physician, and call 911.

Diabetic Emergencies

Patients who have diabetes are seen in clinics for diabetic monitoring and care as well as for the same variety of problems that bring other patients to the physician. The diabetic patient can often be identified by means of the Medic-Alert bracelet. Persons with other medical conditions that may require emergency treatment also wear these bracelets. The nature of the problem is stated on the bracelet and can be of great help in providing an appropriate emergency response.

Diabetes is a disease characterized by the body's inability to metabolize blood glucose. Insulin is an enzyme normally produced in the pancreas in response to food intake. Insufficient insulin prevents the use of glucose by the muscles. When the muscles cannot use glucose, the liver forms ketone bodies to supply the energy for muscle contraction. When excess ketone bodies appear in the blood, they cause ketoacidosis, a change in the pH of the blood. The body attempts to compensate for the acidosis by **hyperventilation** (air hunger with rapid respirations) and excretion of minerals and water in the urine. When the blood glucose level is very high, sugar also "spills over" into the urine. The individual who is terribly thirsty, urinates copious amounts frequently, and has fruity smelling breath (ketones excreted via the respiratory tract) may be approaching **diabetic coma.** This condition is characterized by a relatively slow onset. The patient is diagnosed through blood and urine tests and treated with diet, exercise, and medication such as insulin or oral hypoglycemic agents.

The diabetic patient who has taken insulin but no food may develop **hypoglycemia,** or low blood sugar. Unlike the slow onset of diabetic coma, hypoglycemia is characterized by a *sudden* onset of weakness, sweating, tremors, hunger, and, finally, loss of consciousness. While the patient is still alert and cooperative, hypoglycemia can be quickly remedied by administration of a small amount of candy or sweet fruit juice. Squeeze tubes containing a measured amount of glucose may be stored with the emergency medications. These prepackaged tubes are useful because the gel-like material can be placed inside the patient's cheek. This decreases the chance that a semiconscious or confused patient will aspirate it, as might be the case with candy or juice.

Report the occurrence of hypoglycemia to the physician. You must help these patients to sit or lie down until the sugar takes effect. Occasionally, individuals with the same symptoms do not have diabetes. They may have hypoglycemia without diabetes; the treatment is the same. Table 22-2 summarizes the physical findings associated with both high blood sugar, which is indicative of approaching diabetic coma, and low blood sugar, which may signify an impending insulin reaction.

Cerebrovascular Accident

A **cerebrovascular accident,** also called a **CVA** or a "stroke," occurs most frequently in the elderly, but can occur at any age. Rupture of a cerebral artery can cause a hemorrhage into the brain tissue, or an artery may become occluded and cause an interruption in the blood supply to the area beyond the occlusion. The symptoms may occur very suddenly or may develop over a period of hours. Warning signs may include the following:

- Slurred or difficult speech
- Extreme dizziness
- Severe headache
- Muscle weakness on one or both sides
- Difficulty in vision or deviation in one eye
- Temporary loss of consciousness

These symptoms may be only temporary, but should be reported immediately to a physician. The patient should be helped to a recumbent position with the head elevated. Do not leave the patient, but summon assistance and have the emergency supplies and oxygen at hand. Monitor vital signs every 5 minutes or as ordered by the physician. **Transient ischemic attacks (TIAs)** present similar symptoms but usually last only minutes, or at most a few hours. These temporary attacks should not be ignored, as they are frequently precursors to more permanent damage.

TABLE 22-2
Diabetic Crises

Crisis	Cause	Symptom	Treatment
Hyperglycemia; impending diabetic coma	Food consumption over dietary allowance; fever, infection, stress; insufficient insulin	Increased thirst; increased urinary output; decreased appetite; nausea, vomiting; weakness, confusion; coma	Inform physician immediately; administer sugar-free liquids if conscious
Hypoglycemia; insulin reaction	Insufficient food; excessive exercise	Headache; hunger; diaphoresis; tremors; tachycardia; impaired vision; personality change; loss of consciousness	Administer food with high sugar content; administer glucagon if conscious; inform physician immediately

Seizure Disorder

Seizures occur as a result of a focal or generalized disturbance of brain function and are accompanied by a change in the level of consciousness. A major motor (grand mal) seizure may be preceded by an aura, or premonitory sign. The patient may say, "I'm going to have a spell," and should be assisted to a supine position as rapidly as possible. Frequently the seizure is signaled by a hoarse cry when air is forced past the vocal cords by a sudden contraction of all the abdominal and chest muscles. In the event of a seizure, your first duty is to keep the patient as safe as possible. Notify the physician immediately, request assistance, and *do not leave the patient.* If the patient is on the x-ray table, your first concern is to prevent a fall. Remove any objects that might be hazardous and place padding under the patient's head. Do not attempt to restrain the patient and do not attempt to force objects into the patient's mouth. If a padded tongue blade is available and can be easily inserted, it may help avoid a laceration of the tongue. Loss of consciousness and a rigid arching of the back are followed by alternate relaxation and rigidity of the muscles until the seizure passes and the patient slowly regains consciousness. Have emergency medication (diazepam) ready for administration in the event of prolonged or repeated seizures *(status epilepticus).* While the patient is unconscious, involuntary voiding and defecation may occur. As the seizure passes, turn the patient to a lateral recumbent position to prevent aspiration of secretions, and remain with the patient to provide reassurance and assistance. In the immediate period following the seizure (postictal period), the patient may be somewhat irritable or confused and wish only to sleep.

Less intense partial (focal) seizures may cause severe, uncontrollable tremors. This condition often causes extreme anxiety and hyperventilation in a conscious patient. These seizures are exhausting to the patient and may persist for over an hour without treatment. Instruct the patient to breathe slowly, and place a paper bag over nose and mouth if hyperventilation is otherwise uncontrollable.

Another type of seizure is characterized by a brief loss of consciousness (absence) during which the patient stares, or may lose balance and fall. Many patients are not aware that they undergo this loss of consciousness.

Patients taking anticonvulsant medication may not have seizures for long periods. Most of these medications have a relatively slow excretion rate, which allows the patient to miss a dose or two without precipitating an attack. On the other hand, fatigue or apprehension may initiate a seizure in a previously stable patient.

Realize that the seizure will run its course. The most important actions you can take are to protect the patient from harm and to be an accurate observer. Note when the seizure began and how long it lasted. Did it involve both sides of the body equally, and did the contractions start in one area and progress from one extremity to another? These observations can be helpful to the physician in reaching an accurate diagnosis.

Not all seizure-prone individuals have the same diagnosis. Seizures may be a response to drug sensitivity, infection, epilepsy, tumor, or fever. Only recently have the old superstitions and myths about seizure disorders begun to be dispelled. No direct, consistent correlation exists between seizures and mental acuity, emotional instability, or heredity.

Hyperventilation

The anxious patient who breathes too deeply and too often (hyperventilates) may complain of feeling faint or dizzy and note tingling and numbness in the extremities. These patients have breathed in too much oxygen and have exhaled too much carbon dioxide, which disturbs the chemical balance of the blood. Try to persuade them to breathe more slowly or to breathe into a paper bag, which will help to return their carbon dioxide level to normal.

Vertigo and Postural Hypotension

A "lightheaded" or dizzy sensation is not unusual when patients sit up suddenly from a recumbent position. This is especially common after prolonged periods of bed rest. Postural hypotension is the usual cause. This results from the same basic mechanism that causes syncope, or fainting, which is discussed later in this chapter. Blood pools in the extremities when the torso is elevated and causes a transient cerebral **anoxia** (lack of oxygen). This condition can usually be avoided by having the patient sit up gradually. This sensation frequently affects elderly patients, so remain close to them and provide support when a sudden change in position is necessary.

Vertigo has a different cause. The patient does not feel lightheaded, but feels as if the room is moving or whirling. These patients frequently cling to the table and will fall if not assisted to lie down. They may experience violent nausea. This sensation is usually attributed to either a middle ear disturbance or to a lesion in the brain or spinal cord. A sudden onset in a patient who does not have a previous history of vertigo should be reported immediately to the physician, as this may be associated with a TIA or CVA. Alcohol or the administration of certain drugs may affect individuals in a similar manner and should also be called to the attention of the physician.

Epistaxis

A nosebleed, or **epistaxis,** is rather frightening to the patient but is usually not serious. Remove eyeglasses when necessary, and provide an ample supply of tissues. Instruct the patient to breathe through the mouth and to squeeze firmly against the nasal septum for 10 minutes. The patient should not lie down, blow the nose, or talk. Provide an emesis basin, instructing the patient to spit out blood that runs down the nasopharynx rather than swallow it. If bleeding lasts more than a few minutes, inform the physician, who may want to apply more direct treatment.

Nausea and Vomiting

Nausea and vomiting are not uncommon in health care settings, and a well-prepared radiographer learns to cope easily with this situation. Vomiting can often be prevented by the radiographer's reassuring presence and by instructions that focus the patient's attention on breathing. "Breathe through your mouth, taking short, rapid, panting breaths," or "Take some long, slow deep breaths through your mouth," are both effective instructions and each has strong adherents. On the other hand, if a patient expresses a need for an emesis basin, offer it immediately. Bring the patient a clean emesis basin before removing the soiled one. Provide tissues and water to rinse the mouth. It is especially important to support the patient in a sitting or lateral recumbent position to avoid aspiration of emesis. If the patient loses consciousness, turn the head to the side and clear the airway.

Shock

Shock is a general term used to describe a failure of circulation in which blood pressure is inadequate to support oxygen perfusion of vital tissues and is unable to remove the by-products of metabolism.

Fainting, also called **syncope,** is a very mild form of shock that sometimes occurs when fright, pain, or unpleasant events are beyond the coping ability of the patient's nervous system. Blood pressure falls as the diameter of the blood vessels increases and the heart rate slows. When the blood pressure is too low to supply the brain with oxygen, the patient faints. Placing the patient in a dorsal recumbent position with the feet elevated usually relieves this type of shock. Patients who have not eaten for 12 hours and are feeling anxious and stressed may undergo syncope. Patients who feel faint should be assisted into a sitting or recumbent position. If a chair is not within reach, ease the patient to the floor. If the patient does not rouse immediately, spirits of ammonia held under the nose usually cause a rapid return to consciousness. Small, crushable vials of ammonia are usually stocked with the emergency supplies. A physician's order is not required for their use. Anyone who has more than a momentary loss of consciousness should be evaluated by a physician before the examination is resumed.

Shock other than syncope is a dangerous, potentially fatal condition. It may be caused by blood loss, severe infection, head trauma, heart failure, or a severe allergic response. Early signs of shock are pallor, increased heart rate and respirations, and restlessness or confusion.

The following symptoms indicate some degree of shock in any or all combinations:

- Restlessness and a sense of apprehension
- Increased pulse rate
- Pallor accompanied by weakness or a change in thinking ability
- Cool, clammy skin
- A fall in blood pressure of 30 mm Hg below the baseline systolic pressure

Your responsibility to patients with any type of shock is to recognize its symptoms, to know the location of emergency medical supplies, and to be thoroughly familiar with the emergency protocols of your facility. The physician may call on your knowledge of medications and your medication administration skills during treatment. The radiographer's role in suspected shock is as follows:

- Stop the procedure.
- Assist the patient to a dorsal recumbent position to avoid a fall.
- Obtain help. Notify the physician. If in doubt, call 911. It is much better to be mistaken than to have a patient die because of inadequate treatment.
- Check blood pressure.
- Assist the dyspneic patient with oxygen.
- Be ready to perform CPR.
- Assist the physician or emergency medical team as necessary.
- Chart the occurrence, the treatment administered, and the patient's response on an incident report form and/or in the chart.

SUMMARY

Quality patient care requires that radiographers assess and respond to patients' needs, both emotional and physiological. Assessment of physical status involves trained observation and touch. In addition, physical condition is evaluated by taking vital signs: temperature, pulse, respirations, and blood pressure. The radiographer who can take accurate vital signs and recognize abnormal readings can use these skills both to assist with routine patient evaluation and to identify changes in patient status in an emergency. Shock, respiratory arrest, heart attack, and cardiac arrest are life-threatening emergencies. Radiographers must recognize the signs of these conditions and be capable of responding quickly and appropriately. Other medical emergencies such as syncope, vomiting, hypoglycemia, and epistaxis require appropriate responses as well. When the radiographer is alert and well prepared, any emergency is more likely to have a positive outcome.

▪ Review Questions ▪

1. List the items you will need to assist a patient with a bedpan.
2. List pertinent questions to obtain a history on a patient who has wrist pain. List the questions to ask regarding chest pain.
3. Observe the physical status of a classmate or family member. Record your observations using appropriate terminology.
4. List the normal values for temperature, pulse, respiration, and blood pressure. Suggest possible reasons for abnormal values for these vital signs.
5. What should be the oxygen flow rate for a patient who experiences sudden dyspnea?
6. What are the signs of shock? Which sign is most likely to be noticeable in the early stages?
7. What steps should you take to determine whether an unconscious patient is in a state of cardiac arrest?
8. List precautions for handling a fractured extremity and one to which a cast has just been applied.
9. Compare and contrast syncope, vertigo, and postural hypotension.
10. A patient has just fainted and you note that he is wearing a Medic-Alert bracelet indicating diabetes. What is the most likely cause of the faint? What should you do?

PART V

Ancillary Clinical Skills

23

Medications and Their Administration

Learning Objectives

At the conclusion of this chapter, the student will be able to:

- Explain the role of the radiographer with respect to medication administration in outpatient facilities
- Differentiate between chemical, generic, and trade names for medications
- Look up medication information in standard references and on medication package inserts
- List and describe common routes of medication administration
- Name medications commonly needed in emergencies and describe their effects
- Demonstrate the steps used in the administration of oral medication
- Demonstrate the steps used in the parenteral administration of medication using intradermal, subcutaneous, and intramuscular routes
- Identify the equipment and supplies used for parenteral injections and select appropriate supplies for each route of administration
- Identify sites used for intramuscular injections
- Describe the precautions required to ensure safety when injecting medications
- Chart medications accurately

Key Terms

agonist	intramuscular (IM)
allergen	intravenous (IV)
antagonist	opiate
antidote	opioid
dehydration	parenteral
efficacy	potency
extravasation	proprietary
generic	standing order
hematoma	subcutaneous (SC)
hydration	synergistic
idiosyncratic	topical
infiltration	toxic
intradermal	

THE RADIOGRAPHER'S ROLE IN MEDICATION ADMINISTRATION

Each state has regulations regarding the qualifications of individuals to administer medications. You must be familiar with the rules governing medication administration in your state as well as in your facility. As stated in Chapter 21, health care facilities use written policies and procedures to ensure patient safety and to reduce the risk of liability for errors. Medication administration is an area of patient care with a high potential for both error and medicolegal problems. For this reason, there will be an established policy with respect to who may administer medication in the facility. There may also be standard procedures for medication administration. *Never administer any medication unless you are certain beyond doubt that you are authorized to do so.* If regulations and established policies allow it, your job description may involve medical assisting duties that include medication administration. If you are not permitted to administer medications, your duties may include checking the allergic history of patients, preparing medication for administration, verifying patient identification, assisting the physician, and monitoring the patient after the medication has been given.

Even when your job description is limited primarily to radiography, a knowledge of medications and their administration is valuable in an emergency. The radiographer must be familiar with the location of medications that might be required if there is a sudden change in a patient's status. As discussed in Chapter 22, acute attacks of angina, sudden asthmatic episodes, or insulin reactions are typical of the types of emergencies seen by radiographers where prompt administration of medication may be essential.

Whenever medications are given in a physician's office or clinic, the physician selects the drug, determines the route of administration, and prescribes the exact dosage. *No medication is ever given without a physician's order.* Orders may be verbal or written. Verbal orders should be written or countersigned by the physician before he or she leaves the area or within the following 24-hour period. In some states radiographers are not allowed to accept medication orders by telephone. These matters should also be addressed in the Policy and Procedures manual.

A **standing order** consists of written directions for a specific medication or procedure, signed by a physician, and used only under the specific conditions stated in the order. For example, a standing order might allow a nurse to administer a specific dose of nitroglycerin to a patient experiencing angina when the physician is not present to issue a specific order. Such orders are found in a Policy and Procedures or Standing Orders book available for immediate reference.

Although a comprehensive knowledge of drugs is not usually essential for a radiographer, you should become familiar with the names, dosages, and routes of administration for medications frequently used and those most likely to be needed in emergencies. If this seems intimidating, be reassured that only a limited number of drugs, with a few standard dosages of those drugs, are used with any regularity. Knowledge of these medications greatly facilitates the task of assisting the physician and also aids in determining whether departmental stocks of medications and medication supplies are adequate and up to date. Checking expiration dates on medication supplies is especially important in doctors' offices and clinics where such supplies are used infrequently. Medication knowledge also enables the alert radiographer to prevent errors by questioning and double-checking any medication orders or records that seem unusual or inappropriate.

Any drug may produce side effects in certain patients. Radiographers use this knowledge to anticipate possible adverse drug reactions and to recognize and report signs and symptoms of side effects as they occur. This awareness is very important since radiographers may be the first to observe the onset of medication responses that could have serious consequences.

MEDICATION NOMENCLATURE AND INFORMATION RESOURCES

The terms *drugs* and *medications* are often used interchangeably. As a working definition, medications are substances that are prescribed for treatment and that produce therapeutically useful effects. The more general term *drugs* denotes substances used in diagnosis, treatment, or prevention of disease or as a component of a medication. This term is sometimes confusing because it is also used to denote chemicals such as narcotics or hallucinogens that affect the central nervous system, causing behavioral changes and possible addiction. Drugs may replace a missing substance in the body such as estrogen or insulin. Some medications such as digitalis are made from plants; others, like heparin, come from animal sources; and some such as penicillin are produced by microorganisms. Many drugs today are manufactured from synthetic materials. Drug synthesis and the rapidly expanding application of genetic engineering promise vast possibilities for the future.

Each medication has a **generic** name that identifies its chemical family. If the drug consists principally of one chemical, it may also be referred to by its chemical name. For example, *acetylsalicylic acid* is the chemical name for the generic drug aspirin. Manufacturers give their products "brand names" that are also called **proprietary** or *trade names*. The same generic substance may be manufactured by several different companies and given a different trade name by each. For example, a synthetic antibacterial containing trimethoprim and sulfamethoxazole is produced by Roche Pharmaceuticals under the name *Bactrim* and by Glaxo Wellcome Inc. as *Septra.* The generic and trade names of some drugs are used interchangeably. For instance, the generic term *epinephrine* is used just as frequently as the trade name *Adrenalin* for this common emergency drug. Since medications may be ordered by either generic or trade names, you should be familiar with both terms. When the physician calls for epinephrine, the knowledgeable radiographer will reach for the Adrenalin without having to read the small print on each container in the emergency drug box.

Setting the standards for control of drugs is part of the role of the U.S. Food and Drug Administration (FDA). This role includes strict rules concerning **efficacy** (effectiveness), purity, **potency** (strength), safety, and toxicity of both prescription and nonprescription (over-the-counter or OTC) medications.

The study of drugs is an ongoing process since new medications are constantly being added. No single textbook can provide information to cover all situations. For these reasons, you should be acquainted with other methods of obtaining medication facts on a continuing basis. One useful resource is the information sheet enclosed in each drug package. The FDA requires that all drug packages include the following data: trade name, generic name, chemical composition, chemical strength, usual dosage, indications, contraindications, and reported side effects. Package inserts from frequently used drugs may be kept on file to avoid the necessity of opening a package when this information is needed. If you collect and study inserts from the drugs used most frequently, you will soon develop a working knowledge and useful base of information about medications that are important in your clinical setting.

Another useful resource is a reference work of medication information such as the *Physicians' Desk Reference* (PDR) or *Mosby's GenRx.* The libraries of most health care facilities include a drug reference manual, and the radiographer should become familiar with its use. These references are especially useful in clinic and office situations where medication is given infrequently. They are published annually and list drugs alphabetically by their generic names and their trade names and according to their uses. A separate section indexes the products made by each manufacturer. In the product description you will find information similar to that found in the package inserts. In-service classes and college courses may also provide useful information on medications. Formal instruction may be especially important to you if your state law and job description permit you to administer medications and/or contrast media.

ROUTES OF MEDICATION ADMINISTRATION

The most common routes for medication administration are oral, topical, and parenteral.

Oral medications are swallowed. They are dissolved or digested in the stomach and then pass into the small intestine where the majority of the absorption takes place.

With the **topical** route of administration, a drug is applied to the surface of the skin. This route may be used for a local effect such as when calamine lotion is used to relieve the itch of poison ivy. Some topical medications are used for a systemic effect. These medications are absorbed through the skin into the bloodstream and are applied to the skin in a paste form or on small adhesive disks ("patches"). One such topical drug, nifedipine, is used by patients with heart conditions to increase vascular dilation. Another example is scopolamine, which is applied on adhesive discs to treat vertigo and help prevent motion sickness.

Some medications are applied to the mucous membrane of the mouth and may be classified as oral and/or topical medications. They are placed under the tongue (sublingual) or inside the cheek (buccal) and are absorbed directly into the blood through the mucous membrane. Sublingual or buccal administration makes certain drugs work rapidly because they are immediately available without having to be digested and absorbed through the stomach or bowel. Examples of these medications include nitroglycerin, which is administered sublingually for angina pectoris, and glucose paste, which is administered by buccal application to treat hypoglycemia (see Chapter 22).

While patients may prefer to take medications orally, some drugs cause irritation of the gastrointestinal tract, cannot be absorbed by this route, or must be given by a route that will produce a more rapid response. In the **parenteral** route, medications are injected directly into the body. Parenteral injections may be given in several ways and are classified according to the depth and/or location of injection: **intradermal** injections are shallow injections between the skin layers; **subcutaneous (SC)** injections are deeper, beneath the skin; the **intramuscular (IM)** route involves injection directly into muscle tissue; and **intravenous (IV)** injections are made directly into a vein. IV administration is the parenteral route that offers the most immediate results in terms of effect. As the medication is dispersed by the bloodstream, the time needed to reach the site of action is very short. IV administration is outside the scope of practice for limited radiographers in most circumstances and is beyond the scope of this text.

Procedures for both oral and parenteral medication administration are discussed later in this chapter.

FREQUENTLY USED MEDICATIONS

The following medications are used regularly in many outpatient health care facilities. These general descriptions illustrate how such medications are used but are not meant to be exclusive. The specific drugs used in your facility may be different while meeting the same needs. See Table 23-1 for a more extensive list of medications that you may encounter.

Antiallergic Medications

Antihistamines are drugs used to treat mild to moderate allergic reactions. Diphenhydramine (Benadryl) is the most frequently used antihistamine. It also has sedative and anticholinergic (drying) side effects. For adults the usual oral dose is 25 to 50 mg, and for children weighing over 20 pounds, the dosage is 12.5 to 25 mg. Benadryl may also be given IM or IV if the patient has an allergic reaction. The dosage is 10 to 50 mg

TABLE 23-1

Common Medications by Category

Category	Effect	Example
Adrenergics (vasoconstrictors)	Stimulate the sympathetic nervous system, causing constriction of smooth muscle lining of the respiratory tract	Epinephrine (Adrenalin), ephedrine, isoproterenol (Isuprel), metaraminol (Aramine), phenylephrine hydrochloride (Neo-Synephrine)
Adrenergic blocking agents	Block the production of epinephrine in the body, causing dilation of blood vessels and decreased cardiac output; used as an antihypertensive	Methyldopa (Aldomet), clonidine (Catapres), prazosin (Minipress)
Analgesics	Relieve pain	Acetaminophen (Tylenol), aspirin, codeine, meperidine (Demerol), methadone, morphine, phenacetin
Anesthetics	Promote loss of feeling or sensation	General—sodium pentothal, halothane (Fluothane), nitrous oxide; Local—lidocaine (Xylocaine)
Antiarrhythmics	Prevent or relieve cardiac arrhythmias (dysrhythmias)	Procainamide hydrochloride (Pronestyl)
Antimicrobials	Suppress the growth of microbes	Internal—penicillin, tetracyclines, sulfadiazine, erythromycin, cephalosporins (Keflex, Keflin); External—sulfonamides, thimerosal (Merthiolate), hexachlorophene (pHisoHex)
Anticholinergics	Depress the parasympathetic nervous system and act as antispasmodics of smooth muscle tissue; decrease contractions, saliva, bronchial mucus, digestive secretions, and perspiration; used as preparation for surgery and endoscopy to suppress secretions	Atropine, belladonna, propantheline bromide (Pro-Banthine), scopolamine (Hyoscine)
Anticoagulants	Inhibit the clotting mechanism of the blood; used to keep IV lines and arterial catheters open during diagnostic procedures	Heparin, warfarin (Coumadin)
Anticonvulsants	Inhibit convulsions	Phenytoin (Dilantin), trimethadione (Tridione)
Antidepressants	Relieve or prevent depression	Amitriptyline (Elavil), imipramine (Tofranil), fluoxetine (Prozac), paroxetine (Paxil)
Antiemetics	Relieve or prevent vomiting	Trimethobenzamide hydrochloride (Tigan), prochlorperazine (Compazine)
Antifungals	Treat or prevent fungal infections	Systemic—griseofulvin; Topical—Tinactin
Antihistamines	Relieve the symptoms of allergic reactions	Diphenhydramine (Benadryl), chlorpheniramine maleate (Chlor-Trimeton)
Antiperistaltics	Slow peristalsis of the gastrointestinal tract	Tincture of opium (Paregoric), loperamide (Imodium)

IV or IM and may be increased to 100 mg as necessary. The maximum safe dosage in a 24-hour period is 400 mg. Cortisone is also prescribed for these purposes.

For patients with an acute allergic reaction, epinephrine (Adrenalin) is administered SC, IM, or IV. To control hives, shock, or respiratory distress, the physician administers a small dose of Adrenalin (0.2 to 1 ml of 1:1000 solution) and increases the dosage if required.

When patients with a severe or incapacitating allergic response do not appear to respond to the treatment just described, methylprednisolone (Solu-Medrol) may be administered IV. This is a corticosteroid that acts as an antiinflammatory agent, preventing or reducing edema of the trachea and bronchi. This treatment minimizes the possibility of respiratory arrest. Solu-Medrol is provided in a two-compartment vial with the diluting fluid and soluble powder separated

TABLE 23-1
Common Medications by Category—cont'd

Category	Effect	Example
Antipsychotics	Treat psychoses, schizophrenia	Haloperidol (Haldol), fluphenazine (Prolixin)
Antipyretics	Reduce fever	Aspirin, acetaminophen (Tylenol)
Antitussives	Reduce coughing	Dextromethorphan (Romilar), codeine
Antivirals	Prevent or treat viral diseases	Acyclovir (Zovirax), amantadine zidovudine (AZT)
Barbiturates	Depress the central nervous system, respirations, and blood pressure; induce sleep	Pentobarbital sodium (Nembutal), secobarbital (Seconal), phenobarbital
Bronchodilators	Dilate smooth muscle	Theophylline (Theo-Dur), aminophylline
Cardiac depressants	Restrain or slow heart activity	Quinidine procainamide (Pronestyl)
Cardiac stimulants	Strengthen and tone the heart; increase cardiac output	Digitalis, gitalin (Gitaligin), lanatoside C (Cedilanid)
Cathartics	Stimulate peristalsis; promote defecation	Bisacodyl (Dulcolax), castor oil, magnesium sulfate
Diuretics	Stimulate the flow of urine	Chlorothiazide (Diuril), furosemide (Lasix), acetazolamide (Diamox)
Emetics	Induce vomiting	Ipecac
Hypoglycemics	Lower blood sugar level	Insulin, chlorpropamide (Diabinese), tolbutamide (Orinase)
Opioids (narcotics)	Analgesic sedatives with a potential for addiction; classified as controlled substances under the Harrison Act	Morphine, meperidine (Demerol), codeine
Opioid antagonists	Prevent or counteract respiratory depression and depressive effects of morphine and related drugs	Naloxone hydrochloride (Narcan), nalorphine (Nalline)
Radioactive isotopes	Radioactive forms of elements used for diagnosis and treatment	Iodine 131, cobalt 60
Sedatives	Depress and relax the central nervous system and reduce mental activity	Barbiturates, paraldehyde, chloral hydrate
Skeletal muscle relaxants	Relax skeletal and striated muscle tissue	Succinylcholine chloride (Anectine)
Stimulants	Stimulate the central nervous system	Caffeine, sodium benzoate amphetamines (Benzedrine, Dexedrine)
Tranquilizers	Reduce anxiety	Minor—diazepam (Valium), chlordiazepoxide (Librium); major— chlorpromazine (Thorazine), trifluoperazine (Stelazine)
Vasodilators	Relax the walls of blood vessels, permitting a greater flow of blood	Isosorbide dinitrate (Sorbitrate), trinitroglycerol, hydralazine (Apresoline), papaverine

by a plunger/stopper. The directions for mixing are provided, but you should become familiar with the preparation of this and all common medications before the need arises.

Antimicrobial Medications

The category of antimicrobial medications includes disinfectants such as alcohol and Betadine, an iodine compound commonly used for skin preparation prior to sterile procedures.

Also classed as antimicrobials are the medications, commonly referred to as *antibiotics,* that are given to treat wound infections and infectious diseases. Antibiotics can be subclassified as antibacterial or antifungal according to the type of organisms against which they are most effective. Some antibiotics treat a very narrow range of microorganisms. Trimethoprim/sulfamethoxazole (Bactrim), for example, is used to treat specific infections of the urinary tract. Others are referred to as *broad-spectrum* antibiotics and are effective against a wide variety of pathogens.

The use of some medications has come full circle. As newer medications such as vancomycin came into use, many of the earlier antibiotics fell into disuse. Since more and more infectious organisms have become resistant to the modern antibiotics once used to treat them, some of the old medications are being used in new forms and combinations.

Analgesics

Analgesics are defined as drugs that can relieve pain without causing loss of consciousness. As a group, **opioids** are the most effective analgesics. The term *opioid* describes any drug, natural or synthetic, whose actions are similar to the actions of morphine. The opioid family, whose name derives from opium, includes morphine, codeine, and meperidine (Demerol). **Opiate** is the more specific term applied only to natural opium derivatives. The term *narcotic* is no longer used with precision as a synonym for *analgesic* because it has also come to stand for those analgesic central nervous system (CNS) depressants that may lead to addiction. In a *legal* sense, cocaine, marijuana, and lysergic acid diethylamide (LSD) are classed with opiates as narcotics.

Controlled substances are drugs with a high potential for abuse and misuse that are therefore kept in a locked container. They must be counted daily, and any medication given must be listed on forms that give the date, the patient's name, the dose, and the name and title of the person administering the medication. Use of these drugs is monitored by the U.S. Drug Enforcement Administration (DEA), and they can be prescribed only by persons who hold a DEA license.

Opioids act by depressing the CNS, relieving pain, and producing drowsiness. Excessive doses, however,

can result in depressed respirations, coma, and possible death. Among the most frequently used are morphine sulfate (MS) and Demerol. These injectable medications are given in a dosage of 10 to 30 mg of morphine or 50 to 100 mg of Demerol. Fentanyl (Sublimaze), another highly potent opioid analgesic, is given to patients who are sensitive to other analgesics or who are not responding to such medications with adequate pain relief. The physician determines the dosage of this medication based on age, weight, use of other drugs, and the procedure involved. The action of Sublimaze is almost immediate and lasts 30 to 60 minutes after IV administration. It is supplied in a strength of 50 micrograms per milliliter (mcg/ml), and the usual dose is 1 to 2 ml. Respiratory depression peaks 5 to 10 minutes after injection and may last for several hours, depending on the dosage.

Patients who have received any CNS depressant must be monitored closely. Respiratory depression is a life-threatening side effect. Resuscitation equipment and emergency drugs to counteract these effects should be immediately available. Since patients may be some distance from the radiographer, a pulse oximeter is attached to the patient's finger, toe, or earlobe (Fig. 23-1). A digital readout of the pulse and oxygen saturation of the blood can be viewed from a monitor screen. If the oxygen saturation drops below 90%, the patient is asked to respond and take a few deep breaths, then is observed closely. If the oxygen saturation continues to drop or the patient does not respond adequately, a physician is notified immediately.

Analgesics may be given to help the patient cope with painful procedures or to lessen preexisting pain. Analgesics with a low potential for side effects such as aspirin, ibuprofen (Advil), acetaminophen (Tylenol), and naproxen sodium (Aleve) are frequently used to

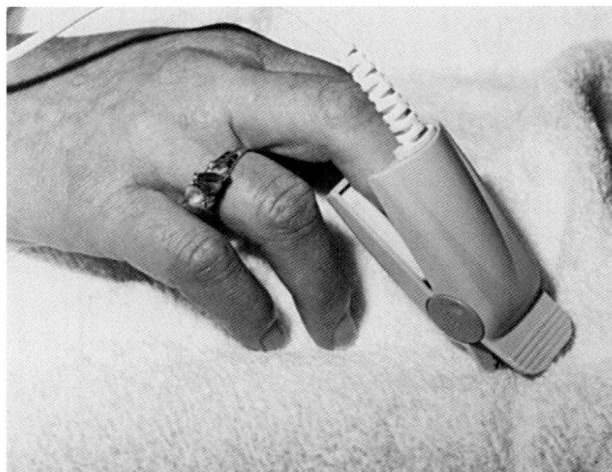

Fig. 23-1 Pulse oximeter probe in place on finger.

alleviate discomfort. With the exception of acetaminophen, these products are all categorized as nonsteroidal antiinflammatory drugs (NSAIDs). Although they are all analgesics, these drugs are not controlled substances. All are available without prescription. Self-medication with OTC analgesics is very common, especially in the elderly. Patients should be encouraged to keep a record of these medications and the frequency of use so that physicians can identify possible incompatibilities among their medications. As a general rule, patients should not be given OTC medications except under a physician's order or under standing orders. At home, children should be given analgesics other than Tylenol only under the direct instruction of a physician. Even Tylenol may be hazardous if recommended dosages are not followed.

Sedatives and Tranquilizers

Sedatives or tranquilizers exert a quieting effect, often inducing sleep. They are not analgesics but may provide relief from pain by promoting muscle relaxation. Tranquilizers reduce anxiety and mental tension more effectively than sedatives and often provide some sedation as well. At low doses tranquilizers do not impair mental acuity, but as the dosage increases patients tend to feel drowsy and speech may become slow and slurred. Some patients experience a brief loss of inhibition, similar to the effect of alcohol, which causes them to talk and act inappropriately. Individuals taking tranquilizers may have slowed reaction times and thus should not drive or operate machinery. Diazepam (Valium) is a tranquilizer commonly prescribed as premedication for various diagnostic and therapeutic procedures.

Phenobarbital and other barbiturates are sedatives and were formerly used as preoperative medications. Their use for this purpose has been largely supplanted by Valium. Phenobarbital is still used with other medications to treat patients with seizures. Both Valium and midazolam (Versed) have a tranquilizing effect and may be given with morphine to patients who are highly anxious and uncomfortable. The premedication dosage of Valium for anxiety ranges from 2 to 20 mg given IM or IV. The physician may administer larger doses as needed to achieve relaxation. Very large doses are sometimes given to control grand mal seizures. When Valium is injected IV, it is administered slowly, taking at least 1 minute for each 5 mg (1 ml) given. The small veins of the hand and wrist should not be used because Valium is irritating to blood vessels and may cause phlebitis and damage to the vein. **Extravasation** of Valium (infiltration into surrounding tissues) can be painful and cause irritation and swelling.

Versed is sometimes given when a previous administration of an analgesic or Valium has not achieved the desired state of pain relief and relaxation. The initial dose of Versed in this instance is 1 mg and can be increased to a usual maximum of 10 mg at the physician's discretion.

Antagonists

The most common **antagonists** encountered by radiographers are those formulated to counteract the effects of sedatives and analgesics.

Valium and Versed are benzodiazepine drugs that may be given to produce relaxation and/or sedation, as previously described. An overdose produces respiratory depression and loss of psychomotor function as a toxic (poisonous) effect. Flumazenil (Mazicon) is a medication developed to counteract the effect of these drugs. Mazicon can antagonize the sedation and reverse the impairment of recall and psychomotor function produced by benzodiazepines. Patients who receive Mazicon should be monitored for re-sedation, respiratory depression, and other residual effects for up to 2 hours, based on the dosage and duration of the benzodiazepine used. For the reversal of conscious sedation by benzodiazepines, the recommended initial dose of Mazicon is 0.2 mg (2 ml) administered IV over 15 seconds. If the desired level of consciousness is not obtained after waiting an additional 45 seconds, additional doses may be given until the desired effect is achieved, to a maximum of 1 mg (10 ml). To minimize the possibility of pain or inflammation, Mazicon should be administered through a freely flowing IV line into a large vein. The use of Mazicon is known to be associated with seizures in some patients who have been taking benzodiazepines over a long period.

Naloxone (Narcan) counteracts the effects of opiates such as morphine and prevents or reverses respiratory depression, sedation, and hypotension. A rapid reversal of opiate depression can cause nausea, vomiting, tachycardia, and nervousness. Although Narcan can be administered SC or IM, the most rapid onset of action is obtained with a dilution of Narcan in saline or 5% dextrose in water administered IV. To reverse respiratory depression, 0.1 to 0.2 mg of Narcan should be administered IV at 2- to 3-minute intervals until adequate ventilation is achieved.

Local Anesthetics

Lidocaine (Xylocaine) is a local anesthetic used to eliminate sensation in a specific area before a painful procedure. You may have received such an injection before having dental work or when having stitches placed to close a wound. Xylocaine is provided in a variety of strengths and is available both with or without epinephrine. The addition of epinephrine causes constriction of adjacent blood vessels, preventing bleeding and localizing the anesthetic effect to the immediate area. If your facility stocks more than one type, be sure you understand clearly which one the physician requires.

MEDICATION PROPERTIES

Drugs differ in the ways they enter the body, are absorbed, reach their site of action, are metabolized, and exit the body. The study of these properties is called *pharmacokinetics.* These processes are important since they affect the ways in which individuals respond to medications. Individual response can vary greatly depending on age, physical condition, sex, weight, or immune status. Response may also depend on the body's state of **hydration,** that is, the water content of the tissues. Insufficient fluid intake causes a state of **dehydration,** which has a negative effect on the body's general state and may alter the effects of medications.

Absorption

Absorption is the process by which the drug enters the systemic circulation in order to provide a desired effect. The method of absorption varies with the route of administration. Medications may be absorbed through the walls of the stomach or intestine; through the skin, fat, or muscle layers of the body; or through mucous membranes.

Distribution

Distribution is the means by which drugs travel from the site of absorption to the site of action, usually through the bloodstream. This process depends on an adequate circulation. Drugs act most rapidly in organs that have an abundant blood supply such as the liver, heart, brain, and kidneys.

Metabolism

Metabolism is the process by which the body transforms drugs into an inactive form that can be excreted from the body. Most drug metabolism occurs in the liver, where enzymes transform drugs into substances that can be excreted by way of the intestinal tract or the kidneys.

Excretion

Excretion refers to the elimination of drugs from the body after they have been metabolized. Drugs may be excreted by way of the kidneys, intestines, lungs, or exocrine glands such as the sweat glands. The kidneys are the chief organs of excretion, but the route depends largely on the chemical makeup of the drug.

Portions of some drugs may escape metabolism and be excreted unchanged in the urine. Volatile substances such as alcohol and certain anesthetics are excreted through the lungs. For this reason, postoperative patients are encouraged to cough and breathe deeply to help clear their bodies of the anesthetic agent. Other drugs are metabolized in the liver, excreted into the bile, and then routed through the intestines for elimination. Some medications are metabolized in the liver and their metabolites (substances produced in the process of metabolism) are transported by the bloodstream to the kidneys for excretion. If kidney function is impaired or if the patient is dehydrated, drugs can be retained in the body and a toxic effect can occur. Therefore adequate fluid intake is very important for most patients on medication.

MEDICATION EFFECTS

The study of the effects of drugs on the normal physiologic functions of the body is called *pharmacodynamics.* After a drug reaches the site of action, it exerts specific effects on the cells. The effect of a drug on specific cells is called the *therapeutic action.* Therapeutic action results in anticipated outcomes such as diuresis (fluid elimination), increased cardiac output, or relief from pain.

The most common mechanism of drug action is the binding of drugs to receptor sites on a cell. The drugs and receptors fit together much like the fit of a lock and key. When receptors and drugs lock together, the therapeutic effects occur. Each cell in the body contains specific, unique receptors. For example, Benadryl blocks the receptor sites of histamine cells and reduces the itching and swelling caused by an allergic reaction.

A drug that produces such a specific action and promotes the desired result is referred to as an **agonist.** Analgesics, for example, are agonists. A drug which attaches itself to the receptor, preventing the agonist from acting, is called an antagonist. Common antagonists include those used to reverse the effects of opioids or benzodiazepines as discussed in the previous section.

Drugs are administered to produce a predictable physiologic response, the therapeutic effect. In addition to the desired outcome, side effects may occur that may or may not be harmless. If the side effects are severe enough to outweigh the benefits of the medication, the physician may choose to discontinue the drug.

Knowing the effects of common medications helps you evaluate changes in the condition of patients in your care. Reference to the medication record in the patient's chart may help you determine whether a change in status is caused by medication or by deterioration in the patient's condition. For example, an anticholinergic medication such as atropine may cause a dry mouth. This is a specific side effect that has nothing to do with the patient's state of hydration. Opiates may slow the respiratory rate, and vasodilators may cause the blood pressure to drop. Such effects are the usual consequence of the specific medication and are taken into account when the drug is prescribed.

Adverse side effects are effects that are not expected as a usual consequence of prescribed medications. They may range from mild nausea, flushing, or diarrhea to critical situations, including cardiac arrest or other life-threatening states. An allergic reaction is an adverse side effect that occurs when a patient has been sensitized to the initial dose of a medication and develops an allergic response to the **allergen** (allergy-causing substance) and related drugs. Drug allergies may be slight or severe, and the extent of the reaction is unpredictable. Urticaria, respiratory distress, or abrupt changes in blood pressures are all symptoms demanding a physician's immediate intervention. Chapter 22 provides information on appropriate responses to acute allergic reactions and other changes in patient condition that may occur as adverse effects of medications.

Toxic effects are poisonous consequences that develop when a drug accumulates in the body because of inadequate excretion, impaired metabolism, overdose, or sensitivity to the drug. Elderly patients are more likely to have poor heart, kidney, or liver function. These conditions increase the possibility that toxic effects might occur. A specific drug that treats a toxic effect is called an **antidote.**

An **idiosyncratic** (unusual or peculiar) reaction occurs when a patient over- or under-reacts to a drug or has an unusual reaction. For example, when phenobarbital (a sedative) is administered, some individuals become very agitated rather than sedated.

There are many occasions when drugs taken together have a **synergistic** (additive) effect that may go far beyond the desired outcome. For example, a patient who is taking a prescribed medication for high blood pressure and then takes an OTC diuretic may become hypotensive and feel weak and faint. Since many drugs interact when taken together, the physician who orders a new drug should have the patient's chart and should question the patient about taking OTC medications or drugs prescribed by a different physician.

MEDICATION ADMINISTRATION

The information in this section provides a basis for assisting the physician in medication administration but is not intended as a substitute for directions from the physician.

In preparation for medication administration, the first step is to check the order. If you have questions about dosage or method of administration, check with the physician or your supervisor. Second, make certain that you have the correct medication in the correct strength and check the expiration date. The medication label indicates not only the name of the medication and the strength (amount per ml), but also gives an expiration date past which the drug should not be used. Third, check for allergies in the chart or obtain an allergy history from the patient. Last but not least, remember to wash your hands before preparing the medication.

A common rule of thumb is to read the label three times: first when selecting the container, second while preparing the dose, and a third time just before administration. This is essential to be absolutely certain that you have the correct drug and the proper strength. Be sure to keep the container until the medication is charted.

When medications are not used frequently (as in an emergency kit), they should be checked often and out-of-date supplies discarded and replaced. When preparing medications, the memory device in Box 23-1 will help you to avoid errors.

Dosage

The metric system is usually used for measuring medications. Liquids are measured in units from liters (L, slightly more than a quart) down to milliliters (ml), which are thousandths of a liter. One milliliter is equal to 1 cubic centimeter (cc), and 1 ounce equals 30 ml.

Since liquid agents are often diluted for use, the strength is expressed as a ratio of the amount of the drug to the total volume of solution. For example, 1:1000 indicates a dilution of one part drug to 1000 parts of water or other solvent.

If the active ingredient is a solid, it is measured by weight in grams (g), milligrams (mg), or micrograms (mcg or µg). The strength of solids dissolved in a liquid is designated in terms of weight per volume, often

---

BOX 23-1

Five Rights of Medication Administration

- The right dose
- Of the right medication
- To the right patient
- At the right time
- By the right route

mg/ml. You will often need to determine how much liquid will provide a given dose of a solid:

$$\frac{\text{Dose (Desired quantity of solid)}}{\text{Strength (Quantity of solid per volume of liquid)}} = \text{Volume (Required quantity of liquid)}$$

Example: If the drug is supplied in a strength of 4 mg/ml, and you want to administer 10 mg, you will need 2.5 ml:

$$\frac{10 \text{ mg}}{4 \text{ mg/ml}} = 2.5 \text{ ml}$$

Conversely, you may need to know how much of a solid is delivered in a given volume of liquid:

$$\text{Strength} \times \text{Volume} = \text{Dose}$$

Example: If 2 ml of solution is given and the strength is 4 mg/ml, the dose is 8 mg:

$$4 \text{ mg/ml} \times 2 \text{ ml} = 8 \text{ mg}$$

Practice these calculations so that you can do them quickly without error whenever you are required to prepare a parenteral medication. Additional examples and practice problems are included in Chapter 26.

Oral Administration

The oral route of medication administration is the easiest and most familiar. The charting abbreviation for oral administration is PO for *per os,* by mouth. When medication is taken orally it must dissolve in the stomach and then pass into the small intestine, where the majority of the absorption takes place.

Oral medications are supplied in a variety of forms, including tablets, capsules, granules, and liquids. Some oral medications are irritating to the stomach and are provided in a coated form that allows tablets to pass through the stomach before dissolving. These medications should not be chewed or broken since this would negate the use of the enteric coating. Some tablets are chewable, but almost all medication should be swallowed with varying amounts of liquid, usually water. Liquid medications are usually taken with water as well. Granules are mixed with a specified amount of liquid. Follow the directions on the package insert.

One way to minimize errors in the administration of drugs is to establish a set routine and follow it unfailingly. The steps in Box 23-2 serve as a guide to establishing a procedure for the administration of oral medications.

Parenteral Equipment

Hypodermic needles are supplied in various diameters and lengths. The gauge of a needle indicates its diameter, and the gauge increases as the diameter of

---

BOX 23-2

Procedure for Oral Administration

- Wash your hands.
- Obtain the proper medication and read the label.
- Prepare the medication tray with a medicine cup and a glass of water (if appropriate). Read the label again.
- If the physician is present, show the physician the label.
- Pour the correct amount of medication directly into the medicine cup. When pouring liquids, hold the label against the palm of your hand so that it will stay clean and legible.
- If requested by the physician to administer the medication, check the patient's identification, stay with the patient while the medication is swallowed, and offer water if permitted.
- Return the tray and discard the remaining water and medicine cup.
- Wash your hands again.
- Chart the medication.
- Discard packaging from single-dose packages or return medication container to storage.

---

the bore decreases. An 18-gauge needle is larger around than a 22-gauge needle and delivers a given volume of fluid more rapidly. A 22-gauge needle can be used for much smaller veins since it makes a smaller hole and excessive bleeding or **hematoma** is less likely when it is removed. This size is also used for IM injections.

The length of hypodermic needles is measured in inches and may vary from ½ inch, used for accessing IV line ports and for intradermal injections, to 4½ inches, needed for intrathecal (spinal canal) injections. A 2½-inch length is typical for IV needles, and the usual gauge ranges from 18 to 22 for adults. Fig. 23-2 shows a hypodermic needle, together with other injection equipment that is discussed later in this section.

Disposable plastic syringes are supplied in individual sterile paper wraps, sometimes with a needle attached. Syringes consist of a barrel with a measurement scale on the outside and a plunger to force the contents through the needle. The tip of the plunger is usually made of latex.

In recent years an increasing number of individuals have become sensitized to latex. This has caused a return to the use of glass syringes for such patients. Glass syringes (Fig. 23-3) are not disposable and are sterilized before each use. Since the plunger is ground to fit the barrel precisely, both barrel and plunger are marked with code numbers that must match in order for the syringe to work. When you open the wrap you will need to assemble the two parts. To avoid contaminating the syringe, pick up the plunger by the handle only and insert it into the barrel. Needles are supplied separately. After selecting a needle suitable for the administration, twist the needle hub firmly onto

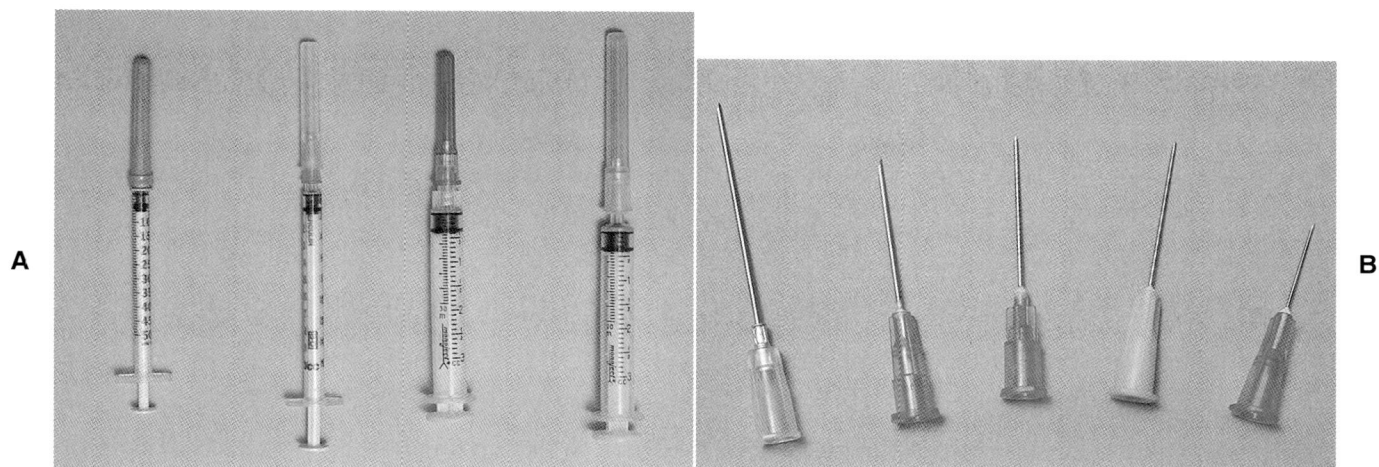

Fig. 23-2 Injection equipment. **A,** Syringes. Left to right: disposable 3-ml syringe for intramuscular injections, 3-ml hypodermic syringe for subcutaneous injection, tuberculin syringe, and insulin syringe. **B,** Hypodermic needles. Left to right: 19-gauge, 20-gauge, 21-gauge, 23-gauge, and 25-gauge.

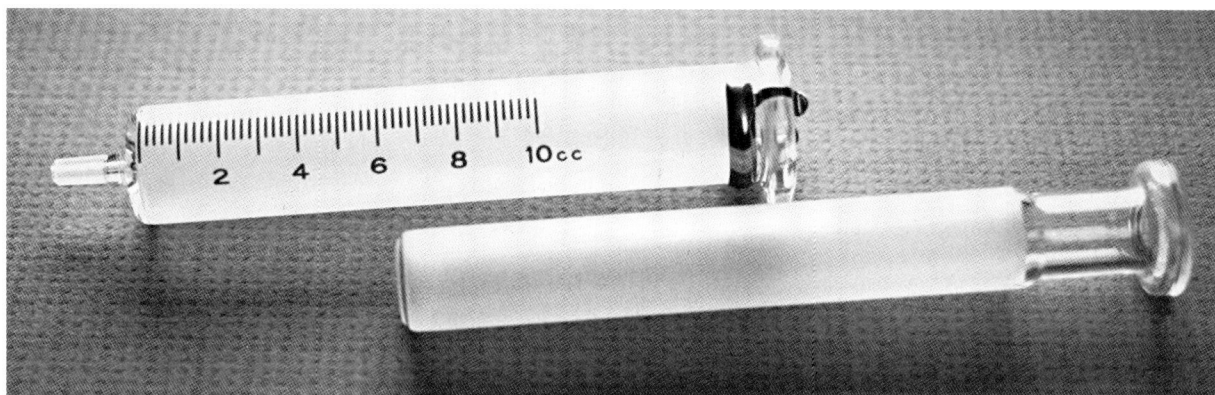

Fig. 23-3 Glass syringe.

the syringe and fill the syringe in the same manner as a disposable syringe. After use, discard the needle into the contaminated sharps container. The barrel and plunger must be separated and placed in a container with disinfectant until they can be sterilized.

Preparation for Injection

To prepare for parenteral injections, specifically intradermal, SC, and IM injections, assemble the proper syringe and needle and an alcohol wipe for cleansing the skin. Next, obtain the medication and read the label carefully. Draw up the medication into the syringe.

If a drug is supplied in ampule form, a small file is needed to nick the neck of the ampule. The top then snaps off easily. Use a 2 × 2 inch gauze sponge to protect your fingers when opening the ampule, since the glass may break unevenly and cut your hand (Fig. 23-4). Next, attach the needle to the syringe securely and remove the needle cover, taking care not to contaminate the needle. Place the needle tip in the

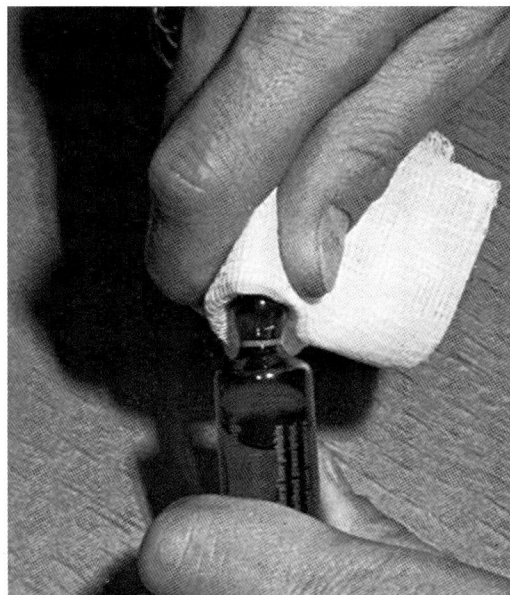

Fig. 23-4 Protect your hand when opening an ampule.

ampule below the solution level and withdraw the required amount of the medication into the syringe. Hold the syringe to the light to check for air bubbles. If bubbles appear, hold the syringe with the needle pointing up, and tap the side of the syringe. As the bubbles rise, they can be ejected with gentle pressure on the plunger. The removal of air is essential to accurate dosage measurement and may also affect patient safety. Now, read the label again. The ampule is retained until after the drug has been administered and charted. If the physician is present when the drug is administered, show the physician the ampule and the syringe while stating aloud what has been done. After administration, the medication is charted and the ampule, together with any remaining medication, is discarded.

If the medication is supplied in a vial, there are several variations in the preparation procedure. First, pull off the vial's protective cap, exposing the rubber stopper and taking care not to contaminate the underlying surface. Since this is a closed system, you must inject a volume of air equal to the amount of fluid you wish to remove. Remove the needle cover and pull down the plunger of the syringe to the desired reading. Insert the needle through the stopper and inject the air into the bottle. Invert the bottle and make sure the needle tip is below the fluid level. Then, pull down the plunger to the desired reading and check for bubbles. If there are any, dislodge them, inject them into the vial, and withdraw the plunger again until the dosage is correct (Fig. 23-5). Then, remove the needle from the vial, replace the needle cover, and proceed as previously described.

The vial was originally designed as a multiple-dose container, but in the past the incidence of contamination was so common that multiple use is now restricted to use for the same patient on the same day. Vials of local anesthetic are sometimes used repeatedly for the same patient during a radiographic procedure. When using a vial after the first time, clean the stopper with an alcohol wipe. (Alcohol is not needed for the first injection because the stopper is sterile on opening the vial.) When the procedure is completed, discard the vial and any remaining medication. Some vials are not meant for multiple use and are so marked. For medications dispensed in relatively small quantities (10 ml or less), vials have largely been replaced by preloaded syringes that are discarded after use.

Parenteral Injection Procedure

Injection techniques are practical skills that cannot be learned through study alone. You will want to observe these techniques performed competently and have an experienced observer present until you gain confidence.

The basic procedure for intradermal, SC, and IM injections is stated in Box 23-3. The loading of the syringe is the same for all injections and is illustrated in Fig. 23-5. The injection sites are also different for these three routes. These variations are summarized in Table 23-2.

Intradermal injections involve a shallow puncture between the layers of the skin. The anterior surface of the forearm is a typical site for intradermal injections. Only very small quantities may be injected intradermally. A tuberculin syringe is used; it is finely calibrated and comes with a very small (26-gauge) needle. The tuberculin skin test discussed in Chapter 21 involves an intradermal injection to the anterior forearm. This route is also sometimes used to test allergic sensitivity.

For SC injections, a 23- to 25-gauge needle with a ⅝-inch length is used. It is directed through the skin

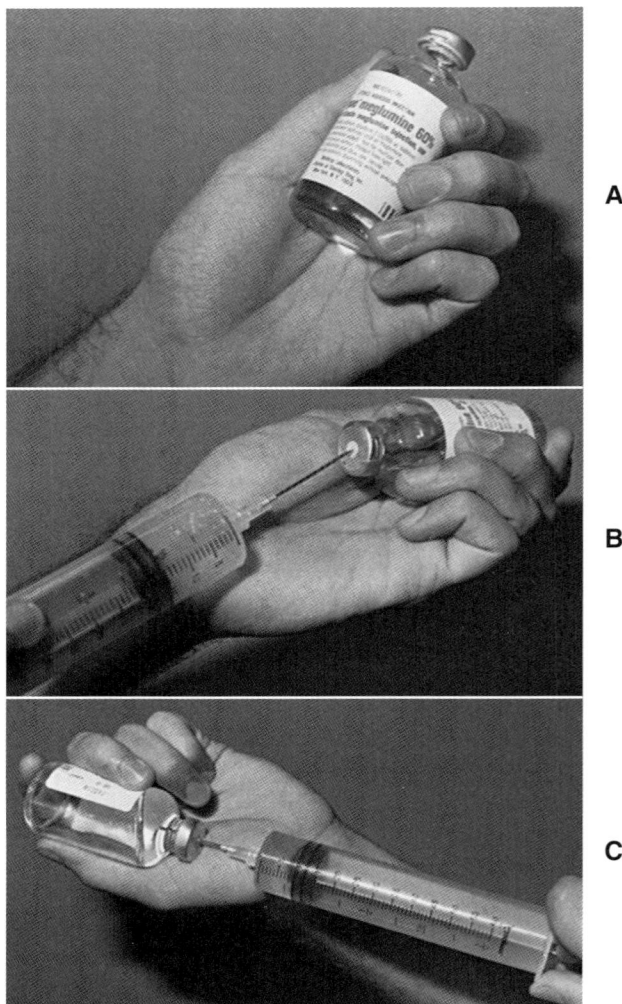

Fig. 23-5 Loading a syringe from a vial. **A,** Check the label for drug name, correct strength, and expiration date. **B,** Pull back on the plunger to the desired dose reading. Inject air into the air space in the vial. **C,** Tip vial downward to withdraw the solution.

at a 45° angle. Syringes used for SC injections are usually 2 ml or smaller in size, since it is painful to inject a large quantity beneath the skin. The most convenient areas for SC injections are on the upper arm and on the outer aspect of the thigh.

IM injections are sometimes given in larger amounts. The syringe may be larger than that used for SC injection (up to 5 ml) and the needle size is also larger, usually 22 gauge. The injection is given into the deltoid muscle of the upper arm (Fig. 23-6), the gluteal muscles in the hip area (Fig. 23-7), or the vastus lateralis muscle of the lateral thigh (Fig. 23-8). For children under 5 years of age, the vastus lateralis site is preferred to the gluteus site because the gluteus maximus muscle is not fully developed and there is risk of damage to the sciatic nerve. Injections are not usually given into the anterior thigh because this site

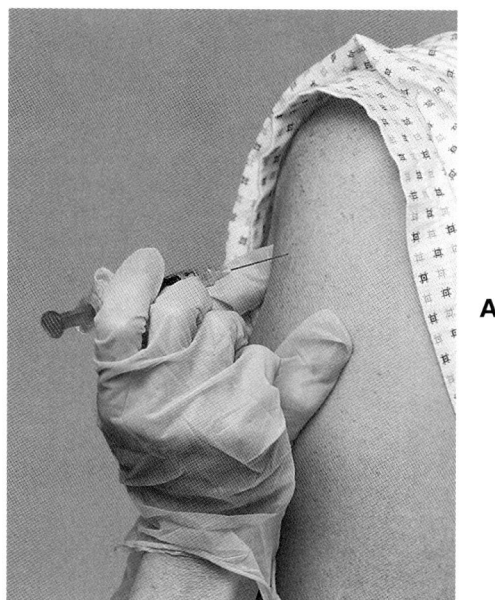

BOX 23-3

**Procedure for Parenteral Administration
(Intradermal, Subcutaneous,
and Intramuscular)**

- Explain to the patient what you are about to do.
- Don clean gloves.
- Select the appropriate site.
- Cleanse the selected area with an alcohol wipe.
- Hold the skin taut with your non-dominant hand.
- Insert the needle at the correct angle and pull back slightly on the plunger.
- If no blood is drawn into the syringe, inject the medication.
- Withdraw the needle quickly and wipe the injection site.
- Dispose of needle and syringe.
- See to the patient's comfort.
- Remove your gloves and wash your hands.
- Chart the medication.
- Dispose of the medication container.

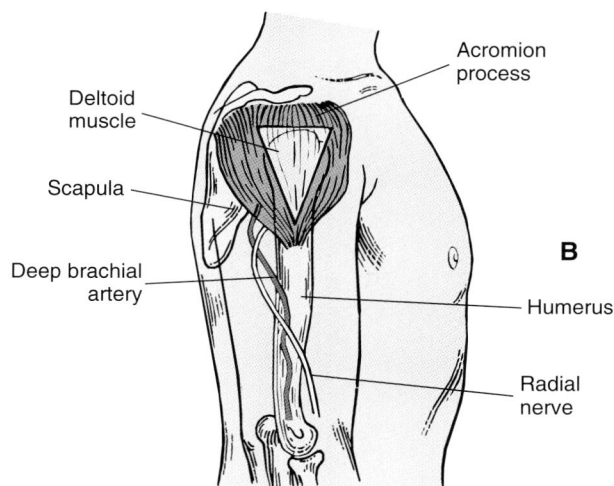

Fig. 23-6 A, Deltoid muscle site for IM injection. **B,** Anatomical view of deltoid muscle injection site.

TABLE 23-2

Parental Injection Summary

Route	Injection Volume	Needle Size	Injection Site(s)	Angle of Injection
Intradermal	<1 ml	⅝ in 26 gauge	Anterior forearm	15°
Subcutaneous (SC)	<2 ml	⅝ in 23-25 gauge	Upper arm, outer aspect of thigh	45°
Intramuscular (IM)	<5 ml	1 to 1½ in 22 gauge	Deltoid muscle, vastus lateralis muscle, gluteal muscle (over age 5)	90°

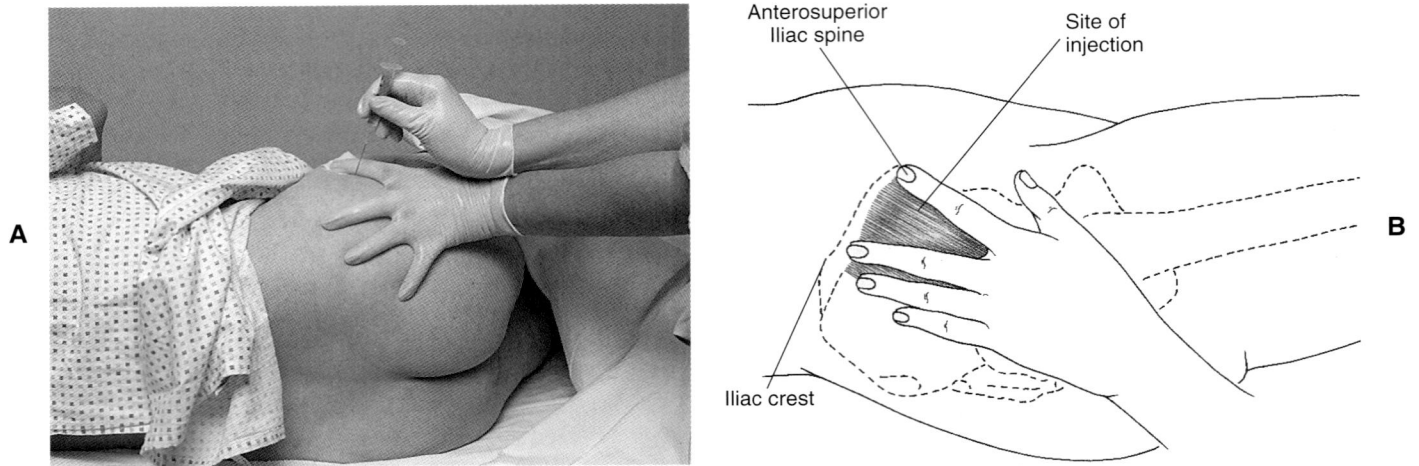

Fig. 23-7 Gluteal muscle site for IM injection. **A,** The injection site into the ventrogluteal muscle avoids major nerves and blood vessels. **B,** Anatomical view of ventrogluteal muscle injection site.

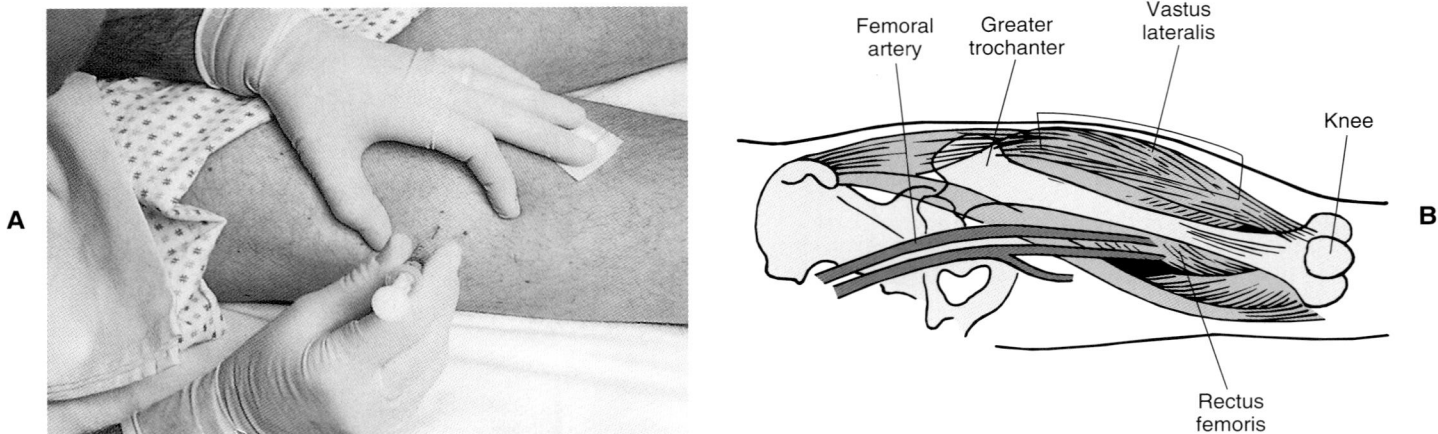

Fig. 23-8 Vastus lateralis muscle for IM injection. **A,** Injection site into the vastus lateralis muscle. **B,** Anatomical view of vastus lateralis muscle injection site.

is extremely painful and the discomfort may persist for several days.

Intravenous Route

IV fluids and medications are administered to meet specific needs. Patients respond rapidly to medication administration via this route. The IV injection is used for delivering most emergency medications when an immediate response is critical. The IV route may serve to transport medications, parenteral nutrition, or chemotherapy. Dehydrated patients may need fluid and electrolyte replacement by means of IV infusion. This is also the route used to inject contrast media for

radiographic examinations of the urinary tract and for some computed tomography (CT) studies, and to provide sedation during invasive procedures and magnetic resonance imaging (MRI) examinations.

As previously stated, starting IV lines and administering medication by the IV route is beyond the limited radiographer's scope of practice, but you may need to assist the physician with these procedures or monitor the infusion of IV fluids after an IV line has been established. If your work involves the IV administration of medication or contrast media, consult other resources that provide specific information on these products and the procedures and precautions for their use.

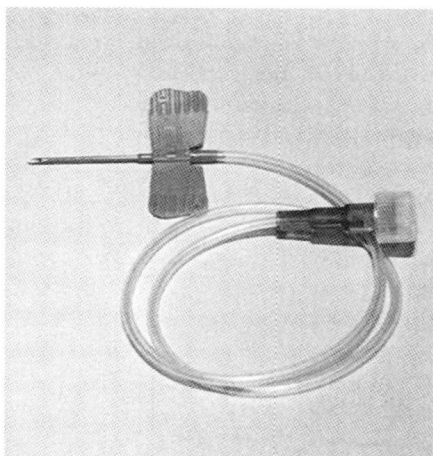

Fig. 23-9 Butterfly set.

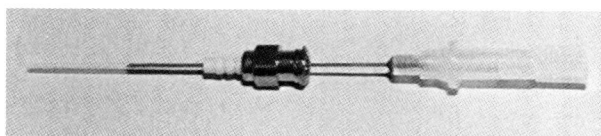

Fig. 23-10 IV catheter set.

Venipuncture may be accomplished with a hypodermic needle, a butterfly set, or an IV catheter. The use of hypodermic needles is generally restricted to phlebotomy for obtaining laboratory samples (Chapter 24) and for single, small injections. A butterfly set (Fig. 23-9) is preferable to a conventional hypodermic needle for most IV injections and is often used for direct injections with a syringe. This apparatus consists of a needle with plastic projections on either side that aid in holding the needle during venipuncture and may be taped to the patient's skin after the needle is in place. This prevents movement of the needle in the vein. Attached to the needle is a short length of tubing with a hub that connects to a syringe. The syringe is filled from a vial or ampule with a needle, as in preparing for other parenteral injections. The needle is then discarded and the syringe is attached to the butterfly tubing. Before the butterfly needle is inserted into the vein, the tubing is filled with liquid from the syringe to avoid injecting air into the vein.

IV catheters are used instead of needles or butterfly sets when repeated or continuous IV injections or infusions are administered. The IV catheter is a two-part system consisting of a needle that fits inside a flexible plastic catheter (Fig. 23-10). The catheter hub has wing-shaped plastic projections similar to the butterfly set. Once the flexible catheter is properly situated in the vein, it is connected to the syringe or supply system. Fluid can be administered by syringe or through tubing from a hanging bottle or bag. Medication can also be injected from a syringe into an injection port on the IV fluid tubing.

Large volumes of solution are intravenously administered by IV infusion. The most common replacement fluids are normal saline or a 5% solution of dextrose in water (D5W). These IV solutions are provided in bottles and plastic bags. They may be stocked with the emergency supplies, or may be stocked for general use if your facility performs procedures that require an IV line. If you are assisting with the starting or replacing of IV fluids, *be certain that the solution is correct.* Less common solutions, including some that contain medication, may be stocked for special purposes. Checking labels for IV fluids is just as important as checking medication labels.

The setup for IV fluid administration requires the bottle or bag of fluid, an infusion set, and an IV pole. The infusion set consists of a drip chamber that attaches to the bag or bottle with tubing that attaches to the needle or catheter. It includes an injection port for adding medication to the infusion.

The bags have a cap over the sterile port through which the drip chamber is inserted. The drip chamber is removed from its wrappings and inserted into the sterile port. Care must be taken not to contaminate either component or both must be discarded. Solutions supplied in bottles have a removable cap and sometimes a rubber diaphragm covering a rubber stopper. The cap is removed and the diaphragm is pulled off without touching the stopper. The drip chamber is then inserted through the stopper, with the practitioner taking care that the clamp on the IV tubing is closed. The bottle or bag is then inverted and hung on the IV pole. When it is in place, the cover at the other end of the IV tubing is removed, the clamp is opened, and the fluid is allowed to run into a basin until the tubing is free of bubbles. The clamp is then closed and the tip covered to keep it sterile. The procedure for setting up an intravenous infusion is illustrated in Fig. 23-11. The physician or nurse will attach the tubing to the needle or catheter and adjust the fluid flow rate.

Monitoring Infusions

If you are caring for a patient who is receiving an IV infusion, it is your duty to monitor the flow rate and the condition of the injection site.

The flow rate is measured in units of drops per minute. Note the flow rate when you first encounter the patient, and then check at regular intervals to ensure that this rate is maintained. Most patients easily tolerate 15 to 20 drops/minute from a standard IV set. At this rate the patient receives approximately 60 ml/hour. The drip rate is controlled by a clamp below

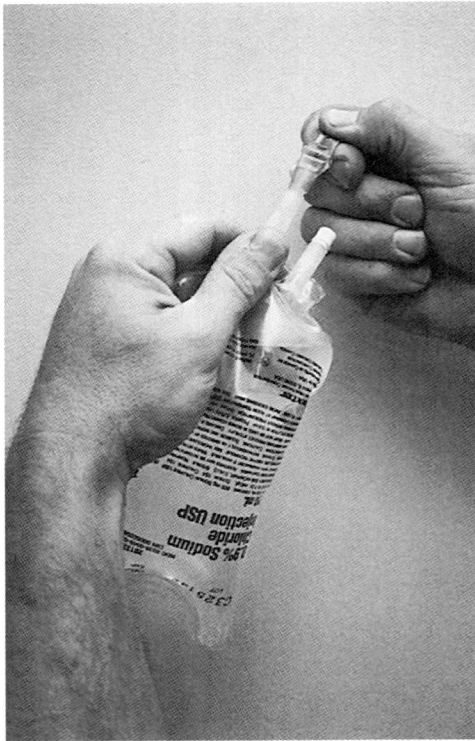

A, Remove protective cover from access port. Avoid contamination.

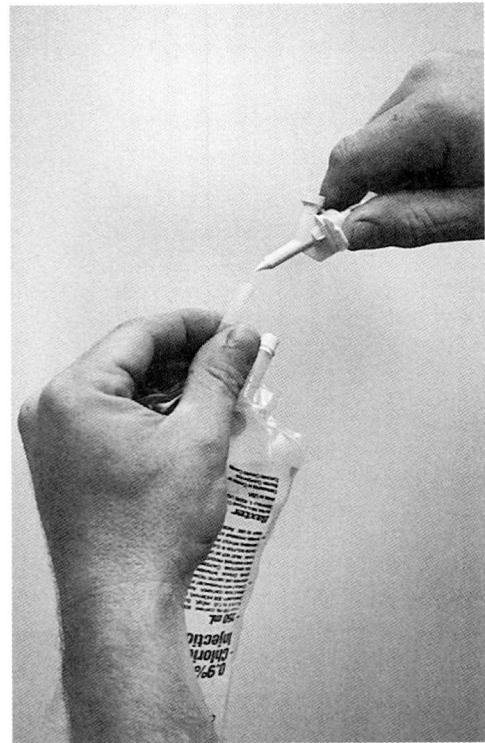

B, With tubing clamped off, insert drip chamber firmly into access port.

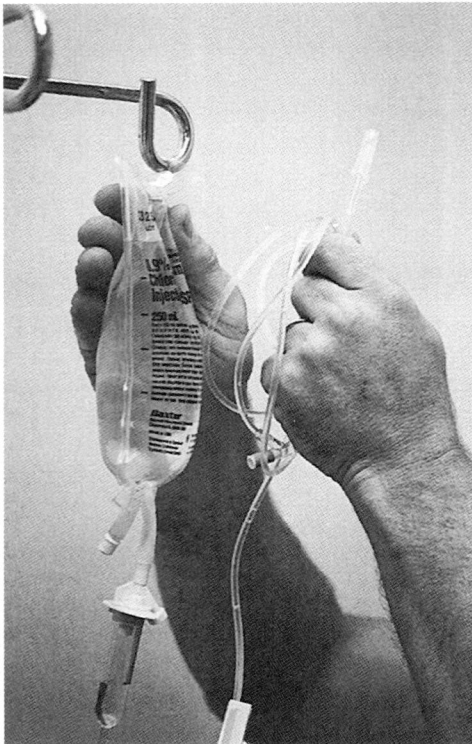

C, Insert bag or bottle and suspend from pole.

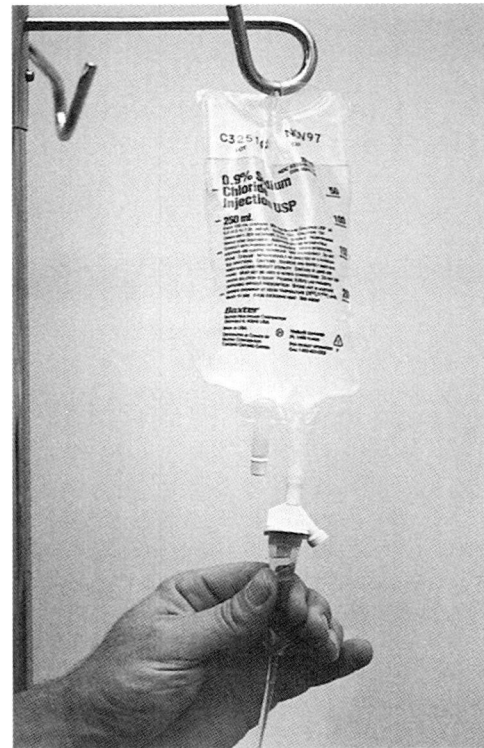

D, Pinch drip chamber to draw fluid into chamber. Fill chamber about half full. Unclamp to fill tubing and reclamp. Setup is ready for attachment to IV catheter.

Fig. 23-11 Intravenous infusion procedure.

the drip meter which can be opened or closed to control the rate of flow.

If an IV infusion runs too fast, a patient with a condition such as chronic obstructive pulmonary disease or congestive heart failure may receive more fluid than can be readily assimilated, causing fluid to accumulate in the lungs (pulmonary edema). Since an IV infusion may also contain medication, the patient could suffer a toxic effect or an overdose. On the other hand, too slow a flow rate might prevent effective treatment.

The height of the bottle or bag affects the flow rate and should always be 18 to 20 inches above the level of the vein. If the bottle is inadvertently placed lower than the vein, blood will flow back into the catheter or tubing and may clot, causing the fluid to stop flowing. This frequently necessitates restarting the IV line at a new site. On the other hand, an IV solution that is too high may cause fluid to infiltrate into the surrounding tissues because of the increased pressure.

Occasionally IV fluids or medications may leak or be accidentally injected into the tissues surrounding a vein. This extravasation, also called **infiltration,** may be both painful and dangerous. The patient is likely to complain of discomfort and you may observe swelling at the site. Remember to check the area around the injection site frequently. If it is cool, swollen, and boggy, the IV solution may have infiltrated.

When extravasation occurs, shut off the flow of fluid immediately and notify a nurse or the physician. The needle must be removed and the problem attended to before proceeding with an injection at another site. When the needle or catheter has been removed and the bleeding has stopped, a cold pack is applied to the affected area to help alleviate the pain. This will also cause constriction of the blood vessels in the area and help to keep the infiltration localized. If cold packs are not readily available, a terry towel can be wrapped around ice cubes, or ice in a plastic bag can be applied. A dry towel may be wrapped around the bag to hold it in place. Replace the cold pack with another as soon as it melts. Ice packs are applied to the extravasation site for 20 to 60 minutes until swelling is diminished. If the patient is allowed to leave before the swelling has completely disappeared, the physician may order continued cold pack treatments at home and instruct the patient to return or go to the emergency room if inflammation or discomfort persists. It is recommended that an incident report be completed for any extravasation involving a potentially irritating medication or contrast medium.

Previously, hot packs were recommended for the treatment of IV infiltration in the belief that the increased circulation that results with heat would result in the body absorbing the extravasated fluid more rapidly. Recent research confirms that the safest and most effective treatment is the cold pack.

Precautions for All Injections

Most needles and syringes are provided in sterile wraps, used once, and then destroyed. One common on-the-job injury in the past has been an accidental skin puncture by a contaminated needle. As discussed in Chapter 21, standard precautions are essential for your safety as well as that of the patient. For this reason use caution and apply the following rules when dealing with parenteral equipment:

- Wear gloves when performing injections or when dealing with any object contaminated by blood.
- Dispose of all syringes and needles directly into a puncture-proof container without recapping.

Several additional points of critical importance must be remembered in all medication administration involving injections:

- Always follow established rules of aseptic technique.
- Read the label three times: before drawing up the medication, after drawing it up, and with the physician before administration.
- Check patient identification before administration.
- Monitor the patient carefully for side effects.

CHARTING MEDICATIONS

When a medication is given by a physician or by a radiographer under the physician's supervision, it is always recorded in the patient's chart. The notation is made in the appropriate section of the chart and includes the time of day, the name of the drug, the dosage, and the route of administration. A typical entry in the medication record might read, "6/21/01, 10:50 AM, Benadryl, 50 mg, PO." Each entry must include the identification of the person who charted it. Initials alone are not considered to be adequate identification. If a medication record calls for initials, there is another place in the chart (often on the same page) where each set of initials is identified with the signer's full name. For legal reasons, the radiographer who charts medication must use the exact procedure established by the health care facility.

If an emergency prevents the charting of medications at the time they are given, make a written notation of the time, drug, and dosage so that accurate information will be available when the charting is completed. If this is not done, the pressure of the situation may lead to confusion of the facts, and the time, sequence, and dosage of several medications may be forgotten or charted incorrectly. Any medication prescribed or administered by a physician should be charted by the physician; if charted by the radiographer, it should be countersigned by the physician. The legal significance of complete accountability in such situations cannot be overemphasized.

Physicians enter an account of the procedure and any medications given in the progress notes. When involved in medication administration, the radiographer is responsible for checking that the drug, time, dose, and route of administration are noted and clearly expressed. This avoids the possibility of duplication or omission. With most charting systems, the radiographer may chart directly on progress notes. Some facilities have a medication administration sheet on which all medications are charted. Charting routines vary. Familiarize yourself with the routine of your facility.

SUMMARY

Depending on the facility and the job description, radiographers may be called upon to play an important role in the administration of medications. Knowledge of common medications and their proper administration will enable you to fulfill this role.

Medications are commonly administered by the oral, topical, or parenteral route, and each route requires specialized knowledge and equipment. IV access systems may also be used in your facility, depending on the procedures performed. The monitoring of these systems requires that the radiographer have a high degree of knowledge and awareness. Supervised clinical practice and familiarity with institutional procedures are required to implement this knowledge.

The charting of medication and the monitoring of patients are significant aspects of medication administration and must not be overlooked. The radiographer must recognize the potential harm and legal complications that could result from medication administration errors and strive for error-free performance.

▪ Review Questions ▪

1. Using the *Physicians' Desk Reference* or *Mosby's GenRx,* find two trade names for acetaminophen and the generic names for Valium and Benadryl.
2. List two pairs of medication names that might be confused because they look or sound alike.
3. Under what circumstances might medication be given parenterally rather than orally?
4. Which route of medication administration has the highest potential for life-threatening side effects? Why?
5. Under what circumstances might a radiographer accept a verbal order from a physician? How should the order be validated before the medication is administered?
6. State the "five rights" of medication administration.
7. A physician has ordered 8 mg of Valium IM. Your supply of this medication has a strength of 2 mg/ml. How much Valium should be drawn into the syringe to administer the correct dose?
8. Compare subcutaneous and intramuscular injections with respect to injection site, needle size, and angle of injection.
9. State the average rate of flow for IV fluids expressed in drops per minute.
10. Describe what is meant by "extravasation." How would you recognize it? What would you do?
11. Write the record of medication administration that should be charted for the Valium order from question 7.

24 *Medical Laboratory Skills*

Learning Objectives

At the conclusion of this chapter, the student will be able to:

- Demonstrate general knowledge of laboratory safety (Standard Precautions)
- Identify basic venipuncture equipment, state its purpose, and demonstrate correct use
- Provide appropriate patient instructions for venipuncture
- Identify an appropriate puncture site for venipuncture
- Demonstrate correct procedures for handling blood specimens
- Dispose of contaminated items correctly and safely
- Solve common problems encountered when collecting blood specimens
- Identify basic equipment for urinalysis, state its purpose, and demonstrate correct use
- Provide appropriate patient instructions for urine specimen collection
- Explain correct timing and technique of specimen collection
- Demonstrate correct step-by-step performance of urinalysis
- Accurately record urinalysis test results
- Demonstrate proper disposal of urine and contaminated items
- Solve common problems encountered with urinalysis

Key Terms

antecubital fossa
biohazardous waste
clean-catch midstream specimen
glans penis
reagent strip

sharps
tourniquet
urethral meatus
urinalysis
venipuncture

Radiographers may be expected to be able to perform basic laboratory procedures in addition to their work in the radiology department, especially if they work in a clinic or medical office. The purpose of this chapter is to provide basic information regarding the skills of performing routine **venipuncture** and **reagent strip** (dipstick) **urinalysis.** On-the-job instruction and supervision is necessary, followed by hours of practice, to become proficient in these skills.

STANDARD PRECAUTIONS

Standard precautions were developed to protect the health care worker from infection with blood-borne pathogens such as hepatitis B virus (HBV), hepatitis C virus (HCV), and human immunodeficiency virus (HIV). The essence of standard precautions is embodied in the statement that *all patients' body fluids are potentially infectious.* More details are available in Chapter 21. The essentials of standard precautions, as they relate to venipuncture and urinalysis, include hand washing, barrier techniques, and proper disposal of biohazardous waste, which is waste that is hazardous to living organisms.

Hand Washing

Hands should always be washed before and after all patient care. Antibacterial foam can be used if soap and water are not available.

Barrier Techniques

Protective gloves, most often latex, should be worn whenever handling body fluids or touching mucous membranes, open wounds, and contaminated skin, equipment, or counter surfaces. Vinyl gloves may be used if the worker or patient has a latex allergy. Liquid-resistant gowns or laboratory coats with knit cuffs are required if there is a risk of splashing. These should then be properly laundered. Disposable gowns are also available. Masks, protective eyewear, and face shields are needed if splashing is anticipated.

Biohazardous Waste Disposal

Biohazardous waste is any refuse that is poisonous or dangerous to living creatures. **Sharps** (anything that can puncture the skin such as needles, glass tubes, glass slides, and finger lancets) must be disposed into an appropriate sharps container (Fig. 24-1). Non-sharps (items that cannot puncture the skin such as a gauze sponge) must be disposed into a biohazard bag. When full, the containers and bags should be autoclaved or taken away by a commer-

Fig. 24-1 Various types of sharps containers.

cial biohazard waste company. Be sure not to fill beyond the line indicated by the manufacturer.

VENIPUNCTURE

Venipuncture is the most common method of blood collection. Any vein can be used for venipuncture. However, the veins in the **antecubital fossa** (the front side of the elbow) are most commonly used and, therefore, are discussed in this chapter.

Equipment

Evacuated Blood Collection Tubes The evacuated tube system (Fig. 24-2) is most commonly used in routine venipuncture. Tubes are either glass or plastic. They come in a variety of sizes to accommodate the amount of specimen required. Specimen volumes commonly range from 2 ml to 15 ml. The tube stoppers are either entirely rubber or rubber inserted into a protective plastic cap. The stoppers are color-coded, indicating the presence or absence of a specific additive. The colors of the stoppers are universal for all manufacturers. However, a tube with the protective plastic cap may have a different color than its counterpart with a plain rubber stopper. Tubes with additives must be gently inverted after filling to ensure adequate mixing. Tubes without additives must be filled before tubes with additives. The vacuum in the tube is expended once the stopper is punctured. Therefore a tube cannot be used a second time following an unsuccessful venipuncture. Tubes are marked with expiration dates. Outdated tubes must not be used. The laboratory facility can provide information regarding tubes and specimen requirements. Table 24-1 summarizes the most commonly used stopper colors, additives, types of specimens, and associated tests.

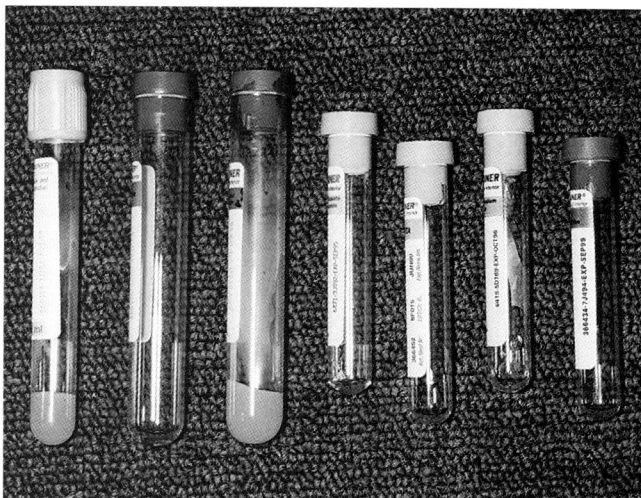

Fig. 24-2 Evacuated blood collection tubes.

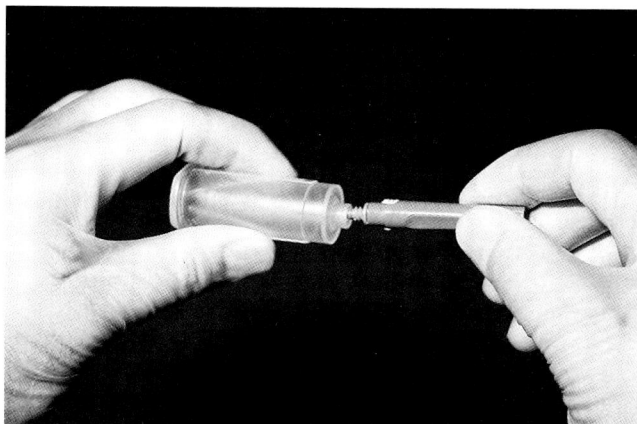

Fig. 24-3 Screwing a venipuncture needle into a needle holder.

TABLE 24-1
Commonly Used Evacuated Tube Guide

Stopper Color	Additive	Type of Specimen	Common Use
Red	None	Serum	Chemistry, serology
Red/gray mottled	Clot activator	Serum	Chemistry, serology
Gold	Clot activator	Serum	Chemistry, serology
Green*	Heparin	Plasma	Chemistry
Gray*	Sodium fluoride	Serum	Glucose
Lavender*	EDTA	Whole blood	Hematology
Light blue*	Sodium citrate	Plasma	Coagulation

Italics indicates tube with plastic protective cap.
*Same color for plain rubber stopper and plastic protective cap.

Needles Standard venipuncture needles are 21-gauge and either 1 or 1½ inches long. One-inch-long needles are adequate for most patients. Smaller gauge needles are available for patients with thinner than average veins. Needles are sterile and can be used only once. They are packaged in protective plastic containers and are sealed with a paper seal. A needle should be discarded if the seal appears broken. A standard venipuncture needle is actually two needles attached to a threaded plastic hub. The threaded hub screws into the needle holder. The needle mounted to the threaded hub is designed to puncture the stopper of the evacuated tube. The other needle mounted to the non-threaded end of the hub is for puncturing the skin.

Needle Holder The needle holder is a reusable plastic holder which is sometimes referred to as a *barrel.* Holders come in different sizes to accommodate dif-

ferent size tubes. As described previously, the needle is screwed into the holder and gently tightened without excessive force (Fig. 24-3). Holders do not have to be sterile but should be kept clean. A freshly prepared 10% bleach solution can be used to clean blood from a holder. Some facilities soak the holders overnight in this solution.

Tourniquet A **tourniquet** is used to facilitate distention of the vein. Types of tourniquets include simple rubber straps, 1-inch-wide latex strips, or Penrose tubing. These tourniquets must be applied with a special slip loop for easy one-handed release. Latex tourniquets should not be used on patients with a latex allergy. Rubber strap tourniquets with Velcro closures are also available. For venipuncture in the antecubital fossa, the tourniquet is applied approximately a hand-width above the crease. The tourniquet should be applied tight enough to cause distention of the veins without causing the patient discomfort or turning the skin white around the tourniquet. Additional information about tourniquet application and removal will be provided later in this chapter.

Alcohol Preps   Alcohol preps are used to cleanse the venipuncture site. Routine venipuncture is not a sterile procedure. Therefore 70% isopropyl alcohol preps are adequate. Betadine or povidone preps must be used if blood is being collected for blood cultures or for blood alcohol testing. Cleansing should begin at the proposed puncture site and move outward in a circular fashion. Once cleansed, the puncture site should not be touched again.

Gauze or Cotton Balls   Once the puncture site is cleansed, the alcohol should be allowed to completely dry, or it can be wiped dry with a gauze sponge or cotton ball. Residual alcohol on the skin can cause discomfort during the puncture. Alcohol can also enter the needle and can lead to hemolysis (rupture of red blood cells). The gauze or cotton ball can also be used to cover the puncture site after the needle is removed from the arm following completion of the venipuncture.

Bandage   A plastic strip bandage is used to cover the puncture. Alternatively, the gauze or cotton ball can be used as a bandage by taping it in place. Care must be taken in case a patient has an allergy to tape. A pressure bandage may be needed if the patient is taking anticoagulant medications.

Biohazard Waste Receptacles   The used needle is disposed into a sharps container. There are various types of sharps containers (see Fig. 24-1). Some automatically unscrew the needle from the holder. Others require manual unscrewing. However, they are all designed to receive the needle with a minimum amount of handling. Non-sharps contaminated items are disposed into a biohazard bag.

Splash Shield   A splash shield is required when uncapping filled evacuated tubes. Even when fully filled, a small amount of vacuum remains in the tube. An aerosol of the blood is created as the cap is removed and air rushes into the tube. Therefore a protective barrier must be placed between the tube and the face when the tube is uncapped. Most commonly a simple plastic shield or a plastic face mask is used. As mentioned earlier, some manufacturers now produce evacuated tubes with a stopper covered by a plastic cap. The plastic cap acts to minimize aerosols produced, but does not obviate the need for a shield.

Patient and Equipment Preparation

Appropriately identify the patient. Providing a basic explanation of the procedure often helps the patient to remain calm. All foreign materials in the mouth such as candy or chewing gum are removed. Most ambulatory patients comfortably tolerate routine venipuncture in a seated position. Standard phlebotomy chairs are best for this purpose. The patient should lie supine if fainting or lightheadedness was experienced during previous venipunctures. Wash hands and apply protective gloves. Examine both arms. Do not perform venipuncture above the site where IV fluids are being infused, or if excess scarring is evident near the proposed puncture site, or on the side of a mastectomy. Patients often know which arm has the best veins. The arm should be fully extended at the elbow and supported. The tourniquet is applied, and the veins palpated.

In the antecubital fossa the median cubital vein is most commonly used. However, the cephalic or basilic veins also can be used (Fig. 24-4). Choice of a vein is based upon palpation, not visualization. Veins palpate with a spongy feel. They do not pulsate. Tendons are hard. Locate the vein best anchored within the subcutaneous tissues. Palpation assesses the depth and course of the vein.

Venipuncture is performed from an ergonomically safe standing position. All of the equipment is organized within easy reach. Open the needle by breaking the paper seal with a twist, and pull it apart. Screw the

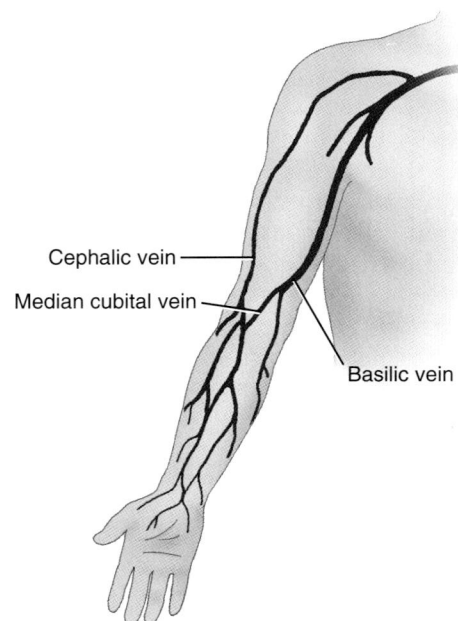

Cephalic vein
Median cubital vein
Basilic vein

Fig. 24-4 Antecubital fossa and veins.

needle into the holder. Discard the empty cap. The first tube can be placed into the holder, but do not engage the stopper onto the internal needle.

Step-by-Step Venipuncture Procedure

With hands washed and protective gloves applied, perform the following steps:

1. Apply the tourniquet; it should not be applied for more than 1 minute, in which case it should be released for at least 3 minutes before continuing.
2. With the arm extended at the elbow, have the patient clench the fist.
3. Palpate the veins and determine the puncture site.
4. Cleanse with alcohol and dry; do not touch the site after cleansing.
5. Grasp the holder with the thumb on top and the second and third fingers below.
6. Uncap the needle by pulling the cap straight off without twisting; the beveled edge of the needle should point upward.
7. With the other hand, anchor the vein by placing the second finger above and the thumb below the puncture site, stretch the skin, and apply gentle downward pressure (Fig. 24-5).
8. Tell the patient to expect a "stick"; allow the patient to watch if desired.
9. Puncture the skin at approximately a 40° angle; use a shallower angle if the vein is superficial.
10. Thread approximately half of the needle into the vein.
11. Anchor the holder and needle in place by applying gentle downward pressure; the fifth finger can be wrapped around the elbow.
12. Engage the first tube onto the internal needle, taking care to not allow the needle in the arm to move; the tube should begin to fill (Fig. 24-6).
13. Have the patient relax the fist, and remove the tourniquet; the tube will fill until the vacuum is expended.
14. Remove the filled tube, again taking care to not allow the needle in the arm to move. If the tube contains an additive, it can be gently inverted at this time or as the next tube fills.
15. Fill additional tubes (if required) in the same manner. (Remember: fill tubes without additives first, followed by tubes with additives.)
16. Remove the last tube from the holder before removing the needle from the arm.
17. Quickly withdraw the needle from the arm, immediately place gauze or a cotton ball over the puncture site, and apply pressure. Do not allow the patient to bend the arm; the patient can be asked to apply the pressure.
18. Immediately dispose of the needle in the appropriate sharps container. WARNING: *Never* recap needles.
19. Label the tubes according to the standards of the laboratory.
20. After 5 minutes, check the puncture site to ensure bleeding has stopped.
21. Apply a bandage. Provide appropriate instructions should the puncture site start to bleed after the patient is dismissed.
22. Dismiss the patient.
23. Appropriately dispose of biohazardous material.
24. Clean the work area.
25. Remove gloves and wash hands.
26. Initiate transport of the specimens to the laboratory.

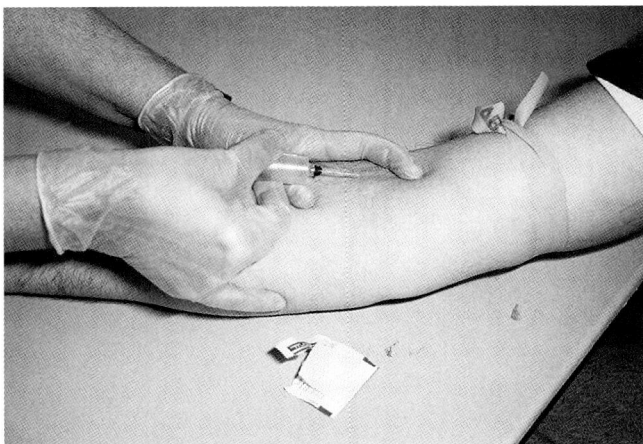

Fig. 24-5 Vein anchored with needle about to puncture.

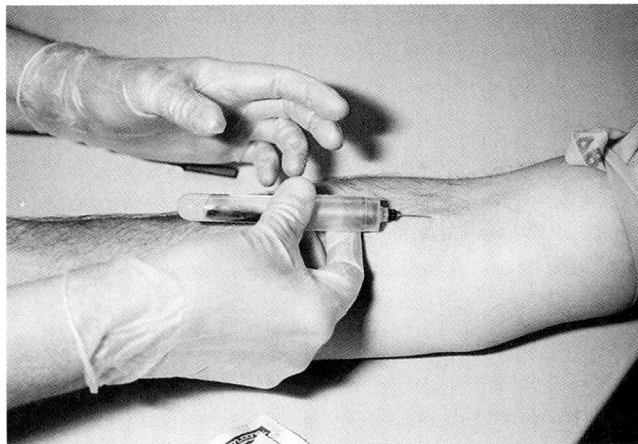

Fig. 24-6 Engaged evacuated tube filling with blood.

Common Problems

Tube Not Filling with Blood A tube not filling with blood means the needle is not in the lumen (channel) of the vein. The needle can be under the skin but not in the vein. The needle may have gone too deep and punctured all the way through the vein. Most commonly the needle lies on either side of the vein under the skin. Gently palpate the position of the tip of the needle. The vein can be anchored again, and the needle redirected into the vein. Gently withdraw the needle slightly if it is determined the initial puncture went too deep. This skill takes practice and experience. Always be sensitive to the patient. Some individuals will not tolerate this procedure. At times the venipuncture needs to be discontinued and attempted again on the other arm.

Patient Feeling Faint Some patients become faint during a routine veni-puncture. Their complexion may turn pale and appear slightly sweaty. At this point they may not be able to respond to verbal questions. If the patient is seated, it is best to discontinue the procedure. Allow the patient to lie down. Once the patient has recovered, retry the procedure, preferably using the other arm with the patient lying supine. If the faint patient was initially supine, the procedure may be attempted again. With experience one can judge whether continuing the procedure is safe for the patient.

URINALYSIS

Urinalysis is the physical, microscopic, or chemical examination of urine. A routine urinalysis consists of three components: (1) macroscopic examination of physical characteristics, (2) chemical analysis performed with a urine reagent strip, and (3) microscopic examination of the urine sediment.

Equipment

Reagent Strips Urine reagent strips (Fig. 24-7) are used to perform from one to ten tests. Small pads of an absorbent material are impregnated with chemical reagents and attached to plastic strips. Once dipped into urine, a pad will change color. The intensity of the color is proportionate to the concentration of the analyte being tested. Reagent strips have outdates and should be discarded if they are too old. To protect the strips from moisture and light, the bottle should be capped immediately after a strip is removed. Also, the bottle should be stored at room temperature. A radiographer gaining initial experience with reagent strips would benefit by thoroughly reading the manufacturer's information provided with each new bottle. Quality assurance using commercially prepared urine controls should

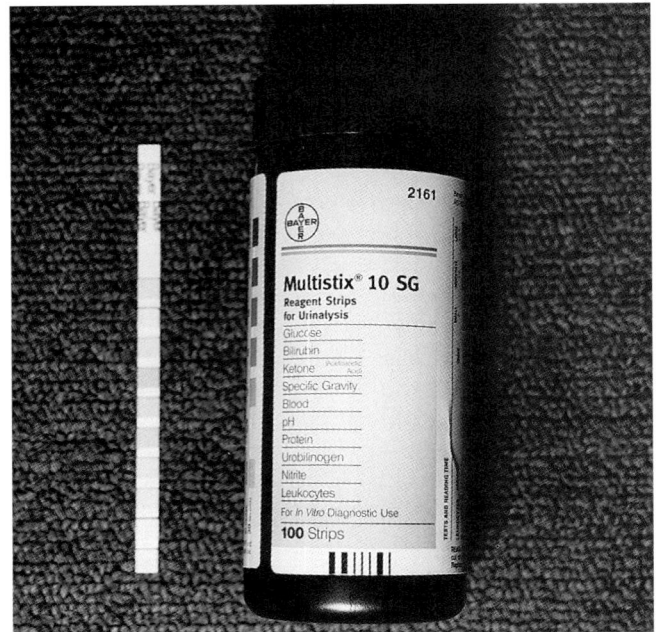

Fig. 24-7 Multiple test urine reagent strips.

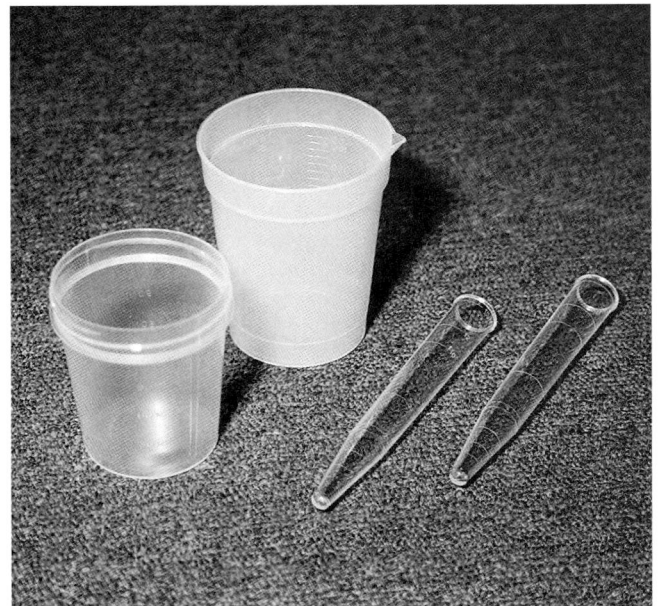

Fig. 24-8 Urine collection cups and urinalysis tubes.

be considered according to the facility's quality control standards.

Collection Containers (Cups) A variety of collection containers is available (Fig. 24-8). Some have caps that should be used if the specimen will not be analyzed immediately following collection. Some have pour spouts that facilitate filling the urinalysis tubes. The containers should be clean and dry, and are used only once.

Urinalysis Tubes Urinalysis tubes are generally made of plastic and must be clear (see Fig. 24-8). They can hold up to 15 mL of urine. Tubes may come with optional caps that should be used if the specimen will not be analyzed immediately.

Gauze Sponges Standard 2 × 2 inch gauze sponges are adequate. Gauze sponges moistened with an appropriate cleansing solution are used to cleanse the patient prior to urine collection.

Cleansing Solution A variety of appropriate cleansing solutions are available. Towelettes premoistened with an appropriate cleansing solution are also available, and can be used in place of gauze sponges. These are used to cleanse the patient prior to urine collection.

Specimen Collection

Appropriate care in collecting urine is often ignored because of the ease of collection. Improperly collected urine may yield incorrect test results.

Timing The greatest amount of diagnostic information is obtained when a first-morning specimen is collected. This urine has incubated in the bladder during the night and is collected as soon as the patient awakens in the morning. However, it is often not practical to obtain a first-morning specimen. Urine collected regardless of the time of day is termed a *random specimen.* Some diagnostic information may be sacrificed using a random specimen. The ordering physician must determine which specimen best suits the clinical situation.

Technique of Collection Urine should always be collected via the **clean-catch midstream specimen** (CCMS) technique. The tissues adjacent to the **urethral meatus** are cleansed with an appropriate cleansing solution. The initial portion of the urine stream is discarded. The middle portion of the urine stream is collected for analysis. The last portion of the urine stream is also discarded.

The pre-collection cleansing technique varies from female to male. A female patient spreads the labia and cleanses in an anterior to posterior direction with three separate gauze sponges moistened with cleansing solution (or premoistened towelettes). A separate sponge is used to cleanse each side of the urinary meatus, and a third sponge is wiped directly over the urethral meatus, the external opening to the urethra. A dry sponge is then wiped directly over the urinary meatus. Keeping the labia spread, a small amount of urine is passed into the toilet. At least 15 ml of urine is collected into an appropriate collection container. The remaining urine in the bladder is then passed into the toilet.

A circumcised male simply cleanses the **glans penis,** the conical tip of the penis, using sponges moistened with cleansing solution by wiping from the center outward. An uncircumcised male must first retract the foreskin and keep it retracted during the cleansing and collection. A dry sponge is used to wipe directly over the meatus. As described previously for the female patient, at least 15 ml of urine is collected into a collection cup from the middle portion of the urine stream.

Specimen Testing

The urine should be analyzed as quickly as possible, but at most within 1 hour after collection. Some analytes may deteriorate if analysis is delayed. If a delay is unavoidable, the specimen should be capped, protected from the light, and refrigerated until the analysis is performed. If refrigerated, the urine should be allowed to warm up to room temperature prior to analysis. After being gently remixed, the urine is transferred into a clear plastic urinalysis tube.

Macroscopic Examination Color and appearance (clarity) are judged by visual examination of urine in a clear plastic urinalysis tube. These characteristics should not be assessed with the urine in the collection cup. Urine color normally ranges from colorless to yellow. Shades of yellow should be reported with modifiers such as *light yellow* or *dark yellow.* Appearance of urine normally ranges from clear to slightly hazy. Degrees of haziness are reported as *slightly hazy, moderately hazy, markedly hazy, cloudy,* or *turbid.* Haziness greater than slight may indicate the need for a microscopic examination of the urine sediment. See the following guidelines for contacting the physician in such an instance.

Chemical Analysis Chemical analysis is performed with a urine reagent strip. Timing is critical, and the test result must be read in the time indicated by the manufacturer. The end-point color is compared with a color chart attached to the reagent strip bottle, and the result is recorded. Some manufacturers provide report forms that directly correlate with the strip (Fig. 24-9). Most test results are reported as *negative* (normal, or none detected), *trace, 1+, 2+, 3+,* or *4+.* Alternatively, test results may be recorded as negative, trace, small, moderate, or large. Specific gravity and pH are re-corded as specific numerical values. Physicians are trained to interpret the results regardless of the reporting system employed.

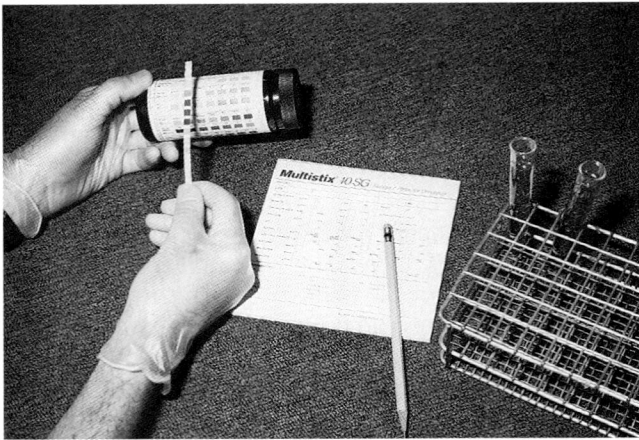

Fig. 24-9 Reading reagent strip and recording the results.

Step-by-Step Reagent Strip Urinalysis

1. Remove a strip and immediately recap the bottle. Be sure the strip is not outdated. Do not touch the test pads.
2. Immerse the strip, taking care that all pads contact the urine. Timing begins once the strip is immersed.
3. Immediately remove the strip and blot the edge of the strip on a paper towel to remove excess urine.
4. Carefully time and read the results.
5. Record the results.

Once testing is completed, the remaining urine can be flushed down the toilet or washed down a sink drain with cold water. The used reagent strip, contaminated paper towel, urinalysis tube, and collection cup are discarded into a biohazard bag.

Common Problems

Collection and Handling Problems   The most common problems encountered in routine urinalysis are usually associated with specimen collection and handling. Urine collected without using the CCMS technique may result in bacterial contamination. Analysis of urine delayed beyond 1 hour after collection, especially if the urine is not refrigerated, may produce erroneous results.

Unmatchable End-Point Color   At times an end-point color not matching the reference chart cannot be explained. It may be due to a faulty strip. The specimen should be retested. If the problem persists, try a different bottle of strips. If this does not resolve the problem, the urine may contain a substance interfering with the determination of that analyte. The spec-

imen should be sent to the laboratory for analysis by a different method.

Occasionally the urine color is abnormal (most often due to a medication). The abnormal urine color may interfere with appropriate color development on the strip. An alternative method of analysis of the chemical analytes is required in this situation.

Multiple End-Point Colors   Multiple end-point colors may occur if excess urine on the strip is not properly blotted (see step 3 described previously). This allows reagents from one test pad to run into another.

A unique occurrence of multiple end-point colors is found with the test area for blood. It may occasionally reveal a green speckled pattern overlying an orange background. This results from intact red blood cells present in the urine. This is reported as "positive for non-hemolyzed blood."

Menstrual Blood Contamination   One of the most common causes of hematuria (blood in the urine) occurs when urine is collected during menstrual bleeding. Often a routine urinalysis can be delayed until the menstrual bleeding has stopped. If it is necessary to perform a urinalysis during menstrual bleeding, the CCMS technique is employed immediately following the insertion of a fresh tampon.

Indications for Microscopic Examination

The microscopic examination of the urine sediment is beyond the scope of this text and should be performed by a trained medical technologist or experienced physician. However, a radiographer performing the first two components of the routine urinalysis needs to be aware that the urine sediment should be examined if there are certain abnormal results from the chemical and physical characteristics analyses. Most commonly, the following test results, either individually or in combination, should be brought to the physician's attention *prior to* discarding the urine specimen so an appropriate decision can be made about performing a sediment examination:

- Hazy urine (more than slightly hazy)
- Positive glucose, protein, blood, nitrite, or leukocyte esterase

At this point, the urine will be **centrifuged** while still in the urinalysis tube. The sediment that accumulates at the bottom of the tube will then be examined under a microscope by an appropriately trained medical technologist or physician.

Miscellaneous Laboratory Tests

A variety of simple laboratory tests are available that are used as diagnostic tools for various conditions, including pregnancy, infectious mononucleosis, rheumatoid arthritis, and colorectal cancer (stool guaiac testing for occult blood). A radiographer may be required to perform such procedures and perhaps instruct patients regarding specimen collection. The manufacturers of these kits provide detailed instructions regarding required specimens, essential equipment, testing procedure, expected test result parameters, and safety precautions. This information should be read with great care. Consult with laboratory personnel if you have any questions.

SUMMARY

Laboratory skills involve contact with potentially infectious body fluids. Therefore Standard Precautions are employed for safety.

Venipuncture is used to obtain blood specimens for a variety of tests. After the procedure has been explained to the patient and the equipment has been prepared, a suitable vein is selected, and the vein is entered using a sterile needle attached to a needle holder. One or more evacuated tubes are then filled with blood. It is important that the correct tubes be selected and correctly handled for accurate test results.

Urinalysis involves proper specimen collection technique, prompt and correct specimen handling, and accurate evaluation. The radiographer may perform macroscopic and chemical analyses. When microscopic analysis is required, qualified personnel perform it. The radiographer must identify the need for this procedure and provide appropriate notification.

■ *Review Questions* ■

1. Which body fluids are considered potentially infectious with blood-borne pathogens?
2. List appropriate barrier techniques used during routine venipuncture.
3. What should be done with filled sharps containers and biohazard bags?
4. Name the color of the stopper of the evacuated tubes used for hematology, chemistry, and coagulation testing.
5. Name the most common vein and location used in routine venipuncture.
6. What size needle is most commonly used in routine venipuncture?
7. How long should pressure be applied to the puncture site following venipuncture?
8. Which evacuated tubes should be inverted after filling?
9. What are the three major components to a routine urinalysis?
10. Which tests involve simple observation of the physical characteristics of urine?
11. How long should the reagent strip be placed into the urine?
12. Describe the basic steps involved in the CCMS collection technique.
13. Urine should be analyzed within how much time after collection?
14. List indications for a microscopic examination of the urine sediment.

25

Additional Procedures for Assessment and Diagnosis

Learning Objectives

At the conclusion of this chapter, the student will be able to:

- Demonstrate the correct procedure for weighing and measuring a patient
- Conduct a Snellen test for distance vision acuity and properly record the results
- Conduct a reading test for near distance vision acuity and properly record the results
- Conduct an Ishihara test for color vision perception and record the results
- Instruct and prepare a patient for electrocardiography and perform the test
- List the leads recorded for a standard 12-lead electrocardiogram and code them correctly
- Recognize common artifacts seen on electrocardiograms and list steps to be taken to correct each
- State the purpose of forced expiration spirometry and two significant measurements involved in this test
- Instruct and prepare the patient for a forced expiration spirometry test and conduct the test
- Discriminate between acceptable and unacceptable forced expiration maneuvers

Key Terms

ECG leads
electrocardiogram (ECG, EKG)
electrocardiograph
electrodes
exercise tolerance test (stress test)
forced expiration test
forced expiratory volume (FEV)
forced vital capacity (FVC)
glottic closure
hemoptysis
hyperopia

Ishihara test
myopia
presbyopia
Snellen alphabet chart
Snellen E chart
spirogram
spirometer
spirometry
standardization mark
stylus
visual acuity

n small clinics and physicians' offices it is often the case that radiographers must be available to perform other functions when there is not enough radiography to keep them occupied. In these circumstances it is common for ra-diographers to be cross-trained in a number of diagnostic procedures and patient assessment skills. Some patient assessment skills have been described in other chapters. Vital signs, for example, are explained in Chapter 22. Obtaining blood and urine samples is covered in Chapter 24. This chapter provides an introduction to additional skills that will enhance the radiographer's value in the clinical setting. It covers weighing and measuring patients, vision screening tests, electrocardiography, and forced expiration spirometry.

This chapter provides a general overview of these procedures. They are not difficult to learn and are often taught in the employment setting without formal instruction. You will need a clinical orientation to the equipment used in your facility and practical instruction in its use to become competent. Additional information about these procedures, including variations for pediatric patients, may be found in nursing and medical assisting texts. Texts devoted to electrocardiography and pulmonary functioning testing, including spirometry, are also available.

MEASURING WEIGHT AND HEIGHT

Measuring patients' weight is an important aspect of patient assessment. Weight is sometimes used to determine medication dosage. In addition, any sudden weight gain or weight loss may be a significant diagnostic sign. Patients whose treatment involves a dietary regimen for weight loss or weight gain will be monitored to determine the effectiveness of the treatment. Infants and children are weighed and measured to monitor their growth and development. For these reasons, many facilities routinely weigh patients on each visit.

Since the height of adult patients is not subject to sudden change, height is usually measured only on the first visit or as part of a comprehensive physical examination.

Measuring Weight

Most facilities use upright balance scales to obtain accurate weight measurements. These scales have two calibration bars, upper and lower. The calibration bars are connected on the ends to form a rectangle (Fig. 25-1). This rectangle tips when the scale is not in balance. An indicator on the right end shows when the rectangle is level and the scale is balanced. The lower calibration bar is divided into 50-pound increments with grooves for placement of the weight at each setting. The upper calibration bar is divided into

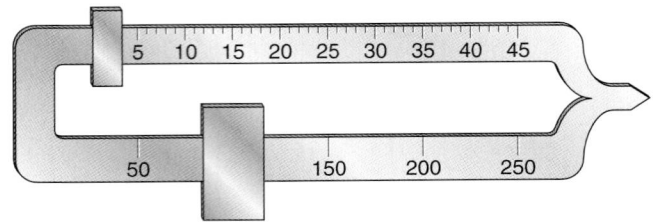

Fig. 25-1 Balance scale has two calibration bars. Lower weight is set in grooves indicating 50-pound increments. Upper weight slides smoothly over 50-pound range in ¼-pound increments.

pounds and quarter pounds with a range of 0 to 50. The upper calibration bar has no grooves and the weight slides smoothly along it. The patient's weight is determined by adding the weight measurements from the two calibration scales.

Patients are usually weighed with their normal clothing. Any heavy outer clothing, such as a jacket, is removed, as are the patient's shoes. Paper may be placed on the scale platform to prevent transmission of disease. The scale should be located in a non-public area, as many patients are self-conscious about their weight. For this reason, take care not to make any comments that might embarrass or upset patients who are sensitive about their weight.

Before the patient steps onto the scale, check to see that the weights on both bars are set at 0 and that the scale is in balance. If the scale is not in balance, a screw at the left end is adjusted until balance is achieved. The scale platform moves slightly, so the patient may feel insecure while stepping onto it. Assist the patient onto the scale platform and ensure that the patient position is stable. The patient must not touch or lean on you or on any object to maintain balance during weight measurement, as this will result in an inaccurate reading. All of the patient's weight must be centered over the scale. Instruct the patient to stand very still, as it is not possible to balance the scale when the platform is in motion.

To obtain the weight measurement, move the weight on the lower calibration bar to the highest groove on the bar that does not cause the scale to tip. Be certain that the weight is properly situated in the groove. Then, slowly move the upper weight until the scale is in balance (Fig. 25-2). Note the readings on both calibration bars and add them together. For example, if the lower weight is set at 150 pounds and the upper weight is at 11½ pounds, the patient's weight is 161½ pounds. Record the patient's weight to the nearest quarter pound.

Some facilities use a digital electronic scale. These battery-powered scales consist of a platform with a digital readout and a foot lever that activates the scale. When the patient is ready to be weighed, activate the scale with your foot. The scale should read

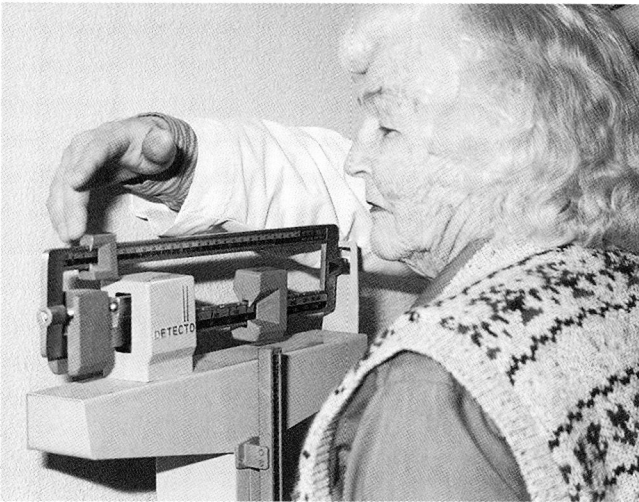

Fig. 25-2 Weighing patient on balance scale.

Fig. 25-3 Measuring patient on balance scale.

000.0. A dial on the bottom of the scale is used to calibrate the scale to zero, when necessary. Assist the patient onto the scale. Within a few seconds, the scale will indicate the patient's weight. If the reading keeps changing or does not appear promptly, this may be due to patient motion. The procedure must be repeated. Assist the patient to step off the scale and provide instruction to hold very still after stepping onto the scale. Reactivate the scale and weigh the patient again.

Measuring Height

Most upright balance scales include a calibration rod for measuring height. The calibration rod extends for height adjustment and has a measuring bar hinged to the top that unfolds into a horizontal position. Use caution and follow the procedure in the proper order so that movement of the measuring bar does not injure the patient.

Extend the calibration rod upward until its top is well above the patient's head and unfold the measuring bar so that it is horizontal. Instruct the patient to step onto the platform with his or her back to the scale. Assist the patient into this position, if required. Instruct the patient to stand erect and to look straight ahead. Then, slowly lower the measuring bar until it rests gently on top of the patient's head (Fig. 25-3). Hold the bar in position while the patient steps off the scale.

After the patient has stepped down, read the height measurement. This measurement is read at the junction of the moveable and stationary portions of the calibration rod (Fig. 25-4) unless the patient is shorter than the stationary portion, in which case it is read directly from the stationary portion. Record the patient's height to the nearest quarter inch.

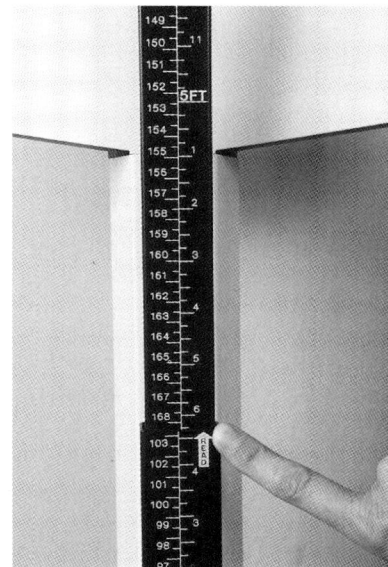

Fig. 25-4 Taking height measurement at junction of stationary and extending calibration rods.

VISION SCREENING TESTS

Vision tests to determine the correction needed for eyeglasses or contact lenses are usually performed by an optometrist. Screening tests, however, may be performed in the physician's office and are commonly used to determine whether **visual acuity** (sharpness of perception) is normal or whether a referral to an optometrist is needed. These tests are usually included in a complete physical examination. Conditions identified by simple screening tests include myopia, hyperopia, presbyopia, and defects in color perception.

Myopia is a condition in which the patient is unable to see normally at a distance and is referred to in lay terms as "nearsightedness." **Hyperopia** or "farsightedness," on the other hand, refers to a vision defect that prevents focus at close range. **Presbyopia** is a type of farsightedness that commonly occurs with advancing age, starting between the ages of 40 and 45. Although few people are totally "colorblind," or completely unable to perceive colors, defects in color perception are not uncommon, especially in males.

Distance Vision Assessment

Distance vision is usually evaluated using a **Snellen alphabet chart,** which has various letters of the alphabet arranged in numbered lines by size (Fig. 25-5). For children who have not learned the alphabet or patients who are unfamiliar with the English

alphabet, the **Snellen E chart** may be used. This chart contains the block letter E in various positions on numbered lines by size (Fig. 25-6).

When using the E chart, the patient must first be instructed how to indicate to the examiner the position of the E. The patient may state the direction of the open end of the letter, left, right, up, or down. For some patients, especially children, it is best to have the patient extend the three middle fingers and demonstrate the position of the E. Practice using a card with a large E held in various positions at close range will ensure that the accuracy of the test is not compromised by confusion in reporting.

Distance vision assessment is usually made at a distance of 20 feet. This distance from the chart may be

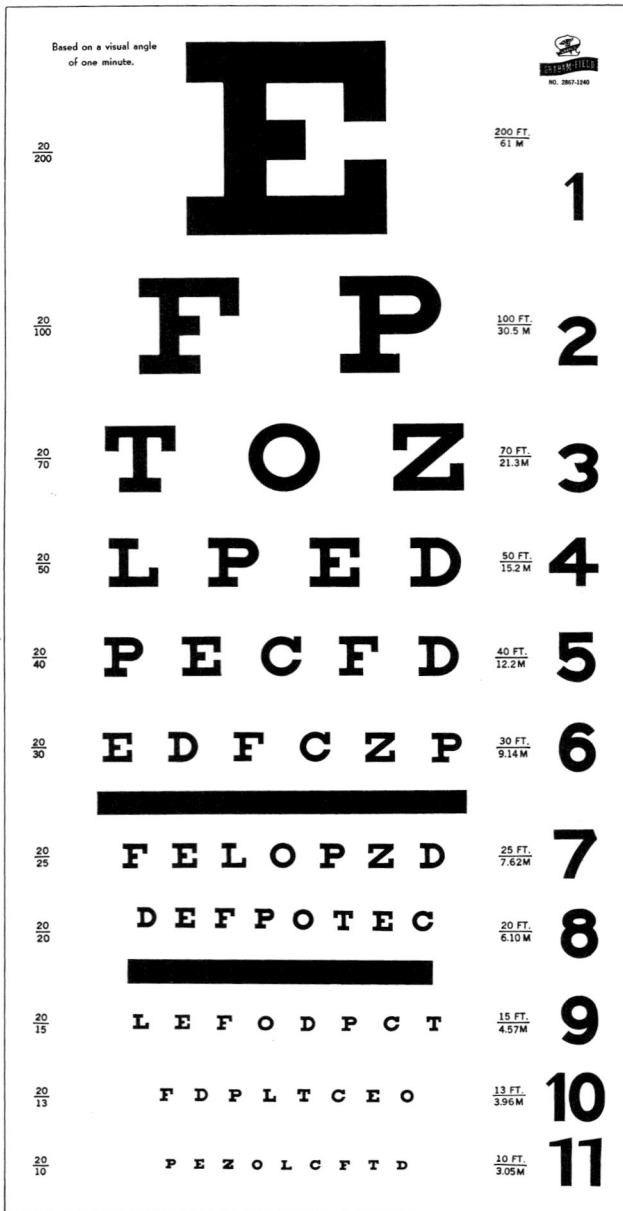

Fig. 25-5 Snellen alphabet eye chart.

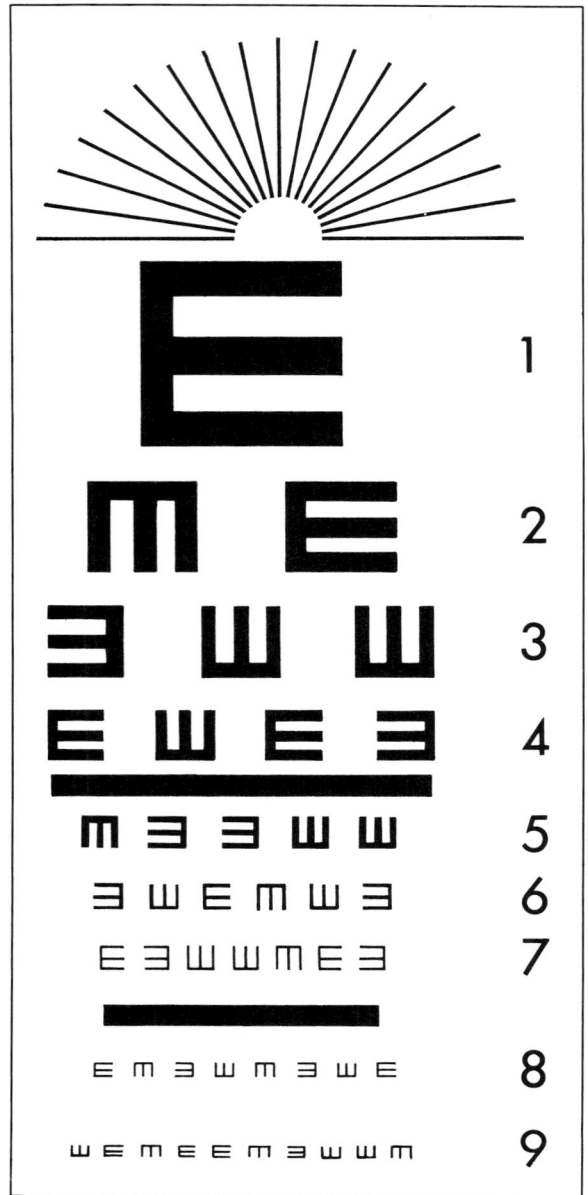

Fig. 25-6 Snellen E chart.

marked on the floor with paint or tape to avoid the need to measure it for each test. The chart should be at eye level in a well-lit area. The patient may be standing or seated for the test.

Each eye is examined separately while the opposite eye is covered. An optical occluder that resembles a plastic cup with a straight handle may be used to cover the eye, or the patient may cover the eye with one hand. The same standard procedure should always be followed to minimize error. The right eye is examined first.

The examiner stands near the chart and points to each line, starting at the top (Fig. 25-7). The patient is instructed to read the indicated line without squinting. The examiner notes the lowest line on the chart that the patient can read. If one or two errors are made in any line, the error(s) are noted and the patient is asked to continue. If more than two errors are made, the previous line is recorded as the lowest line read.

Beside each row of letters on the chart are two numbers with a line between them. The upper number indicates the distance in feet at which the test is conducted, usually 20. The lower number indicates the distance in feet at which a person with normal visual acuity can read the line. This format is used to record measurements of visual acuity. When a patient can read the line marked 20/20, this means that the patient can read at 20 feet what a person with normal vision can read at 20 feet. This indicates normal distance visual acuity, or 20/20 vision. If the smallest line the patient can read is marked 20/40, this indicates that the patient can read at 20 feet what a person with normal acuity can read at 40 feet and therefore indicates that the patient has less than normal distance acuity.

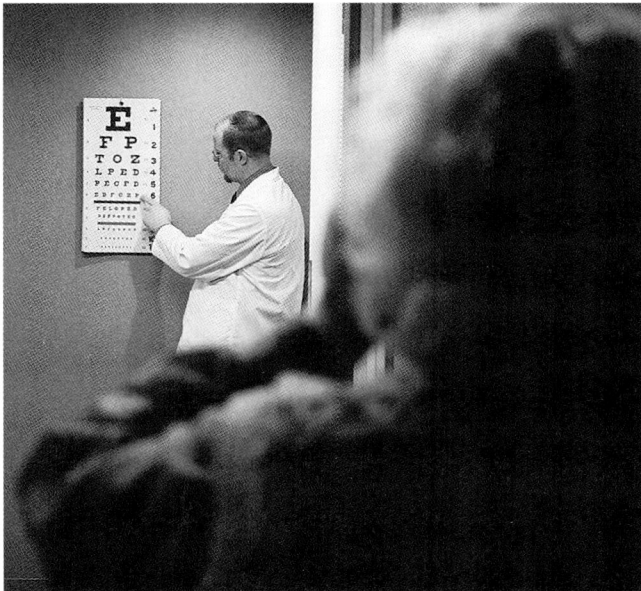

Fig. 25-7 Patient taking distance vision test.

To chart the test results, state the date and time, the name of the test, and the visual acuity measurement for the smallest line the patient was able to read with each eye. If there were errors in the reading of this line, the measurement is followed by a minus sign and the number of errors. Since distance vision is sometimes tested both with and without corrective lenses, the record must indicate how the test was done by adding the abbreviation s̄c (without correction) or c̄c (with correction). If the patient squinted or blinked excessively or the eyes watered during the test, these observations should also be noted. An example of a record of a distance visual acuity test is: "10/21/2001, 10:30 AM, Snellen test: OD 20/20-1. OS 20/30, s̄c. Exhibited squinting while reading 20/25 line OD."

Near Vision Assessment

Several different types of cards are available for the assessment of near vision, but the method for using them is similar in each case. The patient holds a card with lines or paragraphs of print that range in size from the height of newspaper headlines to the type used in telephone books or on road maps. An example is shown in Fig. 25-8. Cards using pictures and Es are also available. The card is held at eye level at a distance of 14 to 16 inches.

The test is conducted in a quiet, well-lit room. The eyes are tested separately, right eye first. The patient may cover one eye during the test, or simply close it. If the patient normally wears reading glasses, they are usually also worn for the test.

This test may be scored in several different ways, depending on the chart used. A common method is similar to the Snellen chart measurement, with 14/14 indicating normal vision when the test is conducted at a reading distance of 14 inches. Results of near vision tests are charted in the same manner as the Snellen test.

Color Perception Assessment

The classic method of evaluating color perception is the **Ishihara test.** It utilizes a special book with a multicolor plate on each page. Each plate shows a circle containing dots that are arranged to form a number. The background also consists of dots, but the color is in contrast to the dots that form the number. The number is clearly visible to those with normal color perception. The first plate in the book is similar to the others, but the number can be read by anyone, even patients who are totally unable to perceive color. This plate is used as a teaching tool to explain the test to the patient.

It is best to conduct this test in a quiet room, well lit by daylight, as color is perceived most accurately in natural light. The book is held approximately 30

60
Nothing can take the place of "the only pair of eyes you will ever have." That is why you are exercising such good judgment in taking care of them as you are now doing.

50
For this reason, you will welcome the suggestion about lenses which are designed and made to give you "greater comfort and better appearance." In man's earliest days he had little use for glasses. He used his eyes chiefly for long distance.

40
He worked by daylight and at tasks with little detail. But now, you use your eyes for much close work—reading, writing, sewing and many other uses which the eyes of primitive man did not know. Now your eyes meet all sorts of lighting conditions, artificial and natural.

30
Many of these conditions produce "overbrightness" or glare. Sometimes it is the direct or reflected glare of sunlight; often it is direct or reflected from artificial light. And very often this glare is uncomfortable—impairs your efficiency. But special lenses, developed by America's leading optical scientists, combat this glare.

25
These lenses give you more comfortable vision and blend harmoniously with your complexion. These lenses are less conspicuous. We are glad to rec- ommend them because they will give you greater comfort and better appearance. Thousands of satisfied wearers testify to their real benefits.

20
You are wise in taking good care of "the only pair of eyes you will ever have." You know how valuable they are, that you can never have another pair. For this reason, you will welcome the suggestion about lenses which are designed and made to give you "greater comfort and better appear- ance." In man's earliest days he had little use for glasses.

The above letters subtend the visual angle of 5' at the designated distance in inches.

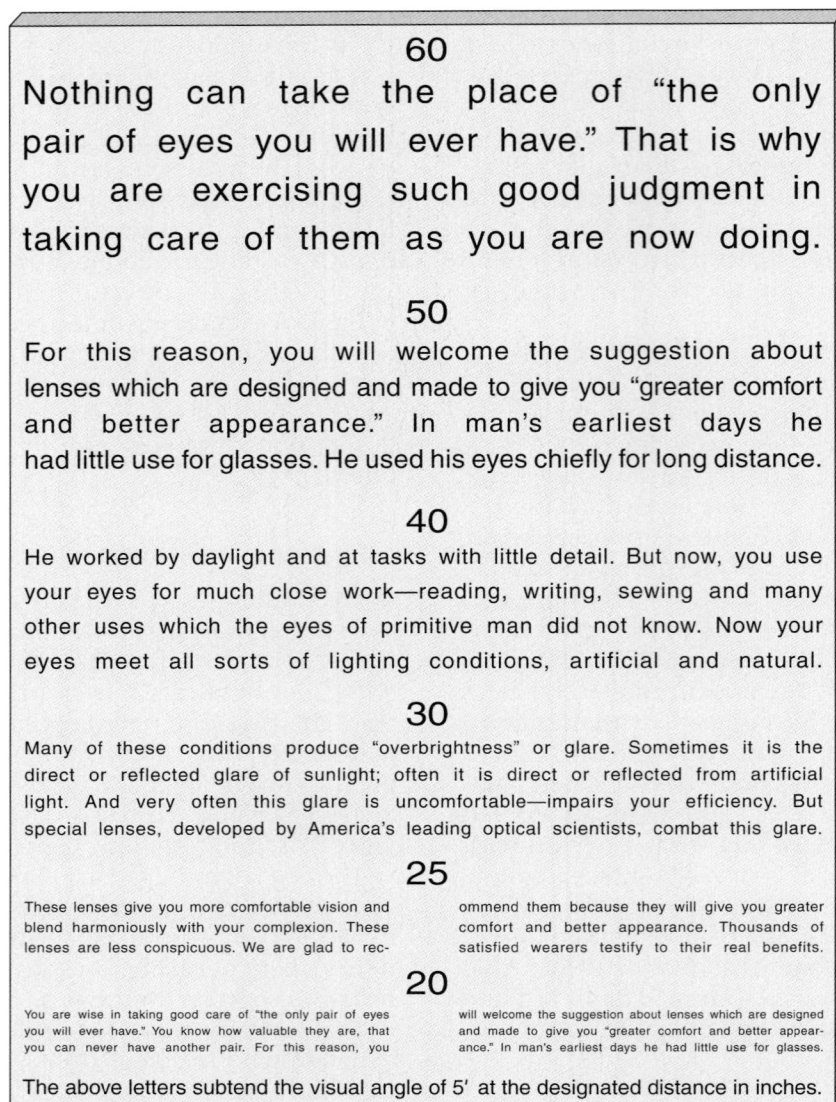

Fig. 25-8 Near vision acuity chart.

inches from the patient. It should be at the patient's eye level and perpendicular to the line of sight. Show the patient the first plate and explain the test. The patient should understand that he or she will have 3 seconds to identify each number.

As the patient identifies the number on each plate, record the response and display the next page. If the patient is unable to identify the number in 3 seconds, record an X for that plate and continue to the next page. The Ishihara test consists of 14 plates. Plates 1 through 11 constitute the basic test. Plates 12, 13, and 14 are used to further evaluate the perception of patients who exhibit red-green color deficiency. If the patient provides normal responses to the first 11 plates, the final 3 plates may be omitted from the test. If the patient reads 10 plates correctly, color perception is considered normal. If the patient can read only 7 or fewer of the plates, the patient is identified as having a color vision deficiency. The test is so structured that a score of 8 or 9 is unlikely. The test is charted by stating the date and time, the test name (Ishihara test), and a listing of the patient's response to each plate presented. Any unusual signs exhibited during the test are also recorded.

ELECTROCARDIOGRAPHY

Electrocardiography is one of the diagnostic tools commonly used in the assessment of heart disease. It is a graphic representation of tiny electrical currents generated within the heart that cause the heart muscle to contract. A machine called an **electrocardiograph** is attached to the patient with cables and **elec-**

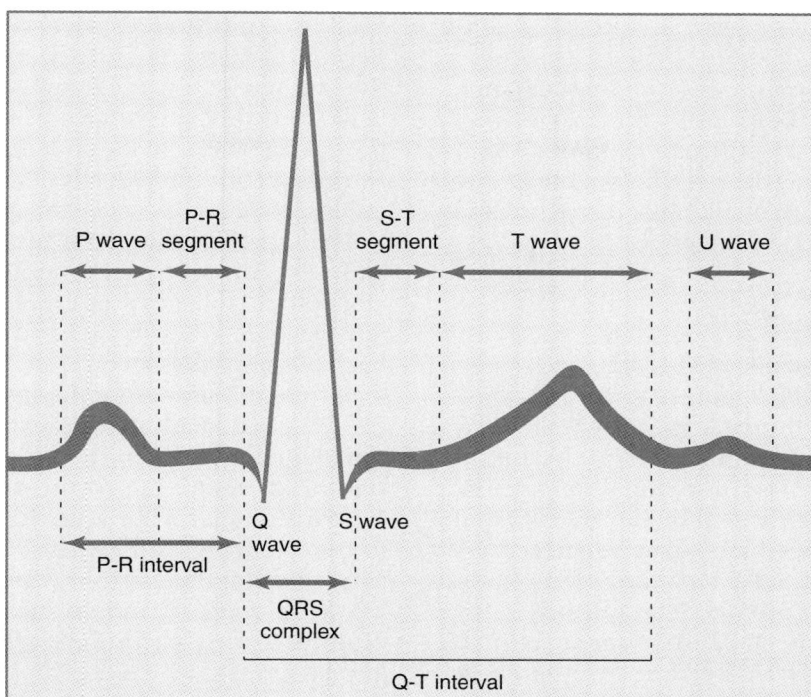

Fig. 25-9 ECG waves indicate each portion of the cardiac cycle.

trodes (contacts that receive electrical signals). Electrical impulses from the patient are transmitted to the machine, where they are amplified and translated into signals that cause movement of a balanced tracing pen called a **stylus.** The pattern of these electrical impulses is traced by the stylus on graph paper. The resulting tracing is called an **electrocardiogram,** abbreviated **ECG** or **EKG.**

The Cardiac Impulse

It may help you to understand this section if you first review the basic anatomy and physiology of the heart in Chapter 16. The myocardium (heart muscle tissue) is of two types: contracting tissue and conducting tissue.

The cardiac impulse is a tiny electrical current that originates at the junction of the vena cava and the right atrium in conducting myocardium called the *sinoatrial (SA) node.* It spreads in circular waves over the atrial walls, causing the atria to contract. The impulse then passes through the atrioventricular (AV) node, a second area of conducting myocardium. From this node, it passes through a band of conducting muscle that connects the atria to the ventricles and is called the *bundle of His.* The bundle of His divides into the left and right bundle branches, conducting the cardiac impulse to the left and right ventricles. The bundle branches further divide into Purkinje fibers, fine strands of conducting muscle

that transmit the impulse to the contracting muscle of the ventricles.

The normal cardiac cycle includes atrial contraction, ventricular contraction, and rest. The transmission of the electrical wave causing contraction of the chamber walls is called *depolarization.* Each contraction is followed by repolarization, an electrical recovery period. Following ventricular repolarization, the heart rests for a moment in a state of polarization, and then the cycle begins again. With normal heart function, this cycle is repeated 50 to 100 times per minute at a regular rate.

The ECG tracing records the electrical impulses as deflections above or below a baseline. These deflections are referred to as *waves* and are labeled P, Q, R, S, and T (Fig. 25-9). Together, the Q, R, and S waves are called the *QRS complex.* The P wave indicates contraction of the atria, the beginning of depolarization. The space between the P wave and the R wave is called the *P-R interval* and represents the time from the beginning of the atrial contraction to the beginning of the ventricular contraction. The QRS complex represents ventricular contraction. The S-T segment indicates the time between ventricular contraction and the beginning of ventricular recovery. The T wave represents ventricular recovery (repolarization). After the T wave the tracing shows a straight line, indicating the period of heart rest. On rare occasions you may observe a small U wave following the T wave. This is an abnormal wave that indicates a low serum potassium level or other metabolic disturbance that affects the conduction of the heart impulses.

The physician observes the morphology (shape), amplitude (height), and duration (graph width) of each wave in relation to the baseline and the other waves. Taken together, these findings enable the physician to detect disturbances in heart rhythm and to identify different types of cardiac disorders.

Electrocardiography Leads

A standard ECG study includes 12 separate recordings called **ECG leads.** Each lead represents cardiac activity as recorded from a different angle. To obtain these various leads, electrodes are placed in 10 locations, one on each arm and each leg, and 6 on the chest. The tracing for each lead must be marked or coded so that the physician will know which angle is represented. The 12 leads are categorized as standard (limb) leads, augmented leads, and precordial (chest) leads.

There are three limb or standard leads, sometimes referred to as *bipolar leads.* They are designated by roman numerals, I, II, and III. Each uses two limb electrodes to record the electrical activity (Fig. 25-10).

Lead I records the electrical potential (voltage) between the right arm and the left arm. Lead II records the voltage between the right arm and the left leg. Lead III records the voltage between the left arm and the left leg. The right leg provides an electrical ground. Normal recordings of leads I, II, and III are illustrated in Fig. 25-11.

The three augmented leads are termed *aVR, aVL,* and *aVF.* The "aV" portion of these terms stands for augmented voltage. R indicates the right arm; L, the left arm; and F stands for foot, indicating the left leg. Each of these leads records the voltage from one limb electrode to the midpoint between two others (Fig. 25-12). Lead aVR records the voltage measured from the midpoint between the left arm and left leg to the right arm. Lead aVL records the voltage measured from the midpoint between the right arm and left leg to the left arm. Lead aVF records voltage measured from the midpoint between the right arm and the left arm to the left leg. Again, the right leg serves as a ground. Normal recordings of the augmented leads are seen in Fig. 25-13.

The six precordial or chest leads are designated V_1 through V_6, each representing a specific location on the chest (Fig. 25-14). These leads indicate the voltage between the chest wall and a point within the heart. Normal recordings of leads V_1 through V_6 are seen in Fig. 25-15.

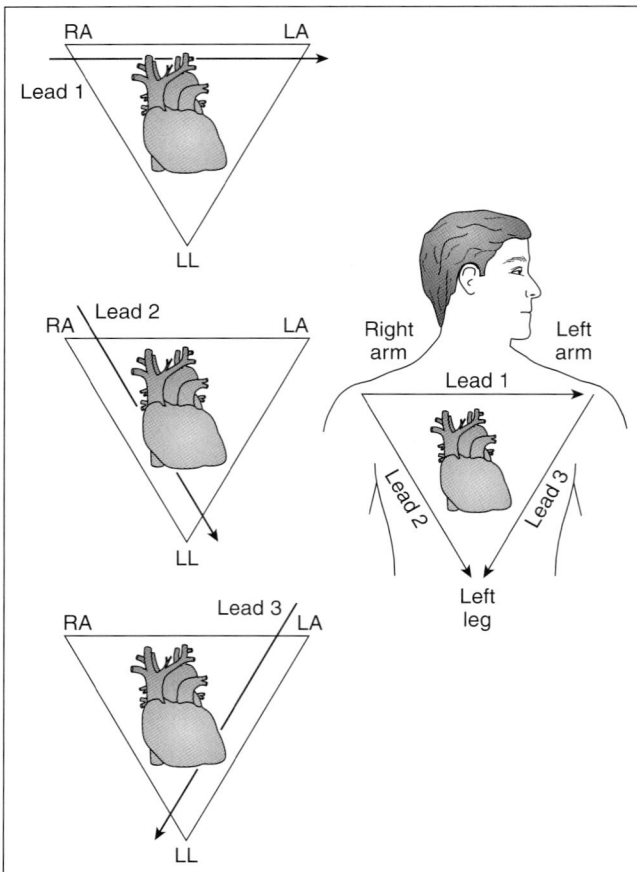

Fig. 25-10 Standard limb leads record activity of the heart from three angles.

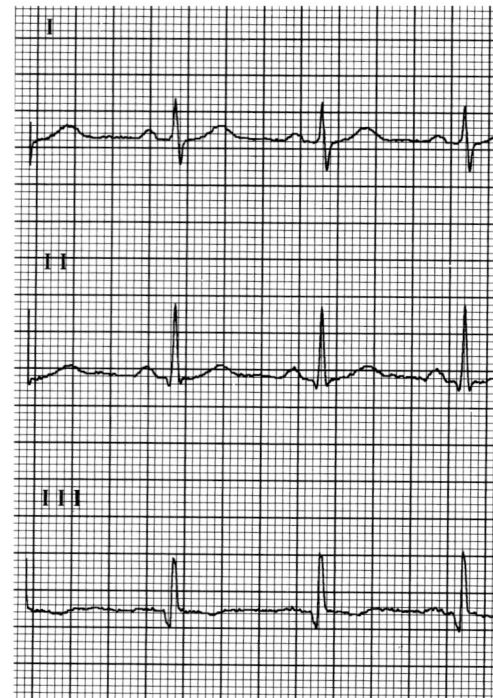

Fig. 25-11 ECG leads I, II, and III.

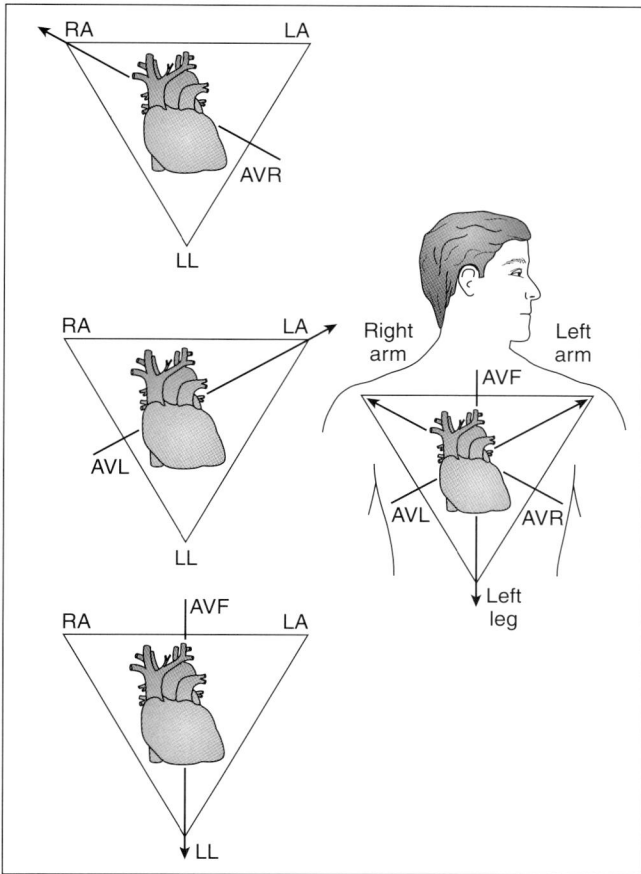

Fig. 25-12 For augmented leads, machine records voltage between one point and the midpoint of two others.

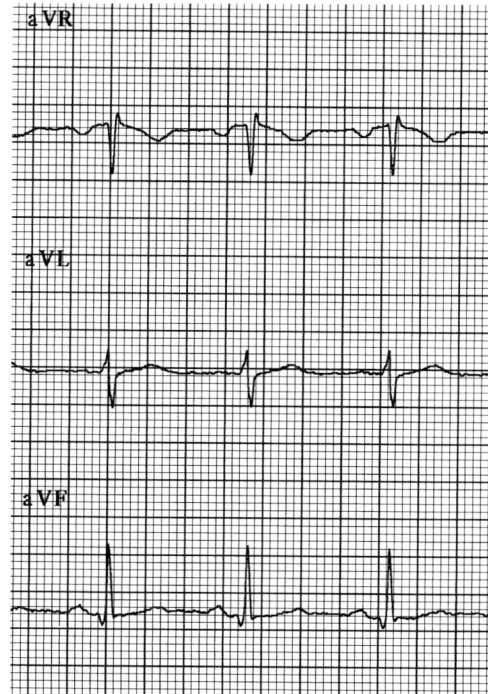

Fig. 25-13 ECG leads AVR, AVL, and AVF.

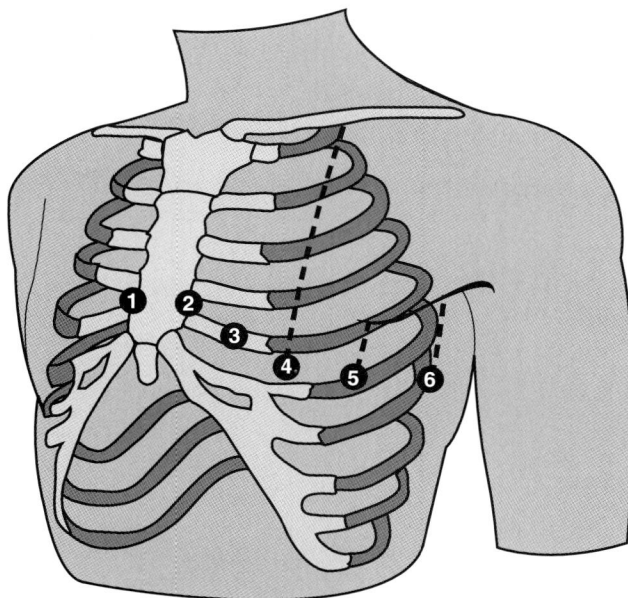

Fig. 25-14 Chest lead locations.

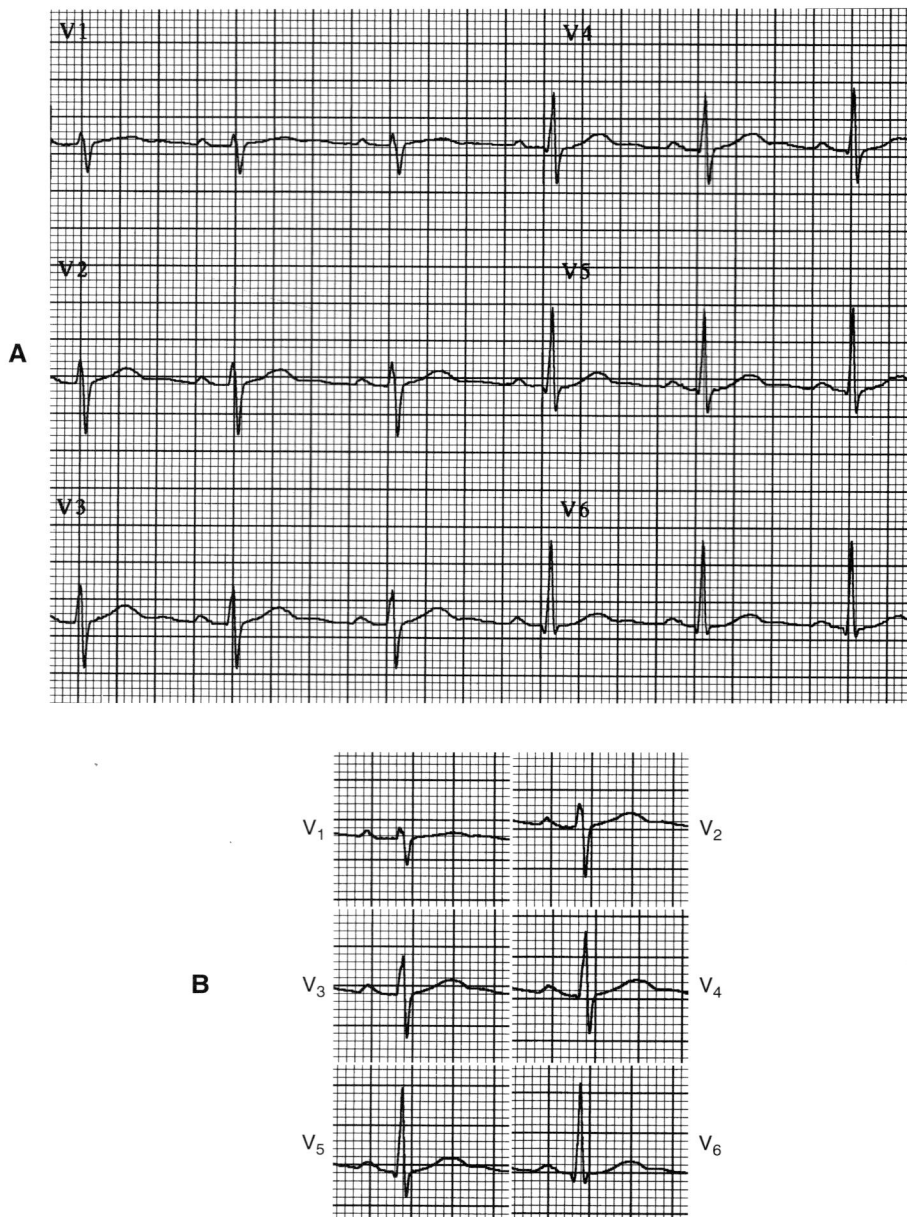

Fig. 25-15 ECG chest leads. **A,** Normal recording. **B,** Individual heartbeats from each of the precordial leads demonstrate progression from the SA node region to the left ventricle. Sequence of precordial wave forms.

Electrocardiograph Paper and Standardization

ECG paper is divided into squares to facilitate measurement of the waves, intervals, and segments by the physician. Each square measures 1 mm in each dimension and every fifth square in each dimension is indicated with a bold line (Fig. 25-16). The bold lines form large squares that are 5 mm by 5 mm in size.

International agreements have established standards for ECGs to ensure that tracings read anywhere in the world will be interpreted in the same way. When the machine is properly calibrated, 1 millivolt (mV) of electrical potential difference will cause the stylus to move vertically 10 mm. This standard ensures that voltage will be interpreted accurately when the graph is read.

To check the calibration of the machine, the standardization button is depressed using a quick, pecking motion of the finger. Depression of the standardization button produces a 1 mV signal that produces a **standardization mark** on the tracing. The mark should be 10 mm high, approximately 2 mm wide, and rectangular in shape. The operation manual for the machine will provide instructions for making adjustments when the standard is not accurate.

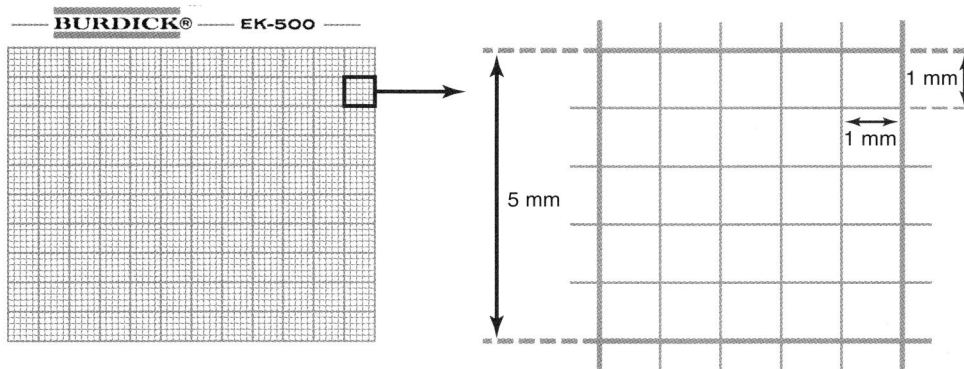

Fig. 25-16 ECG graph paper.

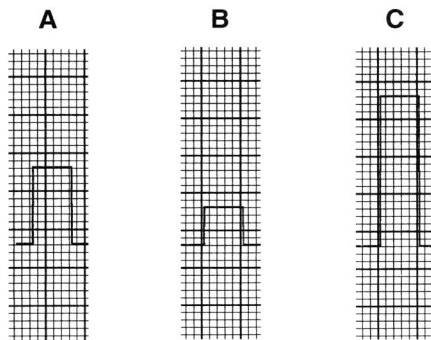

Fig. 25-17 ECG standardization marks. **A,** 1 STD. Normal standardization mark is 10 mm high. **B,** ½ STD. One half standardization mark is 5 mm high. **C,** 2 STD. Double standardization mark is 20 mm high.

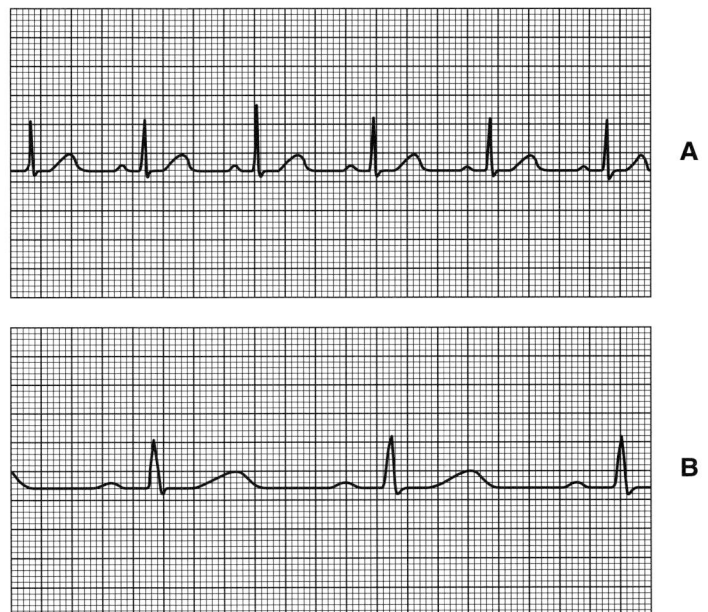

Fig. 25-18 Paper run speed. **A,** Normal recording speed is 25 mm per second. **B,** Double speed is 50 mm per second.

Most machines have three standard (STD) position settings: ½ STD, 1 STD, and 2 STD. The 1 STD setting is usually used. If the amplitude of the QRS complex is so great that it causes the stylus to move off the paper, the ½ STD setting should be used. When recording at the ½ STD setting, 1 mV causes the stylus to deflect only 5 mm. If the amplitude of the QRS complex is very low, the 2 STD setting may be used to increase the readability of the tracing. At the 2 STD setting, the standard mark will be the height of four large squares (20 mm). Fig. 25-17 shows standardization marks for each STD setting.

A standardization mark is usually placed at the beginning of the first lead recording. Some physicians prefer standardization on each of the 12 leads.

The speed of the paper feed must also be standardized for the tracing to be interpreted accurately. The universal recording speed is 25 mm per second. When the patient's heart rate is very rapid or when certain parts of the complex are too close together for accurate assessment, it may be desirable to adjust the machine so that the paper runs at double speed, 50 mm per second. This change in settings will extend the recording to twice its normal length and must be noted on the tracing. The difference in recording between normal and double speed is illustrated in Fig. 25-18.

Types of Electrocardiography Machines

Electrocardiography machines vary depending on age, manufacturer, and level of technical sophistication. An operator's manual should accompany the machine. The radiographer who performs ECGs should be familiar with the features of the equipment and the guidelines provided in the manual.

The least sophisticated machines are single-channel machines that record one lead at a time. The operator

TABLE 25-1

Standard Electrocardiograph Lead Codes

Leads	Electrodes Connected	Marking Code
Standard Limb Leads		
Lead I	LA and RA	.
Lead II	LL and RA	..
Lead III	LL and LA	...
Augmented Limb Leads		
aVR	RA and LA-LL	–
aVL	LA and RA-LL	– –
aVF	LL and RA-LA	– – –
Precordial (Chest) Leads		
V_1	C and LA-RA-LL	– .
V_2	C and LA-RA-LL	– ..
V_3	C and LA-RA-LL	– ...
V_4	C and LA-RA-LL	– ....
V_5	C and LA-RA-LL	– .....
V_6	C and LA-RA-LL	– ......

selects the lead to be recorded and starts and stops the recording manually. There may be one or six lead wires (machine connections) for the chest. If there is only one, it is moved from one location to the next as each of the precordial leads is recorded. The operator may need to enter a code to identify each lead. The standard marking codes are listed in Table 25-1.

Multichannel machines are available that can record three or six leads at once. With these machines, six chest lead wires are used, and all are connected to the patient before the recording begins. A three-channel electrocardiograph is illustrated in Fig. 25-19.

Some machines, both single-channel and multi-channel, are equipped with automatic sequencing capability. Once started, these units proceed automatically from one lead to the next without input from the operator. These machines also identify each lead and place a standardization mark on the tracing. Machines with automatic sequencing usually have a manual option, permitting longer tracings when needed.

Interpretive electrocardiographs incorporate a computer that analyzes the tracing as it is made and

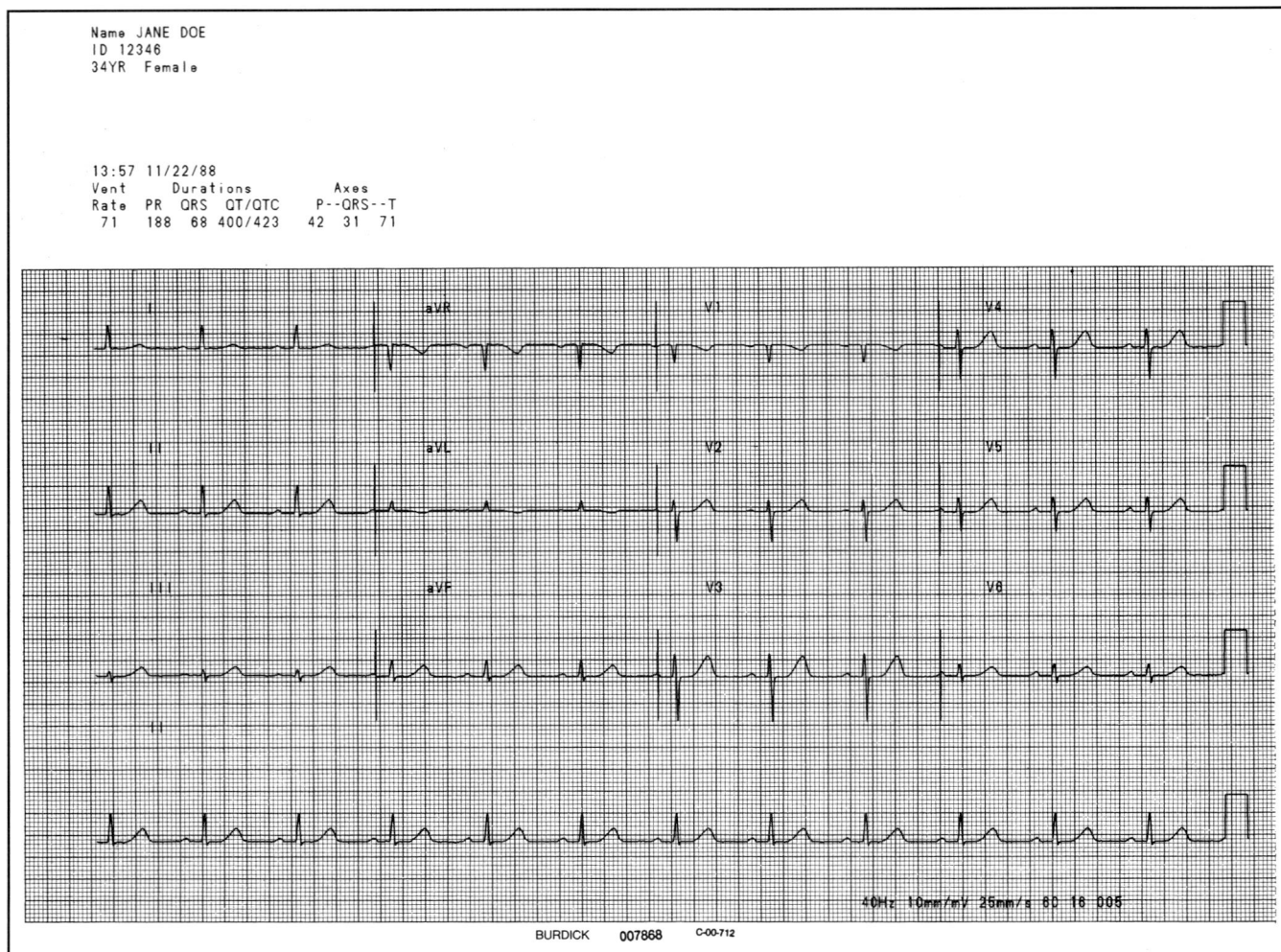

Fig. 25-19 Three-channel recording.

prints the ECG interpretation and the specific reason(s) for the interpretation on the tracing. Patient data is entered into the computer before recording begins, and this data is also printed on the ECG.

Telephone transmission equipment is available for sending ECGs to remote locations for interpretation.

Preparation for Electrocardiography

At the time of scheduling, patients should be informed of what to expect and instructed not to apply any lotion or oil to the arms, legs, or chest. They should be advised that they will need to expose their arms, legs, and chest for the application of the electrodes so that they can dress for convenience in this regard. Patients who have not had an ECG before may assume from the name of the test that electricity will be applied to their bodies. Explain that the test is painless and that only electricity *from* their bodies is being measured. Some facilities provide patient brochures that answer common questions and list instructions.

The room should be quiet and warm with a table for the patient to lie on. The table should be padded and wide enough for the patient to relax comfortably with both arms and legs fully supported. If the table has metal parts, there must be insulation between these parts and the patient. Small pillows under the patient's head and knees will add comfort and make it easier for the patient to lie still for the duration of the test. It is most convenient to position the table so that the operator is on the patient's left side. The ECG machine should be placed as far from other electrical equipment as possible. The power cord should point away from the patient and should not pass under the table.

The patient should disrobe to the waist and wear a gown. Shoes and stockings must be removed and the lower legs exposed. The patient is instructed to lie supine with the legs separated from each other. Any tight clothing should be loosened. Arrange the gown so that the chest is accessible for electrode placement, but modestly covered for the patient's comfort. For the best test results, the patient should rest in this position for about 10 minutes before the start of the test. Some physicians may want you to record any medications that the patient is taking. Explain that the patient should remain in a relaxed position and must not move during the recording of the ECG. Talking during recording is not usually a problem, as long as the patient remains still and relaxed.

Turn on the machine's power switch and allow it to warm up. The operation manual will state the time needed for this process.

The next step is to attach the electrodes to the patient. The skin in the areas where electrodes are applied must be clean, dry, and oil-free. If the patient did not have an opportunity to prepare for the examination, you may have to remove oil or lotion from the skin before applying the electrodes. This can

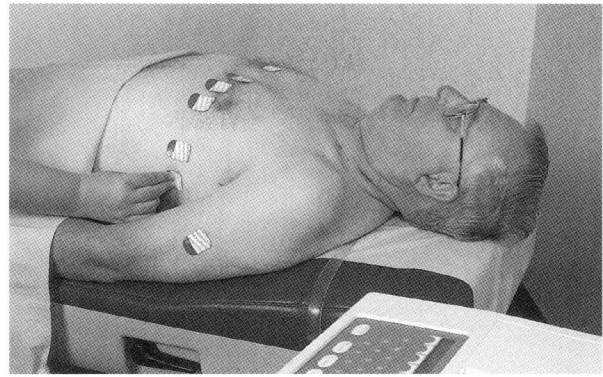

Fig. 25-20 Disposable electrodes in place.

be accomplished by rubbing the area briskly with an alcohol pad and wiping it with a tissue. The electrodes are attached to the patient's skin and have an attachment for the cable ends. Disposable adhesive electrodes are most commonly used, but with some machines nondisposable metal electrodes may be used. Since skin does not conduct electricity well, an electrolyte compound between the skin and the electrode enhances the electrical contact and helps to ensure a reliable tracing. Electrolytes are available in the form of paste, gel, or saturated pads. Electrolytes are incorporated in disposable electrodes.

If disposable electrodes are used, attach one firmly to the fleshy, outer surface of each upper arm and to the inner surface of the calf of each leg. The connection tabs of the arm electrodes should point downward, and the tabs of the leg electrodes should point upward. Attach six electrodes to the chest according to the diagram in Fig. 25-14. The patient in Fig. 25-20 has disposable electrodes properly placed.

If nondisposable electrodes are used, attach an electrode strap and an electrolyte pad to each limb electrode. If using electrolyte gel or paste, use a tongue blade to rub the electrolyte into the electrode site on each arm and leg, continuing to rub until the skin reddens slightly. Wipe off any excess electrolyte, and do not transfer it to the equipment with your fingers. Unequal amounts of electrolyte, or electrolyte contamination of wires or equipment may cause artifacts in the tracing. Place the electrodes on the prepared sites and attach them firmly to the limbs by fastening the straps (Fig. 25-21). The attachment must be securely snug, but not uncomfortably tight. If your machine has six chest lead wires, apply electrolyte to the electrode sites on the chest. Take care that the electrolyte does not spread between the sites. If this should accidentally occur, cleanse the area and repeat the application correctly. Welch electrodes are the usual type of nondisposable electrodes used on the chest. A Welch electrode is bell-shaped with an attached rubber bulb. The bulb is squeezed, the elec-

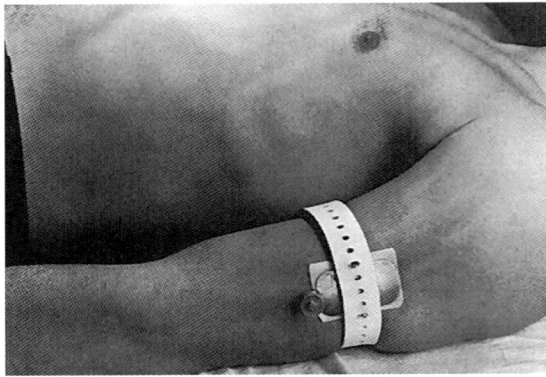

Fig. 25-21 Nondisposable limb electrodes are firmly attached with rubber straps.

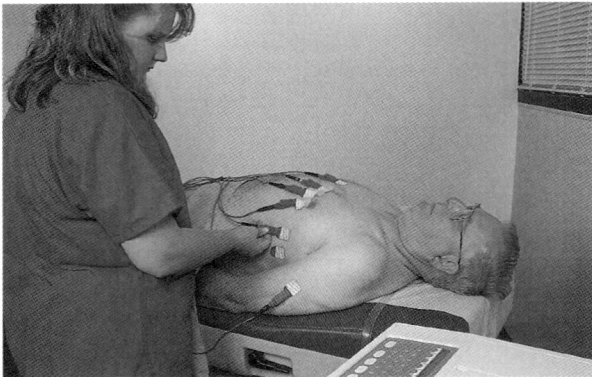

Fig. 25-22 Lead wires are attached to the electrodes, with the operator taking care that the lead wires match the locations.

trode placed on the patient's skin, and the bulb released. The bulb action creates suction between the skin and the electrode, holding the electrode in place. Do not permit the chest electrodes to touch each other.

The patient cable is a branching cable used to connect the electrodes to the machine. The cable has 5 or 10 branches, depending on whether there are one or six precordial connections. Each branch is called a *lead wire* and is labeled for the specific electrode to which it should be attached. Improper connections will result in an inaccurate tracing. To stay focused when making these connections (Fig. 25-22), it is a good practice to read each lead wire label and say to yourself what you are doing while you connect it. For example, "This is the right arm lead wire and I am connecting it to the right arm electrode." When the electrodes are all connected, arrange the cords so that they lie on the patient's body. This practice reduces interference and avoids the possibility that the weight of a hanging cable will loosen a connection.

Obtaining the Tracing

The procedure for making the tracing will vary, depending on the equipment. If you have an electrocardiograph with automated controls, activate the automatic start button and monitor the electrocardiograph while it records all 12 leads. The machine will standardize and code the tracing.

If you have a machine that is manually controlled, set the lead selector control to STD. Turn on the recorder control and activate RUN. The stylus will create a flat line. Use the position control knob to center the baseline on the graph paper. Check the standardization by pressing the standardization button as previously described. The stylus should rise the height of two large squares on the tracing paper (10 mm).

Turn the lead selector control to lead I and code the lead unless your machine does this automatically. Run at least 8 to 10 inches of clean, high-quality tracing with no artifacts. Switch to lead II and then to lead III, repeating the same steps. Run at least 4 to 6 inches of high-quality recording for leads aVR, aVL, and aVF. If there are six chest lead wires, continue recording to obtain 4 to 6 inches of high-quality tracing for leads V_1 through V_6.

If your equipment has a single chest lead wire, turn off the record switch while you prepare to record the chest leads. Apply electrolyte to the six lead locations on the patient's chest, taking care that the electrolyte does not spread between the sites. Beginning with the electrode in the V_1 location and the lead selector dial on V_1 (or V, depending on the machine), turn the switch to RUN and obtain 4 to 6 inches of high-quality recording. Standardize and code the lead. Turn the record switch off while you move the chest electrode to the next position. This prevents excessive movement of the stylus. Repeat these steps for leads V_2, V_3, V_4, V_5, and V_6. When all of the chest leads are complete, slowly turn the lead selector control back to STD, one setting at a time. Turning too rapidly may damage the gears of the dial. Run a straight baseline and then turn off the record switch.

When the ECG is complete, turn off the machine. Disconnect the lead wires and remove the electrodes. It is a good practice to double check that the leads were properly connected as you disconnect them. If electrolyte paste or gel was used, cleanse the skin. Assist the patient off the table and to the dressing room. Discard disposable electrodes or clean nondisposable ones. Return all equipment to its proper place.

Artifacts

ECG artifacts are irregularities in the tracing that do not represent the electrical activity of heart impulses and that interfere with the appearance of the ECG. The operator must monitor all ECG recordings for artifacts. When they occur, the recording should be

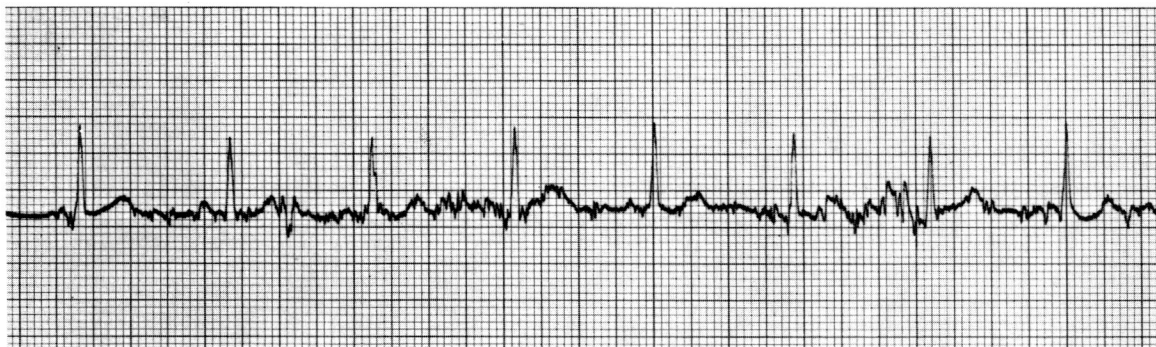

Fig. 25-23 Muscle artifact.

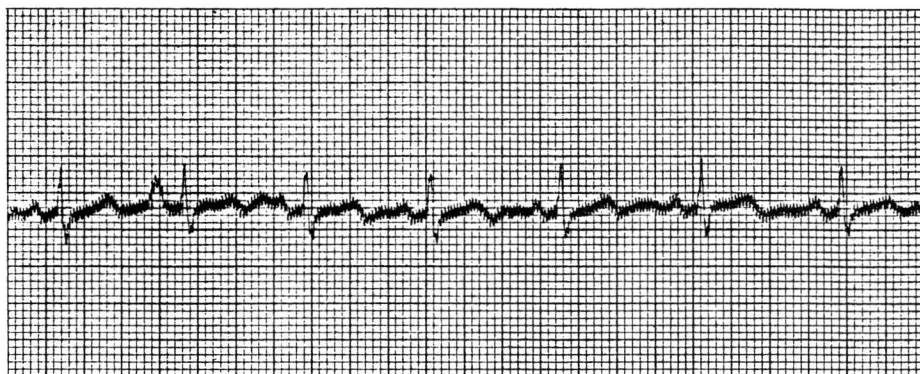

Fig. 25-24 AC artifact.

stopped and the problem corrected. Any leads affected by artifacts should be repeated.

The artifacts most commonly seen are muscle artifact, alternating current artifact, wandering baseline, and interrupted baseline.

Muscle artifact (Fig. 25-23) causes a fuzzy, irregular baseline due to muscle contraction. Muscle artifact is most commonly seen when anxiety or discomfort causes the patient to tense muscles. If the patient is chilly, shivering may cause this artifact. Muscle artifact can usually be avoided by providing reassurance, encouraging the patient to relax and let the muscles go limp, and by ensuring the patient's warmth and comfort. Sometimes muscle artifact occurs as a result of tremors due to nervous system disorders such as Parkinson's disease. In these cases, it may be helpful to have the patient place the hands under the body. If the operator has the freedom to do so, placing gentle pressure on the patient's shoulders may be reassuring and help the patient to relax. The operator must be understanding and try to record when the tremor is at a minimum.

Alternating current (AC) artifact (Fig. 25-24) is caused by electrical interference. AC artifact appears as tiny spiked vertical lines that are very close together and are consistent throughout the tracing. Newer electrocardiographs are electrically shielded and this artifact is seldom seen with modern equipment unless the patient is affected by current leaking from an electric appliance on his or her body, such as a transcutaneous electric nerve stimulation (TENS) unit. With older equipment, electrical interference may be caused by other electrical equipment in the room, fluorescent light fixtures, improper grounding of the electrocardiograph, or wiring in the structure of the room. When AC artifacts appear, turn off all electrical appliances and fluorescent lights in the room and make sure the machine is connected to a properly grounded outlet. Check to be certain that the lead wires follow the body contour and do not dangle from the table, as they are more likely to be influenced by current from other sources when this occurs. If these measures do not solve the problem, try moving the patient's table to another location at a greater distance from the walls.

Subtle wandering baseline (Fig. 25-25) is an indication of poor electrical contact. This artifact is seen when electrodes are not in firm contact with the skin

Fig. 25-25 Subtle wandering baseline.

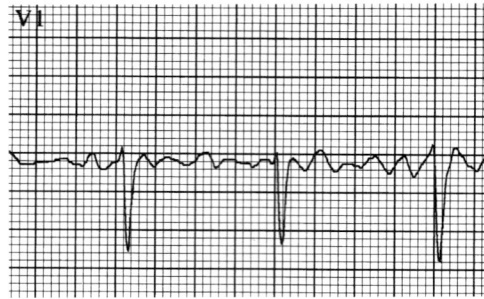

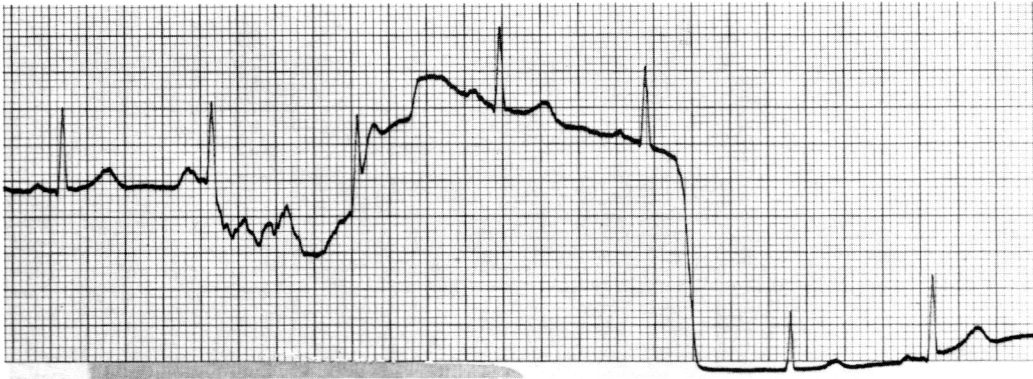

Fig. 25-26 Major wandering baseline.

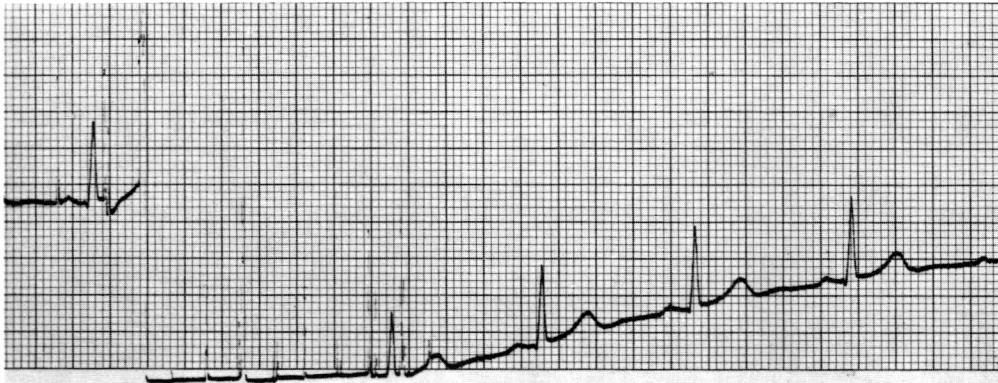

Fig. 25-27 Interrupted baseline.

or when the contacts between the electrodes and the lead wires are loose. Wandering baseline may also be caused when oily skin has not been properly cleansed or when electrolyte is not correctly applied. It is important to identify and correct this artifact because it can sometimes be mistaken for evidence of pathology.

Major wandering baseline (Fig. 25-26) is caused by patient movement or by electrical interference from activity within the room. People moving about near the patient may cause this artifact.

Interrupted baseline (Fig. 25-27) is an artifact that occurs when there is an intermittent interruption of electrical transmission. This may be caused by the detachment of a lead wire or by a broken wire. When a broken wire is determined to be the cause, the patient cable must be repaired or replaced.

Preparing the ECG for Interpretation and Storage

The ECG recording is part of the patient's permanent record. It must be accurately identified with the patient's name and an identifying number such as the birth date or file number. Data should also include the patient's age and sex, the testing date, and any variations from the usual STD or machine speed. If required by the physician, also note any medications taken by the patient prior to the test.

ECG paper is delicate and subject to damage when folded or scratched. Paper clips and staples must not be used to secure ECGs. If clear tape is used, it must be a high-quality tape that will not deteriorate or yellow with age. There are four basic types of graph paper: chemical/thermal, wax-coated, plain paper, and glossy paper. Tape should not be used on chemical/thermal paper because chemicals in the adhesive will darken the tracing.

Depending on the format of the ECG, your facility's protocol may require that it be photocopied and/or mounted in a special mounting folder. Convenient mounts are commercially available that display the entire study on a single surface. When mounting, the tracing is edited to eliminate excess recording and portions that contain artifacts. Select the best portion of each lead when editing. Care must be taken to match each lead to its correct location in the mount and to mount each lead right side up.

After being correctly identified, edited, and prepared, the ECG is presented to the physician for interpretation, together with the patient's chart and any previous ECG studies for comparison.

Exercise Tolerance Testing

An **exercise tolerance test,** also called an *ECG stress test,* is used to detect and/or evaluate cardiac ischemia. The ECG is monitored and recorded while the patient performs exercise. The exercise is usually performed on a treadmill, but a stationary bicycle or stair-steps are sometimes used. The patient's blood pressure and pulse are also monitored during the test. Resting ECGs are usually performed both before and after the stress test.

Exercise tolerance testing is performed only in facilities that have trained staff and special equipment to respond to a cardiac emergency, which could arise as a result of the test.

Various protocols are used for stress testing. If your duties involve assisting with exercise tolerance ECG

tests, you will be instructed in the specific protocols observed in your facility and the procedure for patient instruction and preparation.

SPIROMETRY

The term spirometry refers to the measurement of lung air flow using a machine called a spirometer. While several types of tests may be performed using a spirometer, this discussion covers only the **forced expiration test,** which is the principal test for most purposes. Forced expiration spirometry is a simple and extremely useful form of pulmonary function test (PFT) and is sometimes referred to by that name. It is used to evaluate ventilation in patients for a number of reasons. As a screening test, it may alert the physician to early signs of pulmonary disease. This is especially true when the tests are taken periodically and compared over time. Spirometry is also used to evaluate the effectiveness of treatment for pulmonary conditions. It is particularly helpful in monitoring patients with asthma, chronic bronchitis, cystic fibrosis, and chronic obstructive pulmonary disease (COPD) such as emphysema. Spirometry may be used to assess the risk for surgical procedures that are known to affect lung function such as those requiring a general anesthetic.

In work situations where air quality or inhalation of irritants may cause occupational disease, workers may be required to have a baseline spirometry test when they are hired, and may be monitored with spirometry for signs of pulmonary problems over the course of their work. If your facility serves as a provider of occupational health care, spirometry may be performed as a part of routine pre-employment physical examinations and/or occupational health evaluations. When there is a legal need to establish pulmonary impairment or disability, spirometry provides an objective measurement.

During this test the patient takes a deep breath and exhales forcefully into the spirometer tube. The spirometer measures the air flow, recording how much air is forced out of the lungs and how rapidly this is accomplished. The spirometer provides measurements and mathematical data and also produces a graph of the exhalation, which is called a **spirogram.** Depending on the equipment, data recorded for the test may include a number of parameters and calculations of the relationships between the measurements. All spirometers record at least two significant measurements: FVC and FEV_1. **FVC** stands for **forced vital capacity,** the total amount of air the lungs can hold. **FEV** stands for **forced expiratory volume,** and FEV_1 indicates the quantity of air that can be forcefully exhaled during the first second of the test.

The patient's cooperation and effort are essential to the success of the test, so proper patient instruction is needed. Enthusiastic coaching during the test is equally important.

Equipment

There are two commonly used types of spirometers, the volume-displacement type and the flow-sensing type. Many variations and models of each type exist. Most provide measurements, mathematical data, and a spirogram. An operator's manual should accompany the machine. The radiographer who performs spirometry should be familiar with the features of the equipment and the guidelines provided in the manual.

Recent advances in electronics and microprocessor technology have led to the development of portable spirometers of the flow-sensing type that are ideal for use in physicians' offices and clinics. These units automatically calculate a range of ventilation indices that assist the physician in interpreting the results and eliminate the need for mathematical calculations by the technician.

There are two types of spirograms. One plots flow according to the volume of air and is called a *flow-volume graph* (Fig. 25-28). The other is a time-volume graph in which the volume of air is plotted over the time of the exhalation (Fig. 25-29).

The calibration of equipment is very important. Calibration procedures vary greatly depending on the equipment, and are explained in the operation manual.

Flow-sensing spirometers may have disposable or non-disposable flow sensors. Disposable flow sensors are discarded and replaced after each use, and each new sensor must be calibrated. Non-disposable sensors must be disinfected after each use to prevent the potential spread of disease from patient to patient. The operation manual for the equipment will detail these procedures.

Patient Preparation and Testing

It would be unusual for a physician to order a spirometry test on a patient for which the test is contraindicated. Even so, the radiographer who performs these tests should be aware of the principal contraindications. The test should not be performed on patients who have had recent abdominal surgery, thoracic surgery, or eye surgery, including cataract operations. **Hemoptysis** (coughing up blood) from an unknown cause and pneumothorax (collapsed lung) are also contraindications. Certain cardiovascular conditions may prevent the use of this test. For example, patients with angina or unstable blood pressure may experience an aggravation of their conditions.

Patients taking bronchodilating medications should discontinue their use prior to the test. The physician will specify when the medication should be stopped, depending on the specific drug and the duration of its effectiveness. The effect of bronchodilating medications on the patient's condition is sometimes evaluated with spirometry. A baseline test is given, followed by the administration of bronchodilating medication. After a set interval (usually less than 15 minutes), the test is repeated. The physician will prescribe the medication and the exact procedure for post-bronchodilator testing.

To prepare for the test, the patient should loosen any tight clothing about the throat and the waist. Any

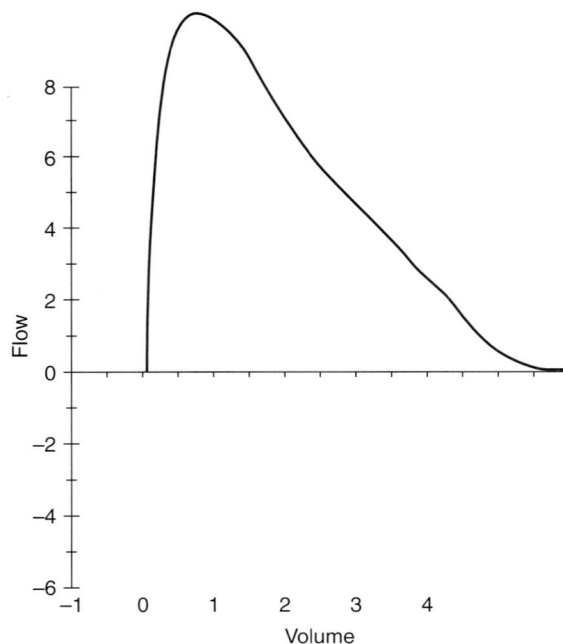

Fig. 25-28 Flow-volume spirogram.

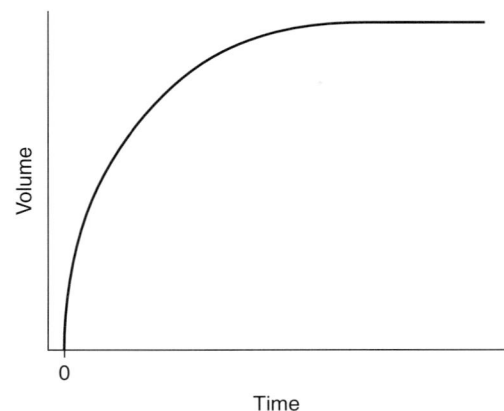

Fig. 25-29 Time-volume spirogram.

clothing that could restrict the patient's ability to breathe should be loosened or removed.

Explain the purpose of the test in simple terms. A good statement is, "This test measures how much air your lungs can hold and how fast you can blow all the air out." Then explain how the patient should perform the test. When instructed to do so, the patient will take the deepest possible breath, seal his or her lips around the mouthpiece, and blow the air out as fast as possible until the lungs are completely empty. It may be helpful for you to demonstrate the maneuver using a disposable mouthpiece that is not connected to the spirometer. The chin should be lifted enough so as not to crowd the throat. The patient should stand or sit erect and should not bend forward during the test (Fig. 25-30). A standing posture is preferred. The lips must seal tightly around the mouthpiece. Disposable nose clamps are not required, but they are sometimes used to ensure that no air escapes from the nose. The patient must make a vigorous effort *from the start* to breathe out hard and fast and must continue to breathe out smoothly until absolutely no more air can be exhaled.

When the equipment and the patient are ready, check the patient's posture and instruct the patient to take the deepest possible breath and then quickly seal

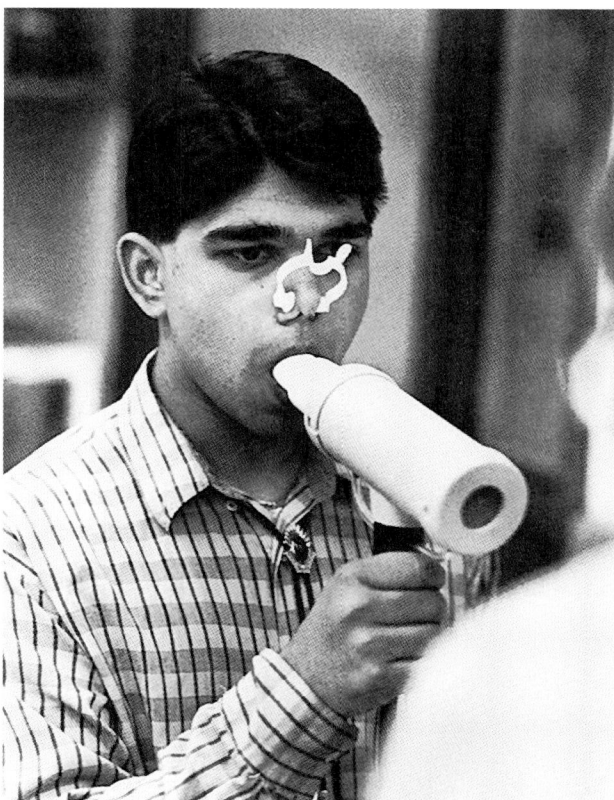

Fig. 25-30 Good posture enhances air flow when taking a spirometry test.

the lips around the mouthpiece. As the patient inhales, encourage deeper and deeper inhalation to obtain a maximum breath. As soon as the mouthpiece is properly situated, instruct the patient to blow. During the maneuver, it is important that you coach the patient, as this will result in a better effort. Raise your voice somewhat and in an urgent tone say, "BLOW, blow hard, keep blowing, keep blowing, don't stop blowing." Continue coaching until the patient has expelled all of the air. At the conclusion of the maneuver, the patient should remove the mouthpiece before taking another breath (unless the equipment is designed for measuring inhaled air). After each maneuver, check the graph and discuss the patient's performance. Point out any aspect of the patient's effort that can be improved and praise the good points. For example, you might say, "That was a great start, but you stopped too soon. Next time, try to keep blowing longer until *all* of the air is exhaled."

The maneuver is repeated until you have obtained three recordings that represent the patient's best effort and are consistent with one another. If this is not accomplished after eight repetitions of the maneuver, stop the test and notify the physician. It is usually not helpful to continue the test, as fatigue will negatively affect the result.

A satisfactory test is characterized by an immediate forceful start, a maximum effort, and a smooth continuous exhalation that does not end abruptly (Fig. 25-31). The most common patient-related problems when performing the forced expiration maneuver are as follows:

- Less than maximum effort (Fig. 25-32)
- Air leaks between the lips and the mouthpiece
- Incomplete inspiration or expiration
- Hesitation at the start of the expiration (Fig. 25-33)
- Coughing, especially during the early part of the expiration (Fig. 25-34)
- **Glottic closure** (sudden cessation of exhalation due to closure of the opening to the trachea) (Fig. 25-35)
- Obstruction of the mouthpiece by the tongue
- Vocalization (voice sounds) during the maneuver
- Poor posture

During the test, be alert for any sign of patient distress. Although this is a very safe test, patients may occasionally experience dizziness, syncope, chest pain, fits of coughing, or difficulty breathing. If any of these symptoms are seen or reported, stop the test and provide appropriate assistance to the patient (see Chapter 22). If the problem does not resolve immediately, notify the physician.

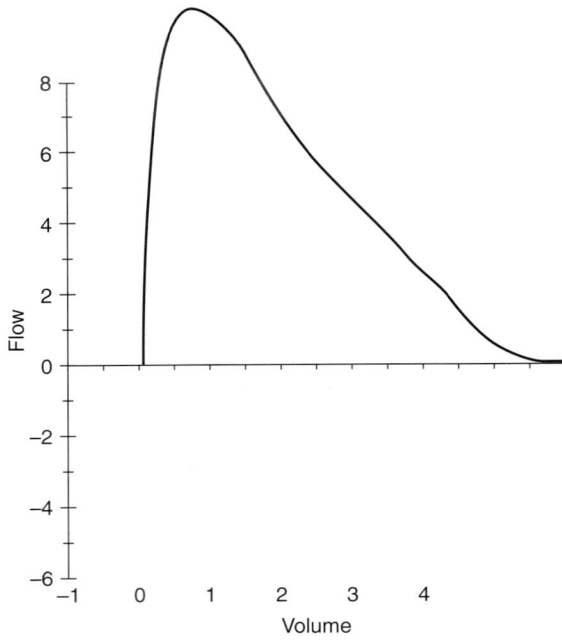

Fig. 25-31 Good effort.

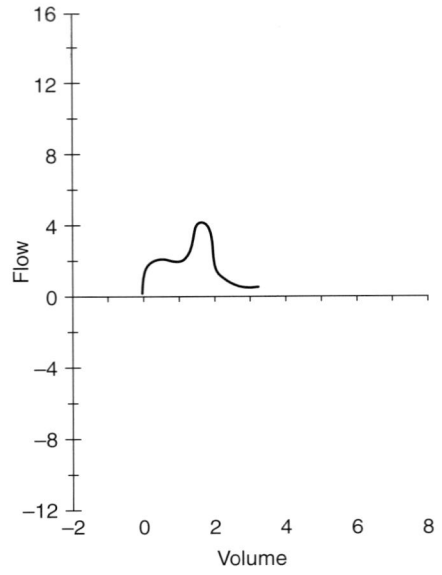

Fig. 25-32 Poor effort.

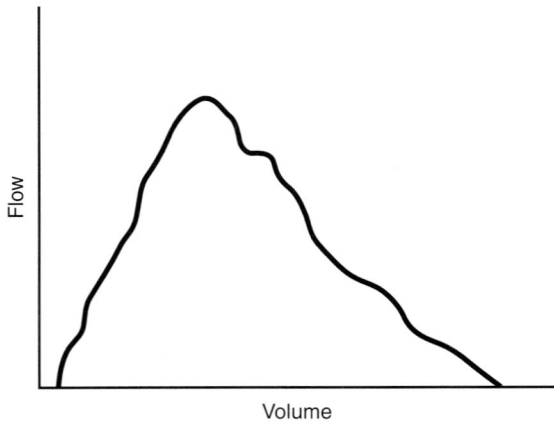

Fig. 25-33 Poor start. Initial exhalation is not forceful enough.

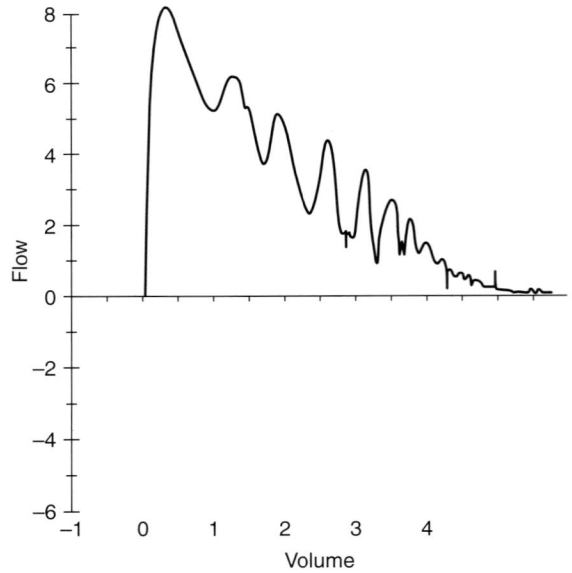

Fig. 25-34 Coughing.

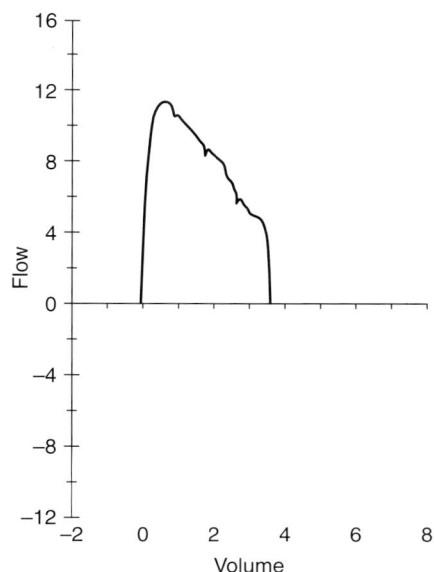

Fig. 25-35 Glottic closure.

Infection Control

There is an obvious and significant danger of infection transmission from body fluids on mouthpieces. For this reason, disposable mouthpieces are usually used. Where non-disposable mouthpieces are used, they must be disinfected with a high-level disinfectant after each use.

Modern spirometers are designed to minimize the possibility of spreading infection. As mentioned before, some spirometers use disposable flow sensors that are discarded after each use for infection control reasons. Unless the spirometer is specifically designed to permit both inhalation and exhalation, the patient should not inhale through the mouthpiece.

Other specific infection control measures vary greatly, depending on the design of the equipment. The operation manual will explain the required procedures, which should be followed explicitly. As with any patient procedure, thorough hand washing before and after the procedure is essential to the patient's safety and to your health.

SUMMARY

Diagnostic assessment skills not directly related to radiography may be a significant asset to the radiographer. This is particularly true when the radiographer is part of the health care team in a small facility. The procedures for weighing and measuring patients and for conducting vision screening tests are commonly performed in physicians' offices and clinics and are easily learned. Electrocardiography and spirometry are more technical and require a higher level of learning for accurate performance. The background knowledge and technical ability learned in radiography are good starting points for learning these complex procedures.

Height and weight are usually measured using a balance scale. Vision screening tests include distance vision acuity measurement using a Snellen chart at a 20-foot distance, near vision evaluation using a reading chart at 14 to 16 inches' distance, and a color perception assessment using the Ishihara color plate test. All must be conducted in a suitable quiet setting with appropriate lighting. Correct recording of results is essential.

Electrocardiography is a heart function test that measures the electrical impulses within the body that cause the heart to beat. The patient is connected to the machine by electrodes and cables, and a tracing is made on graph paper. Radiographers who perform this test must be thoroughly familiar with the equipment used and the procedures for preparing the patient for the test. They must recognize the characteristics of a high-quality tracing and be able to solve artifact problems when they occur.

Forced expiration spirometry is a very useful test for the evaluation of pulmonary health and for monitoring various conditions affecting lung function. To perform this test, the patient blows into the mouthpiece of the spirometer and the machine measures air flow, calculates data, and produces a graph. The radiographer who administers this test should be thoroughly familiar with the equipment, including procedures for its calibration and disinfection. Effective patient teaching and coaching are essential to the success of these tests.

■ Review Questions ■

1. What figures must be added together to calculate the patient's weight using a balance scale?
2. Which direction should the patient face when being measured on a balance scale?
3. What patient instruction should be given at the start of a Snellen test? What, exactly, is meant by the term "20/20 vision"?
4. What is the purpose of an Ishihara test? How is it conducted?
5. Relate the terms "depolarization," "repolarization," and "polarization" to the events that occur in the heart during a heartbeat.
6. If you are taking an ECG tracing and the amplitude of the QRS complex is so great that the stylus moves off the graph paper, what should you do?
7. If you are taking an ECG tracing and the heartbeat is so rapid that the waves cannot be clearly distinguished on the graph, what should you do?
8. How would you recognize muscle artifact on an ECG tracing? What might you do to correct this problem? Describe the appearance of "subtle wandering baseline." How might this problem be solved?
9. List two measurements made using a spirometer.
10. List the characteristics of a good patient effort for a forced expiration spirometry test.
11. What is the minimum number of forced expiration maneuvers involved in a spirometry test? The maximum number?

26

Mathematics for Radiographers

Learning Objectives

At the conclusion of this chapter, the student will be able to:

- Demonstrate calculations using fractions, decimals, percentages, ratios, exponents, and simple algebraic equations
- Identify and use standard measurement units needed by radiographers, state equivalent values for measurements in both the English system and the metric system, and convert measurements from one unit to another
- When given two of the mAs values (mA, time, mAs), calculate the third
- Calculate changes in radiation intensity and required mAs for changes in distance (SID)
- Given a set of exposure factors, calculate the changes needed to change contrast levels using the 15% rule
- Given a set of exposure factors, make appropriate adjustments for changes in patient size using both kVp and mAs
- Given a set of exposure factors, make appropriate adjustments for changes in grid ratio using both kVp and mAs
- Given a set of exposure factors, make appropriate adjustments for changes in the speed of the image receptor system
- Given a transformer ratio and input voltage, calculate the output voltage
- Perform routine medication dose calculations accurately

Key Terms

algebra	improper fraction
base number	lowest terms
common denominator	mixed number
cube	numerator
decimal	percentage
decimal point	power
denominator	proper fraction
dividend	ratio
divisor	remainder
equation	scientific notation
exponent	square
factor	square root

R adiographers often perform mathematic calculations (Fig. 26-1). Measurements, exposure factors, radiation doses, and medication doses require skill in using units of measure and the ability to manipulate these numbers accurately. While the increasing use of computers and computerized equipment has reduced the amount of routine calculation required for radiographers, an understanding of these functions helps radiographers recognize errors when they occur and to make mathematical adjustments when necessary. Certification examinations frequently test these skills by including mathematic problems that must be solved without the aid of a computer or calculator.

It is assumed that the reader can perform basic arithmetic functions: adding, subtracting, multiplying, and dividing whole numbers. This chapter begins with a discussion of basic mathematical principles. Later in the chapter, these skills are applied to practical problems of the types frequently encountered in radiography. Solutions to the practice problems are in Appendix I.

FUNDAMENTAL MATHEMATICAL PRINCIPLES

Terminology

Discussion of calculations is facilitated by naming the various parts of the problems. New terms are introduced throughout this chapter, but defining a few basic terms at the beginning will assist in your understanding of the sections that follow:

- Sum: total, the answer to an addition problem
- Difference: the answer to a subtraction problem
- Product: the answer to a multiplication problem
- **Dividend:** the number divided into in a division problem
- **Divisor:** the number that is divided into the dividend
- Quotient: the answer to a division problem
- **Remainder:** the number that is "left over" when the dividend cannot be evenly divided by the divisor

Fractions

Fractions are parts of whole numbers. They are commonly used in our everyday lives. For example, you can easily relate to the concept of one half (½) of an orange or one quarter (¼) of a dollar. X-ray control panels with synchronous timers have exposure times that are expressed in fractions of seconds.

The lower number of a fraction is called the **denominator.** The denominator indicates how many equal parts the whole has been divided into. The

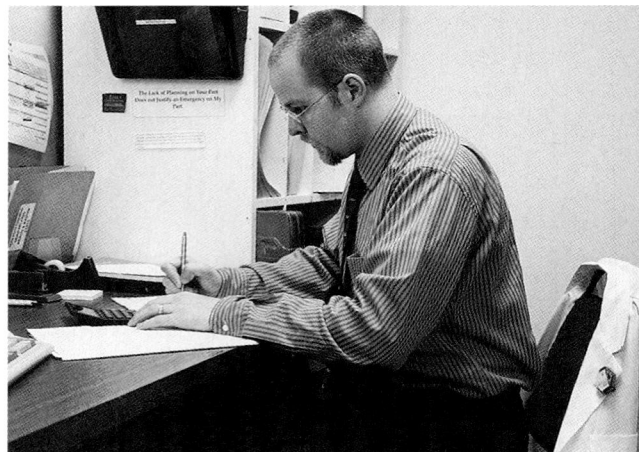

Fig. 26-1 Radiographer struggling with math.

upper number is called the **numerator.** The numerator indicates the number of parts or "pieces" of the divided whole. For example, if you cut a sheet of paper into 8 equal parts, 1 of the parts would be ⅛ of the page. Three parts would be ⅜ of the page.

A **mixed number** consists of a whole number and a fraction. For example, 1½ and 8¾ are mixed numbers.

A **proper fraction** has a value of less than 1. In a proper fraction the numerator is always smaller than the denominator. For example, ⅓, ⅚, and ³⁄₁₆ are proper fractions. An **improper fraction** has a value greater than 1 and the numerator is greater than the denominator. The fractions ³⁄₂, ⁴⁄₃, and ⁷⁄₅ are improper.

Mixed numbers must be changed to improper fractions before they can be multiplied or divided. When a calculation results in an improper fraction, the final result must be converted to a proper fraction. The following rules are used to convert mixed numbers to improper fractions and improper fractions to mixed numbers:

To convert a mixed number to an improper fraction, multiply the whole number times the denominator and add the result to the numerator.

> Example: Convert 2⅔ to an improper fraction.
> Multiply the whole number (2) times the denominator (3): 2 × 3 = 6
> Add the result (6) to the numerator (2): 6 + 2 = 8
> This result (8) is the new numerator. The denominator is unchanged.
> Therefore the result is ⁸⁄₃

To convert an improper fraction to a whole number or mixed number, divide the numerator by the denominator. Any remainder is the numerator of the final fraction. The denominator is unchanged.

Examples: 1. Convert ⁸⁄₄ to a whole number or a mixed number.
Divide the numerator (8) by the denominator (4): $8 \div 4 = 2$
Therefore the result is the whole number 2

2. Convert ⁹⁄₂ into a whole number or a mixed number.
Divide the numerator (9) by the denominator (2): $9 \div 2 = 4$ with a remainder of 1
In this problem, 4 is a whole number and the remainder (1) is the numerator of the fraction. The denominator is unchanged.
Therefore the result is a mixed number: 4½

When both the numerator and the denominator of a fraction are multiplied by the same number, the resulting fraction is equal to the original fraction. This is called "raising the fraction to higher terms." For example, when both the numerator and the denominator of ¾ are multiplied by 2, the resulting fraction is ⁶⁄₈. These two fractions, ¾ and ⁶⁄₈, are equal to each other, and ⁶⁄₈ represents higher terms of ¾.

When both the numerator and the denominator of a fraction can be divided evenly by the same number, the resulting fraction is equal to the original fraction. This is called "reducing the fraction to lower terms." For example, if you divide both the numerator and denominator of ¹⁰⁄₂₀ by 5, the result is ²⁄₄, lower terms of the fraction ¹⁰⁄₂₀ and equal to it. The most usual or common form of a fraction is one in which there is no number except 1 that can be divided evenly into both the numerator and the denominator. Such a fraction is said to be in its **lowest terms.** For example, the fraction ²⁄₄ can be further divided by 2, resulting in ½, the lowest terms of both ¹⁰⁄₂₀ and ²⁄₄.

To reduce a fraction to lowest terms, divide both the numerator and the denominator by the largest number that will divide evenly into both of them.

Example: Reduce ⁹⁄₁₂ to lowest terms.
The largest number that can be divided evenly into both the numerator and the denominator is 3.
Divide the numerator by 3: $9 \div 3 = 3$, the numerator of the new fraction.
Divide the denominator by 3: $12 \div 3 = 4$, the denominator of the new fraction.
Therefore ⁹⁄₁₂ = ¾
Since there is no number except 1 that can be divided evenly into both 3 and 4, ¾ is the lowest terms of ⁹⁄₁₂.

Fractions can be added or subtracted only when they have a **common denominator.** When fractions to be added or subtracted do not have the same denominator, one or more of the fractions is raised to higher terms so the denominators are the same.

The lowest common denominator is the smallest number into which all of the denominators can be divided evenly.

Example: The lowest common denominator of ½, ⅓, and ¼ is 12, the smallest number into which 2, 3, and 4 can all be evenly divided.

To convert fractions to the lowest common denominator, divide the denominator into the lowest common denominator and multiply your answer by the numerator. This number becomes the numerator of the new fraction.

Example: Convert ½ and ⅓ to fractions with a common denominator.
The lowest common denominator is 6.
To convert ½, divide the common denominator (6) by the denominator (3): $6 \div 2 = 3$
Multiply this number (3) by the numerator (1): $3 \times 1 = 3$
Therefore ½ = ³⁄₆. The fraction ½ has been raised to higher terms and its denominator is equal to the common denominator.
To convert ⅓, divide the common denominator (6) by the denominator (3): $6 \div 3 = 2$
Multiply this number (2) by the numerator (1): $2 \times 1 = 2$
Therefore ⅓ = ²⁄₆. The fraction ⅓ has been raised to higher terms and its denominator is equal to the common denominator.

When adding or subtracting fractions, first determine the lowest common denominator. Raise fractions to higher terms, if necessary, so that the denominator of each is the lowest common denominator. Add or subtract the numerators and place the sum or difference over the common denominator.

NOTE: If the answer is an improper fraction, it must be converted to a mixed or whole number. Any fraction in the answer must be reduced to lowest terms.

Example: 1. ⅔ + ¾ + ⅙ = ?
The lowest common denominator is 12.
Raise all fractions to higher terms with 12 as the denominator: ⅔ = ⁸⁄₁₂, ¾ = ⁹⁄₁₂, and ⅙ = ²⁄₁₂
Add the numerators: $8 + 9 + 2 = 19$
Place the sum (19) over the common denominator (12): ¹⁹⁄₁₂
Convert this improper fraction to a mixed number: ¹⁹⁄₁₂ = 1⁷⁄₁₂
Therefore ⅔ + ¾ + ⅙ = 1⁷⁄₁₂

2. ⁷⁄₁₆ − ⅜ = ?
The lowest common denominator is 16.

Raise both fractions to higher terms, if necessary, with 16 as the denominator: $\frac{7}{16}$ does not require conversion, and $\frac{3}{8} = \frac{6}{16}$.
Subtract the second numerator (6) from the first: (7): $7 - 6 = 1$
Place the difference (1) over the common denominator (16): $\frac{1}{16}$
Therefore $\frac{7}{16} - \frac{3}{8} = \frac{1}{16}$

To add mixed numbers, add the whole numbers and the fractions separately. If the sum of the fractions is an improper fraction, convert it to lowest terms and add the whole number to the sum of the other whole numbers.

Examples: 1. $3\frac{1}{3} + 1\frac{1}{2} = ?$
Add the whole numbers: $3 + 1 = 4$
The common denominator of the fractions is 6.
Raise the fractions to higher terms with 6 as the denominator: $\frac{1}{3} = \frac{2}{6}$, $\frac{1}{2} = \frac{3}{6}$
Add the numerators: $2 + 3 = 5$
Place the sum of the numerators (5) over the common denominator (6): $\frac{5}{6}$
Therefore $3\frac{1}{3} + 1\frac{1}{2} = 4\frac{5}{6}$

2. $2\frac{7}{16} + 6\frac{7}{8} = ?$
Add the whole numbers: $2 + 6 = 8$
The lowest common denominator of the fractions is 16.
Raise $\frac{7}{8}$ to higher terms so that the denominator is 16: $\frac{7}{8} = \frac{14}{16}$
Add the fractions: $\frac{7}{16} + \frac{14}{16} = \frac{21}{16}$, an improper fraction.
Convert $\frac{21}{16}$ to a proper fraction: $\frac{21}{16} = 1\frac{5}{16}$
Add the whole number from this calculation (1) to the sum of the other whole numbers (8): $1 + 8 = 9$
Therefore $2\frac{7}{16} + 6\frac{7}{8} = 9\frac{5}{16}$

To subtract mixed numbers, subtract the whole numbers and the fractions separately. If the fraction being subtracted is larger than the fraction from which it is to be subtracted, borrow 1 from the whole number and convert the smaller fraction into an improper fraction.

Example: $3\frac{1}{6} - 1\frac{1}{3} = ?$
Convert fractions to lowest common denominator: $3\frac{1}{6} - 1\frac{2}{6} = ?$
Since the fraction to be subtracted ($\frac{2}{6}$) is larger than the one from which it is to be subtracted ($\frac{1}{6}$), borrow 1 from the whole number (3): $3\frac{1}{6} = 2 + 1\frac{1}{6}$
Convert the borrowed number (1) and the fraction ($\frac{1}{6}$) into an improper fraction: $1\frac{1}{6} = \frac{7}{6}$
The problem now looks like this: $2\frac{7}{6} - 1\frac{2}{6} = ?$

Subtract the whole numbers and the fractions separately: $2 - 1 = 1$; $\frac{7}{6} - \frac{2}{6} = \frac{5}{6}$
Therefore $3\frac{1}{6} - 1\frac{1}{3} = 1\frac{5}{6}$

To multiply fractions, multiply both the numerators and the denominators.

Examples: 1. $\frac{3}{4} \times \frac{1}{5} = ?$
Multiply the numerators: $3 \times 1 = 3$
Multiply the denominators: $4 \times 5 = 20$
Therefore $\frac{3}{4} \times \frac{1}{5} = \frac{3}{20}$

2. $\frac{2}{3} \times \frac{1}{7} \times \frac{2}{5} = ?$
Multiply the numerators: $2 \times 1 \times 2 = 4$
Multiply the denominators: $3 \times 7 \times 5 = 105$
Therefore $\frac{2}{3} \times \frac{1}{7} \times \frac{2}{5} = \frac{4}{105}$

To divide fractions, invert the divisor and multiply. When a fraction is inverted, the numerator becomes the denominator and the denominator becomes the numerator.

Example: $\frac{5}{6} \div \frac{1}{3} = ?$
Invert the divisor and multiply: $\frac{5}{6} \times \frac{3}{1} = \frac{15}{6}$
Convert to a mixed number in lowest terms: $\frac{15}{6} = 2\frac{3}{6} = 2\frac{1}{2}$

To multiply or divide mixed numbers, first convert them to improper fractions. When the problem contains whole numbers, they are converted into fractions with the whole number as the numerator and a denominator of 1. Fractions in the answer must be in the form of proper fractions in lowest terms.

Examples: 1. $1\frac{2}{3} \times 2\frac{1}{2} = ?$
$\frac{5}{3} \times \frac{5}{2} = \frac{25}{6} = 4\frac{1}{6}$

2. $1\frac{3}{4} \div 2 = ?$
Convert mixed numbers and whole numbers to improper fractions: $1\frac{3}{4} = \frac{7}{4}$, $2 = \frac{2}{1}$
Invert the divisor and multiply: $\frac{7}{4} \times \frac{1}{2} = \frac{7}{8}$

When the same number occurs as both a numerator and a denominator in a multiplication or division problem, the two numbers can be canceled. A canceled number becomes the number 1. This often simplifies the calculation.

Example: $\frac{5}{6} \times \frac{2}{5} = ?$
Cancel the 5s: $\frac{1\cancel{5}}{6} \times \frac{2}{\cancel{5}1} = ?$
Multiply to solve: $\frac{1}{6} \times \frac{2}{1} = \frac{2}{6} = \frac{1}{3}$

When *any* numerator and *any* denominator in a multiplication or division problem can be divided evenly by the same number, this reduction simplifies the calculation.

Note: When dividing fractions, cancellation or reduction must occur <u>after</u> the divisor has been inverted.

Example: ½ × 1⅓ = ?

Convert mixed number to improper fraction: ½ × ⁴⁄₃ = ?

Divide the denominator of ½ (2) and the numerator of ⁴⁄₃ (4) by 2, which will divide into both evenly: ¹⁄₂₁ × ²⁴⁄₃

Multiply to solve: ¹⁄₁ × ⅔ = ⅔

Practice Problems Using Fractions

1. Convert the following mixed numbers to improper fractions:
   a. 2½
   b. 5⅚
   c. 3¼
   d. 3⅖
   e. 1⅞

2. Convert the following improper fractions into whole or mixed numbers:
   a. ⁹⁄₄
   b. ¹⁷⁄₆
   c. ¹⁹⁄₂
   d. ²⁴⁄₁₂
   e. ²³⁄₂₀
   f. ²⁷⁄₉
   g. ¹¹⁄₃

3. Reduce the following fractions to lowest terms:
   a. ⁵⁄₁₅
   b. ³⁄₉
   c. ¹²⁄₁₈
   d. ¹⁸⁄₂₀
   e. ¹⁶⁄₂₄
   f. ²⁰⁄₂₅
   g. ⁸⁄₁₀

4. Convert each of the following groups of fractions so that each group has a common denominator:
   a. ⅔, ⅘
   b. ½, ¾, ⁵⁄₁₆
   c. ⅕, ⅔, ³⁄₁₀
   d. ¼, ⅜, ⁵⁄₁₆
   e. ½, ⅜, ⅔, ⁵⁄₁₂, ¾

5. Add each group of fractions in Section 4. Express your answers using proper fractions in lowest terms.

6. Solve the following subtraction problems involving fractions. Express your answers in lowest terms.
   a. ⅜ − ¹⁄₁₆ =
   b. ³⁄₂₀ − ¹⁄₁₀ =
   c. ⅘ − ⅓ =
   d. ⅞ − ¼ =
   e. ⅚ − ⅓ =

7. Add each of the following groups of mixed numbers. Express your answers using proper fractions in lowest terms.
   a. 1½ + 2¼ + 3¾ =

b. 3⅕ + 1³⁄₁₀ + 4³⁄₂₀ =
   c. 2⅗ + 1⅑ =
   d. 4⅓ + 2⅛ + 6⅚ + 10½ =
   e. 21³⁄₁₆ + 10¾ + 5⅛ =

8. Solve the following subtraction problems involving mixed numbers. Express your answers in lowest terms.
   a. 5⅚ − 3⅙ =
   b. 3¾ − 1½ =
   c. 2⅕ − 1³⁄₁₀ =
   d. 4¼ − ⁷⁄₁₆ =
   e. 7½ − 4⅘ =

9. Solve the following multiplication problems involving fractions, mixed numbers, and whole numbers. Cancel or reduce numbers within the fractions, where possible, to simplify your work. Express your answers in proper fractions and in lowest terms.
   a. ¾ × ⁷⁄₁₆ =
   b. 1½ × ⅔ =
   c. 20⅓ × ⅙ × 2 =
   d. ⅖ × ¹⁵⁄₁₆ × 8 =
   e. ⅚ × ⅔ =
   f. 2 ⅔ × ⅛ × ½ =
   g. ⅘ × ²⁄₁₅ × 1¼ =
   h. 6⅕ × ¹⁵⁄₁₆ =
   i. ¼ × ⅕ × ⅛ × ¹⁄₁₀ =
   j. 3⅜ × 1⅗ =

10. Solve the following division problems involving fractions, mixed numbers, and whole numbers. Cancel or reduce numbers within the fractions, where possible, to simplify your work. Express your answers in proper fractions and in lowest terms.
    a. ¹⁄₆₀ ÷ 2 =
    b. 1⅔ ÷ ¹⁄₄ =
    c. 2⅕ ÷ ³⁄₁₀ =
    d. ⅜ ÷ 1⅓ =
    e. 4 ÷ ¾ =

Decimals

A **decimal** is actually a fraction with a denominator of 10, 100, 1000, or any number that consists of a 1 followed by one or more 0s. With these specialized fractions the denominator is not written. Its value is indicated by the position of the numerator in reference to a dot or period called the **decimal point.** Figures to the left of the decimal point are whole numbers. Figures to the right of the decimal point represent the numerator of the fraction. The value of the decimal is determined by the number of figures to the right of the decimal point. The first place to the right of the decimal point is 10ths, the second place 100ths, the third place 1000ths and so on. For example, 0.1 indicates one tenth (¹⁄₁₀) and 2.05 indicates two and five hundredths

(2⁵⁄₁₀₀). You are familiar with the use of decimals to indicate and calculate dollars and cents, with the two figures to the right of the decimal point indicating cents, or hundreths of dollars. Zeros between the decimal point and the figures change the places of the figures and change the value of the decimal. For example, 0.3 equals three tenths (³⁄₁₀), while 0.003 equals three thousandths (³⁄₁₀₀₀). Zeros added to the right of all other figures in a decimal do not change the value of the decimal. These zeros simply raise the decimal to higher terms. For example, 0.08 is equal to 0.0800. The first is read as eight hundredths (⁸⁄₁₀₀) and the second as eight hundred ten thousandths (⁸⁰⁰⁄₁₀₀₀₀). As you know from the preceding section on fractions, ⁸⁰⁰⁄₁₀₀₀₀ can be reduced to ⁸⁄₁₀₀ by dividing both the numerator and the denominator by 100. Therefore, these two fractions are equal and so are the decimals that represent them. For this reason, zeros to the right of the decimal's other figures may be added or dropped without changing the value of the decimal.

To add or subtract decimals, the numbers must be placed so that the decimal points form a vertical line, both in the problem and in the answer.

Examples: 1. 3.4 + 3.04 + 3.004 = ?

$$\begin{array}{r} 3.4 \\ 3.04 \\ +3.004 \\ \hline 9.444 \end{array}$$

2. 6.85 − 2.4315 = ?

For easier subtraction, add zeros as needed, so that both numbers have the same number of decimal places:

$$\begin{array}{r} 6.8500 \\ -2.4315 \\ \hline 4.4185 \end{array}$$

When multiplying decimals, place the decimal point in the product so that the number of decimal places is equal to the total number of decimal places in the numbers being multiplied.

Example: 3.271 × 2.16 = ?

$$\begin{array}{r} 3.271 \text{ (three decimal places)} \\ \times 2.16 \text{ (two decimal places)} \\ \hline 19626 \\ 3271 \\ 6542 \\ \hline 7.06536 \text{ (five decimal places)} \end{array}$$

To divide *into* a decimal, place the decimal point of the quotient in direct alignment with the decimal point of the dividend.

Example: 15.15 ÷ 3 = ?

$$\begin{array}{r} 5.05 \\ 3\overline{)15.15} \end{array}$$

To divide *by* a decimal, move the decimal point to the right of the divisor and move the decimal point of the dividend the same number of spaces to the right. Align the decimal point in the quotient with the new position of the decimal point in the dividend.

Example: 37.3 ÷ .25 = ?

$$\begin{array}{r} 149.2 \\ .25\overline{)37.30.0} \\ \underline{25} \\ 123 \\ \underline{100} \\ 230 \\ \underline{225} \\ 50 \\ \underline{50} \\ 0 \end{array}$$

Note: When calculations involve both decimals and fractions, either the fractions must be converted to decimals, or the decimals must be converted to fractions.

To convert a fraction to a decimal, divide the numerator by the denominator.

Example: Convert ³⁄₄ into a decimal.

$$\begin{array}{r} .75 \\ 4\overline{)3.00} \\ \underline{28} \\ 20 \\ \underline{20} \\ 0 \end{array}$$

When converting a decimal into a fraction or mixed number, the result is a mixed number. Figures to the right of the decimal point become the numerator. The denominator is determined by the number of decimal places in the numerator.

Note: The denominator will be the number 1 followed by the same number of zeros as there are decimal places in the numerator.

Examples: 1. Convert 0.65 into a fraction or mixed number.

$$0.65 = \frac{65}{100} = \frac{13}{20}$$

2. Convert 2.036 into a fraction or mixed number.

$$2.036 = 2\frac{36}{1000} = 2\frac{9}{250}$$

Practice Problems Using Decimals

1. Addition.
   a. $.78 + .01 = ?$
   b. $.24 + .17 + .06 = ?$
   c. $3.2 + 1.04 + .722 = ?$
   d. $31.3 + 5.007 + .516 = ?$
   e. $20 + 1.9 + .838 = ?$
2. Subtraction.
   a. $1.42 - .23 = ?$
   b. $2.008 - .71 = ?$
   c. $4.7 - .528 = ?$
   d. $3 - .54 = ?$
   e. $6.911 - 1.0007 = ?$
3. Multiplication.
   a. $.33 \times .75 = ?$
   b. $.2 \times .934 = ?$
   c. $.03 \times 82 = ?$
   d. $.17 \times 8524 = ?$
   e. $50 \times .7872 = ?$
4. Division.
   a. $.216 \div 3 = ?$
   b. $.12 \div 5 = ?$
   c. $36 \div .09 = ?$
   d. $.49 \div .007 = ?$
   e. $3.44 \div 1.6 = ?$
5. Convert the following fractions and mixed numbers to decimals:
   a. $\frac{3}{5}$
   b. $\frac{1}{8}$
   c. $\frac{1}{20}$
   d. $1\frac{7}{8}$
   e. $3\frac{1}{3}$
6. Convert the following decimals to fractions or mixed numbers. Express your answers in lowest terms.
   a. $.4$
   b. $.625$
   c. 1.8
   d. 2.2
   e. $.008$

Percentage

A **percentage** is a form of fraction with a denominator of 100. The term *percent* means "per hundred," and is indicated by the % sign. For example, the statement that "52% of the population is female," means that out of every 100 people in the population, 52 of them are females. One hundred percent indicates the whole and is equal to the number 1.

Addition or subtraction involving only percentages may be performed as with whole numbers or decimals. A percent sign (%) is added to the answer.

Examples: 1. $4\% + 8\% = 12\%$
2. $35\% - 15\% = 20\%$

When multiplying or dividing percentages or performing calculations that involve a percentage and a whole number or decimal, the percentage must first be converted to a decimal.

To convert a percentage to a decimal, move the decimal point two places to the left.

Examples: 1. $76\% = .76$
2. $4\% = .04$
3. $150\% = 1.5\cancel{0} = 1.5$

To convert a decimal to a percentage, move the decimal point two places to the right and add a percent sign (%).

Examples: 1. $.23 = 23\%$
2. $.08 = 8\%$
3. $1.7 = 170\%$
4. $.177 = 17.7\%$

When performing calculations that involve a percentage and a fraction, either the percentage or the fraction must be converted so that the numbers are expressed in the same form, either decimal or fraction. To convert a percentage to a fraction, the percentage is the numerator of the fraction and 100 is the denominator. If the percentage *contains* a decimal, convert it to an ordinary decimal and then to a fraction.

Examples: 1. $25\% = \frac{25}{100} = \frac{1}{4}$
2. $6.4\% = .064 = \frac{64}{1000} = \frac{8}{125}$

To determine a percentage *of* a number means to multiply the number times the percentage. The percentage is first converted to a decimal.

Example: How much is 25% of 300?
Convert the percentage to a decimal: $25\% = .25$
Multiply the number by the decimal: $300 \times .25 = 75$
Therefore 75 equals 25% of 300.

To determine what percentage one number is in relation to another, divide one number by the other. The dividend is the number whose percentage you are determining, the portion. The divisor is the whole.

Examples: 1. 15 is what percentage of 75?
Divide the portion (15) by the whole (75):
$15 \div 75 = .2$
Convert the decimal into a percentage: $.2 = 20\%$
Therefore 15 equals 20% of 75.

2. 75 is what percentage of 15?
Divide the portion (75) by the whole (15):
$75 \div 15 = 5$
Convert the answer to a percentage: $5 = 500\%$
Therefore 75 equals 500% of 15.

To increase a number by a certain percentage, add the percentage to 100% and multiply the sum times the number to be increased.

Example: Increase 60 by 30%.
Add the percentage increase (30%) to 100%:
    30% + 100% = 130%
Convert the sum of the percentages to a decimal:
    130% = 1.3
Multiply the number to be increased (60) by the converted sum of the percentages (1.3): 60 × 1.3 = 78
Therefore 60 plus 30% of 60 equals 78.

To decrease a number by a certain percentage, subtract the percentage from 100% and multiply the difference times the number to be decreased.

Example: Decrease 60 by 20%.
Subtract the percentage decrease (20%) from 100%:
    100% − 20% = 80%
Convert the difference between the percentages to a decimal: 80% = .8
Multiply the number to be decreased (60) by the converted difference between the percentages (.8): 60 × .8 = 48
Therefore 60 minus 20% of 60 equals 48.

Practice Problems Using Percentages

1. Convert the following percentages to decimals:
   a. 7%
   b. 8.5%
   c. 59%
   d. 99.44%
   e. 140%
   f. 300%
2. Convert the following decimals to percentages:
   a. .2
   b. .19
   c. .91
   d. 1.65
   e. 1.2
   f. 6.0
3. Convert the following percentages to fractions. Convert improper fractions to mixed or whole numbers and reduce fractions to lowest terms.
   a. 60%
   b. 4%
   c. 150%
   d. 12.5%
   e. 62.5%
   f. 200%
4. Perform the following calculations involving only percentages:
   a. 37% + 33% = ?
   b. 100% + 65% = ?
   c. 25% − 13% = ?
   d. 20% × 80% = ?
   e. 150% ÷ 30% = ?
5. Calculatae the value of the following percentages:
   a. 20% of 88
   b. 90% of 200
   c. 50% of 61
   d. 130% of 40
   e. 300% of 12
6. Determine the following percentages. Express your answer to the nearest tenth of a percent.
   a. 13 = ?% of 63
   b. 29 = ?% of 80
   c. 43 = ?% of 200
   d. 40 = ?% of 160
   e. 50 = ?% of 35
7. Calculate the solutions to the following problems that involve increasing and decreasing numbers by a percentage:
   a. Increase 100 by 15%
   b. Increase 30 by 250%
   c. Increase 150 by 25%
   d. Decrease 85 by 10%
   e. Decrease 20 by 12%

Equations

Algebra is a branch of mathematics that provides a method of determining the value of an unknown quantity that has a specific relationship to one or more known quantities. Algebra problems are stated in the form of **equations.** An equation is a mathematical declaration that two mathematical statements (groups of numbers, together with their signs or mathematical functions) are equal to each other. Equations can be very complex, and it is this complexity that causes many students to feel intimidated by the thought of equations. Equations can also be very simple. For example, 3 + 1 = 4 is an equation. The equations that radiographers must solve are relatively simple ones.

The equations used in arithmetic are always stated so that the unknown quantity is to the right of the equal sign. For example, 3 + 1 = ? In algebra, the unknown quantity is usually indicated by a letter, often x, and may be at any position in the equation. The same symbols for mathematical operations used in arithmetic are also used in algebra: plus (+), minus (−), times (×), divided by (÷) and equal (=). For example, 3 + x = 4 is an algebraic equation.

When solving algebraic problems, the equation must be kept in balance. That is, the mathematical statements on both sides of the equal sign must always be equal to each other.

Balance is maintained in an equation by performing the same operation on both sides of the equation.

Example: $3 + x = 4$
Subtract 3 from both sides of the equation:
$3 - 3 + x = 4 - 3$
Perform the mathematical operations:
$0 + x = 1$
Therefore $x = 1$.

In this example, the purpose of subtracting 3 from each side of the equation is to isolate the unknown and determine its value. Other mathematical operations may be performed on both sides of the equation, depending on what is needed to isolate the unknown.

Examples: 1. $3x = 27$
Divide both sides of the equation by 3:

$$\frac{\cancel{3}x}{\cancel{3}} = \frac{\cancel{27}^9}{\cancel{3}}$$

Therefore $x = 9$.

Note that fractions can be canceled and reduced as explained in the section on fractions.

2. $\dfrac{x}{12} = 3$

Multiply both sides of the equation by 12:

$$\cancel{12} \times \frac{x}{\cancel{12}} = 3 \times 12 \quad x = 36$$

Therefore $x = 36$.

3. $x - 10 = 17$
Add 10 to both sides of the equation:
$x - 10 + 10 = 17 + 10$
Perform the mathematical operations:
$-10 + 10 = 0, \ 17 + 10 = 27$
Therefore $x = 27$.

4. $\dfrac{25}{x} \times 3 = 15$

Multiply both sides by x: $\dfrac{25}{x} \times 3 \times x = 15 \times x$

The unknowns can be canceled:
$25 \times 3 = 15x$
Perform calculation: $25 \times 3 = 75, \ 75 = 15x$

Divide both sides by 15: $\dfrac{\cancel{75}^5}{\cancel{15}} = \dfrac{\cancel{15}x}{\cancel{15}}$

Therefore $5 = x$ and $x = 5$.

5. $a + b = c$ (Solve for a)
Subtract b from both sides of the equation:
$a + b - b = c - b$
Therefore, $a = c - b$

6. $a \times b = c$ (Solve for a)
Divide both sides of the equation by b:

$$\frac{a \times \cancel{b}}{\cancel{b}} = \frac{c}{b}$$

Therefore $a = \dfrac{c}{b}$

When an equation consists of two fractions, save steps by eliminating the denominators from consideration by "cross-multiplication." In this operation, each numerator is multiplied by the denominator on the opposite side of the equation.

Example: $\dfrac{16}{4} = \dfrac{8}{x}$

Cross-multiply: $16x = 4 \times 8$
Perform calculation: $16x = 32$
Divide both sides of the equation by 16:

$$\frac{\cancel{16}x}{\cancel{16}} = \frac{\cancel{32}^2}{\cancel{16}}$$

Therefore $x = 2$.

Ratio and Proportion

A **ratio** expresses the operation where one number is divided by another. A ratio may be written using a colon (3:4), a division symbol (3 ÷ 4), a slanted line (3/4), or by placing the dividend over the divisor with a line between them:

$$\frac{3}{4}$$

A proportion is a statement that two ratios are equal to each other. An example of a ratio is 3:4:: 9:12. This is read, "3 is to 4 as 9 is to 12." It could also be written, $\frac{3}{4} = \frac{9}{12}$.

Many mathematical relationships in radiography are proportional to each other, so it is common that the radiographer must solve ratio and proportion problems where one quantity of the proportion is unknown. These problems are set up in the format: a/b = c/d. Numerical values that are known are substituted in the equation and the rules of algebra explained in the previous section are used to solve the problem.

Example: If you walk at the rate of 3 miles per hour, how far can you go in an hour and a half?
The known ratio is 3 mi/1 hr. In the second ratio, the time is known and the distance is not. Therefore, the problem looks like this:

$$\frac{3 \text{ mi}}{1 \text{ hr}} = \frac{x \text{ mi}}{1.5 \text{ hr}}$$

Cross-multiply: $1x = 3 \times 1.5$
Perform calculation: $x = 4.5$ mi

Practice Solving Equations

1. $3x + 6 = 10 + 2$
2. $21/x = 8 - 1$
3. $x - 17 = 13$
4. $20 = 3x - 4$
5. $2x = 18/2$
6. $45 = 9x$
7. $30/x = 6/2$
8. $x/7 = 36/6$
9. $56/8 = 49/x$
10. $12/4 = x/15$

Exponents, Square Roots, and Scientific Notation

Exponents   When a number is multiplied by itself, this operation may be expressed as an **exponent**. An exponent is a small superscript (elevated type position) number that indicates how many times the number is multiplied by itself. For example, the number 64 equals $2 \times 2 \times 2 \times 2 \times 2 \times 2$. The number 2 is multiplied by itself 5 times and the number 2 is used as a multiplication **factor** 6 times. Using an exponent, 64 can be expressed as 2^6. In this case the repeated or **base number** is 2, and the exponent or **power** of the base number is 6. This is read as "2 to the 6th power," or simply as "2 to the 6th." When the exponent is 2 or 3, special terms are commonly used. The number is said to be **squared** when it is multiplied by itself once and the exponent is 2. When the exponent is 3, the number is multiplied by itself twice and is said to be **cubed**.

Examples: 1. What is the square of 12?
$12^2 = 12 \times 12 = 144$
2. What is the value of 2 cubed?
$2^3 = 2 \times 2 \times 2 = 8$
3. Write 1,000,000 as a power of 10.
$1,000,000 = 10 \times 10 \times 10 \times 10 \times 10 \times 10 = 10^6$

NOTE: With powers of 10, the exponent is always equal to the number of zeros in the complete number.
To multiply two powers of the same number, the exponents are added.

Example: $3^2 \times 3^5 = 3^{2+5} = 3^7$

To divide two powers of the same number, the exponents are subtracted.

Example: $10^6 \div 10^4 = 10^{6-4} = 10^2 = 100$

If the power of the divisor is greater than the power of the dividend, the result is a negative power and represents a number that is less than 1.

Example: $10^4 + 10^6 = 10^{4-6} = 10^{-2}$

To determine the value of a negative power, change the sign of the exponent, compute its value, and divide its value into the number 1.

Example: $10^{-2} = 1/10^2 = 1/100 = .01$

NOTE: With negative powers of 10, the number of decimal places is always equal to the number of the negative exponent.

Square Roots   The square root **of a number is that value which, when multiplied by itself, equals the original number. The square root is represented by the radical sign,** $\sqrt{}$.

Examples: 1. $\sqrt{4} = 2$
2. $\sqrt{9} = 3$
3. $\sqrt{16} = 4$
4. $\sqrt{25} = 5$
5. $\sqrt{100} = 10$

Square roots that are small, whole numbers such as those in the examples above can be easily perceived. It is a complex mathematical operation to extract square roots that are not whole numbers. Fortunately today, simple math calculators provide this function.

Scientific Notation

Scientists use a system based on powers of the number 10 to simplify the expression of numbers that are very large or very small. This method is referred to as **scientific notation.**

Large numbers are written with the first figure placed to the left of the decimal point and any other significant (non-zero) figures to the right of the decimal point. This number is then multiplied by the power of 10 that represents the total number of figures in the original number minus 1, that is the total number of figures in the portion of the original number that are to the right of the location where the new decimal point was placed.

Example: Express 48,000,000 using scientific notation.
Place a decimal point after the first figure (4) and drop the insignificant figures (zeros): 4.8.
Count the number of figures in the original number that are to the right of the new position of the decimal point: the number 8 plus 6 zeros equals 7 figures. This number (7) is the power of 10 by which 4.8 is multiplied.
Therefore $48,000,000 = 4.8 \times 10^7$.

To write a very small number in scientific notation, the first significant (non-zero) figure is placed to the left of the decimal point and any other significant figures are placed to its right. The resulting number is multiplied by a *negative* power of 10 that is equal to the number of decimal places between the original location of the decimal point and its new location.

Example: Express .0000000067 using scientific notation.
Place a decimal point after the first significant figure: 6.7.
Count the number of decimal places between the 2 decimal point locations: 8 zeros plus the figure 6 equals 9 decimal places.
This number becomes the negative exponent of 10 by which 6.7 is multiplied.
Therefore $.0000000067 = 6.7 \times 10^{-9}$

Practice Problems Using Exponents, Square Roots, and Scientific Notation

1. Calculate the value of the following exponential terms:
   a. 5 cubed
   b. 2^8
   c. 9^2
   d. 10^4
   e. 10^{-5}
   f. 4^{-3}
2. Solve the following problems involving exponential terms:
   a. $4^2 \times 4^6 = ?$
   b. $10^{10} \times 10^2 = ?$
   c. $3^3 \div 3^2 = ?$
   d. $5^2 \div 5^4 = ?$
   e. $10^6 \div 10^2 = ?$
3. Extract the square roots of the following numbers:
   a. $\sqrt{49} = ?$
   b. $\sqrt{36} = ?$
   c. $\sqrt{64} = ?$
   d. $\sqrt{81} = ?$
   e. $\sqrt{400} = ?$
4. Express the following numbers using scientific notation:
   a. .00075
   b. 23,000
   c. 18,200,000
   d. .0000509
   e. .007365

Measurement Units and Their Conversion

In the United States many measurements are made using the English system, which involves such units as feet, pounds, and gallons. Scientists worldwide prefer measurements made in the metric system because it is based on the number 10, which simplifies many calculations. Radiographers encounter measurements in both systems and must be able to convert measurements readily from one system to the other.

Metric System The basic units of the metric system are the gram for measuring weight, the liter for measuring liquids, and the meter for measuring length. A number of prefixes are used with these units to specify measurements that are larger or smaller than the basic units by factors that are multiples of ten. For example, *kilo* is a prefix meaning 1000. You are already familiar with the fact that 1 kilovolt equals 1000 volts. A kilometer is equal to 1000 meters. The prefix *milli* indicates 1/1000 or .001 times the basic unit. For example, 1 milliliter equals .001 liter. The basic units of the metric system are summarized in Table 26-1. Metric prefixes are summarized in Table 26-2.

English System The relationships between measurements in the English system are far less orderly than those in the metric system. As stated before, the basic units are the pound (weight), the foot (length), and the gallon (liquid), but a number of other units are also used to measure these same parameters in the English system. For example, the ounce and the ton are also weight measurements, the inch and the mile are also used to measure length, and liquids may be measured in ounces and pints. Table 26-3 lists common measurement units of the English system and

TABLE 26-1
Basic Metric Units

Unit	Used to Measure	Abbreviation
Meter	Length	M or m
Liter	Liquid	L or l
Gram	Weight	gm or g

TABLE 26-2
Metric Prefixes

Prefix	Meaning	Abbreviation
Kilo-	1000	k
Hecto-	100	h
Deka-	10	da
Deci-	1/10 (.1)	d
Centi-	1/100 (.01)	c
Milli-	1/1000 (.001)	m
Micro-	1/1,000,000 (.000001)	μ or mc
Nano-	1/1,000,000,000 (.0000000001)	n

TABLE 26-3
English Units

Used to Measure	Unit	Abbreviation	Equivalents
Length	inch	in	
	foot	ft	12 in
	yard	yd	3 ft
	mile	mi	5280 ft
Liquid	ounce	oz	
	pint	pt	16 oz
	quart	qt	2 pt
	gallon	gal	4 qt
Weight	ounce	oz	
	pound	lb	16 oz
	ton		2000 lb

TABLE 26-4
English/Metric Conversion

Type of Unit	English Unit	Metric Equivalent
Length	inch	2.54 cm
Liquid	ounce	30 ml
	quart	0.946 L
Weight	pound	2.2 kg

their relationships to one another. Table 26-4 compares common English measurements with their counterparts in the metric system.

The operation for converting measurements from one unit to another is a form of ratio and proportion problem. It is set up in this format:

$$\frac{\textbf{A units}}{\textbf{B units}} = \frac{\textbf{1 A unit}}{\textbf{B units per unit A}}$$

Examples: 1. Convert 10 km into meters.

$$\frac{10 \text{ km}}{x \text{ m}} = \frac{1 \text{ km}}{1000 \text{ m per km}}$$

Cross-multiply: $x = 10 \times 1000 = 10{,}000$ m.

2. Convert 300 milliamperes (mA) into amperes (A)

$$\frac{300 \text{ mA}}{x \text{ A}} = \frac{1 \text{ mA}}{.001 \text{ mA per A}}$$

Cross-multiply: $x = 300 \times .001 = .3$ amperes

3. Convert 1 foot into centimeters.
1 ft = 12 in
1 in = 2.54 cm

$$\frac{12 \text{ in}}{x \text{ cm}} = \frac{1 \text{ in}}{2.54 \text{ cm per in}}$$

Cross multiply: $x = 12 \times 2.54 = 39.48$ cm

Units of Time Units that measure time are universal and there is only one system, which is a mixture of the English and the metric systems. The base unit is the second. Milliseconds and nanoseconds are short time periods using metric prefixes to indicate thousandths and billionths of seconds, respectively.

The time units greater than a second are probably quite familiar to you. Sixty seconds equal 1 minute, there are 60 minutes in an hour, and 24 hours in a day.

Units of Temperature There are two common scales for measuring temperature. The English system uses the Fahrenheit scales (F). Using this scale, water freezes at 32° F and boils at 212° F.

The Celsius (C) temperature scale is used in the metric system. At one time this scale was called "centigrade," and you may still encounter this term. Each degree on the Celsius scale represents a greater quantity of temperature change than on the Fahrenheit scale. Water freezes at 0° C and boils at 100° C.

The following formulas are used to convert between the Fahrenheit and Celsius temperature scales: F = (C × 1.8) + 32 and C = F − 32/1.8.

Examples: 1. Convert normal body temperature (98.6° F) to the Celsius scale.

$$C = \frac{98.6 - 32}{1.8} = \frac{66.6}{1.8} = 37° \text{ C}$$

2. Convert 25° C to the Fahrenheit scale.
F = (25 × 1.8) + 32
F = 45 + 32 = 77° F

Practice Problems Using Measurement Units
1. Conversions from one metric unit to another.
   a. Convert 75 kilovolts to volts.
   b. Convert 3 meters to centimeters.
   c. Convert 10 milliliters to liters.
   d. Convert 20 grams to kilograms.
   e. Convert 15 centigrams to grams.
2. Conversions from one English unit to another.
   a. Convert 16 inches to yards.
   b. Convert ½ pint to fluid ounces.
   c. Convert 84 inches to feet.
   d. Convert 18 quarts to gallons.
   e. Convert 4.8 pounds to ounces.
3. Conversions between English and metric units.
   a. Convert 3 fl oz to ml.
   b. Convert .25 lb to g.

c. Convert 5 in to m.
d. Convert 30 mm to in.
e. Convert 40 g to oz.
4. Time and temperature conversions.
a. Convert $\frac{1}{20}$ sec to msec.
b. Convert 330 sec to hr.
c. Convert 3.4 days to hr.
d. Convert 19° C to the Fahrenheit scale.
e. Convert 50° F to the Celsius scale.

MATHEMATICS APPLIED TO RADIOGRAPHY

The concept of mAs was introduced in Chapter 4 and further expanded in Chapters 6 and 10. It is the product of mA and exposure time (sec) and indicates the total quantity of radiation involved in an exposure. Several possible combinations of mA and time may be used to obtain a given mAs quantity. The manipulation of these three quantities is an everyday part of the radiographer's work.

The basic formula for determining mAs is: mA × sec = mAs.

Example: Determine the mAs of an exposure made using 300 mA and .4 sec.
300 mA × .4 sec = 120 mAs

According to the principles of algebra explained earlier in this chapter, this formula can be rearranged to isolate any of the three units.

Example: mA × sec = mAs
Divide both sides of the equation by mA to isolate seconds:

$$\frac{\cancel{mA} \times sec}{\cancel{mA}} = \frac{mAs}{mA}$$

Therefore sec = mAs/mA

Example: If the desired mAs is 20 and you wish to use 200 mA, what should be the exposure time?
20 mAs ÷ 200 mA = 1/10 sec or .1 sec.

NOTE: When calculating exposure time for controls that have exposure times in decimals, it is best to divide so as to obtain a decimal. For controls that have exposure times in fractions, form the problem as a fraction and reduce the fraction, if needed, to obtain a fractional exposure time.

The circle in Fig. 26-2 is a handy reminder of the relationship among mA, time, and mAs. When you convert the factor you wish to obtain, the calculation needed is apparent. For example, when mAs is covered, mA and time are separated by a vertical line and should be multiplied to obtain mAs. When mA is covered, time is beneath the horizontal line and mAs is

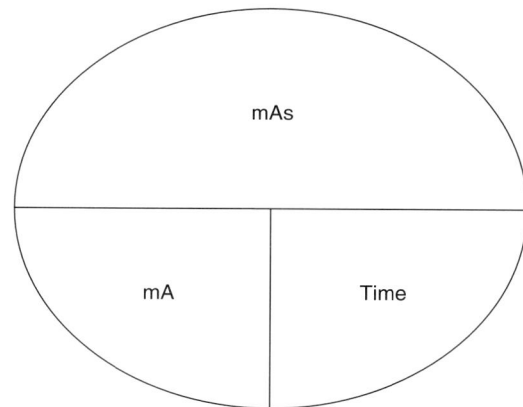

Fig. 26-2 The relationships of mA, time, and mAs.

over it, indicating that mAs is divided by time in seconds to calculate mA.

It is often the case that radiographers calculate values for mA, time, or mAs that are not available on their control panels. It is helpful to remember that variations in exposure of less than 20% are scarcely noticeable on the film. When the ideal exposure factor is not available, simply select the factor that comes closest to the value you want. Most control panels are designed to provide settings within a 20% range of any specific goal you may calculate.

Solving mAs Problems
1. Calculate the mAs for the following exposures. Round any extended decimals to two decimal places.
a. 200 mA, $\frac{1}{40}$ sec
b. 300 mA, $\frac{1}{20}$ sec
c. 100 mA, $\frac{2}{15}$ sec
d. 500 mA, .02 sec
e. 50 mA, .3 sec
f. 150 mA, $1\frac{1}{4}$ sec
g. 400 mA, 2 msec
h. 300 mA, $\frac{1}{120}$ sec
2. Calculate the exposure time for the following exposures. Round any extended decimals to three decimal places.
a. 50 mA, 40 mAs
b. 200 mA, 25 mAs
c. 300 mA, 10 mAs
d. 100 mA, 1 mAs
e. 400 mA, 8 mAs
3. Calculate the mA for the following exposures.
a. 10 mAs, $\frac{1}{30}$ sec
b. 25 mAs, $\frac{1}{2}$ sec
c. 40 mAs, $\frac{2}{15}$ sec
d. 5 mAs, .01 sec
e. 8 mAs, .02 sec

SID and Radiation Intensity

Chapter 6 explains and illustrates the relationship between radiation intensity and source-image distance according to the inverse square law:

Radiation intensity is inversely proportional to the square of the distance.

$$\frac{I_1 \text{ (Original intensity)}}{I_2 \text{ (New intensity)}} = \frac{SID_2^2 \text{ (New distance, squared)}}{SID_1^2 \text{ (Original distance, squared)}}$$

Example: When the SID is changed from 30 inches to 90 inches, what is the relationship between the original radiation intensity and the intensity at the new distance?

Substitute the distances in the formula. The original intensity is assigned a relative value of 1. The new intensity is unknown.

$$\frac{1}{I_2} = \frac{90^2}{30^2}$$

Reduce the fraction: $\dfrac{1}{I_2} = \dfrac{3^2}{1^2}$

Calculate the squares: $\dfrac{1}{I_2} = \dfrac{9}{1}$

Cross multiply: $9 \times I_2 = 1$

To isolate I_2, divide both sides by 9: $\dfrac{9 \times I_2}{9} = \dfrac{1}{9}$

Therefore $I_2 = 1/9$. That is, the new intensity is 1/9 of the original intensity.

While knowing this relationship enhances our understanding, it is not of great practical value to the radiographer. Assuming that the original radiation intensity was satisfactory, the more important question is how to maintain a constant radiation intensity when the distance changes. Mas is used to compensate for changes in distance. The correct change in mAs enables the radiographer to maintain the same radiation intensity when the distance is changed. Since the intensity decreases when the distance increases (an inverse proportion), the mAs must be *increased* when the distance increases (a direct proportion).

The formula for changing mAs to maintain a constant radiation intensity when the distance changes is:

$$\frac{mAs_1}{mAs_2} = \frac{SID_1^2}{SID_2^2}$$

Example: A satisfactory radiograph is made using 20 mAs at 40 in SID. How much mAs is required to produce a similar radiograph at 60 in SID?

Substitute in the formula: $\dfrac{20 \text{ mAs}}{mAs_2} = \dfrac{40^2}{60^2}$

Reduce the fraction: $\dfrac{20 \text{ mAs}}{mAs_2} = \dfrac{\cancel{40^2}}{\cancel{60^2}} = \dfrac{2^2}{3^2}$

Calculate the squares: $\dfrac{20 \text{ mAs}}{mAs_2} = \dfrac{4}{9}$

Cross multiply: $4 \times mAs_2 = 20 \text{ mAs} \times 9$
Multiply: $4 \times mAs_2 = 180 \text{ mAs}$

Divide both sides by 4: $\dfrac{4 \times mAs_2}{4} = \dfrac{\cancel{180} \; 45 \text{ mAs}}{4}$

Therefore the new mAs required at 60 in SID is 45 mAs.

Practice Problems Involving Distance Changes

1. What is the relative change in radiation intensity when the distance changes from 40 in SID to 30 in SID?
2. What is the relative change in radiation intensity when the distance is changed from 72 in SID to 40 in SID?
3. A satisfactory radiograph is made using 15 mAs at 40 in SID. How much mAs is needed to produce a similar radiograph at 48 in SID?
4. A satisfactory radiograph is made using 40 mAs at 72 in SID. How much mAs is needed to produce a similar radiograph at 84 in SID?
5. A satisfactory radiograph is made using 10 mAs at 40 in SID. How much mAs is needed to produce a similar radiograph at 72 in SID?

Exposure Adjustments for Patient/Part Size

Altering kVp for Part Size Change   Adjustments in kilovoltage for variations in part size were discussed in Chapter 10. As stated there, kVp is a useful adjustment only for relatively small variations from normal because large changes in kVp cause significant alteration in the scale of contrast.

Below 85 kVp, an adjustment of 2 kVp per centimeter will compensate for small changes in part size. Above 85 kVp, a change of 3 kVp per centimeter is needed.

Example: A wrist measures 4 cm in PA diameter, and an exposure of 5 mAs at 56 kVp produces a satisfactory radiograph. How much kVp adjustment is needed for the lateral projection, which measures 6 cm?

The kVp range is below 85, so the adjustment is 2 kVp per cm.

The size difference is 2 cm (6 cm − 4 cm).

2 cm × 2 kVp.cm = 4 kVp increase.

The original kVp (56) plus the increase (4) equals the new kVp (60).

The exposure for the lateral wrist projection is 5 mAs at 60 kVp.

Altering mAs for Patient/Part Size Changes   As explained in Chapter 10, mAs is the usual and best choice of factors to adjust when compensating for differences in patient/part size. The mAs is increased by 30% for a 2-cm increase in part size and decreased by 20% for a 2-cm decrease in part size. These percentages may be added or subtracted, respectively, from 100% and the result multiplied by the original mAs. *These changes compound, much like compound interest, and must be applied 2 cm at a time.*

For a 2-cm increase in patient size, increase the mAs by 30% (multiply the mAs by 1.3). For a 2-cm decrease in patient size, decrease the mAs by 20% (multiply the mAs by .8).

Examples: 1. If 20 mAs is a satisfactory exposure for a patient measuring 20 cm, how much mAs is needed for a patient measuring 24 cm?
Multiply the mAs by 1.3 for the first 2 cm increase: 20 × 1.3 = 26 mAs for a 22-cm patient
Multiply the mAs for 22 cm (26) by 1.3 to obtain the mAs for 24 cm: 26 × 1.3 = 33.8 mAs for a 24-cm patient

2. If 60 mAs is a satisfactory exposure for a patient measuring 28 cm, how much mAs is needed for a patient measuring 26 cm?
Multiply the mAs by .8 for a 2-cm decrease: 60 × .8 = 48 mAs for the 26-cm patient.

Practice Problems Involving Patient/Part Size Changes
1. A satisfactory radiograph is made using 96 kVp on a patient measuring 30 cm. Adjust the kVp to compensate for a patient size change to 27 cm.
2. A satisfactory radiograph is made using 74 kVp on a patient measuring 16 cm. Adjust the kVp to compensate for a patient size change to 18 cm.
3. A satisfactory radiograph is made using 50 mAs on a patient measuring 24 cm. Adjust the mAs to compensate for a patient size change to 18 cm.
4. A satisfactory radiograph is made using 30 mAs on a patient measuring 19 cm. Adjust the mAs to compensate for a patient size change to 25 cm.
5. A satisfactory radiograph is made using 10 mAs on a patient measuring 12 cm. Adjust the mAs to compensate for a patient size change to 16 cm.

Altering Contrast with kVp: the 15% Rule

Kilovoltage may be altered to change the scale of contrast. The kVp is increased to lengthen the scale of contrast (decrease contrast and increase latitude) and decreased to shorten the scale, increasing contrast. When kVp is changed, however, the radiographic density is also affected. If the original radiographic density was satisfactory and the radiographer wishes only to change the level of contrast, mAs must be used to compensate for the density change that occurs when kVp is altered. As explained in Chapter 10, the 15% rule provides guidelines for altering the kVp while maintaining a constant radiographic density.

The 15% rule can be used to increase contrast. Decrease the kVp by 15% and multiply the mAs by 2.

Example: A radiograph made using 40 mAs and 90 kVp has satisfactory radiographic density but is lacking in contrast. Suggest a new technique that will provide more contrast with similar radiographic density.
To decrease kVp by 15%, multiply by .85 (100% − 15%): 90 kVp × .85 = 77 kVp

NOTE: When calculating kVp changes, round off to the nearest whole kilovolt.

Multiply the mAs by 2: 40 mAs × 2 = 80 mAs
Therefore the new technique is 80 mAs at 77 kVp.

The 15% rule is also used to decrease contrast, increase latitude, and lower patient dose. Increase the kVp by 15% and divide the mAs by 2.

Example: A satisfactory radiograph is made using 60 mAs and 76 kVp. Suggest a new technique that will provide more latitude and lower the patient dose.
To increase kVp by 15%, multiply by 1.15 (100% + 15%): 76 kVp × 1.15 = 87 kVp
Divide the mAs by 2: 60 mAs ÷ 2 = 30 mAs
Therefore the new technique is 30 mAs at 87 kVp.

NOTE: The 15% rule may be reapplied to the new technique if the first calculation does not produce sufficient change.

Practice Problems Using the 15% Rule
1. An exposure made using 25 mAs and 86 kVp has satisfactory radiographic density. Suggest a new technique that will provide more contrast.
2. An exposure made using 100 mAs and 80 kVp has satisfactory radiographic density. Suggest a new technique that will decrease the patient dose.
3. An exposure made using 10 mAs and 66 kVp has satisfactory radiographic density. Suggest a new technique that will provide more latitude.
4. An exposure made using 40 mAs and 94 kVp has satisfactory radiographic density. Suggest a new technique that will provide more contrast.

5. An exposure made using 60 mAs and 72 kVp has satisfactory radiographic density. Suggest a new technique that will provide less contrast.

Grid Conversions

The use of grids is explained in Chapter 9. Grids absorb both remnant and scatter radiation that would otherwise expose the film. For this reason, grid use results in underexposure unless the exposure is increased. When an established technique is designed for grid use and the exposure is made without a grid, compensation is required to prevent overexposure. The amount of compensation required in each case depends upon the ratio of the grid that is being added or removed. Either mAs or kVp may be used to compensate for adding or removing a grid. The grid conversion table from Chapter 10 is reproduced here as Table 26-5 for convenience in solving grid conversion problems.

Examples: 1. A satisfactory technique is 25 mAs and 78 kVp with a 12:1 grid.
Change the kVp to compensate for taking this film without a grid.
The kVp is reduced when a grid is removed. The quantity of kVp needed to compensate for a 12:1 grid is 16 kVp (from Table 26-5).
78 kVp − 16 kVp = 62 kVp. The mAs is unchanged.

2. A satisfactory technique is 10 mAs and 65 kVp without a grid. Change the mAs to compensate for taking this film with an 8:1 grid.
According to Table 26-5, mAs is multiplied by 2 when 8:1 grid is added.
10 mAs × 2 = 20 mAs. The kVp is unchanged.

NOTE: *When changing grid ratios and compensating with kVp, the amount of the kVp change is equal to the difference in grid ratios. When compensating with mAs, change to a non-grid technique and convert the non-grid technique to a grid technique for the new grid ratio.*

Practice Problems Involving Grid Conversions
1. A satisfactory film is produced using 30 mAs with a 10:1 grid. Adjust this mAs for the removal of the grid.
2. A satisfactory film is produced using 5 mAs without a grid. Adjust this mAs for the addition of a 12:1 grid.
3. A satisfactory film is produced using 75 kVp with an 8:1 grid. Adjust this kVp for the removal of the grid.
4. A satisfactory film is produced using 60 kVp without a grid. Adjust this kVp for the addition of a 10:1 grid.
5. A satisfactory film is produced using 80 kVp and a 12:1 grid. Adjust this kVp for the use of an 8:1 grid.
6. A satisfactory film is produced using 10 mAs and a 10:1 grid. Adjust this mAs for the use of a 6:1 grid.

Calculations Involving Transformer Ratios

Radiographers have no need to calculate transformer output or construction in the course of their work. They are sometimes expected to do so, however, to demonstrate an understanding of transformer function. As explained in Chapter 3, transformers are designed with a certain ratio of windings between the primary and secondary sides. This is called the *transformer ratio*. A step-up transformer has more windings on the secondary side, and a step-down transformer has more windings on the primary side. The voltage on the two sides of the transformer is proportional to the ratio of the windings. When any three of these four factors is known, the fourth can be determined

---

TABLE 26-5

*Grid Conversion Table**

Grid Ratio	From Non-Grid to Grid, Add kVp; from Grid to Non-Grid, Subtract kVp	OR	From Non-Grid to Grid, Multiply mAs by Grid Factor; from Grid to Non-Grid, Divide mAs by Grid Factor
6:1	10		1.5
8:1	12		2
10:1	14		2.5
12:1	16		3
16:1	20		4

*This table is useful for body part sizes in the 8 to 18 cm range.

using the method for ratios and proportions discussed earlier in this chapter.

The transformer ratio is equal to the number of primary windings divided by the number of secondary windings. The input and output voltages of a transformer are proportional to the transformer ratio:

$$\frac{\text{primary windings}}{\text{secondary windings}} = \frac{\text{input voltage}}{\text{output voltage}}$$

Examples: 1. A high-voltage transformer has 100 windings on the primary side and 50,000 windings on the secondary side. What is the transformer ratio?

Transformer ratio =

$$\frac{\text{primary windings}}{\text{secondary windings}} = \frac{100}{50000} = \frac{1}{500}$$

Therefore the transformer ratio = 1:500.

2. If the input voltage for this transformer is 200 V, what is the output voltage?

Substitute in the formula:

$$\frac{1}{500} = \frac{200}{x}$$

Cross-multiply: $x = 200 \times 500 = 100,000\,\text{V} = 100\,\text{kV}$

Practice Problems Involving Transformers

1. A transformer has 90 windings on the primary side and 30 windings on the secondary side. What is its ratio? Is this a step-up or step-down transformer?
2. The transformer in problem 1 has input voltage of 120 V. What is its output?
3. A transformer has 100 windings on the primary side. The input voltage is 120 V, and the output voltage is 12 V. How many windings are on the secondary side?
4. A transformer with a ratio of 1:400 has an output voltage of 96 kV. What is the input voltage?
5. A transformer has 150 windings on the primary side, and 450 on the secondary side. What is the ratio? Is this a step-up or step-down transformer?

Medication Dosage Calculations

Calculation of medication dosage was discussed in Chapter 23. This section reviews that information and provides practice problems.

Medications are provided in many forms, but the most common forms are tablets and liquids. Regardless of form, each medication is provided in specific strengths. The strength of the medication indicates the amount of active drug contained in a certain volume of the medication. For example, a liquid medication may have 5 micrograms of solid dissolved in each milliliter of liquid and the strength would be stated as 5 mcg/ml. A tablet may contain 2 grams of the active drug and would be labeled simply as 2 g/tablet. The basic formula for determining the correct quantity of any drug is as follows:

$$\frac{\textbf{Dose}}{\textbf{Strength}} = \textbf{Volume}$$

In this formula, *dose* refers to the prescribed amount of the active ingredient in the medication. *Strength* refers to the amount of active ingredient per unit of volume as described in the preceding paragraph. *Volume* refers to the total quantity of medication that is administered.

Example: The physician has prescribed a dose of 50 mg of Demerol IM. The available stock has a strength of 20 mg/ml. How much of this solution should you draw up into the syringe?

$$\frac{50\,\text{mg (dose)}}{20\,\text{mg/ml (strength)}} = \frac{2.5\,\text{ml to be injected}}{\text{(volume)}}$$

Practice Problems Using Medication Dosage

1. The prescribed dose is 100 mg. The available stock is in the form of 25 mg tablets. How many should be given?
2. The prescribed dose is 250 mg. The available stock has a strength of 50 mg/ml. How much should be given?
3. The prescribed dose is 40 mcg. The available stock has a strength of 80 mcg/tablet. How much should be given?
4. The prescribed dose is 5 mg. The available stock has a strength of 1 mg/ml. How much should be given?
5. Patient reports taking 8 tablets of Advil a day. The tablets have a strength of 200 mg. What is the patient's daily dose?

State Licensing Agencies for Radiography

As of 1999, the most recent data available, 34 states and Puerto Rico had licensing laws for radiographers in effect. Arkansas and South Carolina have enacted licensure laws, and their rules are forthcoming. The laws and regulations vary widely from state to state.

Students or radiographers desiring information should contact the appropriate agency for the particular state. Although addresses and phone numbers sometimes change, the most current available listings for each of the licensing states and Puerto Rico are listed below.

ARIZONA

State of Arizona, Medical Radiologic Technology Board of Examiners, 4814 South 40th Street, Phoenix, AZ 85040. (602) 255-4845.

CALIFORNIA

State of California, Radiologic Health Branch, 714 P Street, Sacramento, CA 95814. (916) 445-6695.

COLORADO

Colorado State Medical Board, 1560 Broadway, Suite 1300, Denver, CO 80202-5140. (303) 894-7714.

ITEP Exam Processing Center, PO Box 7871, Colorado Springs, CO 80933. (719) 392-2452.

CONNECTICUT

Department of Public Health, Bureau of Health Systems Regulation, Capitol Avenue, MS #12APP, PO Box 340308, Hartford, CT 06134. (860) 509-7562, extension 410.

DELAWARE

State of Delaware, Office of Radiation Control, Robbins Building, PO Box 637, Dover, DE 19903. (302) 736-4731.

FLORIDA

State of Florida, HRS Radiation Control, (850) 487-1004 and (850) 487-3451.

HAWAII

State of Hawaii, Radiologic Technology Board, Department of Health, Noise and Radiation Branch, 591 Ala Moana Boulevard, Honolulu, HI 96813-2498. (808) 548-4383.

ILLINOIS

State of Illinois, Division of Radiologic Technologist Certification, Illinois Department of Nuclear Safety, 1035 Outer Park Drive, Springfield, IL 62704. (217) 785-9915.

INDIANA

State of Indiana, Radiological Health Section, PO Box 1964, Indianapolis, IN 46206-1964. (317) 633-0150.

IOWA

State of Iowa, Department of Health, Lucas State Office Building, Des Moines, IA 50319-0075. (515) 281-3478.

KENTUCKY

State of Kentucky, Radiation Control Branch, 275 East Main Street, Frankfort, KY 40621. (502) 564-3700.

LOUISIANA

Louisiana State Radiologic Technology Board of Examiners, 3108 Cleary Avenue, Suite 207, Metairie, LA 70002. (504) 838-5231.

MAINE

State of Maine, Radiologic Technology Board of Examiners, State House Station #35, Augusta, ME 04333. (207) 582-8723.

MARYLAND

State of Maryland, Public Health Engineer, 2500 Broening Highway, Baltimore, MD 21224. (301) 631-3300.

MASSACHUSETTS

State of Massachusetts, Radiation Control Program, 150 Tremont Street, 11th Floor, Boston, MA 02111. (617) 727-6214.

MINNESOTA

*State of Minnesota, Department of Health—Radiation Control Section, 121 East Seventh Place / PO Box 64975, St. Paul, MN 55164. (612) 215-0941.

MISSISSIPPI

*State Department of Health Professional Licensure, PO Box 1700, Jackson, MS 39215-1700. (601) 987-4153.

MONTANA

State of Montana, Department of Commerce, Board of Radiologic Technologists, 1424 Ninth Avenue, Helena, MT 59620. (406) 444-4288.

NEBRASKA

State of Nebraska, Division of Radiological Health, 301 Centennial Mall South, Lincoln, NE 68509. (402) 471-2168.

*State licensure requirements are pending.

NEW JERSEY

State of New Jersey, Department of Environmental Protection, Bureau of Radiological Health, CN 415, Trenton, NJ 08625-0415. (609) 987-2022.

NEW MEXICO

State of New Mexico, Radiation Protection Bureau, PO Box 968, Santa Fe, NM 87504-0968. (505) 827-2773. (505) 827-2941.

NEW YORK

Bureau of Environment, Radiation Protection, New York State Department of Health, Room 325, 2 University Place, Albany, NY 12203. (518) 458-6482.

OHIO

Ohio Department of Health. Ms. Margaret Cipkala Wanachick, Chief, Radiologic Technology Section, 246 North High Street, PO Box 118, Columbus, OH 43266-0118. (614) 752-4319.

OREGON

Oregon Board of Radiologic Technology, Oregon State Health Division, PO Box 231, Portland, OR 97207. (503) 229-5054.

PENNSYLVANIA

Bureau of Professional & Occupational Affairs, State Board of Medicine, 124 Pine Street, PO Box 2649, Harrisburg, PA 17105-2649. (717) 783-4858.

PUERTO RICO

Puerto Rico Examination Board for Radiology and Radiotherapy Technologists. Physical address: 800 Roberto H. Todd Avenue, Suite 202, Santurce, Puerto Rico. Postal address: Call Box 10200, San Juan, PR 00908-0200. (787) 725-8161. Present President: Mr. Eduardo Brito, RT, MPH.

RHODE ISLAND

Rhode Island Department of Health, Division of Professional Regulations, 3 Capitol Hill, Providence, RI 02908. (401) 277-2827.

TENNESSEE

State of Tennessee, Board of Medical Examiners, 283 Plus Park Boulevard, Nashville, TN 37217. (615) 367-6231.

TEXAS

Texas Department of Health, Medical Radiologic Technology Program, 1100 West 49th Street, Austin, TX 78756-3183. (512) 459-2960.

UTAH

Utah Department of Commerce, Division of Occupational and Professional Licensing, Heber M. Wells Building, 160 East 300 South, Salt Lake City, Utah 84145-0805. (801) 530-6403.

VERMONT

State of Vermont, Board of Radiologic Technology, Division of Licensing and Registration, Office of the Secretary of State, Pavilion Office Building, Montpelier, VT 05609-1101. (802) 828-2886.

VIRGINIA

Commonwealth of Virginia, Department of Health Professions, 6606 West Broad Street, 4th Floor, Richmond, VA 23230. (804) 662-7664.

WASHINGTON

Washington State Office of Radiation, Olympic Building, South 220, 217 Pine Street, Seattle, WA 98101-1549. (206) 464-6840.

WEST VIRGINIA

West Virginia Radiologic Technology Board of Examiners, 1715 Flat Top Road, Cool Ridge, WV 25825. (304) 787-4398.

WYOMING

State of Wyoming, Board of Radiologic Technologist Examiners, 1312 Monroe Avenue, Cheyenne, WY 82001. (307) 778-7319.

Sample X-ray Technique Charts

	EXTREMITIES									PELVIS AND HIP		
	SMALL HAND, WRIST, FOOT			MEDIUM ELBOW, ANKLE, LEG			LARGE SHOULDER, KNEE					
	DETAIL SCREENS XYZ FILM			DETAIL SCREENS XYZ FILM			RAPID SCREENS XYZ FILM			RAPID SCREENS XYZ FILM		
	40″ SID, NON-GRID			40″ SID, NON-GRID			40″ SID, 12:1 GRID			40″ SID, 12:1 GRID		
cm	mA	sec	kVp	mA	sec	kVp	mA	sec	kVp	mA	sec	kVp
2	100	.02	56									
3	100	.03	56									
4	100	.04	56									
5	100	.05	56									
6	100	.06	56	100	.35	62						
7	100	.07	56	100	.04	62						
8	100	.08	56	100	.05	62	100	.04	76			
9				100	.06	62	100	.05	76			
10				100	.07	62	100	.06	76			
11				100	.08	62	100	.07	76			
12				100	.1	62	100	.08	76			
13				100	.11	62	100	.09	76			
14				100	.12	62	100	.1	76	200	.04	80
15							100	.11	76	200	.05	80
16							100	.12	76	200	.06	80
17							100	.13	76	200	.07	80
18							100	.15	76	200	.08	80
19							100	.16	76	200	.1	80
20							100	.18	76	200	.12	80
21										200	.15	80
22										200	.18	80
23										200	.2	80
24										200	.25	80
25										200	.3	80
26										200	.35	80
27										200	.4	80
28										200	.5	80
29										200	.6	80
30										200	.75	80
31										200	.8	80
32										200	1	80

NOTE: Shield gonads.

CERVICAL SPINE								
LATERAL FLEXION, EXTENSION, NON-GRID OBLIQUE			AP LOWER CERVICAL NON-GRID OBLIQUE			AP UPPER CERVICAL (OPEN-MOUTH)		
RAPID SCREENS XYZ FILM			RAPID SCREENS XYZ FILM			RAPID SCREENS XYZ FILM		
72" SID, NON-GRID			40" SID, 12:1 GRID			40" SID, 12:1 GRID		
cm	mA	sec	kVp	mA	sec	kVp	mA	sec	kVp
8	100	.35	76	100	.04	76	100	.04	80
9	100	.04	76	100	.05	76	100	.05	80
10	100	.05	76	100	.06	76	100	.06	80
11	100	.06	76	100	.07	76	100	.07	80
12	100	.07	76	100	.08	76	100	.08	80
13	100	.08	76	100	.1	76	100	.1	80
14	100	.1	76	100	.11	76	100	.11	80
15	100	.11	76	100	.12	76	100	.12	80
16	100	.12	76	100	.15	76	100	.15	80

NOTE: Shield gonads and eyes on all views. Shield thyroid on all views except AP lower cervical. Measure through path of central ray except for APOM. For APOM, measure as for AP lower cervical.

THORACIC SPINE						
	AP			LATERAL		
	RAPID SCREENS XYZ FILM			RAPID SCREENS XYZ FILM		
	40" SID, 12:1 GRID			40" SID, 12:1 GRID		
cm	mA	sec	kVp	mA	sec	kVp
16	200	.04	86			
17	200	.05	86			
18	200	.05	86			
19	200	.06	86			
20	200	.07	86			
21	200	.08	86			
22	200	.1	86			
23	200	.11	86			
24	200	.12	86	25	.6	86
25	200	.13	86	25	.7	86
26	200	.15	86	25	.8	86
27	200	.18	86	25	.8	86
28	200	.2	86	25	.9	86
29	200	.22	86	25	1	86
30	200	.25	86	25	1.2	86
31	200	.27	86	25	1.2	86
32	200	.3	86	25	1.5	86
33	200	.3	86	25	1.8	86
34	200	.35	86	25	2	86
35	200	.4	86	25	2	86
36	200	.5	86	50	1.2	86
37				50	1.5	86
38				50	1.8	86
39				50	2	86
40				50	2.2	86

NOTE: Shield gonads.

	LUMBAR SPINE					
	AP			LATERAL		
	RAPID SCREENS XYZ FILM			RAPID SCREENS XYZ FILM		
	40″ SID, 12:1 GRID			40″ SID, 12:1 GRID		
cm	mA	sec	kVp	mA	sec	kVp
16	200	.06	82			
17	200	.07	82			
18	200	.08	82			
19	200	.1	82			
20	200	.12	82			
21	200	.15	82			
22	200	.18	82			
23	200	.2	82			
24	200	.25	82	200	.22	90
25	200	.3	82	200	.25	90
26	200	.35	82	200	.27	90
27	200	.4	82	200	.3	90
28	200	.5	82	200	.35	90
29	200	.6	82	200	.35	90
30	200	.7	82	200	.4	90
31	200	.8	82	200	.4	90
32	200	.9	82	200	.5	90
33	200	1	82	200	.5	90
34	200	1.2	82	200	.6	90
35	200	1.5	82	200	.7	90
36	200	1.7	82	200	.8	90
37		1.7	82	200	1	90
38		2	82	200	1.2	90
39		2	82	200	1.2	90
40		2.2	82	200	1.5	90

NOTE: Shield gonads.

	CHEST						RIBS					
	PA AND OBLIQUE			LATERAL			ABOVE DIAPHRAGM			BELOW DIAPHRAGM		
							AP, PA, OBLIQUE			AP, PA, OBLIQUE		
	RAPID SCREENS XYZ FILM			RAPID SCREENS XYZ FILM			RAPID SCREENS XYZ FILM			RAPID SCREENS XYZ FILM		
	72″ SID, 10:1 GRID			72″ SID, 10:1 GRID			40″ SID, 12:1 GRID			40″ SID, 12:1 GRID		
cm	mA	sec	kVp	mA	sec	kVp	mA	sec	kVp	mA	sec	kVp
16	300	.004	120				200	.04	72	200	.08	78
17	300	.005	120				200	.05	72	200	.09	78
18	300	.005	120				200	.05	72	200	.1	78
19	300	.006	120				200	.06	72	200	.12	78
20	300	.007	120				200	.06	72	200	.13	78
21	300	.008	120				200	.07	72	200	.15	78
22	300	.008	120				200	.07	72	200	.18	78
23	300	.01	120				200	.08	72	200	.18	78
24	300	.01	120	300	.015	120	200	.08	72	200	.2	78
25	300	.012	120	300	.018	120	200	.1	72	200	.25	78
26	300	.012	120	300	.018	120	200	.1	72	200	.27	78
27	300	.015	120	300	.02	120	200	.12	72	200	.3	78
28	300	.018	120	300	.022	120	200	.12	72	200	.35	78
29	300	.02	120	300	.025	120	200	.15	72	200	.4	78
30	300	.02	120	300	.03	120	200	.15	72	200	.45	78
31	300	.022	120	300	.035	120	200	.18	72	200	.5	78
32	300	.025	120	300	.035	120	200	.18	72	200	.6	78
33	300	.027	120	300	.04	120	200	.2	72	200	.7	78
34	300	.03	120	300	.04	120	200	.2	72	200	.75	78
35	300	.035	120	300	.05	120	200	.22	72	200	.8	78
36	300	.04	120	300	.05	120	200	.25	72	200	.9	78
37	300	.04	120	300	.06	120	200	.27	72	200	1	78
38	300	.05	120	300	.07	120	200	.3	72	200	1.2	78
39	300	.06	120	300	.08	120	200	.33	72	200	1.2	78
40	300	.07	120	300	.1	120	200	.25	72	200	1.5	78
41				300	.12	120	200	.37	72	200	1.5	78
42				300	.05	120	200	.4	72	200	1.7	78

NOTE: Shield gonads. Measure through path of central ray.

ABDOMEN			
AP, OBLIQUE (UPRIGHT OR RECUMBENT)			
RAPID SCREENS XYZ FILM			
40″ SID, 12:1 GRID			
cm	mA	sec	kVp
16	200	.1	70
17	200	.12	70
18	200	.15	70
19	200	.2	70
20	200	.25	70
21	200	.3	70
22	200	.35	70
23	200	.4	70
24	200	.5	70
25	200	.6	70
26	200	.7	70
27	200	.8	70
28	200	1	70
29	200	1.2	70
30	200	1.5	70
31	200	1.7	70
32	200	2	70
33	200	2	70
34	200	2.2	70
35	200	2.5	70
36	200	2.7	70
37	200	3	70
38	200	2.5	76
39	200	3	76
40	200	3	78

	SKULL									SINUSES FACIAL BONES		
	AP, PA PA AXIAL (CALDWELL'S)			LATERAL			AP AXIAL (TOWNE'S), SUBMENTOVERTEX (SMV)					
	RAPID SCREENS XYZ FILM			RAPID SCREENS XYZ FILM			RAPID SCREENS XYZ FILM			RAPID SCREENS XYZ FILM		
	40″ SID, 12:1 GRID			40″ SID, 12:1 GRID			40″ SID, 12:1 GRID			40″ SID, 12:1 GRID		
cm	mA	sec	kVp	mA	sec	kVp	mA	sec	kVp	mA	sec	kVp
10				200	.018	80				100	.06	70
11				200	.022	80				100	.07	70
12				200	.03	80				100	.1	70
13				200	.035	80				100	.11	70
14				200	.045	80				100	.12	70
15	200	.05	80	200	.05	80				100	.15	70
16	200	.07	80	200	.06	80	200	.07	86	100	.18	70
17	200	.08	80	200	.07	80	200	.08	86	100	.2	70
18	200	.09	80	200	.08	80	200	.08	86	100	.25	70
19	200	.1	80	200	.09	80	200	.1	86	100	.3	70
20	200	.12	80	200	.1	80	200	.12	86	100	.35	70
21	200	.15	80	200	.11	80	200	.15	86	100	.4	70
22	200	.18	80	200	.12	80	200	.15	86	100	.4	70
23	200	.18	80				200	.18	86	100	.5	70
24	200	.2	80				200	.18	86	100	.6	70
25	200	.2	80				200	.2	86	100	.7	70
26	200	.22	80				200	.2	86	100	.8	70
27							200	.25	86			
28							200	.25	86			

Optimum Kilovoltage Ranges

NON-GRID/NON-BUCKY

Extremities

Small (fingers, hand, wrist, toes, foot)	50 to 60 kVp
Medium (forearm, elbow, ankle)	55 to 65 kVp
Large (lower leg, knee*, humerus, shoulder*)	60 to 70 kVp

Chest

PA* (heart, lungs, mediastinum)	75 to 90 kVp
Lateral* (heart, lungs, mediastinum)	80 to 90 kVp
Ribs*, PA, AP	60 to 68 kVp

Spine

Cervical, lateral	70 to 80 kVp

Skull

Sinuses* (all views)	60 to 70 kVp
Facial bones* (Waters', Caldwell's, lateral, tangential)	60 to 70 kVp
Nasal bones (axial, lateral)	52 to 60 kVp

WITH GRID/BUCKY

Extremities

Humerus, shoulder	72 to 82 kVp
Knee	70 to 84 kVp

Chest

PA, lateral (heart, lungs, mediastinum)	100 to 130 kVp

*Grid/Bucky techniques preferred.

Ribs (above diaphragm)	66 to 74 kVp
Ribs (below diaphragm)	70 to 78 kVp
Sternum, oblique	60 to 72 kVp

Spine

Cervical, AP lower	70 to 80 kVp
Cervical, AP upper (open-mouth)	74 to 86 kVp
Cervical, lateral, oblique	70 to 80 kVp
Thoracic, AP	80 to 90 kVp
Thoracic, lateral	76 to 86 kVp
Thoracic, swimmer's	75 to 86 kVp
Lumbar, AP	74 to 88 kVp
Lumbar, oblique	80 to 90 kVp
Lumbar, lateral	85 to 100 kVp
Lumbar, lateral spot, L5-S1	90 to 110 kVp
Sacrum, AP	75 to 85 kVp
Sacrum, lateral	90 to 105 kVp
Coccyx, AP	66 to 78 kVp
Lateral	74 to 84 kVp

Pelvis/Hips

AP	72 to 82 kVp
Frog-leg lateral	72 to 82 kVp
Cross-table lateral	76 to 86 kVp

Skull

AP, PA, Caldwell's, Waters'	74 to 86 kVp
Occipital (Towne's), axial (SMV/VSM)	80 to 90 kVp
Sinuses, Caldwell's, PA	70 to 80 kVp
Facial bones, all views except lateral	66 to 76 kVp
Sinuses/facial bones, lateral	60 to 70 kVp

mAs Table

TIME		MILLIAMPERES							
Seconds		25	50	100	150	200	300	400	500
0.008	1/120	0.21	0.42	0.83	1.25	1.67	2.5	3.33	4.17
0.017	1/60	0.42	0.83	1.67	2.5	3.33	5	6.67	8.33
0.025	1/40	0.625	1.25	2.5	3.75	5	7.5	10	12.5
0.033	1/30	0.83	1.67	3.33	5	6.67	10	13.33	16.67
0.042	1/24	1.04	2.08	4.17	6.25	8.33	12.5	16.67	20.83
0.05	1/20	1.25	2.5	5	7.5	10	15	20	25
0.067	1/15	1.67	3.33	6.67	10	13.33	20	26.67	33.33
0.083	1/12	2.08	4.17	8.33	12.5	16.67	25	33.33	41.67
0.1	1/10	2.5	5	10	15	20	30	40	50
0.133	2/15	3.33	6.67	13.33	20	26.67	40	53.33	66.67
0.2	1/5	5	10	20	30	40	60	80	100
0.25	1/4	6.25	12.5	25	37.5	50	75	100	125
0.3	3/10	7.5	15	30	45	60	90	120	150
0.333	1/3	8.33	16.67	33.33	50	66.67	100	133.33	166.67
0.4	2/5	10	20	40	60	80	120	160	200
0.5	1/2	12.5	25	50	75	100	150	200	250
0.6	3/5	15	30	60	90	120	180	240	300
0.8	4/5	20	40	80	120	160	240	320	400
1	1	25	50	100	150	200	300	400	500
1.25	1 1/4	31.25	62.5	125	187.5	250	375	500	625
1.5	1 1/2	37.5	75	150	225	300	450	600	750
1.75	1 3/4	43.75	87.5	175	262.5	350	525	700	875
2	2	50	100	200	300	400	600	800	1000
2.5	2 1/2	62.5	125	250	375	500	750	1000	1250
3	3	75	150	300	450	600	900	1200	1500
3.5	3 1/2	87.5	175	350	525	700	1050	1400	1750
4	4	100	200	400	600	800	1200	1600	2000
5	5	125	250	500	750	1000	1500	2000	2500

Nomogram for Determining Patient Skin Dose from X-ray Exposure*

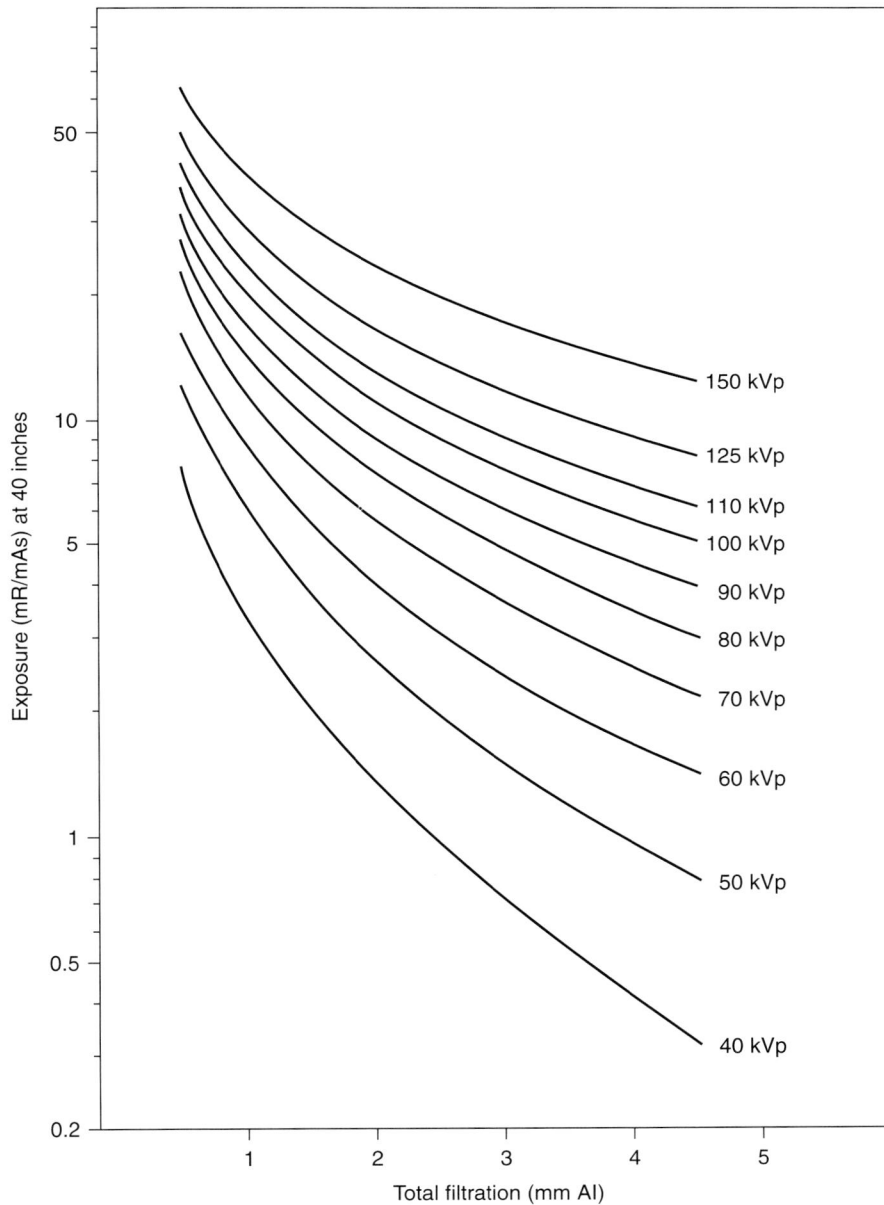

Exposure (mR/mAs) at 40 inches

- 150 kVp
- 125 kVp
- 110 kVp
- 100 kVp
- 90 kVp
- 80 kVp
- 70 kVp
- 60 kVp
- 50 kVp
- 40 kVp

Total filtration (mm Al)

*Courtesy Edward McCullough.

Usual Minimum Views for Routine Examinations

UPPER EXTREMITY

Hand	PA, oblique (anterolateral), lateral
Finger	PA, oblique (anterolateral), lateral
Thumb	AP, oblique, lateral
Wrist	PA, oblique (anterolateral), lateral
Forearm	AP, lateral
Elbow	AP, lateral
Humerus	AP, lateral
Shoulder	AP with external rotation, AP with internal rotation
Shoulder (trauma)	AP without rotation, transthoracic lateral
Clavicle	PA, axial
Scapula	AP, lateral
Acromioclavicular joints	Bilateral AP, with and without weight bearing

LOWER EXTREMITY

Foot	AP, medial oblique, lateral
Toe	AP, medial oblique, lateral
Calcaneus	Axial, lateral
Ankle	AP, medial oblique, lateral
Lower leg	AP, lateral
Knee	AP, lateral
Distal femur	AP, lateral
Proximal femur	AP, lateral
Hip	AP, frog-leg lateral

SPINE

Cervical	AP (lower cervical), AP open-mouth (upper cervical), lateral
Thoracic	AP, lateral
Lumbar	AP, lateral
Sacrum	AP, lateral
Coccyx	AP, lateral
Sacroiliac joints	Axial, bilateral obliques

BONY THORAX AND CHEST

Right upper posterior ribs	AP, right posterior oblique
Left upper posterior ribs	AP, left posterior oblique
Right upper anterior ribs	PA, left anterior oblique
Left upper anterior ribs	PA, right anterior oblique
Right lower posterior ribs	AP, right posterior oblique
Left lower posterior ribs	AP, left posterior oblique
Right lower anterior ribs	PA, left anterior oblique
Left lower anterior ribs	PA, right anterior oblique
Sternum	Lateral, right anterior oblique
Chest	PA, left lateral

ABDOMEN

Abdomen (non-acute)	Variable, depending on diagnostic goal
Acute abdomen	AP supine (KUB), AP upright, PA chest

SKULL, FACIAL BONES, AND PARANASAL SINUSES

Skull	PA, AP axial (Towne's), lateral
Facial bones	PA axial (Caldwell's), Waters', lateral
Paranasal sinuses	PA axial (Caldwell's), Waters', lateral, axial (SMV or VSM)

Critique of Sample Films from Chapter 19

REVIEW FILM #1, FIG. 19-27

This lateral skull radiograph shows objectionable artifacts in the form of hairpins that should have been removed. Screen artifacts are seen over the orbits. An abrasion artifact is seen over the mandible. The lower margin exhibits fog, probably due to cassette leak. Poor centering resulted in failure to show the contour of the cranium at the skull vertex and position of the identification area over the frontal bone.

REVIEW FILM #2, FIG. 19-28

Double exposure. An AP abdomen radiograph is superimposed on a lateral chest view. In addition, there is light fog in the upper left corner due to unlatching of the cassette in daylight. The film identification was not printed in the reserved area. The light artifacts to the viewer's left of L1 represent surgical clips and are within the patient's body.

REVIEW FILM #3, FIG. 19-29

Lateral projection of the tibia and fibula is too light. The exposure should be increased by approximately 200%. No joint is included on the film. The anatomy is not aligned to the film. The collimation is not aligned to the film, producing irregular margins. A small screen artifact is visible in the dark area on the viewer's right near the bottom of the exposed area. The side marker was placed upside down.

REVIEW FILM #4, FIG. 19-30

AP projection of the lower cervical spine is too dark. At least a 60% reduction in mAs is needed. Positioning would be improved by extension of the neck to avoid superimposing the mandible over the lower body of C2.

REVIEW FILM #5, FIG. 19-31

AP projection of the shoulder and upper humerus is too light due to a large, rounded tissue density that covers most of the pertinent anatomy. This shadow represents the patient's breast. While this radiograph could be improved by increasing the mAs by approximately 150%, the best solution would be to position the patient upright so that the breast is not projected over the area of interest.

Charting Terms and Abbreviations*

ABBREVIATIONS TYPICALLY USED IN CHARTING

Abbreviation	Word or Phrase	Abbreviation	Word or Phrase
abd.	abdomen	gyn.	gynecology
a.c.	before meals	(H)	hypodermically
ad lib	freely, as desired	H. or hrs. or hr	hour, hours
amt.	amount	H_2O	water
AP	apical pulse	HA	headache
aq.	water	Hb	hemoglobin
b.i.d. (2 i.d.)	2 times a day	H.S.	bedtime
BP	blood pressure	I & O	intake and output
B.R.P.	bathroom privileges	IM	intramuscular
C or cent	centigrade	IV	intravenous
c̄	with	Kg. or kg.	kilogram
caps.	capsule	KUB	kidneys, ureters, and bladder
cc.	cubic centimeter	L	left
CHF	congestive heart failure	l.	liter
cm.	centimeter	lab.	laboratory
D.C.	discontinue	LBP	low back pain
ECG	electrocardiogram	LLQ	left lower quadrant—abdomen
ED	emergency department	LP	lumbar puncture
EEG	electroencephalogram	LUQ	left upper quadrant—abdomen
ENT	ear, nose, and throat	MI	myocardial infarction
ER	emergency room	mcg	microgram
fld.	fluid	mg.	milligram
G.B.	gallbladder	ml	milliliter
GI	gastrointestinal	MVA	motor vehicle accident
Gm. or gm. or g	gram	NKI	no known injury
gtt.	drop, drops	noct.	at night
GU	genitourinary	N.P.O.	nothing by mouth
		N.S.	normal saline solution

---

*From Ehrlich RA, Daly JA, McCloskey ED: *Patient care in radiography with an introduction to medical imaging,* ed 5, St Louis, 1999, Mosby.

Abbreviation	Word or Phrase
OB, obs.	obstetrics
OD	right eye
O.J.	orange juice
O.P.C.	outpatient clinic
OR	operating room
OS	left eye
P or p̄	after
p.c.	after meals
pH	hydrogen ion concentration
P.O.	by mouth
P.P.	postprandial, after meals
p.r.n.	when necessary, as needed
q.2.h.	every 2 hours
q.h.	every hour
q.i.d. (4 i.d.)	4 times a day
q.s.	sufficient quantity
RBC	red blood count
RUQ	right upper quadrant—abdomen
Rx	therapy
s̄	without
SOB	short of breath
spec	specimen
SQ	subcutaneous
s̄s̄	one half
stat.	at once
t.i.d. (3 i.d.)	3 times a day
TPR	temperature, pulse, respiration
URI	upper respiratory infection
UTI	urinary tract infection
WBC	white blood count
W.C.	wheelchair
wt.	weight
x	times

DESCRIPTIVE TERMS TYPICALLY USED IN CHARTING

Area of Concern to Use	Factor to be Charted	Suggested Terms to Use
Amounts	Large amount	Excessive, profuse, copious
	Moderate amount	Moderate, usual
	Small amount	Scanty, slight
General	Thin and under-nourished	Emaciated
	Fat, greatly overweight	Obese
	Seems very sick	Acutely ill
Appetite	Loss of appetite	Anorexia
	Refuses to eat	Refused food (state reason)
Mental attitude	Has "don't care" attitude	Apathetic
	Afraid, worried	Anxious, apprehensive
	Feeling blue, sad	Depressed
		Other characteristic terms: anxiety, defiance, anger, pain, boredom, worry, happiness, dissatisfaction, irritability
Areas of the back	Small of the back	Lumbar region
	End of spine	Sacral region
	Buttocks	Gluteal area
Bleeding	Very little	Oozing
	Nosebleed	Epistaxis
	Blood in vomitus	Hematemesis
	Blood in urine	Hematuria
	Coughing or spitting up blood	Hemoptysis
	Bleeding stopped	Hemorrhage controlled
Breathing	Breathing	Respiration
	Act of inhaling	Inspiration
	Act of exhaling	Expiration
	Difficulty in breathing	Dyspnea, dyspneic

Area of Concern to Use	Factor to be Charted	Suggested Terms to Use	Area of Concern to Use	Factor to be Charted	Suggested Terms to Use
Breathing	Unable to breathe lying down	Orthopnea	Chill	Blanket applied to keep warm	External heat applied
	Cessation of breathing for short periods	Apnea		Severity (degree of)	Severe, moderate, slight
				Duration	Persistent or short duration
	Rapid breathing	Hyperpnea		Came on suddenly	Sudden onset
	Increasing dyspnea with periods of apnea	Cheyne-Stokes respiration	Level of consciousness	Fully conscious, aware of surroundings	Alert, fully conscious
	Large amount of air inspired or expired	Deep breathing		Only partly conscious	Stuporous
	Small amount of air inspired or expired	Shallow breathing		Unconscious, but can be aroused	Semicomatose
	Abnormal variations in rhythm	Irregular respiration		Unconscious, cannot be aroused	Comatose

Solutions to Practice Problems in Chapter 26

PRACTICE PROBLEMS USING FRACTIONS

1. Converting mixed numbers to improper fractions.
   a. $2\frac{1}{2} = \frac{5}{2}$
   b. $5\frac{5}{6} = \frac{35}{6}$
   c. $3\frac{1}{4} = \frac{13}{4}$
   d. $3\frac{2}{5} = \frac{17}{5}$
   e. $1\frac{7}{8} = \frac{15}{8}$

2. Converting improper fractions to whole or mixed numbers.
   a. $\frac{9}{4} = 2\frac{1}{4}$
   b. $\frac{17}{6} = 2\frac{5}{6}$
   c. $\frac{19}{2} = 9\frac{1}{2}$
   d. $\frac{24}{12} = 2$
   e. $\frac{23}{20} = 1\frac{3}{20}$
   f. $\frac{27}{9} = 3$
   g. $\frac{11}{3} = 3\frac{2}{3}$

3. Reducing fractions to lowest terms.
   a. $\frac{5}{15} = \frac{1}{3}$
   b. $\frac{3}{9} = \frac{1}{3}$
   c. $\frac{12}{18} = \frac{2}{3}$
   d. $\frac{18}{20} = \frac{9}{10}$
   e. $\frac{16}{24} = \frac{2}{3}$
   f. $\frac{20}{25} = \frac{4}{5}$
   g. $\frac{8}{10} = \frac{4}{5}$

4. Converting groups of fractions to a common denominator.
   a. $\frac{2}{3} = \frac{10}{15}$ $\qquad$ $\frac{4}{5} = \frac{12}{15}$
   b. $\frac{1}{2} = \frac{6}{12}$ $\qquad$ $\frac{3}{4} = \frac{9}{12}$ $\qquad$ $\frac{5}{6} = \frac{10}{12}$
   c. $\frac{1}{5} = \frac{6}{30}$ $\qquad$ $\frac{2}{3} = \frac{20}{30}$ $\qquad$ $\frac{3}{10} = \frac{9}{30}$
   d. $\frac{1}{4} = \frac{4}{16}$ $\qquad$ $\frac{3}{8} = \frac{6}{16}$ $\qquad$ $\frac{5}{16}$
   e. $\frac{1}{2} = \frac{12}{14}$ $\qquad$ $\frac{3}{8} = \frac{9}{24}$ $\qquad$ $\frac{2}{3} = \frac{16}{24}$
      $\frac{5}{12} = \frac{10}{24}$ $\qquad$ $\frac{3}{4} = \frac{18}{24}$

5. Adding fractions.
   a. $\frac{10}{15} + \frac{12}{15} = \frac{22}{15} = 1\frac{7}{15}$
   b. $\frac{6}{12} + \frac{9}{12} + \frac{10}{12} = \frac{25}{12} = 2\frac{1}{12}$
   c. $\frac{6}{30} + \frac{20}{30} + \frac{9}{30} = \frac{35}{30} = 1\frac{5}{30} = 1\frac{1}{6}$
   d. $\frac{4}{16} + \frac{6}{16} + \frac{5}{16} = \frac{15}{16}$
   e. $\frac{12}{24} + \frac{9}{24} + \frac{16}{24} + \frac{10}{24} + \frac{18}{24} = \frac{65}{24} = 2\frac{17}{24}$

6. Subtracting fractions.
   a. $\frac{3}{8} - \frac{1}{16} \rightarrow \frac{6}{16} - \frac{1}{16} = \frac{5}{16}$
   b. $\frac{3}{20} - \frac{1}{10} \rightarrow \frac{3}{20} - \frac{2}{20} = \frac{1}{20}$
   c. $\frac{4}{5} - \frac{1}{3} \rightarrow \frac{12}{15} - \frac{5}{15} = \frac{7}{15}$
   d. $\frac{7}{8} - \frac{1}{4} \rightarrow \frac{7}{8} - \frac{2}{8} = \frac{5}{8}$
   e. $\frac{5}{6} - \frac{1}{3} \rightarrow \frac{5}{6} - \frac{2}{6} = \frac{3}{6} = \frac{1}{2}$

7. Adding mixed numbers.
   a. $1\frac{1}{2} + 2\frac{1}{4} + 3\frac{3}{4} \rightarrow 1\frac{2}{4} + 2\frac{1}{4} + 3\frac{3}{4} = 6\frac{6}{4} = 7\frac{2}{4} = 7\frac{1}{2}$
   b. $3\frac{1}{5} + 1\frac{3}{10} + 4\frac{3}{20} \rightarrow 3\frac{4}{20} + 1\frac{6}{20} + 4\frac{3}{20} = 8\frac{13}{20}$
   c. $2\frac{3}{5} + 1\frac{1}{9} \rightarrow 2\frac{27}{45} + 1\frac{5}{45} = 3\frac{32}{45}$
   d. $4\frac{1}{3} + 2\frac{1}{8} + 6\frac{5}{6} + 10\frac{1}{2} \rightarrow 4\frac{8}{24} + 2\frac{3}{24} + 6\frac{20}{24} + 10\frac{12}{24} = 22\frac{43}{24} = 23\frac{19}{24}$
   e. $21\frac{3}{16} + 10\frac{3}{4} + 5\frac{1}{8} \rightarrow 21\frac{3}{16} + 10\frac{12}{16} + 5\frac{2}{16} = 36\frac{17}{16} = 37\frac{1}{16}$

8. Subtracting mixed numbers.
   a. $5\frac{5}{6} - 3\frac{1}{6} = 2\frac{4}{6} = 2\frac{2}{3}$
   b. $3\frac{3}{4} - 1\frac{1}{2} \rightarrow 3\frac{3}{4} - 1\frac{2}{4} = 2\frac{1}{4}$
   c. $2\frac{1}{5} - 1\frac{3}{10} \rightarrow 2\frac{2}{10} - 1\frac{3}{10} \rightarrow 1\frac{12}{10} - 1\frac{3}{10} = \frac{9}{10}$
   d. $4\frac{1}{4} - \frac{7}{16} \rightarrow 4\frac{4}{16} - \frac{7}{16} \rightarrow 3\frac{20}{16} - \frac{7}{16} = 3\frac{13}{16}$
   e. $7\frac{1}{2} - 4\frac{4}{5} \rightarrow 7\frac{5}{10} - 4\frac{8}{10} \rightarrow 6\frac{15}{10} - 4\frac{8}{10} = 2\frac{7}{10}$

9. Multiplying fractions.
   a. $\frac{3}{4} \times \frac{7}{16} = \frac{21}{64}$
   b. $1\frac{1}{2} \times \frac{2}{3} \rightarrow \frac{1\cancel{3}}{\cancel{2}_1} \times \frac{1\cancel{2}}{\cancel{3}_1} \rightarrow \frac{1}{1} \times \frac{1}{1} = \frac{1}{1} = 1$
   c. $20\frac{1}{3} \times \frac{1}{6} \times 2 \rightarrow \frac{61}{3} \times \frac{1}{6} \times \frac{2}{1} = \frac{122}{18} = 6\frac{14}{18} = 6\frac{7}{9}$
   d. $\frac{2}{5} \times \frac{15}{16} \times 8 \rightarrow \frac{1\cancel{2}}{\cancel{5}_1} \times \frac{3\cancel{15}}{\cancel{16}_8} \times \frac{8}{1} \rightarrow \frac{1\cancel{2}}{\cancel{5}_1} \times \frac{3\cancel{15}}{\cancel{16}_8{}_1} \times \frac{1\cancel{8}}{1} \rightarrow \frac{1}{1} \times \frac{3}{1} \times \frac{1}{1} = \frac{3}{1} = 3$
   e. $\frac{5}{6} \times \frac{2}{3} \rightarrow \frac{5}{\cancel{6}_3} \times \frac{1\cancel{2}}{3} \rightarrow \frac{5}{3} \times \frac{1}{3} = \frac{5}{9}$
   f. $2\frac{2}{3} \times \frac{1}{8} \times \frac{1}{2} \rightarrow \frac{1\cancel{8}}{3} \times \frac{1}{\cancel{8}_1} \times \frac{1}{2} \rightarrow \frac{1}{3} \times \frac{1}{1} \times \frac{1}{2} = \frac{1}{6}$
   g. $\frac{4}{5} \times \frac{2}{15} \times 1\frac{1}{4} \rightarrow \frac{4}{5} \times \frac{2}{15} \times \frac{5}{4} \rightarrow \frac{1\cancel{4}}{5_1} \times \frac{2}{15} \times \frac{1\cancel{5}}{\cancel{4}_1} \rightarrow \frac{1}{1} \times \frac{2}{15} \times \frac{1}{1} = \frac{2}{15}$
   h. $6\frac{1}{5} \times \frac{15}{16} \rightarrow \frac{31}{\cancel{5}_1} \times \frac{3\cancel{15}}{16} \rightarrow \frac{31}{1} \times \frac{3}{16} = \frac{93}{16} = 5\frac{13}{16}$
   i. $\frac{1}{4} \times \frac{1}{5} \times \frac{1}{8} \times \frac{1}{10} = \frac{1}{1600}$
   j. $3\frac{3}{8} \times 1\frac{3}{5} \rightarrow \frac{27}{8} \times \frac{8}{5} \rightarrow \frac{27}{\cancel{8}} \times \frac{\cancel{8}}{5} \rightarrow \frac{27}{1} \times \frac{1}{5} = \frac{27}{5} = 5\frac{2}{5}$

10. Dividing fractions.
   a. $\frac{1}{60} \div 2 \rightarrow \frac{1}{60} \div \frac{2}{1} \rightarrow \frac{1}{60} \times \frac{1}{2} = \frac{1}{120}$
   b. $1\frac{2}{3} \div \frac{1}{4} \rightarrow \frac{5}{3} \div \frac{1}{4} \rightarrow \frac{5}{3} \times \frac{4}{1} = \frac{20}{3} = 6\frac{2}{3}$
   c. $2\frac{1}{5} \div \frac{3}{10} \rightarrow \frac{11}{5} \div \frac{3}{10} \rightarrow \frac{11}{5} \times \frac{10}{3} \rightarrow$
      $\frac{11}{5_1} \times \frac{\cancel{10}^2}{3} \rightarrow \frac{11}{1} \times \frac{2}{3} = \frac{22}{3} = 7\frac{1}{3}$
   d. $\frac{3}{8} \div 1\frac{1}{3} \rightarrow \frac{3}{8} \div \frac{4}{3} \rightarrow \frac{3}{8} \times \frac{3}{4} = \frac{9}{32}$
   e. $4 \div \frac{3}{4} \rightarrow \frac{4}{1} \div \frac{3}{4} \rightarrow \frac{4}{1} \times \frac{4}{3} = \frac{16}{3} = 5\frac{1}{3}$

PRACTICE PROBLEMS USING DECIMALS

1. Addition.
   a. $\begin{array}{r} .78 \\ +.01 \\ \hline .79 \end{array}$ d. $\begin{array}{r} 31.3 \\ 5.007 \\ +.516 \\ \hline 36.823 \end{array}$

   b. $\begin{array}{r} .24 \\ .17 \\ +.06 \\ \hline .47 \end{array}$ e. $\begin{array}{r} 20 \\ 1.9 \\ +.838 \\ \hline 22.738 \end{array}$

   c. $\begin{array}{r} 3.2 \\ 1.04 \\ +.722 \\ \hline 4.962 \end{array}$

2. Subtraction.
   a. $\begin{array}{r} 1.42 \\ -.23 \\ \hline 1.19 \end{array}$ d. $\begin{array}{r} 3.00 \\ -.54 \\ \hline 2.46 \end{array}$

   b. $\begin{array}{r} 2.008 \\ -.71 \\ \hline 1.298 \end{array}$ e. $\begin{array}{r} 6.9110 \\ -1.0007 \\ \hline 5.9103 \end{array}$

   c. $\begin{array}{r} 4.700 \\ -.528 \\ \hline 4.172 \end{array}$

3. Multiplication.
   a. $\begin{array}{r} .75 \\ \times.33 \\ \hline 225 \\ .225 \\ \hline .2475 \end{array}$ d. $\begin{array}{r} 8524 \\ \times.17 \\ \hline 1449.08 \end{array}$

   b. $\begin{array}{r} .934 \\ \times.2 \\ \hline .1868 \end{array}$ e. $\begin{array}{r} .7872 \\ \times50 \\ \hline 39.3600 \end{array}$

   c. $\begin{array}{r} 82 \\ \times.03 \\ \hline 2.46 \end{array}$

4. Division.
   a. $3\overline{).216} = .072$ d. $.007\overline{).490.} = 70$

   b. $5\overline{).120} = .024$ e. $1.6\overline{)3.4.40} = 2.15$
   $\begin{array}{r} \underline{32} \\ 24 \\ \underline{16} \\ 80 \\ \underline{80} \\ 0 \end{array}$

   c. $.09\overline{)36.00.} = 400$

5. Converting fractions and mixed numbers to decimals.
   a. $\frac{3}{5} = .6$
   b. $\frac{1}{8} = .125$
   c. $\frac{1}{20} = .05$
   d. $1\frac{7}{8} = 1.875$
   e. $3\frac{1}{3} = 3.333\ldots$ (Recurring decimal. Round off to desired number of decimal places.)

6. Converting decimals to fractions or mixed numbers.
   a. $0.4 = \frac{4}{10} = \frac{2}{5}$
   b. $.625 = \frac{625}{1000} = \frac{5}{8}$
   c. $1.8 = 1\frac{8}{10} = 1\frac{4}{5}$
   d. $2.2 = 2\frac{2}{10} = 2\frac{1}{5}$
   e. $.008 = \frac{8}{1000} = \frac{1}{125}$

PRACTICE PROBLEMS USING PERCENTAGES

1. Converting percentages to decimals.
   a. $7\% = .07$
   b. $8.5\% = .085$
   c. $59\% = .59$
   d. $99.44\% = .9944$
   e. $140\% = 1.4$
   f. $300\% = 3.0$

2. Converting decimals to percentages.
   a. $0.2 = 20\%$
   b. $.19 = 19\%$
   c. $.91 = 91\%$
   d. $1.65 = 165\%$
   e. $1.2 = 120\%$
   f. $6.0 = 600\%$

3. Converting percentages into fractions, mixed numbers, or whole numbers.
   a. $60\% = \frac{60}{100} = \frac{3}{5}$
   b. $4\% = \frac{4}{100} = \frac{1}{25}$
   c. $150\% = \frac{150}{100} = 1\frac{1}{2}$
   d. $12.5\% = .125 = \frac{125}{1000} = \frac{1}{8}$
   e. $62.5\% = .625 = \frac{625}{1000} = \frac{5}{8}$
   f. $200\% = \frac{200}{100} = 2$

4. Performing calculations involving only percentages.
   a. $37\% + 33\% = 70\%$
   b. $100\% + 65\% = 165\%$
   c. $25\% - 13\% = 12\%$
   d. $20\% \times 80\% \rightarrow .2 \times .8 = .16 = 16\%$
   e. $150\% \div 30\% \rightarrow 1.5 \div .3 \rightarrow 15 \div 3 = 5\%$

5. Calculating the value of percentages.
   a. 20% of $88 \rightarrow .2 \times 88 = 17.6$
   b. 90% of $200 \rightarrow .9 \times 200 = 180$
   c. 50% of $61 \rightarrow .5 \times 61 = 30.5$
   d. 130% of $40 \rightarrow 1.3 \times 40 = 52$
   e. 300% of $12 \rightarrow 3 \times 12 = 36$

6. Determining percentages.
   a. $13/63 = .206 = 20.6\%$
   b. $29/80 = .363 = 36.3\%$
   c. $43/200 = .215 = 21.5\%$
   d. $40/160 = .25 = 25\%$
   e. $50/35 = 1.429 = 142.9\%$
7. Calculating the increase or decrease of numbers by a percentage.
   a. $100 \times 1.15 = 115$
   b. $30 \times 3.5 = 105$
   c. $150 \times 1.25 = 187.5$
   d. $85 \times .9 = 76.5$
   e. $20 \times .88 = 17.6$

PRACTICE SOLVING EQUATIONS

1. $3x + 6 = 10 + 2$
   $3x + 6 = 12$
   $\quad 3x = 12 - 6$
   $\quad 3x = 6$
   $\quad\quad x = {}^{6}/_{3}$
   $\quad\quad x = 2$
2. $21/x = 8 - 1$
   $21/x = 7$
   $\quad 21 = 7x$
   $\quad {}^{21}/_{7} = x$
   $\quad\quad 3 = x$
   $\quad\quad x = 3$
3. $x - 17 = 13$
   $x - 17 + 17 = 13 + 17$
   $\quad\quad x = 30$
4. $20 = 3x - 4$
   $20 + 4 = 3x - 4 + 4$
   $\quad 24 = 3x$
   $\quad {}^{24}/_{3} = x$
   $\quad\quad x = 8$
5. $2x = {}^{18}/_{2}$
   $2x = 9$
   $\quad x = {}^{9}/_{2}$
   $\quad x = 4^{1}/_{2} \ (4.5)$
6. $45 = 9x$
   ${}^{45}/_{9} = x$
   $\quad x = 5$
7. $30/x = {}^{6}/_{2}$
   $\quad 6x = 60$
   $\quad\quad x = 10$
8. $x/7 = {}^{36}/_{6}$
   $\quad 6x = 36 \times 7$
   $\quad\quad x = 6 \times 7$
   $\quad\quad x = 42$
9. ${}^{56}/_{8} = 49/x$
   $56x = 49 \times 8$
   $\quad 8x = 7 \times 8$
   $\quad\quad x = 7$
10. ${}^{12}/_{4} = x/15$
    $\quad 4x = 12 \times 15$
    $\quad\quad x = 3 \times 15$
    $\quad\quad x = 45$

PRACTICE PROBLEMS USING EXPONENTS, SQUARE ROOTS, AND SCIENTIFIC NOTATION

1. Calculating the value of exponential terms.
   a. 5 cubed $= 5^3 = 5 \times 5 \times 5 = 125$
   b. $2^8 = 2 \times 2 \times 2 \times 2 \times 2 \times 2 \times 2 \times 2 = 256$
   c. $9^2 = 9 \times 9 = 81$
   d. $10^4 = 10 \times 10 \times 10 \times 10 = 10,000$
   e. $10^{-5} = .00001$
   f. $4^{-3} = \dfrac{1}{4 \times 4 \times 4} = \dfrac{1}{64} = .015625$
2. Solving problems involving exponential terms.
   a. $4^2 \times 4^6 = 4^8$
   b. $10^{10} \times 10^2 = 10^{12}$
   c. $3^3 \div 3^2 = 3^1 = 3$
   d. $5^2 \div 5^4 = 5^{-2}$
   e. $10^6 \div 10^2 = 10^4$
3. Extracting square roots.
   a. $\sqrt{49} = 7$
   b. $\sqrt{36} = 6$
   c. $\sqrt{64} = 8$
   d. $\sqrt{81} = 9$
   e. $\sqrt{400} = 20$
4. Expressing numbers using scientific notation.
   a. $.00075 = 7.5 \times 10^{-4}$
   b. $23,000 = 2.3 \times 10^4$
   c. $18,200,000 = 1.82 \times 10^7$
   d. $.0000509 = 5.09 \times 10^{-5}$
   e. $.007365 = 7.365 \times 10^{-3}$

PRACTICE PROBLEMS USING MEASUREMENT UNITS

1. Conversions from one metric unit to another.
   a. $75 \text{ kV} \times 1000 \text{ V per kV} = 75,000 \text{ V}$
   b. $3 \text{ m}/x \text{ cm} = 1 \text{ m}/100 \text{ cm per m} = 300 \text{ cm}$
   c. $10 \text{ ml}/x \text{ L} = 1000 \text{ ml}/1 \text{ L} = .01 \text{ L}$
   d. $20 \text{ g}/x \text{ kg} = 1000 \text{ g}/1 \text{ kg} = .02 \text{ kg}$
   e. $15 \text{ cg}/x \text{ g} = 100 \text{ cg}/1 \text{ g} = .15 \text{ g}$
2. Conversions from one English unit to another.
   a. $16 \text{ in} \div 26 \text{ in per yd} = .4444 \text{ yd}$
   b. $^{1}/_{2} \text{ pt} \times 16 \text{ oz/pt} = 8 \text{ oz}$
   c. $84 \text{ in} \div 12 \text{ in/ft} = 7 \text{ ft}$
   d. $18 \text{ qt} \div 4 \text{ qt/gal} = 4.5 \text{ gal}$
   e. $4.8 \text{ lb} \times 16 \text{ oz/lb} = 76.8 \text{ oz}$
3. Conversions between English and metric units.
   a. $3 \text{ oz} \times 30 \text{ ml/oz} = 90 \text{ ml}$
   b. $.25 \text{ lb} \div 2.2 \text{ lb/kg} = .1136 \text{ kg}$
      $.1136 \text{ kg} \times 1000 \text{ g/kg} = 113.6 \text{ g}$
   c. $5 \text{ in} \times 2.54 \text{ cm/in} = 12.7 \text{ cm}$ $12.7 \text{ cm} \div 100 \text{ cm/m} = .127 \text{ m}$
   d. $30 \text{ mm} \div 10 \text{ mm/cm} = 3 \text{ cm}$
      $3 \text{ cm} \div 2.54 \text{ cm/in} = 1.18 \text{ in}$
   e. $40 \text{ g} \div 1000 = .04 \text{ kg}$
      $.04 \text{ kg} \times 2.2 \text{ lb/kg} = .088 \text{ lb}$
      $.088 \text{ lb} \times 16 \text{ oz/lb} = 1.408 \text{ oz}$

4. Time and temperature conversions.
   a. $\frac{1}{20}$ sec × 1000 msec/sec = 50 msec
   b. 330 sec ÷ 60 sec/min = 5.5 min
      5.5 min ÷ 60 min/hr = .0917 hr
   c. 3.4 days × 24 hr/day = 81.6 hr
   d. (19° C × 1.8) + 32 = 34.2 + 32 = 66.2° F
   e. 50° F − 32 ÷ 1.8 = 18 ÷ 1.8 = 10° C

SOLVING mAs PROBLEMS

1. Calculating mAs, given mA and time.
   a. 200 mA × $\frac{1}{40}$ sec = 5 mAs
   b. 300 mA × $\frac{1}{20}$ sec = 15 mAs
   c. 100 mA × $\frac{2}{15}$ sec = 13.33 mAs
   d. 500 mA × .02 sec = 10 mAs
   e. 50 mA × .3 sec = 15 mAs
   f. 150 mA × $1\frac{1}{4}$ sec = 187.5 mAs
   g. 400 mA × 2 msec = 400 × .002 sec = .8 mAs
   h. 300 mA × $\frac{1}{120}$ sec = 2.5 mAs
2. Calculating exposure time, given mA and mAs.
   a. 40 mAs ÷ 50 mA = .8 sec or $\frac{4}{5}$ sec
   b. 25 mAs ÷ 200 mA = .125 sec or $\frac{1}{8}$ sec
   c. 10 mAs ÷ 300 mA = .033 sec
   d. 1 mAs ÷ 100 mA = .01 sec
   e. 8 mAs ÷ 400 mA = .02 sec
3. Calculating mA, given mAs and time.
   a. 10 mAs ÷ $\frac{1}{30}$ sec = 10 mAs × 30 = 300 mA
   b. 25 mAs ÷ $\frac{1}{2}$ sec = 25 mAs × 2 = 50 mA
   c. 40 mAs ÷ $\frac{2}{15}$ sec = 40 × $1\frac{5}{2}$ = $\frac{600}{2}$ = 300 mA
   d. 5 mAs ÷ .01 sec = 500 mA
   e. 8 mAs ÷ .02 sec = 400 mA

PRACTICE PROBLEMS INVOLVING DISTANCE CHANGES

1. $\frac{1}{x} = \frac{30^2}{40^2}$

   $\frac{1}{x} = \frac{3^2}{4^2}$

   $\frac{1}{x} = \frac{9}{16}$

   $9x = 16$

   $x = \frac{16}{9}$

   $x = 1\frac{7}{9}$ (1.77) times I_1

2. $\frac{1}{x} = \frac{40^2}{72^2}$

   $\frac{1}{x} = \frac{5^2}{9^2}$

$\frac{1}{x} = \frac{25}{81}$

$25x = 81$

$x = \frac{81}{25}$

$x = 3.24$ times I_1

3. $\frac{15}{x} = \frac{40^2}{48^2}$

   $\frac{15}{x} = \frac{5^2}{6^2}$

   $\frac{15}{x} = \frac{25}{36}$

   $25x = 15 \times 36$

   $25x = 540$

   $x = 21.6$ mAs

4. $\frac{40}{x} = \frac{72^2}{84^2}$

   $\frac{40}{x} = \frac{6^2}{7^2}$

   $\frac{40}{x} = \frac{36}{49}$

   $36x = 40 \times 49$

   $36x = 1960$

   $x = 54.4$ mAs

5. $\frac{10}{x} = \frac{40^2}{72^2}$

   $\frac{10}{x} = \frac{5^2}{9^2}$

   $\frac{10}{x} = \frac{25}{81}$

   $25x = 10 \times 81$

   $25x = 810$

   $x = 32.4$ mAs

PRACTICE PROBLEMS INVOLVING PATIENT/PART SIZE CHANGES

1. 3 kVp/cm × 3 cm = 9 kVp change.
   96 kVp − 9 kVp = 85 kVp for 27 cm.
2. 2 kVp/cm × 2 cm = 4 kVp change.
   74 kVp + 4 kVp = 78 kVp for 18 cm.
3. 50 mAs × .8 = 40 mAs for 22 cm.
   40 mAs × .8 = 32 mAs for 20 cm.
   32 mAs × .8 = 25.6 mAs for 18 cm.
4. 30 mAs × 1.3 = 39 mAs for 21 cm.
   39 mAs × 1.3 = 50.7 mAs for 23 cm.
   50.7 mAs × 1.3 = 65.9 mAs for 25 cm.
5. 10 mAs × 1.3 = 13 mAs for 14 cm.
   13 mAs × 1.3 = 16.9 mAs for 16 cm.

PRACTICE PROBLEMS USING THE 15% RULE

1. 86 kVp × .85 = 73 kVp
   25 mAs × 2 = 50 mAs
2. 8 kVp × 1.15 = 92 kVp
   100 mAs ÷ 2 = 50 mAs
3. 66 kVp × 1.15 = 76 kVp
   10 mAs ÷ 2 = 5 mAs
4. 94 kVp × .85 = 80 kVp
   40 mAs × 2 = 80 mAs
5. 72 kVp × 1.15 = 83 kVp
   60 mAs ÷ 2 = 30 mAs

PRACTICE PROBLEMS INVOLVING GRID CONVERSIONS

1. 30 mAs ÷ 2.5 = 12 mAs
2. 5 mAs × 3 = 15 mAs
3. 75 kVp − 12 kVp = 63 kVp
4. 60 kVp ÷ 14 kVp = 74 kVp
5. 12:1 − 8:1 = 4 (difference in ratios)
   80 kVp − 4 kVp = 76 kVp
6. 10 mAs ÷ 2.5 = 4 mAs (non-grid)
   4 mAs × 1.5 = 6 mAs (6:1 grid)

PRACTICE PROBLEMS INVOLVING TRANSFORMERS

1. $^{90}/_{30} = 3/1 = 3{:}1$ ratio   (step-down)
2. $^{3}/_{1} = 120/x$
   $3x = 120$
   $x = 40$ V
3. $100/x = 120/12$
   $120x = 1200$
   $x = 10$ windings
4. $^{1}/_{400} = x/96{,}000$
   $400x = 96{,}000$
   $x = 96{,}000 ÷ 400 = 240$ V
5. $^{150}/_{450} = ^{1}/_{3} = 1{:}3$ ratio   (step-up)

PRACTICE PROBLEMS USING MEDICATION DOSAGE

1. 100 mg/25 mg/tablet = 4 tablets
2. 250 mg/50 mg/ml = 5 ml
3. 40 mcg/80 μg/tablet = $^{1}/_{2}$ tablet
4. 5 mg/1 mg/ml = 5 ml
5. Dose ÷ 200 mg/tablet = 8 tablets
   Dose = 8 × 200 mg
   Dose = 1600 mg

Photo Credits

Adler AM, Carlton RR: *Introduction to radiography and patient care,* **ed 2, Philadelphia, 1999, WB Saunders.**

Figs. 18-14*AB,* 18-17*A,* 21-16

Atlas RM: *Principles of microbiology,* **St Louis, 1999, Mosby.**

Fig. 21-18

Ballinger PW, Frank ED: *Merrill's atlas of radiographic positions and radiologic procedures,* **ed 9, St Louis, 1999, Mosby.**

Figs. 11-4*B,* 12-35, 15-24, 16-33, 16-34, 17-17, 18-16, 18-19, 18-24, 18-30, 18-31, 18-32, Abdomen 6, Elbow 8, Facial Bones 7, Femur 4, Fullspine 3, Fullspine 5, Hip 6, Hip 7, Shoulder 14, Skull 2, Skull 4, Skull 12

Bonewit-West K: *Clinical procedures for medical assistants,* **ed 5, Philadelphia, 2000, WB Saunders.**

Figs. 25-4, 25-5, 25-6, 25-9, 25-14, 25-16, 25-20, 25-21

Bontrager KL: *Textbook of radiographic positioning and related anatomy,* **ed 5, St Louis, 2001, Mosby.**

Fig. Abdomen 4

Bontrager KL: *Textbook of radiographic positioning and related anatomy,* **St Louis, 1987, Mosby.**

Fig. Shoulder 12

Bushong SC: *Radiologic science for technologists,* **ed 7, St Louis, 2001, Mosby.**

Fig. 3-7

Cipollaro AC: The earliest roentgen demonstration of a pathological lesion in America, *Radiology* **45:555, 1945.**

Fig. 1-2

Deltoff MN: *The portable skeletal x-ray library,* **St Louis, 1997, Mosby.**

Fig. 18-33

Dibner B: *The new rays of Professor Röntgen,* **Norwalk, CT, 1963, Burndy Library.**

Fig. 1-1

Dolan K, Jacoby C, Smoker W: The radiology of facial fractures, *Radiographics* **4:576, 1984.**

Fig. 17-18

Ehrlich RA, Daly JA, McCloskey ED: *Patient care in radiography,* **ed 5, St Louis, 1999, Mosby.**

Figs. 2-10, 2-11, 2-12, 2-13, 2-16, 11-7, 11-8, 11-10, 11-11, 18-1, 18-2, 18-5, 18-11, 18-13*AB,* 18-17*B,* 18-21, 18-37, 18-38, 21-1, 21-2, 21-3, 21-4*ABC,* 21-5*AB,* 21-6, 21-7, 21-8, 21-9, 21-13*AB,* 21-14*AB,* 21-15, 21-17, 21-22*ABCDEFG,* 21-23, 21-24*ABCD,* 21-25*AB,* 21-26*AB,* 21-27*AB,* 21-28*ABCDEFGH,* 21-29*AB,* 22-1, 22-5, 22-6, 22-7, 22-8, 22-9, 22-10, 22-11, 22-12, 22-13, 22-14, 23-1, 23-3, 23-4, 23-5, 23-9, 23-10, 23-11, 23-12

Eisenberg RL: *Atlas of signs in radiology,* **Philadelphia, 1984, JB Lippincott.**

Fig. 16-24*C*

Eisenberg RL, Dennis CA: *Comprehensive radiographic pathology,* **ed 2, St Louis, 1995, Mosby.**

Figs. 12-32, 13-14, 13-16, 13-20*B,* 13-21, 13-23, 13-24, 13-25, 14-12, 14-20, 14-22, 14-25, 15-22, 15-25, 15-29, 15-31, 15-32, 15-34, 15-36, 16-23, 16-24*ABD,* 16-25, 16-26*ACD,* 16-27, 16-28, 16-29, 16-31, 17-13, 17-15, 17-16, 17-20, 17-21, 17-22, 17-23, 17-25, 17-26

Forbes BA, Sahm DF, Weissfeld AS: *Bailey & Scott's diagnostic microbiology,* **ed 4, St Louis, 1998, Mosby.**

Figs. 21-20*AB,* 21-21*AB*

Guebert GM, Yochum TR: *Essentials of diagnostic imaging,* **St Louis, 1995, Mosby.**

Fig. 2-8

Kleinman PK: *Diagnostic imaging of child abuse,* ed 2, St Louis, 1998, Mosby.

Fig. 18-36*ABC*

Linn-Watson TA: *Radiographic pathology,* Philadelphia, 1996, WB Saunders.

Figs. 13-12, 13-15, 14-13, 14-14, 14-15, 14-19, 14-21, 14-23, 16-24*E,* 16-26*BE,* 16-30, 17-14, 17-24

McQuillen-Martensen K: *Radiographic critique,* Philadelphia, 1996, WB Saunders.

Fig. 8-28

Osborn AG: Head trauma. In Eisenberg RL, Amberg JR, editors: *Critical diagnostic pathways in radiology,* Philadelphia, 1981, JB Lippincott.

Fig. 15-30

Perry AG, Potter PA: *Basic nursing: a critical-thinking approach,* ed 4, St Louis, 1999, Mosby.

Figs. 23-2, 23-7

Perry AG, Potter PA: *Clinical nursing skills & techniques,* ed 4, St Louis, 1998, Mosby.

Figs. 22-3, 22-4

Perry AG, Potter PA: *Fundamentals of nursing concepts: process and practice,* ed 5, St Louis, 2000, Mosby.

Figs. 23-6, 23-8

Rogers LF: *Radiology of skeletal trauma,* New York, 1982, Churchill Livingstone.

Fig. 17-19

Silverman FN, Kuhn JP: *Caffey's pediatric x-ray diagnosis,* ed 9, St Louis, 1993, Mosby.

Fig. 13-13

Taylor JAM, Resnick D: *Skeletal imaging,* Philadelphia, 2000, WB Saunders.

Figs. 13-22, 14-16, 14-17, 15-27, 15-28

Thibodeau GA, Patton KT: *Structure and function of the body,* ed 11, St Louis, 2000, Mosby.

Figs. 12-1, 12-2, 12-3, 12-4, 12-5, 12-6, 12-7, 12-8, 12-9, 12-10, 12-11, 12-12, 12-13, 12-14, 16-3, 16-4, 16-5, 16-6, 16-9, 16-10, 18-28

Woods MA: *The clinical medical assistant,* ed 8, Philadelphia, 1999, WB Saunders.

Fig. 25-30

COURTESIES

E.E. Bonic, MD

Fig. 16-20

Eastman Kodak, Rochester, NY

Fig. 4-1

Hologic Systems Division, Newark, DE

Fig. 2-6

R. Kerr, MD

Fig. 16-21

Olympic Medical, Seattle, WA

Fig. 18-15

Spacelabs Medical, Deerfield, WI

Figs. 25-19, 25-21, 25-23, 25-24, 25-26, 25-27

Varian Medical Systems, North Charleston, SC

Fig. 5-18

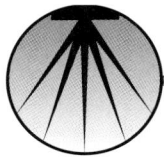

Glossary

acanthion The superior prominence formed at the junction of the two maxillary bones. The positioning landmark where the nose meets the upper lip.

acromion process A large, rounded projection that can be felt on the superior surface of the scapula.

aggressive Describes an attitude characterized by the angry or hostile and forceful expression of feelings or opinions.

agonist A drug that produces a specific action and promotes a desired result.

airborne contamination Dust containing either endospores or droplet nuclei circulating through the air.

algebra A mathematical method for determining the value of an unknown quantity that has a specific relationship to one or more known quantities.

allergen An allergy-causing substance.

Alzheimer's disease A specific type of brain tissue deterioration that causes memory loss and gradual loss of mental function.

anaphylaxis A severe allergic reaction.

angina Chest pain that occurs when the coronary arteries are unable to supply the heart with sufficient oxygen to meet current needs.

anoxia Lack of oxygen.

antagonist A drug, such as Mazicon, used to counteract the effects of sedatives or analgesics.

antecubital fossa The depression on the anterior aspect of the elbow.

antidote A drug that treats a toxic effect.

aorta The largest artery of the body, passing from the heart through the chest and the abdomen.

asepsis The process of reducing the probability of infectious organisms being transmitted to a susceptible individual.

aspiration The process of inhaling a foreign body or substance into the trachea or a bronchus.

assault The threat of touching in an injurious way.

assertive Describes an attitude characterized by the calm, firm expression of feelings or opinions.

asthma A difficulty in breathing caused by constriction of the bronchi.

atelectasis Lung collapse.

atlas The C1 vertebra; a ringlike structure with no vertebral body and a very short spinous process.

autoclave An electric steam chamber that achieves high temperatures under pressure for the purpose of sterilization.

axilla The armpit.

axis The C2 vertebra; the vertebra on which the atlas rotates so the head can turn from side to side.

battered child syndrome The characteristics of child abuse.

battery An unlawful touching of a person without consent.

biohazard symbol A warning mark that indicates a substance that may cause harm chemically or biologically.

biohazardous waste Garbage that is a threat to the environment and the organisms within it.

blow-out fracture Traumatic opening between the orbital floor and the maxillary sinus, usually caused by a blow to the eye.

bronchus An air passageway that connects the trachea to a lung or portion of a lung.

bursitis Inflammation of a bursa.

calcaneus The heel bone.

cardiophrenic angles The inferior medial corners of the lungs.

carpal digits The fingers.

carpus, carpal The eight short bones of the wrist.

carrier An individual infected with an infectious organism who does not contract the disease but is capable of spreading it to others.

cerebral concussion A blow to the head that causes brief unconsciousness or disorientation.

cerebrovascular accident (CVA) A stroke.

cervical spine The most superior section of the vertebral column; consists of seven vertebrae with a lordotic curve.

chart An extensive compilation of information; the medical record of a patient's care.

charting The process of adding records to a document or medical record.

clavicle The collar bone.

clean-catch midstream specimen A urine specimen obtained in such a way as to avoid bacterial contamination of the specimen.

coccyx The most inferior section of the vertebral column, consisting of three to five vertebral segments; also called the *tailbone*.

colon The large intestine.

common denominator When two or more fractions have the same denominator.

contrecoup injury Brain injury due to movement of the brain within the cranium that results from a severe blow to the head on the opposite side.

costophrenic angles The inferior lateral corners of the lungs.

cranium The eight bones that surround the brain; the brain case.

cube To multiply a number by itself two times.

cyanotic Having a bluish coloration in the skin that indicates a lack of sufficient oxygen in tissues.

decimal A fraction with a denominator of 10, 100, 1000, or any number that consists of a 1 followed by one or more zeros.

decimal point The dot at the left of a decimal fraction.

decubitus ulcer A lesion that develops over bony prominences when pressure is exerted over time.

defamation of character Disclosing information that reflects negatively on the person's reputation.

dehydration Lack of fluid in the tissues due to insufficient fluid intake.

demineralization Calcium loss in bones that causes them to become more radiolucent.

denominator The lower number of a fraction.

dens A tooth or toothlike structure or process; the odontoid process of the atlas.

diabetic coma Abnormally high blood glucose levels, characterized by a relatively slow onset.

diaphoretic Perspiring.

diaphragm The large sheath of muscle between the chest and the abdomen that expands and contracts with breathing.

diastolic Referring to the point of least pressure in the arterial system.

disinfection The destruction of pathogens by chemical agents.

diverticulitis A degenerative inflammatory disease of the colon.

droplet contamination Droplets from an infectious individual's nose or mouth.

duodenum The proximal portion of the small intestine.

dyspnea Difficult breathing.

efficacy Effectiveness.

electrocardiogram (ECG, EKG) The tracing made on paper by an electrocardiograph.

electrocardiograph A machine used to assess heart function and diagnose heart disease.

electrode A contact that receives electrical signals, such as the contact between the patient and a lead wire of an electrocardiograph.

emesis Vomit.

empathy A sensitivity to the needs of others that allows you to meet those needs constructively.

emphysema A chronic lung condition characterized by obstruction and destruction of the small airways and alveoli of the lungs, resulting in the inability to effectively exhale stale air.

endospore A bacterial form that is generated to survive harsh environmental conditions.

epistaxis Nosebleed.

equation A mathematical sentence that uses numbers and symbols to show that the value on the left side of the equal sign is equivalent to that on the right side.

esophagus The part of the digestive system that connects the throat to the stomach.

ethics Rules that apply values and moral standards to activities within a profession to define professional behavior.

exercise tolerance test (stress test) An electrocardiograph test taken during physical activity and used to evaluate cardiac ischemia.

exponent A superscript number that indicates how many times the number is multiplied by itself.

external auditory (acoustic) meatus (EAM) The opening to the ear canal.

external occipital protuberance (EOP) The palpable bony prominence in the approximate center of the occipital bone.

extravasation Infiltration of a drug into the tissue surrounding an intravenous injection site.

fabella The normal variation of a small sesamoid bone located posterior to the knee.

facet The articular surface of intervertebral joints. Facets are located on each of the four articular processes that extend superiorly and inferiorly from the vertebral arch.

false imprisonment The unjustifiable detention of a person against his or her will.

fat pad sign Radiographic evidence of displacement of the fat pad in the joint region of the elbow that indicates a fracture involving the elbow joint.

femur The long bone of the thigh.

fibrillation A rapid, weak and inefficient heartbeat.

fibula The thin, shorter long bone located laterally in the lower leg.

fomite An object that has been in contact with pathogenic organisms.

foramen magnum The large round hole in the anterior portion of the occipital bone through which the spinal cord passes.

forced expiration test A common spirometry test used to measure forced expiratory volume and forced vital capacity; a type of pulmonary function test (PFT).

forced expiratory volume (FEV) The quantity of air that can be forcefully exhaled during a specific time period.

forced vital capacity (FVC) The amount of air that can be forced from the lungs following a deep breath.

geriatrics The care of elder adults.

glabella The bony prominence on the frontal bone between the eyebrows.

glans penis The conical tip of the penis.

glenoid process The lateral portion of the scapula that forms the socket of the shoulder joint.

glottic closure Sudden cessation of exhalation due to closure of the opening to the trachea.

gonion The angle of the mandible.

hemoptysis Coughing up of blood.

hemorrhage The continuous abnormal flow of blood.

hilum A depression or pit at that part of the organ where vessels and nerves enter. The hila of the lungs are located on either side of the carina.

humerus The long bone of the upper arm.

hydration The water content of tissues.

hyperopia Farsightedness.

hypertension Abnormally high blood pressure.

hyperventilation Air hunger with rapid respirations.

hypoglycemia Low blood sugar.

hypotension Abnormally low blood pressure.

idiosyncratic Unusual or peculiar.

ileum The distal portion of the small intestine that is connected to the large intestine.

ilium The upper portion of the innominate bone.

improper fraction A fraction with a value greater than 1 where the numerator is greater than the denominator.

incontinence Loss of bowel or bladder control.

informed consent The written acceptance of any procedure that is considered experimental or that involves substantial risk.

innominate bone One of the two composite bones that make up the pelvis.

intentional misconduct An action that violates the law or a professional code of ethics and was committed knowingly.

intervertebral disk A pad of fibrocartilage between vertebral bodies that cushions vertebral motion and absorbs shock.

intradermal Shallow injections made between the skin layers.

intramuscular (IM) Injections made directly into muscle tissue.

intravenous (IV) Injections made directly into a vein.

invasion of privacy Failure to maintain confidentiality of information or the improper exposure or touching of the patient's body.

ischium The posterior inferior portion of the innominate bone.

Ishihara test A test used to measure color perception.

jejunum The second section of the small intestine.

joint effusion Increased fluid in the joint capsule.

KUB The AP supine projection of the abdomen. It stands for *k*idneys, *u*reters, and *b*ladder.

kyphotic curve, kyphosis A posterior convex curvature of the spine.

lacrimal bone One of the smallest and most fragile bones of the face, located at the anterior part of the medial wall of the orbit.

lamina Any thin, flat layer of membrane or other bulkier tissue; right and left posterior portions of the vertebral arch.

lead (ECG) A specific recording on an ECG study.

libel The malicious spreading of information that results in defamation of character or loss of reputation.

lordotic curve, lordosis An anterior convex curvature of the spine.

lowest terms The form of a fraction in which no number except 1 can be divided into the numerator and denominator.

lumbar spine The five vertebrae located inferior to the thoracic spine.

malpractice An act of negligence in the context of the relationship between a professional person and a patient or client.

mandible The lower jaw.

maxilla The upper jaw.

mediastinum The part of the thoracic cavity that encompasses the space between the lungs.

meniscus The C-shaped cartilage that cushions the articular surface of each femoral condyle.

mental protuberance (point) The prominence in the center of the mandible's lower margin.

metacarpal A bone of the hand.

metatarsal A bone of the forefoot.

microbe, microorganism A living organism too small to see with the naked human eye.

microbial dilution Reduction of the total number of microorganisms in an area.

mixed number A number consisting of a whole number and a fraction.

morals Right actions based on familial and societal standards.

multiple myeloma A malignant bone disease that may involve many bones of the body.

myopia Nearsightedness.

nasal conchae Six thin, curved, bony projections that divide the nasal cavities.

nasion The anterior depression in the midline of the skull between the orbits.

negligence The omission of reasonable care or caution.

nonaccidental trauma (NAT) Another name for battered child syndrome.

normal flora Microorganisms that live on or within the body without causing disease.

numerator The upper number of a fraction.

olecranon process A posterior projection at the proximal end of the ulna; the "funny bone" or "crazy bone."

opiate A drug derived from opium.

opioid Any drug whose action is similar to that of morphine.

opportunistic infection An infection that occurs when certain microorganisms take advantage of the opportunity to invade in the absence of an immune response.

orbit Eye socket.

organic brain syndrome A large group of disorders associated with brain damage or impaired cerebral function.

orthopnea Inability to breathe lying down.

orthostatic hypotension A temporary state of low blood pressure that causes patients to feel lightheaded or faint when first sitting up.

osteoarthritis A degenerative joint disease.

osteoblastic A term for a disease that results in increased bone formation.

osteolytic A term for a disease that causes bone destruction.

osteoma A benign bone tumor.

osteomyelitis The inflammation of the bone caused by a pathogenic organism.

osteopenia Calcium loss in bones, causing them to become more radiolucent; demineralization.

osteophytes Enlarged, deformed portions of the bone, usually caused by arthritis.

osteoporosis Calcium loss in bones, often through aging, that results in bones becoming porous, brittle, and more radiolucent.

Paget's disease A disease of the elderly that results in the softening and destruction of the bone, followed by thickening and irregular calcification as the bone is repaired.

palatine bone A part of the roof of the mouth.

parenteral Injected into the body through the skin.

parietal membrane The lining of the body's cavities.

Parkinson's disease A degenerative condition of the nervous system that causes tremors.

PASS The acronym for the operating instructions of the fire extinguisher. *P*ull the pin. *A*im the nozzle. *S*queeze the handle. *S*weep.

patella The kneecap.

pathogen A microorganism capable of causing disease.

pedal digits The toes.

pediatrics The care of children.

pedicle A narrow stalk, stem, or tube of tissue attached to a tumor, skin flap, bone, or organ. The right and left anterior portions of the vertebral arch on either side of the vertebral body.

percentage A form of a fraction with a denominator of 100.

peritoneum The double-walled serous membrane sac that contains the abdominal organs.

phalanx A long bone of the finger or toe.

pleura Membrane that covers the lungs and lines the pleural cavities.

pleural effusion An abnormal collection of fluid in the pleural space.

pneumoconiosis A group of chronic occupational lung diseases caused by the inhalation of irritating dust.

pneumonia Inflammatory disease of the lung.

pneumothorax A collection of air or gas in the pleural space associated with lung collapse.

potency Strength.

presbyopia Farsightedness that occurs with advancing age.

proper fraction A fraction with a value less than 1.

prosthesis Anatomical replacement.

pubis The anterior inferior portion of the innominate bone.

radial flexion Movement of the hand in the lateral direction or away from the body when in anatomical position.

radius The thick, shorter long bone located laterally in the forearm.

ratio The quotient that results from dividing one number into another.

reagent strip A paper strip containing small pads of absorbent material impregnated with chemical reagents; used for urinalysis and other laboratory evaluations.

reasonably prudent person The standard that requires a person to perform as any reasonable person would perform under similar circumstances.

remainder The number that is left over when a number is divided unequally into another.

respondeat superior The legal doctrine that states the employer is liable for employees' negligent acts that occur in the course of their work.

rule of personal responsibility The rule that states that each individual is liable for his or her own negligent conduct.

sacrum The five vertebral segments inferior to the lumbar vertebrae, which fuse together in adulthood to form a solid bony structure.

scapula The shoulder blade.

scientific notation A system based on powers of the number 10 to simplify the expression of numbers that are very large or very small.

scoliosis An abnormal lateral curvature of the spine.

sella turcica The rounded fossa in the center of the anterior superior surface of the sphenoid bone and the location of the pituitary gland.

sesamoid bone A small, flat, oval bone within a tendon that is not counted among the bones of the body.

sharps Anything that can puncture skin such as needles, glass tubes, glass slides, and finger lancets.

sharps container A receptacle for needles, syringes, and other contaminated items capable of puncturing the skin.

slander The malicious spreading of information that results in defamation of character or loss of reputation.

Snellen alphabet chart A chart for testing distance vision.

Snellen E chart A chart for testing distance vision that uses the block letter E in various positions instead of letters of the alphabet.

sphincter A round muscle that opens and closes the opening of an organ.

spirogram The graph that results from a spirometer test.

spontaneous combustion The occurrence of a chemical reaction in or near a flammable material that causes enough heat to generate a fire.

square To multiply a number by itself.

square root That portion of a number which, when multiplied by itself, equals the original number.

Standard Precautions The use of barriers in anticipation of contact with certain bodily fluids and extracts.

standardization mark The mark on an electrocardiograph that results from deflection of the stylus by a 1 mV stimulus. The mark indicates that the machine is recording accurately.

standing order Written directions, signed by a physician, for a specific medication or procedure and under certain circumstances, that may be applied to any patient without further order.

stenosis Narrowing of a passageway, such as an intervertebral foramen.

sterilization The destruction of all microorganisms; surgical asepsis.

sternum The breastbone.

stridor A harsh, high-pitched sound during respiration caused by obstruction of bronchi.

stylus The tracing pen on an electrocardiograph.

subcutaneous (SC) Injections into the fat layer beneath the skin.

suture Synarthroidal joint that connects the bones of the cranium.

syncope Fainting.

synergistic The effects of a combined action.

systolic Referring to maximum blood pressure, occurring during contraction of the left ventricle.

tachycardia A rapid pulse.

talus The ankle bone.

tarsal bones The five short bones of the midfoot.

tendinitis Inflammation of a tendon.

thoracic spine The dorsal spine, which consists of 12 vertebrae with a kyphotic curve; located inferior to the cervical spine.

thorax The upper portion of the trunk. The chest.

thready pulse A weak and rapid pulse.

tibia The thick, longer long bone located medially in the lower leg.

topical Type of medication that is applied to the skin.

tourniquet A rubber strap or tube tied around the arm to facilitate blood collection.

toxic Poisonous.

trachea The windpipe that connects the throat to the bronchi.

transient ischemic attack (TIA) An attack with similar symptoms to a stroke but of short duration.

ulna The thin, longer long bone located medially in the forearm.

ulnar flexion Movement of the hand in the medial direction or toward the body when in anatomical position.

urethral meatus The external opening to the urethra.

urinalysis The physical, microscopic, or chemical examination of urine.

urticaria Hives.

valid choice A selection of alternatives, all of which are acceptable.

values The priorities that are placed on the significance of moral concepts.

vector An arthropod in whose body an infectious organism develops or multiplies before becoming infectious to a new host.

vehicle A medium that transports microorganisms.

vena cava The large vein that brings oxygen-depleted blood from the body to the right atrium of the heart.

venipuncture Collecting blood by inserting a needle into a vein.

vertebra Any one of the 33 bones (26 in the adult) of the spinal column.

vertigo A condition where the patient feels as if the room is moving or whirling.

visceral membrane The covering of the organs.

visual acuity Sharpness of perception.

vomer The bone that forms the inferior portion of the nasal septum.

zygoma The cheek bone.

Index*

*Page numbers followed by b indicate boxes;
f, figures; t, tables.